S0-AEB-631

ENVIRONMENTAL CHEMISTRY

FOURTH EDITION

STANLEY E. MANAHAN
UNIVERSITY OF MISSOURI

WILLARD GRANT PRESS
BOSTON

PWS PUBLISHERS
Prindle, Weber & Schmidt • Willard Grant Press • Duxbury Press •
Statler Office Building • 20 Providence Street • Boston, Massachusetts 02116

Library of Congress Cataloging in Publication Data

Manahan, Stanley E.
Environmental chemistry.

Bibliography: p.
Includes index.
1. Environmental chemistry. I. Title.
QD31.2.M35 1983 628.5'01'54 83-16303
ISBN 0-87150-764-1

ISBN 0-87150-764-1

Cover and text designed by David Foss. Art for this edition drawn by Ryan M. Cooper. Composed in Korinna and English Times by Caron LaVallie of the PWS Composition Department. Cover screenprinted by Lehigh Auto Screen. Text printed and bound by Halliday Lithograph. Printed in the United States of America.

88 87 86 85 84 — 10 9 8 7 6 5 4 3 2 1

PREFACE TO THE FOURTH EDITION

Environmental chemistry is now a well-established chemical subdiscipline. Environmental chemists occupy academic posts and are being employed by industry, utilities, research institutes, and regulatory agencies. It is generally recognized that a knowledge of this subject can be of enormous help in avoiding pollution problems, as well as in solving them after they develop.

Environmental problems and our perceptions of them change with time. When the first edition of this book was published in 1972, detergent phosphates and eutrophication of lakes were burning issues. Publication of the third edition in 1979 saw the "energy crisis" as an overwhelming issue; there was also considerable concern over the effects of chlorofluorocarbons on the ozone layer. Major environmental issues at present include hazardous wastes (particularly "dioxin"), acid rain, and the possible climatic effects of a steady buildup of atmospheric carbon dioxide.

Every effort has been made in this book to present the basic science of environmental chemistry and to keep up to date with the most important issues. Chapter 15, "Organic Pollutants in the Atmosphere," has been added to strengthen the atmospheric chemistry section. The addition of Chapter 19, "The Nature and Sources of Hazardous Wastes," and Chapter 20, "Environmental Chemistry of Hazardous Wastes," was a must because of rapidly growing awareness of long-term problems from improperly discarded hazardous wastes. A minor rearrangement of material has been made in placing Chapter 13, "Particulate Matter in the Atmosphere," before the discussion of photochemical smog. The former topic can be discussed before smog-forming processes and is useful in understanding smog products. Other chapters have been thoroughly revised; obsolete material has been deleted and new material added. Overall, the revision is a substantial one and an even more comprehensive treatment of environmental chemistry than the previous edition.

This text was written for students who have completed both semesters of beginning chemistry, a course in organic chemistry, and a course in analytical chemistry. In practice, it has been the author's experience that students (particularly those strongly motivated to learn the subject) can handle most of the material with as little chemistry background as one semester of beginning chemistry. The chemical equilibrium concepts in Chapters 2–4, though challenging, are presented starting with basic equilibrium principles, and a student with a reasonable aptitude for mathematics can understand them. A brief explanation of structural organic chemistry and organic functional groups is included to enable a student without a course in organic chemistry to handle the many organic examples given in the text.

The author wishes to express his appreciation for the excellent manuscript reviews of this edition by Rudolph Bottei, University of Notre Dame; Samuel Shipp Butcher, Bowdoin College; James Carr, University of Nebraska; David Keeling, University of Hawaii; and Vakula S. Srinivasan, Bowling Green State University. In addition, he would like to acknowledge the efforts of several of the staff of Willard Grant Press: Bruce Thrasher for his steady, highly competent management through the years that the three previous editions of this book have been in print; Edward Murphy for his dynamic, innovative work as Chemistry Editor during recent years; and David Foss for his meticulous, quality-conscious copyediting and production.

Stanley Manahan
University of Missouri–Columbia, 1983

CONTENTS

CHAPTER 1
ENVIRONMENTAL CHEMISTRY AND CHEMICAL HAZARDS

1.1 OUR FRAGILE WORLD

Anyone who has had an opportunity to observe even a small fraction of the natural wonders of our planet would agree that we live in a marvelous world. The majesty of the Rocky Mountains, the beauty of the Maine coast on a clear summer day, and the colors of a desert sunset are all impressive sights. On a more mundane level, we are reminded of the Earth's bounties by the smell and feel of freshly plowed Iowa farmland, by a breath of fresh air on a brisk fall day, or the taste of fresh, clean water on a hot summer day.

It goes without saying that all of these things are threatened by a broad range of human activities. In an effort to increase short-term agricultural production, farmland in many areas is cultivated improperly, resulting in extreme erosion that threatens its very existence. Many urban dwellers would agree that a breath of fresh air is a rare commodity in the city. In many industrialized areas, ground water sources of drinking water are being threatened by the insidious movement of hazardous waste chemical leachates through ground water aquifers. The protection of our environment must be given the highest priority because on it depends the preservation of humankind, itself.

In order to combat threats to our environment, it is necessary to understand the nature and magnitude of the problems involved. Before discussing these problems further, it is essential to recognize the fact that science and technology must play key roles in solving environmental problems. Only through the proper application of science and technology, under the direction of people with a strong environmental consciousness and a basic knowledge of the environmental sciences, can humankind survive on the limited resources of this planet.

In attempting to preserve our environment, we face enormous challenges. These have been documented in a publication issued in 1980 by the U.S. Council on Environmental Quality [1]. This report outlines a number of pressing environmental problems that will likely be present in the year 2000, as summarized in an analysis of the report [2]. One of the most acute problems is population growth, which is expected to increase world population from a current level of a little over 4 billion to 6.35 billion by 2000, with most of the growth in less

---

[1] Barney, G.O. 1980. *The global 2000 report to the President.* 2 vols. Washington, D.C.: U.S. Government Printing Office.

[2] 1980. Environmental outlook grim for the year 2000. *Chemical and Engineering News* 28 July 1980, pp. 21–22.

developed countries. It is also predicted that the amount of arable land will increase by only 4% during this period, a substantial net decline of arable land per capita. Although food production is expected to increase, most of the increase will come in more developed countries. By this time, the worldwide production of a number of commodities, including crude oil, will have already peaked. It is thought that the world fish catch, long a source of protein for much of the world population, will most likely be declining. Another serious problem has to do with world forests, 40% of which may have disappeared by the year 2000. This will mean a decline of 50% in commercial timber supplies and a demand for firewood far exceeding supply. Land, water, and air will all be under severe strain. In nearly half of the world, water requirements will have doubled. Agricultural soils, pushed to their limits by demand for increased food production, will be subject to deterioration from erosion, desertification, loss of organic matter, waterlogging, and the effects of excess salinity and alkali. The atmosphere will be threatened by a significant increase in CO_2 level, which may cause global warming and other marked effects on climate, and by the buildup of ozone-destroying chemicals, particularly chlorofluorocarbons (freons) and N_2O. The net result could be the extinction of up to 20% of current plant and animal species.

Rather than as a collection of frightening predictions, the somewhat grim scenario outlined above should be viewed as conditions that are likely to develop unless there are changes in public policy around the world. Implicit in this discussion is a call for action to thinking citizens in all countries. Only through a knowledge of environmental sciences in general and environmental chemistry in particular can effective action be taken to avert the ecological disaster outlined by the Global 2000 report.

1.2 CHEMICAL HAZARDS

During the last fifteen years the nature and emphasis of environmental concerns have changed appreciably. In 1970, issues such as contamination of water by detergent phosphates received a great deal of publicity. Some of these issues now seem rather insignificant compared to the threats posed to human health by hazardous waste chemicals and their improper disposal. Such threats are exemplified by the Kepone disaster of the mid-1970s, in which approximately 53,000 kilograms of this toxic, persistent pesticide were dumped into the Hopewell, Virginia sewage system, contaminating the James River and perhaps even Chesapeake Bay (see Chapter 7).

Love Canal has come to symbolize the national problem of hazardous waste chemicals. Used during the 1940s as a dump site for about 20,000 metric tons of chemical wastes, Love Canal became a repository for such chemicals as acid chlorides, chlorinated phenols, chlorobenzenes, metal chlorides, and many others. Later, after the dump site had been filled in and covered with soil, a school was built adjacent to the canal, and homes were built around it. During the late 1970s, noxious chemicals were observed oozing from the old dump site, and chemical leachate was found in yards and basements of homes around the canal. Elevated levels of miscarriages and birth defects were reported among nearby residents, and people around the canal complained of various maladies such as chloracne, a skin disorder that can result from exposure to noxious chemicals. As a result, some of the area was evacuated, and a massive cleanup effort was undertaken. As of 1983, more than $100 million in federal, state, and local money had been spent on the Love Canal problem.

According to the Environmental Protection Agency, Love Canal ranks only twenty-fifth among U.S. hazardous waste sites. This means there are many other sites that are even worse and that may pose even greater threats to human welfare. Some of the older waste dumps have been in existence for a very long time. According to the E.P.A., one of the worst sites is the forty-square-mile Tar Creek drainage basin in northeastern Oklahoma, where extensive zinc and lead mining operations were conducted from the early 1900s until the mid-1960s. A *New York Times* article [3] described pollution in this area as follows: ". . . acid waters spurt from abandoned drill holes, a pair of once-clean prairie streams run a rust-laden red, gaseous wisps rise from swampy ponds, and murky pools stand where land has caved into the underground evacuations." The problem became acute in the late 1970s as old mines in the area filled with water. In addition to high levels of zinc and lead, the acidic mine waters contain excessive levels of iron, manganese, cadmium, and other metals. The greatest pollution threat posed by the site is contamination of ground water aquifers.

Another one of the E.P.A.'s ten worst hazardous waste sites is located in a sixty-acre area near Woburn, Massachusetts. Waste chemicals were first dumped at this site by a chemical company in 1853! Later, glue factories and tanneries were constructed near the site, and they, too, dumped hazardous wastes in the area. Among these wastes were toxic heavy elements, including arsenic, chromium, and lead.

Thus, the U.S., along with other industrialized nations, has a long legacy of improperly discarded hazardous wastes. The major threat posed by these chemicals is the pollution of drinking water supplies, particularly ground water. In addition, there are potential dangers to humans and wildlife through direct exposure. New regulations have largely stopped the improper practices that resulted in so many dangerous waste chemical dumps. However, those already in existence, perhaps including some that have not yet been found, will continue to cause health problems and require costly cleanup for many decades. In combating this and other environmental problems, environmental chemistry has a major role to play. These problems are chemical problems, and as such they require chemical solutions. The purpose of this book is to define environmental chemistry and to show how it may play a key part in combating the many environmental problems that face humankind.

1.3 WHAT IS ENVIRONMENTAL CHEMISTRY?

What is environmental chemistry? This question is a little difficult to answer because environmental chemistry encompasses many different topics. It may involve a study of Freon reactions in the stratosphere or an analysis of toxic Kepone deposits in ocean sediments. It also covers the basic chemistry of toxic trace element species synthesized during the manufacture of synthetic natural gas from coal. A reasonably all-inclusive definition is: "Environmental chemistry is the study of the sources, reactions, transport, effects, and fates of chemical species in the water, soil, and air environments, as well as the influence of human activity upon these processes." More simply, **environmental chemistry** is the science of chemical phenomena in the environment.

---

[3] 1981. No. 1 toxic waste site is not a town's no. 1 gripe. *New York Times,* 9 December 1981.

Environmental chemistry is not a new discipline. Excellent work has been done in this field for more than half a century. Until about 1970, most of this work was done in academic departments or industrial groups other than chemistry groups. Much of it was done by people whose primary training was not in chemistry. Thus, when pesticides were synthesized, biologists observed first-hand some of the less desirable consequences of their use. When detergents were formulated, sanitary engineers were startled to see sewage treatment plant aeration tanks vanish under meter-thick blankets of foam, while limnologists wondered why previously normal lakes suddenly became choked with stinking blue-green algae. But until very recently, few chemists were exposed to material dealing with environmental chemistry as part of their training.

At an accelerating rate in recent years, however, many chemists have become deeply involved with the investigation of environmental problems. Academic chemistry departments have found that environmental chemistry courses appeal to students, and many graduate students are attracted to environmental chemistry research. Help-wanted ads have included increasing numbers of openings for environmental chemists among those of the more traditional chemical subdisciplines. Industries have found that well-trained environmental chemists at least help avoid difficulties with regulatory agencies and at best are instrumental in developing profitable pollution-control products and processes.

Some background in environmental chemistry should be part of the training of every chemistry student. The ecologically illiterate chemist can be a very dangerous species. Chemists must be aware of the possible effects their products and processes might have upon the environment. Furthermore, any serious attempt to solve environmental problems must involve the extensive use of chemicals and chemical processes.

There are some things that environmental chemistry is not. It is not just the same old chemistry with a different cover and title. Because it deals with natural systems, it is more complicated and difficult than "pure" chemistry. Students sometimes find this hard to grasp. Accustomed to the clear-cut concepts of relatively simple, well-defined systems, they may find environmental chemistry to be poorly defined, vague, and confusing. More often than not, it is impossible to come up with a simple answer to an environmental chemistry problem. But, building on an ever-increasing body of knowledge, the environmental chemist can make educated guesses as to how environmental systems will behave.

One of environmental chemistry's major challenges is the determination of the nature and quantity of specific pollutants in the environment. Thus, chemical analysis is a vital first step in environmental chemistry research. The difficulty of analyzing for many environmental pollutants can be awesome. Significant levels of air pollutants may consist of less than a microgram per cubic meter of air. For many water pollutants, one part per million by weight (essentially 1 milligram per liter) is a very high value. Environmentally significant levels of some pollutants may be only a few parts per trillion. Thus, it is obvious that the analytical chemistry used to understand some environmental systems requires a very low limit of detection.

However, environmental chemistry is not the same as analytical chemistry, which is only one of the many subdisciplines that are involved in the study of the chemistry of the environment. A "brute-force" approach to environmental control, involving the monitoring of each environmental niche for every possible pollutant, would increase employment for chemists and raise sales of chemical instruments, but it would be a wasteful, unwise way to detect and solve environmental problems. We can be smarter than that. In order for chemistry to make a maximum contribution to the solution of environmental problems, the chemist must work toward an understanding of the nature, reactions, and transport of chemical species in the environment. Analytical chemistry is a fundamental and crucial part of that endeavor but is by no means all of it.

1.4 SOME DEFINITIONS

In some cases pollution is a clear-cut phenomenon, whereas in others it lies largely in the eyes of the beholder. Toxic organochlorine compounds leached into water supplies from a hazardous waste chemical dump are pollutants in anybody's view. However, loud rock music amplified to a high decibel level by the sometimes questionable miracle of modern electronics is pleasant to some people and a very definite form of noise pollution to others. Frequently, time and place determine what may be called a pollutant. The phosphate that the sewage treatment plant operator has to remove from wastewater is chemically the same as the phosphate that the farmer a few miles away has to buy at high prices for fertilizer. Most pollutants are, in fact, resources gone to waste; as resources become more scarce and expensive, economic pressure will almost automatically force solutions to many pollution problems.

A reasonable definition of a **pollutant** is a substance present in greater than natural concentration as a result of human activity and having a net detrimental effect upon its environment or upon something of value in that environment. A **contaminant** is something that causes a deviation from the normal composition of an environment. Contaminants are not classified as pollutants unless they have some detrimental effect.

Every pollutant originates from a source. The source is particularly important, because it is generally the logical place to eliminate pollution. After a pollutant is released from a source, it may act upon a receptor. The **receptor** is anything that is affected by the pollutant. Humans whose eyes smart from oxidants in the atmosphere are receptors. Trout fingerlings that may die after exposure to dieldrin in water are also receptors. Eventually, if the pollutant is long-lived, it may be deposited in a **sink**, a long-time repository of the pollutant. Here it will remain for a long time, though not necessarily permanently. Thus, a limestone wall may be a sink for atmospheric sulfuric acid, through the reaction

$$H_2SO_4 + CaCO_3 \rightarrow CaSO_4 + H_2O + CO_2(g) \qquad (1.4.1)$$

which fixes the sulfate as part of the wall composition.

1.5 WATER, AIR, SOIL, AND LIFE

It is convenient to subdivide environmental chemistry into areas involving the chemistry of the hydrosphere, the lithosphere, the atmosphere, and the biosphere. All matter, from minerals in the outer layers of the Earth's crust to relatively stable ions in the upper reaches of the atmosphere, may be included in one of these categories.

The **hydrosphere** refers to water in its many forms. It includes the oceans, lakes, streams, reservoirs, snowpack, glaciers, the polar ice caps, and water under the ground (ground water). For the study of environmental chemistry, however, liquid water and the reactions of the chemical species in it are of predominant importance.

The **lithosphere** includes the outer parts of the solid Earth. In general, the term refers to minerals encountered in the Earth's crust and to the complex and variable mixture of minerals, organic matter, water, and air making up soil. Insofar as environmental chemistry is concerned, the soil is the most significant part of the lithosphere.

The **atmosphere** is the envelope of gases surrounding the Earth. The atmosphere is subdivided into different regions depending on altitude. Atmospheric chemistry varies a great deal with altitude, exposure to solar radiation, pollution load, and other factors.

The term **biosphere** refers to life. It includes living organisms and their immediate surroundings. The biosphere is influenced tremendously by the chemistry of the environment and, in turn, exerts a powerful influence upon the chemistry of most environments, particularly the lithosphere and hydrosphere. Moreover, as will be discussed in Chapter 11, biological activity is responsible for the present composition of the atmosphere (specifically, high oxygen level and low carbon dioxide level), and plants still influence the atmosphere, for example, by emitting terpenes which form a sort of smog, such as that observed in the Great Smoky or Blue Ridge Mountains.

This book deals primarily with the hydrosphere, lithosphere, and atmosphere. As living beings, we all must be concerned about the preservation and enrichment of life on this planet. Whenever a chemical species is introduced into the environment, the question of its ultimate effect upon life must always be raised. In trying to study and upgrade the environment, priorities must be kept in mind, with life given top priority. Thus, the millions of pounds of glass bottles that accumulate in garbage dumps each year are of relatively less importance than the smaller quantities of heavy metals and toxic organic compounds that get into drinking water sources from hazardous waste chemical dumps. The glass may be an unsightly nuisance, but chemically and biologically it is one of the most innocuous substances known. Heavy metals and toxic organic compounds, on the other hand, can be carcinogenic, mutagenic, and otherwise damaging to life.

The importance of toxic substances and their effects upon living organisms, including humans, has led to the inclusion in this book of Chapter 18, which deals with environmental biochemistry and chemical toxicology. Because of the health effects of toxic substances, it is important for anybody working with chemicals to have a knowledge of this topic.

SUPPLEMENTARY REFERENCES

American Chemical Society. 1978. *Cleaning our environment: a chemical perspective.* Washington, D.C.

Bailey, R.A. 1978. *Chemistry of the environment.* New York: Academic Press.

Bailey, R.A.; Clarke, H.M.; Ferris, J.P.; Krause, S.; and Strong, R.L. 1978. *Chemistry of the environment.* New York: Academic Press.

Barton, R. 1980. *The oceans.* New York: Facts on File.

Bettman, O.L. 1974. *The good old days—they were terrible.* New York: Random House, Inc.

Bhatt, J. J. 1978. *Oceanography.* New York: Van Nostrand.

Bockris, J.O'M., ed. 1977. *Environmental chemistry.* New York: Plenum Publishing Corp.

Caglioti, L. 1983. *The two faces of chemistry.* Cambridge, Mass.: MIT Press.

Chanlett, E.T. 1979. *Environmental protection.* 2nd ed. New York: McGraw-Hill.

Council on Environmental Quality. 1982. *Environmental quality.* Published annually. Washington, D.C.: Superintendent of Documents.

De Bell, G.; Wallace, A.; and Gancher, D., eds. 1980. *The new environmental handbook.* San Francisco: Friends of the Earth.

Dix, H.M. 1980. *Environmental pollution.* New York: John Wiley and Sons, Inc.

Eisenbud, M. 1979. *Environment, technology, and health.* New York: New York University Press.

Garrels, R.M.; MacKenzie, F.T.; and Hunt, C. 1975. *Chemical cycles and the global environment.* Los Altos, Calif.: William Kaufmann, Inc.

1977. *Global chemical cycles and their alteration by man.* Old Greenwich, Conn.: Koehn and Schneider.

Hammond, K.A.; Macinko, G.; and Fairchild, W.B. 1978. *Sourcebook on the environment.* Chicago: University of Chicago Press.

Horne, R.A. 1978. *The chemistry of our environment.* New York: John Wiley and Sons, Inc.

Hutzinger, O., ed. 1980. *The handbook of environmental chemistry.* New York: Springer-Verlag New York, Inc.

Johnston, R.F.; Frank, P.W.; and Michener, C.D., eds. *Ecology and systematics.* Vol. 13. Palo Alto, Calif.: Annual Reviews. Published annually.

Kormondy, E.J. 1977. *Concepts of ecology.* 2nd ed. Englewood Cliffs, N.J.: Prentice-Hall, Inc.

Lenihan, J., and Fletcher, W.W., eds. 1977. *The chemical environment.* New York: Academic Press, Inc.

Lippmann, M., and Schlesinger, R.B. 1979. *Chemical contamination in the human environment.* New York: Oxford University Press.

Moore, J.W., and Moore, E.A. 1976. *Environmental chemistry.* New York: Academic Press, Inc.

Nader, R.; Brownstein, R.; and Richard, J., eds. 1981. *Who's poisoning America?* San Francisco: Sierra Club Books.

Nriagu, J.R., ed. 1976. *Environmental biogeochemistry.* 2 vols. Ann Arbor, Mich.: Ann Arbor Science Publishers, Inc.

Ott, W.R., 1978. *Environmental indices.* Ann Arbor, Mich.: Ann Arbor Science Publishers, Inc.

Parker, S.P., ed. 1980. *McGraw-Hill encyclopedia of environmental science.* New York: McGraw-Hill.

Perry, A.H., and Walker, J.M. 1977. *The ocean-atmosphere system.* New York: Longman.

Pitts, J.N.; Metcalf, R.L.; and Grosjean, D. 1982. *Advances in environmental science and technology.* Published annually. New York: Wiley-Interscience.

Purdom, P.W., and Anderson, S.H. 1983. *Environmental science.* 2nd ed. Columbus, Ohio: Charles E. Merrill Publishing Co.

Schnaiberg, A. 1980. *The environment.* New York: Oxford University Press.

Spiro, T.G., and Stigliani, W.G. 1980. *Environmental issues in chemical perspective.* Baltimore, Md.: State University of New York Press.

Stoker, H.S., and Seager, S.L. 1976. *Environmental chemistry: air and water pollution.* 2nd ed. Glenview, Ill.: Scott, Foresman and Co.

Suffet, I.H., ed. 1977. *Fate of pollutants in the water and air environments.* New York: Wiley-Interscience.

Teja, A.S., ed. 1981. *Chemical engineering and the environment.* New York: Halsted (John Wiley and Sons, Inc.).

Tinsley, I.J. 1979. *Chemical concepts in pollutant behavior.* New York: Wiley-Interscience.

Turk, A.; Wittes, J.T.; Turk, J.; and Wittes, R.E. 1978. *Environmental science.* 2nd ed. Philadelphia: W.B. Saunders and Co.

Vowles, P.D., and Connell, D.W. 1980. *Experiments in environmental chemistry.* Elmsford, N.Y.: Pergamon Press, Inc.

Watkins, J.S.; Bottino, M.L.; and Morisawa, M. 1978. *Our geological environment.* Philadelphia: W.B. Saunders and Co.

Watt, K.E.F. 1982. *Understanding the environment.* Boston: Allyn & Bacon, Inc.

CHAPTER 2
THE PROPERTIES AND COMPOSITION OF NATURAL WATERS

2.1 WATER: QUALITY, QUANTITY, AND CHEMISTRY

Throughout history, the quality and quantity of water available to humans have been vital factors in determining their well-being. Whole civilizations have disappeared because of water shortages resulting from changes in climate. Even in temperate climates, fluctuations in precipitation cause problems. Most of the U.S. suffered from a drought during 1980, and the cracked, muddy bottoms of dry reservoirs provided a common source of newspaper and magazine photographs. The drought was broken in some areas by torrential rains and heavy snowfall in 1981; devastating floods hit the San Francisco area in January, 1982; and a series of rain-laden Pacific storms caused a repeat performance on a larger scale along the West Coast early in 1983.

Although cholera and typhoid are now controlled in most areas of the world, waterborne diseases killed millions of people in the past and still cause great misery in less developed countries. Ambitious programs of dam and dike construction have reduced flood damage, but they have had a number of undesirable side effects in some areas, such as inundation of farmland by reservoirs, and unsafe dams prone to breakage. Problems with water supply quantity and quality remain and in some respects are becoming more serious. These problems include increased water use due to population growth, contamination of drinking water by improperly discarded hazardous wastes (see Chapters 19 and 20), and destruction of wildlife by water pollution.

This chapter is concerned with **aquatic chemistry**, the science dealing with water in rivers, lakes, estuaries, and oceans, as well as ground water, soil water, and water treatment systems. It involves phenomena determining the distribution and circulation of chemical species in natural waters. Any consideration of aquatic chemistry requires some understanding of the sources, transport, characteristics, and composition of water. The chemical reactions that occur in water and the chemical species found in it are strongly influenced by the environment in which the water is found. The chemistry of water exposed to the atmosphere is quite different from that of water at the bottom of a lake. Microorganisms play an essential role in determining the chemical composition of water. Thus, in discussing water chemistry, it is necessary to consider the many general factors that influence this chemistry.

2.2 SOURCES AND USES OF WATER: THE HYDROLOGIC CYCLE

The world's water supply is found in the five parts of the **hydrologic cycle** (Figure 2.1). A large portion of the water is found in the oceans. Another fraction is present as water vapor in the atmosphere (clouds). Some water is contained in the solid state as ice and snow in snowpacks, glaciers, and the polar ice caps. Surface water is found in lakes, streams, and reservoirs. Ground water is located underground.

There is a strong connection between the *hydrosphere,* where water is found, and the *geosphere,* or land; human activities affect both. For example, disturbance of land by conversion of grasslands or forests to agricultural land or intensification of agricultural production may reduce vegetation cover, decreasing **transpiration** (loss of water vapor by plants) and affecting the microclimate. The result is increased rain runoff, erosion, and accumulation of silt in bodies of water. The nutrient cycles may be accelerated, leading to nutrient enrichment of surface waters. This, in turn, can profoundedly affect the chemical and biological characteristics of bodies of water.

The water that humans use is primarily fresh surface water and ground water. In arid regions, a small fraction of the water supply comes from the ocean, a source that is likely to become more important as the world's supply of fresh water dwindles relative to demand. Saline or brackish ground waters may also be utilized in some areas.

Ground water and surface water have appreciably different characteristics. Many substances either dissolve in surface water or become suspended in it on its way to the ocean. Surface water in a lake or reservoir that contains the mineral nutrients essential for algal growth may support a heavy growth of algae. Surface water with a high level of biodegradable organic material, used as food by bacteria, normally contains a large population of bacteria. All these factors have a profound effect upon the quality of surface water.

Ground water may dissolve minerals from the formations through which it passes. Most microorganisms originally present in ground water are gradually

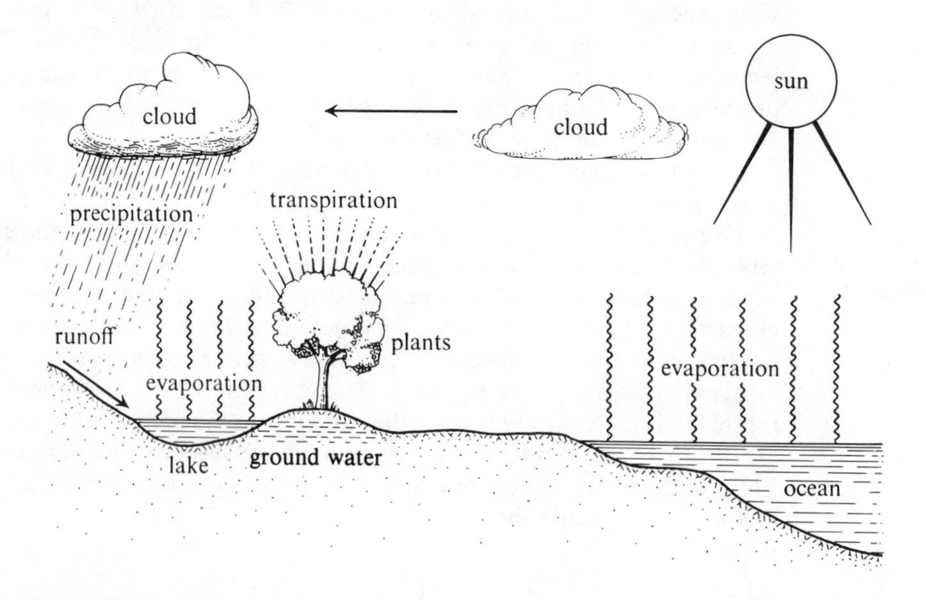

FIGURE 2.1 The hydrologic cycle.

filtered out as it seeps through the ground. Occasionally, the content of undesirable salts may become excessively high in ground water, although it is generally superior to surface water as a domestic water source.

In the continental United States, an average of approximately 1.48×10^{13} liters of water fall as precipitation each day. Dividing this volume of precipitation by the area of the U.S., we obtain an average precipitation of 76 cm per year. Of that amount, approximately 1.02×10^{13} liters per day, or 53 cm per year, are lost by evaporation and transpiration. Thus, the water theoretically available for use is approximately 4.4×10^{12} liters per day, or only 23 centimeters per year. At present, the U.S. uses 1.6×10^{12} liters per day, or 8 centimeters of the average annual precipitation. This amounts to an almost 10-fold increase from a usage of 1.66×10^{11} liters per day at the turn of the century. Even more striking is the per capita increase from 38 liters per day in 1900 to 636 liters per day in 1983. Much of this increase is accounted for by greater agricultural and industrial use, each of which accounts for approximately 46% of total consumption. Municipal uses consume the remaining 8%.

As shown in Figure 2.2, precipitation in the U.S. is unevenly distributed. This is a problem because people in areas with low precipitation often consume more water than people in regions with more rainfall. For example, Arizona, the state ranking next-to-last in precipitation at only one-third the national average, uses twice as much water per person as the national average. Rapid population growth in the more arid states during the last three decades (197% growth for Nevada, 91% for Colorado, and 75% for Utah) has further aggravated the problem. Water shortages are likely to become acute in southwestern U.S. [1], which contains six of the nation's eleven largest cities (Los Angeles, Houston, Dallas, San Diego, Phoenix, and San Antonio). Other problem areas include Florida,

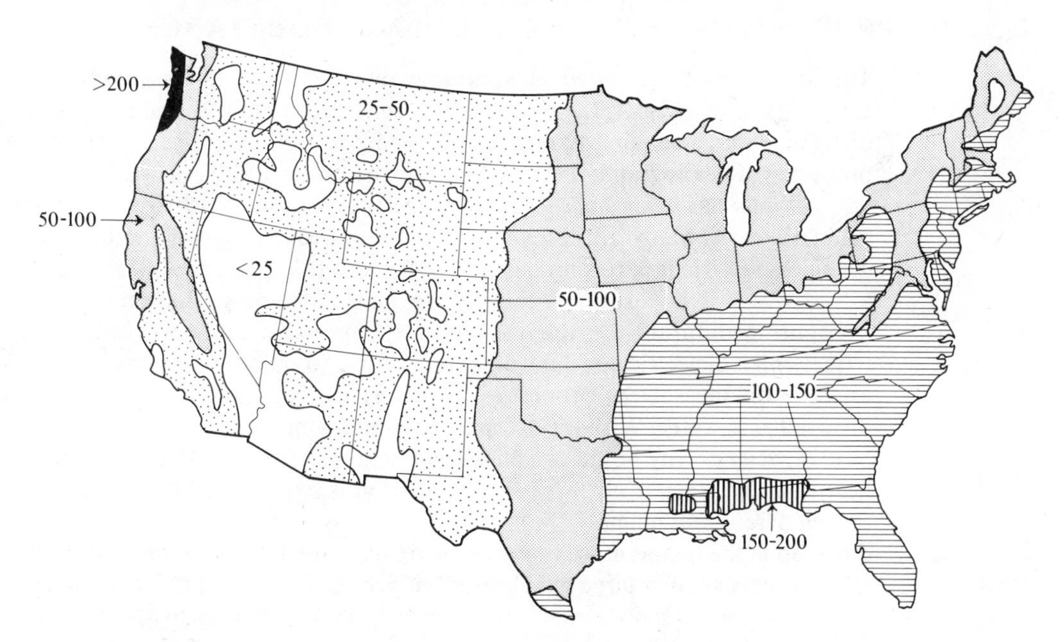

FIGURE 2.2 Distribution of precipitation in the continental U.S., showing average annual rainfall in centimeters.

---

[1] 1981. Rapid population and farm growth may strain southwest's water supply. *New York Times,* August 12, 1981.

where overdevelopment of coastal areas threatens Lake Okeechobee; the Northeast, plagued by deteriorating water systems; and the High Plains, ranging from the Texas panhandle to Nebraska, where irrigation demands on the Ogalla Aquifer are dropping the water table up to two meters per year, with no hope of recharge.

Given present trends, water demand will exceed supply in the U.S. by the year 2000. Despite rigid conservation measures and water recycling, some drastic changes in the pattern of water use appear inevitable. The impact may be particularly severe upon agriculture. In southwestern U.S., for example, agriculture accounts for the bulk of total water usage—85 percent in California, 90 percent in New Mexico, 89 percent in Arizona, and 68 percent in Texas [1]. In some areas, industries and municipalities are willing to buy their water at prices up to ten times that paid for irrigation water. Particularly in Wyoming, Utah, and Colorado, new energy industries are likely to put additional pressure on water prices. The increased cost of water could have marked effects on food prices and availability in the U.S.

Water continues to be the subject of heated disputes among land owners and governmental agencies. One of the latest battles has been over the sale by South Dakota of Missouri River water from the reservoir behind Oahe Dam near Pierre, South Dakota, for use in a pipeline designed to transport coal as a slurry from the Powder River Basin near Gillette, Wyoming to markets in Arkansas and Louisiana. (Slurry is a mixture of equal masses of ground coal and water.) This interbasin water transfer has resulted in considerable objection from states downstream on the Missouri River.

2.3 THE PROPERTIES OF WATER, A UNIQUE SUBSTANCE

The study of water is known as **hydrology**, and is divided into a number of subcategories. **Limnology** is the branch of the science dealing with the characteristics of fresh water, including biological properties as well as chemical and physical properties. **Oceanography** is the science of the ocean and its properties.

Water has a number of unique properties, without which life as we know it could not exist. Many of these properties are due to hydrogen bonding in water. These characteristics are summarized in Table 2.1.

Water is an excellent solvent for many materials; thus it is the basic transport medium for the nutrients and waste products in life processes. The extremely high dielectric constant of water relative to other liquids has a profound effect upon the solvent properties of water, in that most ionic materials are dissociated in water. With the exception of liquid ammonia, water has the highest heat capacity of any liquid or solid, 1 cal $\times$ g^{-1} $\times$ deg^{-1}. Because of this high heat capacity, a relatively large amount of heat is required to change appreciably the temperature of a mass of water; hence, a body of water can have a stabilizing effect upon the temperature of nearby geographic regions. In addition, this property prevents sudden large changes of temperature in large bodies of water and thereby protects aquatic organisms from the shock of abrupt temperature variations. The extremely high heat of vaporization of water, 585 cal/g at 20 °C, likewise stabilizes the temperature of bodies of water and the surrounding geographic regions. It also influences the transfer of heat and water vapor between bodies of water and the atmosphere. Water has its maximum density at a

TABLE 2.1 Important Properties of Water

Property	*Effects and significance*
Excellent solvent.	Transport of nutrients and waste products, making biological processes possible in an aqueous medium.
Highest dielectric constant of any pure liquid.	High solubility of ionic substances and their ionization in solution.
Higher surface tension than any other liquid.	Controlling factor in physiology; governs drop and surface phenomena.
Transparent to visible and longer-wavelength fraction of ultraviolet light.	Colorless, allowing light required for photosynthesis to reach considerable depths in bodies of water.
Maximum density as a liquid at 4 °C.	Ice floats; vertical circulation restricted in stratified bodies of water.
Higher heat of evaporation than any other material.	Determines transfer of heat and water molecules between the atmosphere and bodies of water.
Higher latent heat of fusion than any other liquid except ammonia.	Temperature stabilized at the freezing point.
Higher heat capacity than any other liquid except ammonia.	Stabilization of temperatures of organisms and geographical regions.

temperature (4 °C) above its freezing point. The fortunate consequence of this fact is that ice floats, so that few large bodies of water ever freeze solid. Furthermore, the pattern of vertical circulation of water in lakes, a determining factor in their chemistry and biology, is governed largely by the unique temperature–density relationship of water.

2.4 THE CHARACTERISTICS OF BODIES OF WATER

Surface water occurs primarily in streams, lakes, and reservoirs. As streams mature, they pass through four stages classified as birth, youth, maturity, and old age. Lakes may be classified as oligotrophic, eutrophic, or dystrophic, an order that often parallels the life of the lake. **Oligotrophic** lakes are deep, generally clear, deficient in nutrients, and without much biological activity. **Eutrophic** lakes have more nutrients, support more life, and are more turbid. **Dystrophic** lakes are shallow, clogged with plant life, and normally contain colored water with a low pH.

Some constructed reservoirs are very similar to lakes, while others differ a great deal from them. Reservoirs with a large volume relative to their inflow and outflow are called **storage reservoirs**. Reservoirs with a large rate of flow-through compared to their volume are called **run-of-the-river reservoirs**. The physical, chemical, and biological properties of water in the two types of reservoirs may vary appreciably. Water in storage reservoirs more closely resembles lake water, whereas water in run-of-the-river reservoirs is much like river water.

Impounding water in reservoirs may have some profound effects upon water quality. These changes result from factors such as different velocities, changed detention time, and altered surface-to-volume ratios relative to the streams that were impounded. Some resulting beneficial changes due to impoundment are a decrease in the level of organic matter, a reduction in turbidity, and a decrease in hardness (calcium and magnesium content). Some detrimental changes are lower oxygen levels due to decreased reaeration, decreased mixing, accumulation of pollutants, lack of bottom scour produced by flowing water scrubbing a stream bottom, and increased growth of algae. Algal growth may be enhanced when suspended solids settle from impounded water, causing increased exposure of the algae to sunlight. Stagnant water in the bottom of a reservoir may be of low quality. Oxygen levels frequently go to almost zero near the bottom, and odorous hydrogen sulfide is produced by the reduction of sulfur compounds in the low-oxygen environment. Insoluble iron(III) and manganese(IV) species are reduced to soluble iron(II) and manganese(II) ions which must be removed prior to using the water.

Estuaries constitute another type of body of water, consisting of arms of the ocean into which streams flow. The mixing of fresh and salt water gives estuaries unique chemical and biological properties. Estuaries are the breeding grounds of much marine life, which makes their preservation very important.

Water's unique temperature-density relationship results in the formation of distinct layers within nonflowing bodies of water, as shown in Figure 2.3. During the summer a surface layer (**epilimnion**) is heated by solar radiation and, because of its lower density, floats upon the bottom layer, or **hypolimnion**. This phenomenon is called **thermal stratification**. When an appreciable temperature difference exists between the two layers, they do not mix but behave independently and have very different chemical and biological properties. The epilimnion, which is exposed to light, may have a heavy growth of algae. As a result of exposure to the atmosphere and (during daylight hours) because of the photosynthetic activity of algae, the epilimnion contains relatively higher levels of dissolved oxygen and generally is aerobic. In the hypolimnion, bacterial action on biodegradable organic material may cause the water to become anaerobic. As a consequence, chemical species in a relatively reduced form tend to predominate in the hypolimnion.

The shear-plane, or layer between epilimnion and hypolimnion, is called the **thermocline**. During the autumn, when the epilimnion cools, a point is reached at which the temperatures of the epilimnion and hypolimnion are equal. This disappearance of thermal stratification causes the entire body of water to behave as a hydrological unit, and the resultant mixing is known as **overturn**. An overturn also generally occurs in the spring. During the overturn, the chemical and physical characteristics of the body of water become much more uniform, and a number of chemical, physical, and biological changes may result. Biological activity may increase from the mixing of nutrients. Changes in water composition during overturn may cause disruption in water-treatment processes.

The chemistry and biology of the Earth's vast oceans are unique because of the ocean's high salt content, great depth, and other factors. Oceanographic chemistry is a discipline in its own right. The environmental problems of the oceans have increased greatly in recent years because of ocean dumping of pollutants, oil spills, and increased utilization of natural resources from the oceans.

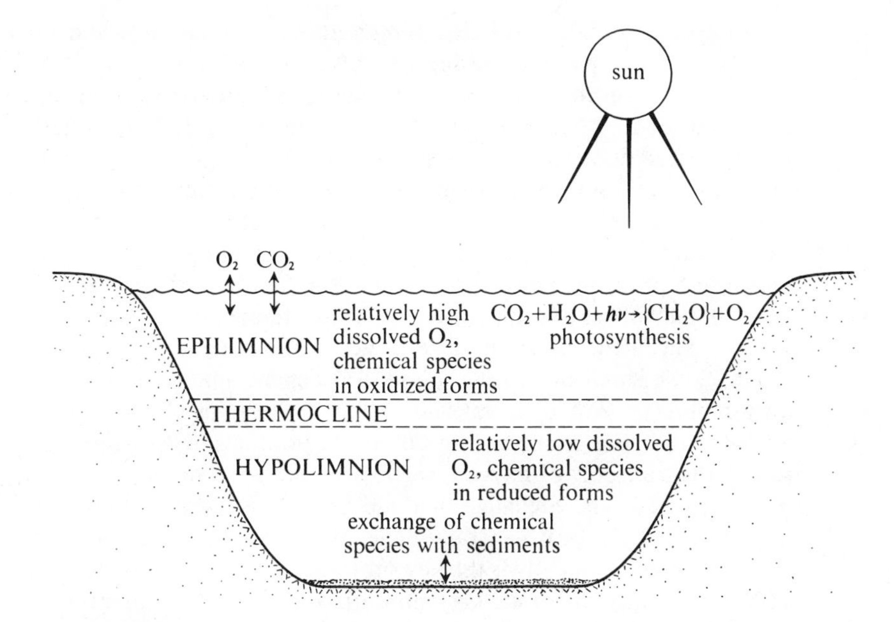

FIGURE 2.3 Stratification of a lake.

2.5 AQUATIC LIFE

The living organisms in aquatic ecosystems may be classified as either autotrophic or heterotrophic. **Autotrophic** organisms utilize solar or chemical energy to fix elements from simple, nonliving inorganic material into complex life molecules that compose living organisms. Algae are typical autotrophic aquatic organisms. Generally, CO_2, NO_3^-, and PO_4^{3-} are sources of C, N, and P, respectively, for autotrophic organisms. Organisms that utilize solar energy to synthesize organic matter from inorganic materials are called **producers**.

Heterotrophic organisms utilize the organic substances produced by autotrophic organisms as energy sources and as the raw materials for the synthesis of their own biomass. **Decomposers** (or **reducers**) are a subclass of the heterotrophic organisms and consist of chiefly bacteria and fungi, which ultimately break down material of biological origin to the simple compounds originally fixed by the autotrophic organisms.

The ability of a body of water to produce living material is known as its **productivity**. Productivity results from a combination of physical and chemical factors. Water of low productivity generally is desirable for water supply or for swimming. Relatively high productivity is required for the support of fish. Excessive productivity can result in choking by weeds and can cause odor problems. The growth of algae may become quite high in very productive waters, with the result that the concurrent decomposition of dead algae reduces oxygen levels in the water to very low values. This set of conditions is commonly called **eutrophication**.

Life forms higher than algae and bacteria—fish for example—comprise a comparatively small fraction of the biomass in most aquatic systems. The influence of these higher life forms upon aquatic chemistry is minimal. However, aquatic life is strongly influenced by the physical and chemical properties of the body of water in which it lives. *Temperature, transparency,* and *turbulence* are the three main physical properties affecting aquatic life. Very low water temperatures result in very slow biological processes, whereas very high temperatures are fatal to most organisms. A difference of only a few degrees can produce large differences in the kinds of organisms present. Thermal discharges of hot water from power plants (*cooling water*) frequently kill temperature-sensitive fish while increasing the growth of other species adapted to higher temperatures. The transparency of water is particularly important in determining the growth of algae. Thus, turbid water may not be very productive of biomass, even though it has the nutrients, optimum temperature, and other conditions needed. Turbulence is an important factor in mixing and transport processes in water. Some small organisms **(plankton)** depend upon water currents for their own mobility. Water turbulence is largely responsible for the transport of nutrients to living organisms and of waste products away from them. It plays a role in the transport of oxygen, carbon dioxide, and other gases through a body of water and in the exchange of these gases at the water-atmosphere interface. Moderate turbulence is generally beneficial to aquatic life.

Oxygen frequently is the key substance in determining the extent and kinds of life in a body of water. Oxygen deficiency is fatal to many aquatic animals such as fish. The presence of oxygen can be equally fatal to many kinds of anaerobic bacteria. The **concentration of dissolved oxygen, DO**, is always one of the first things measured in determining the biological characteristics of a stream or lake.

Biochemical oxygen demand, BOD, is another important water-quality parameter. It refers to the amount of oxygen utilized when the organic matter in a given volume of water is degraded biologically. A body of water with a high biochemical oxygen demand, and no means of rapidly replenishing the oxygen, obviously cannot sustain organisms that require oxygen. The importance of BOD will be discussed in more detail in Chapter 7.

Carbon dioxide is produced by respiratory processes in waters and sediments and can also enter water from the atmosphere. Carbon dioxide is required for the photosynthetic production of biomass by algae and in some cases is a limiting factor. High levels of carbon dioxide produced by the degradation of organic matter in water can cause excessive algal growth and productivity.

The levels of nutrients in water frequently determine its productivity. Aquatic plant life requires an adequate supply of carbon (CO_2), nitrogen (nitrate), phosphorus (orthophosphate), and trace elements such as iron. In many cases, phosphorus is the limiting nutrient and is generally controlled in attempts to limit excess productivity.

The salinity of water also determines the kinds of **biota** (life forms) present. Irrigation waters may pick up harmful levels of salt. Marine life obviously requires or tolerates salt water, whereas many fresh-water organisms are intolerant of salt.

2.6 CHEMICAL MODELS OF NATURAL WATER SYSTEMS

This section introduces some topics that will be expanded upon in later chapters.

Even under carefully controlled conditions in a laboratory, the investigation of chemical species in water can be very difficult. Generally, the media used in laboratory investigations are made up to constant ionic strength with a relatively inert electrolyte. This minimizes effects arising from differences in factors such as activity coefficients or, when potentiometric measurements are made, liquid junction potentials at the solution-reference electrode interface. The temperature of the medium may be regulated to one hundredth of a degree, or even more accurately if necessary. If an equilibrium constant is to be measured, the system may be allowed to reach equilibrium over a very long period. However, the wide variation among values given for the same constants in the chemical literature attests to the difficulty of describing even relatively simple chemical systems under the most carefully controlled conditions.

The difficulties encountered in trying to describe the chemical nature of a natural aquatic system should be obvious. Such systems are very complex and a description of their chemistry must take many variables into consideration. In addition to water, these systems contain mineral phases, gas phases, and organisms. As open, dynamic systems, they have variable inputs and outputs of energy and mass. Therefore, except under unusual circumstances, a true equilibrium condition is not obtained, although an approximately steady-state aquatic system frequently exists. Most metals found in natural waters do not exist as simple hydrated cations in the water, and oxyanions often are found as polynuclear species, rather than as simple monomers. The nature of chemical species in water containing bacteria or algae is strongly influenced by the action of these organisms. Thus, an exact description of the chemistry of a natural water system based upon acid-base, solubility, and complexation equilibrium constants, redox potential, pH, and other chemical parameters is not possible. Therefore, the systems must be described by simplified **models**, often based around equilibrium chemical concepts. Though not exact, nor entirely realistic, such models can yield useful generalizations and insights pertaining to the nature of aquatic chemical processes and provide guidelines for the description and measurement of natural water systems. Though greatly simplified, such models are very helpful in visualizing the conditions that determine chemical species and their reactions in natural waters and wastewaters.

2.7 THE NATURE OF METAL IONS IN WATER

The formula of a metal ion in aqueous solution usually is written M^{n+}, which signifies the simple hydrated metal cation $M(H_2O)_x^{n+}$. A bare metal ion, Mg^{2+} for example, cannot exist as a separate entity in water. Such metal ions are bonded to electron-donor partners to secure the highest stability of their outer electron shells. Metal ions in water are bonded, or *coordinated,* to water molecules or other stronger bases that might be present.

Metal ions in aqueous solution seek to reach a state of maximum stability through chemical reactions. Acid-base, precipitation, complexation, and oxidation-reduction reactions all provide means through which metal ions in water are transformed to more stable forms. The ways in which these reactions affect water chemistry are discussed in later sections.

2.8 HYDRATED METAL IONS AS ACIDS

Hydrated metal ions, particularly those with a charge of +3 or more, tend to lose protons in aqueous solution, and fit the definition of Brönsted acids. (Recall that according to the Brönsted definition, acids are proton donors and bases are proton acceptors.) The acidity of a metal ion increases with charge and decreases with increasing radius. In Equation 2.8.1, the hydrated iron(III) ion is a relatively

$$Fe(H_2O)_6^{3+} + H_2O \rightleftharpoons Fe(H_2O)_5OH^{2+} + H_3O^+ \qquad (2.8.1)$$

strong acid, with K_{a1} of 8.9×10^{-4}. Hydrated trivalent metal ions, such as iron(III), generally are minus at least one hydrogen ion at neutral pH values or above. For tetravalent metal ions, the completely protonated form, $M(H_2O)_x^{4+}$, is rare even at very low pH values. Commonly, O^{2-} is coordinated to tetravalent metal ions; an example is the vanadium(IV) species, VO^{2+}. Generally, divalent metal ions do not lose a hydrogen ion at pH values below 6, whereas monovalent metal ions, such as Na^+, do not act as acids at all in this pH range and exist in water solution as simple hydrated ions.

The tendency of hydrated metal ions to behave as acids may have a profound effect upon the aquatic environment. A good example is *acid mine water* (see Section 5.25), which derives part of its acidic character from the tendency of hydrated iron(III) to lose protons.

$$Fe^{3+} + 3\,H_2O \rightarrow Fe(OH)_3(s) + 3\,H^+ \qquad (2.8.2)$$

Hydroxide, OH^-, bonded to a metal ion, may function as a bridging group to join two or more metals together through the following dehydration-dimerization:

$$2\,Fe(H_2O)_5OH^{2+} \rightarrow (H_2O)_4Fe\underset{\underset{H}{|}}{\overset{\overset{H}{|}}{\langle O \rangle}}Fe(H_2O)_4^{4+} + 2\,H_2O \qquad (2.8.3)$$

Among the metals other than iron(III) forming polymeric species with OH^- as a bridging group are Al(III), Be(II), Bi(III), Ce(IV), Co(III), Cu(II), Ga(III), Mo(V), Pb(II), Sc(II), Sn(II), Sn(IV), and U(VI). Additional hydrogen ions may be lost from water molecules bonded to the dimers, furnishing OH^- groups for further bonding and leading to the formation of polymeric hydrolytic species. If the process continues, colloidal hydroxy polymers are formed, and finally precipitates are produced. This process is thought to be the general process by which hydrated iron(III) oxide, $Fe_2O_3\ x\,H_2O$, or ferric hydroxide, $Fe(OH)_3$, is precipitated from solutions containing iron(III).

It is interesting to calculate the concentrations of the soluble iron species at pH 7, a reasonable pH value for a natural aquatic system. The values of the equilibrium constants are:

$$\frac{[FeOH^{2+}][H^+]}{[Fe^{3+}]} = 8.9 \times 10^{-4} \qquad (2.8.4)$$

$$\frac{[Fe(OH)_2^+][H^+]^2}{[Fe^{3+}]} = 4.9 \times 10^{-7} \qquad (2.8.5)$$

$$\frac{[Fe_2(OH)_2^{4+}][H^+]^2}{[Fe^{3+}]^2} = 1.23 \times 10^{-3} \qquad (2.8.6)$$

Assuming that solid $Fe(OH)_3$ is present, the following relationships also hold:

$$Fe(OH)_3(s) + 3\,H^+ \rightleftharpoons Fe^{3+} + 3\,H_2O \qquad (2.8.7)$$

$$\frac{[Fe^{3+}]}{[H^+]^3} = 9.1 \times 10^3 \qquad (2.8.8)$$

At pH 7, $[Fe^{3+}] = 9.1 \times 10^3(1.0 \times 10^{-7})^3 = 9.1 \times 10^{-18}M$. Using this value for the hydrated ferric ion concentration and substituting into the given equations gives the following concentrations: $[FeOH^{2+}] = 8.1 \times 10^{-14}M$, $[Fe(OH)_2^+] = 4.5 \times 10^{-10}M$, and $[Fe_2(OH)_2^{4+}] = 1.02 \times 10^{-23}M$. Although this treatment represents a simplified picture, it is apparent that the concentration of hydrated ferric ion is negligible in the nearly neutral solutions encountered in most natural waters. Soluble iron at high levels in many ground waters is in the form of iron(II). When exposed to the atmosphere, the iron(II) is oxidized slowly to iron(III) which precipitates from solution, often forming unsightly red-orange deposits.

2.9 SOLUBILITY OF GASES IN WATER

Gases dissolved in water are crucial to the welfare of living species in water. Fish require dissolved oxygen and give off carbon dioxide, while for algae performing photosynthesis the situation is reversed. Dissolved nitrogen can cause severe problems when it forms bubbles in fish blood that cause "bends" and even death in fish. This was responsible for one of the more severe fish kills of recent years, taking a toll of about 400,000 fish on a 40-mile stretch of the Osage River in Missouri in June 1978. Hydrostatic pressure built up in water spilling from the newly constructed Truman dam into a 15-meter-deep basin caused the water to become supersaturated with nitrogen, which killed fish downstream. This is another example of a situation in which the application of rudimentary environmental science in the planning stages of a project would have averted a serious problem.

The equilibrium between molecules of gas in the atmosphere and the corresponding molecules of the same gas in solution,

$$G\,(g) \rightleftharpoons G\,(aq) \qquad (2.9.1)$$

is given in terms of **Henry's Law**: *the solubility of a gas in a liquid is proportional to the partial pressure of that gas in contact with the liquid.* Note that Henry's Law does not account for further chemical reactions of the gas in solution, such as

$$CO_2 + H_2O \rightleftharpoons H^+ + HCO_3^- \qquad (2.9.2)$$

$$SO_2 + H_2O \rightleftharpoons H^+ + HSO_3^- \qquad (2.9.3)$$

Because of this fact, the actual amount of gas taken up by water may be vastly higher than that indicated from Henry's Law.

The solubility of a gas G in water may be expressed by an equilibrium expression such as

$$[G(aq)] = K \times P_G \qquad (2.9.4)$$

TABLE 2.2 Henry's Law Constants for Some Gases in Water at 25 °C

Gas	K, *moles* $\times$ L^{-1} $\times$ *atm*$^{-1}$
O_2	1.28×10^{-3}
CO_2	3.38×10^{-2}
H_2	7.90×10^{-4}
N_2	6.48×10^{-4}
CH_4	1.34×10^{-3}
NO	2.0×10^{-3}

where K is a constant for a specific gas at a given temperature and P_G is the partial pressure of the gas. Some values of K for several gases of interest in water are given in Table 2.2.

Water vapor itself exerts a partial pressure which requires a correction (slight at lower temperatures) in calculating gas solubilities. The partial pressure of water at several temperatures is given in Table 2.3.

The solubility of oxygen in water saturated with air at 1.00 atmosphere pressure may be calculated as an example of a simple gas solubility calculation. From Table 2.3, we see that the partial pressure of water vapor at 25 °C is 0.0313 atm. Since dry air is 20.95% oxygen, the partial pressure of oxygen is given by

$$P_{O_2} = (1.0000\ \text{atm} - 0.0313\ \text{atm}) \times 0.2095 = 0.2029\ \text{atm} \qquad (2.9.5)$$

The molar concentration of oxygen in water is given by Henry's Law:

$$[O_2(aq)] = K \times P_{O_2} = 1.28 \times 10^{-3}\ \text{mole} \times L^{-1} \times \text{atm}^{-1} \times 0.2029\ \text{atm}$$
$$= 2.60 \times 10^{-4} M \qquad (2.9.6)$$

Since the molecular weight of oxygen is 32, the solubility is 8.32 mg/L, or 8.32 parts per million (ppm).

The solubility of gases is affected by temperature, decreasing with increasing temperature. This effect is shown by the **Clausius–Clapeyron equation,**

TABLE 2.3 Partial Pressure of Water at Different Temperatures

T, °C	P_{H_2O}, torr	P_{H_2O}, atm
0	4.579	0.00603
5	6.543	0.00861
10	9.209	0.01212
15	12.788	0.01683
20	17.535	0.02307
25	23.756	0.03126
30	31.824	0.04187
35	42.175	0.05549
40	55.324	0.07279
45	71.88	0.09458
50	92.51	0.12172
100	760.00	1.00000

$$\log \frac{C_2}{C_1} = \frac{\Delta H}{2.303R} \left\{ \frac{1}{T_1} - \frac{1}{T_2} \right\} \quad (2.9.7)$$

where C_1 and C_2 denote the gas concentrations in water at absolute temperatures of T_1 and T_2, respectively; ΔH is the heat of solution in cal/mole; and R is the gas constant (1.987 cal $\times$ deg^{-1} $\times$ $mole^{-1}$).

2.10 OXYGEN IN WATER

Without an appreciable level of dissolved oxygen, many kinds of aquatic organisms cannot exist in water. Dissolved oxygen is consumed by the degradation of organic matter in water. Many fish kills result not from the direct toxicity of pollutants but from oxygen deficiency as a result of its consumption in the biodegradation of pollutants.

Oxygen comprises 20.95% of dry air, and most elemental oxygen comes from the atmosphere. Therefore, the ability of a body of water to reoxygenate itself by contact with the atmosphere is an important characteristic. Oxygen is produced by the photosynthetic action of algae, but this process is really not an efficient means of oxygenating water because some of the oxygen formed by photosynthesis during the daylight hours is lost at night when the algae consume oxygen as part of their metabolic processes. When the algae die, the degradation of their biomass also consumes oxygen.

The solubility of oxygen in water depends upon water temperature (Eq. 2.9.7); the partial pressure of oxygen in the atmosphere; and the salt content of the water. The calculation of oxygen solubility as a function of partial pressure is discussed in Section 2.9, where it is shown that the concentration of oxygen in water at 25 °C in equilibrium with air at atmospheric pressure is only 8.32 mg/L. Thus, water in equilibrium with air cannot contain a high level of dissolved oxygen compared to many other solute species. If oxygen-consuming processes are occurring in the water, the dissolved oxygen level may rapidly approach zero unless some efficient mechanism for the reaeration of water is operative, such as turbulent flow in a shallow stream or air pumped into the aeration tank of an activated sludge secondary waste treatment facility (see Chapter 8). The problem becomes largely one of kinetics, in which there is a limit to the rate at which oxygen is transferred across the air–water interface. This rate depends upon turbulence, air bubble size, temperature, and other factors. It is important to distinguish between solubility, which is the maximum dissolved oxygen concentration at equilibrium, and dissolved oxygen concentration, which is generally not the equilibrium concentration and is limited by the rate at which oxygen dissolves.

If organic matter of biological origin is represented by the formula $\{CH_2O\}$, the consumption of oxygen in water by the degradation of organic matter may be expressed by the following reaction:

$$\{CH_2O\} + O_2 \rightarrow CO_2 + H_2O \quad (2.10.1)$$

The weight of organic material required to consume the 8.3 mg of O_2 in a liter of water in equilibrium with the atmosphere at 25 °C is given by using these figures in a simple stoichiometric calculation based on Equation 2.10.1, which yields a value of 7.8 mg of $\{CH_2O\}$. Thus, the microorganism-mediated degradation of only 7 or 8 mg of organic material can completely consume the oxygen in one liter of water initially saturated with air at 25 °C. The depletion of oxygen to levels below those that will sustain aerobic organisms requires the degradation of even less organic matter at higher temperatures—where the solubility of oxygen is less—or in water not initially saturated with atmospheric oxygen. Furthermore, there is no *chemical sink* (see Section 1.4) for oxygen, unlike the case of CO_2, which can be obtained by algae from dissolved bicarbonate ion by the reaction

$$HCO_3^- \rightleftharpoons CO_2 + OH^- \qquad (2.10.2)$$

There are no chemical reactions that replenish dissolved oxygen; except for oxygen provided by photosynthesis, it must come from the atmosphere.

The temperature effect on the solubility of gases in water (Eq. 2.9.7) is especially important in the case of oxygen. The solubility of oxygen in water decreases from 14.74 mg/L at 0 °C to 7.03 mg/L at 35 °C. With increased water temperature, the decreased solubility of oxygen, combined with the increased respiration rate of aquatic organisms, frequently causes a condition in which an increased demand for oxygen is accompanied by decreased solubility of the gas in water.

It has generally been believed that dissolved oxygen (DO) levels in ground water are always low because of consumption of oxygen by terrestrial microorganisms and by chemical reactions with minerals. However, ground waters have been found in Nevada, Arizona, and the hot springs of Appalachia and Arkansas containing 8 to 10 mg/L of dissolved oxygen [2]. The waters were recovered from depths of 100 to 1000 meters and ranged up to 10,000 years in age. Some of these waters were as far as 80 kilometers from their **recharge points**, points where surface water entered the aquifer. The source of dissolved oxygen in deep ground waters remains largely a mystery. In some cases, highly oxidized minerals, such as iron-titanium minerals, may serve to oxidize ground water.

2.11 CARBON DIOXIDE AND CARBONATE SPECIES

Because of carbon dioxide's acidic character, it is much more complicated to calculate the solubility of CO_2 in water than to calculate the solubility of a nonreactive gas like O_2 or N_2. Consideration of the chemical interactions of carbon dioxide in water is required.

Carbon dioxide, bicarbonate ion, and carbonate ion have an extremely important influence upon the chemistry of water. Many minerals are deposited as salts of the carbonate ion, CO_3^{2-}. Algae in water utilize dissolved CO_2 in the synthesis of biomass. The equilibrium of dissolved CO_2 with the atmosphere,

$$CO_2(\text{water}) \rightleftharpoons CO_2(\text{atmosphere}) \qquad (2.11.1)$$

and the equilibrium of CO_3^{2-} ion between aquatic solution and solid carbonate minerals,

$$MCO_3 \text{ (slightly soluble carbonate salt)} \rightleftharpoons M^{2+} + CO_3^{2-}(\text{water}) \qquad (2.11.2)$$

have a strong buffering effect upon the pH of water.

---

[2] Winograd, I.J., and Robertson, F.N. 1982. Deep oxygenated ground water: anomaly or common occurrence? *Science* 216: 1227–30.

Carbon dioxide is only a very small component of normal dry air. For the calculations in this chapter, we will assume the atmospheric CO_2 content to be 0.0314% by volume. However, as explained in Chapter 11, the atmospheric CO_2 level is increasing and now exceeds 0.0314%. As a consequence of the low level of atmospheric CO_2, water totally lacking in alkalinity in equilibrium with the atmosphere contains only a very low level of carbon dioxide. However, the formation of HCO_3^- and CO_3^{2-} greatly increases the solubility of carbon dioxide. High concentrations of free carbon dioxide in water may adversely affect respiration and gas exchange of aquatic animals. It may even cause death and should not exceed levels of 25 mg/L in water.

A large share of the carbon dioxide found in water is a product of the breakdown of organic matter by bacteria. Even algae, which utilize CO_2 in photosynthesis, produce CO_2 through their metabolic processes in the absence of light. As water seeps through layers of decaying organic matter while infiltrating the ground, it may dissolve a great deal of CO_2 produced by the respiration of organisms in the soil. Later, as water goes through limestone formations, it dissolves calcium carbonate because of the presence of the dissolved CO_2:

$$CaCO_3(s) + CO_2(aq) + H_2O \rightleftharpoons Ca^{2+} + 2\,HCO_3^- \qquad (2.11.3)$$

This process is the one by which limestone caves are formed.

Although CO_2 in water is often represented as H_2CO_3, the equilibrium constant for the reaction

$$CO_2(aq) + H_2O \rightleftharpoons H_2CO_3 \qquad (2.11.4)$$

is only around 2×10^{-3} at 25 °C, so just a small fraction of the dissolved carbon dioxide is actually present as the species H_2CO_3. In this text, nonionized carbon dioxide in water will be designated simply as CO_2, which in subsequent discussions will stand for the total of dissolved molecular CO_2 and undissociated H_2CO_3.

The CO_2-HCO_3^--CO_3^{2-} system in water may be described by the following reactions and equilibrium constants:

$$CO_2 + H_2O \rightleftharpoons H^+ + HCO_3^- \qquad (2.11.5)$$

$$K_1 = \frac{[H^+][HCO_3^-]}{[CO_2]} = 4.45 \times 10^{-7}; \qquad pK_1 = 6.35 \qquad (2.11.6)$$

$$HCO_3^- \rightleftharpoons H^+ + CO_3^{2-} \qquad (2.11.7)$$

$$K_2 = \frac{[H^+][CO_3^{2-}]}{[HCO_3^-]} = 4.69 \times 10^{-11}; \qquad pK_2 = 10.33 \qquad (2.11.8)$$

Given the values of K_1 and K_2, a distribution of species diagram for the CO_2-HCO_3^--CO_3^{2-} system may be prepared with pH as the master variable. Such a diagram shows the major species present in solution as a function of pH. For CO_2 in aqueous solution, the diagram is a series of plots of the fractions present as CO_2, HCO_3^-, and CO_3^{2-} as a function of pH. These fractions, designated as α_x, are given by the following expressions:

$$\alpha_{CO_2} = \frac{[CO_2]}{[CO_2] + [HCO_3^-] + [CO_3^{2-}]} \qquad (2.11.9)$$

$$\alpha_{HCO_3^-} = \frac{[HCO_3^-]}{[CO_2] + [HCO_3^-] + [CO_3^{2-}]} \tag{2.11.10}$$

$$\alpha_{CO_3^{2-}} = \frac{[CO_3^{2-}]}{[CO_2] + [HCO_3^-] + [CO_3^{2-}]} \tag{2.11.11}$$

Substitution of the expressions for K_1 and K_2 into the expressions for α_x gives the fractions of species as a function of acid dissociation constants and hydrogen ion concentration:

$$\alpha_{CO_2} = \frac{[H^+]^2}{[H^+]^2 + K_1[H^+] + K_1K_2} \tag{2.11.12}$$

$$\alpha_{HCO_3^-} = \frac{K_1[H^+]}{[H^+]^2 + K_1[H^+] + K_1K_2} \tag{2.11.13}$$

$$\alpha_{CO_3^{2-}} = \frac{K_1K_2}{[H^+]^2 + K_1[H^+] + K_1K_2} \tag{2.11.14}$$

Examination of the above expressions for fractions of species reveals some basic characteristics of the distribution of species diagram. There are several "landmark points" which make the construction of the diagram relatively easy; they are found at pH = pK_1, pH = ½(pK_1 + pK_2), and pH = pK_2. The significance of these points is shown in Table 2.4.

The distribution of species diagram is shown in Figure 2.4. Examination of the figure shows that bicarbonate ion is the predominant species in the pH range found in most waters, with CO_2 predominating in more acidic waters.

The question of carbon dioxide solubility may now be considered. Assume that pure air is allowed to come to equilibrium with pure water at 25 °C. What is the concentration of CO_2, HCO_3^-, and H^+ in the water? The value of $[CO_2]$ in the water is readily calculated from Henry's Law, given that dry air is 0.0314% CO_2 by volume, the vapor pressure of water is 0.0313 atm at 25 °C, and Henry's Law constant for CO_2 is 3.38×10^{-2} mole × L^{-1} × atm^{-1} at 25 °C.

$$P_{CO_2} = (1.0000 \text{ atm} - 0.0313 \text{ atm}) \times 3.14 \times 10^{-4} \tag{2.11.15}$$
$$= 3.04 \times 10^{-4} \text{ atm}$$

TABLE 2.4 Significant Points on the Distribution of Species Diagram for the CO_2–HCO_3^-–CO_3^{2-} System

pH	α_{CO_2}	$\alpha_{HCO_3^-}$	$\alpha_{CO_3^{2-}}$
$\ll pK_1$	1.00	essentially 0	essentially 0
pK_1 *	0.50	0.50	essentially 0
½(pK_1 + pK_2)†	0.01	0.98	0.01
pK_2 **	essentially 0	0.50	0.50
$\gg pK_2$	essentially 0	essentially 0	1.00

*pH at which $\alpha_{CO_2} = \alpha_{HCO_3^-}$

†pH at which $\alpha_{HCO_3^-}$ has its maximum value

**pH at which $\alpha_{HCO_3^-} = \alpha_{CO_3^{2-}}$

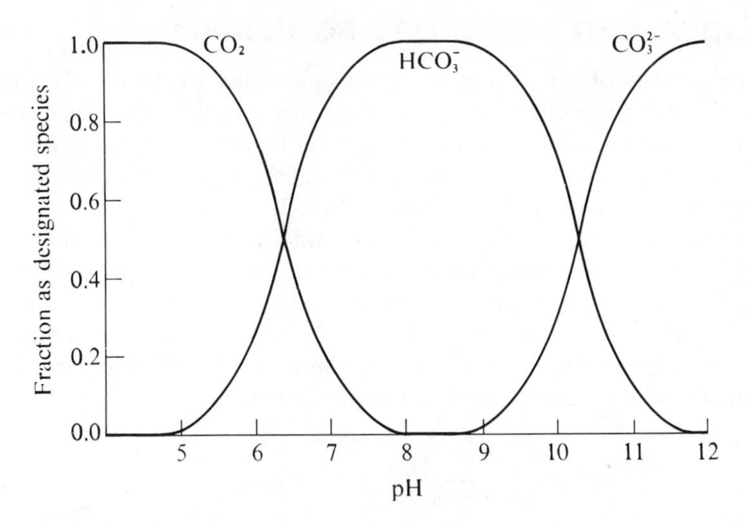

FIGURE 2.4 Distribution of species for the CO_2–HCO_3^-–CO_3^{2-} system in water.

$$[CO_2] = 3.38 \times 10^{-2}\ \text{mole} \times \text{L}^{-1} \times \text{atm}^{-1} \times 3.04 \times 10^{-4}\ \text{atm} \tag{2.11.16}$$

$$= 1.028 \times 10^{-5} M$$

The carbon dioxide dissociates partially in water to produce equal concentrations of H^+ and HCO_3^-:

$$CO_2(aq) + H_2O \rightleftharpoons H^+ + HCO_3^- \tag{2.11.17}$$

The concentrations of H^+ and HCO_3^- are calculated from K_1, which is the acid dissociation constant of CO_2:

$$[H^+] = [HCO_3^-] \tag{2.11.18}$$

$$\frac{[H^+]^2}{[CO_2]} = K_1 = 4.45 \times 10^{-7} \tag{2.11.19}$$

$$[H^+] = (1.028 \times 10^{-5} \times 4.45 \times 10^{-7})^{1/2} \tag{2.11.20}$$

$$[H^+] = [HCO_3^-] = 2.14 \times 10^{-6}$$

$$\text{pH} = 5.67$$

The total amount of CO_2 from air dissolving in 1.00 liter of pure water is the sum of $[CO_2] + [HCO_3^-]$:

$$\begin{pmatrix}\text{total atmospheric } CO_2\\ \text{dissolving in 1.00 liter}\\ \text{of water}\end{pmatrix} = [CO_2] + [HCO_3^-] = 1.242 \times 10^{-5}\ \frac{\text{mole}}{\text{liter}} \tag{2.11.21}$$

The solubility of carbon dioxide is greatly increased by the presence of a base, as will be seen after the following discussion of alkalinity.

2.12 ALKALINITY AND CARBONATE EQUILIBRIA

The capacity of water to accept protons is called **alkalinity**. Alkalinity is important in water treatment and in the chemistry and biology of natural waters. Frequently, the alkalinity of water must be known to calculate the quantities of chemicals to be added in treating the water. Highly alkaline water often has a high pH and generally contains elevated levels of dissolved solids. These characteristics may be detrimental for water to be used in boilers, food processing, and municipal water systems. Alkalinity serves as a pH buffer and reservoir for inorganic carbon, thus helping to determine the ability of a water to support algal growth and other aquatic life. It is used by biologists as a measure of water fertility. Generally, the basic species responsible for alkalinity in water are bicarbonate ion, carbonate ion, and hydroxide ion:

$$HCO_3^- + H^+ \rightarrow CO_2 + H_2O \tag{2.12.1}$$

$$CO_3^{2-} + H^+ \rightarrow HCO_3^- \tag{2.12.2}$$

$$OH^- + H^+ \rightarrow H_2O \tag{2.12.3}$$

Other, usually minor, contributors to alkalinity are ammonia and the conjugate bases of phosphoric, silicic, boric, and organic acids.

Alkalinity generally is expressed as *phenolphthalein alkalinity,* corresponding to titration with acid to the pH at which HCO_3^- is the predominant carbonate species (pH 8.3), or *total alkalinity,* corresponding to titration with acid to the methyl orange endpoint (pH 4.3), where both bicarbonate and carbonate species have been converted to CO_2.

It is important to distinguish between high *basicity*, manifested by a high pH, and high *alkalinity*, a high proton-accepting capability. Whereas pH is an *intensity* factor, alkalinity is a *capacity* factor. This may be illustrated by comparing a solution of $1.00 \times 10^{-3} M$ NaOH with a solution of $0.100 M$ $NaHCO_3$. The sodium hydroxide solution is quite basic, with a pH of 11, but a liter of it will neutralize only 1.00×10^{-3} mole of acid. The pH of the sodium bicarbonate solution is 8.34 ($pH = \frac{1}{2}pK_1 + \frac{1}{2}pK_2$), much lower than that of the NaOH. However, a liter of the sodium bicarbonate solution will neutralize 0.100 mole of acid; therefore, its alkalinity is 100 times that of the more basic NaOH solution.

As an example of a water-treatment process in which water alkalinity is important, consider the use of *filter alum,* $Al_2(SO_4)_3 \cdot 18\,H_2O$, as a coagulant. The hydrated aluminum ion is acidic, and when added to water it reacts with base to form gelatinous aluminum hydroxide.

$$Al(H_2O)_6^{3+} + 3\,OH^- \rightarrow Al(OH)_3(s) + 6\,H_2O \tag{2.12.4}$$

(As this gelatinous material settles, it carries suspended matter with it.) This reaction removes alkalinity from the water. Sometimes the addition of more alkalinity is required to prevent the water from becoming too acidic.

In engineering terms, alkalinity frequently is expressed in units of mg/L of $CaCO_3$. Since the acid-neutralizing reaction of $CaCO_3$ is

$$CaCO_3 + 2\,H^+ \rightarrow Ca^{2+} + CO_2(g) + H_2O \tag{2.12.5}$$

the equivalent weight of calcium carbonate is one-half its formula weight, and it is easy to convert alkalinity to units of equivalents per liter. Expressing alkalinity in terms of mg/L of $CaCO_3$ can, however, lead to confusion, and equivalents/L is preferable notation for the chemist.

The concentrations of CO_2, HCO_3^-, CO_3^{2-}, and OH^- in water may be calculated with a knowledge of the pH, alkalinity, and the appropriate equilibrium constants. Such calculations assume that the contributions to alkalinity from phosphate anions and other miscellaneous species are negligible. A good example is the calculation of species concentrations in a pH 7.00 water having an alkalinity of 1.00×10^{-3}eq/L. Figure 2.4 shows that at pH 7.00 the concentration of CO_3^{2-} is negligible compared to that of HCO_3^-. Furthermore, the concentration of OH^- is only $1.00 \times 10^{-7}M$. Therefore, all of the alkalinity is due to HCO_3^-:

$$[HCO_3^-] = [alk] = 1.00 \times 10^{-3} \tag{2.12.6}$$

where [alk] is the alkalinity. This enables us to calculate $[CO_2]$ from the expression for the acid dissociation constant of this species:

$$[CO_2] = \frac{[H^+][HCO_3^-]}{K_1} = \frac{1.00 \times 10^{-7} \times 1.00 \times 10^{-3}}{4.45 \times 10^{-7}} \tag{2.12.7}$$

$$= 2.25 \times 10^{-4}$$

The value of $[CO_3^{2-}]$ is calculated by substituting into the expression for K_2:

$$[CO_3^{2-}] = \frac{K_2[HCO_3^-]}{[H^+]} = \frac{4.69 \times 10^{-11} \times 1.00 \times 10^{-3}}{1.00 \times 10^{-7}} \tag{2.12.8}$$

$$= 4.69 \times 10^{-7}$$

These concentrations may be compared to those in water having a pH of 10.00 and an alkalinity of 1.00×10^{-3}eq/L. In this case, both CO_3^{2-} and OH^- ions contribute appreciably to the alkalinity. The total alkalinity is given by the equation

$$[alk] = [HCO_3^-] + 2\,[CO_3^{2-}] + [OH^-] \tag{2.12.9}$$

where it should be noted that CO_3^{2-} contributes twice its concentration to the alkalinity of the solution. Substituting the expressions

$$[OH^-] = 1.00 \times 10^{-4} \tag{2.12.10}$$

and

$$[CO_3^{2-}] = \frac{K_2[HCO_3^-]}{[H^+]} \tag{2.12.11}$$

into Equation 2.12.9 gives a value of 4.64×10^{-4} for $[HCO_3^-]$. Solution of Equation 2.12.11 yields $2.18 \times 10^{-4}M$ for $[CO_3^{2-}]$. Thus the total alkalinity is made up of the following contributions:

$$\begin{array}{rl} & 4.64 \times 10^{-4} \text{ eq/L from } HCO_3^- \\ 2 \times 2.18 \times 10^{-4} = & 4.36 \times 10^{-4} \text{ eq/L from } CO_3^{2-} \\ & \underline{1.00 \times 10^{-4} \text{ eq/L from } OH^-} \\ & 1.00 \times 10^{-3} \text{ eq/L total alkalinity} \end{array}$$

These results may be used to show the relationship of the alkalinity of a water to its capacity for production of biomass through algal activity. From the simplified equations for the photochemical synthesis of biomass,

$$CO_2 + H_2O + h\nu \rightarrow \{CH_2O\} + O_2 \qquad (2.12.12)$$

and

$$HCO_3^- + H_2O + h\nu \rightarrow \{CH_2O\} + OH^- + O_2 \qquad (2.12.13)$$

we see that the pH of a solution in which algae are growing without an external source of CO_2 increases as the inorganic carbon is converted to biomass. Such a situation is approximated in cases of rapid algal growth, called *algal blooms,* where the consumption of inorganic carbon is so rapid that equilibrium with atmospheric CO_2 cannot be maintained. In such cases the pH of the water frequently rises to 10 or higher. As an example, consider water with an initial pH of 7.00 and an alkalinity of 1.00×10^{-3} eq/L. How many mg/L of biomass can be produced by the CO_2 consumed when the pH rises from 7.00 to 10.00? Initially, all of the alkalinity is due to bicarbonate ion, so that the concentration of HCO_3^- is $1.00 \times 10^{-3}M$. Substitution into K_1 shows that the concentration of CO_2 at pH 7.00 is $2.25 \times 10^{-4}M$. Therefore, at pH 7.00 the concentration of total inorganic carbon is given by:

$$\begin{pmatrix}\text{total inorganic} \\ \text{carbon conc.}\end{pmatrix} = [CO_2] + [HCO_3^-] = 1.225 \times 10^{-3}M \qquad (2.12.14)$$

At pH 10.00, the concentration of HCO_3^- is $4.64 \times 10^{-4}M$; the concentration of CO_3^{2-} is $2.18 \times 10^{-4}M$; and the concentration of CO_2 is negligible. Therefore, at pH 10,

$$\begin{pmatrix}\text{total inorganic} \\ \text{carbon conc.}\end{pmatrix} = [HCO_3^-] + [CO_3^{2-}] = 0.682 \times 10^{-3}M \qquad (2.12.15)$$

which means that the total inorganic carbon concentration changes from 1.225×10^{-3} moles/L to 0.682×10^{-3} moles/L, a decrease of 0.543×10^{-3} moles/L. This translates to an increase of 0.543×10^{-3} moles/liter of biomass, $\{CH_2O\}$. Since the formula weight of $\{CH_2O\}$ is 30, the weight of biomass produced is 16.3 mg/L. If the initial alkalinity of the water is higher, more biomass can be produced as the pH rises without additional input of CO_2. For this reason, alkalinity is used by biologists as an approximate measure of water fertility.

Since alkalinity in most natural water is due to the presence of HCO_3^- and to a lesser extent CO_3^{2-}, water with high alkalinity has a high concentration of inorganic carbon. This may be illustrated by calculating the solubility of atmospheric CO_2 in water containing initially $1.00 \times 10^{-3}M$ NaOH and having, therefore, an alkalinity of 1.00×10^{-3} equivalents/L. If this water is allowed to come to equilibrium with air, the value of $[CO_2]$ in the water is $1.028 \times 10^{-5}M$, as given by Equation 2.11.16. Reaction of atmospheric CO_2 with hydroxide ion occurs so that the concentration of HCO_3^- is $1.00 \times 10^{-3}M$.

$$CO_2 + OH^- \rightarrow HCO_3^- \qquad (2.12.16)$$

The value of $[H^+]$ is obtained from K_1:

$$[H^+] = K_1 \frac{[CO_2]}{[HCO_3^-]} = 4.45 \times 10^{-7} \frac{1.028 \times 10^{-5}}{1.00 \times 10^{-3}} \qquad (2.12.17)$$

$$= 4.57 \times 10^{-9}$$

$$pH = 8.34$$

At this pH, only small fractions of the total inorganic carbon are present as CO_2 or CO_3^{2-}. The total inorganic carbon concentration is essentially equal to $[HCO_3^-]$, or $1.00 \times 10^{-3}M$. At pH values below 7, H^+ in water detracts significantly from alkalinity, and therefore its concentration must be subtracted in computing the total alkalinity. Therefore, the following equation is the complete equation for alkalinity in a medium where the only contributors to it are HCO_3^-, CO_3^{2-}, and OH^-.

$$[alk] = [HCO_3^-] + 2\,[CO_3^{2-}] + [OH^-] - [H^+] \qquad (2.12.18)$$

2.13 ACIDITY

Acidity as applied to natural water systems is the capacity of the water to neutralize OH^-. Acidic water is not frequently encountered, except in cases of severe pollution. Acidity generally results from the presence of weak acids such as $H_2PO_4^-$, CO_2, H_2S, proteins, fatty acids, and acidic metal ions, particularly Fe^{3+}. Acidity is more difficult to determine than is alkalinity. One reason for the difficulty in determining acidity is that two of the major contributors are CO_2 and H_2S, both volatile solutes which are readily lost from the sample.

$$CO_2 + OH^- \rightarrow HCO_3^- \qquad (2.13.1)$$

$$H_2S + OH^- \rightarrow HS^- + H_2O \qquad (2.13.2)$$

The acquisition and preservation of representative samples of water to be analyzed for these gases is difficult.

The term *free mineral acid* is applied to strong acids such as H_2SO_4 and HCl in water. Acid mine water contains an appreciable concentration of free mineral acid. Whereas total acidity is determined by titration with base to the phenolphthalein endpoint (pH 8.2), free mineral acid is determined by titration with base to the methyl orange endpoint (pH 4.3).

The acidic character of some hydrated metal ions may contribute to acidity, for example:

$$Al(H_2O)_6^{3+} \rightleftharpoons Al(H_2O)_5OH^{2+} + H^+ \qquad (2.13.3)$$

Some industrial wastes contain acidic metal ions and often some excess strong acid. For such wastes the determination of acidity is important in calculating the amount of lime, or other chemicals, that must be added to neutralize the acid.

2.14 CALCIUM IN WATER

Of the cations found in most fresh-water systems, calcium generally has the highest concentration. The chemistry of calcium, although complicated enough, is simpler than that of the transition-metal ions found in water. Calcium is a key element in many geochemical processes, and minerals constitute the primary sources of calcium ion in waters. Among the primary contributing minerals are gypsum, $CaSO_4 \cdot 2\,H_2O$; anhydrite, $CaSO_4$; dolomite, $CaMg(CO_3)_2$; and calcite and aragonite, which are different modifications of $CaCO_3$.

Water containing a high level of carbon dioxide readily dissolves calcium from its carbonate minerals:

$$CaCO_3(s) + CO_2(aq) + H_2O \rightleftharpoons Ca^{2+} + 2\,HCO_3^- \qquad (2.14.1)$$

When Equation 2.14.1 is reversed and CO_2 is lost from the water, calcium carbonate deposits are formed. The concentration of CO_2 in water determines the extent of dissolution of calcium carbonate. The carbon dioxide that water may gain by equilibration with the atmosphere is not sufficient to account for the levels of calcium dissolved in natural waters, especially ground waters. Rather, the respiration of microorganisms degrading organic matter in water, sediments, and soil accounts for the very high levels of CO_2 and HCO_3^- observed in water. This is an extremely important factor in aquatic chemical processes and geochemical transformations.

For fresh water, the typical figures quoted for the concentrations of both HCO_3^- and Ca^{2+} are $1.00 \times 10^{-3}M$. It may be shown that these are reasonable values when the water is in equilibrium with limestone, $CaCO_3$, and with atmospheric CO_2. The concentration of CO_2 in water in equilibrium with air has already been calculated ($[CO_2] = 1.028 \times 10^{-5}M$). The other constants needed to calculate $[HCO_3^-]$ and $[Ca^{2+}]$ are the acid dissociation constant for CO_2:

$$K_1 = \frac{[H^+][HCO_3^-]}{[CO_2]} = 4.45 \times 10^{-7} \qquad (2.14.2)$$

the acid dissociation constant of HCO_3^-:

$$K_2 = \frac{[H^+][CO_3^{2-}]}{[HCO_3^-]} = 4.69 \times 10^{-11} \qquad (2.14.3)$$

and the solubility product of calcium carbonate (calcite):

$$K_{sp} = [Ca^{2+}][CO_3^{2-}] = 4.47 \times 10^{-9} \qquad (2.14.4)$$

The reaction between calcium carbonate and dissolved CO_2 is

$$CaCO_3(s) + CO_2(aq) + H_2O \rightleftharpoons Ca^{2+} + 2\,HCO_3^- \qquad (2.14.5)$$

and has the following equilibrium expression:

$$K' = \frac{[Ca^{2+}][HCO_3^-]^2}{[CO_2]} = \frac{K_{sp}K_1}{K_2} = 4.24 \times 10^{-5} \qquad (2.14.6)$$

The stoichiometry of Equation 2.14.5 gives a bicarbonate ion concentration that is twice that of calcium. Substitution of the value of $[CO_2]$ into the expression for K' yields values of $4.78 \times 10^{-4}M$ for $[Ca^{2+}]$ and $9.56 \times 10^{-4}M$ for $[HCO_3^-]$. Substitution into the expression for K_s yields $9.35 \times 10^{-6}M$ for $[CO_3^{2-}]$. When known concentrations are substituted into the product K_1K_2,

$$K_1K_2 = \frac{[H^+]^2[CO_3^{2-}]}{[CO_2]} = 2.09 \times 10^{-17} \qquad (2.14.7)$$

a value of $4.79 \times 10^{-9}M$ is obtained for $[H^+]$ and the pH is 8.32. The alkalinity is essentially equal to the bicarbonate ion concentration, which is much higher than that of CO_3^{2-} or OH^-.

To summarize, for water in equilibrium with solid calcium carbonate and atmospheric CO_2, the following concentrations are calculated:

$$[CO_2] = 1.028 \times 10^{-5}M \qquad [Ca^{2+}] = 4.78 \times 10^{-4}M$$

$$[HCO_3^-] = 9.56 \times 10^{-4}M \qquad [H^+] = 4.79 \times 10^{-9}M$$

$$[CO_3^{2-}] = 9.35 \times 10^{-6}M \qquad pH = 8.32$$

Factors such as nonequilibrium conditions, high CO_2 concentrations in bottom regions, and increased pH due to algal uptake of CO_2 cause deviations from these values. Nevertheless, they are close to the values found in a large number of natural water bodies.

Calcium ion, along with magnesium and sometimes ferrous ion, accounts for water hardness. The most common manifestation of water hardness is the curdy precipitate formed by soap in hard water. *Temporary hardness* is due to the presence of calcium and bicarbonate ions in water and may be eliminated by boiling the water, thus causing the reversal of Equation 2.14.1:

$$Ca^{2+} + 2\ HCO_3^- \rightleftharpoons CaCO_3(s) + CO_2(g) + H_2O \qquad (2.14.8)$$

Most cooks have observed (and have been disturbed by) the white precipitate of calcium carbonate formed in boiling water having temporary hardness.

The ionic radius of calcium, 0.99 Å, is relatively large compared to some other divalent metal ions found in water. For example, the ionic radius of Mn^{2+} is 0.80 Å and that of Fe^{2+} is 0.75 Å. Thus, the charge density of the Ca^{2+} ion is less than that of these other divalent metal ions. As a consequence, the waters of coordination around Ca^{2+} are less strongly bound, and there is less of a tendency for hydrated Ca^{2+} ions to lose protons. Furthermore, there is relatively little tendency for Ca^{2+} to form complex ions. Under the conditions obtaining in most fresh-water systems, the primary soluble calcium species present is Ca^{2+}. However, at very high levels of HCO_3^-, the ion pair $Ca^{2+}HCO_3^-$ may be present in appreciable amounts. Similarly, in waters of high sulfate content the soluble ion pair $Ca^{2+}SO_4^{2-}$ is present.

2.15 OTHER CHEMICAL SPECIES IN WATER

Much of the preceding discussion has dealt with dissolved species that are particularly abundant or important in water. A number of other chemical species are present naturally, and are significant, in water. Some of these can also be pollutants and are discussed in later chapters. These chemical species and their sources, behavior, and significance in natural waters are summarized in Table 2.5.

TABLE 2.5 Chemical Species Commonly Occurring in Water

Substance	*Sources*	*Behavior and significance in water*
Aluminum	Aluminum-containing minerals	Occurs as $Al(H_2O)_6^{3+}$ below pH 4.0; loses H^+ to yield $Al(OH)(H_2O)_5^{2+}$ from pH 4.5 to 6.5; forms hydroxy-bridged polymers (see Section 2.8); precipitates as gibbsite, $Al_2O_3 \cdot 3\ H_2O$; amphoteric, forming $Al(OH)_4^-$ above pH 10; forms strong complexes with F^-; precipitates with silica and orthophosphate ions.
Chloride, Cl^-	Minerals, pollution	Does not react chemically with many species in water; harmless at relatively low concentrations; major anion associated with excess salinity at higher levels.
Fluoride, F^-	Minerals, water additive	Forms HF, $pK_a = 3.13$, at low pH; forms insoluble salts with Ca(II), Ba(II), Sr(II), Pb(II); commonly substitutes for OH^-; harmful to bones and teeth above approximately 10 mg/L; prevents tooth decay at levels around 1 mg/L, and is commonly added to water for that purpose.

Substance	*Sources*	*Behavior and significance in water*
Iron	Minerals, acid mine water	Occurs as soluble Fe^{2+} under reducing conditions, such as occur in ground water or lake bottom waters; iron(III) must be present as particulate matter or organically bound iron at normal pH's because of the very low solubility of $Fe(OH)_3$ (see Section 2.8); very undesirable solute in water because of formation of $Fe(OH)_3$ deposits; commonly found at levels of 1–10 mg/L in ground waters.
Magnesium	Minerals, such as dolomite, $CaMg(CO_3)_2$	Occurs as Mg^{2+} ion; properties similar to Ca^{2+}, except Mg^{2+} has a much smaller atomic radius of 0.65 Å, holding waters of hydration more strongly; concentrations usually lower than Ca^{2+}, typically 10 mg/L.
Manganese	Minerals	Present as MnO_2 in the presence of oxygen; reduced to soluble Mn^{2+} in ground water and other oxygen-deficient water; low toxicity, but staining tendency of MnO_2 formed by oxidation of Mn^{2+} requires very low levels in municipal water; often precipitates as $MnCO_3$.
Nitrogen	Minerals, decay of nitrogenous organic matter, pollution	Nitrogen species are among the most important species in water. Inorganic nitrogen is present as NO_3^- in the presence, and NH_4^+ in the absence, of oxygen, whereas toxic nitrite, NO_2^-, may be an intermediate form. Nitrate is an algal nutrient that may contribute to excess algal growth. NH_4^+ ion is a weak acid with $pK_a = 9.26$. Unlike NO_3^-, it is strongly bound to soil. Microorganisms catalyze interconversions among various oxidation states of N in water. Organic nitrogen in water is bound to various pollutant organic compounds and biological compounds. (Aspects of nitrogen in water are discussed in Sections 3.12, 4.20, 5.14–5.18, 6.12, 7.11, 7.15, 8.10, 8.19, 10.8, 10.12, and elsewhere in the text.)
Potassium, K^+	Mineral matter, fertilizer runoff, forest fire runoff	May be leached from minerals such as feldspar, $KAlSi_3O_8$; essential plant nutrient; usually occurs at levels of several mg/L, at which it is not a pollutant.
Phosphorus	Minerals, fertilizer runoff, domestic wastes (from detergents)	Occurs in natural waters as anions of orthophosphoric acid, H_3PO_4, $pK_{a1} = 2.17$, $pK_{a2} = 7.31$, $pK_{a3} = 12.36$; the anions $H_2PO_4^-$ and HPO_4^{2-} are predominant in normal water pH ranges; may be present as organic phosphorus; algal nutrient often contributing to excessive algal growth; occurs in natural waters at levels of a few hundredths or tenths of a mg/L.
Silicon	Minerals, such as sodium feldspar albite, $NaAlSi_3O_8$, pollutants	Present in water at normal levels of 1–30 mg/L; occurs as colloidal SiO_2, polynuclear silicate species, such as $Si_4O_6(OH)_6^{2-}$, or silicic acid, H_4SiO_4, $pK_a = 9.46$
Sulfur	Minerals, pollutants, acid mine water, acid rain	Sulfate ion, SO_4^{2-}, predominates under aerobic conditions; hydrogen sulfide, H_2S, is produced in anaerobic waters; H_2S is toxic, but SO_4^{2-} is harmless at moderate levels.
Sodium, Na^+	Minerals, pollution	There are very few reactions by which Na^+ is precipitated or absorbed; no direct harm at lower levels, but higher levels are associated with salt-water pollution, which kills plants; normal levels are several mg/L.

SUPPLEMENTARY REFERENCES

Bowen, R. 1980. *Ground water.* New York: Halsted (John Wiley and Sons, Inc.).

——. 1982. *Surface water.* New York: Wiley-Interscience.

Butler, J.N. 1964. *Ionic equilibrium—a mathematical approach.* Reading, Mass.: Addison-Wesley Publishing Co.

Drever, J.I. 1982. The geochemistry of natural waters. Englewood Cliffs, N.J.: Prentice-Hall, Inc.

Faust, S.D., and Osman, M.A. 1981. *Chemistry of natural waters.* Ann Arbor, Mich.: Ann Arbor Science Publishers, Inc.

Hem, J.D. 1970. *Study and interpretation of the chemical characteristics of natural water.* 2nd ed. U.S. Geological Survey Paper 1473. Washington, D.C.: U.S. Geological Survey.

Höll, K. 1972. *Water: examination, assessment, conditioning, chemistry, bacteriology, biology.* Berlin and New York: Walter de Gruyter, Inc.

Jenne, E.A., ed. 1979. *Chemical modeling in aqueous systems.* ACS Symposium Series No. 93. Washington, D.C.: American Chemical Society.

Lenihan, J., and Fletcher, W.W., eds. 1977. *The marine environment.* New York: Academic Press.

Lerman, A., ed. 1978. *Lakes.* New York: Springer-Verlag.

Moss, B. 1980. *Ecology of fresh waters.* New York: Halsted (John Wiley and Sons, Inc.).

Riley, J.P., and Chester, R., eds. 1978. *Chemical oceanography.* Volumes issued annually. New York: Academic Press.

Ross, D.A. 1977. *Introduction to oceanography.* 2nd ed. Englewood Cliffs, N.J.: Prentice-Hall, Inc.

Rubin, A.J., ed. 1974. *Aqueous environmental chemistry of metals.* Ann Arbor, Mich.: Ann Arbor Science Publishers, Inc.

Smith, F.G.W., ed. 1977. *Handbook of marine science, section I, oceanography, vol. 1, physical chemistry, physics, geology, engineering.* West Palm Beach, Fla.: CRC Press, Inc.

Snoeyink, V.L., and Jenkins, D. 1980. *Water chemistry.* New York: John Wiley and Sons, Inc.

Stumm, W. 1967. Metal ions in aqueous solutions. *Principles and applications of water chemistry,* pp. 520–560. S.D. Faust and J.V. Hunter, eds. New York: John Wiley and Sons, Inc.

Stumm, W., and Morgan, J.J. 1981. *Aquatic chemistry.* 2nd ed. New York: Wiley-Interscience.

Turekian, K.K. 1976. *Oceans.* Englewood Cliffs, N.J.: Prentice-Hall, Inc.

Wetzel, R.G. 1975. *Limnology.* Philadelphia: W.B. Saunders and Co.

QUESTIONS AND PROBLEMS

1. A sample of ground water heavily contaminated with soluble inorganic iron is brought to the surface and the alkalinity is determined without exposing the sample to the atmosphere. A portion of the sample is then allowed to equilibrate with the atmosphere and shows a lower alkalinity. Why?

2. What is the indirect role played by bacteria in the formation of limestone caves?

3. Alkalinity is determined by titration with standard acid. The alkalinity is often expressed as mg/L of $CaCO_3$. If V_p mL of acid of normality N are required to titrate V_s mL of sample to the phenolphthalein endpoint, what is the formula for the phenolphthalein alkalinity as mg/L of $CaCO_3$?

4. Exactly 100 pounds of cane sugar (dextrose), $C_{12}H_{22}O_{11}$, were accidentally discharged into a small stream saturated with oxygen from the air at 25 °C. How many liters of this water could be contaminated to the extent of removing all the dissolved oxygen by biodegradation?

5. Nitrate ion substitutes for oxygen as an electron receptor in the biodegradation of organic matter, undergoing the half-reaction

$$2\,NO_3^- + 12\,H^+ + 10\,e^- \rightarrow N_2\uparrow + 6\,H_2O$$

What must be the levels in mg/L of N in the nitrate form in the water so that the electron-accepting capacity is equivalent to the oxygen in water in equilibrium with the atmosphere at a total pressure of 1.00 atm at 25 °C?

6. Calculate the concentrations of Fe^{3+}, $FeOH^{2+}$, $Fe(OH)_2^+$, and $Fe_2(OH)_2^{4+}$ in a pH-4.00 acidic wastewater in equilibrium with solid $Fe(OH)_3$.

7. Considering formation of $FeOH^{2+}$ and $Fe(OH)_2^+$, but not $Fe_2(OH)_2^{4+}$, calculate $[Fe^{3+}]$, $[FeOH^{2+}]$, $[Fe(OH)_2^+]$, $[H^+]$, and pH in a solution formed by dissolving 1.40×10^{-4} moles of $Fe(NO_3)_3$ in a liter of water having the minimum $[H^+]$ value required to prevent precipitation of solid $Fe(OH)_3$.

8. Exactly 5.60 kg of Fe in the form of ferric nitrate was placed in a waste tank holding initially 100 liters of water. Considering only the formation of insoluble $Fe(OH)_3$ and neglecting all soluble iron(III) species other than Fe^{3+} in solution, what was the final pH and value of $[Fe^{3+}]$?

9. What is the value of $[O_2(aq)]$ for water saturated with a mixture of 50% O_2, 50% N_2 by volume at 25 °C and a total pressure of 1.00 atm?

10. The solubility of oxygen in water is 14.74 mg/L at 0 °C and 7.03 mg/L at 35 °C. Estimate the solubility at 50 °C.

11. Water with an alkalinity of 2.00×10^{-3} equivalents/liter has a pH of 7.00. Calculate $[CO_2]$, $[HCO_3^-]$, $[CO_3^{2-}]$, and $[OH^-]$.

12. Through the photosynthetic activity of algae, the pH of the water in problem 11 was changed to 10.00. Calculate all the preceding concentrations and the weight of biomass, $\{CH_2O\}$, produced. Assume no input of atmospheric CO_2.

13. Calcium chloride is quite soluble, whereas the solubility product of calcium flouride, CaF_2, is only 3.9×10^{-11}. A waste stream of $1.00 \times 10^{-3} M$ HCl is injected into a formation of limestone, $CaCO_3$, where it comes into equilibrium. Give the chemical reaction that occurs and calculate the hardness and alkalinity of the water at equilibrium. Do the same for a waste stream of $1.00 \times 10^{-3} M$ HF.

14. For a solution having 1.00×10^{-3} equivalents/liter total alkalinity (contributions from HCO_3^-, CO_3^{2-}, and OH^-) at $[H^+] = 4.69 \times 10^{-11}$, what is the percentage contribution to alkalinity from CO_3^{2-}?

15. Over the long term, irrigation must be carried out so that there is an appreciable amount of runoff, although a much smaller quantity of water would be sufficient to wet the ground. Why must there be some runoff?

16. The specific species H_2CO_3:
    (a) is the predominant form of CO_2 dissolved in water;
    (b) exists only at pH values above 9;
    (c) makes up only a small fraction of CO_2 dissolved in water, even at a low pH;
    (d) is not known to exist at all;
    (e) is formed by the reaction between CO_2 and OH^-.

17. A wastewater disposal well for carrying various wastes at different times is drilled into a formation of limestone ($CaCO_3$), and the wastewater has time to come to complete equilibrium with the calcium carbonate before leaving the formation through an underground aquifer. Of the following components in the wastewater, the one that would *not* cause an increase in alkalinity due either to the component, itself, or to its reaction with limestone, is:
    (a) NaOH
    (b) CO_2
    (c) HF
    (d) HCl
    (e) all of the preceding would cause an increase in alkalinity.

18. Air is 20.95% oxygen by volume. If air at 1.0000 atm pressure is bubbled through water at 25 °C, what is the partial pressure of O_2 in the water?

19. Alkalinity is *not*:
    (a) a measure of the degree to which water can support algal growth;
    (b) the capacity of water to neutralize acid;
    (c) a measure of the capacity of water to resist a decrease in pH;
    (d) a measure of pH;
    (e) important in considerations of water treatment.

20. Why may oxygen levels become rather low at night in water supporting a heavy growth of oxygen-producing algae?

21. The volume percentage of CO_2 in a mixture of that gas with N_2 was determined by bubbling the mixture at 1.00 atm and 25 °C through a solution of 0.0100*M* $NaHCO_3$ and measuring the pH. If the equilibrium pH was 6.50, what was the volume percentage of CO_2?

CHAPTER 3
REDOX EQUILIBRIA IN NATURAL WATERS

3.1 THE SIGNIFICANCE OF REDOX EQUILIBRIA IN NATURAL WATERS AND WASTEWATERS

Oxidation-reduction reactions are highly significant in the environmental chemistry of natural waters and wastewaters. The reduction of oxygen by organic matter in a lake results in oxygen depletion which can be fatal to fish. The rate at which sewage is oxidized is crucial to the operation of a waste-treatment plant. Reduction of insoluble iron(III) to soluble iron(II) in a reservoir causes problems with iron removal in the water-treatment plant. Oxidation of NH_4^+ to NO_3^- in water is essential for getting the ammonium nitrogen into a form assimilable by algae in the water. It can be readily seen, therefore, that the types, rates, and equilibria of redox reactions largely determine the nature of important solute species in water.

Two important points should be stressed regarding redox reactions in natural waters and wastewaters. First, many of the most important redox reactions are catalyzed by microorganisms. Bacteria are the catalysts by which molecular oxygen reacts with organic matter, iron(III) is reduced to iron (II), and ammonia is oxidized to nitrate ion. The role of microorganisms in catalyzing such reactions is so important that Chapter 5 is devoted to it.

The second important point regarding redox reactions in the hydrosphere involves an analogy that can be drawn between them and acid-base reactions. The activity of the hydrogen ion, H^+, is used to express the degree to which water is acidic or basic. Water with a high hydrogen-ion activity, such as acid mine water, is said to be acidic. Water with a low hydrogen-ion activity, such as an alkaline seep from an alkaline soil, is basic. By analogy, water with a high *electron* activity, such as that in the anaerobic digester of a sewage-treatment plant, is said to be *reducing*. Water with a low electron activity, highly chlorinated water for example, is said to be *oxidizing*. Actually, neither free electrons nor free protons as such are found dissolved in aquatic solution; they are always strongly associated with solvent or solute species. However, the concept of electron activity, like that of hydrogen-ion activity, remains a very useful one to the aquatic chemist.

Examples of some reactions will help illustrate the analogy between redox and acid-base reactions. For example, bicarbonate ion in water reacts with hydrogen ion to produce gaseous carbon dioxide:

$$HCO_3^- + H^+ \rightarrow CO_2(g) + H_2O \qquad (3.1.1)$$

whereas it reacts with electrons through a series of complicated, bacterially catalyzed reactions to yield methane gas:

$$HCO_3^- + 8\,e^- + 9\,H^+ \rightarrow CH_4(g) + 3\,H_2O \qquad (3.1.2)$$

As another example of the analogy between acid-base reactions and redox reactions, consider hydrated ferrous ion acting as an acid through loss of a proton,

$$Fe(H_2O)_6^{2+} \rightleftharpoons Fe(H_2O)_5OH^+ + H^+ \quad (3.1.3)$$

or acting as a reducing agent when it loses an electron,

$$Fe(H_2O)_6^{2+} \rightleftharpoons Fe(H_2O)_6^{3+} + e^- \quad (3.1.4)$$

Generally, the transfer of electrons in a redox reaction is accompanied by proton transfer, providing a close relationship between redox and acid-base equilibria. For example, if iron(II) loses an electron at pH 7, three hydrogen ions are also lost to form highly insoluble ferric hydroxide.

$$Fe(H_2O)_6^{2+} \rightarrow Fe(OH)_3(s) + 3\,H_2O + 3\,H^+ + e^- \quad (3.1.5)$$

This oxidation reaction, producing hydrogen ions, is responsible for much of the acid in acid mine water.

A stratified lake (shown in Figure 2.3) illustrates redox equilibria or changes in an aquatic system. The anaerobic sediment layer is so reducing that carbon can be produced in the −IV oxidation state as CH_4. If the lake becomes anaerobic, the hypolimnion may contain elements in their reduced states: NH_4^+ for nitrogen, H_2S for sulfur, and soluble Fe^{2+} for iron. The surface water may be saturated with atmospheric oxygen and be a relatively oxidizing medium. If allowed to reach thermodynamic equilibrium, it is characterized by the more oxidized forms of the elements present: CO_2 for carbon, NO_3^- for nitrogen, iron as insoluble $Fe(OH)_3$, and sulfur as SO_4^{2-}. Obviously, such changes are vitally important to aquatic organisms and have tremendous influence on water quality. The following sections show how available thermodynamic data are used to predict and explain the equilibrium concentrations of various species in aquatic media.

It should be pointed out that the systems presented in this chapter are assumed to be at equilibrium, a state almost never achieved in any real natural water or wastewater system. Real systems generally exhibit mixed potentials consisting of several different redox systems; they represent steady-state, or dynamic situations. Nevertheless, the picture of a system at equilibrium is very useful in visualizing trends in natural water and wastewater systems, yet the model is still simple enough to comprehend. It is important to realize the limitations of such a picture, however, especially in trying to make measurements of the redox status of water.

3.2 ELECTRON ACTIVITY EXPRESSED AS pE

Just as pH is defined as

$$pH = -\log(a_{H^+}) \quad (3.2.1)$$

where a_{H^+} is the activity of the hydrogen ion in aqueous solution, pE is defined as

$$pE = -\log(a_{e^-}) \quad (3.2.2)$$

where a_{e^-} is the activity of the electron in aqueous solution. Since hydrogen-ion activities may vary over many orders of magnitude, pH is a convenient way of expressing a_{H^+} in terms of manageable numbers. Similarly, a stable aquatic system may have an electron activity ranging over more than 20 orders of magnitude; therefore, it is convenient to express a_{e^-} as pE.

The rigorous thermodynamic definition of pE has been given by Stumm and Morgan [1]. It is based upon the following reaction:

$$2\,H^+(aq) + 2\,e^- \rightleftharpoons H_2(g) \tag{3.2.3}$$

The free-energy change for this reaction is defined as exactly zero when all components of the reaction are at unit activity. (Recall that for ionic solutes, activity approaches concentration at low concentrations and low ionic strengths. The activity of a gas is equal to its partial pressure. Furthermore, the free energy, G, decreases for spontaneous processes occurring at constant temperature and pressure. Processes for which the free-energy change, ΔG, is zero have no tendency toward spontaneous change and are in a state of equilibrium.) This reaction is the one upon which the free energies of formation of all ions in aqueous solution are based. It also forms the basis for defining the free-energy changes for oxidation-reduction processes in water.

Whereas it is relatively easy to visualize the activities of ions in terms of concentration, it is harder to visualize the activity of the electron, and therefore pE, in similar terms. For example, at 25 °C in pure water, a medium of zero ionic strength, the hydrogen-ion concentration is $1.0 \times 10^{-7} M$, the hydrogen-ion *activity* is 1.0×10^{-7}, and the pH is 7.0. The electron activity, however, must be defined in terms of Equation 3.2.3. When $H^+(aq)$ at unit activity is in equilibrium with hydrogen gas at 1 atmosphere pressure (and likewise at unit activity), the activity of the electron in the medium is exactly 1.00 and the pE is 0.0. If the electron activity were increased by a factor of 10 (as would be the case if $H^+(aq)$ at an activity of 0.100 were in equilibrium with H_2 at an activity of 1.00), the electron activity would be 10 and the pE value would be -1.0.

3.3 ELECTRODE POTENTIALS, pE, AND THE NERNST EQUATION

The tendency of metal surfaces to oxidize is responsible for corrosion and is an important factor in heavy-metal contamination of water. For example, copper is generally found at low (harmless) levels in tap water as a consequence of the reverse of a typical redox half-reaction,

$$Cu^{2+} + 2\,e^- \rightleftharpoons Cu \tag{3.3.1}$$

which illustrates an equilibrium between an oxidized species, cupric ion, and its corresponding reduced form, copper metal. When cupric ion in solution is in equilibrium with copper metal, the electron activity of the solution is determined by the relative tendencies of cupric ion to acquire electrons and of copper metal to give them up. The physical measurement of these relative tendencies may be made with a cell consisting of a standard hydrogen electrode connected to a copper half-cell as shown in Figure 3.1. The measured potential of the right-hand electrode in Figure 3.1 versus the standard hydrogen electrode is called the **electrode potential, E**. If the cupric ion and copper metal are both at unit activity, the potential is the **standard electrode potential** (according to IUPAC convention, the

---

[1] Stumm, W., and Morgan, J.J. 1981. *Aquatic chemistry*. 2nd ed. New York: Wiley-Interscience.

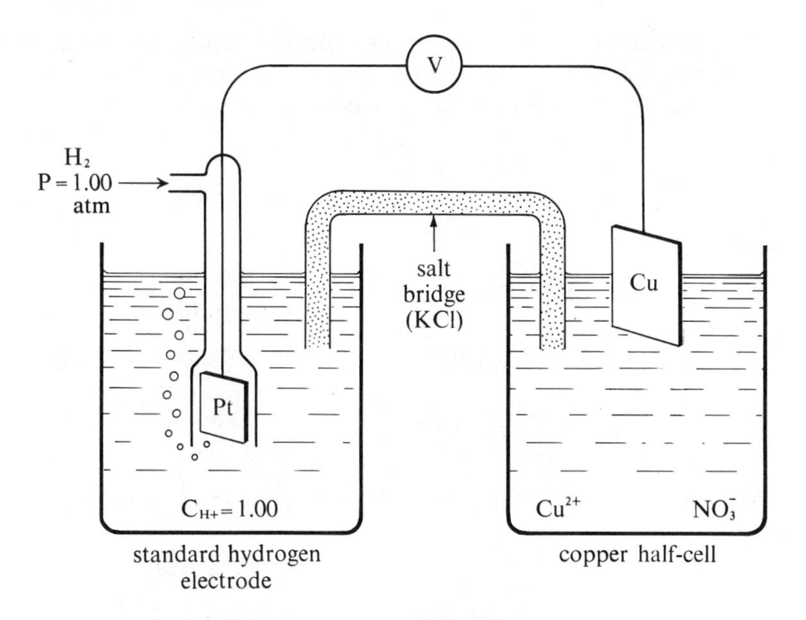

FIGURE 3.1 Cell for the measurement of electrode potentials.

standard **reduction** potential), **E°**. The standard electrode potential for the Cu^{2+}-Cu couple is 0.337 volts. Standard electrode potentials for other systems are presented in Section 3.4 and in Table 3.1 (page 45).

The **Nernst equation** is used to account for the effect of different activities upon electrode potential. Referring to Figure 3.1, if the cupric-ion concentration is increased in the copper half-cell, it is only reasonable to assume that the potential of the copper electrode will become more positive; a higher concentration of electron-deficient cupric ions clustered around the copper electrode tends to draw electrons from the electrode and leaves it more positively charged. Decreased cupric-ion concentration has the opposite effect. The Nernst equation for the half-reaction

$$Cu^{2+} + 2e^- \rightleftharpoons Cu \tag{3.3.2}$$

expresses this effect as:

$$E = E^0 + \frac{2.303RT}{nF} \log [Cu^{2+}] \tag{3.3.3}$$

where E is the potential (in volts) of the copper electrode versus the standard hydrogen electrode; R is the molar gas constant; T is the absolute temperature; n is the number of electrons involved in the reaction; F is the Faraday constant; and the activity of cupric ion has been taken as its concentration (a simplification, valid for more dilute solutions, which will be made throughout this chapter). The activity of copper metal does not appear in the Nernst equation because activities of solid metals are exactly unity by definition. At 25 °C, substitution of numerical values into Equation 3.3.3 yields the following equation:

$$E = 0.337 + \frac{0.0591}{2} \log [Cu^{2+}] \tag{3.3.4}$$

The concept of pE has already been introduced as the negative log of the electron activity. It is now convenient to give a working definition of pE and show how it relates to electrode potential. For a redox reaction involving n electrons, the generalized form of the Nernst equation is

$$E = E^0 + \frac{2.303RT}{nF} \log \frac{[\text{reactants}]}{[\text{products}]} \quad (3.3.5)$$

which at 25 °C is

$$E = E^0 + \frac{0.0591}{n} \log \frac{[\text{reactants}]}{[\text{products}]}$$

It may be shown from thermodynamic arguments that pE is given by

$$pE = \frac{E}{\frac{2.303RT}{F}} = \frac{E}{0.0591} \quad (\text{at } 25\,^\circ\text{C}) \quad (3.3.6)$$

and it follows that pE^0 may be defined in terms of the standard electrode potential:

$$pE^0 = \frac{E^0}{\frac{2.303RT}{F}} = \frac{E^0}{0.0591} \quad (\text{at } 25\,^\circ\text{C}) \quad (3.3.7)$$

Therefore, the Nernst equation in terms of pE takes the following form:

$$pE = pE^0 + \frac{1}{n} \log \frac{[\text{reactants}]}{[\text{products}]} \quad (3.3.8)$$

The Nernst equation in this form is quite simple and offers some advantages in calculating redox equilibria, which will be seen in the remainder of this chapter. At 25 °C, pE is E/0.0591 and pE^0 is $E^0/0.0591$. It is imperative to keep in mind that pE is the negative log of electron activity and *not* the negative log of the electrode potential, E.

3.4 RELATIVE REACTION TENDENCY: WHOLE REACTIONS FROM HALF-REACTIONS

Some typical reduction half-reactions with their standard electrode potentials and pE^0 values are:

$$Hg^{2+} + 2e^- \rightleftharpoons Hg \quad E^0 = +0.789\text{ V} \quad pE^0 = 13.35 \quad (3.4.1)$$

$$Cu^{2+} + 2e^- \rightleftharpoons Cu \quad E^0 = +0.337\text{ V} \quad pE^0 = 5.71 \quad (3.4.2)$$

$$2H^+ + 2e^- \rightleftharpoons H_2 \quad E^0 = 0.000\text{ V} \quad pE^0 = 0.00 \quad (3.4.3)$$

$$Pb^{2+} + 2e^- \rightleftharpoons Pb \quad E^0 = -0.126\text{ V} \quad pE^0 = -2.13 \quad (3.4.4)$$

The more positive the value of a standard potential, or of pE^0, the greater the tendency of the reaction to proceed as written. For example, if a piece of lead foil is dipped into a solution containing cupric ion, the lead acquires a layer of copper metal, through the reaction

$$Cu^{2+} + Pb \rightarrow Cu + Pb^{2+} \quad (3.4.5)$$

which occurs because the cupric ion has a greater tendency to acquire electrons than the lead ion has to retain them. Similarly, copper metal will not cause hydrogen gas to be evolved from solutions of strong acid, because hydrogen ion has less attraction for electrons than does cupric ion. Lead metal, in contrast, will displace hydrogen gas from acidic solutions.

If a half-reaction is written as an oxidation, the sign of E^0 is opposite from that of the measured standard electrode potential. Thus, it is correct to write the reverse of Equation 3.4.2 as:

$$Cu \rightleftharpoons Cu^{2+} + 2\,e^- \qquad E^0 = -0.337\text{ V} \tag{3.4.6}$$

The sign of pE^0 is, of course, also reversed if the reaction is written as an oxidation:

$$Cu \rightleftharpoons Cu^{2+} + 2\,e^- \qquad pE^0 = -5.71 \tag{3.4.7}$$

Regardless of how the reaction is written, or which sign is given to E^0, the copper electrode is still electrostatically positive with respect to the hydrogen electrode, as shown in the electrochemical cell in Figure 3.1.

Half-reactions may be combined to form whole reactions. In principle, half-reactions may be allowed to occur in separate electrochemical half-cells, as shown in Figure 3.2, and are therefore called **cell reactions**. For example, the overall reaction for the reduction of cupric ion by lead metal may be obtained from two half-reactions by subtracting the lead half-reaction, Equation 3.4.4, from the copper half-reaction, Equation 3.4.2:

$$\begin{array}{lll} Cu^{2+} + 2\,e^- \rightleftharpoons Cu & E^0 = +0.337\text{ V} & pE^0 = 5.71 \\ -(Pb^{2+} + 2\,e^- \rightleftharpoons Pb & E^0 = -0.126\text{ V} & pE^0 = -2.13) \\ \hline Cu^{2+} + Pb \rightleftharpoons Cu + Pb^{2+} & E^0 = +0.463\text{ V} & pE^0 = 7.84 \end{array} \tag{3.4.8}$$

The positive values of E^0 and pE^0 for the overall reaction indicate that it tends to go to the right as written. This occurs either as the sum of two separate half-

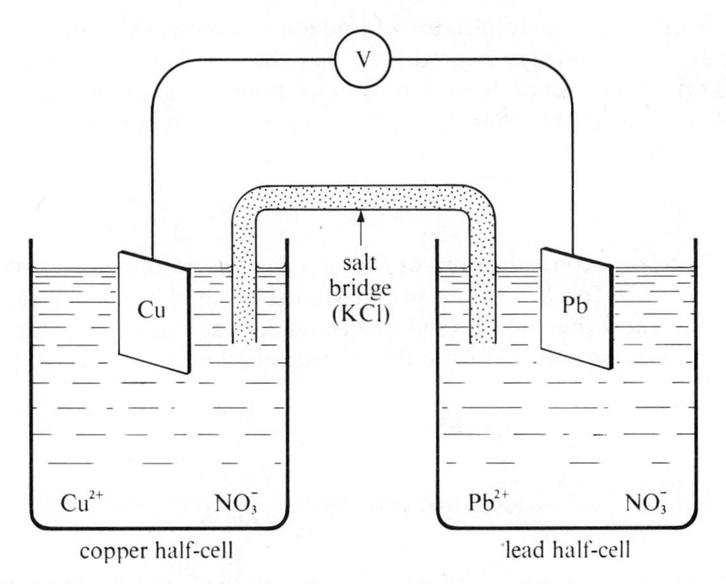

FIGURE 3.2 Cell for the measurement of lead half-cell potential versus copper half-cell potential. The "V" represents a high-resistance voltmeter which does not allow current to flow between the electrodes.

reactions in the two compartments of an electrochemical cell (Figure 3.2), or when lead metal is placed directly in a solution of cupric ion. Therefore, if a waste solution containing cupric ion, a relatively innocuous pollutant, comes into contact with lead in plumbing, toxic lead may go into solution.

What if the activities of Cu^{2+} and Pb^{2+} are not unity? This is accounted for by the Nernst equation. Referring to Figure 3.2, the potential of the copper electrode versus the lead electrode is the E value of Reaction (3.4.9). The effect of concentration upon this E value is given by the Nernst equation,

$$E = 0.463 + \frac{0.0591}{2} \log \frac{[Cu^{2+}]}{[Pb^{2+}]} \tag{3.4.9}$$

and the pE value is given by the equation

$$pE = 7.84 + \frac{1}{2} \log \frac{[Cu^{2+}]}{[Pb^{2+}]} \tag{3.4.10}$$

where 7.84 is the pE^0 value of the overall reaction.

3.5 THE NERNST EQUATION AND CHEMICAL EQUILIBRIUM

Refer again to Figure 3.2. Imagine that instead of the cell being set up to measure the potential between the copper and lead electrodes, the voltmeter, V, were removed and the electrodes directly connected with a wire so that the current might flow between them. The reaction

$$Cu^{2+} + Pb \rightleftharpoons Cu + Pb^{2+} \tag{3.5.1}$$

will occur until the concentration of lead ion becomes so high, and the concentration of copper ion becomes so low, that the reaction stops. The system is at equilibrium and, since current no longer flows, E must be exactly zero. The equilibrium constant *K* for the reaction is given by the equation

$$K = \frac{[Pb^{2+}]}{[Cu^{2+}]} \tag{3.5.2}$$

and the relative concentrations of lead ion and cupric ion must be such that they give the value for *K* when substituted into the equilibrium-constant equation. With the knowledge that E (and pE) equal zero at equilibrium, the equilibrium constant may be obtained from the Nernst equation:

$$E = E^0 - \frac{0.0591}{2} \log \frac{[Pb^{2+}]}{[Cu^{2+}]} \tag{3.5.3}$$

$$0.00 = 0.463 - \frac{0.0591}{2} \log K \tag{3.5.4}$$

Note that products are placed over reactants in the log term, and a minus sign is placed in front to put the equilibrium constant in the correct form (a purely mathematical operation). In terms of pE and pE^0, the corresponding two equations are:

$$pE = pE^0 - \frac{1}{2}\log\frac{[Pb^{2+}]}{[Cu^{2+}]} \tag{3.5.5}$$

$$0.00 = 7.84 - \frac{1}{2}\log K \tag{3.5.6}$$

Solution of either Equation 3.5.4 or 3.5.6 for log K yields a value of 15.7.

In general, therefore, the equilibrium constant for a redox reaction involving n electrons is given by the formula

$$\log K = \frac{nFE^0}{2.303RT} = \frac{nE^0}{0.0591} \quad \text{(at 25 °C)} \tag{3.5.7}$$

where E^0 is the E^0 value for the overall reaction. When the Nernst equation for the overall reaction is expressed in terms of pE and pE^0, the equilibrium constant is given by the equation

$$\log K = n(pE^0) \tag{3.5.8}$$

3.6 THE RELATIONSHIP OF E AND pE TO FREE ENERGY

Aquatic systems and the organisms that inhabit them—just like the steam engine or students hoping to pass physical chemistry—must obey the laws of thermodynamics. Bacteria, fungi, and, indeed, human beings derive their energy from acting as mediators (catalysts) of chemical reactions and extracting a certain percentage of useful energy from them. In predicting or explaining the behavior of an aquatic system, it is helpful to be able to predict the useful energy that can be extracted from chemical reactions in the system, such as microbially mediated oxidation of organic matter to CO_2 and water, or the fermentation of organic matter to methane by anaerobic bacteria in the absence of oxygen. Such information may be obtained by knowing the free-energy change, ΔG, for the redox reaction; ΔG, in turn, may be obtained from E or pE. The free-energy change for a redox reaction involving n electrons is given by either of the two following equations:

$$\Delta G = -nFE \tag{3.6.1}$$

$$\Delta G = -2.303nRT(pE) \tag{3.6.2}$$

The free-energy change in cal/mole is obtained if F is substituted as 23,060 cal(volt-equiv)$^{-1}$. When all reaction components are in their standard states (pure liquids, pure solids, solutes at an activity of 1.00), the following equations apply:

$$\Delta G^0 = -nFE^0 \tag{3.6.3}$$

$$\Delta G^0 = -2.303nRT(pE^0) \tag{3.6.4}$$

3.7 EXPRESSING REACTIONS IN TERMS OF ONE ELECTRON-MOLE

Two typical and important redox reactions that occur in aquatic systems are nitrification

$$NH_4^+ + 2\,O_2 \rightleftharpoons NO_3^- + 2\,H^+ + H_2O \tag{3.7.1}$$

and oxidation of iron(II) to iron(III):

$$4\ Fe^{2+} + O_2 + 10\ H_2O \rightleftharpoons 4\ Fe(OH)_3(s) + 8\ H^+ \tag{3.7.2}$$

What do we really mean when we write reactions such as these? If we are going to make any thermodynamic calculations involving the reactions, we take Reaction 3.7.1 to mean that one mole of ammonium ion reacts with two moles of molecular oxygen to produce one mole of nitrate ion, two moles of hydrogen ion, and one mole of water. Reaction 3.7.2 is taken to mean that four moles of ferrous ion react with one mole of oxygen and ten moles of water to produce four moles of hydrated iron(III) oxide and eight moles of hydrogen ions. The free-energy changes for these reactions may be calculated, but how may a valid comparison of such changes be made?

The most meaningful approach for comparing redox reactions is to write each reaction in terms of one electron-mole. This enables us to compare free-energy changes on a common basis, the transfer of one mole of electrons. The advantage of this approach is illustrated by considering Reaction 3.7.1, which involves an eight-electron change, and Reaction 3.7.2, which involves a four-electron change. Rewriting Equation 3.7.1 for one electron-mole yields

$$\frac{1}{8}NH_4^+ + \frac{1}{4}O_2 \rightleftharpoons \frac{1}{8}NO_3^- + \frac{1}{4}H^+ + \frac{1}{8}H_2O \tag{3.7.3}$$

whereas Reaction 3.7.2, when rewritten for one electron-mole rather than four, yields:

$$Fe^{2+} + \frac{1}{4}O_2 + \frac{5}{2}H_2O \rightleftharpoons Fe(OH)_3(s) + 2\ H^+ \tag{3.7.4}$$

It has already been shown (Eq. 3.5.8) that for a redox reaction involving n electrons, pE° is related to the equilibrium constant by the following equation:

$$pE^\circ = \frac{1}{n}\log K \tag{3.7.5}$$

If the reaction is expressed in terms of a one-electron transfer, the equilibrium-constant equation is simplified even further:

$$pE^\circ = \log K \tag{3.7.6}$$

Reaction 3.7.3, the nitrification reaction written in terms of one electron-mole, has a pE° value of +5.85. The equilibrium-constant expression for this reaction is

$$K = \frac{[NO_3^-]^{1/8}[H^+]^{1/4}}{[NH_4^+]^{1/8}P_{O_2}^{\ 1/4}} \tag{3.7.7}$$

and log K is simply

$$\log K = pE^\circ = 5.85 \tag{3.7.8}$$

3.8 THE DETERMINATION OF E° AND pE° VALUES

Table 3.1 is a compilation of pE° values for redox reactions that are especially important in aquatic systems. How are these values determined? In practice, they may rarely be obtained from a direct potentiometric measurement in an elec-

TABLE 3.1 pE° Values of Redox Reactions Important in Natural Waters (at 25 °C)

Reaction	pE°	pE°(W)[1]
(1) $\frac{1}{4}O_2(g) + H^+(W) + e \rightleftharpoons \frac{1}{2}H_2O$	+20.75	+13.75
(2) $\frac{1}{5}NO_3^- + \frac{6}{5}H^+(W) + e \rightleftharpoons \frac{1}{10}N_2(g) + \frac{3}{5}H_2O$	+21.05	+12.65
(3) $\frac{1}{2}MnO_2(s) + \frac{1}{2}HCO_3^-(10^{-3}) + \frac{3}{2}H^+(W) + e \rightleftharpoons \frac{1}{2}MnCO_3(s) + \frac{3}{8}H_2O$	—	+8.5[2]
(4) $\frac{1}{2}NO_3^- + H^+(W) + e \rightleftharpoons \frac{1}{2}NO_2^- + \frac{1}{2}H_2O$	+14.15	+7.15
(5) $\frac{1}{8}NO_3^- + \frac{5}{4}H^+(W) + e \rightleftharpoons \frac{1}{8}NH_4^+ + \frac{3}{8}H_2O$	+14.90	+6.15
(6) $\frac{1}{6}NO_2^- + \frac{4}{3}H^+(W) + e \rightleftharpoons \frac{1}{6}NH_4^+ + \frac{1}{3}H_2O$	+15.14	+5.82
(7) $\frac{1}{2}CH_3OH + H^+(W) + e \rightleftharpoons \frac{1}{2}CH_4(g) + \frac{1}{2}H_2O$	+9.88	+2.88
(8) $\frac{1}{4}CH_2O + H^+(W) + e \rightleftharpoons \frac{1}{4}CH_4(g) + \frac{1}{4}H_2O$	+6.94	−0.06
(9) $FeOOH(s) + HCO_3^-(10^{-3}) + 2\,H^+(W) + e \rightleftharpoons FeCO_3(s) + 2\,H_2O$	—	−1.67[2]
(10) $\frac{1}{2}CH_2O + H^+(W) + e \rightleftharpoons \frac{1}{2}CH_3OH$	+3.99	−3.01
(11) $\frac{1}{6}SO_4^{2-} + \frac{4}{3}H^+(W) + e \rightleftharpoons \frac{1}{6}S(s) + \frac{2}{3}H_2O$	+6.03	−3.30
(12) $\frac{1}{8}SO_4^{2-} + \frac{5}{4}H^+(W) + e \rightleftharpoons \frac{1}{8}H_2S(g) + \frac{1}{2}H_2O$	+5.75	−3.50
(13) $\frac{1}{8}SO_4^{2-} + \frac{9}{8}H^+ + e \rightleftharpoons \frac{1}{8}HS^- + \frac{1}{2}H_2O$	+4.13	−3.75
(14) $\frac{1}{2}S(s) + H^+(W) + e \rightleftharpoons \frac{1}{2}H_2S(g)$	+2.89	−4.11
(15) $\frac{1}{8}CO_2 + H^+ + e \rightleftharpoons \frac{1}{8}CH_4 + \frac{1}{4}H_2O$	+2.87	−4.13
(16) $\frac{1}{6}N_2(g) + \frac{4}{3}H^+(W) + e \rightleftharpoons \frac{1}{3}NH_4^+$	+4.68	−4.65
(17) $H^+(W) + e \rightleftharpoons \frac{1}{2}H_2(g)$	0.0	−7.00
(18) $\frac{1}{4}CO_2(g) + H^+(W) + e \rightleftharpoons \frac{1}{4}CH_2O + \frac{1}{4}H_2O$	−1.20	−8.20

[1] (W) indicates $a_{H^+} = 1.00 \times 10^{-7} M$ and pE°(W) is a pE° at $a_{H^+} = 1.00 \times 10^{-7} M$.

[2] These data correspond to $a_{HCO_3^-} = 1.00 \times 10^{-3} M$ rather than unity and so are not exactly pE°(W); they represent typical aquatic conditions more nearly than pE° values do.

Source: Werner Stumm and James J. Morgan, *Aquatic Chemistry* (New York: Wiley-Interscience, 1970), p. 318. Reproduced with permission of John Wiley and Sons, Inc.

trochemical cell, as shown in Figure 3.1. Most electrode systems that might be devised do not give potential responses corresponding to the Nernst equation; we say that they do not behave *reversibly*. It is true that we may place a platinum electrode and a reference electrode in water and measure a potential. This potential, referred to the standard hydrogen electrode, is the so-called **E_H value.** Furthermore, the measured potential will be more positive (more oxidizing) in the aerobic surface layers of a lake than in the anaerobic bottom regions. However, attaching any quantitative significance to the E_H value measured directly with an electrode is a very dubious practice. Acid mine waters containing relatively high levels of sulfuric acid and dissolved iron give reasonably accurate E_H values by direct measurement, but most waters do not yield meaningful values of E_H.

Most published E^0 values are calculated from free-energy data (Eq. 3.6.3) or from equilibrium constants (Eq. 3.5.7). As an example, we may calculate E^0 for the reaction

$$Ag^+ + e^- \rightleftharpoons Ag \tag{3.8.1}$$

The standard free energy of formation of aqueous silver ion is 18.43×10^3 cal/mole. Therefore, the standard free energy of Reaction 3.8.1 is -18.43×10^3 cal/mole. Equation 3.6.3 may be written in the form

$$E^0 = -\frac{\Delta G^0}{nF} \tag{3.8.2}$$

where F may be expressed in cal(volt-equiv)$^{-1}$ and has the value 23,060 cal(volt-equiv)$^{-1}$. Substituting into Equation 3.8.2 yields E^0 for the reduction of silver ion:

$$E^0 = -\frac{-18.43 \times 10^3 \text{ cal/mole}}{1\,\dfrac{\text{equiv}}{\text{mole}} \times 23.060 \times 10^3 \text{ cal(volt-equiv)}^{-1}} = 0.799 \text{ volt}$$

3.9 THE LIMITS OF pE IN WATER

Water may be oxidized:

$$2\,H_2O \rightleftharpoons O_2 + 4\,H^+ + 4\,e^- \tag{3.9.1}$$

or it may be reduced:

$$2\,H_2O + 2\,e^- \rightleftharpoons H_2 + 2\,OH^- \tag{3.9.2}$$

These two reactions determine the limits of pE in water. On the oxidizing side (relatively more positive pE values), the pE value is limited by the oxidation of water, Reaction 3.9.1. The evolution of hydrogen, Reaction 3.9.2, limits the pE value on the reducing side. Since these reactions involve hydrogen ion and hydroxide ion, the reactions are pH-dependent.

The boundary condition commonly chosen for the oxidizing limit of water is an oxygen pressure of 1.00 atmosphere, and the boundary condition for the reducing limit of water is a hydrogen pressure of 1.00 atmosphere. These boundary conditions enable us to derive equations relating the stability boundaries of water to pH. Writing the reverse of Reaction 3.9.1 for one electron and setting $P_{O_2} = 1.00$ yield:

$$\frac{1}{4}O_2 + H^+ + e^- \rightleftharpoons \frac{1}{2}H_2O \qquad pE^0 = +20.75 \quad \text{(from Table 3.1)} \tag{3.9.3}$$

$$pE = pE^0 + \log (P_{O_2}^{1/4} [H^+]) \tag{3.9.4}$$

$$pE = 20.75 - pH \tag{3.9.5}$$

Thus, Equation 3.9.5 defines the oxidizing limit of water. At a specified pH, pE values more positive than the one given by Equation 3.9.5 cannot exist at equilibrium in water in contact with the atmosphere.

The pE-pH relationship for the reducing limit of water is given by the following derivation:

$$H^+ + e^- \rightleftharpoons \frac{1}{2}H_2 \qquad pE^0 = 0.00 \tag{3.9.6}$$

$$pE = pE^0 + \log [H^+] \tag{3.9.7}$$

$$pE = -pH \tag{3.9.8}$$

For neutral water (pH = 7.00), substitution into Equations 3.9.8 and 3.9.5 yields −7.00 to 13.75 for the pE range of water. The pE-pH boundaries of stability for water are shown by the dashed lines in Figure 3.3 (Section 3.11).

The decomposition of water is very slow in the absence of a suitable catalyst. Therefore, water may have temporary nonequilibrium pE values more negative than the reducing limit or more positive than the oxidizing limit. An example of the latter is a solution of chlorine in water.

3.10 pE VALUES IN NATURAL AQUATIC SYSTEMS

As discussed previously, it is not generally possible to obtain accurate E_H values by direct potentiometric measurements in natural aquatic systems. If this were possible, it would be very easy to obtain pE in water. However, in principle, pE values may be calculated from the species present in water at equilibrium.

An obviously significant pE value is that of neutral water in thermodynamic equilibrium with the atmosphere. In water under these conditions, P_{O_2} = 0.21 atm and $[H^+] = 1.00 \times 10^{-7}M$. Substitution into Equation 3.9.4 yields:

$$pE = 20.75 + \log (0.21^{1/4} \times 1.00 \times 10^{-7}) \tag{3.10.1}$$

$$pE = 13.58$$

According to this calculation, a pE value of around +13 is to be expected for water in equilibrium with the atmosphere, that is, an aerobic water. What sort of pE value might be encountered in an anaerobic water, however? As an example, consider anaerobic water in which methane and CO_2 are being produced by microorganisms. Assume $P_{CO_2} = P_{CH_4}$ and that pH = 7.00. The relevant half-reaction is

$$\frac{1}{8}CO_2 + H^+ + e^- \rightleftharpoons \frac{1}{8}CH_4 + \frac{1}{4}H_2O \qquad pE^0 = +2.87 \tag{3.10.2}$$

for which the Nernst equation is

$$pE = pE^{\circ} + \log \frac{P_{CO_2}^{1/8}[H^+]}{P_{CH_4}^{1/8}} = 2.87 + \log [H^+] \tag{3.10.3}$$

This equation yields a pE value of −4.13. Note that this pE value does not exceed the reducing limit of water at pH 7.00, which from Equation 3.9.8 is −7.00. It is of interest to calculate the pressure of oxygen in neutral water at a pE value of −4.13. Substitution into Equation 3.9.4 yields

$$-4.13 = 20.75 + \log (P_{O_2}^{1/4} \times 1.00 \times 10^{-7}) \tag{3.10.4}$$

from which the pressure of oxygen is calculated to be 3.0×10^{-72} atm. This incredibly low figure for the pressure of oxygen means that equilibrium with respect to oxygen partial pressure is not achieved under these conditions. Certainly, under any condition approaching equilibrium between comparable levels of CO_2 and CH_4, the partial pressure of oxygen must be very low indeed.

3.11 pE-pH DIAGRAMS

The examples cited so far have shown the close relationship between pE and pH in water. This relationship may be expressed graphically in the form of a **pE-pH diagram**. Such diagrams show the regions of stability and the boundary lines for various species in water. Because of the numerous species that may be formed, such diagrams may become extremely complicated. For example, if a metal is being considered, several different oxidation states of the metal, hydroxy complexes, and different forms of the solid metal oxide or hydroxide may exist in different regions described by the pE-pH diagram. Most waters contain carbonate, and many contain sulfates and sulfides, so that various metal carbonates, sulfates, and sulfides may predominate in different regions of the diagram. In order to illustrate the principles involved, however, we will discuss a simplified pE-pH diagram. The reader is referred to more advanced works on geochemistry and aquatic chemistry for more complicated (and more realistic) pE-pH diagrams [1, 2].

A pE-pH diagram for iron may be constructed assuming a maximum concentration of iron in solution, in this case $1.0 \times 10^{-5}M$. The following equilibria will be considered:

$$Fe^{3+} + e^- \rightleftharpoons Fe^{2+} \qquad pE^{\circ} = +13.2 \tag{3.11.1}$$

$$Fe(OH)_2(s) + 2\,H^+ \rightleftharpoons Fe^{2+} + 2\,H_2O \tag{3.11.2}$$

$$K_{sp} = \frac{[Fe^{2+}]}{[H^+]^2} = 8.0 \times 10^{12} \tag{3.11.3}$$

$$Fe(OH)_3(s) + 3\,H^+ \rightleftharpoons Fe^{3+} + 3\,H_2O \tag{3.11.4}$$

$$K'_{sp} = \frac{[Fe^{3+}]}{[H^+]^3} = 9.1 \times 10^3 \tag{3.11.5}$$

---

[2] Garrels, R.M., and Christ, C.M. 1965. *Solutions, Minerals and Equilibria*. New York: Harper and Row.

(The constants K_{sp} and K_{sp}' are derived from the solubility products of $Fe(OH)_2$ and $Fe(OH)_3$, respectively, and are expressed in terms of $[H^+]$ to facilitate the calculations.) Note that the formation of species such as $Fe(OH)^{2+}$, $Fe(OH)_2^+$, and solid $FeCO_3$, all of which might be of significance in a natural water system, is not considered.

In constructing the pE-pH diagram, several boundaries must be considered. The first two of these are the oxidizing and reducing limits of water (see Section 3.9). At the high pE end, the stability limit of water is defined by Equation 3.9.5:

$$pE = 20.75 - pH \tag{3.9.5}$$

The low pE limit is defined by Equation 3.9.8:

$$pE = -pH \tag{3.9.8}$$

The pE-pH diagram constructed for the iron system must fall between the boundaries defined by Equations 3.9.5 and 3.9.8.

In the high pE, low pH region, Fe^{3+} may exist in equilibrium with Fe^{2+}. The boundary line between these two species is given by the following calculation:

$$pE = 13.2 + \log \frac{[Fe^{3+}]}{[Fe^{2+}]} \tag{3.11.6}$$

$$[Fe^{3+}] = [Fe^{2+}] \quad \text{(by definition of the boundary condition)} \tag{3.11.7}$$

$$pE = 13.2 \quad \text{(independent of pH)} \tag{3.11.8}$$

At pE values exceeding 13.2, as the pH increases from very low values, $Fe(OH)_3$ precipitates from a solution of Fe^{3+}. The pH at which precipitation occurs depends of course upon the concentration of Fe^{3+}. In this example, we have chosen, somewhat arbitrarily, a maximum soluble iron concentration of $1.00 \times 10^{-5} M$. Therefore, at the boundary, $[Fe^{3+}] = 1.00 \times 10^{-5}$. Substitution in Equation 3.11.5 yields:

$$[H^+]^3 = \frac{[Fe^{3+}]}{K_{sp}'} = \frac{1.00 \times 10^{-5}}{9.1 \times 10^3} \tag{3.11.9}$$

$$pH = 2.99 \tag{3.11.10}$$

In a similar manner, the boundary between Fe^{2+} and solid $Fe(OH)_2$ may be defined, assuming $[Fe^{2+}] = 1.00 \times 10^{-5} M$ at the boundary:

$$[H^+]^2 = \frac{[Fe^{2+}]}{K_{sp}} = \frac{1.00 \times 10^{-5}}{8.0 \times 10^{12}} \quad \text{(from Equation 3.11.3)} \tag{3.11.11}$$

$$pH = 8.95 \tag{3.11.12}$$

Throughout a wide pE-pH range, Fe^{2+} is the predominant soluble iron species in equilibrium with the solid hydrated iron(III) oxide, $Fe(OH)_3$. The boundary between these two species depends upon both pE and pH. Substituting Equation 3.11.5 into Equation 3.11.6 yields:

$$pE = 13.2 + \log \frac{K_{sp}'[H^+]^3}{[Fe^{2+}]} \tag{3.11.13}$$

$$pE = 13.2 + \log 9.1 \times 10^3 - \log 1.00 \times 10^{-5} + 3 \times \log [H^+]$$

$$pE = 22.2 - 3\ pH \qquad (3.11.14)$$

The boundary between the solid phases $Fe(OH)_2$ and $Fe(OH)_3$ likewise depends upon both pE and pH, but it does not depend upon an assumed value for total soluble iron. The required relationship is derived from substituting both Equation 3.11.3 and Equation 3.11.5 into Equation 3.11.6:

$$pE = 13.2 + \log \frac{K'_{sp}[H^+]^3}{K_{sp}[H^+]^2} \qquad (3.11.15)$$

$$pE = 13.2 + \log \frac{9.1 \times 10^3}{8.0 \times 10^{12}} + \log [H^+]$$

$$pE = 4.3 - pH \qquad (3.11.16)$$

All of the equations needed to prepare the pE-pH diagram for iron in water have now been derived. To summarize, the equations are (3.9.5), O_2-H_2O boundary; (3.9.8), H_2-H_2O boundary; (3.11.8), Fe^{3+}-Fe^{2+} boundary; (3.11.10), Fe^{3+}-$Fe(OH)_3$ boundary; (3.11.12), Fe^{2+}-$Fe(OH)_2$ boundary; (3.11.14), Fe^{2+}-$Fe(OH)_3$ boundary; and (3.11.16), $Fe(OH)_2$-$Fe(OH)_3$ boundary.

The pE-pH diagram for the iron system in water is shown in Figure 3.3. In this system, at a relatively high hydrogen-ion activity and high electron activity (an acidic reducing medium), ferrous ion, Fe^{2+}, is the predominant species. (In most natural water systems, the solubility range of Fe^{2+} is very narrow because of the precipitation of FeS or $FeCO_3$.) Some ground waters contain appreciable levels of ferrous ion under these conditions. At a very high hydrogen-ion activity and low electron activity (an acidic oxidizing medium), ferric ion, Fe^{3+}, predominates. In an oxidizing medium at lower acidity, solid, hydrated ferric oxide, $Fe(OH)_3$, is the primary iron species present. Finally, in a basic reducing medium, with low hydrogen-ion activity and high electron activity, solid ferrous hydroxide, $Fe(OH)_2$, is stable.

Note that within the pH regions normally encountered in a natural aquatic system, from about 5 to 9, hydrated iron(III) oxide or ferrous ion are the predominant stable iron species. In fact, it is observed that in waters containing dissolved oxygen at any appreciable level, and therefore having a relatively high pE, hydrated iron(III) oxide is essentially the only inorganic iron species found. Such waters may contain a high level of suspended iron, but any truly soluble iron must be in the form of a complex (see Chapter 4).

In highly anaerobic water, which has a low pE, appreciable levels of soluble Fe^{2+} may be encountered. When such water is exposed to atmospheric oxygen, the pE rises and $Fe(OH)_3$ precipitates. The resulting stains from the deposition of hydrated iron(III) oxide have distressed more than one person attempting to achieve a clean wash with water having a high iron content. This phenomenon also explains why red iron oxide stains are found near pumps and springs that bring deep, anaerobic water to the surface. In shallow wells, where the water may become aerobic, solid $Fe(OH)_3$ may precipitate on the well walls, clogging the aquifer outlet. This usually occurs through bacterially-mediated reactions (see Chapter 5).

One species not yet considered is elemental iron. For the half-reaction

$$Fe^{2+} + 2e \rightleftharpoons Fe \qquad pE^0 = -7.45 \qquad (3.11.17)$$

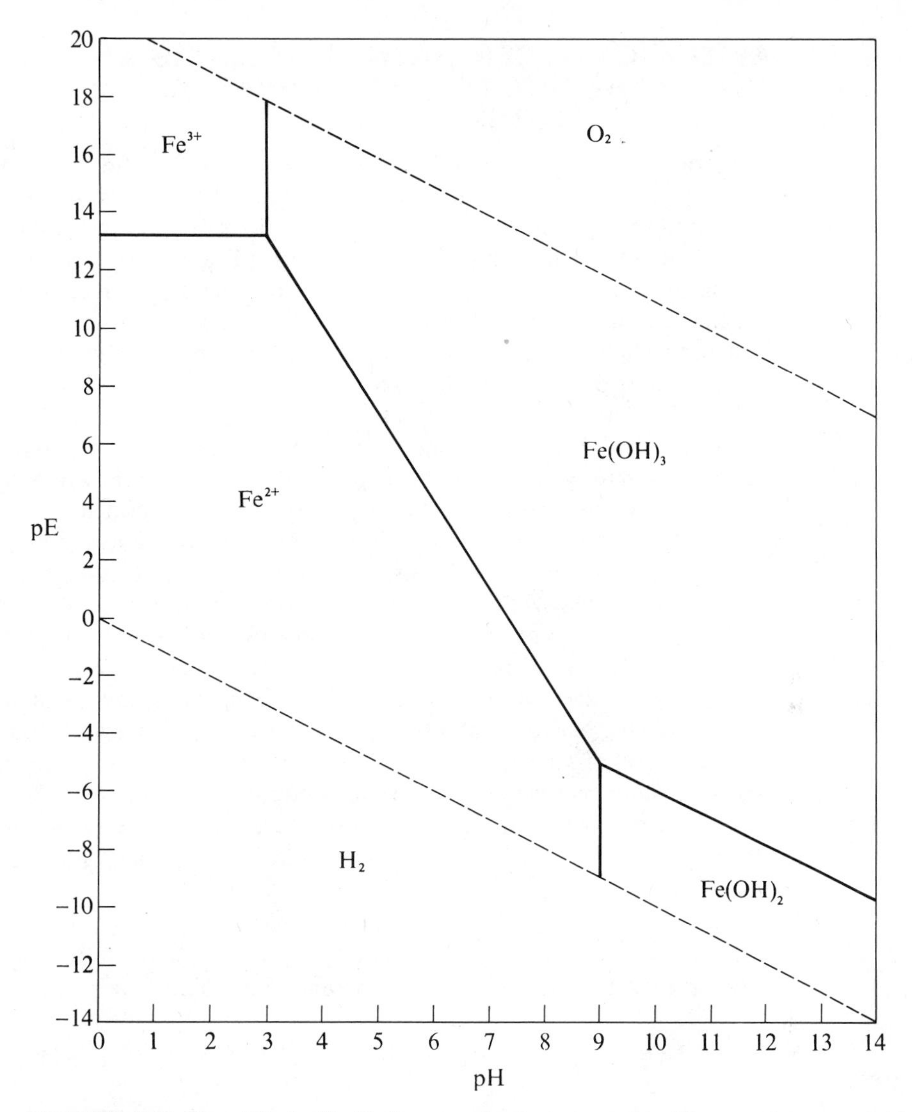

FIGURE 3.3 Simplified pE-pH diagram for iron in water. The maximum soluble iron concentration is $1.00 \times 10^{-5} M$.

the Nernst equation gives pE as a function of $[Fe^{2+}]$:

$$pE = -7.45 + \frac{1}{2} \log [Fe^{2+}] \tag{3.11.18}$$

For iron metal in equilibrium with $1.00 \times 10^{-5} M$ Fe^{2+}, the following pE value is obtained:

$$pE = -7.45 + \frac{1}{2} \log 1.00 \times 10^{-5} = -9.95$$

Examination of Figure 3.3 shows that the pE value for elemental iron in equilibrium with Fe^{2+} is below the reducing limit of water. Iron metal in contact with water, therefore, is thermodynamically unstable, a factor contributing to corrosion.

3.12 LOGARITHMIC CONCENTRATION DIAGRAMS WITH pE AS THE INDEPENDENT VARIABLE: THE NITROGEN SYSTEM IN WATER

Logarithmic concentration diagrams are a very useful way of expressing equilibrium relationships in solution [3]. Such diagrams consist of plots of the logarithms of species concentrations as functions of a master variable such as pH or pE. Redox relationships are shown with pE as an independent variable. In preparing such a diagram, it is necessary to select a formal concentration of an element in solution, then plot the logs of the concentrations of various species involving the element as a function of pE. Since most redox reactions are pH-dependent, a pH value must be specified.

The nitrogen system in water is a useful example for the application of a logarithmic concentration diagram with pE as the master variable, as shown in Figure 3.4. Nitrogen in water exists primarily as ammonium ion, NH_4^+, or nitrate ion, NO_3^-. Under certain conditions, nitrogen in an intermediate oxidation state as nitrite ion may be produced. Nitrite is of concern because of its toxicity. Like so many chemical redox reactions in water, the nitrogen transformation reactions are mediated (catalyzed) by microorganisms. In many cases, microorganisms that have the capability of mediating reactions involving elemental nitrogen, N_2, as either a reactant or product, are not present. Therefore, in preparing the logarithmic concentration diagram involving nitrogen species in water, transitions involving N_2 will be omitted here. A pH value of 7.00 will be chosen. A total nitrogen level of $1.00 \times 10^{-4} M$ will be assumed, which is a reasonable level for a body of water receiving some nitrogen pollution.

At relatively low pE values (pE < 5), NH_4^+ is the predominant species, $[NH_4^+] = 1.00 \times 10^{-4}$, and one equation in the logarithmic concentration diagram is:

$$\log[NH_4^+] = -4.00 \qquad (3.12.1)$$

The half-reaction involving nitrite ion and ammonium ion is

$$\frac{1}{6}NO_2^- + \frac{4}{3}H^+ + e \rightleftharpoons \frac{1}{6}NH_4^+ + \frac{1}{3}H_2O \qquad pE^0 = +15.14 \qquad (3.12.2)$$

which at pH 7.00 yields the following Nernst relationship:

$$pE = 5.82 + \log \frac{[NO_2^-]^{1/6}}{[NH_4^+]^{1/6}} \qquad (3.12.3)$$

Substitution of 1.00×10^{-4} for $[NH_4^+]$ yields the equation relating $\log[NO_2^-]$ to pE in the pE region where ammonium ion is the predominant nitrogen species in solution:

$$\log[NO_2^-] = -38.92 + 6\,pE \qquad (3.12.4)$$

A similar derivation relates $\log[NO_3^-]$ to pE in the region where ammonium ion is the predominant species with a concentration of $1.00 \times 10^{-4} M$:

$$\frac{1}{8}NO_3^- + \frac{5}{4}H^+ + e \rightleftharpoons \frac{1}{8}NH_4^+ + \frac{3}{8}H_2O \qquad pE^0 = +14.90 \qquad (3.12.5)$$

[3] Butler, J.N. 1964. *Ionic equilibrium—a mathematical approach.* Reading, Mass.: Addison-Wesley Publishing Co.

$$pE = 6.15 + \log \frac{[NO_3^-]^{1/8}}{[NH_4^+]^{1/8}} \quad \text{(at pH 7.00)} \qquad (3.12.6)$$

$$\log [NO_3^-] = -53.20 + 8\ pE \qquad (3.12.7)$$

Within a very narrow pE range, around pE 6.50, NO_2^- is the predominant species. Within this pE region, the log of the concentration of nitrite ion is given by the equation

$$\log [NO_2^-] = -4.00 \qquad (3.12.8)$$

and the log of the ammonium-ion concentration is given by substituting 1.00×10^{-4} for $[NO_2^-]$ in Equation 3.12.3:

$$pE = 5.82 + \log \frac{(1.00 \times 10^{-4})^{1/6}}{[NH_4^+]^{1/6}} \qquad (3.12.9)$$

$$\log [NH_4^+] = 30.92 - 6\ pE \qquad (3.12.10)$$

Within the NO_2^- predominance region, the equation for $\log [NO_3^-]$ is given by the following treatment:

$$\frac{1}{2}NO_3^- + H^+ + e \rightleftharpoons \frac{1}{2}NO_2^- + \frac{1}{2}H_2O \qquad pE^0 = +14.15 \qquad (3.12.11)$$

$$pE = 7.15 + \log \frac{[NO_3^-]^{1/2}}{[NO_2^-]^{1/2}} \quad \text{(at pH 7.00)} \qquad (3.12.12)$$

$$\log [NO_3^-] = -18.30 + 2\ pE \quad \text{(when } [NO_2^-] = 1.00 \times 10^{-4}\text{)} \qquad (3.12.13)$$

At pE values appreciably above 7, nitrate ion is the predominant nitrogen species in solution:

$$\log [NO_3^-] = -4.00 \qquad (3.12.14)$$

The value of $\log [NO_2^-]$ may be obtained at pE values greater than 7 by setting $[NO_3^-]$ equal to 1.00×10^{-4} in Equation 3.12.12:

$$pE = 7.15 + \log \frac{(1.00 \times 10^{-4})^{1/2}}{[NO_2^-]^{1/2}} \qquad (3.12.15)$$

$$\log [NO_2^-] = 10.30 - 2\ pE \qquad (3.12.16)$$

A similar substitution into Equation 3.12.6 gives the log of ammonium-ion concentration in the nitrate-predominance region:

$$pE = 6.15 + \log \frac{(1.00 \times 10^{-4})^{1/8}}{[NH_4^+]^{1/8}} \qquad (3.12.17)$$

$$\log [NH_4^+] = 45.20 - 8\ pE \qquad (3.12.18)$$

All of the equations needed to construct the logarithmic concentration diagram for the nitrogen system in water have now been derived. To review, the relevant equations are (3.12.1), (3.12.4), and (3.12.7) in the low pE region where NH_4^+ is the predominant nitrogen species; (3.12.8), (3.12.10), and (3.12.13) in the intermediate pE region where NO_2^- is the predominant species; and (3.12.14), (3.12.16), and (3.12.18) in the high pE region where NO_3^- is the predominant species. The logarithmic concentration diagram is shown in Figure 3.4.

Examination of Figure 3.4 reveals that nitrite ion has only a very limited pE range of stability. In a system at equilibrium with pH 7.00, nitrite ion would

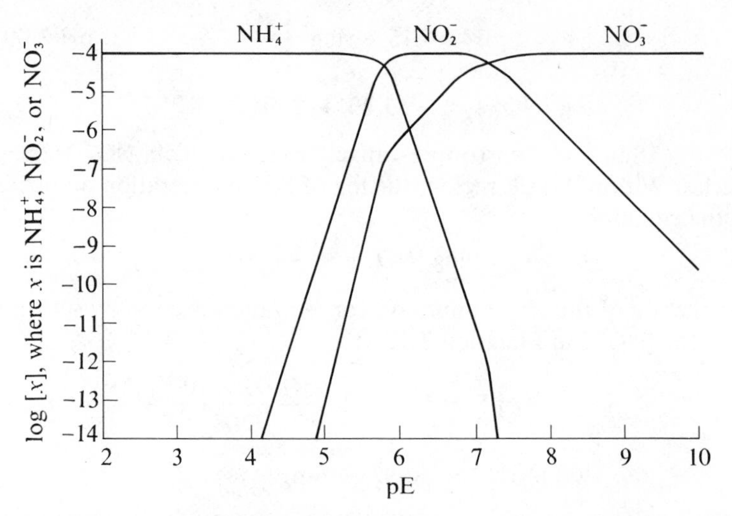

FIGURE 3.4 Logarithmic concentration diagram for the N_4^+–NO_2^-–NO_3^- system in water, with pE as the master variable. The pH is 7.00 and the total nitrogen concentration is $1.00 \times 10^{-4}M$.

occur at appreciable levels only in the pE range of approximately 6–7. The dissolved oxygen level corresponding to the region of nitrite stability may be calculated by substituting into Equation 3.9.4, choosing 6.50 as the approximate pE value for maximum nitrite ion stability at pH 7.00:

$$6.50 = 20.75 + \log (P_{O_2}^{1/4} \times 10^{-7}) \qquad P_{O_2} = 1 \times 10^{-29} \text{ atm} \qquad (3.12.19)$$

Thus, anaerobic conditions generally are required for the production of nitrite ion in water. When nitrite is found in soil, the soil generally is waterlogged. As would be predicted from the narrow range of nitrite-ion stability shown in Figure 3.4, nitrite rarely is found at appreciable levels in water samples.

3.13 CORROSION

One of the most damaging redox phenomena is **corrosion**, defined as the destructive alteration of metal through interactions with its surroundings. In addition to costing billions of dollars each year due to destruction of equipment and structures, corrosion introduces metals into water systems, destroys pollution-control equipment and waste disposal pipes, and is aggravated by water and air pollutants.

Thermodynamically, most metals are in an unstable state relative to their environments. These metals tend toward the more stable forms of ions, salts, oxides, and hydroxides. Fortunately, the rates of corrosion are normally slow, so that metals exposed to air and water may endure for long periods of time. However, protective measures are necessary. These sometimes fail, as is shown, for example, by the gaping holes in automobile bodies, which result from the corrosive action of the salt used to control road ice in many U.S. cities.

Corrosion normally occurs when an electrochemical cell is set up on a metal surface. The area corroded is the anode, where the following oxidation reaction occurs, illustrated for the formation of a divalent ion from a metal, M:

$$M \rightarrow M^{2+} + 2\,e^- \qquad (3.13.1)$$

Several cathodic reactions are possible. One of the most common of these is reduction of H^+ ion (Eq. 3.13.2, next page).

$$2\ H^+ + 2\ e^- \rightarrow H_2(g) \tag{3.13.2}$$

Oxygen may also be involved in cathodic reactions, including reduction to hydroxide, reduction to water, and reduction to hydrogen peroxide:

$$O_2 + 2\ H_2O + 4\ e^- \rightarrow 4\ OH^- \quad \text{(hydroxide)} \tag{3.13.3}$$

$$O_2 + 4\ H^+ + 4\ e^- \rightarrow 2\ H_2O \quad \text{(water)} \tag{3.13.4}$$

$$O_2 + 2\ H_2O + 2\ e^- \rightarrow 2\ OH^- + H_2O_2 \quad \text{(hydrogen peroxide)} \tag{3.13.5}$$

Oxygen may either accelerate corrosion reactions by participating in reactions such as these, or retard them by forming protective oxide films. Bacteria are often involved with corrosion, as discussed in Section 5.27.

Corrosion of pipes and fixtures is especially likely to occur in municipal water distribution systems when the water supply is lacking in hardness (see Section 2.14) and alkalinity (see Section 2.12). The Seattle, Washington, Water Department has such a water supply [4]. This water is surface water from the Cedar and Tolt Rivers in the Cascade Mountains. Because of good watershed protection, the water is so pure that only disinfection with chlorine is required to bring it up to Federal drinking water standards. The corrosiveness of water tends to increase with decreasing pH. In 1970, three changes were made in the treatment of water supplied by the Cedar River which decreased the average pH of the water, so that water entering the treatment system with a pH of around 7.6 left with a pH in the range of 6.8 to 7.2. The first of these changes was an increase in chlorination (see Section 8.7 and Equation 8.7.1) to reduce the occurrence of bacteria in the water distribution system. In addition, at the request of the U.S. Public Health Service, the addition of ammonia to the water system (ammoniation) was stopped in order to maintain a free available chlorine residual (as opposed to combined available chlorine; see Section 8.7) in the water distribution system. Finally, as a result of the 1968 citizens' vote, fluoridation was begun in 1970, using hydrofluosilicic acid, H_2SiF_6. In addition to low pH, other factors that contribute to the corrosivity of Cedar River water are relatively high dissolved oxygen levels and a relatively high (halogen + sulfate): alkalinity ratio. A study of the water supply [4] showed that addition of 1.7 mg CaO per liter of water raised the pH to around 8.3 and greatly decreased the corrosivity of the water.

SUPPLEMENTARY REFERENCES

Brubaker, G.R., and Phipps, P., eds. 1979. *Corrosion chemistry.* Washington, D.C.: American Chemical Society.

Leidheiser, H., Jr., 1979. *Corrosion control by coatings.* Princeton, N.J.: Science Press.

Lingane, J. 1958. *Electroanalytical chemistry.* New York: Wiley-Interscience.

Pourbaix, M., and DeZoubov, N. 1966. *Atlas of electrochemical equilibria in aqueous solutions.* Elmsford, N.Y.: Pergamon Press.

Robinson, J.S. 1979. *Corrosion inhibitors—recent developments.* Park Ridge, N.J.: Noyes Data Corp.

---

[4] Hoyt, B.P.; Herrera, C.E.; and Kirmeyer, G.J. 1982. *Seattle distribution system corrosion control study, Vol. I. Cedar River water pilot plant study.* EPA—600/S2-82-026. Cincinnati, Ohio: U.S. Environmental Protection Agency.

Stumm, W. 1966. *Redox potential as an environmental parameter; conceptual significance and operational limitation.* Third International Conference on Water Pollution Research (Munich, Germany). Washington, D.C.: Water Pollution Control Federation.

West, J.M. 1980. *Basic corrosion and oxidation.* New York: Halsted (John Wiley and Sons, Inc.).

QUESTIONS AND PROBLEMS

1. The acid-base reaction for the dissociation of acetic acid is

$$HOAc + H_2O \rightleftharpoons H_3O^+ + OAc^-$$

with $K_a = 1.75 \times 10^{-5}$. Break this reaction down into two half-reactions involving H^+ ion. Break down the redox reaction

$$Fe^{2+} + H^+ \rightleftharpoons Fe^{3+} + \frac{1}{2}H_2(g)$$

into two half-reactions involving the electron. Discuss the analogies between the acid-base and redox processes.

2. Assuming a bicarbonate ion concentration $[HCO_3^-]$ of $1.00 \times 10^{-3}M$ and a value of 3.5×10^{-11} for the solubility product of $FeCO_3$, what would you expect to be the stable iron species at pH 9.5 and pE -8.0, as shown in Figure 3.3? (You may find it helpful to refer to Section 2.11).

3. Assuming that the partial pressure of oxygen in water is that of atmospheric O_2, 0.21 atm, rather than the 1.00 atm assumed in deriving Equation 3.9.5, derive an equation describing the oxidizing pE limit of water as a function of pH.

4. Plot $\log P_{O_2}$ as a function of pE at pH 7.00.

5. Calculate the pressure of oxygen for a system in equilibrium in which $[NH_4^+] = [NO_3^-]$ at pH 7.00.

6. Calculate the values of $[Fe^{3+}]$, pE, and pH at the point in Figure 3.3 where Fe^{2+} at a concentration of $1.00 \times 10^{-5}M$, $Fe(OH)_2$, and $Fe(OH)_3$ are all in equilibrium.

7. What is the pE value in a solution in equilibrium with air (21% O_2 by volume) at pH 6.00?

8. What is the pE value at the point on the Fe^{2+}-$Fe(OH)_3$ boundary line (see Figure 3.3) in a solution with a soluble iron concentration of $1.00 \times 10^{-4}M$ at pH 6.00?

9. What is the pE value in an acid mine water sample having $[Fe^{3+}] = 7.03 \times 10^{-3}M$ and $[Fe^{2+}] = 3.71 \times 10^{-4}M$?

10. At pH 6.00 and pE 2.58, what is the concentration of Fe^{2+} in equilibrium with $Fe(OH)_3$?

11. What is the calculated value of the partial pressure of O_2 in acid mine water of pH 2.00, in which $[Fe^{3+}] = [Fe^{2+}]$?

12. What is the major advantage of expressing redox reactions and half-reactions in terms of exactly one electron-mole?

13. Why are pE values that are determined by reading the potential of a platinum electrode versus a reference electrode generally not very meaningful?

14. What determines the oxidizing and reducing limits, respectively, for the thermodynamic stability of water?

15. How would you expect pE to vary with depth in a stratified lake?

16. Upon what half-reaction is the rigorous definition of pE based?

CHAPTER 4

COMPLEXATION IN NATURAL WATERS AND WASTEWATERS

4.1 SPECIATION OF METALS

Thus far, we have discussed metals in water primarily in terms of hydrated metal ions, $M(H_2O)_6^{n+}$, and their hydrolysis products. As techniques for the study of metals in water have become more sophisticated, however, there has been increasing recognition of the importance of **speciation** of metals in natural waters and wastewaters. Metals may exist in water reversibly bound to inorganic anions or to organic compounds, or they may be present as **organometallic** compounds containing carbon-to-metal bonds. The properties of these species in areas such as solubility, transport, and biological effects are often vastly different from the properties of the metal ions, themselves. Therefore, it is important to consider the diverse forms in which metals may exist in water.

This chapter considers metal species with an emphasis upon metal complexes. Special attention is given to *chelation*, in which particularly strong metal complexes are formed. Stress is placed upon calculations that serve as models for estimating the form of chelatable metal ions in natural waters and wastewaters.

4.2 COMPLEXATION

A metal ion in water may combine with an electron donor (Lewis base) to form a **complex** or **coordination compound** (or ion). Thus, cadmium ion combines with a ligand, cyanide ion, to form the complex ion $CdCN^+$:

$$Cd^{2+} + CN^- \rightarrow CdCN^+ \tag{4.2.1}$$

Additional cyanide ligands may be added to form the progressively weaker (i.e. more easily dissociated) complexes $Cd(CN)_2$, $Cd(CN)_3^-$, and $Cd(CN)_4^{2-}$.

In this example, the cyanide ion is a **unidentate ligand**, which means that it possesses only one site that bonds to the cadmium metal ion. Complexes of unidentate ligands are of relatively little importance in solution in natural waters. Of considerably more importance are complexes with **chelating agents**. A chelating agent has more than one atom that may be bonded to a central metal ion at one time to form a ring structure. Thus, pyrophosphate ion $P_2O_7^{4-}$ bonds to two sites on a calcium ion to form a chelate:

```
      O     O
      ‖     ‖
   O—P—O—P—O²⁻
      |     |
      O     O
       \Ca/
```

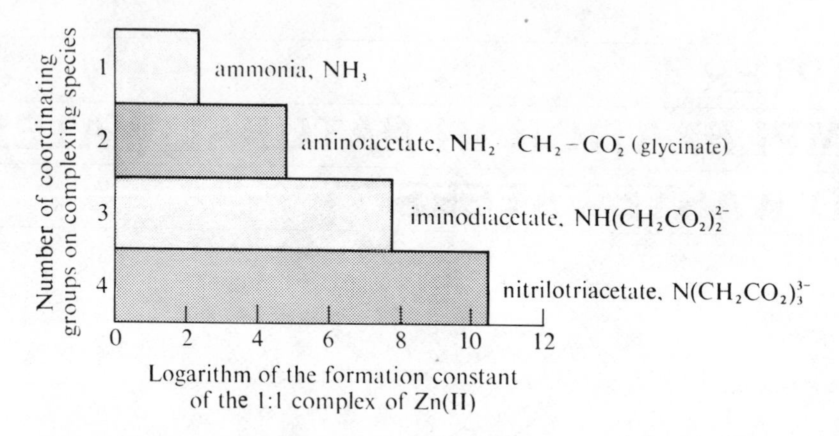

FIGURE 4.1 Effect of chelation upon stability of metal complexes.

In general, since a chelating agent may bond to a metal ion in more than one place simultaneously, chelates are more stable than complexes involving unidentate ligands. The chelate effect upon metal-complex stability is shown in Figure 4.1. It can be seen that as the number of electron-donor binding sites increases from one (NH_3) to four (nitrilotriacetate), the formation constant of the 1:1 complex increases by several orders of magnitude.

Structures of metal chelates take a number of different forms, all characterized by rings in various configurations. The structure of a tetrahedrally coordinated chelate of nitrilotriacetate ion is shown in Figure 4.2.

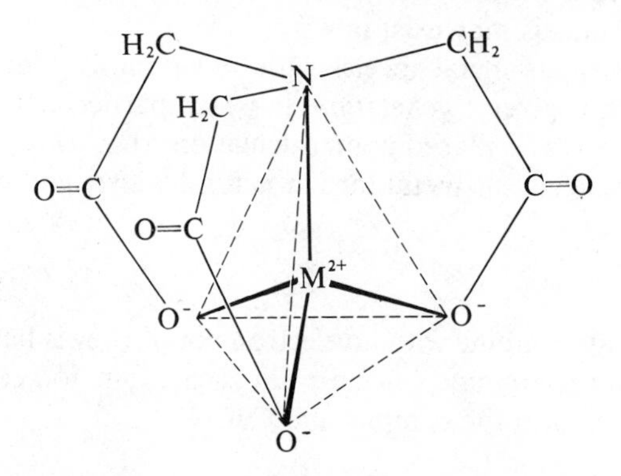

FIGURE 4.2 Nitrilotriacetate chelate of a divalent metal ion in a tetrahedral configuration.

The ligands found in natural waters and wastewaters contain a variety of organic groups which can donate the electrons required to bond the ligand to a metal ion [1]. Among the most common of these groups are:

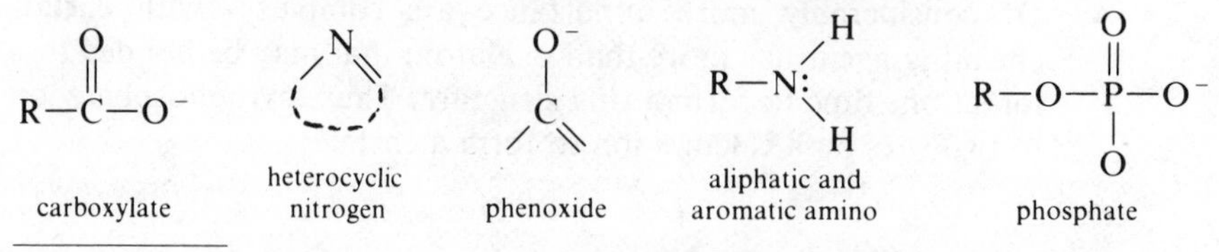

[1] Martell, A.E. 1971. Principles of complex formation. In *Organic compounds in aquatic environments,* S.D. Faust and J.V. Hunter, eds, pp. 262–392. New York: Marcel Dekker, Inc.

These ligands complex most metal ions found in unpolluted waters and biological systems (Mg^{2+}, Ca^{2+}, Mn^{2+}, Fe^{2+}, Fe^{3+}, Cu^{2+}, Zn^{2+}, and VO^{2+}). They also bind to contaminant metal ions such as Co^{2+}, Ni^{2+}, Sr^{2+}, Cd^{2+}, and Ba^{2+}.

Complexation may have a number of effects, including reactions of both ligands and metals. Among the ligand reactions are oxidation-reduction, decarboxylation, and hydrolysis. Complexation may cause changes in oxidation state of the metal and may result in a metal becoming solubilized from an insoluble compound. The formation of insoluble complex compounds may remove metal ions from solution.

Complex compounds of metals such as iron (in hemoglobin) and magnesium (in chlorophyll) are vital to life processes. Naturally occurring chelating agents, such as humic substances and amino acids, are found in water and soil. The high concentration of chloride ion in seawater results in the formation of some chloro- complexes. Synthetic chelating agents such as sodium tripolyphosphate, sodium ethylenediaminetetraacetate (EDTA), sodium nitrilotriacetate (NTA), and sodium citrate are produced in large quantities for use in metal-plating baths, industrial water treatment, detergent formulations, and food preparation. Small quantities of these compounds enter aquatic systems through waste discharges.

Complexing agents in wastewater are of concern primarily because of their ability to solubilize heavy metals from plumbing and from deposits containing heavy metals. Complexation may reduce the efficiency with which heavy metals are removed with sludge in conventional biological waste treatment. Removal of chelated iron is difficult with conventional municipal water-treatment processes. Iron(III) and perhaps several other essential micronutrient metal ions are kept in solution by chelation in algal cultures. The availability of chelating agents may be a factor in determining algal growth. The yellow-brown color of some natural waters is due to naturally occurring chelates of iron.

4.3 OCCURRENCE OF CHELATING AGENTS IN WATER

A great deal of evidence exists suggesting that chelating agents are common water pollutants. For example, two fractions of material of different molecular weight, capable of chelating cupric ion, have been isolated from secondary sewage effluent [2]. The molecular weight of one of these fractions was in the same range as those of common synthetic chelating agents.

Citric acid has the following structure:

$$
\begin{array}{c}
CO_2H \\
| \\
HO_2C{-}CH_2{-}C{-}CH_2{-}CO_2H \\
| \\
OH
\end{array}
$$

It is a component of the biochemical Krebs cycle and is encountered in water that has been exposed to any substantial biological activity. This chelating agent is found in citrus foods, in soft drinks, and as an additive in other foods.

[2] Bender, M.E.; Matson, W.R.; and Jordan, R.A. 1970. On the significance of metal complexing agents in secondary sewage effluents. *Environmental science and technology* 4:520–1.

Chelates formed by the strong chelating agent ethylenediaminetetraacetate, whose structure is illustrated at the beginning of Section 4.7, have been shown to greatly increase the migration rates of radioactive ^{60}Co from pits and trenches used by the Oak Ridge National Laboratory in Oak Ridge, Tennessee for disposal of intermediate-level radioactive wastes [3]. EDTA was used as a cleaning and solubilizing agent for the decontamination of hot cells, equipment, and reactor components. Analysis of water from sample wells in the disposal pits showed EDTA concentrations of $3.4 \times 10^{-7} M$. The presence of EDTA 12–15 years after its burial attests to its low rate of biodegradation. In addition to cobalt, EDTA strongly chelates plutonium and radioisotopes of Am^{3+}, Cm^{3+}, and Th^{4+}. These chelates are much less strongly sorbed by mineral matter and are vastly more mobile than the unchelated metal ions.

Contrary to the above findings, only very low concentrations of chelatable radioactive plutonium were observed in ground water near the Idaho Chemical Processing Plant's low-level waste disposal well [4]. No plutonium was observed in wells at any significant distance from the disposal well. The waste processing procedure used was designed to destroy any chelating agents in the waste prior to disposal, and no chelating agents were found in the water pumped from the test wells.

The determination of binding capacity of natural water has been the subject of a great deal of study in recent years. One such study showed that about 50 percent of the copper binding capacity of lake water was due to soluble organic compounds in solution [5]. The binding capacity was less for filtered lake water, implying that some of the copper binding entities in the water were colloidal materials composed of matter such as clays, silica, or manganese dioxide.

Other examples could be cited of the presence of chelating agents in natural waters and wastewaters. These compounds play a strong role in the chemistry and biology of many waters. Therefore, we will spend the rest of this chapter examining the nature of complexation and its effects upon water chemistry and biology.

4.4 BONDING AND STRUCTURE OF METAL COMPLEXES

A **complex** consists of a central atom to which ligands possessing electron-donor properties are bonded. The ligands may be negatively charged or neutral. The resulting complex may be neutral or may have a positive or negative charge. The ligands are said to be contained within the **coordination sphere** of the central metal atom. Depending upon the type of bonding involved, the ligands within the

---

[3] Means, J.L.; Crerar, D.A.; and Duguid, J.O. 1978. Migration of radioactive wastes: radionuclide mobilization by complexing agents. *Science* 200: 1477–81.

[4] Cleveland, J.M., and Rees, T.F. 1982. Characterization of plutonium in ground water near the Idaho Chemical Processing Plant. *Environmental Science and Technology* 16: 437–439.

[5] Blutstein, H., and Shaw, R.F. 1981. Characterization of copper binding capacity in lake water. *Environmental Science and Technology* 15: 1100–2.

coordination sphere are held in a definite structural pattern. However, in solution, ligands of many complexes exchange rapidly between solution and the coordination sphere of the central metal ion.

The **coordination number** of a metal atom, or ion, is the number of ligand electron-donor groups that are bonded to it. The most common coordination numbers are 2, 4, and 6. Polynuclear complexes contain two or more metal atoms joined together through bridging ligands, frequently OH. An example of such a complex, the dinuclear complex of iron(III), was discussed in Section 2.8.

Several different kinds of bonds are involved in complexation. Many theories, ranging from very simple to very complicated, have been offered to describe this bonding. Two very general pictures are those of ionic bonding and covalent bonding. Ionic bonding, most applicable in the case of central metal ion having a noble-gas structure, involves an electrostatic attraction between the positively charged central metal ion and the negatively charged ligand. The simple picture of covalent bonding as applied to complexation is that electron pairs are donated to the central metal ion by the ligand. These electron pairs fill empty orbitals on the metal ion, thus allowing the electron distribution of the metal ion to approach that of a noble gas.

For additional information on complexation, the reader is referred to any advanced inorganic chemistry textbook. Little is known, however, about the structure and bonding of many important metal complexes in aquatic and soil environments, and more information is needed in this area.

4.5 SPECIFICITY IN CHELATION

Although chelating agents are never entirely specific for a particular metal ion, some complicated chelating agents of biological origin approach almost complete specificity for certain metal ions. One example of such a chelating agent is ferrichrome, which forms extremely stable chelates with iron(III). A number of microorganisms synthesize ferrichrome, and for laboratory use the compound is extracted from fungi. Ferrichromes are complicated peptides in which a trihydroxamate is the actual group binding the iron(III):

$$\text{hydroxamate:}\quad -\overset{\overset{\displaystyle O}{\|}}{C}-\overset{\overset{\displaystyle OH}{|}}{N}-$$

It has been observed that blue-green algae of the *Anabaena* species secrete appreciable quantities of iron-selective hydroxamate chelating agents during periods of heavy algal bloom [6]. These blue-green algae readily take up hydroxamate chelated iron, whereas some competing green algae, e.g., *Scenedesmus,* do not. Thus, the chelating agent serves a dual function of promoting the growth of certain blue-green algae while suppressing the growth of other species, allowing the former to predominate.

---

[6] Murphy, T.P.; Lean, D.R.S.; and Nalewajko, C. 1976. Blue-green algae: Their excretion of iron-selective chelators enables them to dominate other algae. *Science* 192: 900–2.

4.6 CALCULATION OF SPECIES CONCENTRATIONS IN SOLUTIONS OF COMPLEXES

The stability of complex ions in solution is expressed in terms of formation constants. For the species $ZnNH_3^{2+}$, formed by the reaction

$$Zn^{2+} + NH_3 \rightleftharpoons ZnNH_3^{2+} \tag{4.6.1}$$

the formation constant K_1 is given by:

$$K_1 = \frac{[ZnNH_3^{2+}]}{[Zn^{2+}][NH_3]} = 3.9 \times 10^2 \tag{4.6.2}$$

In the previous equations, the waters of hydration have been omitted for simplicity. For the species $Zn(NH_3)_2^{2+}$, formed from $ZnNH_3^{2+}$ by the reaction

$$ZnNH_3^{2+} + NH_3 \rightleftharpoons Zn(NH_3)_2^{2+} \tag{4.6.3}$$

the formation constant K_2 is given by the following expression:

$$K_2 = \frac{[Zn(NH_3)_2^{2+}]}{[ZnNH_3^{2+}][NH_3]} = 2.1 \times 10^2 \tag{4.6.4}$$

Both K_1 and K_2 are called **stepwise formation constants,** in that they describe the addition of NH_3 to the central zinc ion as a stepwise process. An overall formation constant is employed to describe the process of adding several ligands to the central metal ion. For example, the formation of $Zn(NH_3)_2^{2+}$ may be expressed by the reaction

$$Zn^{2+} + 2\,NH_3 \rightleftharpoons Zn(NH_3)_2^{2+} \tag{4.6.5}$$

having the overall formation constant β_2 given by:

$$\beta_2 = \frac{[Zn(NH_3)_2^{2+}]}{[Zn^{2+}][NH_3]^2} = K_1K_2 = 8.2 \times 10^4 \tag{4.6.6}$$

Similarly, for the formation of $Zn(NH_3)_3^{2+}$, $\beta_3 = K_1K_2K_3$, and for the formation of $Zn(NH_3)_4^{2+}$, $\beta_4 = K_1K_2K_3K_4$.

4.7 COMPLEXATION BY DEPROTONATED LIGANDS

In most circumstances, metal ions and hydrogen ions compete for ligands, making the calculation of species concentration more complicated. Before going into such calculations, however, it is instructive to look at an example in which the ligand has lost all ionizable protons. At pH values of 11 or above, EDTA is essentially all in the deprotonated form. This is the tetranegative ion, abbreviated Y^{4-}, whose structural formula is:

```
      O  H                   H  O
      ‖  |                   |  ‖
  ⁻O—C—C                     C—C—O⁻
         |                  /|
         H \    H  H      /  H
            \   |  |     /
             N—C—C—N
            /   |  |     \
         H /    H  H      \  H
         |                  \|
  ⁻O—C—C                     C—C—O⁻
      ‖  |                   |  ‖
      O  H                   H  O
```

Chelating agents in alkaline solution are sometimes used for cleaning metal surfaces by removing the oxide surface layer. Therefore, let us take an example of wastewater from such a cleaning process, containing copper(II) at a total level of 5.0 mg/L and excess uncomplexed EDTA at a level of 200 mg/L (expressed as the disodium salt, $Na_2H_2C_{10}H_{12}O_8N_2 \cdot 2\ H_2O$, formula weight 372). If the wastewater is highly basic, pH 11, the excess EDTA will be present as the tetranegative anion, abbreviated Y^{4-}. In addition, in the absence of chelation the copper would be primarily in the form of insoluble $Cu(OH)_2$ or CuO.

The questions to be asked are: Will most of the copper be present as the EDTA complex? If so, what will be the equilibrium concentration of the hydrated cupric ion, Cu^{2+}? To answer the first question, we must calculate the molar concentration of uncomplexed excess EDTA, $[Y^{4-}]$. Since disodium EDTA with a formula weight of 372 is present at 200 mg/L (ppm), the total molar concentration of EDTA is $5.4 \times 10^{-4} M$. This concentration is essentially equal to that of free Y^{4-} in a solution having a pH of 11. The formation constant K_1 of the copper-EDTA complex CuY^{2-} is given by

$$K_1 = \frac{[CuY^{2-}]}{[Cu^{2+}][Y^{4-}]} = 6.3 \times 10^{18} \tag{4.7.1}$$

The ratio of complexed copper to uncomplexed copper is

$$\frac{[CuY^{2-}]}{[Cu^{2+}]} = [Y^{4-}]K_1 = 5.4 \times 10^{-4} \times 6.3 \times 10^{18}$$

$$= 3.3 \times 10^{15} \tag{4.7.2}$$

and therefore, essentially all of the copper is present as the complex ion. The molar concentration of total copper(II) in a solution containing 5.0 mg/L copper(II) is $7.9 \times 10^{-5} M$. Since we have already shown that essentially all of the copper is present as the EDTA complex, we may say that $[CuY^{2-}] = 7.9 \times 10^{-5}$ mmols/mL. The very low concentration of uncomplexed, hydrated cupric ion is given by

$$[Cu^{2+}] = \frac{[CuY^{2-}]}{K_1[Y^{4-}]} = \frac{7.9 \times 10^{-5}}{6.3 \times 10^{18} \times 5.4 \times 10^{-4}} \tag{4.7.3}$$

$$= 2.3 \times 10^{-20} \text{ mmols/mL}$$

Thus, in the medium described, the concentration of hydrated cupric ion is extremely low compared to total cupric ion. Any phenomenon in solution that depends upon the concentration of the hydrated cupric ion (such as a physiological effect or an electrode response) would be very different in the medium described, as compared to the effect observed if all of the copper at a level of 5.0 mg/L were present as Cu^{2+} in a more acidic solution and in the absence of complexing agent. The phenomenon of reducing the concentration of hydrated metal ion to very low values is one of the most important effects of complexation, especially where strongly chelating ligands are involved.

4.8 COMPLEXATION BY PROTONATED LIGANDS

Generally, complexing agents, particularly chelating compounds, are conjugate bases of Brönsted acids. As examples, NH_3 is the conjugate base of the NH_4^+ acid cation, and glycinate anion, $H_2N\text{-}CH_2\text{-}CO_2^-$, is the conjugate base of glycine,

$^{+}H_3N\text{-}CH_2\text{-}CO_2^-$. Therefore, in many cases hydrogen ion competes with metal ions for a ligand, so that the strength of chelation depends upon pH. In the nearly neutral pH range usually encountered in natural waters, most organic ligands are present in the protonated, or partially protonated, form.

In order to understand the competition between hydrogen ion and metal ion for a ligand, it is useful to know the distribution of ligand species as a function of pH. Consider nitrilotriacetic acid, $N(CH_2CO_2H)_3$, as an example. The trisodium salt of this compound, commonly abbreviated NTA, has been proposed as a detergent phosphate substitute and is a strong chelating agent. Its environmental implications are discussed further in Section 4.15.

Nitrilotriacetic acid, designated H_3T, loses hydrogen ion in three steps to form the nitrilotriacetate anion, T^{3-}. The structure of the T^{3-} anion is

```
       O  H
       ‖  |
   ⁻O—C—C
          |  \        H  O
          H   \       |  ‖
               N — C—C—O⁻
          H   /       |
          |  /        H
   ⁻O—C—C
       ‖  |
       O  H
```

It may coordinate through three CO_2^- groups and through the nitrogen atom, as shown in Figure 4.2. Note the similarity of the NTA structure to that of EDTA, discussed in Section 4.7.

The stepwise ionization of H_3T is given by the following equilibria:

$$H_3T \rightleftharpoons H^+ + H_2T^- \tag{4.8.1}$$

$$K_{a1} = \frac{[H^+][H_2T^-]}{[H_3T]} = 2.18 \times 10^{-2}; \quad pK_{a1} = 1.66 \tag{4.8.2}$$

$$H_2T^- \rightleftharpoons H^+ + HT^{2-} \tag{4.8.3}$$

$$K_{a2} = \frac{[H^+][HT^{2-}]}{[H_2T^-]} = 1.12 \times 10^{-3}; \quad pK_{a2} = 2.95 \tag{4.8.4}$$

$$HT^{2-} \rightleftharpoons H^+ + T^{3-} \tag{4.8.5}$$

$$K_{a3} = \frac{[H^+][T^{3-}]}{[HT^{2-}]} = 5.25 \times 10^{-11}; \quad pK_{a3} = 10.28 \tag{4.8.6}$$

These equilibrium expressions show that uncomplexed NTA may exist in solution as any one of the four species H_3T, H_2T^-, HT^{2-}, or T^{3-}, depending upon the pH of the solution. The fractions of these species depend upon pH, as can be shown graphically by a distribution-of-species diagram, with pH as a master (independent) variable, like that drawn for CO_2 in Figure 2.4. Such a diagram for NTA is shown in Figure 4.3. It is constructed as follows: the fraction of NTA present as the T^{3-} ligand is given by the expression

$$\alpha_{T^{3-}} = \frac{[T^{3-}]}{[H_3T] + [H_2T^-] + [HT^{2-}] + [T^{3-}]} \tag{4.8.7}$$

where $\alpha_{T^{3-}}$ is the fraction of total NTA species present as that anion. The K_a expressions may be combined to express $[H_3T]$, $[H_2T^-]$, and $[HT^{2-}]$ in terms of the K_a's, $[H^+]$, and $[T^{3-}]$. When the resulting expressions are substituted into Equation 4.8.7, $[T^{3-}]$ cancels in the numerator and denominator, and upon rearrangement of terms, $\alpha_{T^{3-}}$ is expressed as

$$\alpha_{T^{3-}} = \frac{K_{a1}K_{a2}K_{a3}}{[H^+]^3 + K_{a1}[H^+]^2 + K_{a1}K_{a2}[H^+] + K_{a1}K_{a2}K_{a3}} \quad (4.8.8)$$

which simplifies to

$$\alpha_{T^{3-}} = \frac{K_{a1}K_{a2}K_{a3}}{G} \quad (4.8.9)$$

where G is the denominator of Equation 4.8.8. The expressions for the fractions of the other species may be derived similarly,

$$\alpha_{HT^{2-}} = \frac{K_{a1}K_{a2}[H^+]}{G} \quad (4.8.10)$$

$$\alpha_{H_2T^-} = \frac{K_{a1}[H^+]^2}{G} \quad (4.8.11)$$

$$\alpha_{H_3T} = \frac{[H^+]^3}{G} \quad (4.8.12)$$

The preceding expressions may be used to draw a distribution-of-species diagram in which the fractions of the four possible species are plotted as a function of pH. Significant points on the plot are given in Table 4.1. The fractions of species present at each of the significant points may be calculated, and these values, along with the fractions of species in highly acidic and highly basic solutions, may be used to construct the diagram.

Plots of the fractions of NTA species as a function of pH are shown in Figure 4.3. The complexing anion T^{3-} is the predominant species only at relatively high pH values, much higher than would be encountered ordinarily in natural waters. The HT^{2-} species has an extremely wide range of predominance, however, spanning the entire normal pH range of ordinary fresh waters.

TABLE 4.1 Fractions of NTA Species at Selected pH Values

pH *value*	α_{H_3T}	$\alpha_{H_2T^-}$	$\alpha_{HT^{2-}}$	$\alpha_{T^{3-}}$
pH below 1.00	1.00	0.00	0.00	0.00
$pH = pK_{a1}$	0.49	0.49	0.02	0.00
$pH = \frac{1}{2}(pK_{a1} + pK_{a2})$	0.16	0.68	0.16	0.00
$pH = pK_{a2}$	0.02	0.49	0.49	0.00
$pH = \frac{1}{2}(pK_{a2} + pK_{a3})$	0.00	0.00	1.00	0.00
$pH = pK_{a3}$	0.00	0.00	0.50	0.50
pH above 12.00	0.00	0.00	0.00	1.00

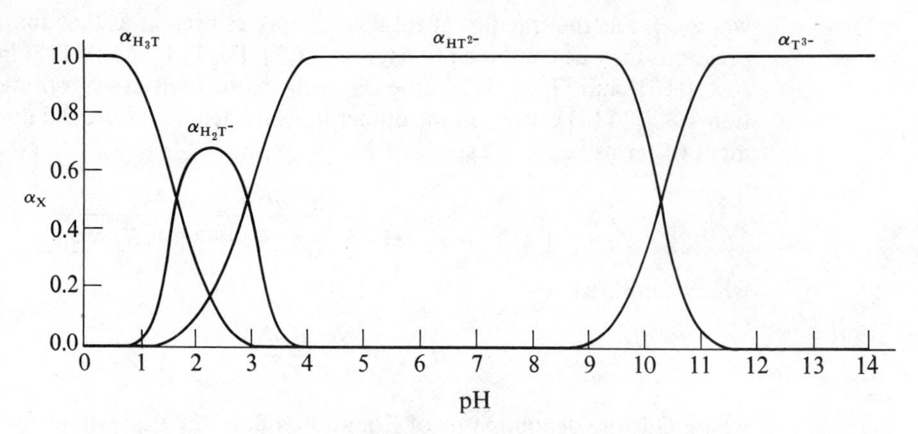

FIGURE 4.3 Plot of fraction of species α_x as a function of pH for NTA species in water.

Since T^{3-} is the actual complexing species, it is informative to calculate the fraction of T^{3-} in a neutral solution, pH 7.00. We may substitute the value of 1.00×10^{-7} for $[H^+]$ into Equation 4.8.9 and calculate $\alpha_{T^{3-}}$. The calculation can be simplified by noting that the first term in the denominator of Equation 4.8.8 comes from $[H_3T]$; the second term comes from $[H_2T^-]$; the third term comes from $[HT^{2-}]$; and the fourth term comes from $[T^{3-}]$. Reference to Figure 4.3 reveals that at pH 7.00, HT^{2-} comprises all but a very small fraction of the total NTA present in solution. Therefore, the only term of significant magnitude in the denominator is the term $K_{a1}K_{a2}[H^+]$, and the expression for the fraction present as T^{3-} at pH 7.00 simplifies to:

$$\alpha_{T^{3-}} = \frac{K_{a1}K_{a2}K_{a3}}{K_{a1}K_{a2}[H^+]} = \frac{K_{a3}}{[H^+]} = \frac{5.25 \times 10^{-11}}{1.00 \times 10^{-7}} = 5.25 \times 10^{-4} \qquad (4.8.13)$$

4.9 COMPLEXATION OF LEAD ION BY NTA

When the complexing agent is the conjugate base of a weak acid, as is the case with NTA, the competition of hydrogen ion for the ligand must be taken into account in calculating the concentration of uncomplexed metal ion. The complexation of lead by NTA at pH 7.00 provides an example of how the competition of hydrogen ion for the ligand may be calculated. Assume a solution 1.00×10^{-2} *M* in *uncomplexed* NTA, and 1.00×10^{-5} *F* in *total* lead(II). At pH 7.00, essentially all of the NTA is present as HT^{2-}, so that the complexation reaction is

$$HT^{2-} + Pb^{2+} \rightleftharpoons PbT^- + H^+. \qquad (4.9.1)$$

This reaction is obtained by adding Reactions 4.9.2 and 4.9.4. Its equilibrium constant is the product of Equations 4.9.3 and 4.9.5:

$$HT^{2-} \rightleftharpoons H^+ + T^{3-} \qquad (4.9.2)$$

$$K_{a3} = \frac{[H^+][T^{3-}]}{[HT^{2-}]} = 5.25 \times 10^{-11} \qquad (4.9.3)$$

$$Pb^{2+} + T^{3-} \rightleftharpoons PbT^- \qquad (4.9.4)$$

$$K_f = \frac{[PbT^-]}{[Pb^{2+}][T^{3-}]} = 2.45 \times 10^{11} \tag{4.9.5}$$

$$HT^{2-} + Pb^{2+} \rightleftharpoons PbT^- + H^+ \tag{4.9.1}$$

$$K = \frac{[PbT^-][H^+]}{[HT^{2-}][Pb^{2+}]} = K_{a3} \times K_f \tag{4.9.6}$$

$$= 5.25 \times 10^{-11} \times 2.45 \times 10^{11} = 12.9$$

It is first necessary to establish whether the lead is present primarily as Pb^{2+} or as PbT^-. The ratio of the concentrations of these two species may be obtained from Equation 4.9.6. The hydrogen-ion concentration is given as $1.00 \times 10^{-7} M$. Since HT^{2-} is present in large excess, its concentration is 1.00×10^{-2} M. Rearranging terms and substituting into Equation 4.9.6 yields

$$\frac{[Pb^{2+}]}{[PbT^-]} = \frac{[H^+]}{[HT^{2-}]K} = \frac{1.00 \times 10^{-7}}{1.00 \times 10^{-2} \times 12.9} = 7.75 \times 10^{-7} \tag{4.9.7}$$

which shows that all but a very small fraction of the lead is present as PbT^-. Therefore, the concentration of PbT^- is essentially the same as that of lead(II), or $1.00 \times 10^{-5} M$. The expression for K may now be solved for $[Pb^{2+}]$:

$$[Pb^{2+}] = \frac{[PbT^-][H^+]}{[HT^{2-}]K} = \frac{1.00 \times 10^{-5} \times 1.00 \times 10^{-7}}{1.00 \times 10^{-2} \times 12.9} \tag{4.9.8}$$

$$= 7.75 \times 10^{-12} M$$

Even at pH 7.00, where T^{3-} constitutes only a very small fraction of the NTA present, it still has a strong chelating ability for lead. The concentration of Pb^{2+} is so low that the solubility product of $Pb(OH)_2$, 1.6×10^{-20}, is not exceeded:

$$[Pb^{2+}][OH^-]^2 = 7.75 \times 10^{-12}(1.00 \times 10^{-7})^2 = 7.75 \times 10^{-26}$$

A completely rigorous treatment of the lead-NTA system would have to take into account the formation of species such as $Pb(OH)^+$, $Pb(OH)T^{2-}$, and PbHT. New metal-containing species continue to be discovered in complex aquatic systems, adding to the difficulty of totally describing such systems by equilibrium calculations.

4.10 CONDITIONAL FORMATION CONSTANTS

Sometimes it is convenient to perform calculations involving complexation using conditional formation constants. A **conditional formation constant** is a constant that is valid under a given set of solution conditions. Specifically, a conditional formation constant may be calculated at a given pH, and thus takes into account the protonation of the ligand. To use the case of the lead-NTA complex as an example, the reaction between Pb^{2+} and uncomplexed NTA in any form may be written as

$$Pb^{2+} + H_xT \rightleftharpoons PbT^- + xH^+ \tag{4.10.1}$$

where H_xT represents uncomplexed NTA, whether it be in the form of T^{3-}, HT^{2-}, H_2T^-, or H_3T. The equilibrium constant for Reaction 4.10.1 is the conditional formation constant for PbT^-, K_f'. It is a pH-dependent term given by the equation

$$K_f' = \frac{[PbT^-]}{[Pb^{2+}]C_T}, \tag{4.10.2}$$

where C_T is the total concentration of uncomplexed NTA. The concentration of the chelating ligand, T^{3-}, may be calculated from the concentration of total uncomplexed NTA if the fraction of NTA present is T^{3-} as known:

$$[T^{3-}] = C_T\alpha_{T^{3-}} \tag{4.10.3}$$

Substituting Equation 4.10.3 into Equation 4.10.2 yields:

$$K_f' = \frac{[PbT^-]}{[Pb^{2+}][T^{3-}]}\alpha_{T^{3-}} \tag{4.10.4}$$

Recalling the expression for K_f, the formation constant of PbT^-,

$$K_f = \frac{[PbT^-]}{[Pb^{2+}][T^{3-}]} = 2.45 \times 10^{11} \tag{4.10.5}$$

and substituting into the expression for K_f' yields the following equation:

$$K_f' = K_f\alpha_{T^{3-}} \tag{4.10.6}$$

Thus, considering protonation of the NTA ligand, the conditional formation constant of the lead-NTA complex is simply the product of K_f and the fraction of NTA present as the chelating ligand T^{3-}.

To calculate the hydrated lead-ion concentration under the conditions of the preceding example, recall that C_T was $1.00 \times 10^{-2} M$ and [Pb(II)] was $1.00 \times 10^{-5} M$. Furthermore, it was shown in Equation 4.8.13 that at pH 7.00 the value of $\alpha_{T^{3-}}$ is 5.25×10^{-4}. For the lead-NTA complex at pH 7.00, the conditional formation constant is

$$K_f' = K_f\alpha_{T^{3-}} = 2.45 \times 10^{11} \times 5.25 \times 10^{-4} = 1.29 \times 10^{8} \tag{4.10.7}$$

and the expression for K_f' is

$$\frac{[PbT^-]}{[Pb^{2+}]C_T} = 1.29 \times 10^{8} \tag{4.10.8}$$

As shown in the preceding example, $[PbT^-]$ is essentially equal to the total lead(II) concentration, $1.00 \times 10^{-5} M$. Substituting $1.00 \times 10^{-2} M$ for C_T yields $7.75 \times 10^{-12} M$ for the value of $[Pb^{2+}]$. Note that the value of $[Pb^{2+}]$ obtained from the conditional formation constant is the same as that obtained from the special constant already derived, which contains a term in $[H^+]$.

Conditional formation constants are also used to account for complexation of the metal ion by another complexing agent forming weaker complexes. In a seawater medium, for example, cadmium(II) is present partially as Cd^{2+} and partially as the $CdCl^+$ ion pair. If the extent of cadmium binding by NTA in seawater is to be calculated, a conditional formation constant can be derived and calculated, taking into consideration the fact that the fraction of the metal present as the hydrated ion is less than 1.00, even before any of the stronger chelating agent is added.

4.11 REACTION OF CHELATING AGENTS WITH METAL HYDROXIDES

A major question regarding the widespread introduction of strong chelating agents into aquatic ecosystems is that of possible solubilization of toxic heavy metals from sediments through the action of chelating agents. Experimentation is required to answer such a question, but calculations are helpful in predicting probable effects. The extent of solubilization of heavy metals depends upon a number of factors, including the stability of the metal chelates; the concentration of the complexing agent in the water; pH; and the nature of the insoluble metal deposit.

As an example of the kind of calculation that might be made, consider possible solubilization of lead from solid $Pb(OH)_2$ by NTA at pH 8.00. As illustrated in Figure 4.3, essentially all NTA is present as HT^{2-} ion at pH 8.00. Therefore, the solubilization reaction is

$$Pb(OH)_2(s) + HT^{2-} \rightleftharpoons PbT^- + OH^- + H_2O \qquad (4.11.1)$$

which may be obtained by adding the following reactions:

$$Pb(OH)_2(s) \rightleftharpoons Pb^{2+} + 2\,OH^- \qquad (4.11.2)$$

$$K_s = [Pb^{2+}][OH^-]^2 = 1.61 \times 10^{-20} \qquad (4.11.3)$$

$$HT^{2-} \rightleftharpoons H^+ + T^{3-} \qquad (4.11.4)$$

$$K_{a3} = \frac{[H^+][T^{3-}]}{[HT^{2-}]} = 5.25 \times 10^{-11} \qquad (4.11.5)$$

$$Pb^{2+} + T^{3-} \rightleftharpoons PbT^- \qquad (4.11.6)$$

$$K_f = \frac{[PbT^-]}{[Pb^{2+}][T^{3-}]} = 2.45 \times 10^{11} \qquad (4.11.7)$$

$$H^+ + OH^- \rightleftharpoons H_2O \qquad (4.11.8)$$

$$\frac{1}{K_w} = \frac{1}{[H^+][OH^-]} = \frac{1}{1.00 \times 10^{-14}} \qquad (4.11.9)$$

$$Pb(OH)_2(s) + HT^{2-} \rightleftharpoons PbT^- + OH^- + H_2O \qquad (4.11.1)$$

$$K = \frac{[PbT^-][OH^-]}{[HT^{2-}]} = \frac{K_s K_{a3} K_f}{K_w} \qquad (4.11.10)$$

$$= 2.07 \times 10^{-5}$$

Assume that a sample of water contains 25 mg/L of NTA as the trisodium salt $N(CH_2CO_2Na)_3$, formula weight 257. The NTA may be present as either HT^{2-} or PbT^-. The total concentration of NTA both complexed and uncomplexed is 9.7×10^{-5} mmol/mL. Assuming a system in which NTA at pH 8.00 is in equilibrium with solid $Pb(OH)_2$, the NTA may be primarily in the uncomplexed form, HT^{2-}, or in the lead complex, PbT^-. The predominant species may be determined by calculating the ratio of $[PbT^-]$ to $[HT^{2-}]$ from the expression for K, noting that at pH 8.00, $[OH^-]$ has a value of $1.00 \times 10^{-6} M$:

$$\frac{[PbT^-]}{[HT^{2-}]} = \frac{K}{[OH^-]} = \frac{2.07 \times 10^{-5}}{1.00 \times 10^{-6}} = 20.7 \tag{4.11.11}$$

Since the ratio of [NTA chelated to lead(II)] to [unchelated NTA] is approximately 20 to 1, most of the NTA in solution is present as the lead chelate. The concentration of PbT^- on a molar basis is just slightly less than the 9.7×10^{-5} mmols/mL total NTA present. The atomic weight of lead is 207 and the concentration of lead in solution is approximately 20 mg/L. This reaction is pH-dependent such that the fraction of NTA chelated decreases with increasing pH.

4.12 REACTION OF CHELATING AGENTS WITH METAL CARBONATES

Solid lead carbonate, $PbCO_3$, is stable within the pE–pH region and alkalinity conditions normally found in natural waters. An example similar to one in the preceding section may be worked, assuming that equilibrium is established with $PbCO_3$ rather than with solid $Pb(OH)_2$.

Let us assume that 25 mg/L of trisodium NTA is in equilibrium with $PbCO_3$ at pH 7.00, and then calculate whether the lead will be complexed appreciably by the NTA. The carbonate ion, CO_3^{2-}, reacts with H^+ to form HCO_3^-, which may in turn react with additional hydrogen ion to produce CO_2 and H_2O. As discussed in Chapter 2, the acid-base equilibrium reactions for the species CO_2-HCO_3^--CO_3^{2-} are

$$CO_2 + H_2O \rightleftharpoons H^+ + HCO_3^- \tag{4.12.1}$$

$$K'_{a1} = \frac{[H^+][HCO_3^-]}{[CO_2]} = 4.45 \times 10^{-7} \tag{4.12.2}$$

$$pK'_{a1} = 6.35 \tag{4.12.3}$$

$$HCO_3^- \rightleftharpoons H^+ + CO_3^{2-} \tag{4.12.4}$$

$$K'_{a2} = \frac{[H^+][CO_3^{2-}]}{[HCO_3^-]} = 4.69 \times 10^{-11} \tag{4.12.5}$$

$$pK'_{a2} = 10.33 \tag{4.12.6}$$

where the acid dissociation constants of the carbonate species are designated as K'_a to distinguish them from the acid dissociation constants of NTA. Examination of these pK_a values, or of Figure 2.4, reveals that at pH 7.00, and at higher values within the limits expected in natural waters, the predominant carbonic species is HCO_3^-. Therefore, within the pH range of approximately 7–10, the CO_3^{2-} liberated by the reaction of NTA with $PbCO_3$ will go into solution as HCO_3^-. The reaction of $PbCO_3$ with HT^{2-} in the pH range of 7–10 is

$$PbCO_3(s) + HT^{2-} \rightleftharpoons PbT^- + HCO_3^- \tag{4.12.7}$$

which is obtained by adding the following equations:

$$PbCO_3(s) \rightleftharpoons Pb^{2+} + CO_3^{2-} \tag{4.12.8}$$

$$K_s = [Pb^{2+}][CO_3^{2-}] = 1.48 \times 10^{-13} \tag{4.12.9}$$

$$HT^{2-} \rightleftharpoons H^+ + T^{3-} \tag{4.12.10}$$

$$K_{a3} = \frac{[H^+][T^{3-}]}{[HT^{2-}]} = 5.25 \times 10^{-11} \qquad (4.12.11)$$

$$Pb^{2+} + T^{3-} \rightleftharpoons PbT^- \qquad (4.12.12)$$

$$K_f = \frac{[PbT^-]}{[Pb^{2+}][T^{3-}]} = 2.45 \times 10^{11} \qquad (4.12.13)$$

$$CO_3^{2-} + H^+ \rightleftharpoons HCO_3^- \qquad (4.12.14)$$

$$\frac{1}{K'_{a2}} = \frac{[HCO_3^-]}{[H^+][CO_3^{2-}]} = \frac{1}{4.69 \times 10^{-11}} \qquad (4.12.15)$$

$$PbCO_3(s) + HT^{2-} \rightleftharpoons PbT^- + HCO_3^- \qquad (4.12.7)$$

$$K = \frac{[PbT^-][HCO_3^-]}{[HT^{2-}]} = \frac{K_s K_{a3} K_f}{K'_{a2}} \qquad (4.12.16)$$

$$= 4.06 \times 10^{-2}$$

From the expression for K, Equation 4.12.16, it may be seen that the degree to which $PbCO_3$ is solubilized as PbT^- depends upon the concentration of bicarbonate ion. Although this concentration will vary appreciably, the figure commonly used to describe natural waters is a bicarbonate ion concentration of 1.00 $\times$ $10^{-3}M$, as shown in Section 2.14. Using 1.00 $\times$ $10^{-3}M$ for the value of $[HCO_3^-]$, the distribution of the chelating agent between PbT^- and HT^{2-} may be calculated:

$$\frac{[PbT^-]}{[HT^{2-}]} = \frac{K}{[HCO_3^-]} = \frac{4.06 \times 10^{-2}}{1.00 \times 10^{-3}} = 40.6 \qquad (4.12.17)$$

Thus, under the given conditions, most of the NTA in equilibrium with solid $PbCO_3$ would be present as the lead complex. As in the previous example, at a trisodium NTA level of 25 mg/L the concentration of soluble lead(II) would be approximately 20 mg/L. At relatively higher concentrations of HCO_3^-, the tendency to solubilize lead would be diminished, whereas at lower concentrations of HCO_3^-, NTA would be more effective in solubilizing lead.

4.13 EFFECT OF CALCIUM ION UPON THE REACTION OF CHELATING AGENTS WITH SLIGHTLY SOLUBLE SALTS

Calcium ion, Ca^{2+}, generally is present in natural aquatic systems. Since the ion forms chelates, it competes for the chelating agent with a metal in a slightly soluble salt, such as $PbCO_3$. At pH 7.00, the reaction between calcium ion and NTA is

$$Ca^{2+} + HT^{2-} \rightleftharpoons CaT^- + H^+ \qquad (4.13.1)$$

which has an equilibrium constant K' given by the following expression:

$$K' = \frac{[CaT^-][H^+]}{[Ca^{2+}][HT^{2-}]} = 1.48 \times 10^8 \times 5.25 \times 10^{-11}$$

$$K' = 7.75 \times 10^{-3} \qquad (4.13.2)$$

The value of K' is the product of the formation constant of the calcium-NTA complex, 1.48×10^8, and K_{a3} of NTA, 5.25×10^{-11}. The fraction of NTA bound as the calcium complex, CaT^-, depends upon the concentration of Ca^{2+} and the pH. A moderate level of calcium in natural water is 40 mg/L, or $1.00 \times 10^{-3}M$. Assuming an uncomplexed calcium-ion concentration of $1.00 \times 10^{-3}M$ and pH 7.00, the ratio of NTA present in solution as the calcium complex to that present as HT^{2-} may be calculated:

$$\frac{[CaT^-]}{[HT^{2-}]} = \frac{[Ca^{2+}]}{[H^+]}K' = \frac{1.00 \times 10^{-3}}{1.00 \times 10^{-7}} \times 7.75 \times 10^{-3} \quad (4.13.3)$$

$$\frac{[CaT^-]}{[HT^{2-}]} = 77.5$$

Therefore, most of the NTA in equilibrium with 40 mg/L Ca^{2+} would be present as the calcium complex, CaT^-.

The reaction of lead carbonate with CaT^- is:

$$PbCO_3(s) + CaT^- + H^+ \rightleftharpoons Ca^{2+} + HCO_3^- + PbT^- \quad (4.13.4)$$

$$K'' = \frac{[Ca^{2+}][HCO_3^-][PbT^-]}{[CaT^-][H^+]} \quad (4.13.5)$$

Reaction 4.13.4 may be obtained by subtracting Reaction 4.13.1 from Reaction 4.12.7, and its equilibrium constant may be obtained by dividing the equilibrium constant of Reaction 4.13.1 by that of Reaction 4.12.7:

$$PbCO_3(s) + HT^{2-} \rightleftharpoons PbT^- + HCO_3^- \quad (4.12.7)$$

$$K = \frac{[PbT^-][HCO_3^-]}{[HT^{2-}]} = 4.06 \times 10^{-2} \quad (4.12.16)$$

$$Ca^{2+} + HT^{2-} \rightleftharpoons CaT^- + H^+ \quad (4.13.1)$$

$$K' = \frac{[CaT^-][H^+]}{[Ca^{2+}][HT^{2-}]} = 7.75 \times 10^{-3} \quad (4.13.2)$$

$$PbCO_3(s) + CaT^- + H^+ \rightleftharpoons Ca^{2+} + HCO_3^- + PbT^- \quad (4.13.4)$$

$$K'' = \frac{K}{K'} = \frac{4.06 \times 10^{-2}}{7.75 \times 10^{-3}} = 5.24 \quad (4.13.5)$$

Having obtained the value of K'', it is now possible to determine the distribution of NTA between PbT^- and CaT^-. Thus, for water containing NTA chelated to calcium at pH 7.00, a concentration of HCO_3^- of $1.00 \times 10^{-3}M$, a concentration of Ca^{2+} of $1.00 \times 10^{-3}M$, and in equilibrium with solid $PbCO_3$, the distribution of NTA between the lead complex and the calcium complex is:

$$\frac{[PbT^-]}{[CaT^-]} = \frac{[H^+]K''}{[Ca^{2+}][HCO_3^-]} = \frac{1.00 \times 10^{-7} \times 5.24}{1.00 \times 10^{-3} \times 1.00 \times 10^{-3}} \quad (4.13.6)$$

$$= 0.524$$

We see that only about 1/3 of the NTA would be present as the lead chelate. Note that under the identical conditions but in the absence of Ca^{2+}, approximately all

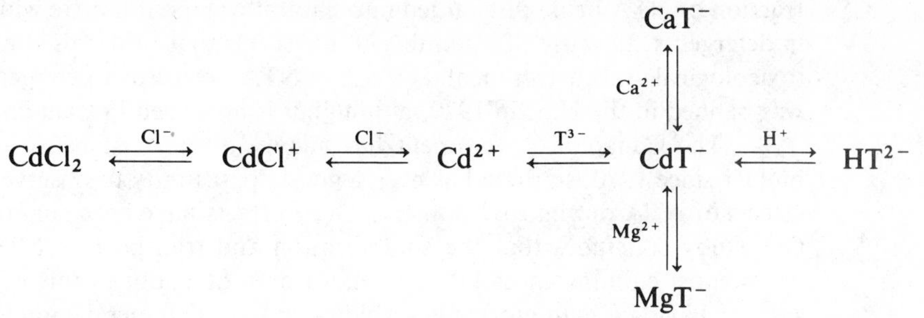

$$
\begin{array}{ccccccccc}
 & & & & & & CaT^- & & \\
 & & & & & & \uparrow\downarrow Ca^{2+} & & \\
CdCl_2 & \overset{Cl^-}{\rightleftharpoons} & CdCl^+ & \overset{Cl^-}{\rightleftharpoons} & Cd^{2+} & \overset{T^{3-}}{\rightleftharpoons} & CdT^- & \overset{H^+}{\rightleftharpoons} & HT^{2-} \\
 & & & & & & \uparrow\downarrow Mg^{2+} & & \\
 & & & & & & MgT^- & &
\end{array}
$$

FIGURE 4.4 Chelation scheme of cadmium in seawater.

of the NTA in equilibrium with solid $PbCO_3$ was chelated to NTA (Section 4.12). Since the fraction of NTA present as the lead chelate is directly proportional to the solubilization of $PbCO_3$, differences in calcium concentration will affect the degree to which NTA solubilizes lead from lead carbonate.

As noted in Section 4.10, the formation of chloro complexes is an additional consideration when chelation of metal ions is considered in seawater. The chelation of cadmium(II) with NTA has been considered as a model for chelation in sea water [7]. The overall scheme is shown in Figure 4.4. This scheme illustrates the complicated interactions involved in determining the degree of chelation of a heavy metal ion (cadmium) in seawater. It can be seen that the presence of excess chloride results in the formation of chloro-complexes; higher concentrations of H^+ break up the cadmium-NTA chelate (CdT^-) by protonation of the ligand; and both Ca^{2+} and Mg^{2+} compete for the NTA ligands.

4.14 LIMITATIONS OF THEORETICAL CALCULATIONS

The calculations in the preceding sections illustrate very clearly the complexity involved in calculating concentrations of species in natural aquatic systems. It is obvious that there is no simple answer to a question such as, "Will NTA solubilize heavy metals from sediment deposits?" Factors such as pH, bicarbonate-ion concentration, calcium-ion concentration, and the nature of the sediment must be taken into account. Not the least of the problems involved in such calculations is the lack of accurately known values of equilibrium constants to be used under the conditions being considered. Furthermore, we have not even considered the kinetic factors, which are quite important. Such calculations can be used only as general guidelines to determine areas in which more data should be obtained.

4.15 NTA IN THE AQUATIC ENVIRONMENT

As mentioned in Section 4.8, the trisodium NTA salt can be used as a detergent builder, substituting for phosphate builders. Although NTA is biodegradable, the present inadequacy of many sewage disposal systems could lead to an appreciable

---

[7] Raspor, B.; Valenta, P.; Nurnberg, H.W.; and Branica, M. 1977. The chelation of cadmium with NTA in seawater as a model for the typical behavior of trace metal chelates in natural waters. *The science of the total environment* 9: 87–109.

fraction of NTA being discharged into natural waters, if it were widely employed in detergents. Because of a number of questions that were raised about possible toxicological and environmental effects of NTA, its use as a detergent component was banned in the U.S. in 1970, although it is now used in Canada and Sweden.

The Canadian government has allowed the use of NTA as a detergent builder since 1970, so there has been a good opportunity to observe the chemical effects of NTA on the environment. These effects have been summarized [8, 9]. One study concludes that the solubilization and transport of NTA is minimal under most circumstances [8]. The major area of doubt in this regard involves systems having a high inorganic turbidity and/or high metal content. Other conclusions regarding NTA in water and wastewater include the following: (1) NTA normally degrades rapidly in aerobic biological wastewater treatment systems and fresh water systems. (2) The degradation of NTA under anaerobic conditions is relatively uncertain. (3) There is some possibility of retention of a fraction of NTA by sludge, such as that from the primary treatment of sewage. (4) The removal of NTA in physical-chemical waste treatment processes is relatively uncertain and should be studied further. Particular attention should be given to the removal of NTA by lime, alum, and iron coagulation and carbon adsorption. (5) Investigations are needed of the degradation of NTA at extremes of pH and temperature. (6) NTA is not expected to increase eutrophication by providing significant levels of algal nutrients. (7) In general, NTA does not pose acute or chronic toxicity hazards to aquatic organisms.

4.16 POLYPHOSPHATES IN WATER

Phosphorus occurs in many anionic forms in combination with oxygen. Some of these oxoanions are strong complexing agents. Since about 1930, salts of polymeric phosphorus oxoanions have been used increasingly for water treatment, for water softening, and as detergent builders. When used for water treatment, polyphosphates "sequester" calcium ion in a soluble or suspended form. The effect is to reduce the equilibrium concentration of calcium ion and prevent the precipitation of calcium carbonate in water pipes, boilers, and so on. Furthermore, when water is softened properly with polyphosphates, calcium does not form precipitates with soaps or interact detrimentally with detergents.

The simplest form of phosphate is orthophosphate, PO_4^{3-}:

$$\left[\begin{array}{ccc} & O & \\ & | & \\ & P & \\ O\diagup & | & \diagdown O \\ & O & \end{array}\right]^{3-}$$

The orthophosphate ion possesses three sites for attachment of H^+. As noted in Table 2.5, orthophosphoric acid, H_3PO_4, has a pK_{a1} of 2.17, a pK_{a2} of 7.31, and a

[8] 1978. *Ecological effects of non-phosphate detergent builders: final report on NTA*. n.p.: Great Lakes Science Advisory Board.

[9] Malaiyandi, M.; Williams, D.T.; and O'Grady, R. 1979. A national survey of nitrilotriacetic acid in Canadian drinking water. *Environmental science and technology* 13: 59–62.

pK_{a3} of 12.36. Much of the orthophosphate in natural waters originates from the hydrolysis of polymeric phosphate species.

Pyrophosphate ion, $P_2O_7^{4-}$, is the first of a series of unbranched chain polyphosphates produced by the condensation of orthophosphate:

$$2\,PO_4^{3-} + H_2O \rightarrow P_2O_7^{4-} + 2\,OH^- \qquad (4.16.1)$$

A long series of linear polyphosphates may be formed, the second of which is triphosphate ion, $P_3O_{10}^{5-}$. These species consist of PO_4 tetrahedra with adjacent tetrahedra sharing a common oxygen atom at one corner. The structural formulas of the acidic forms, $H_4P_2O_7$ and $H_5P_3O_{10}$, are:

$$\begin{array}{ccccc} & OH & & OH & \\ & | & & | & \\ HO— & P & —O— & P & —OH \\ & | & & | & \\ & O & & O & \end{array} \qquad \begin{array}{ccccccc} & OH & & OH & & OH & \\ & | & & | & & | & \\ HO— & P & —O— & P & —O— & P & —OH \\ & | & & | & & | & \\ & O & & O & & O & \end{array}$$

pyrophosphoric (diphosphoric) acid — triphosphoric acid

It is easy to visualize the longer chains composing the higher linear polyphosphates. **Vitreous sodium phosphates** are mixtures consisting of linear phosphate chains with from 4 to approximately 18 phosphorus atoms each. Those with intermediate chain lengths comprise the majority of the species present.

The acid-base behavior of the linear-chain polyphosphoric acids may be explained in terms of their structure by comparing them to orthophosphoric acid. Pyrophosphoric acid, $H_4P_2O_7$, has four ionizable hydrogens. The value of pK_{a1} is quite small (relatively strong acid), whereas pK_{a2} is 2.64, pK_{a3} is 6.76, and pK_{a4} is 9.42 [10]. In the case of triphosphoric acid, $H_5P_3O_{10}$, the first two pK_a values are small, pK_{a3} is 2.30, pK_{a4} is 6.50, and pK_{a5} is 9.24. When linear polyphosphoric acids are titrated with base, the titration curve has an inflection at a pH of approximately 4.5 and another inflection at a pH close to 9.5. Each member of the polyphosphate chain has one readily ionizable hydrogen, and these are removed in titrating to the first equivalence point. The end phosphorus atoms have two OH groups each. One of the OH groups on an end phosphorus atom has a readily ionizable hydrogen, whereas the other loses its hydrogen much less readily. Therefore, one mole of triphosphoric acid, $H_5P_3O_{10}$, loses three moles of hydrogen ion at a relatively low pH (below 4.5), leaving the diprotonated species:

$$\begin{array}{ccccccc} & O^- & & O^- & & O^- & \\ & | & & | & & | & \\ HO— & P & —O— & P & —O— & P & —OH \\ & | & & | & & | & \\ & O & & O & & O & \end{array}$$

At intermediate pH values (below 9.5), an additional two moles of hydrogen atoms ("end hydrogen") are lost to form the following species:

$$\begin{array}{ccccccc} & O^- & & O^- & & O^- & \\ & | & & | & & | & \\ ^-O— & P & —O— & P & —O— & P & —O^- \\ & | & & | & & | & \\ & O & & O & & O & \end{array}$$

---

[10] Van Wazer, J.R., and Callis, C.F. 1958. Metal complexing by phosphates. *Chemical reviews* 58: 1011–45.

Titration of a linear-chain polyphosphoric acid up to pH 4.5 yields the number of moles of phosphorus atoms per mole of acid. Titration from pH 4.5 to pH 9.5 yields the number of end phosphorus atoms. Orthophosphoric acid, H_3PO_4, differs from the other linear polyphosphoric acids in that it has a third ionizable hydrogen, which is removed in only the most basic media.

Ring polyphosphates constitute another type of polyphosphate in which the PO_4 tetrahedron is the basic structural member. The simplest member of this group is trimetaphosphoric acid, $H_3P_3O_9$, which has a six-member ring. The next higher member of the series is tetrametaphosphoric acid, $H_4P_4O_{12}$, possessing an eight-member ring. The structures of these species are:

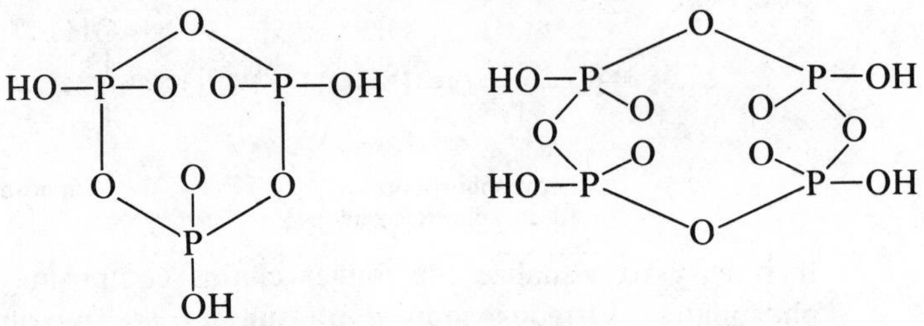

trimetaphosphoric acid

tetrametaphosphoric acid

The cyclic phosphoric acids do not show two breaks in their titration curves because there are no end hydrogens.

4.17 HYDROLYSIS OF POLYPHOSPHATES

All of the polymeric phosphates hydrolyze to simpler products in water. The rate of hydrolysis depends upon a number of factors, including pH, and the ultimate product is always some form of orthophosphate. The simplest hydrolytic reaction of a polyphosphate is that of pyrophosphoric acid to orthophosphoric acid:

$$H_4P_2O_7 + H_2O \rightarrow 2\ H_3PO_4 \qquad (4.17.1)$$

Researchers have found evidence that algae and other microorganisms catalyze the hydrolysis of polyphosphates. Even in the absence of biological activity, polyphosphates hydrolyze at a reasonable rate in water. Because of the chemical hydrolysis of polyphosphates in water there is much less concern about the possibility of their transporting heavy metal ions than is the case with organic chelating agents, which must depend upon microbial degradation for their decomposition.

4.18 COMPLEXATION BY POLYPHOSPHATES

In general, chain phosphates are good complexing agents and even form complexes with alkali-metal ions. Ring phosphates form much weaker complexes than do chain species. The reasons for the different chelating abilities of chain phosphates and ring phosphates become obvious when one considers their structures. For a chain polyphosphate complex at pH values exceeding approximately

4.5 (removal of H^+ from all but end P atoms), it is likely that a metal complex in solution has the following structure [10]:

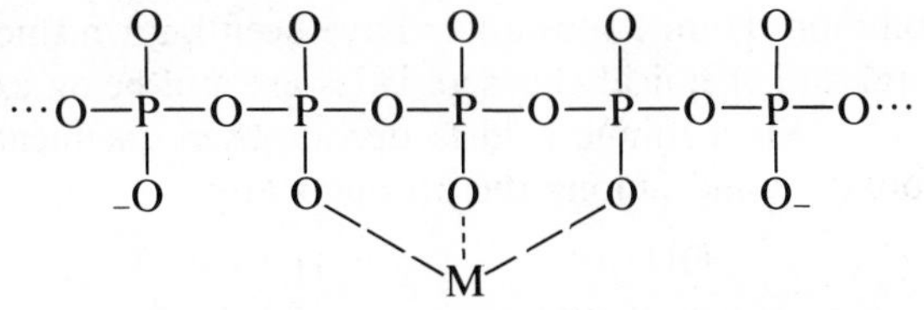

```
      O       O       O       O       O
      |       |       |       |       |
···O—P—O—P—O—P—O—P—O—P—O···
      |       |       |       |       |
     -O       O       O       O       O-
               \      :      /
                 \    :    /
                      M
```

In this structure, the metal ion may assume a bonding configuration with an oxygen atom in three separate, adjacent PO_4 groups. Because of structural restrictions, a ring polyphosphate cannot simultaneously chelate a metal ion at three different sites. This fact probably explains why chain phosphates form much more stable complexes with multivalent cations than do ring phosphates.

The difference in stability between chain phosphate complexes and ring phosphate complexes is illustrated very clearly in the case of Cu^{2+} complex species. The formation constants of the copper(II) complexes of the species discussed previously are: $Cu(P_2O_7)^{2-}$, $10^{8.8}$; $Cu(P_3O_{10})^{3-}$, $10^{8.70}$; $Cu(P_3O_9)^-$, $10^{1.55}$; and $Cu(P_4O_{12})^{2-}$, $10^{3.18}$. The latter two complexes involve ring phosphates and are much weaker.

4.19 HUMIC SUBSTANCES AS COMPLEXING AGENTS

The most important class of complexing agents that occur naturally are the **humic substances**. These are degradation-resistant materials formed during the decomposition of vegetation. They occur as deposits in soil, marsh sediments, peat, coal, lignite, or in almost any location where large quantities of vegetation have decayed. They are best classified on the basis of solubility. If a material containing humic substances is extracted with strong base, and the resulting solution is acidified, the products are (a) a nonextractable plant residue called **humin**; (b) a material that precipitates from the acidified extract, called **humic acid**; and (c) an organic material that remains in the acidified solution, called **fulvic acid**. Because of their acid-base, sorptive, and complexing properties, both the soluble and insoluble humic substances have a strong effect upon the properties of water. In general, fulvic acid dissolves in water and exerts its effect in water solution. Humin and humic acid remain insoluble and affect water quality through exchange of species, such as cations or organic materials, with water.

Humic substances are high-molecular-weight, polyelectrolytic macromolecules. Molecular weights range from a few hundred for fulvic acid to tens of thousands for the humic acid and humin fractions. These substances contain a carbon skeleton with a high degree of aromatic character and with a large percentage of the molecular weight incorporated in functional groups, most of which contain oxygen. They may contain protein-like material and a carbohydrate fraction. These fractions are relatively easy to hydrolyze from the aromatic nucleus, which is resistant to chemical and biochemical attack.

The elementary composition of most humic substances is within the following ranges: C, 45–55%; O, 30–45%; H, 3–6%; N, 1–5%; and S, 0–1%. Typically, humic acids isolated from Okefenokee Swamp in Georgia were found to contain 52% C, 43% O, 4.3% H, 0.56% N, and less than 0.05% ash [11]. The terms *humin, humic acid,* and *fulvic acid* do not refer to single compounds but to a wide range of compounds of generally similar origin, with many properties in common. Humic substances have been known since before 1800, but their structural and chemical characteristics are still being explained.

When humic acid is decomposed chemically, some of the typical compounds found among the products are:

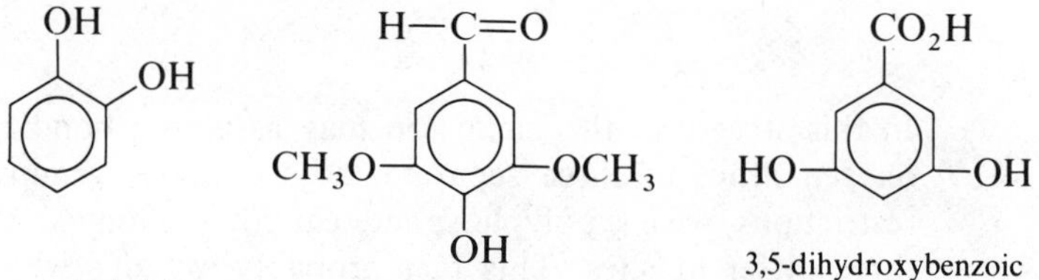

catechol   syringaldehyde   3,5-dihydroxybenzoic acid

It is reasonable to assume, therefore, that the basic structure or skeleton of humic acid consists largely of condensation products of compounds such as these. Furthermore, it is likely that -O- and -N- linkages are contained in the structure:

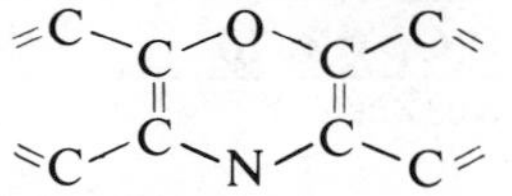

In addition, hydrogen bonds between functional groups may be involved in holding together the aromatic components of the humic acid and fulvic acid molecules.

A hypothetical structural formula for fulvic acid is:

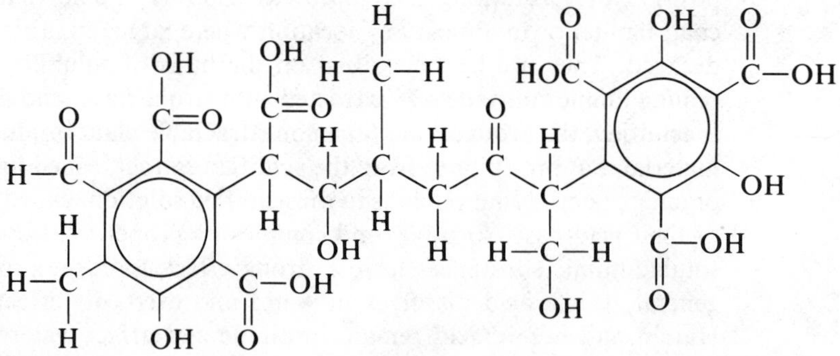

This structure is typical of the type of compound composing fulvic acid. The compound has a formula weight of 666, and its chemical formula may be represented by $C_{20}H_{15}(CO_2H)_6(OH)_5(CO)_2$. As shown in the hypothetical compound, the functional groups that may be present in fulvic acid are carboxyl, phenolic hydroxyl, alcoholic hydroxyl, and carbonyl. The functional groups vary with the particular acid sample. Approximate ranges in units of milliequivalents per gram of acid are: total acidity, 12–14; carboxyl, 8–9; phenolic hydroxyl, 3–6;

---

[11] Josephson, J. 1982. Humic substances. *Environmental science and technology* 16: 20A–24A.

alcoholic hydroxyl, 3–5; and carbonyl, 1–3. In addition, some methoxyl groups, $-OCH_3$, may be encountered at low levels.

The binding of metal ions by humic substances is one of the most important environmental qualities of humic substances. This binding can occur as chelation between a carboxyl group and a phenolic hydroxyl group,

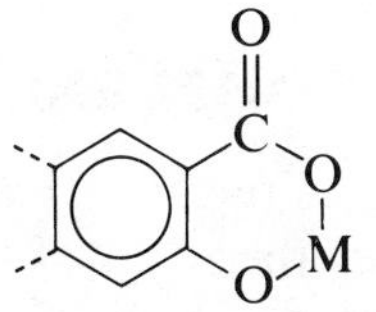

as chelation between two carboxyl groups,

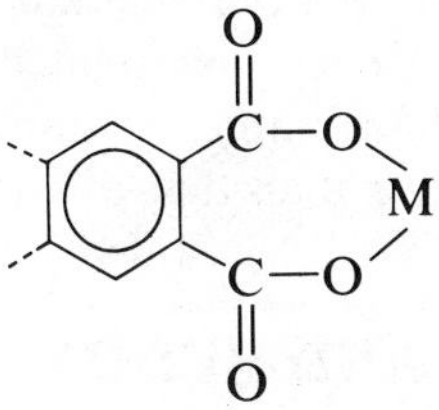

or as complexation with a carboxyl group:

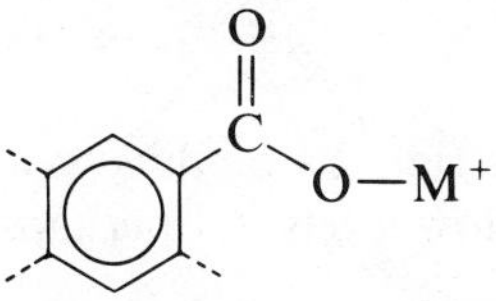

Iron and aluminum are very strongly bound to humic substances, whereas magnesium is rather weakly bound. Other common ions, such as Ni^{2+}, Pb^{2+}, Ca^{2+}, and Zn^{2+}, are intermediate in their binding to humic substances.

The role played by soluble fulvic-acid complexes of metals in natural waters is not well known. They probably keep some of the biologically important transition-metal ions in solution and are particularly involved in iron solubilization and transport. Fulvic acid-type compounds are associated with color in water. These yellow materials, called **Gelbstoffe**, frequently are encountered along with soluble iron.

Insoluble humic substances, the humins and humic acids, effectively exchange cations with water and may accumulate large quantities of metals. Lignite coal, which is largely a humic-acid material, tends to remove some metal ions from water.

Special attention has been given to humic substances since about 1970, following the discovery of **trihalomethanes** (THMs, such as chloroform and dibromochloromethane) in water supplies. It is now generally believed that these suspected carcinogens can be formed in the presence of humic substances during the disinfection of raw municipal drinking water by chlorination (see Chapter 8). The humic substances produce THMs by reaction with chlorine. The formation of THMs can be reduced by removing as much of the humic material as possible, prior to chlorination [11].

For further details on the interesting and significant substances comprising humic materials, the reader is referred to more extensive works on the subject [12, 13, 14].

4.20 AMINO ACIDS AS COMPLEXING AGENTS

Amino acids may be formed from the breakdown of proteins in natural waters. They constitute another class of naturally occurring complexing agents. Amino acids such as glycine, $^{+}H_3NCH_2CO_2^{-}$, form very stable complexes with some metal ions. For example, the overall formation constants of the monoglycinate and diglycinate complexes of copper(II) are 1.3×10^8 and 3×10^{15}, respectively. The role of naturally occurring amino acids in the transport of heavy metal ions in waters has not been investigated extensively, however. They are much more biodegradable than humic substances, so their significance in metal binding is probably much less than that of chelating agents like fulvic acid.

4.21 COMPLEXATION AND REDOX EQUILIBRIA

Complexation may have a strong effect upon equilibria by shifting reactions, such as that for the oxidation of lead,

$$Pb \rightleftharpoons Pb^{2+} + 2e^- \qquad (4.21.1)$$

strongly to the right by binding to the product ion, thus cutting its concentration down to very low levels. Of perhaps more important is the fact that upon oxidation,

$$M + \frac{1}{2}O_2 \rightarrow MO \qquad (4.21.2)$$

many metals form self-protective coatings of oxides, carbonates, or other insoluble species, which prevent further chemical reaction. Copper and aluminum roofing and structural iron are examples of materials which are thus self-protecting. A chelating agent in contact with such metals can result in continual dissolution of the protective coating so that the exposed metal corrodes readily. For example, chelating agents in wastewater may increase the corrosion of metal plumbing, thus adding heavy metals to effluents. Solutions of chelating agents employed to clean metal surfaces in metal-plating operations have a similar effect.

---

[12] Saar, R.A., and Weber, J.H. 1982. Fulvic acid: Modifier of metal-ion chemistry. *Environmental science and technology* 16: 510A–517A.

[13] Schnitzer, M., and Khan, S.U. 1972. *Humic substances in the environment.* New York: Marcel Dekker, Inc.

[14] Gamble, D.S., and Schnitzer, M. 1973. The chemistry of fulvic acid and its reactions with metal ions. In *Trace metals and metal-organic interactions in natural waters,* pp. 265–302. P.C. Singer, ed. Ann Arbor, Mich.: Ann Arbor Science Publishers, Inc.

SUPPLEMENTARY REFERENCES

American Chemical Society. 1968. *Trace inorganics in water.* ACS Advances in Chemistry Series No. 73. Washington, D.C.

Butler, J.N. 1964. *Ionic equilibrium—a mathematical approach.* Reading, Mass.: Addison-Wesley Publishing Co.

Faust, S.D., and Hunter, J.V. 1971. *Organic compounds in aquatic environments.* New York: Marcel Dekker, Inc.

Hem, J.D. 1970. *Study and interpretation of the chemical characteristics of natural water.* 2nd ed. U.S. Geological Survey Paper 1473. Washington, D.C.

Kirk, P.W.W.; Lester, J.N.; and Perry, R. 1983. Investigations into the fate of nitrilotriacetic acid in sewage sludge applied to agricultural land. *Water, Air, and Soil Pollution* 20: 161–70.

Liao, W.; Christman, R.F.; Johnson, J.D.; Millington, D.S.; and Hass, J.R. 1982. Structural characterization of aquatic humic material. *Environmental Science and Technology* 16: 403–410.

Lindberg, S.E., and Harris, R.C. 1974. Mercury-organic matter associations in estuarine sediments and interstitial water. *Environmental Science and Technology* 5: 459.

Morel, F., and Morgan, J. 1972. A numerical method for computing equilibria in aqueous chemical systems. *Environmental Science and Technology* 6: 58–67.

Singer, P.C., ed. 1973. *Trace metals and metal-organic interactions in natural waters.* Ann Arbor, Mich.: Ann Arbor Science Publishers, Inc.

Stevenson, F.J. 1982. Humus chemistry: Genesis, composition, reactions. New York: John Wiley and Sons, Inc.

Stiff, J.J. 1971. The chemical states of copper in polluted fresh water and a scheme of analysis to differentiate them. *Water Research* 5: 585–599.

Stumm, W. 1967. Metal ions in aqueous solutions. In *Principles and applications of water chemistry,* pp. 529–60. S.D. Faust and J.V. Hunter, eds. New York: John Wiley and Sons, Inc.

Theis, T.L., and Singer, P.C. 1974. Complexation of iron(II) by organic matter and its effect on iron(II) oxygenation. *Environmental Science and Technology* 8: 569–73.

QUESTIONS AND PROBLEMS

1. A proposed method for the analysis of total heavy metal complexing agents in water is based upon the solubilization of complexed copper from a freshly precipitated basic copper carbonate, in a solution adjusted to pH 10 with sodium carbonate. Look up the formation constants of some typical metal complexes and show why copper was chosen as a "tracer metal." Do you think the method would work for the analysis of EDTA in the presence of iron(III)? Discuss some possible interferences with the method.

2. Why do chelating agents generally form stronger complexes than do unidentate ligands, given comparable complexing groups?

3. How would you determine whether or not a metal in solution is in the form of a complex ion? Are there any cases in which the influence of the complexing agent upon redox equilibrium would be helpful in determining if a metal were complexed?

4. Try to visualize the structures of some polyphosphate complexes in three dimensions. Explain why ring polyphosphates generally form weaker complexes than linear-chain polyphosphates.

5. In the example given in Section 4.7, the possibility of cupric hydroxide precipitating from the solution was mentioned. Given a value of 3×10^{-20} for the solubility product of $Cu(OH)_2$, show that it is soluble in a medium containing 200 mg/L of disodium EDTA at pH 11.

6. Work out the example in Section 4.9, assuming pH 12.50. In what form would the NTA be present? What would be the ratio of $[Pb^{2+}]$ to $[PbT^-]$? What would be the concentration of Pb^{2+}?

7. For the example discussed in Section 4.11, as the pH is lowered the solubility of lead increases because the following equilibrium is shifted to the right:

$$Pb(OH)_2 \rightleftharpoons Pb^{2+} + 2\,OH^-$$

For water containing 25 mg/L of trisodium NTA in equilibrium with $Pb(OH)_2$, at what pH is $[PbT^-]$ equal to $[Pb^{2+}]$?

8. Calculate the ratio $[PbT^-]/[HT^{2-}]$ for NTA in equilibrium with $PbCO_3$ in a medium having $[HCO_3^-] = 3.00 \times 10^{-3}M$.

9. If the medium in problem 8 contained excess calcium such that the concentration of uncomplexed calcium, $[Ca^{2+}]$, were $5.00 \times 10^{-4}M$, what would be the ratio $[PbT^-]/[CaT^-]$ at pH 7?

10. A wastewater stream containing $1.00 \times 10^{-3}M$ disodium NTA, Na_2HT, as the only solute is injected into a limestone ($CaCO_3$) formation through a waste disposal well. After going through this aquifer for some distance and reaching equilibrium, the water is sampled through a sampling well. What is the reaction between NTA species and $CaCO_3$? What is the equilibrium constant for the reaction? What are the equilibrium concentrations of CaT^-, HCO_3^-, and HT^{2-}? (The appropriate constants may be looked up in Chapters 2 and 4.)

11. If the wastewater stream in problem 10 were 0.100*M* in NTA and contained other solutes that exerted a buffering action such that the final pH were 9.00, what would be the equilibrium value of HT^{2-} concentration in moles/liter?

12. Exactly 1.00×10^{-3} mole of $CaCl_2$, 0.100 mole of NaOH, and 0.100 mole of Na_3T were mixed and diluted to 1.00 liter. What was the concentration of Ca^{2+} in the resulting mixture?

13. How does chelation influence corrosion?

14. The following ligand has more than one site for binding to a metal ion. How many such sites does it have?

```
   O  H  H  H  O
   ‖  |     |  ‖
⁻O—C—C—N—C—C—O⁻
      |     |
      H     H
```

15. If a solution containing initially 25 mg/L trisodium NTA is allowed to come to equilibrium with solid $PbCO_3$ at pH 8.50 in a medium that contains $1.76 \times 10^{-3}M$ HCO_3^- at equilibrium, what is the value of the ratio $[PbT^-]/[HT^{2-}]$?

16. After a low concentration of NTA has equilibrated with $PbCO_3$ at pH 7.00, in a medium having $[HCO_3^-] = 7.50 \times 10^{-4}M$, what is the ratio $[PbT^-]/[HT^{2-}]$?

17. What detrimental effect may dissolved chelating agents have upon conventional biological waste treatment?

18. Why is chelating agent usually added to artificial algal growth media?

19. What common complex compound of magnesium is essential to certain life processes?

20. What is always the ultimate product of polyphosphate hydrolysis?

21. Water containing $1.00 \times 10^{-6} M$ Na_2HT (disodium NTA) as its only impurity is brought to equilibrium with solid $CaCO_3$ and dry air containing 0.0314% CO_2 at 1.00 atm total pressure. What is the value of $[HT^{2-}]$ in solution at equilibrium?

22. A solution containing initially $1.00 \times 10^{-5} M$ CaT^- is brought to equilibrium with solid $PbCO_3$. At equilibrium, pH = 7.00, $[Ca^{2+}] = 1.50 \times 10^{-3} M$, and $[HCO_3^-] = 1.10 \times 10^{-3} M$. At equilibrium, what is the fraction of total NTA in solution as PbT^-?

23. What is the fraction of NTA present as HT^{2-} after HT^{2-} has been brought to equilibrium with solid $PbCO_3$ at pH 7.00, in a medium in which $[HCO_3^-] = 1.25 \times 10^{-3} M$?

CHAPTER 5
MICROORGANISMS: THE CATALYSTS OF AQUATIC CHEMICAL REACTIONS

5.1 MICROORGANISMS AND VIRUSES

Microorganisms—bacteria, fungi, and algae—are living catalysts that enable a vast number of chemical processes to occur in water and soil. A majority of the important chemical reactions occurring in water, particularly those involving organic matter and oxidation-reduction processes, occur through bacterial intermediaries. Algae are the primary producers of biological organic matter (biomass) in water. Microorganisms are responsible for the formation of many sediment and mineral deposits; they also play the dominant role in secondary waste treatment.

Pathogenic microorganisms must be eliminated from water purified for domestic use. In the past, major epidemics of typhoid, cholera, and other waterborne diseases resulted from pathogenic microorganisms in water supplies. Even today, constant vigilance is required to ensure that water for domestic use is free of pathogens.

Although this chapter considers primarily the role played by microorganisms in aquatic chemical transformations, special mention should be made of **viruses** in water. Viruses cannot grow by themselves, but reproduce in the cells of host organisms. They are only about 1/30–1/20 the size of bacterial cells, and they cause a number of diseases, such as polio, viral hepatitis, and perhaps cancer. It is thought that many of these diseases are waterborne.

Because of their small size ($0.025–0.100\mu$), and biological instability, viruses are difficult to isolate and culture. They often survive municipal water treatment, including chlorination. Thus, although viruses have no effect upon the overall environmental chemistry of water, they are an important consideration in the treatment and use of water.

5.2 TYPES OF MICROORGANISMS IN WATER

Microorganisms important in aquatic chemistry may be divided among three categories: bacteria, fungi, and algae. Fungi and bacteria (with the exception of photosynthetic bacteria) are classified as *reducers*. Reducers break down chemical compounds to more simple species and thereby extract the energy needed for their growth and metabolism. Since reducers can utilize only chemical energy, any chemical transformation mediated by them must involve a net loss of free energy. However, compared to higher organisms, the energy utilization of bacteria and fungi is very efficient.

Algae are classified as *producers*, because they utilize light energy and store it as chemical energy. In the absence of sunlight, however, algae utilize

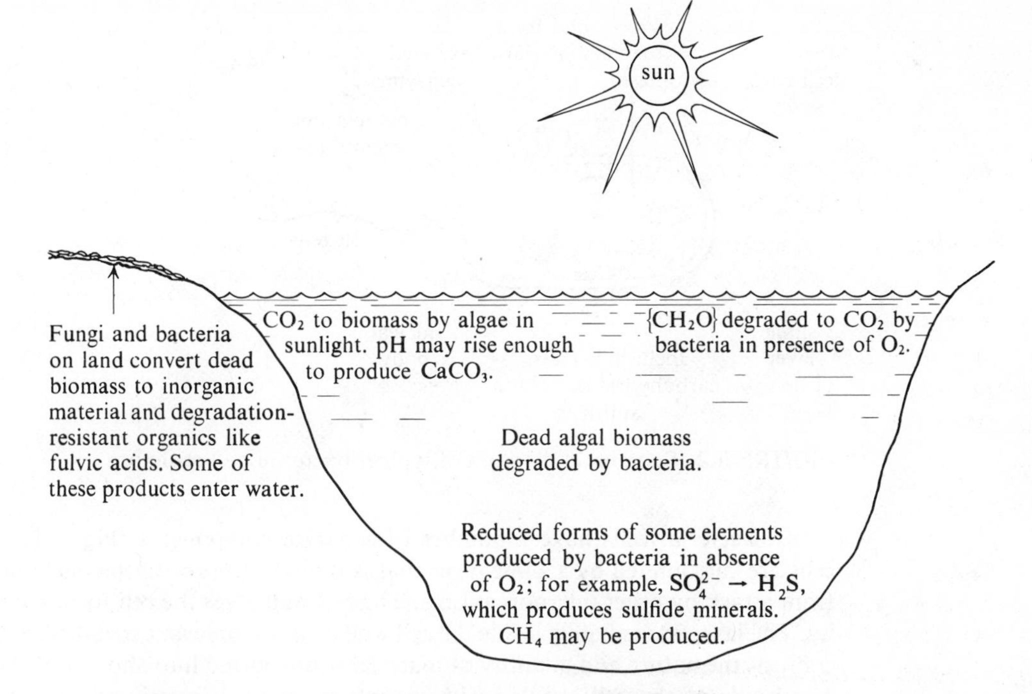
sun
Fungi and bacteria on land convert dead biomass to inorganic material and degradation-resistant organics like fulvic acids. Some of these products enter water.
CO_2 to biomass by algae in sunlight. pH may rise enough to produce $CaCO_3$.
{CH_2O} degraded to CO_2 by bacteria in presence of O_2.
Dead algal biomass degraded by bacteria.
Reduced forms of some elements produced by bacteria in absence of O_2; for example, $SO_4^{2-} \rightarrow H_2S$, which produces sulfide minerals. CH_4 may be produced.

FIGURE 5.1 Effects of microorganisms on the chemistry of water in nature.

chemical energy for their metabolic needs. In a sense, therefore, bacteria and fungi may be looked upon as environmental catalysts, whereas algae function as aquatic solar fuel cells. Some of the effects of microorganisms on the chemistry of water in nature are illustrated in Figure 5.1.

5.3 BACTERIA

Bacteria may be shaped as **rods, spheres, or spirals**. They may occur individually or grow as groups ranging from two to millions of individual cells. Individual bacteria cells are very small and may be observed only through a microscope. Most bacteria fall into the size range of 0.5–3.0 microns. However, considering all species, a size range of 0.3–50 microns is observed. In general, it is assumed that a filter with 0.45-micron pores will remove all bacteria from water passing through it.

The metabolic activity of bacteria is greatly influenced by their small size. Their surface-to-volume ratio is extremely large, so that the inside of a bacterial cell is highly accessible to a chemical substance in the surrounding medium. Thus, for the same reason that a finely divided catalyst is more efficient than a more coarsely divided one, bacteria may bring about very rapid chemical reactions compared to those mediated by larger organisms. Bacteria excrete **exoenzymes** (see Chapter 18) that break down solid food material to soluble components which can penetrate bacterial cell walls, where the digestion process is completed [1].

---

[1] Grutsch, J.F. 1980. The α-β-γ's of wastewater treatment. *Environmental Science and Technology* 14: 276–81.

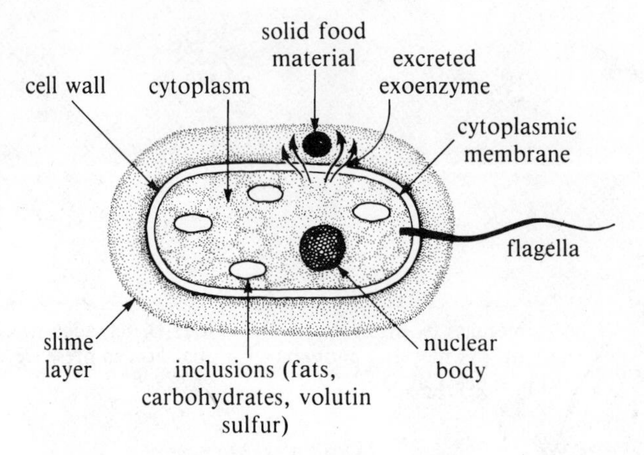

FIGURE 5.2 Schematic diagram of typical bacterial cell structure

Bacterial cells have a number of separate components (Fig. 5.2). Some cells are surrounded by a **slime layer** that is thought to protect the bacterial cells from attack by other microorganisms. The **cell wall** gives the cell form and rigidity. The layer immediately inside the cell wall is the **cytoplasmic membrane**, which controls the nature and quantity of materials transported into and out of the cell. The inside of the cell is filled with **cytoplasm**, the medium in which the cell's metabolic processes are carried out. The **nuclear body** (generally not considered to be a true nucleus) controls metabolic processes and reproduction. In addition to these features, the cell may contain **inclusions** of reserve food material, consisting of fats, carbohydrates, and even elemental sulfur. Some bacteria possess moveable **flagella**, hair-like appendages that give the bacteria mobility. Like fungi and algae, bacteria produce **spores**, metabolically inactive bodies that form and survive under adverse conditions in a "resting" state until conditions favorable for growth occur.

Although individual bacterial cells cannot be seen by the naked eye, bacterial colonies arising from individual cells are readily visible. A common method of counting individual bacterial cells in water consists of spreading a measured volume of water on a plate of agar gel containing bacterial nutrients. Wherever a viable bacterial cell adheres to the plate, a bacterial colony consisting of many cells will grow. These visible colonies are counted and related to the number of cells present initially. In most natural water samples it is necessary to dilute the sample with a sterile buffer solution to prevent the formation of too many colonies.

5.4 FUNGI

Fungi are nonphotosynthetic organisms. Most frequently they possess a filamentous structure. The morphology (structure) of fungi covers a wide range. Some fungi are as simple as the microscopic unicellular yeasts, whereas other fungi form large, intricate toadstools. The microscopic filamentous structures of fungi generally are much larger than bacteria and usually are 5–10 microns in width. Fungi are aerobic (oxygen-requiring) organisms and generally can thrive in more acidic media than can bacteria. They are also more tolerant of higher concentrations of heavy metal ions than are bacteria.

Perhaps the most important function of fungi in the environment is the breakdown of cellulose in wood and other plant materials. To accomplish this, fungal cells secrete a biological catalyst (enzyme) called **cellulase**. This enzyme breaks insoluble cellulose down to soluble carbohydrates that can be absorbed by the fungal cell. Because it acts outside the organism, it is called an **extracellular enzyme** or **exoenzyme**.

Although fungi do not grow well in water, they play an important role in determining the composition of natural waters and wastewaters because of the large amounts of their decomposition products that enter water. An example of such a product is humic material, which interacts with hydrogen ions and metals (see Section 4.19).

5.5 ALGAE

The term **algae** is somewhat difficult to define. For the purposes of our discussion, algae may be considered as generally microscopic organisms that subsist on inorganic nutrients and produce organic matter from carbon dioxide by photosynthesis. The general nutrient requirements of algae are carbon (from CO_2 or HCO_3^-), nitrogen (generally as NO_3^-), phosphorus (as some form of orthophosphate), sulfur (as SO_4^{2-}), and trace elements including sodium, potassium, calcium, magnesium, iron, cobalt, and molybdenum.

In a highly simplified form, the production of organic matter by algal photosynthesis is described by the reaction

$$CO_2 + H_2O \xrightarrow{h\nu} \{CH_2O\} + O_2(g) \tag{5.5.1}$$

where $\{CH_2O\}$ represents a unit of carbohydrate and $h\nu$ stands for the energy of a quantum of light. Fogg [2] has represented the overall formula of the algae *Chlorella* as $C_{5.7}H_{9.8}O_{2.3}N$ (including phosphorus, the formula would be $C_{5.7}H_{9.8}O_{2.3}NP_{0.06}$). Using Fogg's formula for algal biomass, the overall reaction for photosynthesis is:

$$5.7\ CO_2 + 3.4\ H_2O + NH_3 \xrightarrow{h\nu} C_{5.7}H_{9.8}O_{2.3}N + 6.25\ O_2 \tag{5.5.2}$$

In the absence of light, algae metabolize organic matter in the same manner as do nonphotosynthetic organisms. Thus, algae may satisfy their metabolic demands by utilizing chemical energy from the degradation of stored starches or oils, or from the consumption of algal protoplasm itself. In the absence of photosynthesis, the metabolic process consumes oxygen, so during the hours of darkness an aquatic system with a heavy growth of algae may become depleted in oxygen.

5.6 AUTOTROPHIC AND HETEROTROPHIC ORGANISMS

Bacteria may be divided into two main categories, **autotrophic** and **heterotrophic**. **Autotrophic bacteria** are not dependent upon organic matter for growth and

[2] Fogg, G.E. 1953. *The metabolism of algae.* New York: John Wiley and Sons, Inc.

thrive in a completely inorganic medium; they use carbon dioxide or other carbonate species as a carbon source. A number of sources of energy may be used, depending upon the species of bacteria; however, a biologically mediated chemical reaction always supplies the energy.

An example of autotrophic bacteria is *Gallionella*. In the presence of oxygen, these bacteria grow in a medium consisting of NH_4Cl, phosphates, mineral salts, CO_2 (as a carbon source), and solid FeS (as an energy source). It is believed that the following is the energy-yielding reaction for this species:

$$4\ FeS(s) + 9\ O_2 + 10\ H_2O \rightarrow 4\ Fe(OH)_3(s) + 4\ SO_4^{2-} + 8\ H^+ \quad (5.6.1)$$

Starting with the simplest inorganic materials, autotrophic bacteria must synthesize all of the complicated proteins, enzymes, and other materials needed for life processes. It follows, therefore, that the biochemistry of autotrophic bacteria is quite complicated. Because of their consumption and production of a wide range of minerals, autotrophic bacteria are involved in many geochemical transformations.

Heterotrophic bacteria depend upon organic compounds, both for their energy and for the carbon required to build their biomass. They are much more common in occurrence than autotrophic bacteria. Heterotrophic bacteria are the microorganisms primarily responsible for the breakdown of pollutant organic matter in waters and of organic wastes in biological waste-treatment processes.

Algae are autotrophic organisms, using CO_2 as a carbon source and light as an energy source. Fungi are entirely heterotrophic, deriving carbon and energy from the degradation of organic matter.

5.7 AEROBIC AND ANAEROBIC BACTERIA

Another classification system for bacteria depends upon their requirement for molecular oxygen. **Aerobic bacteria** require oxygen as an electron receptor:

$$O_2 + 4\ H^+ + 4\ e^- \rightarrow 2\ H_2O \quad (5.7.1)$$

Anaerobic bacteria function only in the complete absence of molecular oxygen. Frequently, molecular oxygen is quite toxic to anaerobic bacteria.

A third class of bacteria, **facultative bacteria**, utilize free oxygen when it is available and use other substances as electron receptors (oxidants) when molecular oxygen is not available. Common oxygen substitutes in water are nitrate ion (see Section 5.17) and sulfate ion (see Section 5.21).

5.8 KINETICS OF BACTERIAL GROWTH

The population size of bacteria and unicellular algae as a function of time in a growth culture is illustrated by Figure 5.3, which shows a population curve for a bacterial culture. Such a culture is started by inoculating a rich nutrient medium with a small number of bacterial cells. The population curve consists of four regions. The first region is characterized by little bacterial reproduction and is called the **lag phase**. The lag phase occurs because the bacteria must become acclimated to the new medium. Following the lag phase comes a period of very rapid bacterial growth. This is the **log phase**, or **exponential phase**, during which the population doubles over a regular time interval called the **generation time**.

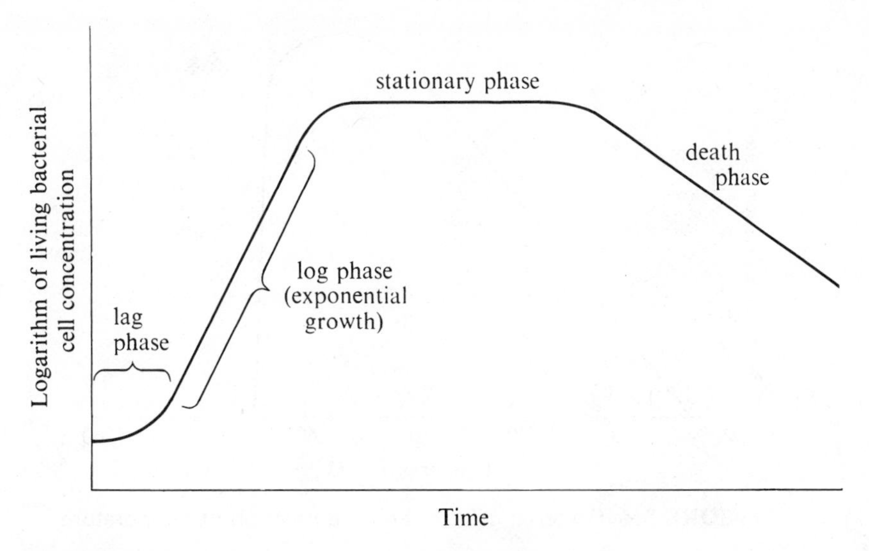

FIGURE 5.3 Population curve for a bacterial culture.

This behavior can be described by a mathematical model in which growth rate is proportional to the number of individuals present and there are no limiting factors such as death or lack of food:

$$\frac{dN}{dt} = kN \tag{5.8.1}$$

This equation can be integrated to give

$$\ln \frac{N}{N_0} = kt \quad \text{or} \quad N = N_0 e^{kt} \tag{5.8.2}$$

where N is the population at time t and N_0 is the population at time $t = 0$. Thus, another way of describing population growth during the log phase is to say that the logarithm of bacterial population increases linearly with time. The generation time, or doubling time, is $(\ln 2)/k$, analogous to the half-life of radioactive decay. Fast growth during the log phase can cause very rapid microbial transformations of chemical species in water.

The log phase terminates and the **stationary phase** begins when a limiting factor is encountered. Typical factors limiting growth are depletion of an essential nutrient; build-up of toxic material; and exhaustion of oxygen. During the stationary phase, the number of viable cells remains virtually constant. Depending upon the bacterial species and other circumstances, the stationary phase may be either very long or very short in duration. After the stationary phase, the bacteria begin to die faster than they reproduce, and the population enters the **death phase**.

Temperature has a strong effect upon bacterial growth. However, a given temperature does not affect different kinds of bacteria in the same way, since they have different optimum temperatures for growth. **Psychrophilic bacteria** are bacteria having temperature optima below approximately 20 °C. The temperature optima of **mesophilic bacteria** lie between 20 °C and 45 °C. Bacteria having

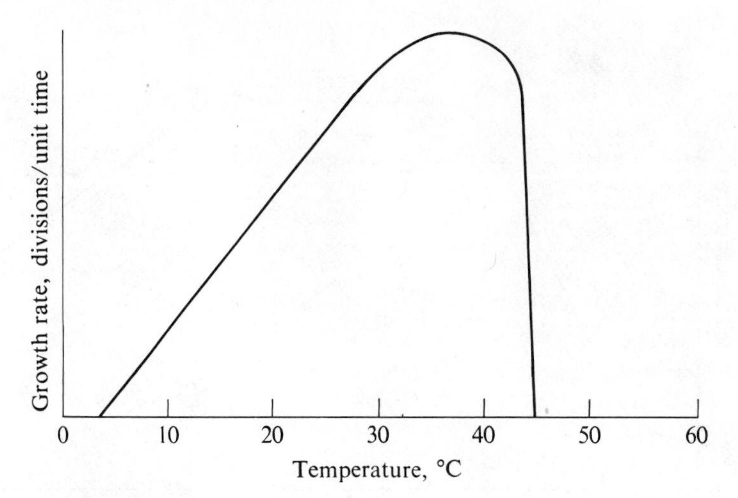

FIGURE 5.4 Bacterial growth rate as a function of temperature.

temperature optima above 45 °C are called **thermophilic bacteria**. The temperature range for optimum growth of bacteria is remarkably wide, with some bacteria being able to grow at 0 °C, and some thermophilic bacteria existing at temperatures as high as 80 °C.

Figure 5.4 shows the temperature dependence of the growth rate of a hypothetical species of bacteria with a temperature optimum of approximately 36 °C. Typically, such curves show an abrupt drop in growth rate beyond the temperature optimum; that is, the temperature optimum is much closer to the temperature maximum than it is to the temperature minimum. Presumably, vital enzymes are destroyed at temperatures not far above the optimum.

Over a relatively short range of temperature, a plot of bacterial growth rate as a function of the reciprocal of the absolute temperature, $1/K$, is linear. Students of physical chemistry will recognize such a relationship as an *Arrhenius plot*, which relates a kinetic rate constant to temperature.

As an interesting sidelight, some bacteria grow better, or even exclusively, at very high pressures in the oceans' depths, where pressures may be around 600 bars (a bar is 1.01 atm). Such bacteria are called **barophilic bacteria**.

5.9 MICROBIALLY MEDIATED OXIDATION AND REDUCTION REACTIONS

Bacteria obtain the energy needed for their metabolic processes and reproduction by mediating redox reactions. Nature provides a large number of such reactions, and bacterial species have evolved that utilize many of these. The most environmentally important redox reactions occurring in water and soil through the action of bacteria are summarized in Table 5.1. As a consequence of their participation in such reactions, bacteria are involved in many biogeochemical processes in water and soil. Bacteria are essential participants in many important elemental cycles in nature, including those of nitrogen, carbon, and sulfur. They are responsible for the formation of many mineral deposits, including some of iron and manganese. On a smaller scale, some of these deposits form through bacterial action in natural water systems and even in pipes used to transport water.

Much of the remainder of this chapter is devoted to a discussion of important redox reactions mediated by bacteria, including those summarized in Table 5.1.

5.10 MICROBIAL TRANSFORMATIONS OF CARBON

Carbon is an essential life element and composes a high percentage of the dry weight of microorganisms. For most microorganisms, the bulk of net energy-yielding or energy-consuming metabolic processes involve changes in the oxidation state of carbon. These chemical transformations of carbon have important environmental implications. For example, when algae and other plants fix CO_2 as carbohydrate, represented as $\{CH_2O\}$, carbon changes from the $+4$ to the 0 oxidation state.

$$CO_2 + H_2O \xrightarrow{h\nu} \{CH_2O\} + O_2(g) \tag{5.10.1}$$

Energy from sunlight is stored as chemical energy in organic compounds. However, when the algae die, bacterial decomposition results in the reverse of Reaction 5.10.1, energy is released, and oxygen is consumed.

In the presence of oxygen, the principal energy-yielding reaction of bacteria is the oxidation of organic matter. Since it is generally more meaningful to compare reactions on the basis of the reaction of one electron-mole, the aerobic degradation of organic matter is conveniently written as

$$\frac{1}{4}O_2 + \frac{1}{4}\{CH_2O\} \rightarrow \frac{1}{4}CO_2 + \frac{1}{4}H_2O \tag{5.10.2}$$

for which the free-energy change is -29.9 kcal (see *aerobic respiration,* Table 5.1). From this general type of reaction, bacteria and other microorganisms extract the energy needed to carry out their metabolic processes; to synthesize new cell material; for reproduction; and for locomotion.

Partial microbial decomposition of organic matter is a major step in the production of peat, lignite, coal, oil shale, and petroleum. Under reducing conditions, particularly below water, the oxygen content of the original plant material (approximate empirical formula, $\{CH_2O\}$) is lowered, leaving materials with relatively higher carbon contents.

5.11 METHANE-FORMING BACTERIA

The production of methane in **anoxic** (oxygen-less) sediments is favored by high organic levels and low nitrate and sulfate levels [3]. Methane production plays a key role in local and global carbon cycles as the final step in the anaerobic decomposition of organic matter. This process is the source of about 80% of the methane entering the atmosphere.

---

[3] Sansone, F.J., and Martens, C.S. 1981. Methane production from acetate and associated methane fluxes from anoxic coastal sediments. *Science* 211: 707–709.

TABLE 5.1 Principal Microbially Mediated Oxidation and Reduction Reactions

	Oxidation	$pE^0(w)$[1]		*Reduction*	$pE^0(w)$[1]
(1)	$\frac{1}{4}\{CH_2O\} + \frac{1}{4}H_2O \rightleftharpoons \frac{1}{4}CO_2(g) + H^+(w) + e$	−8.20	(A)	$\frac{1}{4}O_2(g) + H^+(w) + e \rightleftharpoons \frac{1}{2}H_2O$	−13.75
(1a)	$\frac{1}{2}HCOO^- \rightleftharpoons \frac{1}{2}CO_2(g) + \frac{1}{2}H^+(w) + e$	−8.73	(B)	$\frac{1}{5}NO_3^- + \frac{6}{5}H^+(w) + e \rightleftharpoons \frac{1}{10}N_2(g) + \frac{3}{5}H_2O$	+12.65
(1b)	$\frac{1}{2}\{CH_2O\} + \frac{1}{2}H_2O \rightleftharpoons \frac{1}{2}HCOO^- + \frac{3}{2}H^+(w) + e$	−7.68	(C)	$\frac{1}{8}NO_3^- + \frac{5}{4}H^+(w) + e \rightleftharpoons \frac{1}{8}NH_4^+ + \frac{3}{8}H_2O$	+6.15
(1c)	$\frac{1}{2}CH_3OH \rightleftharpoons \frac{1}{2}\{CH_2O\} + H^+(w) + e$	−3.01	(D)	$\frac{1}{2}\{CH_2O\} + H^+(w) + e \rightleftharpoons \frac{1}{2}CH_3OH$	−3.01
(1d)	$\frac{1}{2}CH_4(g) + \frac{1}{2}H_2O \rightleftharpoons \frac{1}{2}CH_3OH + H^+(w) + e$	+2.88	(E)	$\frac{1}{8}SO_4^{2-} + \frac{9}{8}H^+(w) + e \rightleftharpoons \frac{1}{8}HS^- + \frac{1}{2}H_2O$	−3.75
(2)	$\frac{1}{8}HS^- + \frac{1}{2}H_2O \rightleftharpoons \frac{1}{8}SO_4^{2-} + \frac{9}{8}H^+(w) + e$	−3.75	(F)	$\frac{1}{8}CO_2(g) + H^+(w) + e \rightleftharpoons \frac{1}{8}CH_4(g) + \frac{1}{4}H_2O$	−4.13
(3)	$\frac{1}{8}NH_4^+ + \frac{3}{8}H_2O \rightleftharpoons \frac{1}{8}NO_3^- + \frac{5}{4}H^+(w) + e$	+6.16	(G)	$\frac{1}{6}N_2 + \frac{4}{3}H^+(w) + e \rightleftharpoons \frac{1}{3}NH_4^+$	−4.68
(4)[1]	$FeCO_3(s) + 2\,H_2O \rightleftharpoons FeOOH(s) + HCO_3^-\ (10^{-3}) + 2\,H^+(w) + e$	−1.67			
(5)[1]	$\frac{1}{2}MnCO_3(s) + H_2O \rightleftharpoons \frac{1}{2}MnO_2(s) + \frac{1}{2}HCO_3^-\ (10^{-3}) + \frac{3}{2}H^+(w) + e$	−8.5			

Sequence of Microbial Mediation

Model 1: Excess organic material (water initially contains O_2, NO_3^-, SO_4^{2-} and HCO_3^-). Examples: Hypolimniaan of a eutrophic lake, sediments, sewage treatment plant digester.

	Combination	$pE^0(w)^2$	$\Delta G^0(w)$, kcal
Aerobic respiration	(1) + (A)	21.95	−29.9
Denitrification	(1) + (B)	20.85	−28.4
Nitrate reduction	(1) + (C)	14.36	−19.6
Fermentation[3]	(1b) + (D)	4.67	−6.4
Sulfate reduction	(1) + (E)	4.45	−5.9
Methane fermentation	(1) + (F)	4.07	−5.6
N-fixation	(1) + (G)	3.52	−4.8

Model 2: Excess O_2 (water initially contains organic material, SH^-, NH_4^+, and possibly Fe(II) and Mn(II)). Examples: aerobic waste treatment, self-purification in streams, epilimnion of lake.

	Combination	$pE^0(w)^2$	$\Delta G^0(w)$, kcal
Aerobic respiration	(A) + (1)	21.95	−29.9
Sulfide oxidation	(A) + (2)	17.50	−23.8
Nitrification	(A) + (3)	7.59	−10.3
Ferrous oxidation[4]	(A) + (4)	15.42	−21.0
Mn(II) oxidation[4]	(A) + (5)	5.75	−7.2

[1] pE° at H^+ ion activity of 1.00×10^{-7}; H^+(w) designates $[H^+] = 1.00 \times 10^{-7} M$. pE°(w) values in the left column are given for reduction, although the reaction is written as an oxidation.

[2] pE°(w) = log K(w) for a reaction written for a one-electron transfer. The term K(w) is the equilibrium constant for the reaction in which the activity of the hydrogen ion has been set at 1.00×10^{-7} and incorporated into the equilibrium constant.

[3] Fermentation is interpreted as an organic redox reaction where one organic substance is reduced by oxidizing another organic substance (for example, alcoholic fermentation; the products are metastable thermodynamically with respect to CO_2 and CH_4).

[4] The data for pE°(w) or ΔG°(w) of these reactions correspond to $a_{HCO_3^-} = 1.00 \times 10^{-3} M$ rather than to unity.

Source: W. Stumm and J.J. Morgan, *Aquatic Chemistry,* New York: Wiley-Interscience, 1970, pp. 336–7. Reproduced with permission of John Wiley and Sons, Inc.

The carbon from microbially produced methane can come from either the reduction of CO_2 or the fermentation of organic matter, particularly acetate. The anoxic production of methane can be represented in the following simplified manner. When carbon dioxide acts as an electron receptor in the absence of oxygen, methane gas is produced:

$$\frac{1}{8}CO_2 + H^+ + e^- \rightarrow \frac{1}{8}CH_4 + \frac{1}{4}H_2O \qquad (5.11.1)$$

This reaction is mediated by **methane-forming bacteria**. When organic matter is degraded microbially, the half-reaction for one electron-mole of $\{CH_2O\}$ is

$$\frac{1}{4}\{CH_2O\} + \frac{1}{4}H_2O \rightarrow \frac{1}{4}CO_2 + H^+ + e^- \qquad (5.11.2)$$

Adding Reactions 5.11.1 and 5.11.2 yields the overall reaction for the anaerobic degradation of organic matter by methane-forming bacteria, which involves a free-energy change of −5.55 kcal.

$$\frac{1}{4}\{CH_2O\} \rightarrow \frac{1}{8}CH_4 + \frac{1}{8}CO_2 \qquad (5.11.3)$$

This reaction, in reality a series of complicated processes, is a **fermentation reaction**, a redox process in which both the oxidizing agent and reducing agent are organic substances. It may be seen that only about one-fifth as much free energy is obtained from the reaction of one electron-mole using the methane-forming reaction, 5.11.3, as from a one electron-mole reaction involving complete oxidation of one electron-mole of the organic matter, 5.10.2.

There are four main categories of methane-producing bacteria. These bacteria, differentiated largely by morphology, are *Methanobacterium, Methanobacillus, Methanococcus,* and *Methanosarcina*. The methane-forming bacteria are *obligately* anaerobic; that is, they cannot tolerate the presence of molecular oxygen. The necessity of avoiding any exposure to oxygen makes the laboratory culture of these bacteria very difficult.

Methane formation is a valuable process responsible for the degradation of large quantities of organic wastes, both in biological waste-treatment processes (see Chapter 8) and in nature. Methane production is used in biological waste-treatment plants to further degrade excess sludge from the activated sludge process. In the bottom regions of natural waters, methane-forming bacteria degrade organic matter in the absence of oxygen. This eliminates organic matter which would otherwise require oxygen for its biodegradation. If this organic matter were transported to aerobic water containing dissolved O_2, it would exert a **biological oxygen demand (BOD)**. Methane production is a very efficient means for the removal of BOD. The reaction

$$CH_4 + 2\,O_2 \rightarrow CO_2 + 2\,H_2O \qquad (5.11.4)$$

shows that 1 mole of methane requires 2 moles of oxygen for its oxidation to CO_2. Therefore, the production of 1 mole of methane and its subsequent evolution from water are equivalent to the removal of 2 moles of oxygen demand. In a sense, therefore, the removal of 16 grams (1 mole) of methane is equivalent to the addition of 64 grams (2 moles) of available oxygen to the water.

There is some potential for producing methane fuel from anaerobic digestion of organic wastes. Some installations use cattle feedlot wastes. Methane is routinely generated by the action of anaerobic bacteria and is used for heat and engine fuel at sewage treatment plants (see Section 8.10). Methane produced underground in old garbage dumps is being tapped by some municipalities; however, methane seeping into basements of buildings constructed on landfill containing garbage has caused serious explosions and fires.

5.12 BACTERIAL UTILIZATION OF HYDROCARBONS

Methane is oxidized under aerobic conditions by a number of strains of bacteria. One of these, *Methanomonas,* is a highly specialized organism that cannot use any material other than methane as an energy source. Methanol, formaldehyde, and formic acid are intermediates in the microbial oxidation of methane to carbon dioxide.

Several types of bacteria can degrade higher hydrocarbons and use them as energy and carbon sources. These bacteria include *Micrococcus, Pseudomonas, Mycobacterium,* and *Nocardia.* The microbial degradation of hydrocarbons is an important environmental process because it is the primary means by which petroleum wastes are eliminated from water and soil.

The most common initial step in the microbial oxidation of alkanes involves conversion of a terminal $-CH_3$ group to a $-CO_2H$ group. More rarely, the initial enzymic attack involves addition of an oxygen atom to a nonterminal carbon, forming a ketone. After formation of a carboxylic acid from the alkane, further oxidation normally occurs by a process illustrated by the following reaction, a β-oxidation:

$$CH_3CH_2CH_2CH_2CH_2CO_2H + 3\ O_2 \rightarrow CH_3CH_2CH_2CO_2H + 2\ CO_2 + 2\ H_2O \quad (5.12.1)$$

Since 1904, it has been known that the oxidation of fatty acids involves oxidation of the β-carbon atom, followed by removal of two-carbon fragments. A complicated cycle involving a number of steps is involved. The residue at the end of each cycle is an organic acid with two fewer carbon atoms than its precursor at the beginning of the cycle.

Hydrocarbon degradability varies. Microorganisms show a strong preference for straight-chain hydrocarbons. One reason for this is that branching inhibits β-oxidation at the site of the branch. The presence of a quaternary carbon, such as

$$\cdots CH_2-\overset{\displaystyle CH_3}{\underset{\displaystyle CH_3}{\overset{|}{\underset{|}{C}}}}-CH_3$$

particularly inhibits alkane degradation.

Despite their chemical stability, aromatic rings are susceptible to microbial oxidation. The overall process leading to ring cleavage is

$$C_6H_6 \xrightarrow{O_2} C_6H_4(OH)_2 \xrightarrow{O_2} HO_2C-CH=CH-CH=CH-CO_2H \quad (5.12.2)$$

in which cleavage is preceded by addition of $-OH$ to adjacent carbon atoms. Among the microorganisms that attack aromatic rings is the fungus *Cunninghamella elegans* [4]. It metabolizes a wide range of hydrocarbons, including: C_3–C_{32} alkanes; alkenes; and aromatics, including toluene, naphthalene, anthracene, biphenyl, and phenanthrene. A study of the metabolism of naphthalene by this organism led to the isolation of the following metabolites (the percentage yields are given in parentheses):

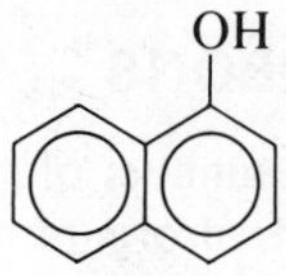

1-naphthol (67.9%) 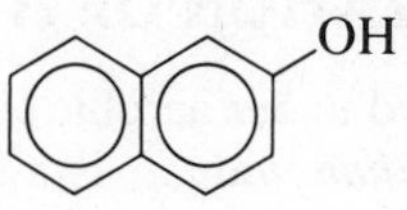 2-naphthol (6.3%) 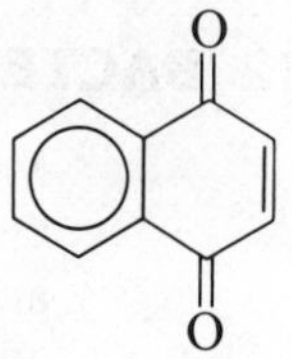 1,4-naphthoquinone (2.8%)

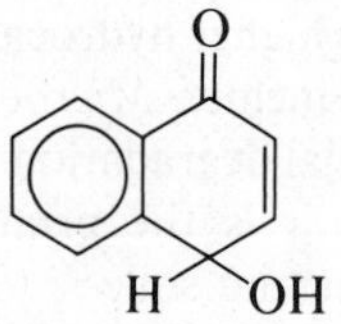

4-hydroxy-1-tetralone (16.7%) 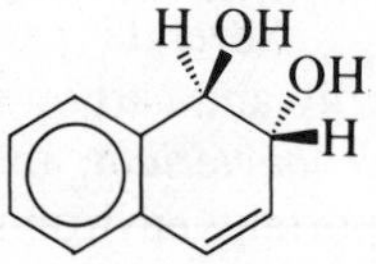 *trans*-1,2-dihydroxy-1,2-dihydronaphthalene (5.3%)

Biochemists believe that 1,2-naphthalene oxide,

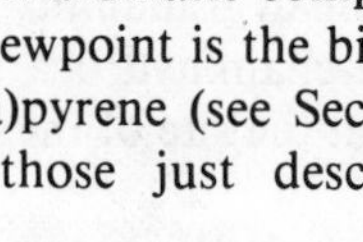

is formed by the initial attack of oxygen on naphthalene, leading to the other products shown.

The biodegradation of an aromatic compound that is particularly important from the environmental viewpoint is the biodegradation of the polynuclear aromatic hydrocarbon benzo(a)pyrene (see Section 13.16). It proceeds by oxidation processes similar to those just described for aromatic rings and naphthalene.

The biodegradation of petroleum is especially important in the elimination of oil spills (about 5×10^6 metric tons per year). This oil is degraded by both marine bacteria and filamentous fungi. In some cases, the rate of degradation is limited by available nitrate and phosphate.

The physical form of crude oil makes a large difference in its degradability. Degradation in water occurs at the water-oil interface. Therefore, thick layers of crude oil prevent contact with bacterial enzymes and O_2. Apparently, bacteria synthesize an emulsifier that keeps the oil dispersed in the water as a fine colloid and therefore accessible to the bacterial cells.

---

[4] Cerniglia, C.E., and Gibson, D.T. 1977. Metabolism of naphthalene by *Cunninghamella elegans*. *Applied and Environmental Microbiology* 34: 363–70.

During the 1950s and 1960s, biodegradation of petroleum for the synthesis of protein was investigated extensively. Increasing petroleum prices and potential contamination of the product by polynuclear aromatic hydrocarbons and other petroleum residues resulted in cancellation of several projects in this area. However, the use of methanol as a food source in an aerated medium containing water, ammonia, and essential mineral micronutrients has been employed to produce protein at the pilot-plant level [5]. The type of bacteria used is *Methylomonas clara.* A dried product, tradename *Probion,* consists of approximately 70% protein, 8% fats, 10% nucleic acids, 7% mineral ash, and 5% water, and is suitable for animal feed. A more extensively refined product containing more than 90% protein along with minerals and water is expected to be suitable for human consumption. Developments such as this may result in the utilization of waste organic materials for food by bacterial conversion to food products.

5.13 MICROBIAL UTILIZATION OF CARBON MONOXIDE

Strong evidence exists that soil microorganisms remove carbon monoxide from the environment [6]. It has been found that carbon monoxide is removed rapidly from air in contact with soil. Air containing 120 parts per million of carbon monoxide is completely freed of the contaminant after only three hours of contact with 2.8 kg of soil. Neither sterilized soil nor green plants grown under sterile conditions show any capacity to remove carbon monoxide from air.

Investigators have isolated from soil 16 different kinds of fungi capable of removing atmospheric CO. These fungi include some commonly occurring strains of the ubiquitous *Penicillium* and *Aspergillus.* In addition to the fungi isolated, it is possible that some bacteria are involved in CO removal. Although some microorganisms metabolize CO, other aquatic and terrestrial organisms produce this gas.

5.14 NITROGEN TRANSFORMATIONS BY BACTERIA

Some of the most important microorganism-mediated chemical reactions in aquatic and soil environments are those involving nitrogen compounds. Among these chemical transformations are **nitrogen fixation**, whereby molecular nitrogen is fixed as organic nitrogen; **nitrification**, the process of oxidizing ammonia to nitrate; **nitrate reduction**, the process by which nitrogen in nitrate ion is reduced to form compounds having nitrogen in a lower oxidation state; and **denitrification**, the reduction of nitrate and nitrite to N_2, with a resultant net loss of nitrogen gas to the atmosphere. Each of these important chemical processes will be discussed separately. They are summarized in the nitrogen cycle shown in Figure 5.5. This cycle describes the dynamic processes through which nitrogen is

---

[5] 1978. Single cell protein pilot unit starts up. *Chemical and Engineering News* 15 May 1978, p. 20.

[6] Inman, R.E. 1971. Soil bugs eat CO. *Chemical and Engineering News* 10 May 1971, p. 24.

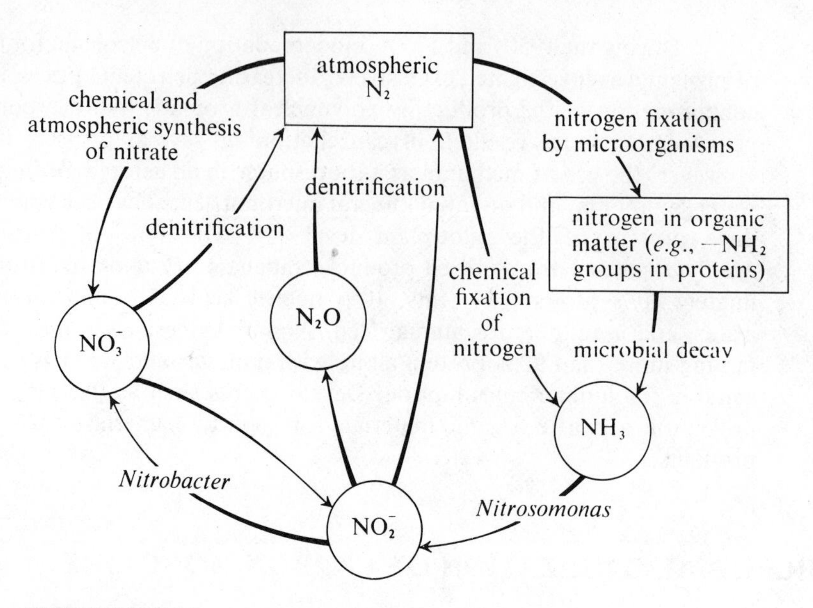

FIGURE 5.5 The nitrogen cycle.

interchanged among the atmosphere, organic matter, and inorganic compounds. It is one of nature's most vital dynamic processes.

The degradation of organic nitrogen compounds is an important type of microbial process. Nitrilotriacetic acid (NTA), the chelating detergent builder discussed in Chapter 4, is an example of such a nitrogen-containing organic compound whose environmental degradation is important. Although readily degraded in fresh waters, sewage-treatment systems, and soil, NTA does not biodegrade well in ocean or estuary waters [7]. This may be due to inhibition of the NTA monooxygenase enzyme by NaCl or by some other unknown inhibitory factors in estuary and sea water.

5.15 NITROGEN FIXATION

The overall microbial process for the fixation of atmospheric nitrogen in a chemically bound form,

$$3\ \{CH_2O\} + 2\ N_2 + 3\ H_2O + 4\ H^+ \rightarrow 3\ CO_2 + 4\ NH_4^+ \qquad (5.15.1)$$

is actually a complicated process and is not completely understood. Biological nitrogen fixation is a key biochemical process in the environment and is essential for plant growth in the absence of synthetic fertilizers.

Only a few species of aquatic microorganisms have the ability to fix atmospheric nitrogen. Among the aquatic bacteria having this capability are photosynthetic bacteria, *Azotobacter*, and several species of *Clostridium*. Among the

[7] Bourquin, A.W., and Przybyszewski, V.A. 1977. Distribution of bacteria with nitrilotriacetate—degrading potential in an estuarine environment. *Applied and Environmental Microbiology* 34: 411–8.

algae, blue-green algae can fix atmospheric nitrogen. In most natural fresh water systems, however, the fraction of nitrogen fixed by organisms in the water, relative to that originating from the decay of organic material, fertilizer runoff, and other external sources, is quite low.

The best-known and most important form of nitrogen-fixing bacteria is *Rhizobium*, which enjoys a *symbiotic* (mutually advantageous) relationship with leguminous plants such as clover or alfalfa. The *Rhizobium* bacteria are found in **root nodules**, special structures attached to the roots of legumes (see Figure 10.5). The nodules develop as a result of the bacteria irritating the root hairs of the developing legume plants. The nodules are connected directly to the vascular (circulatory) system of the plant, enabling the bacteria to derive photosynthetically-produced energy directly from the plant. Thus, the plant provides the energy required to break the strong triple bonds in the dinitrogen molecule, converting the nitrogen to a reduced form which is directly assimilated by the plant [8]. When the legumes die and decay, NH_4^+ ion is released and is converted by microorganisms to nitrate ion which is assimilable by other plants. Some of the ammonium ion and nitrate released may be carried into natural water systems.

Some nonlegume angiosperms fix nitrogen through the action of actinomycetes bacteria contained in root nodules. Shrubs and trees in this nitrogen-fixing category are abundant in fields, forests, and wetlands throughout the world. Their rate of nitrogen fixation is comparable to that of legumes.

Free-living bacteria associated with some grasses are stimulated by the grasses to fix nitrogen [8]. One such bacterium is *Spirillum lipoferum*. In tropical surroundings, the amount of reduced nitrogen fixed by such bacteria can amount to the order of 100 kg per hectare per year.

Because of the cost of energy required to fix nitrogen synthetically, efforts are underway to increase the efficiency of natural means of nitrogen fixation. One approach uses recombinant DNA methodologies (see Chapter 18) in attempts to transfer the nitrogen-fixing capabilities of nitrogen-fixing bacteria directly to plant cells. Though a fascinating possibility, this transfer has not yet been achieved on a practical basis. The other approach uses more conventional plant breeding and biological techniques in attempts to increase the range and effectiveness of the symbiotic relationship existing between some plants and nitrogen-fixing bacteria.

One matter of concern is that successful efforts to increase nitrogen fixation may upset the global nitrogen balance. Total global fixation of nitrogen was estimated at 237 million metric tons in 1974 [9]. This is up from an estimated 174 million metric tons in 1950 and 150 million metric tons in 1850. Potential accumulation of excess fixed nitrogen is the subject of some concern because of aquatic nitrate pollution and microbial production of N_2O gas. Some atmospheric scientists fear that excess N_2O gas may be involved in depletion of the protective atmospheric ozone layer (see Sections 11.9 and 12.11). There has been speculation that continuation of the "agricultural revolution," accompanied by increased evolution of N_2O as an indirect result of greater production of

---

[8] Rawls, R. 1980. Nitrogen fixation research advances. *Chemical and Engineering News* 20 December 1980, pp. 14–16.

[9] 1976. Worldwide nitrogen fixation estimated. *Chemical and Engineering News* 16 February 1976.

TABLE 5.2 Major Sources of Fixed Nitrogen[1]

Process[2]	*Percent of total N fixed*
Natural processes on agricultural land	38
Natural processes on forested and uncultivated land	25
Synthetic fixation	24
Combustion	9
Lightning	4

[1]Based upon estimates in Reference [9].
[2]An estimated 1 million metric tons of nitrogen, 0.4% of the total, is fixed in the oceans each year.

chemically and biologically fixed nitrogen, could seriously threaten the ozone layer. Table 5.2 lists the major sources of fixed nitrogen.

5.16 NITRIFICATION

Nitrification, the conversion of N(−III) to N(V), is a very common and extremely important process in water and in soil. As we have seen, aquatic nitrogen in thermodynamic equilibrium with air is in the +5 oxidation state, NO_3^-, whereas in most biological compounds, nitrogen is present as N(−III), such as $-NH_2$ in amino acids. The equilibrium constant for the overall nitrification reaction, written for one electron-mole,

$$\frac{1}{4}O_2 + \frac{1}{8}NH_4^+ \rightarrow \frac{1}{8}NO_3^- + \frac{1}{4}H^+ + \frac{1}{8}H_2O \quad (5.16.1)$$

is $10^{7.59}$ (Table 5.1), so the reaction is highly favored from a thermodynamic viewpoint.

Nitrification is especially important in nature because nitrogen is absorbed by plants primarily as nitrate. When fertilizers are applied in the form of ammonium salts or anhydrous ammonia, a microbial transformation to nitrate enables maximum assimilation of nitrogen by the plants.

Nitrification, the conversion of ammoniacal nitrogen to nitrate ion, takes place if extensive aeration is allowed to occur in the activated sludge sewage-treatment process (see Section 8.10). As the sewage sludge settles out in the settler, the bacteria in the sludge uses this nitrate as an oxygen source, producing N_2 gas (see denitrification, Section 5.18). The bubbles of nitrogen gas cause the sludge to rise, so that it does not settle properly. This can hinder the proper treatment of sewage through carryover of sludge into effluent water.

In nature, nitrification is catalyzed by two groups of bacteria, *Nitrosomonas* and *Nitrobacter*. *Nitrosomonas* bacteria bring about the transition of ammonia to nitrite,

$$NH_3 + \frac{3}{2}O_2 \rightarrow H^+ + NO_2^- + H_2O \quad (5.16.2)$$

whereas *Nitrobacter* mediate the oxidation of nitrite to nitrate:

$$NO_2^- + \frac{1}{2}O_2 \rightarrow NO_3^- \qquad (5.16.3)$$

Both of these highly specialized types of bacteria are **obligate aerobes**; that is, they function only in the presence of molecular O_2. These bacteria are also **chemolithotrophic**, meaning that they can utilize oxidizable inorganic materials as electron donors in oxidation reactions to yield needed energy for metabolic processes.

For the aerobic conversion of one electron-mole of ammoniacal nitrogen to nitrite ion at pH 7.00,

$$\frac{1}{6}NH_4^+ + \frac{1}{4}O_2(g) \rightarrow \frac{1}{6}NO_2^- + \frac{1}{3}H^+ + \frac{1}{6}H_2O \qquad (5.16.4)$$

the free-energy change is -10.8 kcal. The free-energy change for the aerobic oxidation of one electron-mole of nitrite ion to nitrate ion,

$$\frac{1}{2}NO_2^- + \frac{1}{4}O_2(g) \rightarrow \frac{1}{2}NO_3^- \qquad (5.16.5)$$

is -9.0 kcal. Both steps of the nitrification process involve an appreciable yield of free energy. It is interesting to note that the free-energy yield per electron-mole is approximately the same for the conversion of NH_4^+ to NO_2^- as it is for the conversion of NO_2^- to NO_3^-, about 10 kcal/electron-mole.

5.17 NITRATE REDUCTION

The general term **nitrate reduction** refers to microbial processes by which nitrogen in chemical compounds is reduced to lower oxidation states. In the absence of free oxygen, nitrate may be used by some bacteria as an alternate electron receptor. The most complete possible reduction of nitrogen in nitrate ion involves the acceptance of 8 electrons by the nitrogen atom, with the consequent conversion of nitrate to ammonia (+V to −III oxidation state). Nitrogen is an essential component of protein, and any organism that utilizes nitrogen from nitrate for the synthesis of protein must first reduce the nitrogen to the −III oxidation state (ammoniacal form). However, incorporation of nitrogen into protein generally is a relatively minor use of the nitrate undergoing microbially mediated reactions and is more properly termed **nitrate assimilation**.

Generally, when nitrate ion functions as an electron receptor, the product is NO_2^-:

$$\frac{1}{2}NO_3^- + \frac{1}{4}\{CH_2O\} \rightarrow \frac{1}{2}NO_2^- + \frac{1}{4}H_2O + \frac{1}{4}CO_2 \qquad (5.17.1)$$

The free-energy yield per electron-mole is only about 2/3 of the yield when oxygen is the electron receptor; however, nitrate ion is a good electron receptor in the absence of molecular oxygen. One of the factors limiting the use of nitrate ion in this function is the relatively low concentration of NO_3^- in most waters. Furthermore, nitrite, NO_2^-, is relatively toxic and tends to inhibit the growth of many bacteria after building up to a certain level. Sodium nitrate is sometimes used as a "first-aid" treatment in sewage lagoons that have become oxygen-deficient. It provides an emergency source of oxygen to reestablish normal bacterial growth.

5.18 DENITRIFICATION

An important special case of nitrate reduction is **denitrification**, in which the reduced nitrogen product is a nitrogen-containing gas. At pH 7.00, the free-energy change per electron-mole of reaction,

$$\frac{1}{5}NO_3^- + \frac{1}{4}\{CH_2O\} + \frac{1}{5}H^+ \rightarrow \frac{1}{10}N_2(g) + \frac{1}{4}CO_2(g) + \frac{7}{20}H_2O \qquad (5.18.1)$$

is -2.84 kcal. The free-energy yield per mole of nitrate reduced to N_2 (5 electron-moles) is lower than that for the reduction of the same quantity of nitrate to nitrite. More important, however, the reduction of a nitrate ion to N_2 gas consumes 5 electrons, compared to only 2 electrons for the reduction of NO_3^- to NO_2^-.

Denitrification is an important process in nature. It is the mechanism by which fixed nitrogen is returned to the atmosphere. Denitrification is also useful in advanced water treatment for the removal of nutrient nitrogen (see Section 8.19). Because nitrogen gas is a nontoxic volatile substance that does not inhibit microbial growth, and since nitrate ion is a very efficient electron acceptor, denitrification allows the extensive growth of bacteria under anaerobic conditions.

Loss of nitrogen to the atmosphere may also occur through the formation of N_2O and NO by bacterial action on nitrate and nitrite, catalyzed by the action of several types of bacteria. Production of N_2O relative to N_2 is enhanced during denitrification in soils by increased concentrations of NO_3^-, NO_2^-, and O_2 [10]. Nitrous oxide production is influenced by the synthesis and presence of nitrogenous oxide reductase enzymes (see Section 18.2). Although it was once believed that N_2O was formed only during the reduction of nitrate and nitrite, it has been demonstrated [11] that nitrification of ammonium ion and urea (a widely used fertilizer discussed in Section 10.12) results in the production of N_2O by the action of several types of bacteria, including *Nitrosomonas europaea*. As discussed in Chapter 12, N_2O and NO participate in atmospheric chemical processes, and their increased production as a result of greater use of chemically fixed nitrogen fertilizer may be of some concern.

5.19 COMPETITIVE OXIDATION OF ORGANIC MATTER BY NITRATE ION AND OTHER OXIDIZING AGENTS

The successive oxidation of organic matter by dissolved O_2, NO_3^-, and SO_4^{2-} brings about an interesting sequence of nitrate-ion levels in sediments and hypolimnion waters initially containing O_2 but lacking a mechanism for reaeration [12]. This is shown in Figure 5.6, where concentrations of dissolved O_2,

---

[10] Firestone, M.K.; Firestone, R.B.; and Tiedje, J.M. 1980. Nitrous oxide from soil denitrification: factors controlling its biological production. *Science* 208: 749–51.

[11] Bremner, J.M., and Blackmer, A.M. 1978. Nitrous oxide: emission from soils during nitrification of fertilizer nitrogen. *Science* 199: 295–6.

[12] Bender, M.L.; Fanning, K.A.; Froelich, P.N.; and Maynard, V. 1977. Interstitial nitrate profiles and oxidation of sedimentary organic matter in eastern equatorial Atlantic. *Science* 198: 605–8.

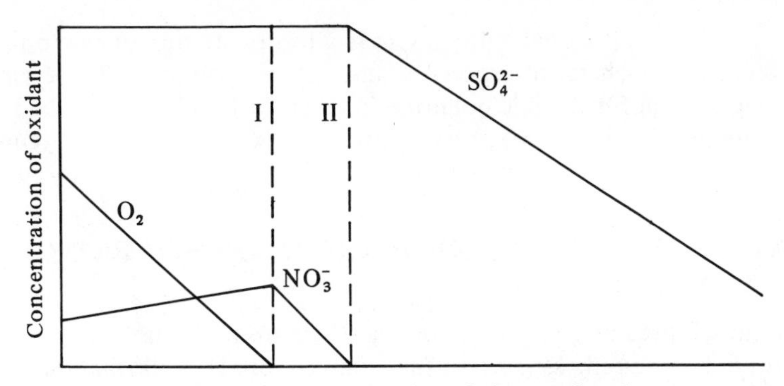

FIGURE 5.6 Oxidation of organic matter by O_2, NO_3^-, and SO_4^{2-}.

NO_3^-, and SO_4^{2-} are plotted as a function of total organic matter metabolized. This behavior is explained by the following sequence of (unbalanced) chemical reactions:

$$\{(CH_2O)_{106}(NH_2)_{16}P\} + O_2 \rightarrow CO_2 + NO_3^- + H_2PO_4^- + H_2O + H^+ \quad (5.19.1)$$

$$\{(CH_2O)_{106}(NH_2)_{16}P\} + NO_3^- + H^+ \rightarrow CO_2 + N_2 + H_2PO_4^- + H_2O \quad (5.19.2)$$

$$\{(CH_2O)_{106}(NH_2)_{16}P\} + SO_4^{2-} + H^+ \rightarrow CO_2 + NH_4^+ + H_2PO_4^- + H_2S + H_2O \quad (5.19.3)$$

So long as some O_2 is present, some nitrate may be produced from organic matter. After exhaustion of molecular oxygen, nitrate is the favored oxidizing agent, and its concentration falls from a maximum value (I) to zero (II). Sulfate, which is usually present in a large excess over the other two oxidants, then becomes the favored electron receptor, enabling biodegradation of organic matter to continue.

5.20 MICROBIAL TRANSFORMATIONS OF SULFUR

Sulfur compounds are very common in water. Sulfate ion, SO_4^{2-}, is found in varying concentrations in practically all natural waters. Organic sulfur compounds, both those of natural origin and pollutant species, are very common in natural aquatic systems, and the degradation of these compounds is an important microbial process. Sometimes the degradation products, such as the odiferous and toxic H_2S, cause serious problems with water quality.

There is a strong analogy between sulfur in the environment and nitrogen in the environment. Sulfur in living material is present primarily in its most reduced state, for example, as the hydrosulfide group, $-SH$. Nitrogen in living material is present in the (−III) oxidation state, for example, as $-NH_2$. When organic sulfur compounds are decomposed by bacteria, the initial sulfur product is generally the reduced form, H_2S. When organic nitrogen compounds are decomposed by microorganisms, the reduced form of nitrogen, NH_4^+, is produced. Just as some microorganisms can produce elemental nitrogen from nitrogen compounds, some bacteria produce and store elemental sulfur from sulfur compounds. In the presence of oxygen, some bacteria convert reduced forms of sulfur to the oxidized form in SO_4^{2-} ion, whereas other bacteria catalyze the oxidation of reduced nitrogen compounds to nitrate ion.

The total global sulfur cycle is an intricate mix of microbially mediated processes, geochemical processes, and SO_2 air pollution. The cycle is discussed along with sulfur dioxide pollution in Chapter 12, because of the role played by pollutant SO_2. The atmospheric sulfur cycle is summarized in Figure 12.1.

5.21 OXIDATION OF H_2S AND REDUCTION OF SULFATE BY BACTERIA

Although organic sulfur compounds often are the source of H_2S in water, they are not required as the sulfur source for H_2S formation. The bacteria *Desulfovibrio* can reduce sulfate ion to H_2S. In so doing, they utilize sulfate as an electron acceptor in the oxidation of organic matter. The overall reaction is:

$$SO_4^{2-} + 2\{CH_2O\} + 2H^+ \rightarrow H_2S + 2CO_2 + 2H_2O \quad (5.21.1)$$

Actually, other bacteria besides *Desulfovibrio* are required to oxidize organic matter completely to CO_2. The oxidation of organic matter by *Desulfovibrio* generally terminates with acetic acid, and accumulation of acetic acid is characteristic of these bacteria. The activities of these bacteria are particularly evident in bottom waters. Because of the high concentration of sulfate ion in seawater, bacterially mediated formation of H_2S causes pollution problems in some coastal areas and is a major source of atmospheric sulfur. In waters where sulfide formation occurs, the sediment is often black in color due to the formation of FeS.

Bacterially-mediated reduction of sulfur in calcium sulfate deposits produces elemental sulfur interspersed in the pores of the limestone product. The highly generalized chemical reaction for this process is

$$2\,CaSO_4 + 3\,\{CH_2O\} \xrightarrow[\text{(bacteria)}]{} 2\,CaCO_3 + 2\,S + CO_2 + 3\,H_2O \quad (5.21.2)$$

although the stoichiometric amount of free sulfur is never found in these deposits due to the formation of volatile H_2S, which escapes.

Whereas some bacteria can reduce sulfate ion to H_2S, others can oxidize hydrogen sulfide to higher oxidation states. The purple sulfur bacteria and green sulfur bacteria derive energy for their metabolic processes through the oxidation of H_2S. These bacteria utilize CO_2 as a carbon source and are strictly anaerobic. The aerobic colorless sulfur bacteria may use molecular oxygen to oxidize H_2S,

$$2\,H_2S + O_2 \rightarrow 2\,S + 2\,H_2O \quad (5.21.3)$$

elemental sulfur,

$$2\,S + 2\,H_2O + 3\,O_2 \rightarrow 4\,H^+ + 2\,SO_4^{2-} \quad (5.21.4)$$

or thiosulfate ion:

$$S_2O_3^{2-} + H_2O + 2\,O_2 \rightarrow 2\,H^+ + 2\,SO_4^{2-} \quad (5.21.5)$$

Oxidation of sulfur in a low oxidation state to sulfate ion produces sulfuric acid, a strong acid. One of the colorless sulfur bacteria, *Thiobacillus thiooxidans,* is tolerant of 1*N* acid solutions, a remarkable acid tolerance. When elemental sulfur is added to excessively alkaline soils, the acidity is increased because of a microorganism-mediated reaction (5.21.4), which produces sulfuric acid. Elemental sulfur may be deposited as granules in the cells of purple sulfur bacteria and colorless sulfur bacteria. Such processes are important sources of elemental sulfur deposits.

5.22 MICROORGANISM-MEDIATED DEGRADATION OF ORGANIC SULFUR COMPOUNDS

Sulfur occurs in many types of biological compounds. As a consequence, organic sulfur compounds of natural and pollutant origin are very common in water. The degradation of these compounds is an important microbial process having a strong effect upon water quality.

Among some of the common sulfur-containing functional groups found in aquatic organic compounds are hydrosulfide (—SH), disulfide (—S—S—), sulfide (—S—), sulfoxide ($-\overset{O}{\overset{\|}{S}}-$), sulfonic acid ($-SO_2OH$), thioketone ($-\overset{S}{\overset{\|}{C}}-$), and thiazole (a heterocyclic sulfur group). Protein contains some amino acids with sulfur functional groups, whose breakdown is important in natural waters. Among the more common sulfur-containing amino acids are:

$$\underset{\text{cysteine}}{HO_2C{-}CH(NH_2){-}CH_2{-}SH}$$

$$\underset{\text{cystine}}{HO_2C{-}CH(NH_2){-}CH_2{-}S{-}S{-}CH_2{-}CH(NH_2){-}CO_2H}$$

$$\underset{\text{methionine}}{HO_2C{-}CH(NH_2){-}CH_2{-}CH_2{-}S{-}CH_3}$$

These amino acids are readily broken down by bacteria and fungi.

The biodegradation of sulfur-containing amino acids can result in production of volatile organic sulfur compounds such as methyl thiol, CH_3SH, and dimethyl disulfide, CH_3SSCH_3. These compounds have strong, unpleasant odors. Their formation, in addition to that of H_2S, accounts for much of the odor associated with the biodegradation of sulfur-containing organic compounds.

Hydrogen sulfide is formed from a large variety of organic compounds through the action of a number of different kinds of microorganisms. A typical sulfur-cleavage reaction producing H_2S is the conversion of cysteine to pyruvic acid through the action of cysteine desulfhydrase enzyme in bacteria:

$$HS{-}\underset{H}{\overset{H}{C}}{-}\underset{NH_2}{\overset{H}{C}}{-}CO_2H + H_2O \xrightarrow[\text{cysteine desulfhydrase}]{\text{bacteria}} CH_3{-}\overset{O}{\overset{\|}{C}}{-}CO_2H + H_2S + NH_3 \quad (5.22.1)$$

Because of the numerous forms in which organic sulfur may exist, a variety of sulfur products and biochemical reaction paths must be associated with the biodegradation of organic sulfur compounds. Much remains to be learned in this area.

5.23 MICROBIAL CONVERSIONS OF SELENIUM

Directly below sulfur in the periodic table, **selenium** is subject to bacterial oxidation and reduction [13]. These transitions are important because selenium is a crucial element in nutrition, particularly of livestock. Diseases related to either selenium excesses or deficiency have been reported in at least half of the states of the U.S. and in 20 other countries, including the major livestock-producing countries. Livestock in New Zealand, in particular, suffer from selenium deficiency.

Microorganisms are closely involved with the selenium cycle, and microbial reduction of oxidized forms of selenium has been known for some time [13]. More recently, a soil-dwelling strain of *Bacillus megaterium* has been found to be capable of oxidizing elemental selenium to selenite, SeO_3^{2-} [13]. This observation provides a heretofore missing link in the selenium cycle.

5.24 IRON AND MANGANESE BACTERIA

Some bacteria, including *Ferrobacillus, Gallionella,* and some forms of *Sphaerotilus,* utilize iron compounds in obtaining energy for their metabolic needs. These bacteria catalyze the oxidation of iron(II) to iron(III) by molecular oxygen:

$$4\ Fe(II) + 4\ H^+ + O_2 \rightarrow 4\ Fe(III) + 2\ H_2O \tag{5.24.1}$$

The carbon source for some of these bacteria is CO_2. Since they do not require organic matter for carbon, and because they derive energy from the oxidation of inorganic matter, these bacteria may thrive in environments where organic matter is absent.

The microorganism-mediated oxidation of iron(II) is not a particularly efficient means of obtaining energy for metabolic processes. For the reaction

$$FeCO_3(s) + \frac{1}{4}O_2 + \frac{3}{2}H_2O \rightarrow Fe(OH)_3(s) + CO_2 \tag{5.24.2}$$

the change in free energy is approximately 10 kcal/electron-mole. Approximately 220 g of iron(II) must be oxidized to produce 1.0 g of cell carbon. The calculation assumes CO_2 as a carbon source and a biological efficiency of 5%. The production of only 1.0 g of cell carbon would produce approximately 430 g of solid $Fe(OH)_3$. It follows that large deposits of hydrated iron(III) oxide form in areas where iron-oxidizing bacteria thrive.

Some of the iron bacteria, notably *Gallionella,* secrete large quantities of hydrated iron(III) oxide in the form of intricately branched structures. The bacterial cell grows at the end of a twisted stalk of the iron oxide. Individual cells of *Gallionella,* photographed through an electron microscope have shown [2] that the stalks consist of a number of strands of iron oxide secreted from one side of the cell (Fig. 5.7).

---

[13] Sarathchandra, S.U., and Watkinson, J.H. 1981. Oxidation of elemental selenium to selenite by *Bacillus megaterium. Science* 211: 600–601.

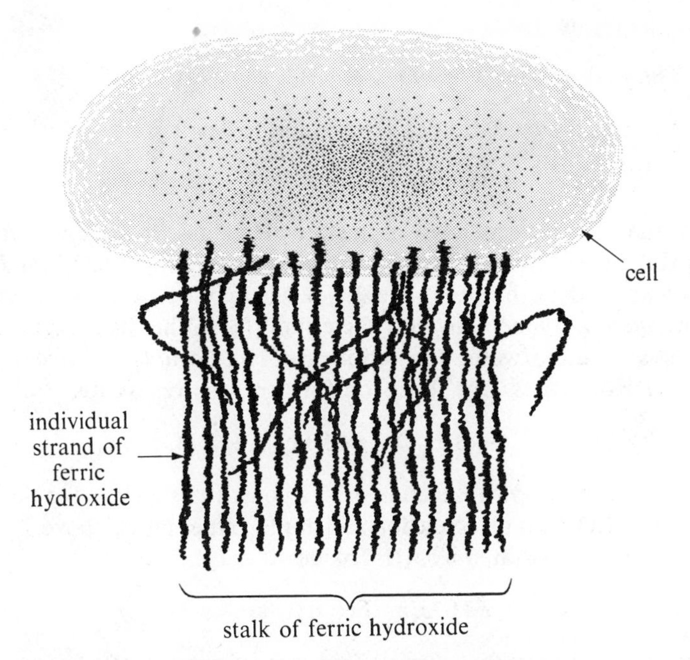

FIGURE 5.7 Sketch of a cell of *Gallionella,* showing ferric hydroxide secretion.

At nearly neutral pH values, bacteria deriving energy by mediating the air oxidation of iron(II) must compete with direct chemical oxidation of iron(II) by O_2. The latter process is relatively rapid at pH 7. As a consequence, these bacteria tend to grow in a narrow layer in the region between the oxygen source and the source of iron(II). Therefore, iron bacteria are sometimes called *gradient organisms,* and they grow at intermediate pE values.

Bacteria are strongly involved in the oceanic manganese cycle. Manganese nodules, a potentially important source of manganese, copper, nickel, and cobalt (see Section 17.2) yield different species of bacteria which mediate both the oxidation and reduction of manganese. These reactions are enzymatic in nature and are influenced by seawater cations, particularly Ca^{2+} and Mg^{2+} [14].

5.25 ACID MINE WATERS

Acid mine drainage is one of the most common and damaging problems in the aquatic environment. Many waters flowing from coal mines and draining from the "gob piles" left over from coal processing and washing are practically sterile due to high acidity.

Acid mine water results from the presence of sulfuric acid produced by the oxidation of pyrite, FeS_2. Microorganisms are closely involved in the overall process, which consists of several reactions [15]. The first of these reactions is the

---

[14] Ghiorse, W.C., and Ehrlich, H.L. 1974. Effects of seawater cations and temperature on manganese dioxide-reductase activity in a marine bacillus. *Applied Microbiology* 28: 785–792.

[15] Singer, P. and Stumm, W. 1970. *Science* 167: 121.

oxidation of pyrite:

$$2\ FeS_2(s) + 2\ H_2O + 7\ O_2 \rightarrow 4\ H^+ + 4\ SO_4^{2-} + 2\ Fe^{2+} \quad (5.25.1)$$

The next step is the oxidation of ferrous ion to ferric ion,

$$4\ Fe^{2+} + O_2 + 4\ H^+ \rightarrow 4\ Fe^{3+} + 2\ H_2O \quad (5.25.2)$$

a process that occurs very slowly at the low pH values found in acid mine waters. Below pH 3.5, the iron oxidation is catalyzed by the iron bacterium *Thiobacillus ferrooxidans,* and in the pH range 3.5–4.5 it may be catalyzed by a variety of *Metallogenium,* a filamentous iron bacterium [16]. Other bacteria that may be involved in acid mine water formation are *Thiobacillus thiooxidans* and *Ferrobacillus ferrooxidans.* The ferric ion further dissolves pyrite,

$$FeS_2(s) + 14\ Fe^{3+} + 8\ H_2O \rightarrow 15\ Fe^{2+} + 2\ SO_4^{2-} + 16\ H^+ \quad (5.25.3)$$

which in conjunction with Reaction 5.25.2 constitutes a cycle for the dissolution of pyrite. $Fe(H_2O)_6^{3+}$ is an acidic ion and at pH values much above 3, the iron(III) precipitates as the hydrated iron(III) oxide

$$Fe^{3+} + 3\ H_2O \rightleftharpoons Fe(OH)_3(s) + 3\ H^+ \quad (5.25.4)$$

The beds of streams afflicted with acid mine drainage often are covered with "yellowboy," an unsightly deposit of amorphous, semigelatinous $Fe(OH)_3$. The most damaging component of acid mine water, however, is sulfuric acid. It is directly toxic and has other undesirable effects.

The prevention and cure of acid mine water is one of the major challenges facing the environmental chemist. One approach to eliminating excess acidity involves the use of carbonate rocks. When acid mine water is treated with limestone, the following reaction occurs:

$$CaCO_3(s) + 2\ H^+ + SO_4^{2-} \rightarrow Ca^{2+} + SO_4^{2-} + H_2O + CO_2(g) \quad (5.25.5)$$

Unfortunately, because iron(III) is generally present Reaction 5.25.4 occurs as the pH is raised. The hydrated iron(III) oxide that forms as a result of elevated pH soon covers the particles of carbonate rock with a relatively impermeable layer. This armoring effect prevents further neutralization of the acid.

5.26 MICROBIAL DEGRADATION OF PESTICIDES

The major types of pesticides and their uses are discussed in Section 7.18. The biodegradation of pesticides in the aquatic and terrestrial environments is quite important to environmental quality and varies greatly. **Herbicides**, which are designed for plant control, and **insecticides**, which are used to control insects, generally do not have any detrimental effect upon microorganisms. However, effective **fungicides** must be antimicrobial in action. Therefore, in addition to killing harmful fungi, fungicides frequently harm beneficial **saprophytic fungi** (fungi that decompose dead organic matter) and bacteria.

---

[16] Walsh, F., and Mitchell, R. 1972. A pH-dependent succession of iron bacteria. *Environmental Science and Technology* 6: 809–12.

The biodegradation of pesticides by microorganisms occurs by way of a number of stepwise, microbially catalyzed reactions [17]. These reactions will be discussed individually with examples.

Oxidation occurs by the action of *oxygenase enzymes* (see Chapter 18 for a discussion of biochemical terms). The microbially catalyzed conversion of aldrin to dieldrin is an example of epoxide formation, a major step in many oxidation mechanisms.

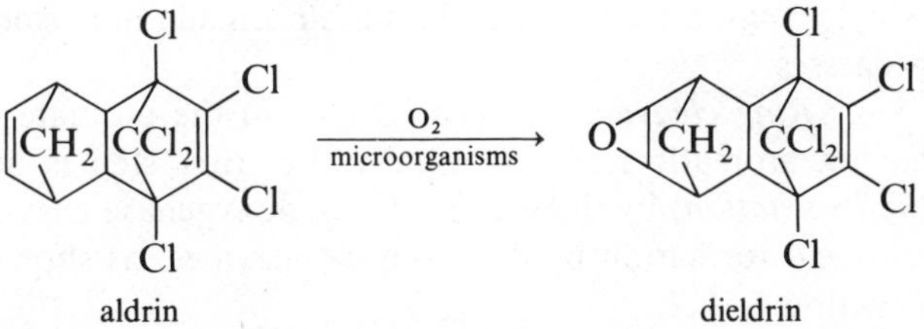

$$\text{aldrin} \xrightarrow[\text{microorganisms}]{O_2} \text{dieldrin} \quad (5.26.1)$$

Introduction of $-OH$ groups onto aromatic rings is another major oxidation step. Hydrocarbon chains are often oxidized by β-cleavage; see Reaction 5.12.1.

Reduction of pesticides is largely confined to conversion of $-NO_2$ groups to $-NH_2$ groups. Reduction of quinones to phenols is also possible.

Hydrolysis is a third major step in microbial degradation of pesticides. Esters and amides are normally hydrolyzed. At least one species of *Pseudomonas* hydrolyzes malathion in a type of hydrolysis reaction typical of those by which pesticides are degraded:

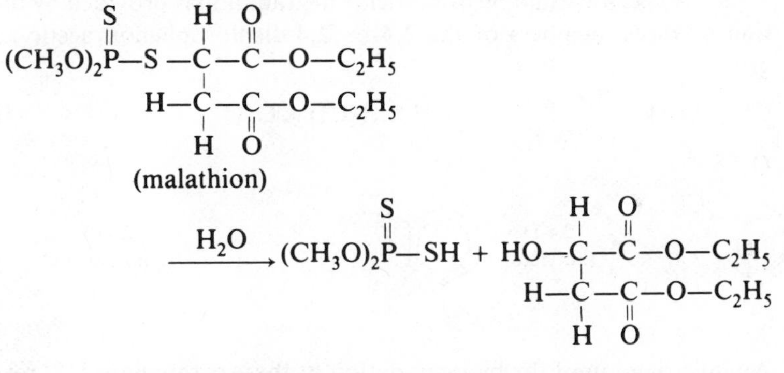

$$(CH_3O)_2P(=S)\text{—S—CH(—C(=O)—O—}C_2H_5\text{)—CH}_2\text{—C(=O)—O—}C_2H_5 \text{ (malathion)}$$

$$\xrightarrow{H_2O} (CH_3O)_2P(=S)\text{—SH} + \text{HO—CH(—C(=O)—O—}C_2H_5\text{)—CH}_2\text{—C(=O)—O—}C_2H_5 \quad (5.26.2)$$

Dehalogenation reactions involving the replacement of a halogen atom with $-OH$ are mediated by some bacteria:

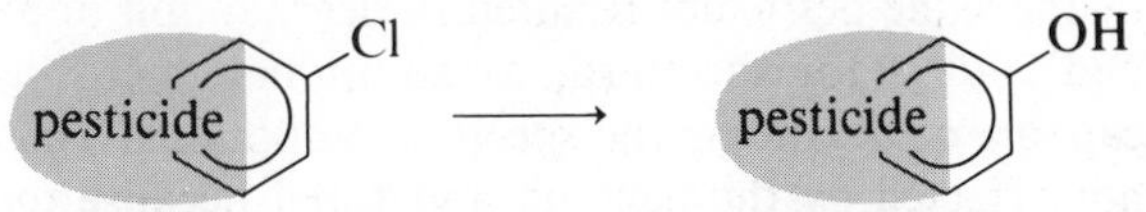

$$\text{pesticide—}C_6H_4\text{—Cl} \longrightarrow \text{pesticide—}C_6H_4\text{—OH}$$

This is a major pathway for the degradation of organochlorine hydrocarbons.

Alkyl groups attached to oxygen, sulfur, or nitrogen atoms can be removed by *dealkylation* reactions, such as the removal of an ethyl group from the herbicide simazine, shown in Equation 5.26.3 on the next page.

---

[17] Higgins, I.J. and Burns, R.G. 1975. *The chemistry and microbiology of pollution.* New York: Academic Press.

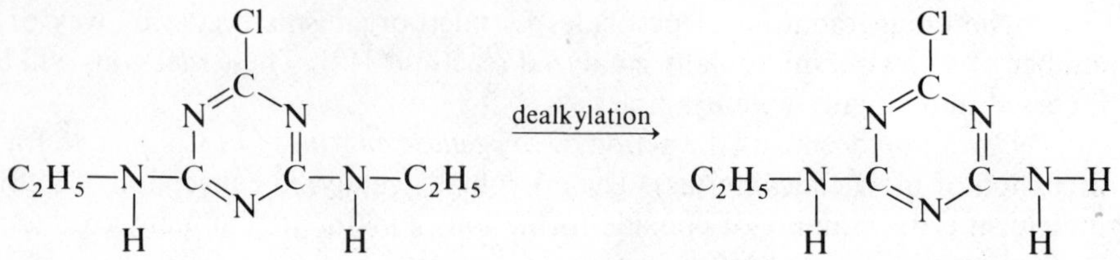

$$C_2H_5{-}NH{-}C_3N_3(Cl){-}NH{-}C_2H_5 \xrightarrow{\text{dealkylation}} C_2H_5{-}NH{-}C_3N_3(Cl){-}NH{-}H \qquad (5.26.3)$$

Alkyl groups attached to carbon are normally not removed directly by microbial processes.

Ring cleavage is a crucial step in the ultimate degradation of pesticides having aromatic rings. Normally, the first step is addition of $-OH$ groups (*hydroxylation*) by the action of monooxygenase enzymes, followed by cleavage and acid formation by dioxygenase enzymes, as shown for an aromatic ring in Reaction 5.12.2.

Condensation reactions may occur that effectively deactivate a pesticide. These reactions involve joining the pesticide molecule to some other organic molecule.

It is fortunate that microorganisms need not utilize a particular pesticide as a sole carbon source in order to affect its degradation. This is due to the phenomenon of **cometabolism**, which results from a lack of specificity in the microbial degradation processes. Thus, bacterial degradation of small amounts of a pesticide may occur while the microorganism involved is metabolizing much larger quantities of another substrate.

A specific example of pesticide degradation is provided by the biodegradation of three members of the 2,4-D (2,4-dichlorophenoxyacetic acid) herbicide group:

2,4-D (acetate): $CH_2CO_2H{-}O{-}C_6H_3Cl_2$ (Cl at 2 and 4 positions)

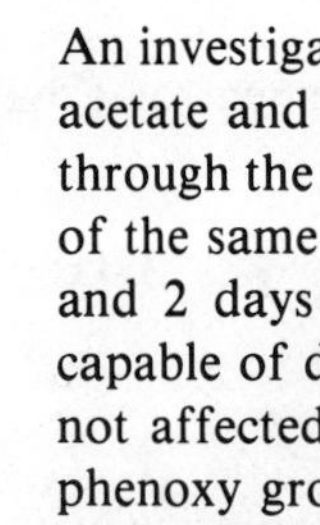

2,4-D (butyrate): $CH_2CH_2CH_2CO_2H{-}O{-}C_6H_3Cl_2$ (Cl at 2 and 4 positions)

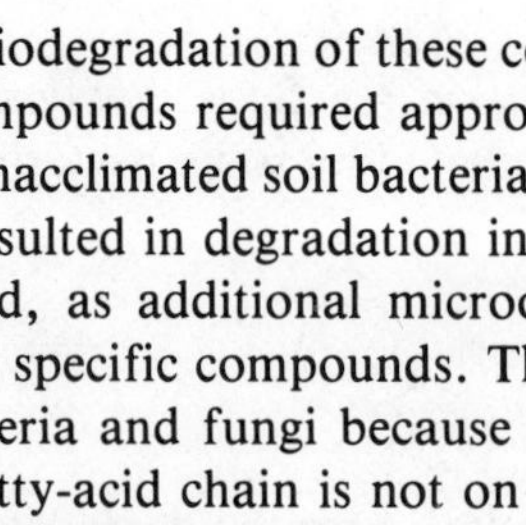

2,4-D (α-propionate): $H_3CCHCO_2H{-}O{-}C_6H_3Cl_2$ (Cl at 2 and 4 positions)

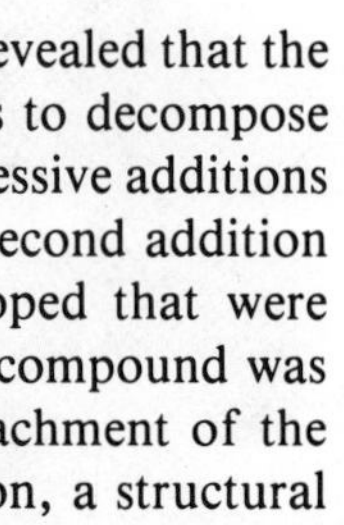

An investigation of the biodegradation of these compounds [18] revealed that the acetate and butyrate compounds required approximately 15 days to decompose through the action of nonacclimated soil bacteria and fungi. Successive additions of the same pesticides resulted in degradation in 5 days for the second addition and 2 days for the third, as additional microorganisms developed that were capable of degrading the specific compounds. The α-propionate compound was not affected by the bacteria and fungi because the point of attachment of the phenoxy group to the fatty-acid chain is not on a terminal carbon, a structural feature that hinders biological action.

Further studies were performed on compounds of the 2,4-D type to determine structural factors inhibiting biodegradation. Two structural features were

---

[18] Alexander, M. 1964. Microbiology of pesticides and related hydrocarbons. In *Principles and applications in aquatic microbiology,* H. Heukelekian and N.C. Dondero, eds. pp. 15–42. New York: John Wiley and Sons, Inc.

found to have a strong influence upon biodegradation. The first is the presence of chlorine on carbon atoms 3 or 5 (*meta*) relative to the linkage to the fatty-acid group; compounds with this feature do not biodegrade at a significant rate. Second, compounds in which the fatty acid is linked to the aromatic ring at the end position opposite the CO_2H group degrade readily, whereas compounds having linkage of the aromatic ring to the α carbon (adjacent to the CO_2H group) are virtually nondegradable.

There appear to be two general reaction paths for the degradation of 2, 4-dichlorophenoxy alkyl carboxylic acids, represented by the following general formula:

$(CH_2)_xCO_2H$ — O — (2,4-dichlorophenyl ring: Cl at 2, Cl at 4) $= Cl_2\phi O(CH_2)_xCO_2H$

In the compounds examined in the previously mentioned studies, the value of x ranged from 2–7. When these compounds were degraded by the bacteria *Nocardia,* the following decomposition products were obtained (the compound preceding the arrow is the parent compound):

$$Cl_2\phi O(CH_2)_3CO_2H \rightarrow Cl_2\phi O(CH_2)CO_2H$$
$$Cl_2\phi O(CH_2)_4CO_2H \rightarrow Cl_2\phi O(CH_2)_2CO_2H \quad (5.26.4)$$
$$Cl_2\phi O(CH_2)_5CO_2H \rightarrow Cl_2\phi O(CH_2)_3CO_2H$$
$$Cl_2\phi O(CH_2)_6CO_2H \rightarrow Cl_2\phi O(CH_2)_4CO_2H$$

The initial attack of *Nocardia* is probably a β-cleavage of the aliphatic chain:

$$Cl_2\phi OCH_2CH_2CH_2CH_2CO_2H + 3\,O_2 \longrightarrow Cl_2\phi OCH_2CH_2CO_2H + 2\,CO_2 + 2\,H_2O \quad (5.26.5)$$

It is likely that the biochemical pathway of this reaction is very similar to that of the well-established β-oxidation of fatty acids.

When the compounds just discussed are degraded by *Flavobacterium,* the products are alkyl carboxylic acids:

$$Cl_2\phi(CH_2)_3CO_2H \rightarrow (CH_2)_3CO_2H \quad (5.26.6)$$
$$Cl_2\phi(CH_2)_4CO_2H \rightarrow (CH_2)_4CO_2H$$

The chlorinated aromatic ring residues generally are not observed. Therefore, a possible reaction path for the degradation of these herbicides by *Flavobacterium* is shown in Equation 5.26.7 on the next page.

$$\text{Cl}_2\text{C}_6\text{H}_3\text{–O–(CH}_2)_x\text{CO}_2\text{H} \longrightarrow \text{Cl}_2\text{C}_6\text{H}_3\text{OH} + \text{H}_3\text{C(CH}_2)_{x-1}\text{CO}_2\text{H}$$

$$\longrightarrow \text{(Cl)} + \text{Cl–C}_6\text{H}_3\text{(OH)}_2 \longrightarrow \text{HO}_2\text{C–C=C–C(Cl)=C–CO}_2\text{H} \longrightarrow \text{degradation products} \quad (5.26.7)$$

Synthetic herbicides such as 2,4,5-trichlorophenoxyacetic acid (identical in structure to 2,4-D, except that it has an additional Cl in the 5 position) generally degrade poorly because they are present at such low levels that they are not used as a sole carbon source by microorganisms. A technique known as *plasmid-assisted molecular breeding* has been utilized to develop bacterial strains capable of the total degradation of 2,4,5-T present as a sole source of carbon at concentrations exceeding 1 mg/mL [19]. Additional advances in the breeding of bacterial strains capable of degrading specific pollutants may be anticipated, particularly as techniques of genetic engineering advance.

One of the most extensively used general insecticides is **parathion** (O, O-diethyl-O-*p*-nitrophenylphosphorothioate), which is a rather toxic organophosphate compound:

$$(\text{C}_2\text{H}_5\text{O})_2\overset{\text{S}}{\overset{\|}{\text{P}}}\text{–O–C}_6\text{H}_4\text{–NO}_2$$

Microbial detoxification of this compound in wastes is an attractive possibility, particularly considering that the standard chemical treatment for it (hydrolysis to *p*-nitrophenol and diethylthiophosphoric acid in alkali) is slow and produces pollutant products. It has been found [20] that mixed microbial cultures can degrade parathion at the rate of 50 $\text{mg/L}^{-1}\text{h}^{-1}$. However, no single bacterial species has been found capable of degrading the pesticide. Thus, the degradation mechanism must require symbiotic action among several microorganisms. It was found that the enzymatic hydrolysis of parathion produces *p*-nitrophenol,

$$\text{HO–C}_6\text{H}_4\text{–NO}_2$$

[19] Kellogg, S.T.; Chatterjee, K.K.; and Chakrabarty, A.M. 1981. Plasmid-assisted molecular breeding: New technique and enhanced biodegradation of persistent toxic chemicals. *Science* 214: 1133–5.

[20] Munnecke, D.M., and Hsieh, D.P.H. 1974. Microbial decontamination of parathion and *p*-nitrophenol in aqueous media. *Applied Microbiology* 28: 212–7.

which further hydrolyzes to nitrous acid, HNO_2, and hydroquinone,

HO–⟨◯⟩–OH

as a metabolic intermediate. An extensive lag period was attributed to inhibition by *p*-nitrophenol metabolite.

The preceding treatment provides only a few examples of pesticide biodegradation. Pesticides in common use are composed of literally hundreds of chemical formulations. Many possible reaction paths are involved in their degradation. One possibility that must be kept in mind is that the degradation intermediates may be more toxic to some organisms than are the original pesticides. An example of this is the bioconversion of DDT to DDD:

$$\text{Cl}-C_6H_4-\underset{\text{H}}{\overset{\text{CCl}_3}{\text{C}}}-C_6H_4-\text{Cl} \longrightarrow \text{Cl}-C_6H_4-\underset{\text{H}}{\overset{\text{H-C(Cl)}_2}{\text{C}}}-C_6H_4-\text{Cl}$$

DDT DDD (5.26.8)

The latter compound is more toxic to some insects than DDT and is even manufactured as a pesticide. (The same situation applies to microbially mediated conversion of aldrin to dieldrin.)

5.27 MICROBIAL CORROSION

Corrosion is a redox phenomenon [21] and was discussed in Section 3.13. Much corrosion is bacterial in nature. Bacteria involved with corrosion set up their own electrochemical cells in which a portion of the surface of the metal being corroded forms the anode of the cell and is oxidized. Structures called **tubercles** form in which bacteria pit and corrode metals as shown in Figure 5.8.

It is beyond the scope of this book to discuss corrosion in detail. However, its significance and effects should be kept in mind by the environmental chemist.

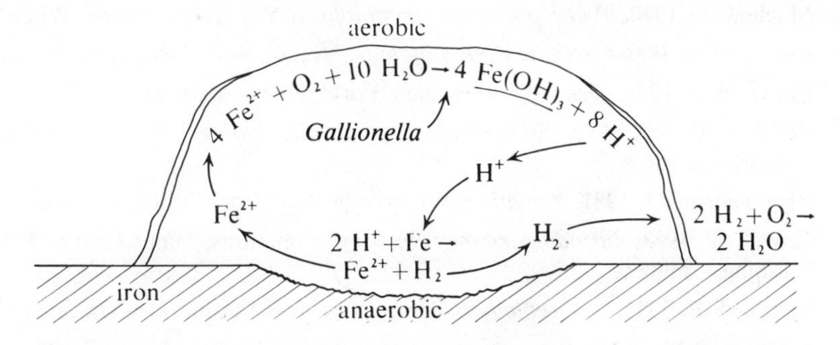

FIGURE 5.8 Tubercle in which the bacterially mediated corrosion of iron occurs through the action of *Gallionella*.

---

[21] Laque, F.L. 1973. Corrosion. In *The encyclopedia of chemistry,* 3rd ed., pp. 309–311. New York: Van Nostrand Reinhold.

SUPPLEMENTARY REFERENCES

Alexander, M. 1974. Microbial formation of environmental pollutants. In *Advances in applied microbiology*. Vol. 18, pp. 1–74. D. Perlman, ed. New York: Academic Press.

Bollag, J-M. 1974. Microbial transformations of pesticides. In *Advances in applied microbiology,* Vol. 18, pp. 75–131. D. Perlman, ed. New York: Academic Press.

Broughton, W.J., ed. 1981. *Nitrogen fixation.* Vol. 1. New York: Clarendon (Oxford University Press).

Burns, R.C., and Hardy, R.W.F. 1975. *Nitrogen fixation in bacteria and higher plants.* New York: Springer-Verlag, Inc.

Dart, R.K., and Stretton, R.J. 1977. *Microbiological aspects of pollution control.* Amsterdam: Elsevier Scientific Publishing Co.

Edmonds, P. 1978. *Microbiology—an environmental perspective.* New York: Macmillan.

Fenchel, T., and Blackburn, T.H. 1979. *Bacteria and mineral cycling.* New York: Academic Press.

Hardy, R.W.F., ed. 1977. *A treatise on dinitrogen fixation.* New York: Wiley-Interscience.

Higgins, I.G., and Burns, R.G. 1975. *The chemistry and microbiology of pollution.* New York: Academic Press.

Hill, I.R., and Wright, S.J.L., eds. 1978. *Pesticide microbiology.* New York: Academic Press.

Krumbein, W.E., ed. 1978. *Environmental biogeochemistry and geomicrobiology.* Ann Arbor, Mich.: Ann Arbor Science Publishers, Inc.

Kuzentsov, S.I. 1975. *The microflora of lakes and its geochemical significance.* C.H. Oppenheimer, ed. Austin, Texas: Univ. of Texas Press.

Laskin, A.I., ed. 1982. *Advances in applied microbiology.* New York: Academic Press, Inc.

Leisinger, T.; Hutter, R.; Cook, A.M.; and Nüesch, J., eds. 1981. *Microbial degradation of xenobiotics and recalcitrant compounds.* New York: Academic Press.

Maki, A.W.; Dickson, K.L.; and Cairns, J., eds. 1980. *Biotransformation and fate of chemicals in the aquatic environment.* Washington, D.C.: American Society for Microbiology.

Matsumura, F., and Krishna Murti, C.R., eds. 1982. *Biodegradation of pesticides.* New York: Plenum Publishing Corp.

McKinney, R.E. 1962. *Microbiology for sanitary engineers.* New York: McGraw-Hill Book Co.

Mitchell, R. 1970. *Water pollution microbiology.* Vol. 1. New York: Wiley-Interscience.

——. 1978. *Water pollution microbiology.* Vol. 2. New York: Wiley-Interscience.

Payne, W.J. 1981. *Denitrification.* New York: Wiley-Interscience.

Perlman, D., ed. 1982. *Advances in applied microbiology.* Issued annually. New York: Academic Press.

Rheinheimer, G. 1981. *Aquatic microbiology.* New York: Wiley-Interscience.

Ross, F.C. 1983. *Introductory microbiology.* Columbus, Ohio: Charles E. Merrill Publishing Company.

Round, F.E. 1981. *The ecology of algae.* New York: Cambridge University Press.

Sharpley, J.M. 1966. *Elementary petroleum microbiology.* Houston, Texas: Gulf Publishing Co.

Stanier, R.Y.; Adelberg, E.A.; Ingraham, J.L.; and Wheelis, M.L. 1979. *The microbial world.* Englewood, N.J.: Prentice-Hall, Inc.

Starr, M.P.; Ingraham, J.L.; and Raffel, S., eds. 1983. *Annual review of microbiology.* Published annually. Palo Alto, Calif.: Annual Reviews.

Umbreit, W.W. 1976. *Essentials of bacterial physiology.* Stroudsburg, Pa.: Dowden, Hutchinson, and Ross, Inc.

QUESTIONS AND PROBLEMS

1. As $CH_3CH_2CH_2CH_2CO_2H$ biodegrades in several steps to carbon dioxide and water, various chemical species are observed. What stable chemical species would be observed as a result of the first step of this degradation process?

2. Which of the following statements is true regarding the production of methane in water?
   (a) It occurs in the presence of oxygen.
   (b) It consumes oxygen.
   (c) It removes biological oxygen demand from the water.
   (d) It is accomplished by aerobic bacteria.
   (e) It produces more energy per electron-mole than does aerobic respiration.

3. At time zero, the cell count of a bacterial species mediating aerobic respiration of wastes was 1×10^6 cells per liter. At 30 minutes it was 2×10^6; at 60 minutes it was 4×10^6; at 90 minutes, 7×10^6; at 120 minutes, 10×10^6; and at 150 minutes, 13×10^6. From this data, which of the following logical conclusions would you draw?
   (a) The culture was entering the log phase at the end of the 150-minute period.
   (b) The culture was in the log phase throughout the 150-minute period.
   (c) The culture was leaving the log phase at the end of the 150-minute period.
   (d) The culture was in the lag phase throughout the 150-minute period.
   (e) The culture was in the death phase throughout the 150-minute period.

4. Which of the following structures is a relatively biodegradable water pollutant?

(a)

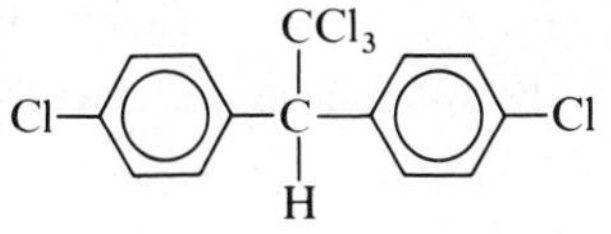

(b)

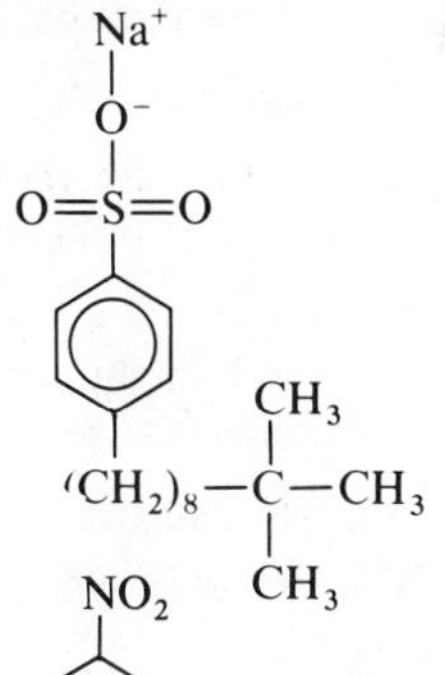

(c)

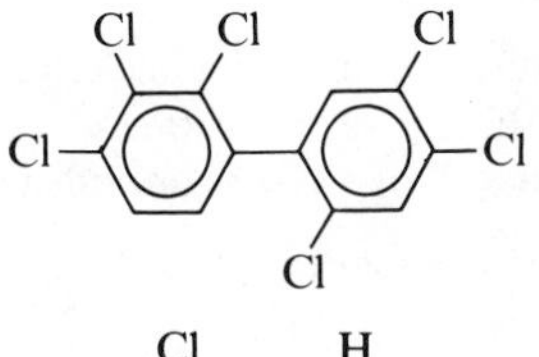

(d)

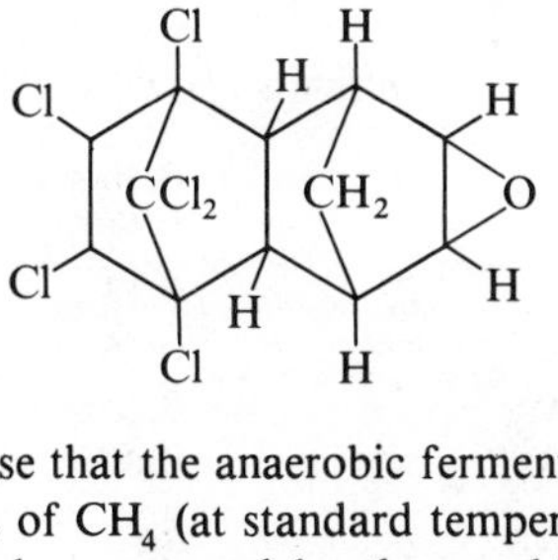

(e)

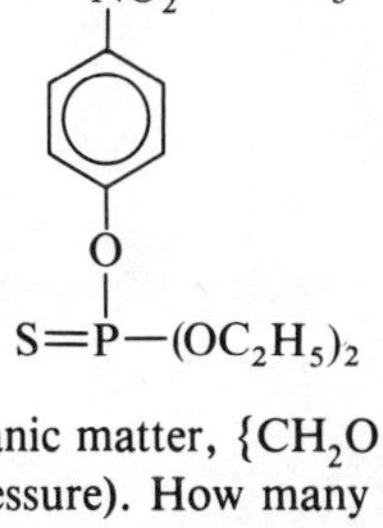

5. Suppose that the anaerobic fermentation of organic matter, $\{CH_2O\}$, in water yields 15.0 L of CH_4 (at standard temperature and pressure). How many grams of oxygen would be consumed by the aerobic respiration of the same quantity of $\{CH_2O\}$? (Recall the significance of 22.4 L in chemical reactions of gases.)

6. What weight of $FeCO_3(s)$, using Reaction 4 in Table 5.1, gives the same free-energy yield as 1.00 g of organic matter, using Reaction 1, when oxidized by oxygen at pH 7.00?

7. How many bacteria would be produced after 10 hours by one bacterial cell, assuming exponential growth with a generation time of 20 minutes?

8. Referring to Reaction 5.16.1, calculate the concentration of ammonium ion in equilibrium with oxygen in the atmosphere and $1.00 \times 10^{-5} M\ NO_3^-$ at pH 7.00.

9. When a bacterial nutrient medium is inoculated with bacteria grown in a markedly different medium, the lag phase (Fig. 5.3) often is quite long, even if the bacteria eventually grow well in the new medium. Can you explain this behavior?

10. Most plants assimilate nitrogen as nitrate ion. However, ammonia (NH_3) is a popular and economical fertilizer. What essential role do bacteria play when ammonia is used as a fertilizer? Do you think any problems might occur when using ammonia in a waterlogged soil lacking oxygen?

11. Why is the growth rate of bacteria as a function of temperature (Fig. 5.4) not a symmetrical curve?

12. Discuss the analogies between bacteria and a finely divided chemical catalyst.

13. Would you expect autotrophic bacteria to be more complex physiologically and biochemically than heterotrophic bacteria? Why?

14. Wastewater containing 8 mg/L O_2 (atomic weight O = 16), $1.00 \times 10^{-3} M\ NO_3^-$, and $1.00 \times 10^{-2} M$ soluble organic matter, $\{CH_2O\}$, is stored isolated from the atmosphere in a container richly seeded with a variety of bacteria. Assume that denitrification is one of the processes which will occur during storage. After the bacteria have had a chance to do their work, which of the following statements will be true?
    (a) No $\{CH_2O\}$ will remain.
    (b) Some O_2 will remain.
    (c) Some NO_3^- will remain.
    (d) Denitrification will have consumed more of the organic matter than aerobic respiration.
    (e) The composition of the water will remain unchanged.

15. Of the four classes of microorganisms—algae, fungi, bacteria, and virus—which has the least influence on water chemistry?

16. Figure 5.2 shows the main structural features of a bacterial cell. Which of these do you think might cause the most trouble in water-treatment processes such as filtration or ion exchange, where the maintenance of a clean, unfouled surface is critical? Explain.

17. The day after a heavy rain washed a great deal of cattle feedlot waste into a farm pond, the following counts of bacteria were obtained:

Time	*Thousands of viable cells per mL*
6:00 A.M.	0.10
7:00 A.M.	0.11
8:00 A.M.	0.13
9:00 A.M.	0.16
10:00 A.M.	0.20
11:00 A.M.	0.40
12:00 noon	0.80
1:00 P.M.	1.60
2:00 P.M.	3.20

To which portion of the bacterial growth curve, Figure 5.3, does this time span correspond?

18. A bacterium capable of degrading 2,4-D herbicide was found to have its maximum growth rate at 32 °C. Its growth rate at 12 °C was only 10% of the maximum. Do you think there is another temperature at which the growth rate would also be 10% of the maximum? If you believe this to be the case, of the following temperatures, choose the one at which it is most plausible for the bacterium to also have a growth rate of 10% of the maximum: 52 °C, 37 °C, 8 °C, 20 °C.

19. Addition of which two half-reactions in Table 5.1 is responsible for:
    (a) Elimination of an algal nutrient in secondary sewage effluent using methanol as a carbon source.
    (b) A process responsible for a bad-smelling pollutant when bacteria grow in the absence of oxygen.
    (c) A process that converts a common form of commercial fertilizer to a form that most crop plants can absorb.
    (d) A process responsible for the elimination of organic matter from wastewater in the aeration tank of an activated sludge sewage-treatment plant.
    (e) A characteristic process that occurs in the anaerobic digester of a sewage treatment plant.

20. What is the surface area in square meters of 1.00 gram of spherical bacterial cells, 1.00 micron in diameter, having a density of 1.00 g/cm^3?

21. What is the purpose of exoenzymes in bacteria?

22. Match each species of bacteria listed in the left column with its function on the right.

(a) *Spirillum lipoferum*	(1) reduces sulfate to H_2S
(b) *Bacillus megaterium*	(2) catalyzes oxidation of Fe^{2+} to Fe^{3+}
(c) *Thiobacillus ferrooxidans*	(3) fixes nitrogen in grasses
(d) *Desulfovibrio*	(4) oxidizes elemental selenium to selenite

23. What factors favor the production of methane in anoxic surroundings?

CHAPTER 6
LIQUID-SOLID-GAS INTERACTIONS IN AQUATIC CHEMISTRY

6.1 CHEMICAL INTERACTIONS INVOLVING SOLIDS, GASES, AND WATER

Homogeneous chemical reactions occurring entirely in aqueous solution are rather rare in natural waters and wastewaters. Instead, most significant chemical and biochemical phenomena in water involve interactions between species in water and in another phase. Some of these important interactions are illustrated in Figure 6.1. Production of solid biomass through the photosynthetic activity of algae (see Section 5.5) occurs within a suspended algal cell and involves exchange of dissolved solids and gases between the surrounding water and the cell. Similar exchanges are involved when bacteria degrade organic matter (often in the form of small particles) in water. Chemical reactions occur that produce solids or gases in water. Iron and many important trace-level elements are transported through aquatic systems as colloidal chemical compounds or are sorbed to solid particles. Pollutant hydrocarbons and some pesticides may be present on the water surface as an immiscible liquid film. Sedimentary material can be washed physically into a body of water.

The solids involved in aquatic chemical interactions may be placed in the two categories of *sediments* (bulk solids) and *suspended colloidal material.* Colloidal material, which may also consist of gases or water-immiscible liquids, is very reactive because of its high surface area per unit weight. Colloids are involved with many significant phenomena in aquatic chemistry and are discussed in some detail in this chapter.

6.2 FORMATION OF SEDIMENTS

Sediments typically consist of mixtures of clay, silt, sand, organic matter, and various minerals. Their composition may range from pure mineral matter to predominantly organic material. Sediments are deposited in the bottom regions of bodies of water through a number of physical, chemical, and biological processes. For example, the sedimentary material may be simply carried into a body of water by erosion or through sloughing (caving-in) of the shore. Thus, clay, sand, organic matter, and other materials may be washed into a lake and settle out as layers of sediment. Simple precipitation reactions can occur, a few examples of which follow. When a phosphate-rich wastewater enters a body of very hard water (high calcium-ion concentration), the following reaction occurs, producing hydroxyapatite:

$$5\,Ca^{2+} + OH^- + 3\,PO_4^{3-} \rightarrow Ca_5OH(PO_4)_3(s) \qquad (6.2.1)$$

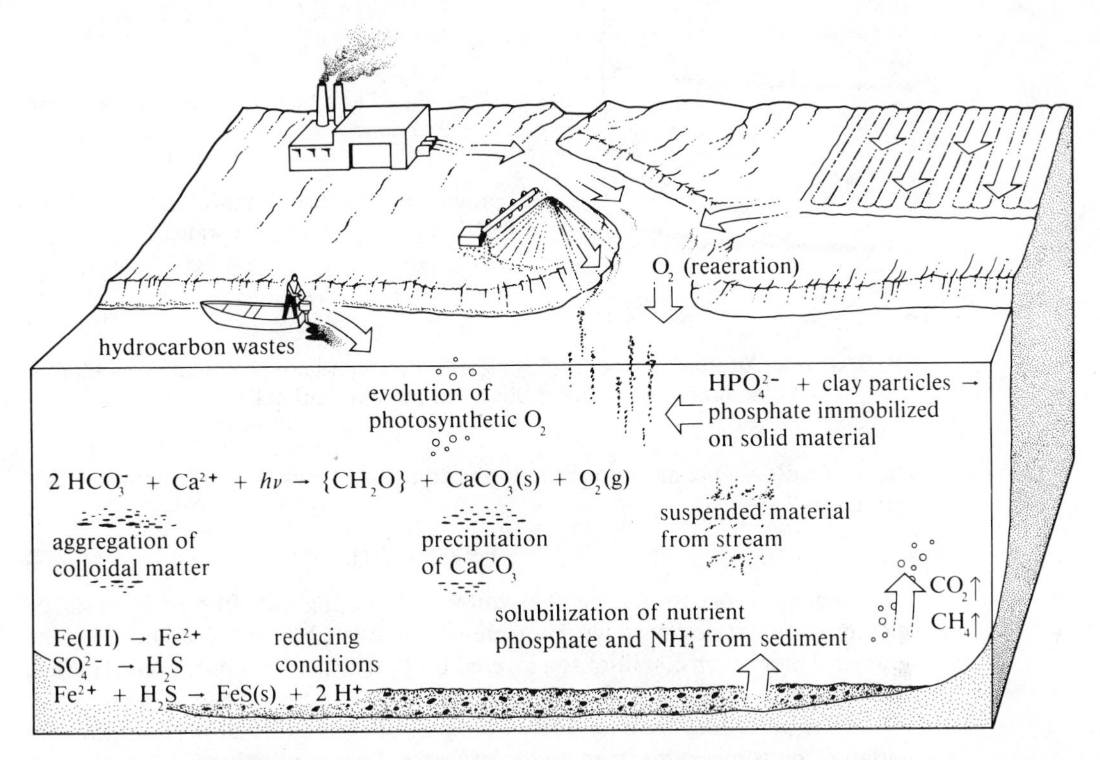

FIGURE 6.1 Many significant chemical, biochemical, and physical processes in water involve interactions among different phases.

Calcium carbonate sediment may form when water rich in carbon dioxide and containing a high level of calcium as temporary hardness (see Section 8.4) loses carbon dioxide to the atmosphere,

$$Ca^{2+} + 2\,HCO_3^- \rightarrow CaCO_3(s) + CO_2 + H_2O \qquad (6.2.2)$$

or when the pH is raised by a photosynthetic reaction:

$$Ca^{2+} + 2\,HCO_3^- + h\nu \rightarrow \{CH_2O\} + CaCO_3(s) + O_2 \qquad (6.2.3)$$

Oxidation of reduced forms of an element can result in its transformation into an insoluble species; this occurs when iron(II) is oxidized to iron(III), producing a precipitate of insoluble ferric hydroxide:

$$4\,Fe^{2+} + 10\,H_2O + O_2 \rightarrow 4\,Fe(OH)_3(s) + 8\,H^+ \qquad (6.2.4)$$

A decrease in pH can result in the production of an insoluble humic-acid sediment from base-soluble organic humic substances in solution (see Section 4.19).

Biological activity is responsible for the formation of some aquatic sediments. Some bacterial species produce large quantities of iron(III) oxide (see Section 5.24) as part of their energy-extracting mediation of the oxidation of iron(II) to iron(III). In anaerobic bottom regions, some bacteria use sulfate ion as an electron acceptor,

$$SO_4^{2-} \rightarrow H_2S \qquad (6.2.5)$$

whereas other bacteria reduce iron(III) to iron(II):

$$Fe(OH)_3(s) \rightarrow Fe^{2+} \qquad (6.2.6)$$

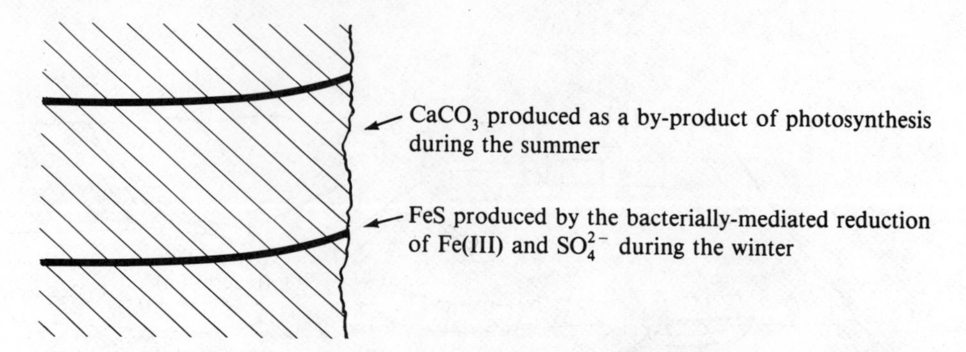
$CaCO_3$ produced as a by-product of photosynthesis during the summer
FeS produced by the bacterially-mediated reduction of Fe(III) and SO_4^{2-} during the winter

FIGURE 6.2 Alternate layers of $CaCO_3$ and FeS in a lake sediment. This kind of sediment structure is observed in Lake Zürich [1].

The net result is a precipitation reaction producing a black layer of ferrous sulfide sediment:

$$Fe^{2+} + H_2S \rightarrow FeS(s) + 2\,H^+ \tag{6.2.7}$$

This frequently occurs during the winter, alternating with the production of calcium carbonate by-product from photosynthesis (Reaction 6.2.3) during the summer. Under such conditions, a layered bottom structure is produced [1] made up of alternate layers of FeS and $CaCO_3$, as shown in Figure 6.2.

The preceding are only a few examples of reactions that result in the formation of bottom sediments in bodies of water. Eventually, these sediments may become covered and form sedimentary minerals.

6.3 SURFACE SORPTION BY SOLIDS

Many of the properties and effects of solids in contact with water have to do with the sorption of solutes by solid surfaces [2]. Surfaces in finely divided solids tend to have excess surface energy. This is the result of an imbalance of chemical forces among surface atoms, ions, and molecules. Surface energy level may be lowered by a reduction in surface area. Normally, this reduction is accomplished by aggregation of particles or by sorption of solute species.

Interactions at surfaces are best understood in the case of metal oxides. Figure 6.3 represents an oxide of a hypothetical metal, MO. Figure 6.3(a) represents a piece of metal oxide. The surface metal ions are coordinatively unsaturated (that is, capable of accepting electron pairs from donor atoms) Lewis acids. The unsaturated condition may be overcome by acceptance of electron pairs from oxygen atoms provided by coordination with water, as shown in Figure 6.3(b). Commonly, the chemisorbed water molecules dissociate, as shown

---

[1] Stumm, W., and Stumm-Zollinger, E. 1972. The role of phosphorus in eutrophication. In *Water pollution microbiology,* pp. 11–42. R. Mitchell, ed. New York: Wiley-Interscience.

[2] Schindler, P.W. 1981. Surface complexes at oxide-water interfaces. In *Adsorption of inorganics at solid-liquid interfaces,* pp. 1–50. M.A. Anderson and A.J. Rubin, eds. Ann Arbor, Mich.: Ann Arbor Science Publishers, Inc.

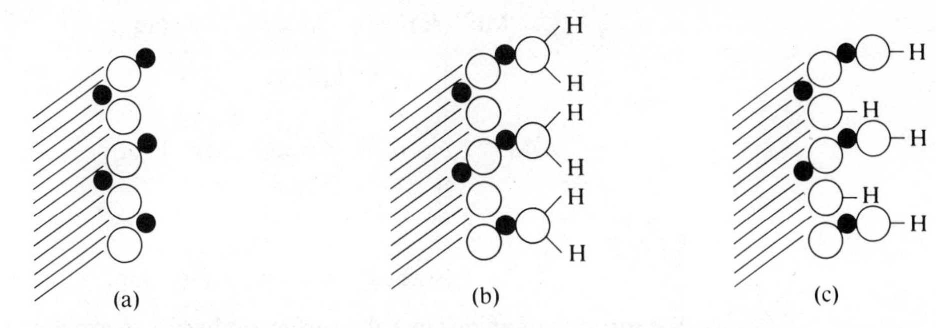

F HEAD

FIGURE 6.3 Representation of the surface of a metal oxide, MO, showing sorption and dissociation of water molecules. Each o represents an oxygen atom; each ●, a metal atom.

in Figure 6.3(c) and represented by the reaction

$$\mathrm{MO} + \mathrm{H_2O} \rightarrow \mathrm{M(OH)_2} \tag{6.3.1}$$

Filling the surface coordination sites (where electron-pair donor groups may be bonded to the surface) with hydroxyl groups yields a **hydroxylated surface**. Solutes in water may interact in a number of ways with hydroxylated surfaces. Predominant among these interactions are acid-base reactions of the type

$$\mathrm{M{-}OH} \rightleftharpoons \mathrm{MO^-} + \mathrm{H^+} \tag{6.3.2}$$

$$\mathrm{M{-}OH} + \mathrm{H^+} \rightleftharpoons \mathrm{M{-}OH_2^+} \tag{6.3.3}$$

where M represents a monovalent unit of the metal oxide. A metal ion, Mt^{z+}, may be complexed at the oxide surface,

$$\mathrm{M{-}OH} + \mathrm{Mt}^{z+} \rightleftharpoons \mathrm{M{-}OMt}^{z-1} + \mathrm{H^+} \tag{6.3.4}$$

or chelation may occur

$$\begin{array}{l}\mathrm{M{-}OH}\\ \\ \mathrm{M{-}OH}\end{array} + \mathrm{Mt}^{z+} \rightleftharpoons \begin{array}{l}\mathrm{M{-}O}\diagdown\\ \qquad\quad \mathrm{Mt}^{z-2}\\ \mathrm{M{-}O}\diagup\end{array} + 2\mathrm{H^+} \tag{6.3.5}$$

A metal ion complexed with a ligand, L, may bond by displacement of either $\mathrm{H^+}$ or $\mathrm{OH^-}$:

$$\mathrm{M{-}OH} + \mathrm{Mt}^{z+}\mathrm{L} \rightleftharpoons \mathrm{MOMtL}^{(z-1)} + \mathrm{H^+} \tag{6.3.6}$$

$$\mathrm{M{-}OH} + \mathrm{Mt}^{z+}\mathrm{L} \rightleftharpoons \mathrm{M{-}MtL}^{(z+1)} + \mathrm{OH^-} \tag{6.3.7}$$

Furthermore, in the presence of a ligand, dissociation of the complex and sorption of metal, complex, and ligand must be considered, as shown by the scheme below in which "(sorbed)" represents sorbed species and "(aq)" represents dissolved species [3].

---

[3] Benjamin, M.M., and Leckie, J.O. 1981. Conceptual model for metal-ligand-surface interactions during adsorption. *Environmental Science and Technology* 15: 1050–1057.

$$
\begin{array}{ccc}
\mathrm{Mt^{z+}(sorbed)} & \rightleftharpoons & \mathrm{Mt^{z+}(aq)} \\
\updownarrow & & \updownarrow \\
\mathrm{Mt^{z+}L(sorbed)} & \rightleftharpoons & \mathrm{Mt^{z+}L(aq)} \\
\updownarrow & & \updownarrow \\
\mathrm{L(sorbed)} & \rightleftharpoons & \mathrm{L(aq)}
\end{array}
$$

The sorption of anions by solid surfaces is harder to explain than the sorption of cations. Phosphates may be sorbed on hydroxylated surfaces by displacement of hydroxides (ion exchange):

$$
\begin{array}{l} \mathrm{M{-}OH} \\ \mathrm{M{-}OH} \end{array} + \mathrm{HPO_4^{2-}} \rightleftharpoons \begin{array}{l} \mathrm{M{-}O} \\ \mathrm{M{-}O} \end{array} \!\!> \mathrm{P(OH)({=}O)} + 2\mathrm{OH^-} \qquad (6.3.8)
$$

The degree of anion sorption varies. As with phosphate, sulfate may be sorbed by chemical bonding. Most sulfate sorption occurs below pH 7. Chloride and nitrate are sorbed only by electrostatic attraction, such as occurs with positively charged colloidal particles in soil at a low pH. In addition, there are specific means for sorbing fluoride, molybdate, selenate, selenite, arsenate, and arsenite ions.

6.4 THE NATURE OF COLLOIDAL PARTICLES

Many minerals, some organic pollutants, proteinaceous material, some algae, and some bacteria exist in water as **colloids**. Colloidal particles generally are defined as those ranging in diameter from 0.001 micron to 1 micron. Their high specific area, high interfacial energy, high surface-charge density, and other physical-chemical characteristics strongly influence their properties and behavior. The physical and chemical behavior of colloids is in between that of solutions and that of suspensions of larger particles. A schematic drawing of a colloidal suspension is shown in Figure 6.4.

Colloidal suspensions exhibit a characteristic property known as the **Tyndall effect**. This effect is manifested by a light blue hue observed at right angles to a beam of white light shining through the suspension. Colloidal particles have this light-scattering property because they are of the same general dimensions as the wavelength of visible light.

Colloids may be classified on the basis of their interaction with water. Like most classification systems, however, the classification of colloids is not completely rigorous. The three classes of colloids are *hydrophilic colloids, hydrophobic colloids,* and *association colloids.* The term *hydrophobic colloids* implies that the particles in such colloids are water-repelling, although that is not strictly true for many colloidal materials in this class. Examples of hydrophobic colloids are clay particles, petroleum droplets, and very small gold particles.

Hydrophilic colloids generally consist of macromolecules, such as proteins and synthetic polymers, that are characterized by strong specific interaction with

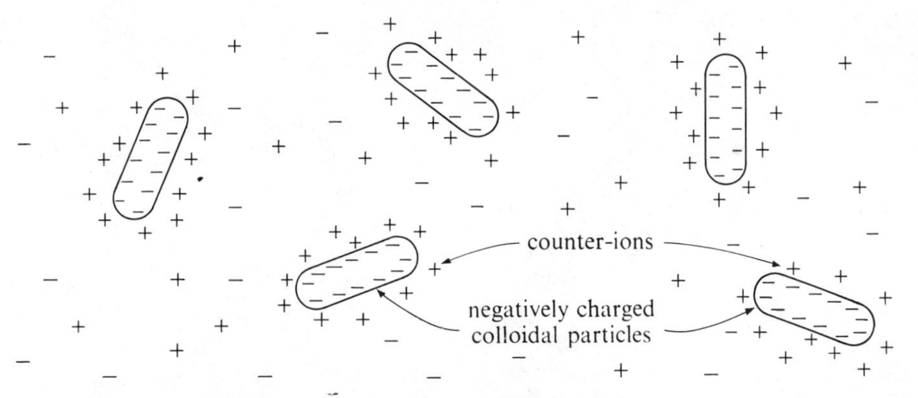

FIGURE 6.4 Schematic drawing of negatively charged colloidal particles, their charges balanced by positive ions in the immediately surrounding solution (counter-ions). Colloidal particles suspended in water may be either negatively or positively charged.

water and spontaneous colloid formation when placed in water. In a sense, hydrophilic colloids are solutions of very large molecules or ions.

Hydrophobic colloids, as illustrated in Figure 6.4, have a positive or negative charge and are surrounded by **counter-ions** of opposite charge. The colloid and its oppositely charged counter-ions compose an **electrical double layer,** which causes the particles to repel each other. Hydrophobic colloids are relatively sensitive to the addition of small amounts of salts, which cause the colloids to aggregate and settle out of suspension. Hydrophilic colloids are generally much less affected by salts.

The particles composing **association colloids** are formed from relatively small individual molecules or ions that are soluble in the solvent medium. The larger suspended particles formed by the aggregation of the smaller soluble units are called **micelles**. Soaps and detergents form association colloids [4]. To understand this phenomenon, consider a typical soap, sodium stearate, which has the following formula:

$$\begin{array}{cccccccccccccccccccc}
 & H & H & H & H & H & H & H & H & H & H & H & H & H & H & H & H & H & O & \\
 & | & | & | & | & | & | & | & | & | & | & | & | & | & | & | & | & | & \| & \\
H- & C- & C- & C- & C- & C- & C- & C- & C- & C- & C- & C- & C- & C- & C- & C- & C- & C- & C- & O^-Na^+ \\
 & | & | & | & | & | & | & | & | & | & | & | & | & | & | & | & | & | & & \\
 & H & H & H & H & H & H & H & H & H & H & H & H & H & H & H & H & H & &
\end{array}$$

It may be seen from the formula that the stearate anion contains an organic portion, which is the $CH_3(CH_2)_{16}$- chain, and an ionic portion, the -CO_2^- group. The aliphatic chain is hydrophobic and the -CO_2^- group is hydrophilic. The aliphatic chain is subject to the **hydrophobic effect**, in which the hydrocarbon portions of the anions cluster together, avoiding contact with the water. Thus, the stearate anions form micelles consisting of as many as 100 anions arranged with their aliphatic chains inside the micelle and their hydrophilic -CO_2^- groups on the

[4] Swisher, R.D. 1970. *Surfactant biodegradation*. New York: Marcel Dekker, Inc.

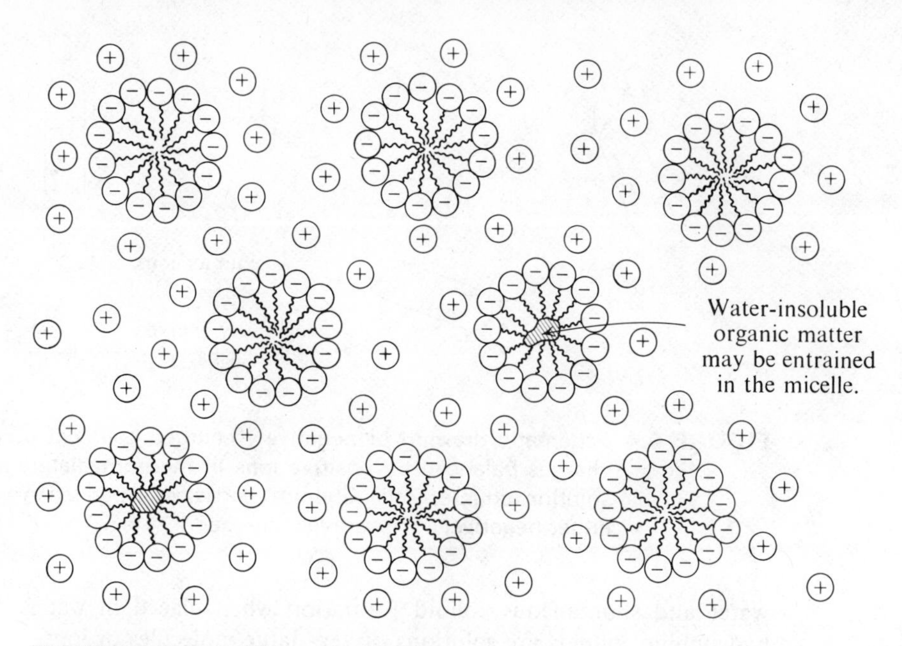
Water-insoluble
organic matter
may be entrained
in the micelle.

FIGURE 6.5 Colloidal soap micelles.

surface of the particle, as shown in Figure 6.5. The Na^+ counter-ions are located on or near the particle surface.

The stability of colloids is a prime consideration in determining their behavior. Colloid stability is involved, for example, with the formation of sediments, dispersion and agglomeration of bacterial cells, and dispersion and removal of pollutants (such as crude oil from an oil spill).

The two main phenomena contributing to the stabilization of colloids are **hydration** and **surface charge**. Hydrated colloidal particles have a coating of water on the surface. The water layer may prevent contact of the particles so that they cannot form larger units. Therefore, hydrophilic colloids may be stabilized by hydration.

A surface charge on colloidal particles may prevent aggregation, since particles with charges of the same sign will repel each other in aquatic suspension. The surface charge frequently is pH-dependent, and around pH 7 most colloidal particles found in natural waters have a negative charge. Negatively charged particles found in natural water include algal cells, bacterial cells, proteins, and colloidal petroleum droplets.

As an example of how a colloidal particle may acquire a charge, consider colloidal MnO_2 in aqueous suspension. Figure 6.6 shows a small unit of MnO_2 without implying anything about its actual structure. We may first visualize anhydrous MnO_2, which, according to its chemical formula, has two O atoms for each Mn atom. In contact with water, MnO_2 is hydrated with H_2O molecules that bind to surface Mn atoms. Loss of H^+ ions from these H_2O molecules yields a colloidal particle with a net negative charge. Gain of H^+ by surface O atoms gives the particle a net positive charge. Thus, the charge is dependent upon the H^+ ion concentration in the surrounding medium and varies with pH.

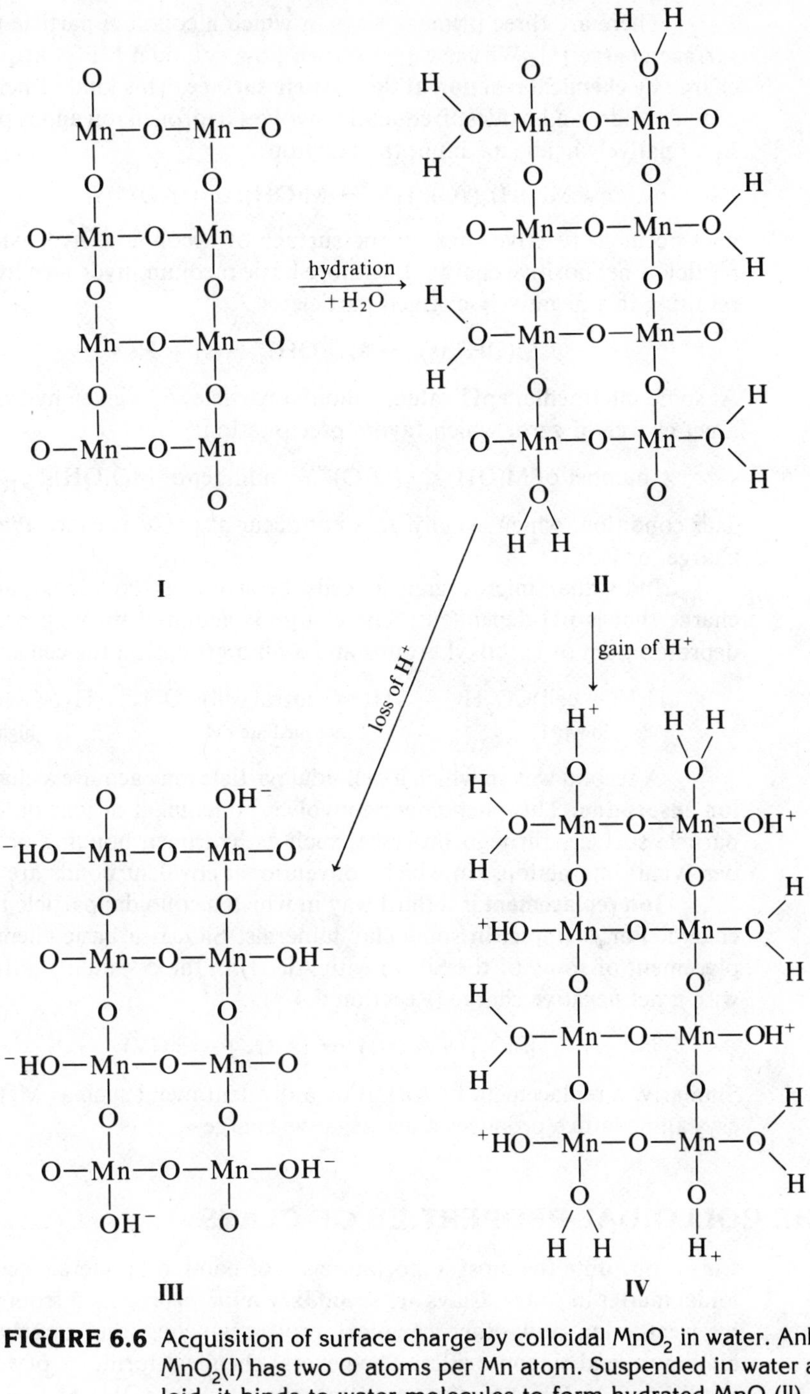

hydration $+H_2O$

I

II

loss of H^+

gain of H^+

III

IV

FIGURE 6.6 Acquisition of surface charge by colloidal MnO_2 in water. Anhydrous MnO_2(I) has two O atoms per Mn atom. Suspended in water as a colloid, it binds to water molecules to form hydrated MnO_2(II). Loss of H^+ from the bound H_2O yields a negatively charged colloidal particle (III). Gain of H^+ ion by surface O atoms yields a positively charged particle (IV). The former process (loss of H^+ ion) predominates for metal oxides.

There are three primary ways in which a colloidal particle may acquire a surface charge [5]. We have just shown how colloidal MnO_2 acquires a surface charge by **chemical reaction at the particle surface**. This kind of behavior, typical of hydroxides and oxides, frequently involves hydrogen ion and is pH-dependent. In a relatively acidic medium, the reaction

$$M(OH)_n(s) + H^+ \rightarrow M(OH)_{n-1}(H_2O)^+(s) \quad (6.4.1)$$

may occur at reactive sites on the surface of a colloidal hydroxide, giving the particle a net positive charge. In a more basic medium, hydrogen ion may be lost, resulting in a negatively charged particle:

$$M(OH)_n(s) \rightarrow MO(OH)^-_{n-1}(s) + H^+ \quad (6.4.2)$$

At some intermediate pH value, colloidal particles of a given hydroxide will have a net charge of zero, which favors precipitation:

$$\text{number of } M(OH)_{n-1}(H_2O)^+ = \text{number of } MO(OH)^-_{n-1} \quad (6.4.3)$$

This condition, which usually does not occur at pH 7, is called the **zero point of charge**, or **ZPC**.

Individual microorganism cells behaving as colloidal particles have a charge that is pH-dependent. The charge is acquired through protonation and deprotonation of carboxyl groups and amino groups on the cell surface:

$$\underset{\text{low pH}}{{}^+H_3N(+\text{ cell})CO_2H} \qquad \underset{\text{intermediate pH}}{{}^+H_3N(\text{neutral cell})CO_2^-} \qquad \underset{\text{high pH}}{H_2N(-\text{ cell})CO_2^-}$$

A second way in which a colloidal particle may acquire a charge is through **ion adsorption**. This phenomenon involves attachment of ions onto the colloidal particle surface, through processes such as hydrogen bonding or London (Van der Waal) interactions, in which conventional covalent bonds are not formed.

Ion replacement is a third way in which a colloidal particle may gain a net charge. For example, in some clay minerals, SiO_2 is a basic chemical unit. Replacement of some of the Si(IV) with Al(III) in the crystalline lattice yields sites with a net negative charge (Reaction 6.4.4).

$$[SiO_2] + Al(III) \rightarrow [AlO_2^-] + Si(IV) \quad (6.4.4)$$

Similarly, a replacement of Al(III) by a divalent metal such as Mg(II) in the clay crystalline lattice produces a net negative charge.

6.5 THE COLLOIDAL PROPERTIES OF CLAYS

Clays constitute the most important class of common minerals occurring as colloidal matter in water. Clays are **secondary minerals** (formed from primary rocks by weathering and other chemical processes) consisting largely of hydrated aluminum and silicon oxides. The general chemical formulas of some clays are: kaolinite, $Al_2(OH)_4Si_2O_5$; montmorillonite, $Al_2(OH)_2Si_4O_{10}$; nontronite, $Fe_2(OH)_2Si_4O_{10}$; and hydrous mica, $KAl_2(OH)_2(AlSi_3)O_{10}$. Iron and magnesium commonly are associated with clay minerals. The most common general classes of

---

[5] Stumm, W., and Morgan, J.J. 1981. *Aquatic chemistry*. 2nd ed. New York: Wiley-Interscience.

clay minerals are illites, montmorillonites, chlorites, and kaolinites. These clay minerals are distinguished from each other by general chemical formula, structure, and chemical and physical properties.

Clays are characterized by a layered structure consisting of sheets of silicon oxide alternating with sheets of aluminum oxide. The silicon oxide sheets are made up of tetrahedra in which each silicon atom is surrounded by four oxygen atoms. Of the four oxygen atoms in each tetrahedron, three are shared with other silicon atoms that are components of other tetrahedra. The sheet thus composed is called the **tetrahedral sheet**.

The aluminum oxide is contained in an **octahedral sheet**, so named because each aluminum atom is surrounded by six oxygen atoms in an octahedral configuration. The structure is such that some of the oxygen atoms are shared between aluminum atoms, and some are shared with the tetrahedral sheet.

Structurally, clays may be classified as either *two-layer clays* or *three-layer clays*. In **two-layer clays**, oxygen atoms are shared between a tetrahedral sheet and an adjacent octahedral sheet. In **three-layer clays**, an octahedral sheet shares oxygen atoms with tetrahedral sheets on either side. These layers composed of either two or three sheets are called **unit layers**. A unit layer of a two-layer clay typically is around 0.7 nanometer (nm) thick, whereas that of a three-layer clay exceeds 0.9 nm in thickness. The structure of the two-layer clay kaolinite is represented in Figure 6.7. Some clays, particularly the montmorillonites, may absorb large quantities of water between unit layers, a process accompanied by swelling of the clay.

As described in Section 6.4, clay minerals may attain a net negative charge by ion replacement, in which Si(IV) ions and Al(III) ions are replaced by metal ions of similar size but lesser charge. Compensation must be made for this negative charge by cations associated with the layer surfaces. Since these cations need not fit specific sites in the crystalline lattice of the clay, they may be relatively large ions, such as K^+, Na^+, or NH_4^+. These cations are called **exchangeable cations** and are exchangeable for other cations in water. The amount of exchangeable cations, expressed as milliequivalents of monovalent cations per 100 g of dry clay, is called the **cation-exchange capacity, CEC,** of the clay. The cation-exchange capacity is an extremely important quantitative characteristic of colloids and sediments having cation-exchange capabilities.

Clay minerals are among the most common suspended matter found in natural waters. Because of their structure and high surface area per unit weight, these minerals have a strong tendency to sorb chemical species from water. Thus, clays play a role in the transport and reactions of biological wastes, organic chemicals, gases, and other pollutant species in water. However, clay minerals also may effectively immobilize dissolved chemicals in water and so exert a purifying action. It is thought that some microbial processes occur at clay particle surfaces. In some cases, sorption of organics by clay inhibits biodegradation. Thus, clay may play a role in the microbial degradation, or nondegradation, of organic wastes.

6.6 AGGREGATION OF PARTICLES

The process by which particles aggregate and precipitate from colloidal suspension are quite important in the aquatic environment. The settling of biomass during biological waste treatment depends upon the aggregation of bacterial cells.

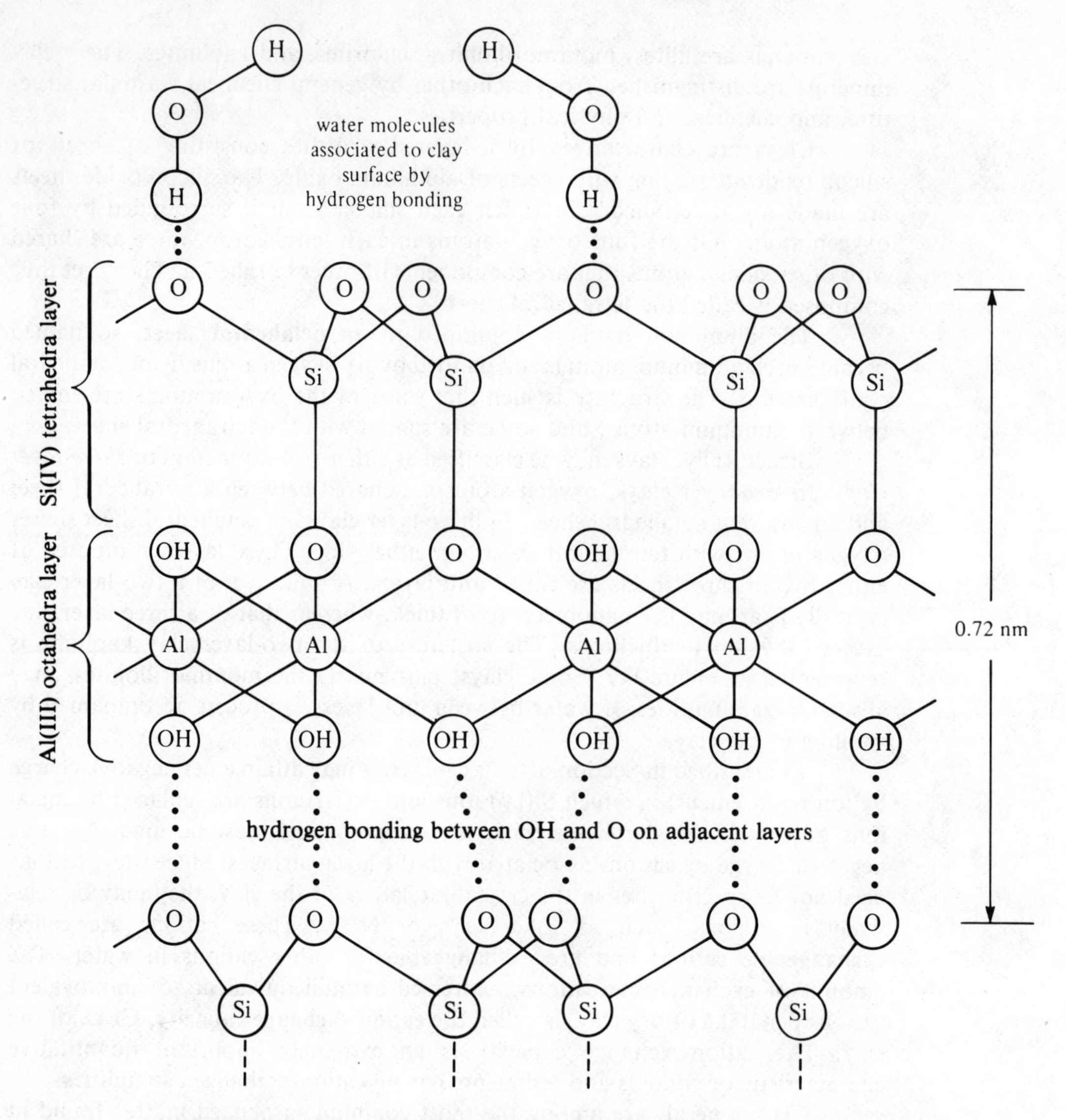

FIGURE 6.7 Structure of kaolinite, a two-layer clay.

Other processes involving the aggregation of colloidal particles are the formation of bottom sediments and the clarification of turbid water for domestic or industrial use.

The processes by which colloidal particles aggregate are quite complicated and have been the subject of considerable investigation. La Mer [6] has divided these processes into two classes, *coagulation* and *flocculation*. Colloidal particles are prevented from aggregating by the electrostatic repulsion between the electrical double layers (adsorbed-ion layer and counter-ion layer). **Coagulation** in-

---

[6] La Mer, V.K., and Healy, T.W. 1963. *Rev. Pure Appl. Chem.* 13: 112.

volves the reduction of this electrostatic repulsion, such that colloidal particles of identical materials may aggregate. **Flocculation** depends upon the presence of **bridging compounds**, which form chemically bonded links between colloidal particles and enmesh the particles in relatively large masses called **floc networks.**

6.7 COAGULATION BY ADDITION OF SALTS

Hydrophobic colloids often are readily coagulated by the addition of small quantities of salts that contribute ions to solution. Such colloids are stabilized by electrostatic repulsion. Therefore, the simple explanation of coagulation by ions in solution is that the ions reduce the electrostatic repulsion between particles to such an extent that the particles aggregate. Because of the double layer of electrical charge surrounding a charged particle, this aggregation mechanism is sometimes called **double-layer compression**. It is particularly noticeable in estuaries where sediment-laden fresh water flows into the sea, and is largely responsible for deltas formed where large rivers enter oceans.

6.8 COAGULATION OF SILICA BY IRON(III)

A typical example of the chemical interactions involved in the aggregation of colloids may be found in an investigation [7] of the iron(III)-induced coagulation of silica, colloidal SiO_2. The charge on colloidal silica particles is pH-dependent, with a zero point of charge (ZPC) at a pH of approximately 2. Excess H^+ ions on the particle surfaces yield a net positive charge at pH values below the pH of the ZPC, whereas excess OH^- groups on the surface yield a net negative charge at pH values above the pH of the ZPC (see Section 6.4).

The investigation of silica aggregation was conducted by measuring light scattered by colloidal particles as iron(III) was added at pH 5 [7]. At this pH, the silica has a negative charge, and the predominant iron species are $Fe(OH)^{2+}$, $Fe(OH)_2^+$, and polynuclear complexes with the general formula

$$Fe_x(OH)_y(H_2O)_z^{3x-y}$$

Coagulation occurred abruptly at a value of log[iron(III)] of -5.26, and the colloidal suspension was abruptly restabilized at a value of log[iron(III)] of -4.48. These results are explained by an initial neutralization of the negative surface charge on the SiO_2 particles by sorption of positive iron(III) ions, allowing coagulation to occur. As more iron(III) is added, the sorption of positive iron(III) ions results in the formation of positive colloidal particles, restabilizing the colloid as a positive colloidal suspension, illustrated in Figure 6.8 (next page).

---

[7] O'Melia, C.R., and Stumm, W. 1967. Aggregation of silica dispersions by iron(III). *J. Colloid and Interface Science* 23: 437–47.

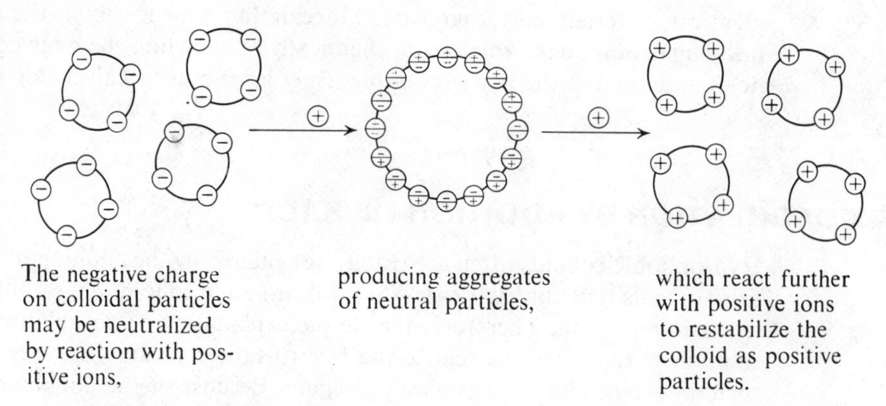

The negative charge on colloidal particles may be neutralized by reaction with positive ions,
producing aggregates of neutral particles,
which react further with positive ions to restabilize the colloid as positive particles.

FIGURE 6.8 Aggregation of negatively charged colloidal particles by reaction with positive ions, followed by restabilization as a positively charged colloid.

6.9 FLOCCULATION OF COLLOIDS BY POLYELECTROLYTES

Polyelectrolytes of both natural and synthetic origin are effective agents in flocculating colloids. Polyelectrolytes are polymers with a high formula weight that normally contain ionizable functional groups. Typical examples of synthetic polyelectrolytes are shown in Table 6.1

It can be seen from Table 6.1 that anionic polyelectrolytes have negatively charged functional groups, such as $—SO_3^-$ and $—CO_2^-$. Cationic polyelectrolytes have positively charged functional groups, normally H^+ bonded to N. Nonionic polymers that serve as flocculants normally do not have charged functional groups.

Somewhat paradoxically, *anionic* polyelectrolytes may flocculate *negatively charged* colloidal particles. The mechanism by which this occurs involves bridging between the colloidal particles by way of the polyelectrolyte anions. Strong chemical bonding has to be involved, since both the particles and the polyelectrolytes are negatively charged. However, the process does occur and is particularly important in biological systems; for example, in the cohesion of tissue cells, clumping of bacterial cells, and antibody-antigen reactions.

The flocculation process induced by anionic polyelectrolytes is greatly facilitated by the presence of a low concentration of a metal ion capable of binding with the functional groups on the polyelectrolyte. A good example is provided by a study of the influence of Ca^{2+} ion on the coagulation of negatively charged AgBr colloidal particles [8]. Addition of three anionic polyelectrolytes, namely, polyacrylic acid, hydrolyzed polyacrylamide, and polystyrenesulfonate, resulted in coagulation of the colloid, followed by restabilization of the colloidal suspension at higher polyelectrolyte concentrations. The reasons for this behavior are summarized in Figure 6.9.

---

[8] Sommerauer, A.; Sussman, D.L.; and Stumm, W. 1968. The role of complex formation in the flocculation of negatively charged sols with anionic polyelectrolytes. *Kolloid-Zeitschrift und Zeitschrift für Polymere* 254: 147–54.

TABLE 6.1 Some Synthetic Polyelectrolytes and Neutral Polymers Used as Flocculants

Anionic polyelectrolytes		*Cationic polyelectrolytes*	
$[-CH_2-CH(C_6H_4SO_3^-)-]_n$ (H H / —C—C— / H / SO_3^-) polystyrene sulfonate	$[-CH_2-CH(CO_2^-)-]_n$ (H H / —C—C— / CO_2^- H) polyacrylate	$[-CH_2-CH(C_5H_4NH^+)-]_n$ (H H / —C—C— / H / N / H^+) polyvinyl pyridinium	$[-CH_2-CH_2-NH_2^+-]_n$ (H H H / —C—C—N± / H H H) polyethylene imine

Nonionic polymers

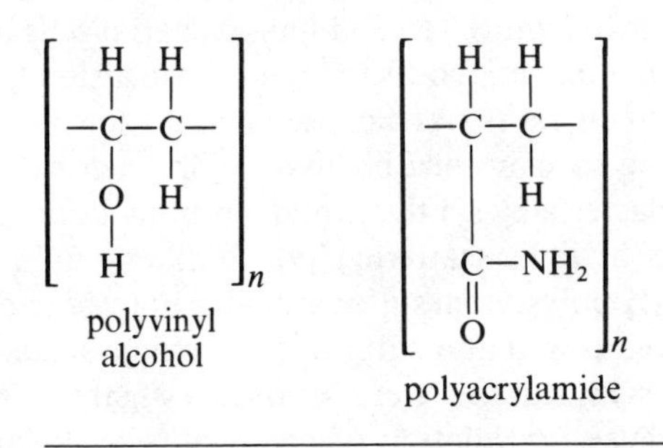

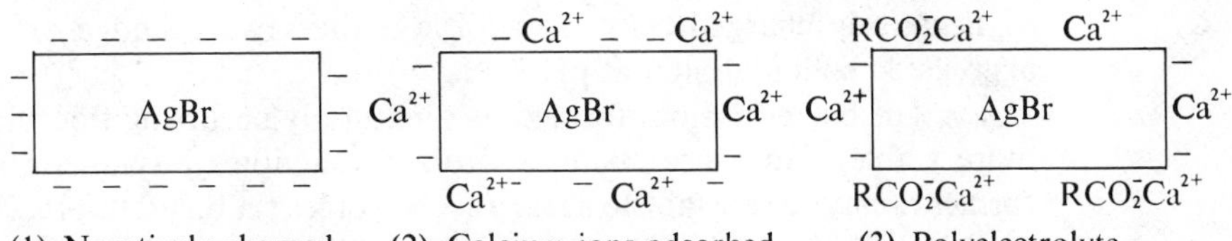

AgBr Ca^{2+} $^-$O$_2$C—R—CO$_2^-$Ca^{2+} AgBr

Ca^{2+} $^-$O$_2$C—R—CO$_2^-$Ca^{2+}

(4) Flocculated sol bridged by polyelectrolyte groups

Ca^{2+} $^-$O$_2$C—R···

···R—CO$_2^-$Ca^{2+} AgBr Ca^{2+} $^-$O$_2$C—R···

···R—CO$_2^-$Ca^{2+} Ca^{2+} $^-$O$_2$C—R···

Ca^{2+} $^-$O$_2$C—R···

(5) Restabilized AgBr sol in excess polyelectrolyte

FIGURE 6.9 Mechanism of flocculation of negatively charged AgBr colloidal particles by anionic polyelectrolytes in the presence of calcium ion.

6.10 FLOCCULATION OF BACTERIA BY POLYMERIC MATERIALS

The aggregation and settling of microorganism cells is a very important process in aquatic systems and is essential to the function of biological waste-treatment systems. In biological waste-treatment processes, such as the activated sludge process, microorganisms utilize carbonaceous solutes in the water to produce biomass (see Section 8.10). The primary objective of biological waste treatment is the removal of carbonaceous material and, consequently, its oxygen demand. Part of the carbon is evolved from the water as CO_2, produced by the energy-yielding metabolic processes of the bacteria. However, a significant fraction of the carbon is removed as **bacterial floc**, consisting of aggregated bacterial cells that have settled from the water. The formation of this floc is obviously an important phenomenon in biological waste treatment. There is evidence that polymeric substances, including polyelectrolytes, that are formed by the bacteria induce bacterial flocculation [9].

Within the pH range of normal natural waters (pH 5–9), bacterial cells are negatively charged. The ZPC of most bacteria is within the pH range 2–3. However, even at the ZPC, stable bacterial suspensions may exist. Therefore, surface charge is not necessarily required to maintain bacterial cells in suspension in water, and it is believed that bacterial cells remain in suspension because of the hydrophilic character of their surfaces. As a consequence, some sort of chemical interaction involving bridging species must be involved in bacterial flocculation.

The flocculation of bacteria by synthetic and naturally occurring polymers has been studied by means of light scattering [9]. Synthetic anionic polymers studied were polyacrylamide, polystyrenesulfonate, and polygalactouronic acid. A nonionic polymer, dextran, as well as a naturally occurring substance extracted from well-flocculating sewage sludge, were also investigated. All of these substances were found to cause flocculation of bacterial cells. Relatively strong chemical interactions are required, because anionic polymers cause flocculation of negatively charged cells. In addition, the sewage-sludge extract effectively aggregated colloidal SiO_2 at pH 7.

The bacterial flocs formed by a naturally occurring flocculating agent are more stable than those resulting from the action of synthetic polymers. The former are more resistant to **shear forces**, forces that tend to break flocs apart by mechanical action; they are also less dependent upon pH and polymer concentration. Therefore, the naturally occurring flocculating agents must have stronger chemical interactions with the bacterial cells than do the synthetic polymers. The chemical composition and structure of the flocculating substances produced by bacteria are not known and are worthy of extensive investigation. It is likely that these bioflocculants are polymeric carbohydrates. They probably contain the same functional groups as those in the flocculating synthetics, for instance, $-CO_2^-$ and $-OH$ groups. It is possible that the efficiency of organic removal from waste-treatment processes could be increased by adding bioflocculants to increase the settling of biomass.

---

[9] Busch, P.L., and Stumm, W. 1968. Chemical interactions in the aggregation of bacteria bioflocculation in waste treatment. *Environmental Science and Technology* 2: 49–53.

6.11 SORPTION OF METALS BY METAL OXIDES—MANGANESE(IV) OXIDE

Some hydrated metal oxides, such as manganese(IV) oxide and iron(III) oxide, efficiently sorb various species from aquatic solution. The sorption ability is especially pronounced for relatively fresh metal oxides. A good example is sorption by colloidal MnO_2 [10]. This oxide may be produced in natural waters by the oxidation of Mn(II) or by the reduction of Mn(VII). Manganese(II) normally is produced in natural waters by the reduction of manganese oxides in anaerobic bottom sediments. Manganese(VII) often is added to water as permanganate salts to diminish taste and odor or to oxidize iron(II).

The formation and precipitation of manganese(IV) oxide is an important process in aquatic chemistry, in limnological transformations, and particularly in water purification. Soluble Mn^{2+} commonly is removed from water by oxidation to insoluble MnO_2 by air or other oxidizing agents such as chlorine or potassium permanganate. The resulting metal oxide suspension must aggregate prior to filtration, a process that normally involves a surface reaction with metal ions, particularly Ca^{2+}.

Freshly precipitated MnO_2 may have a surface area as large as several hundred square meters per gram. The hydrated oxide may undergo either deprotonation, yielding a negatively charged surface, or protonation, yielding a positively charged surface, as shown in Figure 6.6. The ZPC of hydrous MnO_2 occurs when the surface concentration of H^+ is equal to the surface concentration of OH^-, which occurs between the pH values of 2.8 and 4.5. Since the pH of most normal natural waters exceeds 4.5, hydrous MnO_2 colloids are usually negatively charged. The surfaces of these colloids are said to be *hydroxylated* (see Figure 6.3c).

The sorption of metal ions by hydrous MnO_2 may be illustrated by the following reaction for sorption of a divalent metal ion:

$$\begin{array}{l} \text{O—H} \\ \;| \\ \text{Mn—OH} \\ \;| \\ \text{O—H} \\ \;| \\ \text{Mn—OH} \\ \;| \\ \text{O—H} \end{array} + M^{2+} \longrightarrow \begin{array}{l} \text{O—H} \\ \;| \\ \text{Mn—O} \diagdown \\ \;| \\ \text{O—H} \quad M \\ \;| \\ \text{Mn—O} \diagup \\ \;| \\ \text{O—H} \end{array} + 2H^+ \qquad (6.11.1)$$

The sorption of the metal ions Ag^+, Mn^{2+}, Ba^{2+}, Ca^{2+}, Sr^{2+}, and Nd^{3+} has been studied [10] in $1 \times 10^{-3} M$ suspensions of MnO_2 maintained at pH 5.0 in $1 \times 10^{-2} M$ $NaClO_4$. The colloidal hydrous MnO_2 was formed in the presence of the sorbing species. The preparative reaction

$$3\,Mn^{2+} + 2\,MnO_4^- + 2\,H_2O \rightarrow 5\,MnO_2(s) + 4\,H^+ \qquad (6.11.2)$$

involved quantities of Mn^{2+} ion and MnO_4^- ion necessary to produce suspensions equivalent to a $1.00 \times 10^{-3} M$ solution of MnO_2. The metal sorption reactions

---

[10] Posselt, H.S.; Anderson, F.J.; and Weber, W.J., 1968. Cation sorption on colloidal hydrous manganese dioxide. *Environmental Science and Technology* 2: 1087–93.

were allowed to proceed for 30 minutes, after which time the suspensions were filtered through 0.1-micron-pore-diameter filters, and the metals were determined by titration with EDTA (see Sections 4.7 and 9.2).

It was found that the metal ions were sorbed within 5–10 minutes. The equilibrium distribution of the sorbing species between the MnO_2 suspension and solution was in reasonable agreement with the Langmuir equation. The **Langmuir equation** is

$$X = \frac{X_m bC}{1 + bC} \tag{6.11.3}$$

where X is the quantity of material sorbed per unit quantity of sorbing solid; X_m is the limiting value of X, or monolayer capacity; b is a constant related to the energy of sorption; and C is the equilibrium concentration of the sorbed solute remaining in solution. This equation can berearranged into

$$\frac{1}{X} = \frac{1}{X_m} + \frac{1}{X_m bC} \tag{6.11.4}$$

from which X_m and b may be evaluated from plots of $1/X$ versus $1/C$.

The sorptive capacity of MnO_2, as measured by X_m, is reasonably similar for the divalent cations Mg^{2+}, Ca^{2+}, Sr^{2+}, and Ba^{2+}, ranging from 0.100 mole/mole for Mg^{2+} to 0.180 mole/mole for Ba^{2+}. However, the sorptive capacity of hydrous MnO_2 for Mn^{2+} (X_m = 0.284 mole/mole) is much higher than that for any of the other divalent cations. It is possible that the MnO_2 at the surface reacts specifically with Mn^{2+}:

$$MnO_2(s) + Mn^{2+} + 2\,H_2O \rightarrow 2\,Mn(O)OH(s) + 2\,H^+ \tag{6.11.5}$$

The evidence suggests that when cations are sorbed by hydrous MnO_2, hydrogen ion (or sorbed metal ion, if present) is displaced by the metal ion removed from solution. It is possible to follow the sorption of cations on colloidal MnO_2 by titration of the hydrogen ion liberated by the process. The order of degree of sorption of divalent metal ions was found to be $Mg^{2+} < Ca^{2+} < Sr^{2+} < Ba^{2+}$. It is believed that the **ion charge density** (charge/ion radius) is a strong determining factor in the sorption of metal ions within a specific group. The ionic radius of the hydrated metal ion (including the waters of hydration) increases with increasing charge density of the bare ion. The sorptive capacity of the alkaline earths, for example, decreases with increasing charge density; this is because the more strongly hydrated ions are more surrounded by larger clusters of water molecules and are hindered by their size from undergoing the sorption process.

6.12 CATION EXCHANGE WITH BOTTOM SEDIMENTS

Bottom sediments are important sources of inorganic and organic matter in streams, fresh-water impoundments, estuaries, and oceans. It is incorrect to consider bottom sediments simply as wet soil. Normal soils are in contact with the atmosphere and are aerobic, whereas generally, the environment around bottom sediments is anaerobic, and they are subjected to reducing conditions. Bottom sediments undergo continuous leaching, whereas soils do not. The level of organic matter in sediments is generally higher than that in soils.

One of the most important characteristics of bottom sediments is their ability to exchange cations with the surrounding aquatic medium. This characteristic has been investigated [11] using methods similar to those used to study soils. **Cation-exchange capacity (CEC)** and **exchangeable cation status (ECS)** were the two parameters studied in a number of sediment samples. Cation-exchange capacity measures the capacity of a solid to sorb cations (see Sections 6.5 and 10.6). It varies with pH and with salt concentration. Exchangeable cation status refers to the amounts of specific ions bonded to a given amount of sediment. Generally, both CEC and ECS are expressed as milliequivalents per 100 g of solid. Some soil fertility parameters, such as available potassium, are measured in part by ECS.

Because of the generally anaerobic nature of bottom sediments, special care must be exercised in their collection and treatment. Particularly, contact with atmospheric oxygen rapidly oxidizes exchangeable Mn^{2+} and Fe^{2+} to non-exchangeable oxides containing the metals in higher oxidation states (MnO_2 and Fe_2O_3). Therefore, sediment samples must be sealed and frozen as soon as possible after collection.

A common method for the determination of CEC consists of: (1) treating the sediment with a solution of an ammonium salt so that all exchangeable sites are occupied by NH_4^+ ion; (2) displacing the ammonium ion with a solution of NaCl; and (3) determining the quantity of displaced ammonium ion. The CEC values may then be expressed as the number of milliequivalents of ammonium ion exchanged per 100 g of dried sample. Note that the sample must be dried *after* exchange.

The basic method for the determination of ECS consists of stripping all of the exchangeable metal cations from the sediment sample with ammonium acetate. The metal cations Fe^{2+}, Mn^{2+}, Zn^{2+}, Cu^{2+}, Ni^{2+}, Na^+, K^+, Ca^{2+}, and Mg^{2+} are then determined in the leachate. Exchangeable hydrogen ion is very difficult to determine by direct methods. It is generally assumed that the total cation exchange capacity minus the sum of all exchangeable cations except hydrogen ion is equal to the exchangeable hydrogen ion.

A number of sediment samples from rivers, fresh-water impoundments, and bays were analyzed [11] to determine CEC and ECS. The CEC values ranged from 21.5–100 milliequivalents/100 g, although the fresh-water sediments fell generally within the range of 20–30 milliequivalents/100 g. The exchangeable cations, in terms of milliequivalents per 100 g, fell within the following ranges: Fe^{2+}, 0.7–12.4; Mn^{2+}, 0.2–9.7; Na^+, 0.1–32.9 (the latter measurement was from a salt-water sediment); K^+, 0.3–2.7; Ca^{2+}, 1.3–5.6; Mg^{2+}, 1.1–15.3; H^+, 0.0–56.2.

From these values, it is obvious that sediments are an important repository of metal ions that may be exchanged with surrounding waters. Furthermore, because of their capacity to sorb and release hydrogen ions, sediments have an important buffering effect in some waters.

---

[11] Toth, S.J., and Ott, A.N. 1970. Characterization of bottom sediments: cation exchange capacity and exchangeable cation status. *Environmental Science and Technology* 4: 935–9.

6.13 PHOSPHORUS EXCHANGE WITH BOTTOM SEDIMENTS

Phosphorus is one of the key elements in aquatic chemistry (see Table 2.5) and is thought to be the limiting nutrient in the growth of algae under many conditions. Exchange with sediments plays a role in making phosphorus available for algae and contributes, therefore, to eutrophication.

The synthesis of biomass consumes a great deal of phosphorus. Aquatic biomass is only about one percent phosphorus by dry weight, but even at this low percentage, biosynthesis through algal growth rapidly depletes water of its soluble phosphorus. Although phosphorus is recycled from dead biological material, decay and conversion to phosphate ion is necessary to make organic phosphorus available to algae and other plants.

Sedimentary phosphorus may be classified into four types: (1) phosphate minerals; (2) nonoccluded phosphorus; (3) occluded phosphorus; and (4) organic phosphorus. Examples of each type are listed below.

(1) One example of a common, discrete **phosphate sediment mineral** is hydroxyapatite, $Ca_5OH(PO_4)_3$. Several other water-forming, phosphorus-containing minerals include magnesium hydroxyphosphate, $3\ Mg_3(PO_4)_2 \cdot Mg(OH)_2$, and vivianite, $Fe_3(PO_4)_2 \cdot 8\ H_2O$.

(2) Orthophosphate ions may form deposits on the surface of minerals such as SiO_2 and calcite, $CaCO_3$. This **nonoccluded phosphorus** is in direct contact with water and presumably is more rapidly solubilized than other forms of phosphorus.

(3) Amorphous hydrated oxides of iron and aluminum and amorphous aluminosilicates have a particular affinity for orthophosphate. The term **occluded phosphorus** refers to orthophosphate ions contained within the matrix structures of these minerals. It is generally assumed that such phosphorus is not so readily available as nonoccluded phosphorus.

(4) The phosphorus incorporated within aquatic biomass, usually of algal or bacterial origin, along with organic phosphate esters, may be classified as **organic phosphorus**. In some waters receiving heavy loads of domestic or industrial wastes, inorganic polyphosphates (from detergents, for example) may be present in sediments. Runoff from fields where liquid polyphosphate fertilizers have been used might possibly provide polyphosphates sorbed on sediments.

The reaction of orthophosphate species with inorganic sedimentary materials has been studied extensively in recent years. For example, a study of the reaction of phosphate with alumina and kaolinite [12] has shown an initially rapid sorption reaction followed by a slow uptake of phosphate over many days. The rate of phosphorus uptake during the slow part of the process is expressed by the equation

$$-\frac{d[\mathrm{P}]}{dt} = kAS[\mathrm{P}] \qquad (6.13.1)$$

---

[12] Chen, Y.S.R.; Butler, J.N.; and Stumm, W. 1974. Kinetic study of phosphate reaction with aluminum oxide and kaolinite. *Environmental Science and Technology* 7: 327–32.

where [P] is the molar concentration of soluble orthophosphate species; *t* is time in days; *k* is the rate constant for the first-order reaction; *A* is the surface area per unit weight of sediment; and *S* is the number of grams per liter of sediment in suspension. In these studies, *S* was 7.5 g/L for kaolinite and 2.5 g/L for α-Al_2O_3. The value of *A* was 10.7 m^2/g for the kaolinite used and 10 m^2/g for the alumina. At 50 °C and pH 5, the value of *k* was found to be 8×10^{-4} for alumina and 2×10^{-4} for kaolinite when the initial soluble orthophosphate concentration was around $3 \times 10^{-4} M$. The rate was slower by a factor greater than 10 at pH 7.0. The slow uptake process probably involves the formation of a hexagonal $AlPO_4$ phase on the sediment surface.

As indicated in the preceding study, reaction of orthophosphate with aluminum is an important mechanism for phosphate removal from water. A similar reaction occurs beween iron and phosphate. A study [13] of phosphorus extractable from sediment in an estuary has shown a strong correlation between phosphorus extractable with HCl-H_2SO_4 and iron extractable from the sediment with oxalate. The extractable phosphorus decreased from 1.6 mg per gram of sediment in the upper fresh-water region of the estuary to 0.3 mg/g in the lower salt-water region (where the correlation with extractable iron was much weaker). These findings imply that sediment washing into the estuary gradually loses increasing amounts of phosphate to the aquatic phase as the salinity increases. When it is considered, for example, that the flow of the Tar River into the shallow Pamlico Estuary of North Carolina carries an average 343 tons/day of silt-clay sediment, the amount of phosphorus released is appreciable.

6.14 TRACE-LEVEL METALS IN SEDIMENTS AND SUSPENDED MATTER

Sediments and suspended particles are important repositories for trace amounts of metals such as chromium, cadmium, copper, molybdenum, nickel, cobalt, and manganese. These metals may be present in a number of forms, including discrete compounds or ions held by cation-exchanging clays; they may be bound to hydrated oxides of iron or manganese (Section 6.11); or they may be chelated by insoluble humic substances (see Section 4.19). The form of the metals depends upon redox conditions. Examples of specific trace-metal-containing compounds that may be stable in natural waters under oxidizing and reducing conditions are given in Table 6.2. Solubilization of metals from sedimentary or suspended matter is often a function of the complexing agents present. These include amino acids, such as histidine, tyrosine, or cysteine; citrate ion; and, in the presence of seawater, chloride ion.

Suspended particles containing trace elements may be in the submicron size range. Many reported values of "dissolved" trace metals in water are inaccurate because these particles may pass through the 0.45-μ filter commonly used to distinguish between soluble and insoluble matter. Little is known about the availability to aquatic microorganisms of trace metals contained in ultra-fine suspended matter. Certainly, metals in true solution are relatively more available than are those not in solution. However, metals contained in colloidal solids

---

[13] Upchurch, J.B.; Edzwald, J.K.; and O'Melia, C.R. 1974. Phosphates in sediments of Pamlico Estuary. *Environmental Science and Technology* 8: 56–8.

TABLE 6.2 Inorganic Trace-Metal Compounds That May Be Stable under Oxidizing and Reducing Conditions

	Discrete compound that may be present	
Metal	*Oxidizing conditions* (*high* pE)	*Reducing conditions* (*low* pE, S(–II) *present*)
Cadmium	$CdCO_3$	CdS
Copper	$Cu_2(OH)_2CO_3$	CuS
Iron	$Fe_2O_3 \cdot x(H_2O)$	FeS, FeS_2
Mercury	HgO	HgS
Manganese	$MnO_2 \cdot x(H_2O)$	MnS, $MnCO_3$
Nickel	$Ni(OH)_2$, $NiCO_3$	NiS
Lead	$2\ PbCO_3 \cdot Pb(OH)_2$, $PbCO_3$	PbS
Zinc	$ZnCO_3$, $ZnSiO_3$	ZnS

suspended in water are more accessible than those in sediments. Among the factors involved in metal availability are the identity of the metal; its chemical form (type of binding, oxidation state); the nature of the suspended material; the type of organism; and the physical and chemical conditions in the water.

The pattern of trace-metal occurrence in suspended matter in relatively unpolluted water tends to correlate well with that of the parent minerals from which the suspended solids originated; anomalies appear in polluted waters where industrial sources add to the metal content of the stream. The occurrence of mercury from a natural source has been studied in detail for the Kuskokwim River in Alaska [14]. The mercury in the river originates in deposits of cinnabar, HgS, in the river's tributaries. The mean mercury content in this river's water is $0.34 \pm 0.13\ \mu g/L$, while the mercury content in suspended sediment is 3.9 ± 0.52 mg/kg. By comparison, in parts of the river system that are remote from mercury deposits, the mean mercury content of the water is only 0.085–0.19 μg/L, while that of the suspended sediments is only 0.1–1 mg/kg. It was observed that anomalously high mercury concentrations near the mercury deposits are rapidly lowered downstream as the solids containing mercury become mixed with other sediments. The distance that sediment-borne mercury is carried downstream varies inversely with particle size.

6.15 SORPTION OF ORGANIC COMPOUNDS BY SUSPENDED MATERIALS, SEDIMENTS, AND SOIL

Many organic compounds interact with suspended material and sediments in bodies of water. Settling of suspended material containing sorbed organic matter carries organic compounds into the sediment of a stream or lake. For example, this phenomenon is largely responsible for the presence of herbicides in sediments containing contaminated soil particles eroded from crop land. Some organics are carried into sediments by the remains of organisms or by fecal pellets from zooplankton that have accumulated organic contaminants.

---

[14] Nelson, H.; Larsen, B.R.; Jenne, E.A.; and Sorg, D.H. 1977. Mercury dispersal from lode sources in the Kuskokwin River drainage, Alaska. *Science* 198: 820.

Suspended particulate matter affects the mobility of organic compounds sorbed to particles. Furthermore, sorbed organic matter undergoes chemical degradation and biodegradation at different rates and by different pathways compared to organic matter in solution. The interaction of natural colloids with metals has been studied extensively, but comparatively few studies have been performed on organics [15]. Additional studies of organic-colloid interactions are needed to understand the toxicity, transport, and degradation of organic contaminants in natural waters.

There is, of course, a vast variety of organic compounds that get into water. As one would expect, they react with sediments in different ways, the type and strength of binding varying with the type of compound. The most common types of sediments considered for their organic binding ability are clays, organic (humic) substances, and clay-humic substance complexes. Both clays and humic substances act as cation exchangers. Therefore, these materials sorb cationic organic compounds through ion exchange. This is a relatively strong sorption mechanism, greatly reducing the mobility and biological activity of the organic compound. When sorbed by clays, cationic organic compounds are generally held between the layers of the clay mineral structure. Here their biological activity is essentially zero.

Since most sediments lack strong anion exchange sites, negatively charged organics are not held strongly at all. Thus, these compounds are relatively mobile in water. Their biological activity (and biodegradability) remains high in water despite the presence of solids.

The degree of sorption of organic compounds is generally inversely proportional to their water solubility. The more water-insoluble compounds tend to be taken up strongly by lipophilic ("fat-loving") solid materials. Compounds having a relatively high vapor pressure can be lost from water or solids by evaporation. When this happens, photochemical reactions (see Chapter 11) can play an important role in their degradation.

The herbicide 2,4-D (2,4-dichlorophenoxyacetic acid) has been studied extensively in regard to sorption reactions. Most of these studies have dealt with pure clay minerals, however, whereas soils and sediments are likely to have a strong clay-fulvic acid complex component. The sorption of 2,4-D by such a complex can be described using an equation of the Freundlich isotherm type,

$$X = KC^n \tag{6.15.1}$$

where X is the amount sorbed per unit weight of solid, C is the concentration of 2,4-D at equilibrium, and n and K are constants. These values are determined by plotting log X versus log C. If a Freundlich-type equation is obeyed, the plot will be linear with a slope of n and an intercept of log K. For sorption of 2,4-D on an organoclay complex at 5 °C, n was found to be 0.76 and log K was 0.815. At 25 °C, n was 0.83 and log K was 0.716.

Sorption of comparatively nonvolatile hydrocarbons by sediments removes these materials from contact with aquatic organisms but also greatly retards their biodegradation. Aquatic plants produce some of the hydrocarbons that are found in sediments. Photosynthetic organisms, for example, produce quantities of *n*-heptadecane. Pollutant hydrocarbons in sediments are indicated

---

[15] Means, J.C., and Wijayratne, R. 1982. Role of natural colloids in the transport of hydrophobic pollutants. *Science* 215: 968–970.

by a smooth chain-length distribution of *n*-alkanes and thus can be distinguished from hydrocarbons generated photosynthetically in the water. An analysis of sediments in Lake Zug, Switzerland, for example, has shown a predominance of pollutant petroleum hydrocarbons near densely populated areas.

The sorption of neutral species like petroleum obviously cannot be explained by ion-exchange processes. It probably involves phenomena such as Van der Waals forces (induced dipole-dipole interaction involving a neutral molecule) and hydrogen bonding.

Obviously, uptake of organic matter by suspended and sedimentary material in water is an important phenomenon. Were it not for this phenomenon, it is likely that pesticides in water would be much more toxic. Biodegradation is generally slowed down appreciably, however, by sorption by a solid. In certain intensively farmed areas, there is a very high accumulation of pesticides in the sediments of streams, lakes, and reservoirs. The sorption of pesticides by solids and the resulting influence on their biodegradation is an important consideration in the licensing of new pesticides.

A study of organic compounds in water [16] has resulted in the determination of an empirical relationship among the partition coefficients of the distribution of compounds between water and *n*-octanol, their aqueous solubilities, and their bioaccumulation in rainbow trout. The compounds studied included organophosphate pesticides, organochlorine pesticides, polychlorinated biphenyls, aromatic acids, aliphatic hydrocarbons, and aromatic hydrocarbons. The solubility-partition coefficient correlation was found to be accurate to within one order of magnitude for solubilities ranging from 10^{-3}–10^{4} ppm and partition coefficients ranging from 10–10^{7}. Figure 6.10 shows the correlation between partition coefficients and aqueous solubilities of the organic compounds studied. Figure 6.11 shows the correlation between aqueous solubilities and bioconcentration factors.

The transfer of surface water to ground water often results in sorption of some water contaminants by soil and mineral material [17]. To take advantage of this purification effect, some municipal water supplies are drawn from beneath the surface of natural or artificial river banks as a first step in water treatment. The movement of water from waste landfills to aquifers is also an important process (see Chapter 20) in which pollutants in the landfill leachate may be sorbed by solid material through which the water passes.

The sorption of dilute solutions of halogenated and aromatic hydrocarbons by soil and sand has been studied under simulated water infiltration conditions [17]. The relationship between the sorption equilibria observed may be expressed by the formula

$$S = K_p C \tag{6.15.2}$$

where S and C are the concentrations of hydrocarbons in the solid and liquid phases, respectively, and K_p is the partition coefficient. It was found that the two

---

[16] Chiou, C.T.; Freed, V.H.; Schmedding, D.W.; and Kohnert, R.L. 1978. Partition coefficient and bioaccumulation of selected organic chemicals. *Environmental Science and Technology* 11: 475–8.

[17] Schwarzenbach, R.P., and Westall, J. 1982. Transport of nonpolar organic compounds from surface water to groundwater. Laboratory sorption studies. *Environmental Science and Technology* 15: 1360–7.

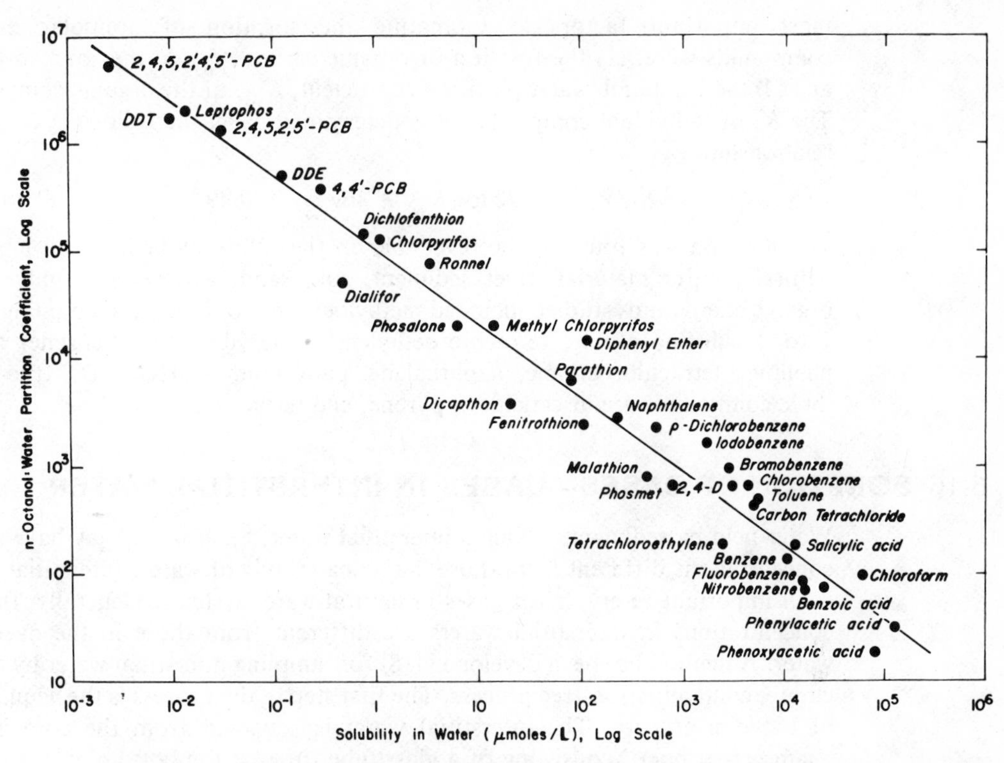

FIGURE 6.10 Partition coefficients versus aqueous solubilities of environmentally significant organic chemicals at room temperature. (From Reference 16. Reprinted with permission from *Environmental Science and Technology*. Copyright by the American Chemical Society.)

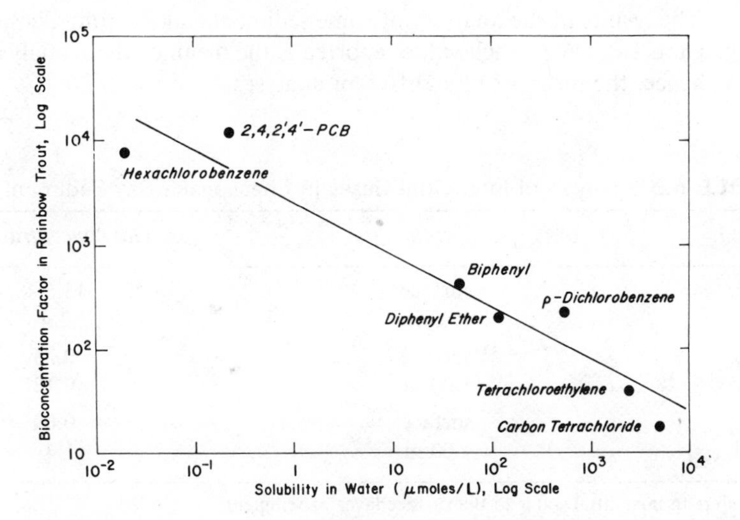

FIGURE 6.11 Aqueous solubilities versus bioconcentration factors of environmentally significant organic chemicals in rainbow trout. (From Reference 16. Reprinted with permission from *Environmental Science and Technology*. Copyright by the American Chemical Society.)

most important factors in estimating the sorption of nonpolar organic compounds were: (1) the fraction of organic carbon, f_{oc}, in the solid sorbents; and (2) the 1-octanol/water partition coefficient, K_{ow}, of the organic compound. The K_p of individual compounds was determined using the following empirical relationship:

$$\log K_p = 0.72 \log K_{ow} + \log f_{oc} + 0.49 \quad (6.15.3)$$

The sorption was found to be reversible on the solids studied, which included natural aquifer material, river sediment, soil, sand, and sewage sludge. The organic compounds studied included methylbenzene compounds containing from 1 to 4 chlorine atoms, tetrachloroethylene, *n*-butylbenzene, benzene, acetophenone, tetrachloroethane, naphthalene, parathion, β-BHC, DDT (the latter three compounds are insecticides), pyrene, and tetracene.

6.16 SORPTION OF GASES—GASES IN INTERSTITIAL WATER

Water held by sediments is called **interstitial water.** Such water may have solute concentrations different from those in the main body of water. Interstitial water is an important reservoir for gases in natural water systems. Generally, the gas concentrations in interstitial waters are different from those in the overlying water. A method has been developed [18] for sampling interstitial water by an accurate, contamination-free process. The first step in the process is the acquisition of a sediment core. The interstitial water is squeezed from the core into a "sampler-stripper" consisting of a glass tube fitted at the bottom with a fritted glass disc. The gas is stripped from the sample by helium dispersed by the glass frit and is then analyzed by gas chromatography. The analytical procedure does not effectively separate O_2 and Ar. Oxygen is eliminated by adding 0.1*N* $CrSO_4$ to the tube from which the gases were stripped:

$$4\,Cr(II) + O_2 + 4\,H^+ \rightarrow 4\,Cr(III) + 2\,H_2O \quad (6.16.1)$$

The results of the analyses of some sediments taken from Chesapeake Bay are given in Table 6.3. Each value reported is the mean of three analyses of three cores, hence, the mean of nine different analyses.

TABLE 6.3 Analysis of Interstitial Gases in Chesapeake Bay Sediment Samples

Gas	*Depth*	*Gas concentration, mL/L*
N_2	surface*	13.5
N_2	1.00 m	2.4
Ar	surface	0.35
Ar	1.00 m	0.12
CH_4	surface	0.00
CH_4	1.00 m	1.4×10^2

* Refers to interstitial water in the surface layer of sediment.

[18] Reeburgh, W.S. 1968. Determination of gases in sediments. *Environmental Science and Technology* 2: 140–1.

Examination of Table 6.3 shows that CH_4 could not be detected at the sediment surface. The equilibrium concentration of methane in air is very low, and it is biodegradable under aerobic conditions. Therefore, it is reasonable that the methane concentration at the sediment surface should be quite low. However, of the gases analyzed, by far the highest concentration at a depth of 1 meter was that of methane. The methane is produced by the anaerobic fermentation of organic matter (see Section 5.11):

$$2\,\{CH_2O\} \rightarrow CH_4(g) + CO_2(g) \qquad (6.16.2)$$

The concentrations of argon and nitrogen are much lower at a depth of 1 meter than they are at the sediment surface. This finding may be explained by the stripping action of the fermentation-produced methane rising to the sediment surface.

A method has been reported for the sampling and determination of interstitial gases in sediments at 20–5,500 m depth [19]. As little as 0.1 μg of Ar, O_2, N_2, and CH_4 were measured in 15-mL samples. Only 14 minutes were required for the analysis using a dual-column gas chromatograph and a thermal conductivity detector. Sample collection at up to 12 meters depth in the sediment was performed with evacuated stainless steel cylinders in which interstitial water was filtered by being forced through membrane filters by hydrostatic pressure. These *in situ* samplers were equipped with one-way poppet valves so that they would seal themselves after filling.

SUPPLEMENTARY REFERENCES

Baker, R.A., ed. 1980. *Contaminants and sediments.* 2 vols. Ann Arbor, Mich.: Ann Arbor Science Publishers, Inc.

Drever, James L. 1982. *The geochemistry of natural waters.* Englewood Cliffs, N.J.: Prentice-Hall, Inc.

Faust, S.D., and Hunter, J.V. 1971. *Organic compounds in aquatic environments.* New York: Marcel Dekker, Inc.

Garrels, R.M., and Christ, C.L. 1965. *Solutions, minerals and equilibria.* New York: Harper and Row, Inc.

Helmke, P.A.; Koons, R.D.; Schomberg, P.J.; and Kiskander, I.K. 1977. Determination of trace element contamination of sediments by multielement analysis of clay-size fraction. *Environmental Science and Technology* 10: 984–989.

Hem, J.D. 1970. *Study and interpretation of the chemical characteristics of natural water.* 2nd ed. U.S. Geological Survey Paper 1473. Washington, D.C.

Hiemenz, P.C. 1977. *Principles of colloid and surface chemistry.* New York: Marcel Dekker, Inc.

Schindler, P.W. 1967. Heterogeneous equilibria involving oxides, hydroxides, carbonates, and hydroxide carbonates. *Equilibrium concepts in natural water systems.* W. Stumm, ed. ACS Advances in Chemistry Series No. 67. Washington, D.C.: American Chemical Society.

Shaw, D.J. 1980. *Introduction to colloid and surface chemistry.* 3rd ed. Woburn, Mass.: Butterworth Publishers, Inc.

---

[19] Whiticar, M.J. 1982. Determination of interstitial gases and fluids in sediment collected with an *in situ* sampler. *Analytical Chemistry* 54: 1796–8.

Stumm, W., and Morgan, J.J. 1981. *Aquatic chemistry.* 2nd ed. New York: Wiley-Interscience.

van Olphen, H. 1977. *An introduction to clay colloid chemistry.* 2nd ed. New York: John Wiley and Sons, Inc.

QUESTIONS AND PROBLEMS

1. A sediment sample was taken from a lignite strip-mine pit containing highly alkaline (pH-10) water. Cations were displaced from the sediment by treatment with HCl. A total analysis of cations in the leachate yielded, on the basis of millimoles per 100 g of dry sediment, 150 millimoles of Na^+, 5 millimoles of K^+, 20 millimoles of Mg^{2+}, and 75 millimoles of Ca^{2+}. What is the cation exchange capacity of the sediment in milliequivalents per 100 g of dry sediment?

2. The Langmuir equation discussed in Section 6.11 is used to describe sorption of solutes by suspended matter. Suppose it was found that for an equilibrium concentration of solute in solution of 3.00×10^{-3} moles/liter, 0.50×10^{-3} moles of the solute were sorbed per gram of solid in suspension in solution. When the equilibrium concentration of solute was lowered to 1.00×10^{-3} moles/liter, 0.250×10^{-3} moles of the solute were sorbed per gram of adsorbent. What is the limiting amount of solute that can be sorbed per gram of adsorbent in units of moles per gram?

3. Of the following, the least likely mode of transport of iron(III) in a normal stream is:
   (a) bound to suspended humic material
   (b) bound to clay particles by cation-exchange processes
   (c) as suspended Fe_2O_3
   (d) as soluble Fe^{3+} ion
   (e) bound to colloidal clay-humic substance complexes.

4. How does freshly precipitated colloidal iron(III) hydroxide interact with many divalent metal ions in solution?

5. Of the following, the one that is *not* a characteristic of clay is:
   (a) layered structure
   (b) aluminum and silicon oxide components
   (c) surface charge through ion replacement
   (d) replacement of Si(IV) or Al(III) ions by ions of similar size but lesser charge
   (e) membership in the class of primary rocks or minerals.

6. What stabilizes colloids made of bacterial cells in water?

7. What is thought to be the mechanism by which bacterial cells aggregate?

8. What is a good method for the production of freshly precipitated MnO_2?

9. A sediment sample was equilibrated with a solution of NH_4^+ ion, and the NH_4^+ was later displaced by Na^+ for analysis. A total of 33.8 milliequivalents of NH_4^+ was bound to the sediment and later displaced by Na^+. After drying, the sediment weighed 87.2 g. What was its CEC in milliequivalents/100 g?

10. A sediment sample with a CEC of 67.4 milliequivalents/100 g was found to contain the following exchangeable cations in milliequivalents/100 g: Ca^{2+}, 21.3; Mg^{2+}, 5.2; Na^+, 4.4; K^+, 0.7. The quantity of hydrogen ion, H^+, was not measured directly. What was the ECS of H^+ in milliequivalents/100 g?

11. What is the meaning of *zero point of charge* as applied to colloids? Is the surface of a colloidal particle totally without charged groups at the ZPC?

12. The concentration of methane in an interstitial water sample was found to be 150 mL/L at STP. Assuming that the methane was produced by the fermentation of organic matter, $\{CH_2O\}$, what weight of organic matter was required to produce the methane in a liter of the interstitial water?

13. What is the difference between CEC and ECS?

14. Match the sedimentary mineral on the left with its conditions of formation on the right:

(a) $FeS(s)$	(1) May be formed when anaerobic water is exposed to O_2.
(b) $Ca_5OH(PO_4)_3$	(2) May be formed when aerobic water becomes anaerobic.
(c) $Fe(OH)_3$	(3) Photosynthesis by-product.
(d) $CaCO_3$	(4) May be formed when wastewater containing a particular kind of contaminant flows into a body of very hard water.

15. In terms of their potential for reactions with species in solution, how might metal atoms, M, on the surface of a metal oxide, MO, be described?

16. For what purpose is a polymer with the following general formula used?

```
 ┌  H   H  ┐
 │  |   |  │
 │ -C---C- │
 │  |   |  │
 │  |   H  │
 │ (C6H4)  │
 │  |      │
 │ SO3-    │
 └         ┘n
```

CHAPTER 7
WATER POLLUTION

7.1 WATER QUALITY

Throughout history, the quality of drinking water has been a factor in determining human welfare. Fecal pollution of drinking water has frequently caused waterborne diseases that have decimated the populations of whole cities. Unwholesome water polluted by natural sources has caused great hardship for people forced to drink it or use it for irrigation.

Today there are still occasional epidemics of bacterial and viral diseases caused by infectious agents carried in drinking water. Some municipal water supplies still have unsafe levels of potential pathogens. You may even have doubts about the safety of the water in a tempting, sparkling mountain brook—there may be a campsite latrine along its banks a few hundred meters upstream. However, waterborne diseases have in general been well controlled, and drinking water in technologically advanced countries in the 1980s is remarkably free of the disease-causing agents that were very common water contaminants only a few decades earlier.

Currently, waterborne toxic chemicals pose the greatest threat to the safety of water supplies in industrialized nations. This is particularly true of ground water in the U.S., which exceeds in volume the flow of all U.S. rivers, lakes, and streams. In some areas, the quality of groundwater is subject to a number of chemical threats. There are many possible sources of chemical contamination. These include wastes from industrial chemical production, metal-plating operations, and pesticide runoff from agricultural lands. Some specific pollutants include industrial chemicals such as chlorinated hydrocarbons; heavy metals, including cadmium, lead, and mercury; saline water; bacteria, particularly coliforms; and general municipal and industrial wastes.

Since World War II there has been a tremendous growth in the manufacture and use of synthetic chemicals. Many of the chemicals have contaminated water supplies. Two examples are insecticide and herbicide runoff from agricultural land, and industrial discharge into surface waters. Most serious, though, is the threat to ground water from waste chemical dumps and landfills, storage lagoons, treating ponds, and other facilities. These threats are discussed in more detail in Chapters 19 and 20.

It is clear that water pollution should be a concern of every citizen. Understanding the sources, interactions, and effects of water pollutants is essential for controlling pollutants in an environmentally safe and economically

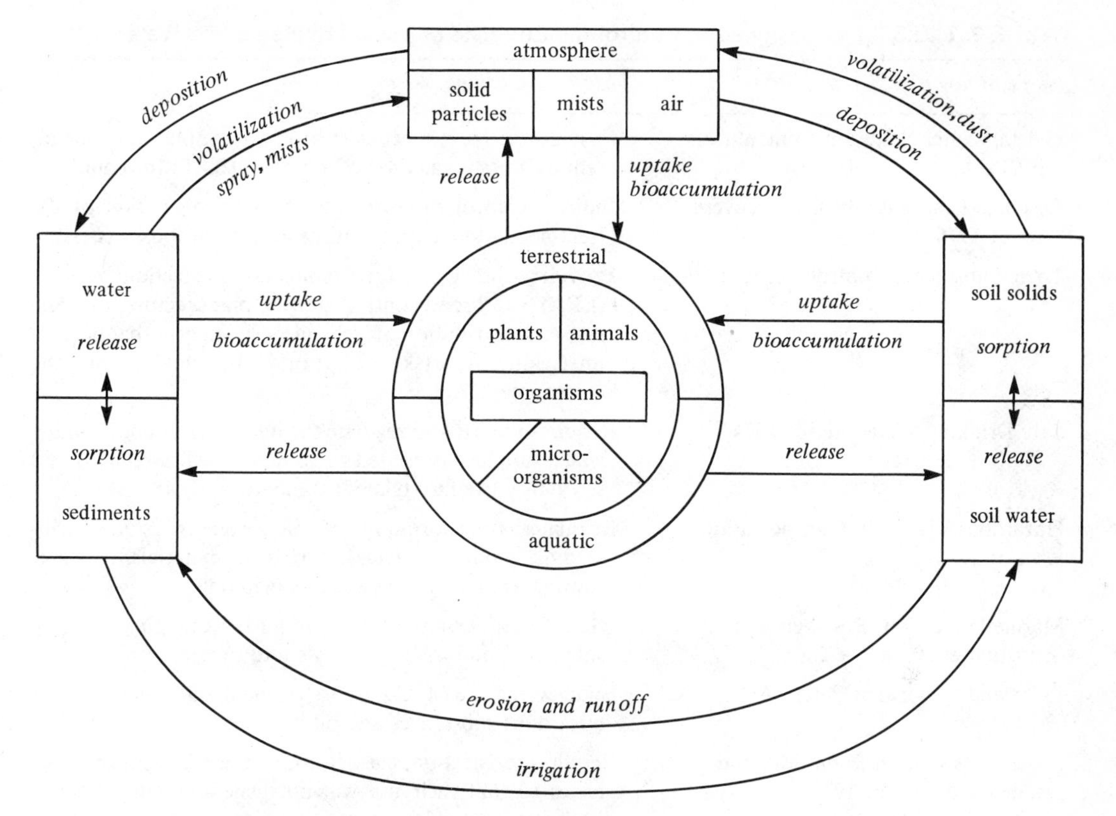

FIGURE 7.1 Pollutant cycles in the environment.

acceptable manner. Above all, an understanding of water pollution and its control depends upon a basic knowledge of aquatic environmental chemistry. That is why we have presented so much material on aquatic chemistry prior to discussing pollution. Water pollution may be studied much more effectively with a sound background in the basic properties of water, aquatic microbial reactions, sediment-water interactions, and other factors involved with the reactions, transport, and effects of these pollutants.

In considering water pollution, it is useful to keep in mind an overall picture of possible pollutant cycles as shown in Figure 7.1. This figure illustrates the major routes of pollutant interchange among the biotic, terrestrial, atmospheric, and aquatic environments.

Federal legislation, most of which was passed during the 1970s, provides regulatory agencies with a number of laws that can be used to prevent water pollutants from getting into U.S. waters [1]. Since it is helpful to consider water pollution from the perspective of laws designed to control it, these laws are summarized in Table 7.1 (see next page).

---

[1] Barrett, B.R. 1978. Controlling the entrance of toxic pollutants into U.S. waters. *Environmental Science and Technology* 12: 154–62.

TABLE 7.1 U.S. Laws Designed to Control the Entrance of Toxic Pollutants into Waters

Name of law	*Major provisions*
Federal Water Pollution Control Acts (FWPCA)	Provide the basic mechanism for water-pollution control; require "best available technology" (BAT) for control.
Resource Conservation and Recovery Act of 1976	Indirect control of water pollution by regulation of the treatment, storage, and disposal of hazardous wastes.
Toxic Substances Control Act of 1976	Provides the U.S. Environmental Protection Agency (EPA) with broad control over the manufacture, processing, and distribution of "chemical substances that pose an unreasonable risk of injury to health or the environment."
Safe Drinking Water Act of 1974	Requires the EPA to regulate the quality of public drinking-water supplies as related to primary (health-related) and secondary (welfare-related) standards.
Hazardous Materials Transportation Act of 1974	Regulates transportation of dangerous substances, including oxidants, acids, corrosive materials, disease-causing agents, and radioactive materials.
Marine Protection, Research and Sanctuaries Act of 1972	"The Ocean Dumping Act", controls dumping of toxic substances and other materials into the oceans.
Ports and Waterways Safety Act of 1972	Enables the Coast Guard to regulate bulk shipment of hazardous substances and oil by water.
Federal Insecticide, Fungicide, and Rodenticide Act of 1972	Requires registration, classification, and regulation of pesticides so that their use will not generally cause adverse environmental effects.
Atomic Energy Act of 1954 (amended)	Regulates environmental release of radioactive wastes.
Clean Water Act of 1977	Regulates quality of effluents discharged from publicly owned treatment works (POTW).
Comprehensive Environmental Response, Compensation, and Liability Act of 1980	Sets up a "Superfund" program to clean up toxic waste dumps threatening water supplies.

7.2 CLASSES OF WATER POLLUTANTS

Water pollutants can be divided among some general classifications, as summarized in Table 7.2. Most of these categories of pollutants, and several subcategories, are discussed in this chapter. An enormous amount of material is published on this subject each year, and it is impossible to cover it all in one chapter. Therefore, many references for additional reading are provided at the end of this chapter.

7.3 POLLUTANT TRACE ELEMENTS IN WATER

The term **trace elements** refers to those elements that occur at very low levels in a given system. The somewhat ambiguous term probably arose from the inadequacy of earlier analytical techniques—before modern methods such as atomic absorption, plasma emission, neutron-activation analysis, gas chromatography,

TABLE 7.2 General Classifications of Water Pollutants

Class of pollutant	*Significance*
Trace elements	Health, aquatic biota
Metal-organic combinations	Metal transport
Inorganic pollutants	Toxicity, aquatic biota
Asbestos	Human health
Algal nutrients	Eutrophication
Radionuclides	Toxicity
Acidity, alkalinity, salinity (pollutants if in excess)	Water quality, aquatic life
Sewage	Water quality, oxygen levels
Biochemical oxygen demand	Water quality, oxygen levels
Trace organic pollutants	Toxicity
Pesticides	Toxicity, aquatic biota, wildlife
Polychlorinated biphenyls	Possible biological effects
Chemical carcinogens	Incidence of cancer
Petroleum wastes	Effect on wildlife, esthetics
Pathogens	Health effects
Detergents	Eutrophication, wildlife, esthetics
Sediments	Water quality, aquatic biota, wildlife
Taste, odor, and color	Esthetics

mass spectrometry, and others extended the limits of detection to the very low levels currently attainable. In many early investigations, it was only possible to detect the *presence* of an element, as it was said to be present at a "trace level." A reasonable definition of a trace element is one that occurs at a level of a few parts per million or less. The term **trace substance** is a more general one applied to both elements and chemical compounds.

Table 7.3 (pages 150–151) summarizes the more important trace elements encountered in natural waters. Some of these are recognized as nutrients required for animal and plant life. Of these, many are essential at low levels but toxic at higher levels. This is typical behavior for many substances in the aquatic environment, a point that must be kept in mind in judging whether a particular element is beneficial or detrimental. Some of these elements, such as lead or mercury, have such toxicological and environmental significance that they are discussed in detail in separate sections. For more detailed information on the significance of trace elements in the aquatic environment, the reader may consult a number of references on the subject (see references 2–9 on page 152).

Some of the heavy metals are among the most harmful of the elemental pollutants. These elements, which are in general the metals in the lower right-hand corner of the periodic table, include essential elements like iron as well as toxic metals like lead, cadmium, and mercury. Most of them have a tremendous affinity for sulfur and attack sulfur bonds in enzymes, thus immobilizing the enzymes. Protein carboxylic acid ($—CO_2H$) and amino ($—NH_2$) groups are also chemically bound by heavy metals. Cadmium, copper, lead and mercury ions bind to cell membranes, hindering transport processes through the cell wall. Heavy metals may also precipitate phosphate biocompounds or catalyze their decomposition. The biochemical effects of metals are discussed in Chapter 18.

TABLE 7.3 Occurrence and Significance of Trace Elements in Natural Waters

Element	*Sources*	*Effects and significance*	*U.S. Public Health Service Limit, mg/liter*[1]	*Occurrence: % of samples, highest and mean concentrations* (μg/liter)[2]
Arsenic	Mining by-product, pesticides, chemical waste	Toxic, possibly carcinogenic	0.05	5.5% (above 5 μg/L), 336, 64
Beryllium	Coal, nuclear power and space industries	Acute and chronic toxicity, possibly carcinogenic	Not given	Not given
Boron	Coal, detergent formulations, industrial wastes	Toxic to some plants	1.0	98% (above 1 μg/L), 5000, 101
Cadmium	Industrial discharge, mining waste, metal plating, water pipes	Replaces zinc biochemically, causes high blood pressure and kidney damage, destroys testicular tissue and red blood cells, toxic to aquatic biota.	0.01	2.5%, not given, 9.5
Chromium	Metal plating, cooling-tower water additive (chromate), normally found as Cr(VI) in polluted water	Essential trace element (glucose tolerance factor), possibly carcinogenic as Cr(VI)	0.05	24.5%, 112, 9.7
Copper	Metal plating, industrial and domestic wastes, mining, mineral leaching	Essential trace element, not very toxic to animals, toxic to plants and algae at moderate levels	1.0	74.4%, 280, 15
Fluorine (fluoride ion)	Natural geological sources, industrial waste, water additive	Prevents tooth decay at about 1 mg/L, causes mottled teeth and bone damage at around 5 mg/L in water	0.8–1.7 depending on temperature	Not given
Iodine (iodide)	Industrial waste, natural brines, seawater intrusion	Prevents goiter	Not given	Rare in fresh water

Iron	Corroded metal, industrial wastes, acid mine drainage, low pE water in contact with iron minerals	Essential nutrient (component of hemoglobin), not very toxic, damages materials (bathroom fixtures and clothing)	0.05	75.6%, 4600, 52
Lead	Industry, mining, plumbing, coal, gasoline	Toxicity (anemia, kidney disease, nervous system), wildlife destruction	0.05	19.3% (above 2 μg/L), 140, 23
Manganese	Mining, industrial waste, acid mine drainage, microbial action on manganese minerals at low pE	Relatively nontoxic to animals, toxic to plants at higher levels, stains materials (bathroom fixtures and clothing)	0.05	51.4% (above 0.3 μg/L), 3230, 58
Mercury	Industrial waste, mining, pesticides, coal	Acute and chronic toxicity	Not given	Not given
Molybdenum	Industrial waste, natural sources, cooling-tower water additive	Possibly toxic to animals, essential for plants	Not given	32.7 (above 2 μg/L), 5400, 120
Selenium	Natural geological sources, sulfur, coal	Essential at low levels, toxic at higher levels, causes "alkali disease" and "blind staggers" in cattle, possibly carcinogenic	0.01	Not given
Silver	Natural geological sources, mining, electroplating, film-processing wastes, disinfection of water	Causes blue-grey discoloration of skin, mucous membranes, eyes	0.05	6.6% (above 0.1 μg/L), 38, 2.6
Zinc	Industrial waste, metal plating, plumbing	Essential element in many metalloenzymes, aids wound healing, toxic to plants at higher levels; major component of sewage sludge, limiting land disposal of sludge	5.0	76.5% (above 2 μg/L), 1180, 64

[1] *Public Health Service Drinking Water Standards,* U.S. Public Health Service, 1962.

[2] J.F. Kopp and R.C. Kroner, *Trace Metals in Waters of the United States,* United States Environmental Protection Agency. The first figure is the percentage of samples showing the element; the second is the highest value found; the third is the mean value in positive samples (samples showing the presence of the metal at detectable levels).

Some of the metalloids, elements on the borderline between metals and nonmetals, are significant water pollutants. Arsenic, selenium, and antimony are of particular interest.

Inorganic chemicals manufacture has the potential to contaminate water with trace elements. There are an estimated 750 inorganic chemical plants in the U.S. [10]. Among the industries regulated for potential trace element pollution of water are those producing chlor-alkali (mercury cells and diaphragm cells), hydrofluoric acid, sodium dichromate (sulfate process and chloride ilmenite process), aluminum fluoride, chrome pigments, copper sulfate, nickel sulfate, sodium bisulfate, sodium hydrosulfate, sodium bisulfite, titanium dioxide, and hydrogen cyanide. In the June 29, 1982 *Federal Register,* limits on pollutants discharged in wastewater from inorganic chemical plants were published [11]. These affect 144 industrial chemical manufacturing plants producing gases, pigments from metal-bearing ores, and inorganic chemicals. It was estimated that these regulations would remove from wastewater 1.4 million kilograms of toxic metals, including chromium, as well as 3.6 million kilograms of chlorine.

The control and treatment measures to be used in cleaning up wastewaters from inorganic chemical plants include the following: (1) equalization, alkaline precipitation, and settling-clarification to precipitate metals and fluorides; (2) dual-media filtration to remove suspended metal compounds; (3) sulfide precipitation of arsenic, cadmium, copper, lead, mercury, and silver; (4) use of coagulants to aid settling and filtration of precipitated compounds; (5) reduction of Cr(VI) to Cr(III) by Fe(II), SO_2, or sulfide; (6) treatment with air to reduce chemical oxygen demand and to oxidize soluble Fe(II) to settleable solid $Fe(OH)_3$; (7) alkaline chlorination at pH 10–11 to oxidize cyanide to cyanate; and (8) dechlorination by catalytic decomposition, chemical treatment, or thermal treatment. Membrane processes, application of xanthates, and activated charcoal adsorption can also be employed.

---

[2] Hemphill, D.D., ed. 1982. *Trace substances in environmental health.* Vols. 1–16. Published annually since 1967. Columbia, Mo.: Univ. of Missouri Environmental Trace Substances Center.

[3] Singer, P.C., ed. 1973. *Trace metals and metal-organic interactions in natural waters.* Ann Arbor, Mich.: Ann Arbor Science Publishers, Inc.

[4] Kothny, E.L., ed. 1973. *Trace elements in the environment.* ACS Advances in Chemistry Series 123. Washington, D.C.: American Chemical Society.

[5] Förstner, U., and Wittman, G.T.W. 1979. *Metal pollution in the aquatic environment.* New York: Springer-Verlag, Inc.

[6] Laws, E.A. 1981. *Aquatic pollution.* New York: Wiley-Interscience.

[7] Kopp, J.F., and Kroner, R.C. 1969. *Trace metals in waters of the United States.* Washington, D.C.: United States Environmental Protection Agency.

[8] National Academy of Sciences. 1974. *Geochemistry and the environment, vol. I, the relationship of selected trace elements to health and disease.* Washington, D.C.

[9] National Academy of Sciences, 1977. *Geochemistry and the environment, vol. II, the relationship of selected trace elements to health and disease.* Washington, D.C.

[10] 1980. Inorganics industry faces new regs. *Chemical and Engineering News* 4 August 1980, p. 6.

[11] 1982. Discharge of inorganic pollutants to be regulated. *Chemical and Engineering News* 12 July 1982, p. 19.

7.4 ARSENIC

Arsenic has been the chemical villain of more than a few muder plots. Acute arsenic poisoning can result from the ingestion of more than about 100 mg of the element. Chronic poisoning occurs with the ingestion of small amounts of arsenic over a long period of time. There is some evidence that this element is also carcinogenic.

Arsenic occurs in the Earth's crust at an average level of 2–5 ppm. The combustion of fossil fuels, particularly coal, introduces large quantities of arsenic into the environment, much of it reaching natural waters. Arsenic occurs with phosphate minerals and enters into the environment along with some phosphorus compounds. Some pesticides, particularly those in wide use before World War II, contain highly toxic arsenic compounds. The most common of these are lead arsenate, $Pb_3(AsO_4)_2$; sodium arsenite, Na_3AsO_3; and Paris Green, $Cu_3(AsO_3)_2$. Another major source of arsenic is mine tailings. Arsenic produced as a by-product of copper, gold, and lead refining greatly exceeds the commercial demand for arsenic, and it accumulates as waste material.

There is some evidence that arsenic, like mercury, may be converted to highly toxic methyl derivatives by bacteria, according to the following reactions:

$$H_3AsO_4 + 2\,H^+ + 2\,e \rightarrow H_3AsO_3 + H_2O \qquad (7.4.1)$$

$$H_3AsO_3 \xrightarrow{\text{methylcobalamin}} \underset{\text{(methyl arsinic acid)}}{CH_3AsO(OH)_2} \qquad (7.4.2)$$

$$CH_3AsO(OH)_2 \xrightarrow{\text{methylcobalamin}} \underset{\text{(dimethyl arsinic acid)}}{(CH_3)_2AsO(OH)} \qquad (7.4.3)$$

$$(CH_3)_2AsO(OH) + 4\,H^+ + 4\,e \rightarrow \underset{\text{(dimethyl arsine)}}{(CH_3)_2AsH} \qquad (7.4.4)$$

7.5 CADMIUM

Pollutant cadmium in water may arise from industrial discharges and mining wastes. Cadmium is widely used in metal plating. Chemically, cadmium is very similar to zinc, and these two metals frequently undergo geochemical processes together. Both metals are found in water in the +2 oxidation state.

The effects of acute cadmium poisoning in humans are very serious. Among them are high blood pressure, kidney damage, destruction of testicular tissue, and destruction of red blood cells. Cadmium poisoning causes the malady known in Japan as "itai, itai," translated as "ouch, ouch." This condition is manifested by bone fracture and causes intense pain. Some people in the Jintsu River basin of Japan have been afflicted with "itai, itai." The source of cadmium was river water polluted by mining operations. The cadmium was ingested with the water used for drinking or from rice irrigated by the contaminated water.

It is believed that much of the physiological action of cadmium arises from its chemical similarity to zinc. Specifically, cadmium may replace zinc in some enzymes, thereby altering the stereostructure of the enzyme and impairing its catalytic activity. Disease symptoms ultimately result. Cadmium is certainly a dangerous water pollutant, constituting a major water-quality problem.

Cadmium and zinc are common water and sediment pollutants in harbors surrounded with industrial installations. The estuarine migration and redistribution of zinc and cadmium from such sources has been the subject of a detailed study [12]. In the particular location studied, concentrations of cadmium as high as 130 ppm dry weight sediment were found in harbor sediments, and concentrations ranging up to 1.9 ppm were found in bay sediments outside the harbor. During the summer, when the harbor water was most stagnant, an interesting variation in cadmium concentration was observed with increasing depth. The aerobic surface layer contained a relatively high dissolved cadmium level, primarily as the soluble ion pair $CdCl^+$. The anaerobic bottom layer of the water was cadmium-poor because microbial reduction of sulfate had produced sulfide,

$$2\,\{CH_2O\} + SO_4^{2-} + H^+ \rightarrow 2\,CO_2 + HS^- + 2\,H_2O \qquad (7.5.1)$$

which had precipitated cadmium as insoluble cadmium sulfide:

$$CdCl^+ + HS^- \rightarrow CdS\,(s) + H^+ + Cl^- \qquad (7.5.2)$$

Mixing of bay and harbor water by high winds during the winter resulted in desorption of cadmium from harbor sediments by aerobic bay water. This dissolved cadmium was carried out into the bay, where it reacted with suspended solid materials, which then became incorporated with the bay sediments. This is only one example of the sort of complicated interaction of hydraulic, chemical, solution–solid, and microbiological factors involved in the transport and distribution of a pollutant in an aquatic system.

7.6 LEAD

Lead occurs in water in the +II oxidation state and arises from a number of industrial and mining sources. Lead from leaded gasoline is a major source of atmospheric and terrestrial lead, and much of this lead eventually enters natural water systems. Under provisions of the Clean Air Act, the Environmental Protection Agency instituted a program in 1973 to gradually decrease the lead content of gasoline. As of early 1982, large refineries were restricted to adding an average of 0.5 grams of lead per gallon to their overall gasoline production; producers of under 50,000 gallons of gasoline per day were allowed to add up to 2.65 g of lead per gallon of gasoline. In July, 1982, EPA proposed that lead content, averaged over both leaded and unleaded gasoline, be limited to 0.5 g/gal with a limit of 1.1 g/gal in leaded gasoline. These measures and the conversion of the U.S. auto fleet to vehicles that require unleaded gasoline should appreciably help the lead contamination problem. In addition to pollutant sources, lead-bearing limestone and galena (PbS) may contribute lead to natural waters.

Despite greatly increased total use of lead by industry, evidence from hair samples and other sources indicates that body burdens of this toxic metal may, in fact, have decreased during recent decades. This may be the result of less lead used in plumbing and other products that come in contact with food or drink. In

[12] Holmes, C.W.; Slade, E.A.; and McLerran, C.J. 1974. Migration and redistribution of zinc and cadmium in a marine estuarine system. *Environmental Science and Technology* 8: 255–9.

fact, some historians believe that the use of lead containers for food and drink poisoned ancient Rome's ruling class and accelerated its downfall.

Acute lead poisoning in humans causes severe dysfunction in the kidneys, reproductive system, liver, and the brain and central nervous system. Sickness or death results. Lead poisoning from environmental exposure is thought to have caused mental retardation in many children. Mild lead poisoning causes anemia. The victim may have headaches and sore muscles and may feel generally fatigued and irritable.

Lead is probably not a major problem in drinking water, except in those cases where old lead pipe is still in use. Lead is a constituent of solder and some pipe-joint formulations, so that household water does have some contact with lead. Water that has stood in household plumbing for some time may accumulate spectacular levels of lead (along with zinc, cadmium, and copper). Next time you return from vacation, let the water run for some time before drinking it!

7.7 MERCURY

Mercury generates the most concern of any of the heavy-metal pollutants. Mercury is found as a trace component of many minerals, with continental rocks containing an average of around 80 parts per billion, or slightly less, of this element. Cinnabar, red mercuric sulfide, is the chief commercial mercury ore. The fossil fuels coal and lignite contain mercury, often at levels of 100 parts per billion or even higher. As the use of these fuels grows in response to the energy shortage, more must be learned about the fate of this mercury, and appropriate steps must be taken to prevent environmental hazards from it.

Metallic mercury is used, for example, in laboratory vacuum apparatus. The primary use of mercury metal is as an electrode in the electrolytic generation of chlorine gas. Large quantities of inorganic mercury(I) and mercury(II) compounds are used annually. Organic mercury compounds find wide application as pesticides, particularly fungicides. These mercury compounds include aryl mercurials such as phenyl mercuric dimethyldithiocarbamate

$$C_6H_5{-}Hg{-}S{-}\overset{\overset{S}{\|}}{C}{-}N(CH_3)_2$$

(used in paper mills as a slimicide and as a mold retardant for paper), and alkylmercurials such as ethylmercuric chloride, C_2H_5HgCl, used as a seed fungicide. The alkyl mercury compounds tend to resist degradation and are generally considered to be more of an environmental threat than either the aryl or inorganic compounds.

Mercury enters the environment from a large number of miscellaneous sources related to human use of the element. These include discarded laboratory chemicals, batteries, broken thermometers, lawn fungicides, amalgam tooth fillings, and pharmaceutical products. Taken individually, each of these sources may not contribute much of the toxic metal, but the total effect can be substantial. Sewage effluent sometimes contains up to 10 times the level of mercury found in typical natural waters.

The toxicity of mercury was tragically illustrated in the Minamata Bay area of Japan during the period 1953–1960. A total of 111 cases of mercury poisoning were reported among people who had consumed mercury-contaminated seafood from the bay. Of those afflicted, 43 died. Congenital defects were observed in 19 babies whose mothers had consumed seafood contaminated with mercury. The level of metal in the contaminated seafood was 5–20 parts per million. The mercury source was waste from a chemical plant that drained into Minamata Bay.

Among the toxicological effects of mercury are neurological damage, including irritability, paralysis, blindness, or insanity; chromosome breakage; and birth defects. The milder symptoms of mercury poisoning, such as depression and irritability, have a psychopathological character. Therefore, mild mercury poisoning may escape detection. Some forms of mercury are relatively nontoxic and have been used as medicines, in the treatment of syphilis for example, for centuries. Other forms of mercury, particularly organic compounds, are highly toxic.

Because there are few major natural sources of mercury, and since most inorganic compounds of this element are relatively insoluble, it was assumed for some time that mercury was not a serious water pollutant in water. Indeed, a survey of trace metals in U.S. waters conducted from 1962 to 1967 did not even list mercury. However, in 1970 alarming mercury levels were discovered in fish in Lake Saint Clair between Michigan and Ontario, Canada. A subsequent survey by the Federal Water Quality Administration revealed a number of other waters contaminated with mercury. It was found that several chemical plants, particularly caustic-chemical plants, were each releasing up to 14 or more kilograms of mercury in wastewaters each day.

The unexpectedly high concentrations of mercury found in water and in fish tissues result from the formation of soluble monomethylmercury ion, CH_3Hg^+, and volatile dimethylmercury, $(CH_3)_2Hg$, by anaerobic bacteria in sediments. Mercury from these compounds becomes concentrated in fish lipid (fat) tissue and the concentration factor from water to fish may exceed 10^3. The methylating agent by which inorganic mercury is converted to methylmercury compounds is methylcobalamin, a vitamin B_{12} analog:

$$HgCl_2 \xrightarrow{\text{methylcobalamin}} CH_3HgCl + Cl^- \quad (7.7.1)$$

It is believed that the bacteria that synthesize methane produce methylcobalamin as an intermediate in the synthesis. Thus, waters and sediments in which anaerobic decay is occurring probably contain all of the necessary ingredients for methylmercury production. In neutral or alkaline waters, the formation of dimethyl mercury, $(CH_3)_2Hg$, is favored. This volatile compound can escape to the atmosphere.

7.8 METAL-ORGANIC COMBINATIONS IN WATER

An appreciation of the strong influence of complexation and chelation on heavy metals' behavior in natural waters and wastewaters may be gained by reading Chapter 4, which deals with that subject. Methylmercury formation is discussed in Section 7.7. Both topics involve the combination of metals and organic entities in water. It must be stressed that the interaction of metals with organic compounds is of utmost importance in determining the role played by the metal in an aquatic system.

There are two major types of metal-organic interactions to be considered in an aquatic system. The first of these is *complexation*, usually chelation when organic ligands are involved. A reasonable definition of complexation by organics applicable to natural water and wastewater systems is a system in which a species is present that *reversibly* dissociates to a metal ion and a complexing species as a function of hydrogen-ion concentration:

$$ML + 2\,H^+ \rightleftharpoons M^{2+} + H_2L \qquad (7.8.1)$$

In this equation, M^{2+} is a metal ion and H_2L is the acidic form of a complexing—frequently chelating—ligand, L^{2-}. *Organometallic compounds*, on the other hand, normally are considered to be those involving organic and metal compounds that do not dissociate reversibly at lower pH or greater dilution. These compounds generally contain a carbon-metal bond. Furthermore, the organic component, and sometimes the particular oxidation state of the metal involved, may not be stable apart from the organometallic compound. There are, of course, many kinds of organometallic compounds, of which methylmercury is only one example. Evidence continues to accumulate on the role of bacteria in producing methylated forms of metals. For example, recent investigations [13, 14] have shown that a species of *Pseudomonas* bacteria isolated from Chesapeake Bay produces organotin compounds such as hydrophobic Me_4Sn, where Me represents the methyl group, by biomethylation and bioreduction of Sn(IV). Also produced under the same conditions are methyl stannates with the general formula, Me_nSnH_{4-n}, where n = 1–3. Many pesticide formulations containing metals are organometallic compounds.

The interaction of trace metals with organic compounds in natural waters is discussed in a collection of works on that subject [3]. It is too vast an area to cover in detail in this chapter; however, it may be noted that metal-organic interactions may involve organic species of both pollutant (such as EDTA) and natural (such as fulvic acids) origin. These interactions are influenced by, and sometimes play a role in, redox equilibria; formation and dissolution of precipitates; colloid formation and stability; acid-base reactions; and microorganism-mediated reactions in water. Metal-organic interactions may increase or decrease the toxicity of metals in aquatic ecosystems, and they have a strong influence on the growth of algae in water.

7.9 CYANIDE AND OTHER INORGANIC CHEMICAL SPECIES IN WATER

Many important inorganic water pollutants were considered in Section 7.3, as part of the discussion on pollutant trace elements. Inorganic pollutants that con-

---

[13] Jackson, J.A.; Blair, W.R.; Brinckman, F.E., and Iverson, W.P. 1982. Gas chromatographic speciation of methylstannates in the Chesapeake Bay using purge and trap sampling with a tin-selective detector. *Environmental Science and Technology* 16: 110–119.

[14] Hallas, L.E., Means, J.C., and Cooney, J.J. 1982. Methylation of tin by estuarine microorganisms. *Science* 215: 1505–7.

tribute acidity, alkalinity, or salinity to water are considered separately in this chapter. Still another class is that of algal nutrients. This leaves unclassified, however, some important inorganic pollutant species, of which cyanide ion, CN^-, is probably the most important. Others include ammonia, carbon dioxide, hydrogen sulfide, nitrite, and sulfite.

Cyanide, a deadly poisonous substance, exists in water as HCN, a weak acid with a K_a of 6×10^{-10}. The cyanide ion has a strong affinity for many metal ions, forming relatively less-toxic ferrocyanide, $Fe(CN)_6^{4-}$, with iron(II), for example. Volatile HCN is very toxic and was used between 1940 and 1967 in gas chamber executions in the U.S.

Cyanide is widely used in industry, especially for metal cleaning and electroplating. It is also one of the main gas and coke scrubber effluent pollutants from gas works and coke ovens. Cyanide is widely used in certain mineral-processing operations. Until about 1970, a particular gold mine in the United States used almost 1-¼ tons of cyanide per day to leach gold from a daily output of more than 5,000 tons of ore. Approximately 75 pounds of the cyanide were dumped into a creek each day. The same creek previously had received approximately 30 pounds of mercury daily, as a result of a leaching process that was replaced by cyanide leaching because of the polluting effect of the mercury. There was a distinct lack of aquatic life in the stream, and it was not recommended as a source of drinking water.

Excessive levels of ammoniacal nitrogen cause water-quality problems. Ammonia is the initial product of the decay of nitrogenous organic wastes, and its presence frequently indicates the presence of such wastes. It is a normal constituent of low-pE ground waters and is sometimes added to drinking water, where it reacts with chlorine to provide residual chlorine (see Section 8.7). Since the pK_a of ammonium ion, NH_4^+, is 9.26, most ammonia in water is present as the protonated form rather than as NH_3.

Hydrogen sulfide, H_2S, is a product of the anaerobic decay of organic matter containing sulfur. It is also produced in the anaerobic reduction of sulfate by microorganisms (see Chapter 5). Hydrogen sulfide is evolved as a gaseous pollutant from geothermal waters (see Section 17.4). Wastes from chemical plants, paper mills, textile mills, and tanneries may also contain H_2S. Its presence is easily detected by its characteristic rotten-egg odor. In water, H_2S is a weak diprotic acid with pK_{a1} of 6.99 and pK_{a2} of 12.92; S^{2-} is not present in normal natural waters. The sulfide ion has tremendous affinity for many heavy metals, and precipitation of metallic sulfides often accompanies production of H_2S (see Figure 6.2).

Free carbon dioxide, CO_2, is frequently present in water at high levels due to decay of organic matter. It is also added to softened water during water treatment as part of a recarbonation process (see Chapter 8). Excessive carbon dioxide levels may make water more corrosive and may be harmful to aquatic life.

Nitrite ion, NO_2^-, occurs in water as an intermediate oxidation state of nitrogen. Its pE range of stability is relatively narrow (see Figure 3.4). Nitrite is added to some industrial process water to inhibit corrosion; it is rarely found in drinking water at levels over 0.1 mg/L.

Sulfite ion, SO_3^{2-}, is found in some industrial wastewaters. Sodium sulfite is commonly added to boiler feedwaters as an oxygen scavenger:

$$2\,SO_3^{2-} + O_2 \rightarrow 2\,SO_4^{2-} \quad (7.9.1)$$

Since pK_{a1} of sulfurous acid is 1.76 and pK_{a2} is 7.20, sulfite exists as either HSO_3^- or SO_3^{2-} in natural waters, depending upon pH. It may be noted that hydrazine, N_2H_4, also functions as an oxygen scavenger:

$$N_2H_4 + O_2 \rightarrow 2\ H_2O + N_2(g) \qquad (7.9.2)$$

7.10 ASBESTOS IN WATER

The toxicity of inhaled asbestos is well established. The fibers scar lung tissue and cancer eventually develops, often 20 or 30 years after exposure. It is not known for sure whether asbestos is toxic in drinking water. This has been a matter of considerable concern because of the dumping of taconite (iron ore tailings) containing asbestos-like fibers into Lake Superior. The fibers have been found in drinking waters of cities around the lake. After having dumped the tailings into Lake Superior since 1952, the Reserve Mining Company at Silver Bay on Lake Superior solved the problem in 1980 [15] by constructing a 6-square-mile containment basin inland from the lake. This $370-million facility keeps the taconite tailings covered with a 3-meter layer of water to prevent escape of fiber dust.

In April, 1982, a lawsuit was settled involving the Reserve Mining Company, the states of Michigan, Wisconsin, and Minnesota, the Federal Government, and 23 other parties. The final agreement called for Reserve to pay $1.84 million to filter the water supplies of Duluth, Minnesota and three other cities that had asbestos-like fibers in their drinking water sources, all pumped from Lake Superior.

7.11 ALGAL NUTRIENTS AND EUTROPHICATION

The term **eutrophication** is derived from the Greek word meaning "well-nourished". It describes a condition of lakes or reservoirs involving excess algal growth, which may eventually lead to severe deterioration of the body of water. The first step in eutrophication of a body of water is an input of nutrients from watershed runoff or sewage. The nutrient-rich body of water then produces a great deal of plant biomass by photosynthesis, along with a smaller amount of animal biomass. Dead biomass accumulates in the bottom of the lake, where it partially decays, recycling nutrient carbon dioxide, phosphorus, nitrogen, and potassium. If the lake is not too deep, bottom-rooted plants begin to grow, accelerating the accumulation of solid material in the basin. Eventually a marsh is formed, which finally fills in to produce a meadow or forest.

Eutrophication is by no means a new phenomenon; for instance, it is basically responsible for the formation of huge deposits of coal and peat. However, human activity can accelerate eutrophication. To understand why this is so, refer to Table 7.4 (see page 160). This table shows the chemical elements needed for plant growth. Most of these are present at a level more than sufficient to support plant life in the average lake or reservoir. Hydrogen and oxygen come

---

[15] 1980. Tailings' end, cleaner days for Silver Bay. *Time* 31 March 1980, p. 45.

TABLE 7.4 Essential Plant Nutrients: Sources and Functions

Nutrient	*Source*	*Function*
Macronutrients		
carbon (CO_2)	atmosphere, decay	biomass constituent
hydrogen	water	biomass constituent
oxygen	water	biomass constituent
nitrogen (NO_3^-)	decay, atmosphere (nitrogen-fixing organisms), pollutants	protein constituent
phosphorus (phosphate)	decay, minerals, pollutants	DNA, RNA constituent
potassium	minerals, pollutants	metabolic function
sulfur (sulfate)	minerals	proteins, enzymes
magnesium	minerals	metabolic function
calcium	minerals	metabolic function
Micronutrients		
B, Cl, Co, Cu, Fe, Mo, Mn, Na, Si, V, Zn	minerals, pollutants	metabolic function and/or constituent of enzymes

from the water itself. Carbon is provided by CO_2 from the atmosphere or from decaying vegetation. Sulfate, magnesium, and calcium are normally present in abundance from mineral strata in contact with the water. The micronutrients are required at only very low levels (for example, approximately 40 ppb for copper). Therefore, the nutrients most likely to be limiting are the "fertilizer" elements: nitrogen, phosphorus, and potassium. These are all present in sewage and are of course found in runoff from heavily fertilized fields. They are also constituents of various kinds of industrial wastes. Each of these elements can also come from natural sources—phosphorus and potassium from mineral formations, and nitrogen fixed by bacteria, blue-green algae, or discharge of lightning in the atmosphere.

Generally, the single plant nutrient most likely to be limiting is phosphorus, and it is generally named as the culprit in excessive eutrophication. Household detergents are a common source of phosphate in wastewater, and eutrophication control has concentrated upon eliminating phosphates from detergents, removing phosphate at the sewage-treatment plant, and preventing phosphate-laden sewage effluents (treated or untreated) from entering bodies of water. (See Chapter 4 for additional details regarding phosphates and detergent phosphate substitutes in water.)

In some cases, nitrogen or even carbon may be limiting nutrients. This is particularly true of nitrogen in seawater.

The whole eutrophication picture is a complex one, and continued research is needed to solve the problem. It is indeed ironic that in a food-poor world, nutrient-rich wastes from over-fertilized fields or sewage are causing excessive plant growth in many lakes and reservoirs. This illustrates again the point that in general there is no such thing as a pollutant—there are only resources (in this case, plant nutrients) gone to waste.

7.12 RADIONUCLIDES IN THE AQUATIC ENVIRONMENT

The massive production of **radionuclides** (radioactive isotopes) by weapons and nuclear reactors since World War II has been accompanied by increasing concern about the effects of radioactivity upon health and the environment. Radionuclides are produced as fission products of heavy nuclei of such elements as uranium or plutonium. They are also produced by the reaction of neutrons with stable nuclei. These phenomena are illustrated in Figure 7.2. Radionuclides are formed in large quantities as waste products in nuclear power generation. Their ultimate disposal is a problem that has caused much controversy regarding the widespread use of nuclear power. Artificially produced radionuclides are also widely used in industrial and medical applications, particularly as "tracers." With so many possible sources of radionuclides, it is impossible to entirely eliminate radioactive contamination of aquatic systems. Furthermore, radionuclides may enter aquatic systems from natural sources. Therefore, the transport, reactions, and biological concentration of radionuclides in aquatic ecosystems are of great importance to the environmental chemist.

Radionuclides differ from other nuclei in that they emit ionizing radiation—alpha particles, beta particles, and gamma rays. The most massive of these emissions is the **alpha particle**, a helium nucleus of atomic weight 4, consisting of two neutrons and two protons. The symbol for an alpha particle is ${}^{4}_{2}\alpha$. An example of alpha production is found in the radioactive decay of uranium-238:

$$ {}^{238}_{92}U \rightarrow {}^{234}_{90}Th + {}^{4}_{2}\alpha + \text{energy} \quad (7.12.1)$$

This transformation consists of a uranium nucleus, atomic number 92 and atomic mass 238, losing an alpha particle, atomic number 2 and atomic mass 4. The product is a thorium nucleus, atomic number 90 and atomic mass 234.

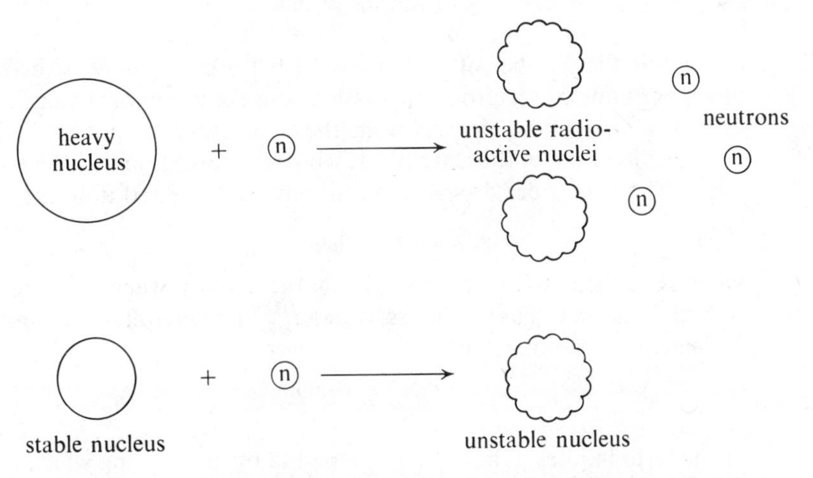

FIGURE 7.2 A heavy nucleus, such as that of U-235, may absorb a neutron and break up, yielding lighter radioactive nuclei. A stable nucleus may absorb a neutron, producing a radioactive nucleus.

Beta radiation consists of either highly energetic, negative electrons, designated ${}_{-1}^{0}\beta$, or positive electrons, called positrons and designated ${}_{1}^{0}\beta$. A typical beta emitter, chlorine-38, may be produced by irradiating chlorine with neutrons. The chlorine-37 nucleus, natural abundance 24.5%, absorbs a neutron to produce chlorine-38 and gamma radiation:

$${}_{17}^{37}\mathrm{Cl} + {}_{0}^{1}\mathrm{n} \rightarrow {}_{17}^{38}\mathrm{Cl} + \gamma \tag{7.12.2}$$

The chlorine-38 nucleus is radioactive and loses a negative beta particle to become an argon-38 nucleus:

$${}_{17}^{38}\mathrm{Cl} \rightarrow {}_{18}^{38}\mathrm{Ar} + {}_{-1}^{0}\beta \tag{7.12.3}$$

Since the negative beta particle has essentially no mass and a -1 charge, the stable product isotope, argon-18, has the same mass and a charge 1 greater than chlorine-38.

Gamma rays are electromagnetic radiation similar to X-rays, though more energetic. Since the energy of gamma radiation is often a well-defined property of the emitting nucleus, it may be used in some cases for the qualitative and quantitative analysis of radionuclides.

The primary effect of alpha particles, beta particles, and gamma rays upon materials is the production of ions; therefore they are called **ionizing radiation**. Due to their large size, alpha particles do not penetrate matter deeply, but cause an enormous amount of ionization along their short path of penetration. Therefore, alpha particles present little hazard outside the body, but are very dangerous when ingested. Although beta particles are more penetrating than alpha particles, they produce much less ionization per unit path length. Gamma rays are much more penetrating than particulate radiation. Their degree of penetration is proportional to their energy.

The **decay** of a specific radionuclide follows first-order kinetics; that is, the number of nuclei disintegrating in a short time interval is directly proportional to the number of radioactive nuclei present. The rate of decay, $-dN/dt$, is given by the equation

$$\text{decay rate} = -\frac{dN}{dt} = \lambda N \tag{7.12.4}$$

where N is the number of radioactive nuclei present and λ is the rate constant, which has units of reciprocal time. Since the exact number of disintegrations per second is difficult to determine in the laboratory, radioactive decay is often described in terms of the **activity**, A, which is proportional to the absolute rate of decay. The first-order decay equation may be expressed in terms of A,

$$A = A_0 e^{-\lambda t} \tag{7.12.5}$$

where A is the activity at time t; A_0 is the activity when t is zero; and e is the natural logarithm base. The **half-life**, $t_{1/2}$, is generally used instead of λ to characterize a radionuclide:

$$t_{1/2} = \frac{0.693}{\lambda} \tag{7.12.6}$$

As the term implies, a half-life is the period of time during which half of a given number of atoms of a specific kind of radionuclide decay. Ten half-lives are required for the loss of 99.9% of the activity of a radionuclide.

Radiation damages living organisms by initiating harmful chemical reactions in tissues. For example, bonds are broken in the macromolecules that carry out life processes. In cases of acute radiation poisoning, bone marrow which produces red blood cells is destroyed and the concentration of red blood cells is diminished. Radiation-induced genetic damage is of great concern. Such damage may not become apparent until many years after exposure. As humans have learned more about the effects of ionizing radiation, the dosage level considered to be safe has steadily diminished. For example, the United States Atomic Energy Commission has dropped the maximum permissible concentration of some radioisotopes to levels of less than one ten-thousandth of those considered safe in the early 1950s. It is possible that even the slightest exposure to ionizing radiation entails some damage. Some radiation is unavoidably received from natural sources. For the majority of the population, exposure to natural radiation exceeds that from artificial sources.

The study of the ecological and health effects of radionuclides involves consideration of many factors. Among these are the type and energy of radiation emitted and the half-live of the source. In addition, the degree to which the particular element is absorbed by living species and the chemical interactions and transport of the element in aquatic ecosystems are important factors. Radionuclides having a very short half-life may be hazardous when produced but decay too rapidly to affect the environment into which they are introduced. Radionuclides with a very long half-life may be quite persistent in the environment but of such low activity that little environmental damage is caused. Therefore, in general, radionuclides with intermediate half-lives are the most dangerous. They persist long enough to enter living systems while still retaining a high activity. Because they may be incorporated within living tissue, radionuclides of "life elements" are particularly dangerous. Much concern has been expressed over strontium-90, a common waste product of nuclear testing. This element is interchangeable with calcium in bone. Strontium-90 fallout drops onto pasture and crop land and is ingested by cattle. Eventually, it enters the bodies of infants and children by way of cow's milk.

Some radionuclides found in water, primarily radium and potassium-40, originate from natural sources, particularly leaching from minerals. Others come from pollutant sources, primarily nuclear-power plants and testing of nuclear weapons. The levels of radionuclides found in water typically are measured in units of picoCuries/liter. A Curie is 3.7×10^{10} disintegrations per second, and a picoCurie is 10^{-12} that amount, or 3.7×10^{-2} disintegrations per second. Therefore, a picoCurie is 2.2 disintegrations per minute.

Public Health Service specifications stipulate that water supplies should not contain more than 3 picoCuries/L of naturally occurring radium-226 and not more than 10 picoCuries/L of strontium-90. Furthermore, the gross beta activity in the absence of Sr-90 and alpha emitters is not to exceed 1000 picoCuries/L. When the gross beta activity is exceeded, the specific radionuclides must be identified and a determination made to see if their ingestion will result in radiation exposure exceeding the maximum allowed limit.

As the use of nuclear power increases, the possible contamination of water by fission-product radioisotopes becomes more of a cause for concern. (If nations continue to refrain from testing nuclear weapons above ground, it is hoped that radioisotopes from this source will contribute only minor amounts of

TABLE 7.5 Radionuclides in Water

Radionuclide	*Half-life*	*Nuclear reaction, description, source*
Naturally occurring and from cosmic reactions		
Carbon-14	5730 years	$^{14}N(n, p)\ ^{14}C$,* thermal neutrons from cosmic or nuclear-weapon sources reacting with N_2
Silicon-32	~300 years	$^{40}Ar(p,x)\ ^{32}Si$, nuclear spallation (splitting of the nucleus) of atmospheric argon by cosmic-ray protons
Potassium-40	~1.4×10^9 years	0.0119% of natural potassium
Naturally occurring from ^{238}U series		
Radium-226	1620 years	Diffusion from sediments, atmosphere
Lead-210	21 years	$^{226}Ra \rightarrow$ 6 steps $\rightarrow ^{210}Pb$
Thorium-230	75,200 years	$^{238}U \rightarrow$ 3 steps $\rightarrow ^{230}Th$ produced *in situ*
Thorium-234	24 days	$^{238}U \rightarrow ^{234}Th$ produced *in situ*
From reactor and weapons fission		
Strontium-90	28 years	These are the fission-product radioisotopes of greatest significance because of their high yields and biological activity.
Iodine-131	8 days	
Cesium-137	30 years	
Barium-140	13 days	The isotopes from barium-140 through krypton-85 are listed in generally decreasing order of fission yield.
Zirconium-95	65 days	
Cerium-141	33 days	
Strontium-89	51 days	
Ruthenium-103	40 days	
Krypton-85	10.3 years	
Cobalt-60	5.25 years	From nonfission neutron reactions in reactors
Manganese-54	310 days	From nonfission neutron reactions in reactors
Iron-55	2.7 years	$^{56}Fe(n, 2n)\ ^{55}Fe$, from high-energy neutrons acting on iron in weapon hardware
Plutonium-239	24,300 years	$^{238}U(n, \gamma)\ ^{239}Pu$, neutron capture by uranium

* This notation denotes the isotope nitrogen-14 reacting with a neutron, n, giving off a proton, p, and forming the isotope carbon-14; other nuclear reactions may be similarly deduced from the notation shown. (Note that x represents nuclear fragments from the spallation reaction.)

radioactivity to water.) Table 7.5 summaries the major natural and artificial radionuclides likely to be encountered in water.

Transuranic elements are of growing concern in the oceanic environment. These alpha emitters are long-lived and highly toxic. As their production increases, so does the risk of environmental contamination. Included among these elements are various isotopes of neptunium, plutonium, americium, and curium. Specific isotopes, with half-lives in years given in parentheses, are: Np-237 (2.14×10^6); Pu-236 (2.85); Pu-238 (87.8); Pu-239 (2.44×10^4); Pu-240 (6.54×10^3); Pu-241 (15); Pu-242 (3.87×10^5); Am-241 (433); Am-243 (7.37×10^3); Cm-242 (0.22); and Cm-244 (17.9).

7.13 ACIDITY, ALKALINITY, AND SALINITY

Aquatic biota are sensitive to extremes of pH. Largely because of osmotic effects, they cannot live in a medium having a salinity to which they are not adapted. Thus, a fresh-water fish soon succumbs in the ocean, and sea fish normally cannot live in fresh water. Excess salinity soon kills plants not adapted to it. There are, of course, ranges in salinity and pH in which organisms live. As shown in Figure 7.3, these ranges frequently may be represented by a reasonably symmetrical curve, along the fringes of which an organism may live without really thriving. These curves do not generally exhibit a sharp cutoff at one end or the other, as does the curve representing the growth of bacteria over a temperature range (see Chapter 5).

The most common source of pollutant acid in water is acid mine drainage. The sulfuric acid in such drainage arises from the microbial oxidation of pyrite or other sulfide minerals as described in Chapter 5. The values of pH encountered in acid-polluted water may fall below 3, a condition deadly to most forms of aquatic life except the culprit bacteria mediating the pyrite and iron(II) oxidation. Industrial wastes frequently contribute strong acid to water. Sulfuric acid produced by the air oxidation of pollutant sulfur dioxide (see Chapter 12) enters natural waters as acidic rainfall. In cases where the water does not have contact with a basic mineral, such as limestone, the water pH may become dangerously low. This condition occurs in some Canadian lakes, for example.

Excess alkalinity, and frequently accompanying high pH, generally are not introduced directly into water by human activity. However, in vast areas of the United States, the soil and mineral strata are alkaline and impart a high alkalinity to water. Human activity can aggravate the situation; for example, the exposing of alkaline overburden from strip mining to surface or ground water. Excess alkalinity in water is manifested by a characteristic fringe of white salts at the edges of a body of water or on the banks of a stream.

Water salinity may be increased by a number of human activities. Water passing through a municipal water system inevitably picks up salt from a number of processes; for example, recharging water softeners with sodium chloride. Salts can leach from spoil piles. One of the major environmental constraints on the production of shale oil, for example, is the high percentage of leachable sodium

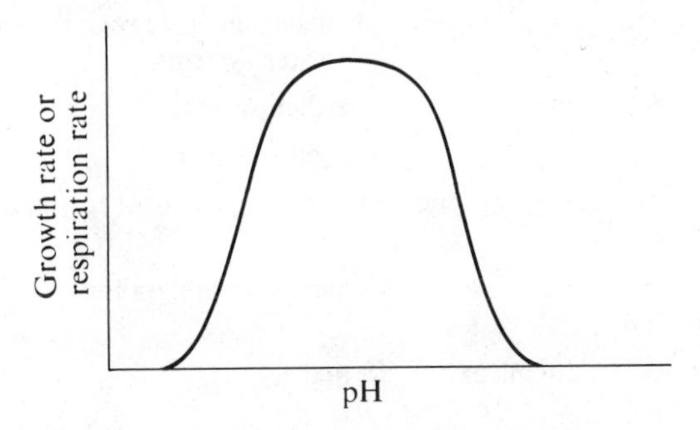

FIGURE 7.3 A generalized plot of pH range of a medium versus growth of an aquatic organism in that medium.

sulfate in piles of spent shale. Careful control of these wastes is necessary to prevent further saline pollution of water in areas where salinity is already a problem. Irrigation adds a great deal of salt to water, a phenomenon responsible for the Salton Sea in California, and is a source of conflict between the United States and Mexico over saline contamination of the Rio Grande and Colorado rivers. Irrigation and intensive agricultural production have caused **saline seeps** in some of the Western states. These occur when water seeps into a slight depression in tilled, sometimes irrigated, fertilized land, carrying salts (particularly sodium, magnesium, and calcium sulfates) along with it. The water evaporates in the dry summer heat, leaving a salt-laden area behind which no longer supports much plant growth. With time, these areas grow, removing productive crop land from production.

7.14 SEWAGE AND WATER POLLUTION

Sewage from domestic, commercial, food-processing, and industrial sources contains an incredible array of pollutants, as shown in Table 7.6. Some of these pollutants, particularly oxygen-demanding substances (see Section 7.15), oil, grease, and solids, are removed by primary and secondary sewage-treatment processes. Others, such as salts, heavy metals, and refactory (degradation-resistant) organics, are not efficiently removed.

Disposal of inadequately treated sewage can cause severe problems. For example, offshore disposal of sewage results in the formation of beds of sewage

TABLE 7.6 Some of the Primary Constitutents of Sewage from a City Sewage System

Constituent	*Sources*	*Effects in water*
oxygen-demanding substances	mostly organic materials, particularly human feces	consume dissolved oxygen
refractory organics	industrial wastes, household products	toxic to aquatic life
viruses	human wastes	cause disease (possibly cancer); major deterrent to sewage recycle through water systems
detergents	household detergents	esthetics; toxic to aquatic life
phosphates	detergents	algal nutrients
grease and oil	cooking, food processing, and industrial wastes	esthetics; harmful to some aquatic life
salts	human wastes, water softeners, industrial wastes	increase water salinity
heavy metals	industrial wastes, chemical laboratories	toxicity
chelating agents	some detergents, industrial wastes	heavy-metal ion transport
solids	all sources	esthetics; harmful to aquatic life

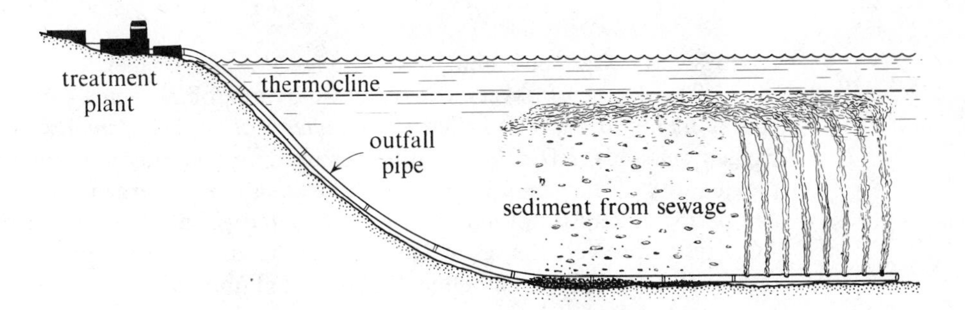

FIGURE 7.4 Settling of solids from an ocean-floor sewage effluent discharge.

residues. Municipal sewage typically contains about 0.1% solids, even after treatment, and these settle out in the ocean in a typical pattern, illustrated in Figure 7.4. The warm sewage water rises in the cold hypolimnion and is carried in one direction or another by tides or currents. It does not rise above the thermocline; instead, it spreads out as a cloud from which the solids rain down on the ocean floor. Aggregation of sewage colloids is aided by dissolved salts in seawater (see Chapter 6), thus promoting the formation of sediment.

Another major disposal problem with sewage is the sludge produced as a product of the sewage treatment process (see Chapter 8). This sludge contains organic material which continues to degrade slowly; refractory organics; and heavy metals. The amounts of sludge produced are truly staggering. For example, the city of Chicago, which until recently had about one-half of the secondary sewage-treatment capacity in the world, produces about 3 million tons of sludge each year. A major deterrent to the safe disposal of such amounts of sludge is the presence of potentially dangerous components such as heavy metals.

Better control of sewage sources is needed to minimize sewage pollution problems. Particularly, if heavy metals and refractory organic compounds could be controlled at the source, sewage, or treated sewage effluents, could be employed for irrigation, recycle to the water system, or ground water recharge.

7.15 OXYGEN-DEMANDING SUBSTANCES IN WATER

Oxygen is a vitally important species in water (see Chapter 2). In water, oxygen is consumed rapidly by the oxidation of organic matter, $\{CH_2O\}$:

$$\{CH_2O\} + O_2 \xrightarrow{\text{microorganisms}} CO_2 + H_2O \qquad (7.15.1)$$

Unless the water is reaerated efficiently, as by turbulent flow in a shallow stream, it rapidly becomes depleted in oxygen and will not support higher forms of aquatic life.

In addition to the microorganism-mediated oxidation of organic matter, oxygen in water may be consumed by the biooxidation of nitrogenous material,

$$NH_4^+ + 2\,O_2 \rightarrow 2\,H^+ + NO_3^- + H_2O \qquad (7.15.2)$$

and by the chemical or biochemical oxidation of chemical reducing agents:

$$4\,Fe^{2+} + O_2 + 10\,H_2O \rightarrow 4\,Fe(OH)_3(s) + 8\,H^+ \qquad (7.15.3)$$

$$2\ SO_3^{2-} + O_2 \rightarrow 2\ SO_4^{2-} \qquad (7.15.4)$$

All these processes contribute to the deoxygenation of water.

The degree of oxygen consumption by microbially-mediated oxidation of contaminants in water is called the *biochemical oxygen demand* (or biological oxygen demand), **BOD**. This parameter is commonly measured by determining the quantity of oxygen utilized by suitable aquatic microorganisms during a five-day period. There is nothing particularly sacred about a five-day period for the BOD test; however, the test originated in England where the maximum stream flow is five days. It was assumed that any contaminant not decomposing in five days would reach the ocean. Despite its somewhat arbitrary nature, a five-day BOD test remains a respectable measure of the short-term oxygen demand exerted by a pollutant.

The addition of oxidizable pollutants to streams produces a typical **oxygen sag curve** as shown in Figure 7.5. Initially, a well-aerated, unpolluted stream is relatively free of oxidizable material; the oxygen level is high; and the bacterial population is relatively low. With the addition of oxidizable pollutant, the oxygen level drops because reaeration cannot keep up with oxygen consumption. In the decomposition zone, the bacterial population rises. The septic zone is characterized by a high bacterial population and very low oxygen levels. The septic zone terminates when the oxidizable pollutant is exhausted, and then the recovery zone begins. In the recovery zone, the bacterial population decreases and the dissolved oxygen level increases until the water regains its original condition.

Although BOD is a reasonably realistic measure of water quality insofar as oxygen is concerned, the test for determining it is time-consuming and cumbersome to perform. **Chemical oxygen demand, COD**, is a much more easily determined parameter. Basically, the test for COD consists of the chemical oxidation of material in water by dichromate ion in 50% H_2SO_4:

$$3\ \{CH_2O\} + 16\,H^+ + 2\ Cr_2O_7^{2-} \rightarrow 4\ Cr^{3+} + 3\ CO_2 + 11\ H_2O \qquad (7.15.5)$$

After oxidation of the oxidizable material in water, the amount of unreacted dichromate is determined by titration with a standard reducing agent. The COD

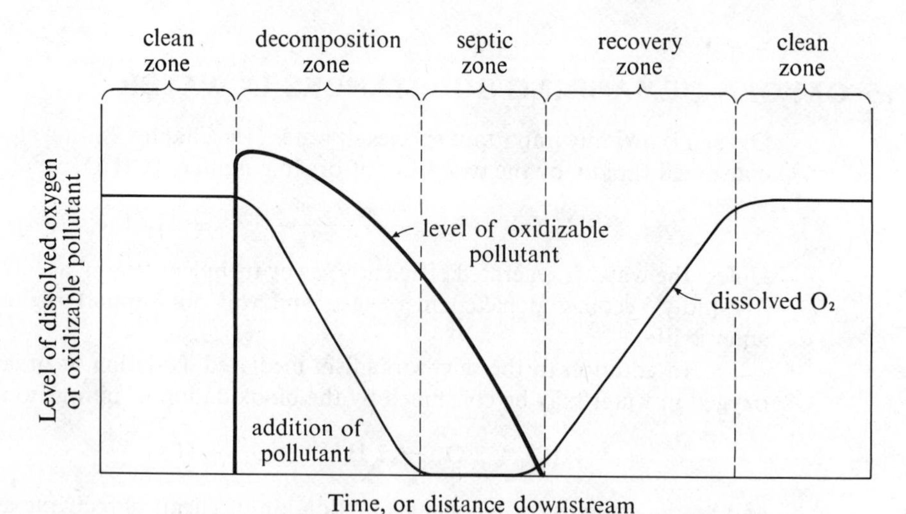

FIGURE 7.5 Oxygen sag curve resulting from the addition of oxidizable pollutant material to a stream.

of water may differ appreciably from the BOD. The presence of poorly biodegradable compounds in water will result in a COD higher than the BOD. The easier and faster COD test often may be substituted for the BOD test, although there is not always a definite correlation between them.

Another parameter, **total organic carbon (TOC)**, is frequently measured by catalytically oxidizing carbon in the water and detecting CO_2. It is gaining in popularity because it is easily measured instrumentally (see Chapter 9).

7.16 SOAPS, DETERGENTS, AND DETERGENT BUILDERS

Soaps are salts of the higher fatty acids. For example, stearic acid, $C_{17}H_{35}COOH$, reacts with sodium hydroxide to form the salt $C_{17}H_{35}COO^-Na^+$. This salt, sodium stearate, is a soap. The other two common components of ordinary soap are sodium palmitate, $C_{17}H_{31}COO^-Na^+$, and sodium oleate, $C_{17}H_{33}COO^-Na^+$.

The cleaning action of soap results largely from its emulsifying power. This concept may be understood by considering the dual nature of the soap anion. An examination of its structure shows that the stearate ion consists of an ionic carboxyl "head" and a long hydrocarbon "tail":

$$\begin{array}{cccccccccccccccccccc}
 & H & H & H & H & H & H & H & H & H & H & H & H & H & H & H & H & H & O & \\
 & | & | & | & | & | & | & | & | & | & | & | & | & | & | & | & | & | & \| & \\
H- & C- & C- & C- & C- & C- & C- & C- & C- & C- & C- & C- & C- & C- & C- & C- & C- & C- & C- & O^- \\
 & | & | & | & | & | & | & | & | & | & | & | & | & | & | & | & | & | & & \\
 & H & H & H & H & H & H & H & H & H & H & H & H & H & H & H & H & H & &
\end{array}$$

In the presence of oils, fats, and other water-insoluble organic materials, the tendency is for the "tail" of the anion to dissolve in the organic matter, whereas the "head" remains in aquatic solution. Thus, the soap **emulsifies**, or suspends, organic material in water. In the process, the anions form colloidal soap micelles, discussed in Section 6.4.

Soap lowers the surface tension of water. At 25 °C, the surface tension of pure water is 71.8 dynes/cm, whereas in the presence of dissolved soap the surface tension is lowered to around 25–30 dynes/cm.

The primary disadvantage of soap as a cleaning agent comes from its reaction with divalent cations to form insoluble salts of fatty acids:

$$2\,C_{17}H_{35}CO_2^-Na^+ + Ca^{2+} \rightarrow Ca(C_{17}H_{35}CO_2)_2(s) + 2\,Na^+ \qquad (7.16.1)$$

These insoluble salts, usually salts of magnesium or calcium, are not at all effective as cleaning agents. In addition, the insoluble "curds" form unsightly deposits on clothing and in washing machines. If sufficient soap is used, all of the divalent cations may be removed by their reaction with soap and the water containing excess soap will have good cleaning qualities. This is the approach commonly used when soap is employed in the bathtub or wash basin, where the insoluble calcium and magnesium salts can be tolerated. However, in applications such as washing clothing, the water must be softened by the removal of calcium and magnesium or their complexation by substances such as polyphosphates (see Chapter 4).

Although the formation of insoluble calcium and magnesium salts has resulted in the essential elimination of soap as a cleaning agent for clothing, dishes, and most other materials, it has distinct advantages from the environmental standpoint. As soon as soap gets into sewage or an aquatic system, it generally precipitates as calcium and magnesium salts. Hence, any effects that soap might have in solution are eliminated. With eventual biodegradation, the soap is completely eliminated from the environment. Therefore, aside from the occasional formation of unsightly scum, soap does not cause any substantial pollution problems.

Synthetic detergents have good cleaning properties and do not form insoluble salts with "hardness ions" such as calcium and magnesium. Such synthetic detergents have the additional advantage of being the salts of relatively strong acids, and, therefore, they do not precipitate out of acidic waters as insoluble acids, an undesirable characteristic of soaps.

Synthetic detergents usually have a surface-active agent, or **surfactant**, added to them that lowers the surface tension of water to which the detergent is added, making the water "wetter." Until the early 1960s, the most common surfactant used was an alkyl benzene sulfonate, ABS, a sulfonation product of an alkyl derivative of benzene:

```
          O                    H    H    H    H    H    H    H    H    H
          ||                   |    |    |    |    |    |    |    |    |
Na+ -O—S—(C6H4)—C—C—C—C—C—C—C—C—C—CH3
          ||                   |    |    |    |    |    |    |    |    |
          O                   CH3   H   CH3   H   CH3   H   CH3   H   CH3
```

ABS, alkyl benzene sulfonate,

ABS suffered the distinct disadvantage of being only very slowly biodegradable because of its branched-chain structure (see Section 5.12). The most objectionable manifestation of the nonbiodegradable detergents, insofar as the average citizen was concerned, was the "head" of foam that began to appear in glasses of drinking water in areas where sewage was recycled through the domestic water supply. Sewage-plant operators were disturbed by spectacular beds of foam which appeared near sewage outflows and in sewage treatment plants. Occasionally, the entire aeration tank of an activated sludge plant would be smothered by a blanket of foam. Among the other undesirable effects of persistent detergents upon waste-treatment processes were: lowered surface tension of water; deflocculation of colloids; flotation of solids; emulsification of grease and oil; and destruction of useful bacteria. Consequently, ABS was replaced by a biodegradable detergent known as LAS.

LAS, α-dodecane benzenesulfonate, has the general structure:

```
   H H H H H H H H H H H H
   | | | | | | | | | | | |
 H—C—C—C—C—C—C—C—C—C—C—C—C—H
   | | | | | | | | | | | |
   H H H H H H H H  (C6H4) H H H
                      |
                   O=S=O   ←—— sulfonate group
                      |
                    O-Na+
```

where the benzene ring may be attached at any point on the alkyl chain except at the ends. LAS is more biodegradable than ABS because the alkyl portion of LAS is not branched and does not contain the tertiary carbon which is so detrimental to biodegradability. Since LAS has replaced ABS in detergents, problems arising from the surface-active agent in the detergents (e.g., toxicity to fish fingerlings) have greatly diminished and the levels of surface-active agents found in water have decreased markedly.

Most of the environmental problems currently attributed to detergents do not arise from the surface-active agents, which basically improve the wetting qualities of water. The **builders** added to detergents continue to cause environmental problems, however. Builders bind to hardness ions, making the detergent solution alkaline and greatly improving the action of the detergent surfactant. A commercial solid detergent contains only 10–30% surfactant. In addition, most detergents still contain polyphosphates (about 6% for detergents manufactured in 1982), added to complex calcium and to function as a builder. Other ingredients include anticorrosive sodium silicates, amide foam stabilizers, soil-suspending carboxymethylcellulose, diluent sodium sulfate, and water absorbed by other components. Of these materials, the polyphosphates have caused the most concern as environmental pollutants. Phosphate detergents have been considered a major source of phosphate in water. Since phosphate is popularly regarded as the essential algal nutrient that can most readily be limited in natural waters, the removal of phosphates from detergents has become a primary goal for environmental improvement, and some local legislation has been enacted limiting detergent phosphates. Finding satisfactory substitutes for polyphosphates has been difficult, however. The precipitating builders, such as sodium carbonate and sodium silicate, precipitate calcium as the carbonate and silicate. These insoluble products may deposit on clothing and make it stiff. In addition, some of the detergent formulations utilizing precipitating builders are so basic that they constitute a toxicological hazard. Other detergent builders that are strong complexing agents pose unresolved questions in regard to possible heavy-metal ion transport.

The problem of finding environmentally acceptable detergent builders is typical of the difficulties encountered in solving environmental problems. It has been extremely hard to find builders with performance qualities equal to those of the polyphosphates. It has not been proven conclusively that the elimination of phosphates from detergents would materially reduce eutrophication. In an efficient, properly operated sewage-treatment plant, phosphate removal is a relatively easy and inexpensive process and will certainly improve in efficiency as it comes into wider use. Phosphate removal at the sewage-treatment plant removes phosphate originating from organic wastes as well as that from detergents. It is possible that phosphate removal by precipitation provides an additional fringe benefit through the coprecipitation of other undesirable impurities in water. The advantages of phosphate use in detergents may well justify the expense of phosphate removal from wastewater.

7.17 TRACE ORGANIC POLLUTANTS AND BIOREFRACTORY COMPOUNDS IN WATER

By 1983, the American Chemical Society's Chemical Abstracts Service had registered a total of well over 4 million chemical compounds, most of which are synthetic organic compounds. Approximately 33,000 of these chemicals are in commercial production. About 60 million metric tons of organic chemicals are manufactured each year in the U.S. Many of these chemicals appear as water

TABLE 7.7 Organic Compounds Found in U.S. Waterways

Fraction	*Total found*	*Most abundant compounds*
Acid-extractable	110	Methyl stearate, methyl palmitate, diethylhexyl phthalate, methyl myristate, C_{15} terpineol
Base-extractable	89	Dibutyl phthalate, diethylhexyl phthalate, C_{15} terpineol, C_{10} terpineol
Purgeable	81	C_1 to C_6 halogenated hydrocarbons, chloroform, trichloroethylene, tetrachloroethylene, 1, 2-dichloromethane, bromodichloromethane, toluene

pollutants. Most of these compounds, particularly the less biodegradable ones, are substances to which living organisms have not been exposed until recent years. Frequently, their effects upon organisms are not known, particularly for long-term exposures at very low levels. The potential of synthetic organics for causing genetic damage, cancer, or other ill effects is uncomfortably high. On the positive side, organic pesticides enable a level of agricultural productivity without which millions would starve. Synthetic organic chemicals are increasingly taking the place of natural products in short supply. Thus it is that organic chemicals are essential to the operation of a modern society. Because of their potential danger, however, acquisition of knowledge about their environmental chemistry must have a high priority.

Examination of water from the Great Lakes and the major industrialized river basins in the U.S. [16] has revealed that approximately 200 organic chemicals are found in these waters within the detection limits of modern analytical methods (which can detect many organics to below 1 ppb). These compounds are summarized in Table 7.7. Although some of the more abundant compounds listed in Table 7.7 occur naturally, it is thought that they are primarily of synthetic origin. A majority of the compounds found are of entirely synthetic origin, many are known to be toxic, and some are suspected carcinogens.

Some organic compounds are environmental contaminants whose effects are not yet completely documented. A major class of these consists of phthalic acid esters used as plasticizers to impart flexibility and increase the workability of plastics such as polyvinyl chloride (PVC). These esters, primarily di-2-ethylhexylphthalate (DEHP), and including others such as di-*n*-butylphthalate (DNBP), are produced at levels of approximately one billion pounds per year. Comprising up to 40% by weight of some plastics, these compounds are encountered in water, blood stored for transfusions, and other fluids that have contacted plastics. Aquatic organisms concentrate the esters from water by factors of up to several thousand.

Harmful effects to living organisms have not been demonstrated for phthalate esters at levels normally encountered in the environment. However, because of the widespread occurrence of these contaminants, continued surveillance and investigation of possible toxic effects has been advised.

---

[16] 1977. EPA finds some 200 chemicals in waterways. *Chemical and Engineering News* 21 Nov. 1977, pp. 6–7.

Biorefractory organics are the organic compounds of most concern in wastewater, particularly in sources of drinking water. These are nonbiodegradable, low-molecular-weight compounds of low volatility. Generally, they are aromatic or chlorinated hydrocarbons, or both. Included in the list of refractory organic industrial wastes are acetone, benzene, bornyl alcohol, bromobenzene, bromochlorobenzene, butylbenzene, camphor chloroethyl ether, chloroform, chloromethyl ethyl ether, chloronitrobenzene, chloropyridine, dibromobenzene, dichlorobenzene, dichloroethyl ether, dinitrotoluene, ethylbenzene, ethylene dichloride, 2-ethylhexanol, isocyanic acid, isopropylbenzene, methylbiphenyl, methyl chloride, nitrobenzene, styrene, tetrachloroethylene, trichloroethane, toluene, and 1,2-dimethoxy benzene. Many of these compounds have been found in drinking water, and some are known to cause taste and odor problems in water. Biorefractory compounds are not completely removed by biological treatment, and other approaches (air stripping, solvent extraction, ozonation, and carbon absorption) are being tried.

A 1976 court settlement between the U.S. Environmental Protection Agency and several plaintiffs consisting of environmentally concerned organizations resulted in the establishment of a list of 115 organic **priority pollutants** that must be analyzed during the development and promulgation phases of environmental regulations. These pollutants are to be analyzed by gas chromatography/mass spectrometry (see Section 9.9) and are divided into four groups: phenolic acids (A); base/neutrals (B/N); pesticides (P); and purgeable organic compounds (VOA). Typical of the acids are phenol, 2-chlorophenol, and 2-nitrophenol. Some base/neutrals are anthracene, benzidine, hexachlorocyclopentadiene, benzo(a)anthracene, nitrobenzene, and *N*-nitrosodimethylamine. Pesticides include aldrin, chlordane, heptachlor, and 2,3,7,8-TCDD (see Section 7.18). The purgeable organic compounds are quite varied and include acrolein, acrylonitrile, benzene, carbon tetrachloride, benzo(a)pyrene, diethyl phthalate, and vinyl chloride. Priority pollutants are those that are likely to be found in industrial wastewater discharges.

One class of organic compounds that pose a unique problem consist of organic liquids that are denser than water and immiscible in it [17]. These are primarily halogenated solvents, such as ethylene dichloride, carbon tetrachloride, and chloroform. In a stream these liquids form localized pools along the undulating stream bottom. Strong turbulence may keep the liquids in suspension until they reach a calm area and settle out. They may travel great distances as globs and pools along channelized river bottoms. One incident involving such a liquid occurred in 1981, when a total of 1,340,000 liters of ethylene dichloride missing from a storage terminal was found as a layer 13 meters below the surface of Lake Ferguson, near Greenville, Mississippi.

7.18 PESTICIDES IN WATER

The introduction of DDT during World War II marked the beginning of a period of very rapid growth in pesticide use. Pesticides are employed for many different

---

[17] Meyer, R.A.; Kirsch, M.; and Marx, L.F. 1981. Detection and mapping of insoluble sinking pollutants. EPA-600/S2-81-198. Cincinnati, Ohio: U.S. Environmental Protection Agency, Municipal Environmental Research Laboratory.

purposes [18]. Chemicals used in the control of invertebrates include insecticides, molluscicides for the control of snails and slugs, and nematicides for the control of microscopic roundworms. Vertebrates are controlled by rodenticides which kill rodents, avicides used to repel birds, and piscicides used in fish control. Herbicides are used to kill plants. Plant growth regulators, defoliants, and plant desiccants are used for various purposes in the cultivation of plants. Fungicides are used against fungi, bactericides against bacteria, and algicides against algae. Annual U.S. pesticide production has risen from about 330 million kilograms of active ingredients in 1962 to about 680 million kilograms of active ingredients in 1982. Although insecticide production has remained about level during the last two decades, herbicide production has increased greatly as chemicals have increasingly replaced cultivation of land in the control of weeds (see no-till agriculture, Section 10.14).

Large quantities of pesticides enter water either directly, in applications such as mosquito control, or indirectly, primarily from drainage of agricultural lands. Major water pollutant pesticides are listed in Table 7.8. These generally belong to the classes of chlorinated hydrocarbons, organic phosphates, and carbamates, which are derived from carbamic acid,

```
   H  O
   |  ||
H—N—C—OH
```

The chlorinated hydrocarbon **DDT** (dichlorodiphenyltrichloroethane or 1,1,1-trichloro-2,2-di-(4-chlorophenyl)-ethane) is probably the best-known pesticide. It was used in massive quantities following World War II. Although its toxicity is generally low, its persistence and accumulation in food chains have led to a ban on DDT use in the United States. It is still employed in other countries. For example, Zimbabwe uses about 1000 metric tons of DDT per year, and high levels of the insecticide have been found in human milk, dairy products, beef, and corn in Zimbabwe [19].

The toxicity of pesticides varies widely. For example, diflubenzuron (see description below) has an LD_{50} (a dose killing half of exposed animals) of 5 grams per kilogram of body weight for mammals [20], compared to 5 mg/kg for methyl parathion. This represents a toxicity ratio of 10^3. Moreover, a few pesticides have little or no toxicity. In May, 1982, an insecticide consisting of soap from potassium salts of fatty acids was cleared for use on food crops up to the day of harvest. This biodegradable, exceptionally safe material is effective against white flies, scale crawlers, mealy bugs, aphids, plant bugs, spider mites, and leafhoppers.

Much of the environmental harm caused by pesticides results from the use of broad-spectrum pesticides, which have composed the bulk of production since World War II. Diflubenzuron, a benzoylphenyl urea with the formula shown at the top of the next page.

---

[18] Ware, G.W. 1978. *The pesticide book*. San Francisco: W.H. Freeman and Company.

[19] Cowell, A. 1981. Wildlife menaced in African valley. *New York Times* 22 October 1981.

[20] Marx, J.L. 1977. Chitin synthesis inhibitors: new class of insecticides. *Science* 197: 1170–2.

$$\text{(2,6-F}_2\text{C}_6\text{H}_3)-\overset{\overset{\displaystyle O}{\|}}{C}-\overset{\overset{\displaystyle H}{|}}{N}-\overset{\overset{\displaystyle O}{\|}}{C}-\overset{\overset{\displaystyle H}{|}}{N}-C_6H_4-Cl$$

is an example of a much more specific pesticide. It inhibits the biosynthesis of the polysaccharide outer covering of insects, which is made up of chitin, possibly by inhibition of the enzyme chitin synthetase. Other animals do not have similar structures made of chitin and are not affected. Still in the experimental stage, diflubenzuron has been shown effective against flies, beetles, gypsy moths, gnats, mites, and the boll weevil at application rates of 0.5–2 ounces per acre. There are many facets to the problem of pesticides in the environment. Some of these are illustrated in Figure 7.6.

A number of water pollution and health problems have been associated with the manufacture of pesticides. For example, degradation-resistant hexa-

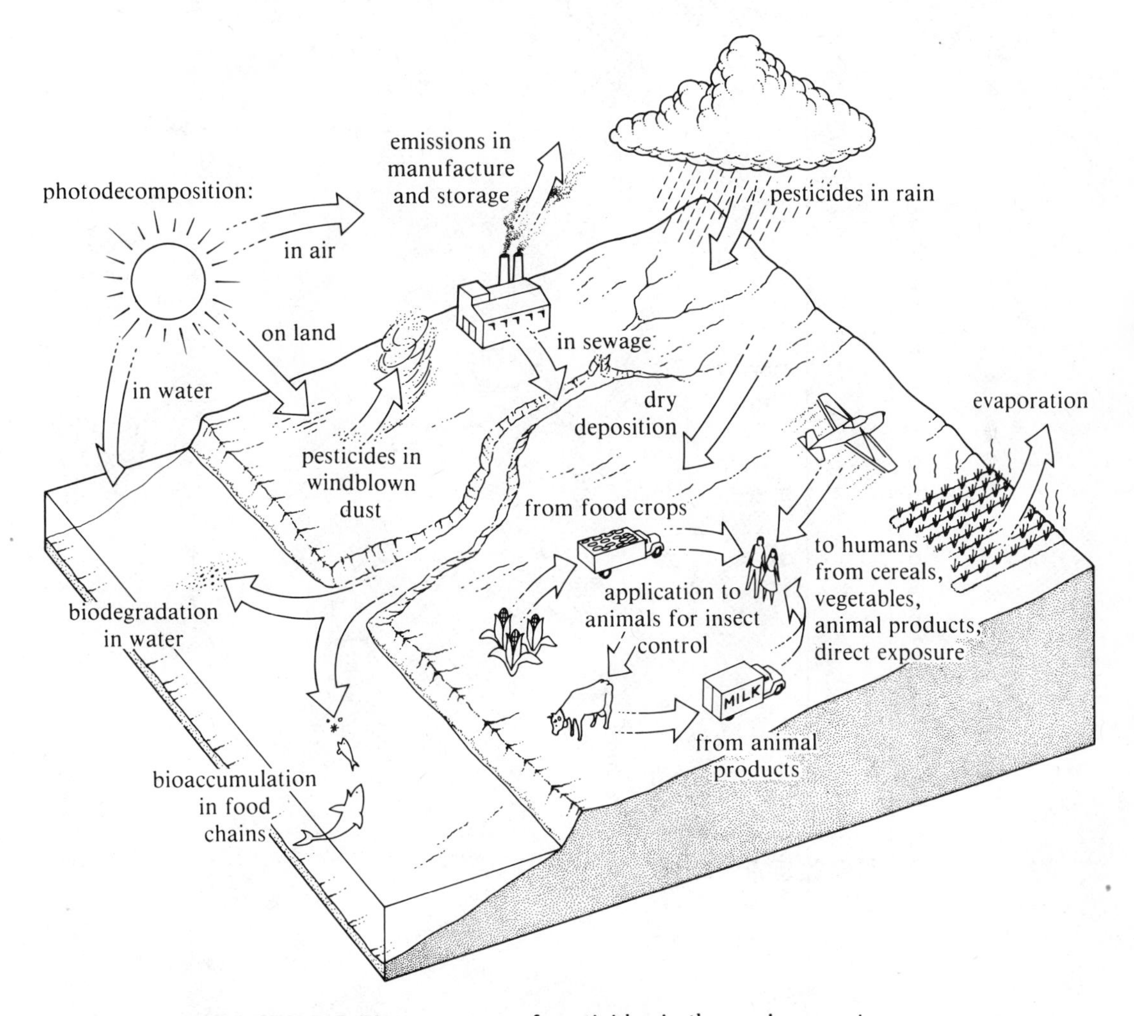

FIGURE 7.6 Major aspects of pesticides in the environment.

TABLE 7.8 Pesticides Commonly Encountered in Water[1]

Pesticide	*Formula*	*Fresh-water quality criteria*[2]	*Uses and characteristics*
Aldicarb Temik®	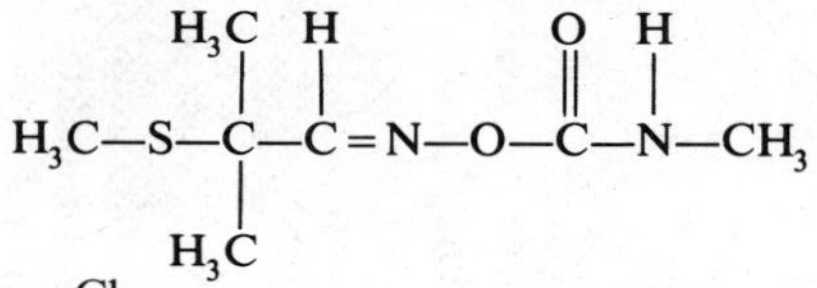 $H_3C{-}S{-}C(CH_3)_2{-}CH{=}N{-}O{-}C(=O){-}NH{-}CH_3$	—	Plant systemic insecticide taken up by roots or leaves; accused of contaminating well water on Long Island and banned in that area in 1982.
Aldrin–Dieldrin*	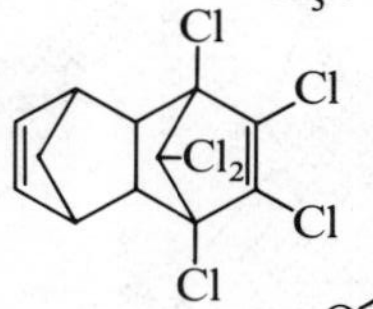 aldrin; 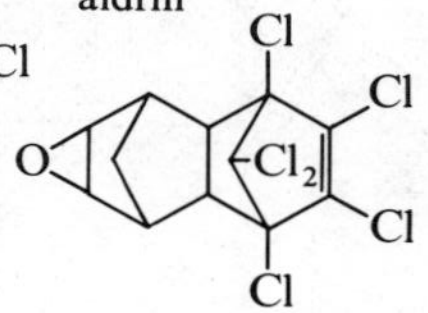 dieldrin (endo–exo)	0.003 $\mu g/L$	Persistent and stable in soil, effective against insects in soil. Organisms convert aldrin to dieldrin, known to be carcinogenic to mice. Banned in the U.S. for most uses in 1975.
Chlordane*	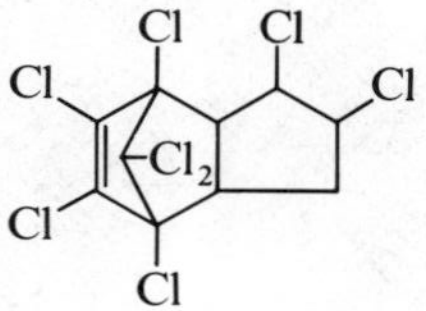	0.01 $\mu g/L$	First of the cyclodiene[3] insecticides to be used, especially effective against termites. Potential carcinogen, use restricted in U.S. in 1978.
Chlorophenoxy herbicides*	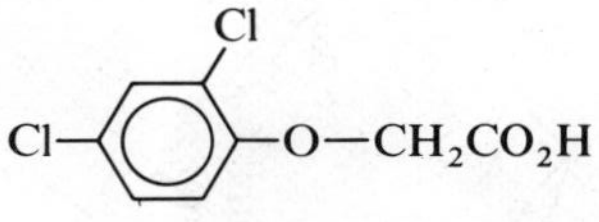 2,4-D: 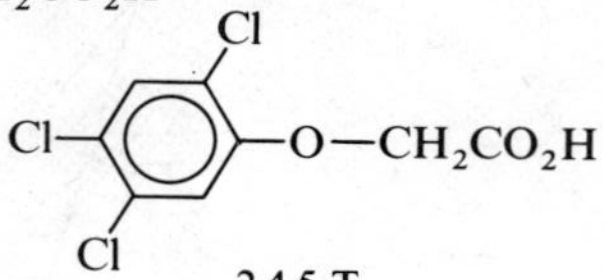 $Cl{-}C_6H_3(Cl){-}O{-}CH_2CO_2H$; 2,4,5-T: $Cl{-}C_6H_2(Cl)_2{-}O{-}CH_2CO_2H$	100 $\mu g/L$ (domestic water supplies)	Highly selective for the control of broad-leaved weeds; relatively nontoxic to animals. May contain highly toxic tetrachlorodioxin (TCDD) as a manufacturing impurity.

DDT*	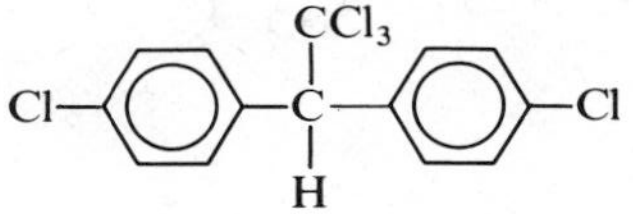	0.001 μg/L	Low acute toxicity to mammals; persistent; accumulates in food chain. Some evidence of carcinogenicity, use banned in the U.S. in 1972.
Demeton*	mixture of $(C_2H_5O)_2P(=O)-S-CH_2CH_2-S-C_2H_5$ and $(C_2H_5O)_2P(=S)-O-CH_2CH_2-S-C_2H_5$	0.1 μg/L	Lower acute toxicity than most organophosphate pesticides; used to control sucking insects. Action lasts 4–6 weeks under field conditions; systemic insecticide (taken up by roots, translocated to stems and leaves, toxic to insects sucking plant juices).
Endosulfan*	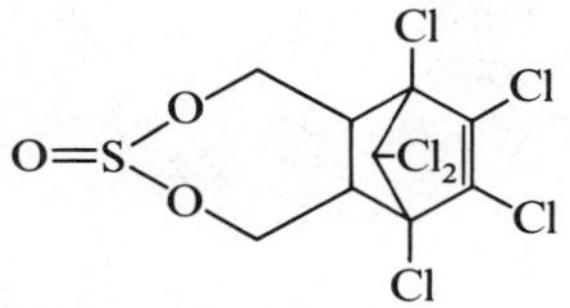	0.003 μg/L	Used on fruit and berries to control aphids, beetles, and caterpillars; lower chronic toxicity to mammals than most cyclodiene pesticides.
Endrin*	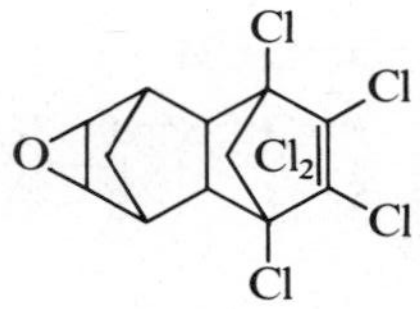 (endo–endo)	0.004 μg/L	Only pesticide effective against black currant mud mite; also used as a zoocide. Special precautions must be used to avoid skin contact during application; readily photolyzes to nontoxic ketone form.
Guthion (azinphosmethyl)*	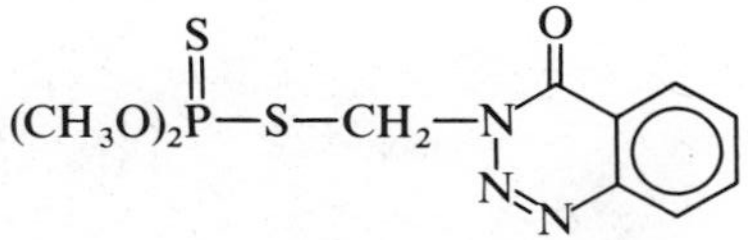	0.01 μg/L	Used for control of pests on fruit, cotton, other crops; effective acaricide (controls mites); relatively toxic to mammals.
Heptachlor*	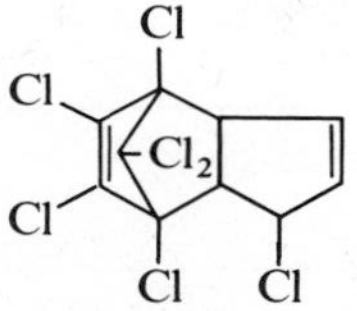	0.001 μg/L	Used to control pests in soil; insecticide in feed. Changes to the more toxic epoxide which persists for a long time in soil; use restricted in U.S. in 1978.

Table 7.8 *(continued)*

Pesticide	*Formula*	*Fresh-water quality criteria*[2]	*Uses and characteristics*
Lindane*	$C_6H_6Cl_6$ (γ-isomer)	0.01 μg/L	Used to control insects, plant pests, animal parasites; widely manufactured because of convenience, lack of odor, minimal residue.
Malathion*	$(CH_3O)_2P(=S)-S-CH(COOC_2H_5)-CH_2-COOC_2H_5$	0.1 μg/L	Safer than most organophosphate pesticides, comparatively little hazard to mammals.
Methoxychlor*	$CH_3O-C_6H_4-CH(CCl_3)-C_6H_4-OCH_3$	0.03 μg/L	Popular DDT substitute, reasonably biodegradable; low toxicity to mammals.
Mirex*	$C_{10}Cl_{12}$ (cage structure, Cl_{12})	0.001 μg/L	Used almost exclusively to control the imported fire ant in the southeastern U.S.
Methyl parathion*	$O_2N-C_6H_4-O-P(=S)-(OCH_3)_2$	—	Used to control many plant pests; ranks second in U.S. pesticide consumption.
Toxaphene*	Cl_x-camphene skeleton with $=CH_2$ and $(CH_3)_2$ (x = average of 8)	5 μg/L	Pesticide most widely used in the U.S.; EPA has proposed banning it because of tests showing it is carcinogenic in mice and rats.

Carbaryl (Sevin®)	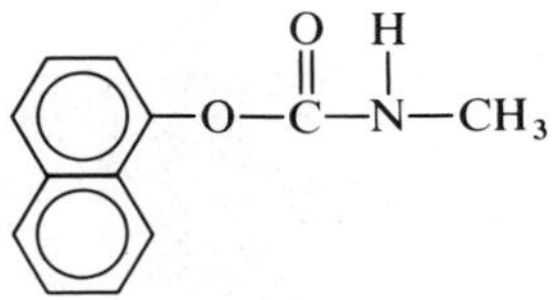 $C_{10}H_7$–O–C(=O)–N(H)–CH_3	—	Carbamate pesticide widely used as a lawn and garden insecticide; low toxicity to mammals.
Diazinon	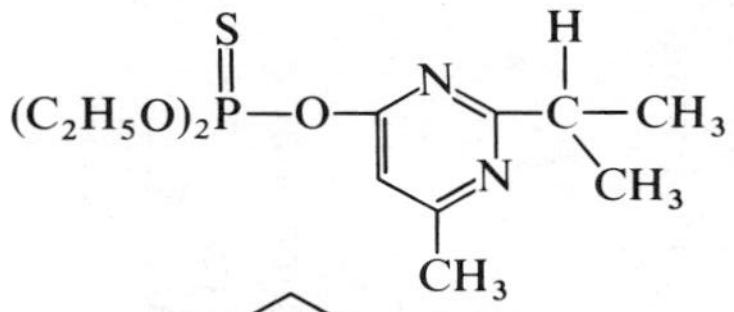 $(C_2H_5O)_2P(=S)$–O–(pyrimidine ring: N, N; CH_3)–C(H)(CH_3)–CH_3	—	Used to control plant pests and animal parasites; relatively high toxicity to mammals.
Carbofuran	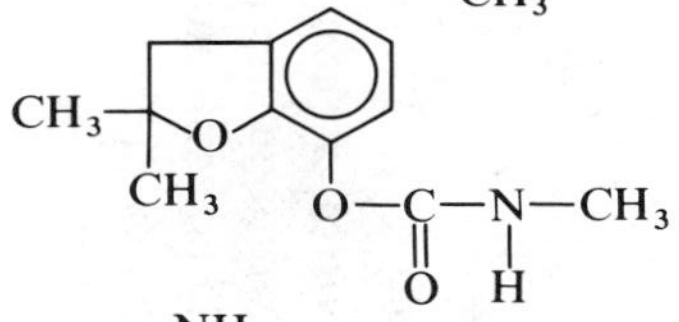 CH_3, CH_3 (dihydrobenzofuran ring, O)–O–C(=O)–N(H)–CH_3	—	Good plant systemic pesticide because of high water-solubility; readily taken up by roots and leaves; subject of litigation by exposed workers in Texas, resulting in a 1982 court order directing EPA to release health and safety information
Picloram	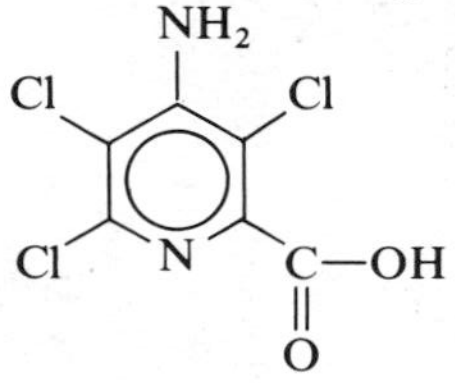 NH_2, Cl, Cl, Cl (pyridine ring, N)–C(=O)–OH	—	Popular herbicide used against broad-leaved and woody plants; taken up by either roots or foliage; accused of causing cancer deaths in timbered areas of North Carolina.
Diquat	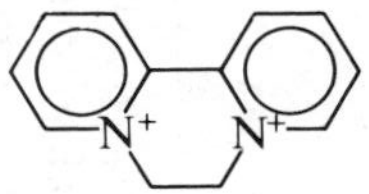 N^+, N^+	—	These are the only important pesticides of the bipyridylium (2 pyridine rings) type. They are contact herbicides applied directly to plant tissue, causing rapid cell membrane destruction and a frost-bitten appearance. Paraquat was used to spray marijuana in the southeastern U.S. in August, 1983, causing considerable controversy.
Paraquat	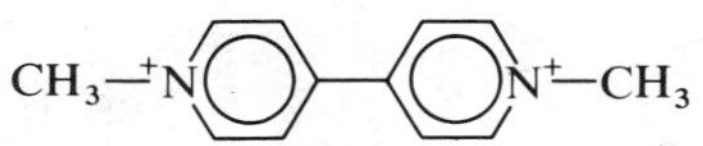 CH_3–^+N(pyridine)–(pyridine)N^+–CH_3	—	

[1]Pesticides designated by an asterisk are those discussed as particularly troublesome water pollutants in *Quality Criteria for Water,* U.S. Environmental Protection Agency, Washington, D.C., 1976. The others are noteworthy for large-scale use in the U.S.

[2]Except where noted, these are criteria for the protection of fresh-water marine life.

[3]Cyclodiene, or "diene," pesticides are those whose synthesis is based upon hexachlorocyclopentadiene.

chlorobenzene (see structure below) is used as a raw material for the synthesis of other pesticides and has often been found in water. Release of by-product 2,3,7,8-tetrachloro-*p*-dioxin (commonly called TCDD or, simply, dioxin; see structure below) in a 1976 pesticide manufacturing plant explosion in Seveso, Italy, exposed many humans to this substance, which is one of the most toxic compounds known [21]. Approximately 3,500 barrels of herbicide wastes, some of them leaking, were found behind a Jacksonville, Arkansas factory that had manufactured 2,4,5-T herbicide in 1979. Subsequent investigations revealed that the wastes contained 40 ppm of TCDD, a very high level. Dioxin was subsequently found in the local sewage treatment plant, several nearby gardens, and in fish exposed to water in a creek flowing through the plant site. Subsequent medical examination of workers in the factory showed a high incidence of chloracne (skin eruptions) and other health problems possibly arising from exposure to toxic substances.

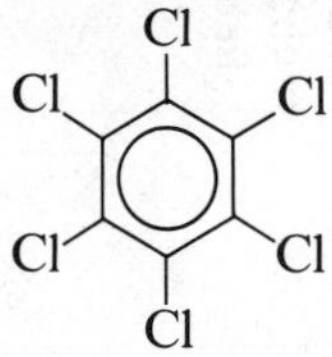

hexachlorobenzene

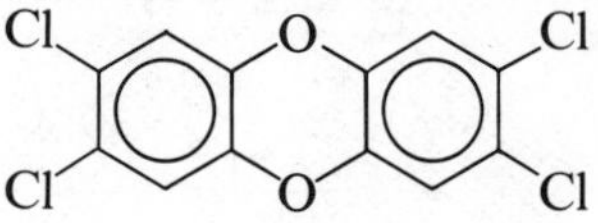

TCDD

The most notable case of TCDD contamination resulted from the spraying of waste oil mixed with TCDD on roads and horse arenas in Missouri in the early 1970s. The oil was used to try to keep dust down in these areas. The extent of contamination was revealed by studies conducted in late 1982 and early 1983. As a result, the EPA offered to buy out the entire TCDD-contaminated town of Times Beach, Missouri, in March, 1983. This case is discussed further in Chapter 20.

One of the greatest environmental disasters ever to result from pesticide manufacture involved the production of Kepone, structural formula

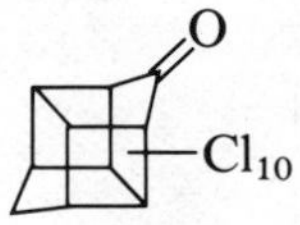

kepone

This pesticide has been used for the control of banana-root borer, tobacco wireworm, ants, and cockroaches [22]. Although the use of Kepone has been banned in the U.S. since 1977, a closely related compound, Kelevan, is employed for the control of the Colorado potato borer in Eastern Europe and Ireland. Kelevan is also used against the banana-root borer in South America and Cameroons, Carribean.

Kepone exhibits acute, delayed, and cumulative toxicity in birds, rodents, and humans, and it causes cancer in rodents. It was manufactured in Hopewell, Virginia, during the mid-1970s. During this time, workers were exposed to Kepone and are alleged to have suffered health problems as a result. The plant

---

[21] Rawls, R.L., and Sullivan, D.A. 1977. Italy seeks answers following toxic substance release. *Chemical and Engineering News* 23 August 1977.

[22] Epstein, S.S. 1978. Kepone hazard evaluation. *The Science of the Total Environment* 9: 1–62.

was connected to the Hopewell sewage system. It has been reported [22] that frequent infiltration of Kepone wastes caused the Hopewell sewage treatment plant to become inoperative at times. Reportedly, as much as 53,000 kg of Kepone may have been dumped into the sewage system during the years that the plant was operated. The sewage effluent was discharged to the James River, resulting in extensive environmental dispersion and toxicity to aquatic organisms. Decontamination of the river would involve dredging 135 million cubic meters of river sediment, which would then have to be detoxified. A prohibitively high price tag of several billion dollars has been estimated for the cleanup cost.

One of the more highly publicized incidents involving insecticides during recent years was that of the Mediterranean fruit fly (Med-fly) infestation of parts of California (Figure 7.7). The trouble started with the capture of several Mediterranean fruit flies in insect traps in Santa Clara County, California, in June, 1980. This was a matter of great concern because of the damage done to fruits by this pest. The female fly injects 300–400 eggs under the skin of the fruit, resulting in the growth of approximately half-inch-long pupa from which the adult flies emerge. The fruit is, of course, ruined. For many months the Governor of California resisted aerial spraying, in part because of citizen opposition. Finally, faced with quarantines of the state's $14.5 billion fruit industry, officials relented and allowed aerial spraying with malathion (see Table 7.8) during the summer of 1981, at a cost of $60 million to the state. Some private concerns used the more powerful diazinon in their orchards. As of 1982, the "Med-fly" problem appeared to be largely solved. However, on June 25, 1982 a fertile "Med-fly" was found in a trap near Stockton, on the edge of the San Joaquin valley. As a result, shipment of fruits and vegetables was banned in the surrounding 80-square-mile area, and a 9.5-square-mile area around Stockton was slated for spraying.

On an encouraging note, some threatened bird species have shown increased populations in the 10 years since DDT was banned. According to the U.S. Department of the Interior [23], threatened species that are now returning to old habitats include the bald eagle, the brown pelican, the peregrine falcon, and the

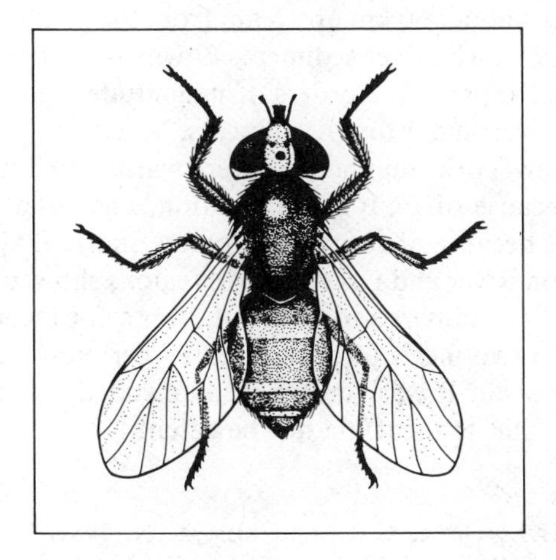

FIGURE 7.7 Mediterranean fruit fly.

[23] 1982. Bird populations show a rise in decade since ban on DDT. *New York Times* 11 March 1982.

osprey. From a low of only about 1,000 breeding pairs along the Atlantic Coast in the 1960s, the once highly threatened brown pelican showed an increase to about 5,000 breeding pairs by 1982. Unfortunately, use of DDT in Central and South America, where the pesticide is not banned, continues to threaten some migratory birds, such as the peregrine falcon.

7.19 PCB's AND PBB's

First discovered as an environmental pollutant in 1966, **polychlorinated biphenyls (PCB's)** have been found throughout the world in water, sediments, bird tissue, and fish tissue. The synthesis of PCB's involves the chlorination of biphenyl,

$$\text{C}_6\text{H}_5\text{–C}_6\text{H}_5 + x\,\text{Cl}_2 \xrightarrow[\text{FeCl}_2]{\text{Fe}} \text{C}_{12}\text{H}_{10-x}(\text{Cl})_x + x\,\text{HCl} \qquad (7.19.1)$$

(ring positions labeled 2, 3, 4, 5, 6 and 2′, 3′, 4′, 5′, 6′)

where x may range from 1 to 10. Up to 210 compounds are possible from this synthesis, and a mixture of compounds is always obtained. Until their manufacture was discontinued in 1977, PCB's were used as coolant-insulation fluids in transformers and capacitors; as plasticizers; for the impregnation of cotton and asbestos; and in some epoxy paints. Their low vapor pressure, high dielectric constants, chemical inertness, and extreme thermal stability are all properties that made PCB's useful in many applications. Unfortunately, these same properties contributed to extreme stability in the environment. Several chemical formulations have been developed to substitute for PCB's in electrical applications. Disposal of PCB's from discarded electrical equipment and other sources remains a problem, particularly since PCB's can survive ordinary incineration by escaping as vapors through the smokestack. However, they can be destroyed by special incineration processes.

PCB's are especially prominent pollutants in the sediments of the Hudson River [24]. This is the result of waste discharges from two capacitor manufacturing plants, which operated about 60 km upstream from the southernmost dam on the river from 1950 to 1976. The river sediments downstream from the plants exhibit PCB levels of about 10 ppm, 1–2 orders of magnitude higher than levels commonly encountered in river and estuary sediments.

During Fall 1981, New York state hunters were warned to limit their consumption of wild ducks because of PCB contamination, and Montana hunters were given similar warnings because of contamination by Endrin [25]. Dissection of 63 ducks from the Hudson River and Lake Ontario regions showed contamination levels of 7.5 ppm PCB's, compared to limits of 3 ppm for chickens. It was suggested that no more than two meals of duck be eaten per month and that the skin and fat should be removed. It was further recommended that if the ducks were cooked with stuffing, that the stuffing not be eaten!

---

[24] Bopp, R.F., Simpson, H.J., Olsen, C.R., and Kostyk, N. 1981. Polychlorinated biphenyls in sediments of the tidal Hudson River. *Environmental Science and Technology,* 15: 210–216.

[25] Faber, H. 1981. New York hunters warned on eating wild ducks. *New York Times* 8 October 1981.

Polybrominated biphenyl (PBB) is a chemical fire retardant which was accidently mistaken for magnesium oxide and mixed with cattle feed distributed in Michigan in 1973 [26]. As a result, over 30,000 cattle, approximately 6000 hogs, 1500 sheep, 1.5 milion chickens, 18,000 pounds of chese, 2700 pounds of butter, 34,000 pounds of dry milk products, and 5 million eggs had to be destroyed. Farm families eating PBB-contaminated foods have ingested the substance, and it has been detected in the blood of many Michigan residents. A 1976 study of Michigan dairy farmers and chemical-plant employees showed that persons who had ingested PBB tended to have less disease resistance and increased incidence of rashes, liver ailments, and headaches [27]. The economic cost of the Michigan PBB incident exceeded $100 million.

SUPPLEMENTARY REFERENCES

Afghan, B.K., and Mackay, D., eds. 1980. *Hydrocarbons and halogenated hydrocarbons in the environment.* New York: Plenum Publishing Corp.

1978. *An assessment of mercury in the environment.* Washington, D.C.: National Academy of Sciences.

1977. *Arsenic.* Washington, D.C.: National Academy of Sciences.

Berry, J.W.; Osgood, D.W.; and St. John, P.A. 1974. *Chemical villains: a biology of pollution.* St. Louis, Mo.: C.V. Mosby Co.

Borchardt, J.A. 1977. *Viruses and trace contaminants in water and wastewater.* Ann Arbor, Mich.: Ann Arbor Science Publishers, Inc.

1978. *Cleaning our environment—a chemical perspective.* Washington, D.C.: American Chemical Society.

Cremlyn, R. 1978. *Pesticides.* New York: John Wiley and Sons, Inc.

Cusine, D.J., and Grant, J.P., eds. 1980. *The impact of marine pollution.* London: Croom Helm.

Davis, E.S., and Wilk, V.A. 1982. *Toxic chemicals.* Washington, D.C.: Farmworker Justice Fund.

Dinges, R. 1981. *Natural systems for water pollution control.* New York: Van Nostrand Reinhold Co.

DiStacio, J.I., ed. 1981. *Surfactants, detergents, and sequestrants.* Park Ridge, N.J.: Noyes Data Corp.

Edwards, C.A. 1973. *Environmental pollution by pesticides.* New York: Plenum Publishing Corp.

Egginton, J. 1980. *The poisoning of Michigan.* New York: W.W. Norton & Co. Inc.

Eisenreich, S.J., ed. 1981. *Atmospheric pollutants in natural waters.* Ann Arbor, Mich.: Ann Arbor Science Publishers, Inc.

Eisler, R. 1981. *Trace metal concentrations in marine organisms.* Elmsford, N.Y.: Pergamon Press, Inc.

Fair, G.M.; Geyer, J.C.; and Okun, D.A. 1968. *Water and wastewater engineering.* Vols. I and II. New York: John Wiley and Sons, Inc.

---

[26] Carter, L.J. 1976. Michigan's PBB incident: chemical mix-up leads to disaster. *Science* 192: 240–3.

[27] 1978. Aftermath of two environmental shocks. *U.S. News and World Report* 13 February 1978, pp. 43–4.

Faust, S.D., and Hunter, J.V. 1971. *Organic compounds in aquatic environments.* New York: Marcel Dekker, Inc.

Förstner, U., and Wittman, G.T.W. 1979. *Metal pollution in the aquatic environment.* New York: Springer-Verlag New York, Inc.

Gehm, H.W., and Bregman, J.I., eds. 1976. *Handbook of water resources and pollution control.* New York: Van Nostrand Reinhold Co.

Gould, R.F., ed. 1972. *Fate of organic pesticides in the aquatic environment.* Advances in Chemistry Series III. Washington, D.C.: American Chemical Society.

Gunther, F.A., and Gunther, J.D., eds. 1980. *Residue reviews.* Series of volumes. New York: Springer-Verlag New York, Inc.

Harrison, R.M., and Laxen, D.P.H. 1981. *Lead pollution.* London: Chapman and Hall.

Hayes, W.J. 1975. *Toxicology of pesticides.* Baltimore, Md.: Williams and Wilkins Co.

Hutzinger, O. 1982. *Chlorinated dioxins and related compounds.* Elmsford, N.Y.: Pergamon Press, Inc.

——, ed. 1982. Anthropogenic compounds. In *The handbook of environmental chemistry,* Vol. 3, Part B. New York: Springer-Verlag New York, Inc.

Hutzinger, O.; Frei, R.W.; Merian, E.; and Reggiani, G. 1983. Proceedings of the Third International Symposium on Chlorinated Dioxins and Related Compounds (*Chemosphere* 12: 4/5). Elmsford, N.Y.: Pergamon Press, Inc.

Kaufman, D.D.; Still, G.G.; Paulson, G.D.; and Bandal, S.K., eds. 1976. *Bound and conjugated pesticide residues.* ACS Symposium Series No. 29. Washington, D.C.: American Chemical Society.

Kavanaugh, M.C., and Leckie, J.O., eds. 1980. *Particulates in water.* Washington, D.C.: American Chemical Society.

Kearney, P.C., and Kaufman, D.D., eds. 1975. *Herbicides: chemistry, degradation, and mode of action.* Vol. I. New York: Marcel Dekker, Inc.

Kennedy, M.V., ed. 1978. *Disposal and decontamination of pesticides.* ACS Symposium Series No. 73. Washington, D.C.: American Chemical Society.

Khan, M.A.Q. 1977. *Pesticides in aquatic environments.* New York: Plenum Publishing Corp.

Laws, E.A. 1981. *Aquatic pollution.* New York: Wiley-Interscience.

McEwan, F.L., and Stephenson, G.R. 1979. *The use and significance of pesticides in the environment.* New York: Wiley-Interscience.

National Research Council, Safe Drinking Water Committee. 1982. *Drinking water and health.* Vol. 4. Washington, D.C.: National Academy Press.

1978. *Nitrates: an environmental assessment.* Washington, D.C.: National Academy of Sciences.

Novotny, V., and Chesters, G. 1981. *Handbook of nonpoint source pollution.* New York: Van Nostrand Reinhold Co.

NRCC/CNRC. 1982. *Chlorinated phenols: criteria for environmental quality.* Ottawa, Ontario, Canada: NRCC/CNRC.

Nriagu, J.O. 1980. *Zinc in the environment.* New York: Wiley-Interscience.

——, ed. 1979. *Copper in the environment.* New York: Wiley-Interscience.

Plimmer, J.R., ed. 1977. *Pesticide chemistry in the 20th century.* ACS Symposium Series No. 37. Washington, D.C.: American Chemical Society.

——, ed. 1982. *Pesticide residues and exposure.* Washington, D.C.: American Chemical Society.

1979. *Polychlorinated biphenyls.* Washington, D.C.: National Academy of Sciences.

1982. Proceedings of the conference on heavy metals in the environment. Edinburgh, United Kingdom: CEP Consultants, Ltd.

Scher, H.B., ed. 1977. *Controlled release pesticides.* ACS Symposium Series No. 53. Washington, D.C.: American Chemical Society.

Sheets, T.J., and Pimentel, D., eds. 1979. *Pesticides: Contemporary roles in agriculture, health, and environment.* Clifton, N.J.: Humana Press, Inc.

Sittig, M. 1979. *Detergent manufacture—including zeolite builders and other new materials.* Park Ridge, N.J.: Noyes Data Corp.

Stiff, M.J., ed. 1980. *River pollution control.* New York: Halsted (John Wiley and Sons, Inc.).

Thibodeaux, L.J. 1979. *Environmental movement of chemicals in air, water, and soil.* New York: John Wiley and Sons, Inc.

QUESTIONS AND PROBLEMS

1. Which of the following statements is true regarding chromium in water?
   (a) Chromium(III) is suspected of being carcinogenic.
   (b) Chromium(III) is less likely to be found in a soluble form than is chromium(VI).
   (c) The toxicity of chromium(III) in electroplating wastewaters is decreased by oxidation to chromium(VI).
   (d) Chromium is not an essential trace element.
   (e) Chromium is known to form methylated species analogous to methylmercury compounds.

2. What do mercury and arsenic have in common in regard to their interactions with bacteria in sediments?

3. What are some characteristics of radionuclides that make them especially hazardous to humans?

4. To what class do pesticides containing the following group belong?

$$-\overset{\displaystyle H}{\overset{|}{N}}-\overset{\displaystyle O}{\overset{\|}{C}}-$$

5. Which of the following characteristics are *not* possessed by the compound

$$Na^+\ {}^-O-\overset{O}{\overset{\|}{\underset{\underset{O}{\|}}{S}}}-C_6H_4-(CH_2)_9-CH_3$$

   (a) one end of the molecule is hydrophilic and the other end is hydrophobic
   (b) surface-active qualities
   (c) the ability to lower surface tension of water
   (d) good biodegradability
   (e) tendency to cause foaming in sewage-treatment plants.

6. A certain pesticide is fatal to fish fingerlings at a level of 0.50 parts per million in water. A leaking metal can containing 5.00 kg of the pesticide was dumped into a stream with a flow of 10.0 liters per second moving at 1 kilometer per hour. The container leaks

pesticide at a constant rate of 5 mg/sec. For what distance (in km) downstream is the water contaminated by fatal levels of the pesticide by the time the container is empty?

7. What are two reasons that Na_3PO_4 is not used as a detergent builder instead of $Na_5P_3O_{10}$?

8. Of the compounds $CH_3(CH_2)_{10}CO_2$, $(CH_3)_3C(CH_2)_2CO_2H$, $CH_3(CH_2)_{10}CH_3$, and ϕ-$(CH_2)_{10}CH_3$ (where ϕ is a benzene ring), which is the most readily biodegradable?

9. A pesticide sprayer got stuck while trying to ford a stream flowing at a rate of 136 liters per second. Pesticide leaked into the stream for exactly 1 hour and at a rate that contaminated the stream at a uniform 0.25 ppm of methoxychlor. How much pesticide was lost from the sprayer during this time?

10. A sample of water contaminated by the accidental discharge of a radionuclide used for medicinal purposes showed an activity of 12,436 counts per second at the time of sampling and 8,966 cps exactly 30 days later. What is the half-life of the radionuclide?

11. What are two reasons that soap is environmentally less harmful than ABS surfactant used in detergents?

12. What is the exact chemical formula of the specific compound designated as PCB?

13. A radioisotope has a nuclear half-life of 24 hours and a *biological* half-life of 16 hours (half of the element is eliminated from the body in 16 hours). A person accidentally swallowed sufficient quantities of this isotope to give an initial "whole body" count rate of 1000 counts per minute. What was the count rate after 16 hours?

14. What is the primary detrimental effect upon organisms of salinity in water arising from dissolved NaCl and Na_2SO_4?

15. Give a specific example of each of the following general classes of water pollutants:
    (a) trace elements; (b) metal-organic combinations
    (c) pesticides

16. Match each compound in the left column with the description corresponding to it in the right columm.

    (a) CdS
    (b) $(CH_3)_2$ AsH
    (c)

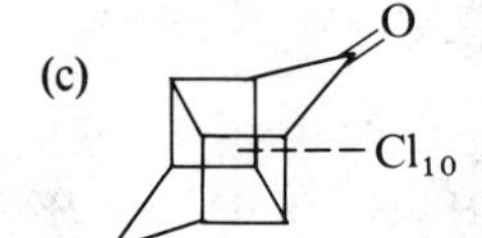

    (d)

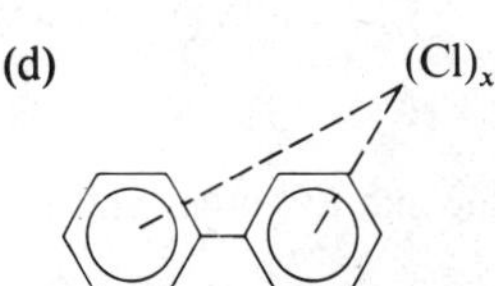

    (1) Pollutant released to a U.S. stream by a poorly controlled manufacturing process.
    (2) Insoluble form of a toxic trace element likely to be found in anaerobic sediments.
    (3) Common environmental pollutant formerly used as a transformer coolant.
    (4) Chemical species thought to be produced by bacterial action.

17. A polluted water sample is suspected of being contaminated with one of the following: soap, ABS surfactant, or LAS surfactant. The sample has a very low BOD relative to its COD or TOC. Which is the contaminant?

CHAPTER 8
WATER TREATMENT

8.1 WATER TREATMENT AND WATER USE

The treatment of water may be divided into three major categories: (1) purification for domestic use; (2) treatment for specialized industrial applications; and (3) treatment of wastewater to make it acceptable for release or reuse. The type and degree of treatment are strongly dependent upon the source and intended use of the water. Water for domestic use must be thoroughly disinfected to eliminate disease-causing microorganisms, but may contain appreciable hardness in the form of dissolved calcium and magnesium. Water to be used in boilers may contain bacteria but must be quite soft to prevent scale formation. Wastewater being discharged into a large river requires less rigorous treatment than water to be reused in an arid region. As world demand for limited water resources grows (see Section 2.2), more sophisticated and extensive means will have to be employed to treat water.

8.2 THE WATER TREATMENT PLANT

The modern **water treatment plant** is often called upon to perform wonders with the water fed to it. The clear, safe, even tasteful water that comes from a faucet may have started as a murky liquid pumped from a polluted river laden with mud and swarming with bacteria. Or, it may have started with well water, much too hard for domestic use and containing high levels of stain-producing dissolved iron and manganese. The water treatment plant operator's job is to make sure that the water plant product presents no hazards to the consumer.

A schematic diagram of a typical municipal water treatment plant is shown in Figure 8.1 (next page). This particular facility treats water containing excessive hardness and a high level of iron. The raw water taken from wells first goes to an **aerator**. Contact of the water with air removes volatile solutes such as hydrogen sulfide, carbon dioxide, methane, and volatile odorous bacterial metabolites. The addition of oxygen also aids iron removal by oxidizing soluble iron(II) to insoluble iron(III). The addition of lime (CaO or $Ca(OH)_2$) after aeration raises the pH and results in the formation of precipitates containing the hardness ions Ca^{2+} and Mg^{2+}. These precipitates settle from the water in a **primary basin**. Much of the solid material remains in suspension and requires the addition of coagulants to settle the colloidal particles. Among the most common coagulants used are ferric and aluminum sulfates, which form gelatinous hydroxides of their respective metals. Activated silica or synthetic polyelectrolytes may

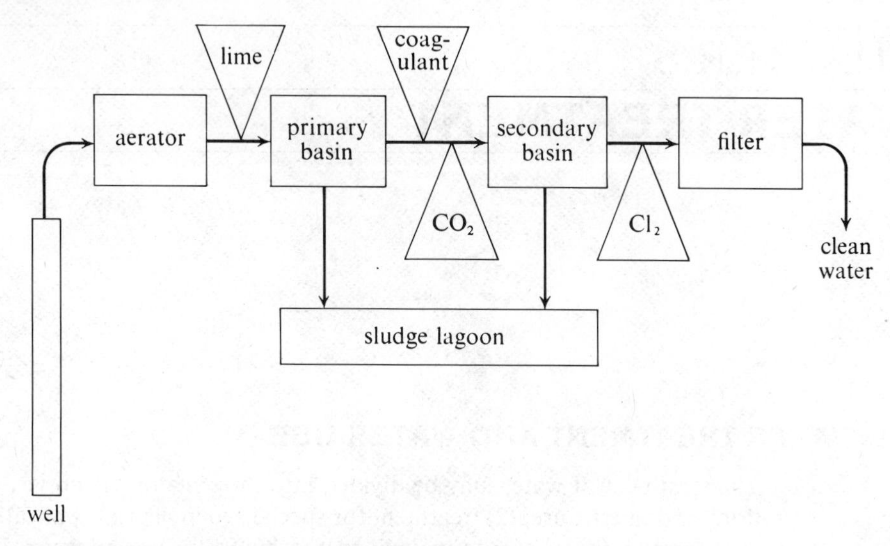

FIGURE 8.1 A schematic diagram of a typical municipal water treatment plant.

also be added to stimulate coagulation or flocculation. The settling occurs in a **secondary basin** after the addition of carbon dioxide to lower the pH. Sludge from both the primary and secondary basins is pumped to a **sludge lagoon**. The water is finally chlorinated, filtered, and pumped to the city water mains.

There are many variations upon the above treatment process. A number of different chemical processes are employed, depending upon what species must be removed from the water. In this chapter, we will discuss some of the main chemical processes involved [1–4].

8.3 REMOVAL OF SOLIDS

Relatively large solid particles are removed from water by simple settling and filtration. Particles in the size range of colloids do not settle readily and are removed by coagulation-flocculation processes, the principles of which are discussed in Section 6.6.

Salts of aluminum and iron are the coagulants most often used in water treatment. Of these, alum or filter alum, a hydrated aluminum sulfate with the formula $Al_2(SO_4)_3 \cdot 18\ H_2O$, is most commonly used. When this salt is added to water, the aluminum ion hydrolyzes by reactions that consume alkalinity in the water, such as:

$$Al(H_2O)_6^{3+} + 3\ HCO_3^- \rightarrow Al(OH)_3(s) + 3\ CO_2 + 6\ H_2O \qquad (8.3.1)$$

The gelatinous hydroxide thus formed carries suspended material with it as it

---

[1] 1971. *Water quality and treatment.* 3rd ed. New York: McGraw-Hill Book Co.

[2] Fair, G.M.; Geyer, J.C.; and Okun, D.A. 1968. *Water and wastewater engineering.* Vols. I and II. New York: John Wiley and Sons, Inc.

[3] Lehr, J.H. 1980. *Domestic water treatment.* New York: McGraw-Hill Book Co.

[4] 1977. *Betz handbook of industrial water conditioning.* 7th ed. Trevose, Pa.: BETZ.

settles. In addition, however, it is likely that positively charged hydroxyl-bridged dimers such as

$$(H_2O)_4Al\overset{\overset{H}{|}}{\underset{\underset{H}{|}}{\langle O \rangle}}Al(H_2O)_4^{4+}$$

and higher polymers are formed which interact specifically with colloidal particles, bringing about coagulation. Metal ions in coagulants also react with virus proteins and destroy up to 99% of the virus in water.

Anhydrous ferric sulfate added to water forms ferric hydroxide in a reaction analogous to (8.3.1). An advantage of ferric sulfate is that it works over a wide pH range of approximately 4–11. Hydrated ferrous sulfate, $FeSO_4 \cdot 7\ H_2O$, or *copperas*, is also commonly used as a coagulant. It forms a gelatinous precipitate of hydrated iron(III) oxide; in order to function, it must be oxidized to iron(III) by dissolved oxygen in the water at a pH higher than 8.5, or by chlorine, which can oxidize iron(II) at lower pH values.

Sodium silicate partially neutralized by acid aids coagulation, particularly when used with alum. The chemical mechanism by which this *activated silica* operates is still largely unknown.

Natural and synthetic polyelectrolytes are used in flocculating particles. Among the natural compounds so used are starch and cellulose derivatives, proteinaceous materials, and gums composed of polysaccharides. More recently, selected synthetic polymers that are effective flocculants have come into use. Neutral polymers and both anionic and cationic polyelectrolytes have been used successfully as flocculants in various applications.

8.4 REMOVAL OF WATER HARDNESS

Calcium and magnesium salts, which generally are present in water as bicarbonates or sulfates, cause **water hardness**. One of the most common manifestations of water hardness is the insoluble "curd" formed by the reaction of soap with calcium or magnesium ions. The formation of these insoluble soap salts is discussed in Section 7.16. Although ions that cause water hardness do not form insoluble products with detergents, they do adversely affect detergent performance. Therefore, calcium and magnesium must be complexed or removed from water for detergents to function properly.

Another problem caused by hard water is the formation of mineral deposits. For example, when water containing calcium and bicarbonate ions is heated, insoluble calcium carbonate is formed:

$$Ca^{2+} + 2\ HCO_3^- \rightarrow CaCO_3(s) + CO_2(g) + H_2O \qquad (8.4.1)$$

This product coats the surfaces of hot water systems, clogging pipes and reducing heating efficiency. Dissolved salts such as calcium and magnesium bicarbonates and sulfates can be especially damaging in boiler feedwater. Clearly, the removal of water hardness is essential for many uses of water.

Several processes are used for softening water. On a large scale, such as in community water-softening operations, the lime-soda process is used. This process involves the treatment of water with lime, $Ca(OH)_2$, and soda ash, Na_2CO_3. Calcium is precipitated as $CaCO_3$ and magnesium as $Mg(OH)_2$. When the calcium is present primarily as "bicarbonate hardness," it can be removed by the addition of $Ca(OH)_2$ alone:

$$Ca^{2+} + 2\ HCO_3^- + Ca(OH)_2 \rightarrow 2\ CaCO_3(s) + 2\ H_2O \qquad (8.4.2)$$

When bicarbonate ion is not present at substantial levels, a source of CO_3^{2-} must be provided at a high enough pH to prevent conversion of most of the carbonate to bicarbonate. These conditions are obtained by the addition of Na_2CO_3. For example, calcium present as the chloride can be removed from water by the addition of soda ash:

$$Ca^{2+} + 2\ Cl^- + 2\ Na^+ + CO_3^{2-} \rightarrow CaCO_3(s) + 2\ Na^+ + 2\ Cl^- \qquad (8.4.3)$$

Note that the removal of bicarbonate hardness results in a net removal of soluble salts from solution, whereas removal of nonbicarbonate hardness involves the addition of at least as many equivalents of ionic material as are removed.

The precipitation of magnesium as the hydroxide requires a higher pH than the precipitation of calcium as the carbonate:

$$Mg^{2+} + 2\ OH^- \rightarrow Mg(OH)_2(s) \qquad (8.4.4)$$

The high pH required may be provided by the basic carbonate ion from soda ash:

$$CO_3^{2-} + H_2O \rightarrow HCO_3^- + OH^- \qquad (8.4.5)$$

Some large-scale, lime-soda softening plants make use of the precipitated calcium carbonate product as a source of additional lime. The calcium carbonate is first heated to at least 825 °C to produce *quicklime,* CaO:

$$CaCO_3 \underset{\Delta}{\rightarrow} CaO + CO_2 \qquad (8.4.6)$$

The quicklime is then slaked with water to produce calcium hydroxide:

$$CaO + H_2O \rightarrow Ca(OH)_2 \qquad (8.4.7)$$

The water softened by lime-soda softening plants usually suffers from two defects. First, because of super-saturation effects, some $CaCO_3$ and $Mg(OH)_2$ usually remain in solution. If not removed, these compounds will precipitate at a later time and cause harmful deposits or undesirable cloudiness in water. The second problem results from the use of highly basic sodium carbonate, which gives the product water an excessively high pH, up to pH 11. To overcome these problems, the water is recarbonated by bubbling CO_2 into it. The carbon dioxide converts the slightly soluble calcium carbonate and magnesium hydroxide to their soluble bicarbonate forms:

$$CaCO_3(s) + CO_2 + H_2O \rightarrow Ca^{2+} + 2\ HCO_3^- \qquad (8.4.8)$$

$$Mg(OH)_2(s) + 2\ CO_2 \rightarrow Mg^{2+} + 2\ HCO_3^- \qquad (8.4.9)$$

The CO_2 also neutralizes excess hydroxide ion:

$$OH^- + CO_2 \rightarrow HCO_3^- \qquad (8.4.10)$$

The pH generally is brought within the range 7.5–8.5 by recarbonation. The source of CO_2 used in the recarbonation process may be from the combustion of carbonaceous fuel. Scrubbed stack gas from a power plant frequently is utilized. Water adjusted to a pH, alkalinity, and Ca^{2+} concentration very close to $CaCO_3$ saturation is labeled *chemically stabilized.* It neither precipitates $CaCO_3$ in water mains, which can clog the pipes, nor dissolves protective $CaCO_3$ coatings from the pipe surfaces. Water with Ca^{2+} concentration much below $CaCO_3$ saturation is called an *aggressive* water.

Calcium may be removed from water very efficiently by the addition of orthophosphate:

$$5\,Ca^{2+} + 3\,PO_4^{3-} + OH^- \rightarrow Ca_5OH(PO_4)_3(s) \qquad (8.4.11)$$

It should be pointed out that the chemical formation of a slightly soluble product for the removal of undesired solutes such as hardness ions, phosphate, iron, and manganese must be followed by sedimentation in a suitable apparatus. Frequently, coagulants must be added, and filtration employed for complete removal of these sediments.

Water may be purified by ion exchange, the reversible transfer of ions between aquatic solution and a solid material capable of bonding ions. The removal of NaCl from solution by two ion-exchange reactions is a good illustration of this process. First the water is passed over a solid *cation* exchanger in the hydrogen form, represented by $H^{+\,-}Cat(solid)$:

$$H^{+\,-}Cat(solid) + Na^+ + Cl^- \rightarrow Na^{+\,-}Cat(solid) + H^+ + Cl^- \qquad (8.4.12)$$

Next the water is passed over an *anion* exchanger in the hydroxide ion form, represented by $OH^{-\,+}An(solid)$:

$$OH^{-\,+}An(solid) + H^+ + Cl^- \rightarrow Cl^{-\,+}An(solid) + H_2O \qquad (8.4.13)$$

Thus, the cations in solution are replaced by hydrogen ion and the anions by hydroxide ion, yielding water as the product.

The softening of water by ion exchange does not require the removal of all ionic solutes, just those cations responsible for water hardness. Generally, therefore, only a cation exchanger is necessary. Furthermore, the sodium rather than the hydrogen form of the cation exchanger is used, and the divalent cations are replaced by sodium ion. Sodium ion at low concentrations is harmless in water to be used foremost purposes, and sodium chloride is a cheap and convenient substance with which to recharge the cation exchangers.

A number of materials have ion-exchanging properties [5]. Among the minerals especially noted for their ion-exchange properties are the aluminum silicate minerals, or **zeolites**. An example of a zeolite which has been used commercially in water softening is glauconite, $K_2(MgFe)_2Al_6(Si_4O_{10})_3(OH)_{12}$. Synthetic zeolites have been prepared by drying and crushing the white gel produced by mixing solutions of sodium silicate and sodium aluminate.

The discovery in the mid-1930s of synthetic ion-exchange resins composed of organic polymers with attached functional groups marked the beginning of modern ion-exchange technology. Structures of typical synthetic ion exchangers

---

[5] Fair, G.M.; Geyer, J.C.; and Okun, D.A. 1968. *Water and wastewater engineering.* Vol. II, Chapter 29. New York: John Wiley and Sons, Inc.

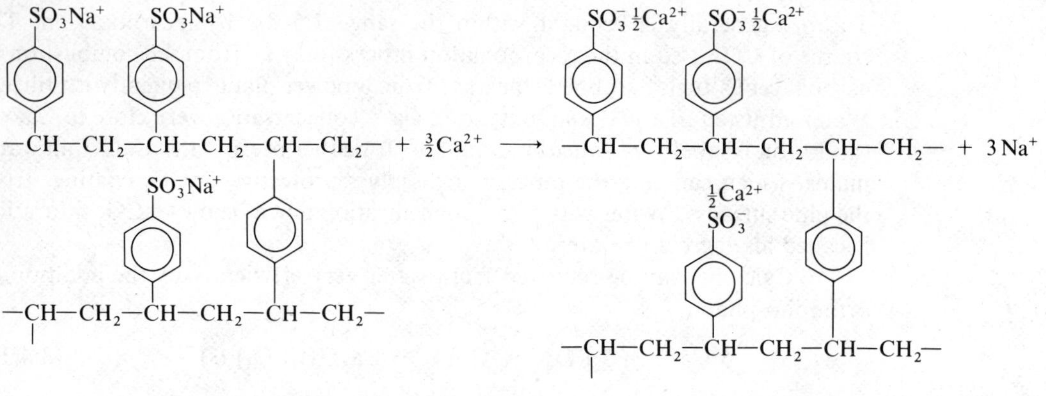

FIGURE 8.2 Strongly acidic cation exchanger. Sodium exchange for calcium in water is shown.

are shown in Figures 8.2 and 8.3. The cation exchanger shown in Figure 8.2 is called a **strongly acidic cation exchanger** because the parent $-SO_3^- H^+$ group is a strong acid. When the functional group binding the cation is the $-CO_2^-$ group, the exchange resin is called a **weakly acidic cation exchanger**, because the $-CO_2H$ group is a weak acid. Figure 8.3 shows a **strongly basic anion exchanger** in which the functional group is a quaternary ammonium group, $-N^+(CH_3)_3$. In the hydroxide form, $-N^+(CH_3)_3OH^-$, the hydroxide ion is readily released; hence the exchanger is classified as strongly basic.

The water-softening capability of a cation exchanger is shown in Figure 8.2, where sodium ion on the exchanger is exchanged for calcium ion in solution. The same reaction occurs with magnesium ion. Water softening by cation exchange is now a widely used, effective, and economical process. In many areas having a low water flow, however, it is not likely that home water softening by ion exchange may be used universally without some deterioration of water quality arising from the contamination of wastewater by sodium chloride. Such contamination results from the periodic need to regenerate a water softener with sodium chloride, in order to displace calcium and magnesium ions from the resin and replace these hardness ions with sodium ions:

$$Ca^{2+}[^-Cat(solid)]_2 + 2\,Na^+ + 2\,Cl^- \rightarrow 2\,Na^{+\,-}Cat(solid) + Ca^{2+} + 2\,Cl^- \qquad (8.4.14)$$

During the regeneration process, a large excess of sodium chloride must be used —several pounds for a home water softener. Appreciable amounts of dissolved sodium chloride can be introduced into sewage by this route.

Strongly acidic cation exchangers are used for the removal of water hardness. Weakly acidic cation exchangers having the $-CO_2H$ group as a functional group are useful for removing alkalinity. Alkalinity, the capacity of a water to

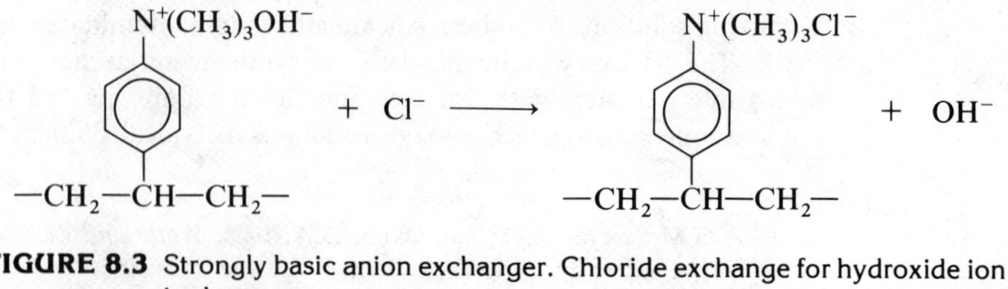

FIGURE 8.3 Strongly basic anion exchanger. Chloride exchange for hydroxide ion is shown.

neutralize an acid (see Section 2.12), generally is manifested by bicarbonate ion. This species is a sufficiently strong base to neutralize the acid of a weak acid cation exchanger:

$$2\,R-CO_2H + Ca^{2+} + 2\,HCO_3^- \rightarrow [R-CO_2^-]_2Ca^{2+} + 2\,H_2O + 2\,CO_2(g) \qquad (8.4.15)$$

However, weak bases such as sulfate ion or chloride ion are not strong enough to remove hydrogen ion from the carboxylic acid exchanger. An additional advantage of these exchangers is that they may be regenerated almost stoichiometrically with dilute strong acids, thus avoiding the potential pollution problem caused by the use of excess sodium chloride to regenerate strongly acidic cation changers.

A convenient approach to eliminating the effects of hardness ions in water is chelation or, as it is sometimes known, **sequestration**. A complexing agent is added which greatly reduces the concentrations of free hydrated cations. Polyphosphate salts, EDTA, and NTA (see Chapter 4) are chelating agents commonly used for water softening. Polysilicates are used to complex iron. To understand the effect of complexing agents in preventing the formation of precipitates arising from hardness ions, consider the following example of calcium carbonate scale formation. In water initially containing 40 mg/L ($1.00 \times 10^{-3}M$) calcium ion, the concentration of carbonate ion required to precipitate calcium carbonate is given by:

$$[Ca^{2+}][CO_3^{2-}] = K_s = 7.2 \times 10^{-9} \qquad (8.4.16)$$

$$[CO_3^{2-}] = \frac{K_s}{[Ca^{2+}]} = \frac{7.2 \times 10^{-9}}{1.00 \times 10^{-3}} = 7.2 \times 10^{-6}M \qquad (8.4.17)$$

In the presence of NTA at a one-fold molar excess over calcium at pH 7.00, the concentration of uncomplexed calcium ion, Ca^{2+}, may be calculated as shown in (8.4.18). Here, HT^{2-} represents the monoprotonated NTA species (the predominant form throughout the pH range of most natural waters—see Figure 4.3) and T^{3-} represents the complexing nonprotonated NTA anion.

$$Ca^{2+} + HT^{2-} \rightleftharpoons CaT^- + H^+ \qquad K = \frac{[CaT^-][H^+]}{[Ca^{2+}][HT^{2-}]} \qquad (8.4.18)$$

Equation 8.4.18 is obtained by adding the following equations:

$$HT^{2-} \rightleftharpoons H^+ + T^{3-} \qquad K_{a3} = 5.25 \times 10^{-11} \qquad (8.4.19)$$

$$Ca^{2+} + T^{3-} \rightleftharpoons CaT^- \qquad K_f = 1.48 \times 10^8 \qquad (8.4.20)$$

$$Ca^{2+} + HT^{2-} \rightleftharpoons CaT^- + H^+ \qquad K = K_{a3}K_f = 7.75 \times 10^{-3}$$

Assuming that most of the calcium is in the complexed form,

$$[CaT^-] = 1.00 \times 10^{-3}M$$

that there is a one-fold excess of NTA over calcium,

$$[HT^{2-}] = [CaT^-] = 1.00 \times 10^{-3}M$$

and that the pH is 7.00, we obtain the following value for the hydrated calcium ion concentration:

$$[Ca^{2+}] = \frac{[CaT^-][H^+]}{K[HT^{2-}]} = \frac{1.00 \times 10^{-3} \times 1.00 \times 10^{-7}}{7.75 \times 10^{-3} \times 1.00 \times 10^{-3}} = 1.29 \times 10^{-5}M$$

The concentration of carbonate ion now required for the precipitation of $CaCO_3$ is

$$[CO_3^{2-}] = \frac{K_s}{[Ca^{2+}]} = \frac{7.2 \times 10^{-9}}{1.29 \times 10^{-5}} = 5.58 \times 10^{-4}M \qquad (8.4.21)$$

which is higher by a factor of almost 100 than the concentration required in the absence of the complexing agent. In the presence of a larger excess of complexing agent over chelatable metal ion, the concentration of carbonate ion required for the precipitation of calcium carbonate would be even higher. Thus, chelation is an effective method of softening water without actually having to remove calcium and magnesium from solution. Because of possible heavy-metal transport (see Chapter 4) this approach should be used only in closed systems.

8.5 REMOVAL OF IRON AND MANGANESE

Soluble iron and manganese are found in many ground waters because of reducing conditions which favor the soluble +2 oxidation state of these metals. Iron is the more commonly encountered of the two metals. In ground water, the level of iron seldom exceeds 10 mg/L, and that of manganese is rarely higher than 2 mg/L. The U.S. Public Health Service recommends maximum levels of 0.3 mg/L for iron and 0.05 mg/L for manganese in drinking water.

The basic method for removing both of these metals depends upon oxidation to higher insoluble oxidation states. The oxidation is generally accomplished by aeration. The rate of oxidation is pH-dependent in both cases, with a high pH favoring more rapid oxidation. In the case of iron, the rate of iron(II) oxidation, $-d[Fe(II)]/dt$, is given by the equation

$$-\frac{d[Fe(II)]}{dt} = k[Fe(II)][O_2][OH^-]^2 \qquad (8.5.1)$$

where k is the rate constant for the reaction [6]. The oxidation of soluble Mn(II) to insoluble MnO_2 is a complicated process. It appears to be catalyzed by solid MnO_2, which is known to adsorb Mn(II). This adsorbed Mn(II) is slowly oxidized on the MnO_2 surface.

Chlorine and potassium permanganate are sometimes employed as oxidizing agents for iron and manganese. There is some evidence that organic chelating agents with reducing properties hold iron(II) in a soluble form in water. In such cases, chlorine is effective because it destroys the organic compounds and enables the oxidation of iron(II).

In water with a high level of carbonate, $FeCO_3$ and $MnCO_3$ may be precipitated directly by raising the pH above 8.5 by the addition of sodium carbonate or lime. This approach is less popular than oxidation, however.

Relatively high levels of insoluble iron(III) and manganese (IV) frequently are found in water as colloidal material which is difficult to remove. These metals may be associated with humic colloids or "peptizing" organic material that binds to colloidal metal oxides, stabilizing the colloid.

---

[6] Stumm, W., and Lee, G.F. 1961. Oxygenation of ferrous iron. *Ind. Eng. Chem.* 53: 143.

8.6 REMOVAL OF ORGANICS FROM DRINKING WATER

Very low levels of exotic organic compounds (synthesized by humans, but not by natural processes) in drinking water are suspected of contributing to cancer and other maladies. Concern over this problem was increased following the Environmental Protection Agency's November 1974 report that at least eight potential carcinogens had been detected in New Orleans drinking water. Some of these are chlorinated organic compounds produced by chlorination of water for disinfection (see Section 8.7).

In January, 1978, the EPA proposed that municipal water systems serving 75,000 or more persons be required to stay within a 100-ppb limit for chloroform and other trihalomethane compounds. In most cases where this limit is exceeded, the standard can be met by treatment with granular activated carbon, a process described in Section 8.13. Although 32 municipal water treatment plants were built during the 1970s with the provision for using activated carbon, only 2 were using that treatment as of 1981 [7]. It is now believed that trihalomethanes originate from chlorination of organics in water, especially humic substances (see Section 4.19). Removal of organics to very low levels prior to chlorination has been found to be effective in preventing trihalomethane formation.

8.7 DISINFECTION OF WATER

Chlorine is the most commonly used disinfectant employed for killing bacteria in water. When chlorine is added to water, it rapidly hydrolyzes according to the reaction

$$Cl_2 + H_2O \rightarrow H^+ + Cl^- + HOCl \tag{8.7.1}$$

which has the following equilibrium constant:

$$K_n = \frac{[H^+][Cl^-][HOCl]}{[Cl_2]} = 4.5 \times 10^{-4} \tag{8.7.2}$$

Hypochlorous acid, HOCl, is a weak acid that dissociates according to the reaction,

$$HOCl \rightleftharpoons H^+ + OCl^- \tag{8.7.3}$$

with an ionization constant of 2.7×10^{-8}. From equation 8.7.3 it can be shown that the concentration of elemental Cl_2 is negligible at equilibrium above pH 3 when chlorine is added to water at levels below 1.0 g/L.

Sometimes, hypochlorite salts are substituted for chlorine gas as a disinfectant. Calcium hypochlorite, $Ca(OCl)_2$, is commonly used. The hypochlorites are safer to handle than gaseous chlorine.

The two chemical species formed by chlorine in water, HOCl and OCl^-, are known as **free available chlorine**. Free available chlorine is very effective in

---

[7] Miller, S. 1981. Activated carbon—the dilemma facing its use. *Environmental Science and Technology* 15: 502.

killing bacteria. In the presence of ammonia, monochloramine, dichloramine, and trichloramine are formed:

$$NH_4^+ + HOCl \rightarrow NH_2Cl + H_2O + H^+ \quad \text{(monochloramine)} \qquad (8.7.4)$$

$$NH_2Cl + HOCl \rightarrow NHCl_2 + H_2O \quad \text{(dichloramine)} \qquad (8.7.5)$$

$$NHCl_2 + HOCl \rightarrow NCl_3 + H_2O \quad \text{(trichloramine)} \qquad (8.7.6)$$

The chloramines are called **combined available chlorine**. Chlorination practice frequently provides for formation of combined available chlorine which, although a weaker disinfectant than free available chlorine, is more readily retained as a disinfectant throughout the water distribution system. Too much ammonia in water causes excess demand for chlorine and is considered undesirable.

At sufficiently high Cl:N molar ratios in water containing ammonia, some HOCl and OCl^- remain unreacted in solution, and a small quantity of NCl_3 is formed. The ratio at which this occurs is called the **breakpoint**. Chlorination beyond the breakpoint ensures disinfection. It has the additional advantage of destroying the more common materials that cause odor and taste in water.

At moderate levels of NH_3-N (approximately 20 mg/L), when the pH is between 5.0 and 8.0, chlorination with a minimum 8:1 weight ratio of Cl to NH_3-nitrogen produces efficient denitrification:

$$NH_4^+ + HOCl \rightarrow NH_2Cl + H_2O + H^+ \qquad (8.7.7)$$

$$2\, NH_2Cl + HOCl \rightarrow N_2(g) + 3\, H^+ + 3\, Cl^- + H_2O \qquad (8.7.8)$$

This reaction is used to remove pollutant ammonia from wastewater. However, recent concern over possible organochlorine compounds produced by chlorination of organic wastes may hinder the development of this approach to ammonia removal. Chlorine produces a number of by-products in the presence of organic matter in water. Some of these may be harmful to health. Typical of such by-products is chloroform, produced by the chlorination of humic substances in water.

Chlorine is also used to treat water other than drinking water. It is employed to disinfect effluent from sewage treatment plants, as an additive to the water in electric power plant cooling towers, and to control microorganisms in food processing [8]. Despite some known harmful effects, chlorine use for disinfection is increasing steadily.

Ozone is sometimes used as a disinfectant in place of chlorine, particularly in Europe. Figure 8.4 shows the basic components of an ozone water-treatment system. Basically, air is filtered, cooled, dried, and pressurized, then subjected to an electrical discharge of approximately 20,000 volts. The ozone produced is then pumped into a contact chamber where water contacts the ozone for 10–15 minutes. The recent concern over possible production of toxic organochlorine compounds by water chlorination processes has increased interest in ozonation. Furthermore, ozone is more destructive to viruses than is chlorine. Unfortunately, the solubilility of ozone in water is relatively low, which limits its disinfective power.

---

[8] Hileman, B. 1982. The chlorination question. *Environmental Science and Technology* 16: 15A–18A.

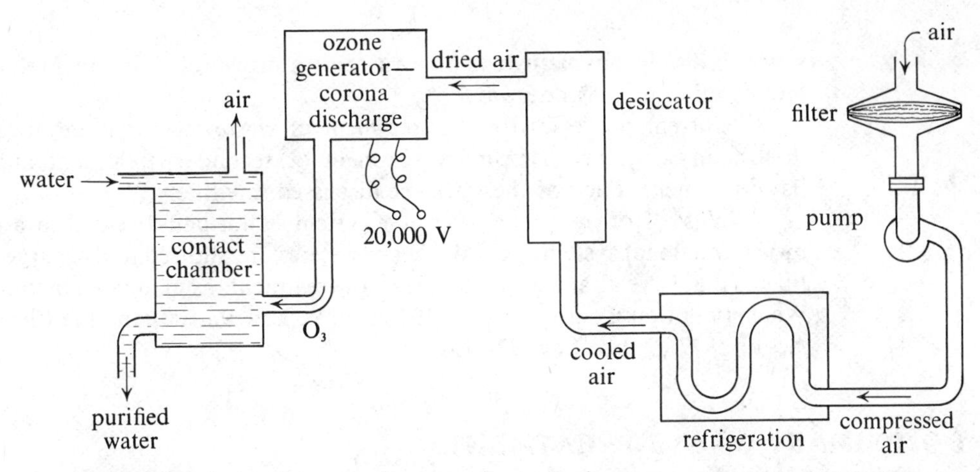

FIGURE 8.4 A schematic diagram of a typical ozone water-treatment system.

A major consideration with ozone is the rate at which it decomposes spontaneously in water, according to the overall reaction,

$$2\ O_3 \rightarrow 3\ O_2 \tag{8.7.9}$$

This rate of decomposition has been found to follow the empirical equation below [9]:

$$-\frac{d[O_3]}{dt} = k_0[OH^-]^{0.55}[O_3]^2 \tag{8.7.10}$$

Because of the decomposition of ozone in water, some chlorine must be added to maintain disinfectant throughout the water distribution system.

Iron(VI) in the form of ferrate ion, FeO_4^{2-}, is a strong oxidizing agent with excellent disinfectant properties. It has the additional advantage of removing heavy metals, viruses, and phosphate. It may well find application in the future.

8.8 SEWAGE TREATMENT

Typical municipal sewage contains oxygen-demanding materials, sediments, grease, oil, scum, pathogenic bacteria, viruses, salts, algal nutrients, pesticides, refractory organic compounds, heavy metals, and an astonishing variety of flotsam ranging from children's socks to sponges. It is the job of the waste-treatment plant to remove as much of this material as possible.

Several characteristics are used to describe sewage. These include **turbidity** (international turbidity units); suspended solids (ppm); total dissolved solids (ppm); acidity (H^+ ion concentration or pH); and dissolved oxygen (in ppm O_2). In addition, the **potassium permanganate value (PV)** may be determined by measuring how much $KMnO_4$ reacts with a given volume of sewage, thus providing a measure of unstable organic compounds. Biological oxygen demand and chemical oxygen demand (see Section 7.15) are used as measures of oxygen-demanding substances. The time required for methylene blue dye added to the

[9] Gurol, M.D., and Singer, P.C. 1982. Kinetics of ozone decomposition: a dynamic approach. *Environmental Science and Technology* 16: 377–383.

water to disappear (**methylene blue value**) is a measure of the time required for fermentation to get underway.

Current processes for the treatment of wastewater may be divided into three main categories: (1) primary treatment, (2) secondary treatment, and (3) tertiary treatment. Each of these will be discussed separately.

Waste from a municipal water system is normally treated in a **publicly owned treatment system, POTW**. These systems are allowed to discharge only effluents that have attained a certain level of treatment, as mandated by the Federal Water Pollution Control Acts of 1972 (Public Law 92-500) and the Clean Water Act of 1977 (Public Law 92-215).

8.9 PRIMARY WASTE TREATMENT

Primary treatment of wastewater consists of the removal of insoluble matter such as grit, grease, and scum from water. The first step in primary treatment normally is **screening** [10]. Screening removes or reduces the size of trash and large solids that get into the sewage system. These solids are collected on screens and scraped off for subsequent disposal. Most screens are cleaned with power rakes. **Comminuting devices** shred and grind solids in the sewage. Particle size may be reduced to the extent that the particles can be returned to the sewage flow.

Grit in wastewater consists of such materials as sand and coffee grounds which do not biodegrade well and generally have a high settling velocity. **Grit removal** is practiced to prevent accumulation of grit in other parts of the treatment system, to reduce clogging of pipes and other parts, and to protect moving parts from abrasion and wear. Grit normally is allowed to settle in a tank under conditions of low flow velocity, and it is then scraped mechanically from the bottom of the tank.

Primary sedimentation removes both settleable and floatable solids. During primary sedimentation there is a tendency for flocculent particles to aggregate for better settling, a process that may be aided by the addition of chemicals. The material that floats in the primary settling basin is known collectively as **grease**. In addition to fatty substances, the grease consists of oils, waxes, free fatty acids, and insoluble soaps containing calcium and magnesium. Normally, some of the grease settles with the sludge and some floats to the surface, where it may be removed by a skimming device.

8.10 SECONDARY WASTE TREATMENT BY BIOLOGICAL PROCESSES

The most obvious harmful effect of biodegradable organic matter in wastewater is BOD, the removal of dissolved oxygen by the microorganism-mediated degradation of the organic matter. Streams receiving organic wastes are capable of self purification through this phenomenon, which is utilized in biological waste treatment. **Secondary treatment** by biological processes takes many forms but consists basically of the following: microorganisms are allowed to degrade

---

[10] 1976. *Operation of wastewater treatment plants.* Manual of Practice No. 11. Washington, D.C.: Water Pollution Control Federation.

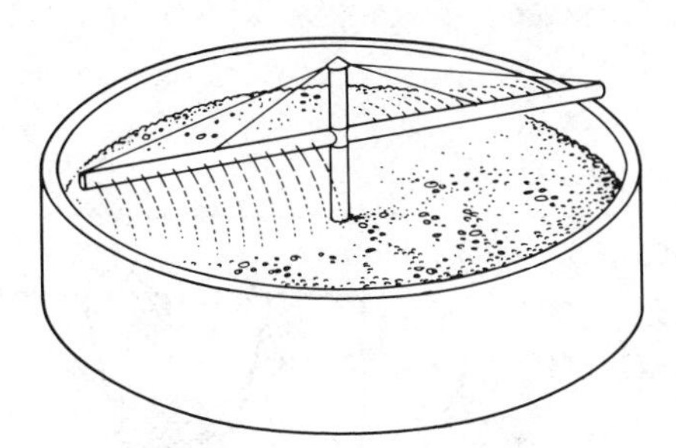

FIGURE 8.5 A trickling filter waste-treatment system.

organic material in solution or in suspension, in the presence of added oxygen, until the BOD of the waste has been reduced to acceptable levels. The waste is oxidized biologically under conditions controlled for optimum bacterial growth and at a site where this growth does not influence the environment.

One of the most commonly used biological waste treatment processes is the **trickling filter** (Fig. 8.5), in which wastewater is sprayed over rocks or other solid support material covered with microorganisms. The structure of the trickling filter is such that contact of the wastewater with air is allowed and degradation of organic matter occurs by the action of the microorganisms.

Rotating biological reactors, another type of treatment system, consist of groups of large plastic discs mounted close together on a shaft. The device is positioned such that half of each disc is immersed in wastewater and half exposed to air. The shaft rotates constantly, so that the submerged portion of the discs is always changing. The discs, usually made of high-density polyethylene or polystyrene, accumulate thin layers of attached biomass, which degrades organic matter in the sewage. Oxygen is absorbed by the biomass and by the layer of wastewater adhering to it during the time that the biomass is exposed to air.

Both trickling filters and rotating biological reactors are examples of **fixed-film biological (FFB) processes**. The greatest advantage of these processes is their low energy consumption. The energy consumption is minimal because it is not necessary to pump air or oxygen into the water, as is the case with the popular *activated sludge process* described below. The trickling filter has long been a standard means of wastewater treatment; for example, a 31-acre trickling filter installation in Baltimore which became operational in 1907 is still being used. As of 1982, 236 municipal wastewater treatment plants were using rotating biological reactors [11].

The **activated sludge process** is probably the most versatile and effective of all waste treatment processes (Figure 8.6 on page 200). Microorganisms in the aeration tank convert organic material in wastewater to microbial biomass and CO_2. Organic nitrogen is converted to ammonium ion or nitrate. Organic phosphorus is converted to orthophosphate. The microbial cell matter formed as part of the waste degradation processes is normally kept in the aeration tank until the microorganisms are past the log phase of growth (Section 5.8), at which point

[11] Josephson, J. 1982. Fixed-film biological processes. *Environmental Science and Technology* 16: 380A–384A.

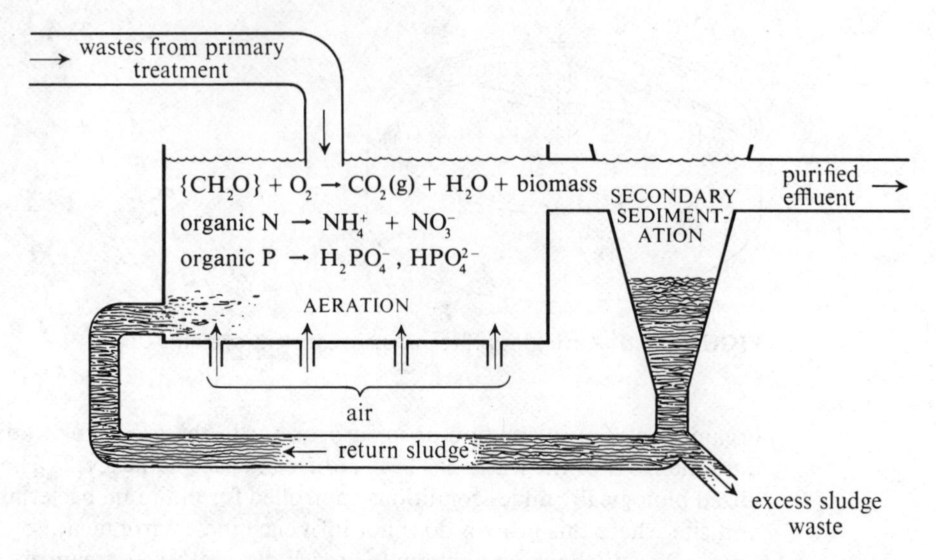

FIGURE 8.6 Schematic diagram of a conventional activated sludge system for secondary biological wastewater treatment.

the cells flocculate relatively well to form settleable solids. These solids settle out in a settler and a fraction of them is discarded. Part of the solids, the return sludge, is recycled to the head of the aeration tank and comes into contact with fresh sewage. The combination of a high concentration of "hungry" cells in the return sludge and a rich food source in the influent sewage provides optimum conditions for the rapid degradation of organic matter.

The degradation of organic matter that occurs in an activated sludge facility also occurs in streams and other aquatic environments. However, in general, when a degradable waste is put into a stream, it encounters only a relatively small population of microorganisms capable of carrying out the degradation process. Thus, several days may be required for the buildup of a sufficient population of organisms to degrade the waste. In the activated sludge process, continual recycling of active organisms provides the optimum conditions for waste degradation, and a waste may be degraded within the very few hours that it is present in the aeration tank.

The activated sludge process provides two pathways for the removal of BOD, as illustrated schematically in Figure 8.7. BOD may be removed by (1) **oxidation** of organic matter to provide energy for the metabolic processes of the microorganisms, and (2) **synthesis**, incorporation of the organic matter into cell mass. In the first pathway, carbon is removed in the gaseous form as CO_2. The second pathway provides for removal of carbon as a solid in biomass. That portion of the carbon converted to CO_2 is vented to the atmosphere and does not present a disposal problem. The disposal of waste sludge, however, is a problem, primarily because it is only about 1% solids and contains many undesirable components. Normally, partial water removal is accomplished by drying on sand filters, vacuum filtration, or centrifugation. The dewatered sludge may be incinerated or used as land fill. To a certain extent, sewage sludge may be digested anaerobically to produce methane and carbon dioxide:

$$2\,\{CH_2O\} \xrightarrow[\text{anaerobic bacteria}]{\text{methane-producing}} CH_4(g) + CO_2(g) \qquad (8.10.1)$$

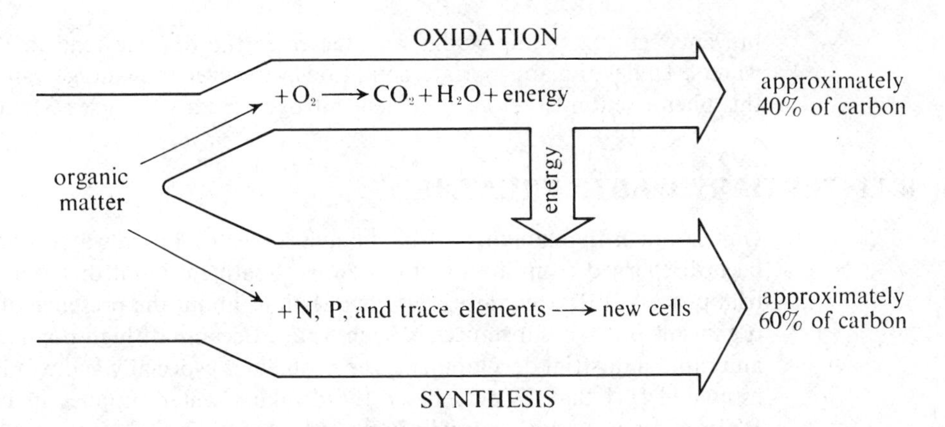

FIGURE 8.7 General metabolic scheme for the aerobic microbial degradation of organic wastes.

This reduces both the volatile-matter content and the volume of the sludge by about 60%. A carefully designed plant may produce enough methane to provide for all of its power needs.

One of the most desirable means of sludge disposal is to use it to fertilize and condition soil. However, care has to be taken that excessive levels of heavy metals are not applied to the soil as part of the sludge [12]. Problems with various kinds of sludges resulting from water treatment are discussed further in Section 8.23.

Nitrification (the microbially mediated conversion of ammonium nitrogen to nitrate; see Section 5.16), is a significant process that occurs during biological waste treatment. Ammonium ion is normally the first inorganic nitrogen species produced in the biodegradation of nitrogenous organic compounds. It is oxidized, under the appropriate conditions, first to nitrite,

$$2\ NH_4^+ + 3\ O_2 \xrightarrow{Nitrosomonas} 4\ H^+ + 2\ NO_2^- + 2\ H_2O \qquad (8.10.2)$$

then to nitrate:

$$2\ NO_2^- + O_2 \xrightarrow{Nitrobacter} 2\ NO_3^- \qquad (8.10.3)$$

These reactions occur in the aeration tank of the activated sludge plant and are favored in general by long retention times, low organic loadings, large amounts of suspended solids, and high temperatures. Nitrification can reduce sludge settling efficiency because the denitrification reaction

$$4\ NO_3^- + 5\ \{CH_2O\} + 4\ H^+ \rightarrow 2\ N_2(g) + 5\ CO_2(g) + 7\ H_2O \qquad (8.10.4)$$

occurring in the oxygen-deficient settler causes bubbles to form on the sludge floc (aggregated sludge particles), making it so buoyant that it floats to the top. This

---

[12] Naylor, L.M., and Loehr, R.C. 1981. Increase in dietary cadmium as a result of application of sewage sludge to agricultural land. *Environmental Science and Technology* 15: 881–886.

prevents settling of the sludge and increases the organic load in the receiving waters. Under the appropriate conditions, however, advantage can be taken of this phenomenon to remove nutrient nitrogen from water (see Section 8.19).

8.11 TERTIARY WASTE TREATMENT

Unpleasant as the thought may be, many people drink used water—water that has been discharged from a municipal sewage treatment plant or from some industrial process [13]. This raises serious questions about the presence of pathogenic organisms or toxic substances in such water. Because of high population density and rapid industrial development, the problem is especially acute in Europe. It is estimated [11] that 32 percent of the drinking water supplies in England and Wales contain "used" water; in Paris and surrounding areas, the extent of reused water ranges from 50 to 70 percent; and in West Germany, the Ruhr River has at times consisted of as much as 40 percent treated wastewater.

Obviously, there is a great need to treat wastewater in a manner that makes it amenable to reuse. This requires treatment beyond the secondary processes previously discussed. Such treatment is discussed in the following sections.

Tertiary waste treatment is a term used to describe a variety of processes performed on the effluent from secondary waste treatment. A broader term, **advanced waste treatment**, applies to any water-treatment process that removes more contaminants from wastewater than do the treatments now in general use. The contaminants removed by tertiary waste treatment fall into the general categories of (1) suspended solids; (2) dissolved organic compounds; and (3) dissolved inorganic materials, including the important class of algal nutrients. Each of these categories presents its own problems with regard to water quality. Suspended solids are primarily responsible for residual biological oxygen demand in secondary sewage effluent waters. The dissolved organics are the most hazardous from the standpoint of potential toxicity. The major problem with dissolved inorganic materials is that presented by algal nutrients, primarily nitrates and phosphates. In addition, potentially hazardous toxic metals may be found among the dissolved inorganics.

In addition to these chemical contaminants, secondary sewage effluent often contains a number of disease-causing microorganisms, requiring disinfection in cases where humans may later come into contact with the water. Among the bacteria that may be found in secondary sewage effluent are organisms causing tuberculosis, dysenteric bacteria *(Bacillus dysenteriae, Shigella dysenteriae, Shigella paradysenteriae, Proteus vulgaris),* cholera bacteria *(Vibrio cholerae),* bacteria causing mud fever *(Leptospira icterohemorrhagiae),* and bacteria causing typhoid fever (*Salmonella typhosa, Salmonella paratyphi*). In addition, viruses causing diarrhea, eye infections, infectious hepatitis, and polio may be encountered. Ingestion of sewage still causes disease, even in the United States.

Disinfection is usually accomplished by chlorination (see Section 8.7). Although chlorination is effective, there is some concern about the products of

---

[13] Miller, S. 1981. Water reuse. *Environmental Science and Technology* 15: 499–501.

the reaction of chlorine with residual organic matter in the water. An alternative process, ozonation of wastewater (Section 8.7), is practiced in Europe.

8.12 REMOVAL OF SUSPENDED SOLIDS FROM WASTEWATER

Suspended solids in secondary sewage effluent arise primarily from sludge that was not removed in the settling process. These solids account for a large part of the BOD in the effluent and may interfere with other aspects of tertiary waste treatment. For example, these solids may clog membranes in electrodialysis or reverse osmosis processes (see Sections 8.15 and 8.17). The quantity of material involved may be rather high. Processes designed to remove suspended solids often will remove 10–20 mg/L of organic material from secondary sewage effluent. In addition, a small amount of the inorganic material is removed as well.

Suspended materials may be removed from water by simple filtration. A special type of filtration procedure known as **microstraining** is especially effective in the removal of very small particles. These filters are woven from stainless steel wire so fine that it is barely visible. This enables preparation of filters with openings only 60–70μm across. These openings may be reduced to 5–15μm across by partial clogging with small particles, such as bacterial cells. Typically, such filters reduce suspended-solid values from about 55 ppm to 7 ppm and BOD values from about 50 ppm to below 15 ppm. The cost of this treatment is normally substantially lower than the costs of competing processes. High flow rates at low back pressures are normally achieved.

Coagulation–filtration is a much more effective procedure than filtration alone for the removal of suspended material from water. As the term implies, the process consists of the addition of coagulants that aggregate the particles into larger size particles, followed by filtration. Either alum or lime, often with added polyelectrolytes, is most commonly employed for coagulation, which proceeds by the mechanisms described in Section 8.3.

The filtration process is usually performed on a medium such as sand or anthracite coal. Often, to reduce clogging, several media with progressively smaller interstitial spaces are used. One example is the **rapid sand filter**, which consists of a layer of sand supported by layers of gravel particles, the particles becoming progressively larger with increasing depth. The substance that actually filters the water is coagulated material that collects in the sand. As more material is removed, the buildup of coagulated material eventually clogs the filter and must be removed by back flushing.

8.13 REMOVAL OF DISSOLVED ORGANICS FROM WASTEWATER

Many organic compounds survive, or are produced by, secondary wastewater treatment. Almost half of these are humic substances (see Section 4.19) with a molecular-weight range of 1000–5000. Among the remainder are found ether-extractable materials, carbohydrates, proteins, detergents, tannins, and lignins. The humic compounds, because of their high molecular weight and anionic character, influence some of the physical and chemical aspects of waste treatment. The ether-extractables contain many of the biorefractory compounds (see Section 7.17) and are of particular concern regarding potential toxicity, carcinogenicity, and mutagenicity. In the ether extract are found many fatty acids [14], including stearic acid, $CH_3(CH_2)_{16}CO_2H$; palmitic acid, $CH_3(CH_2)_{14}CO_2H$; pentadecanoic acid, $CH_3(CH_2)_{13}CO_2H$; and several others. Hydrocarbons of the *n*-alkane class have been identified in the effluent. Some higher aromatic hydrocarbons, including naphthalene, diphenylmethane, diphenyl, and methylnaphthalene also are found. Isopropyl benzene and dodecyl benzene have been identified, along with phenol, dioctylphthalate, and triethylphosphate.

The standard method for the removal of dissolved organic material is adsorption on activated carbon, a product that is produced from a variety of carbonaceous materials, including wood, pulp-mill char, peat, and lignite. The carbon is produced by charring the raw material anaerobically below 600 °C, followed by an activation step consisting of partial oxidation. Carbon dioxide may be employed as an oxidizing agent at 600–700 °C,

$$CO_2 + C \rightarrow 2\,CO \tag{8.13.1}$$

or the carbon may be oxidized by water at 800–900 °C:

$$H_2O + C \rightarrow H_2 + CO \tag{8.13.2}$$

These processes develop porosity, increase the surface area, and leave the C atoms in arrangements that have affinities for organic compounds.

Activated carbon comes in two general types: granulated activated carbon, consisting of particles 0.1–1 mm in diameter, and powdered activated carbon, in which most of the particles are 50–100μm in diameter.

The exact mechanism by which activated carbon holds organic materials is not known. However, one reason for the effectiveness of this material as an adsorbent is its tremendous surface area. A solid cubic foot of carbon particles may have a combined pore and surface area of approximately 10 square miles!

A quantitative expression for the adsorptive capacity of activated carbon has been difficult to define. A favored parameter has been the **phenol value**, defined as the amount of carbon in milligrams required to reduce by 90% the phenol in 1 L of a solution containing 100 μg/L of phenol. The phenol values of commercial activated carbons generally are within the range of 15–30. Unfortunately, other organic compounds are not necessarily removed with equal efficiency, and specific compounds must be tested directly.

Although interest is increasing in the use of powdered activated carbon for water treatment, currently granular carbon is more widely used. It may be employed in a fixed bed, through which water flows downward. Accumulation of particulate matter requires periodic backwashing. An expanded bed in which particles are kept slightly separated by water flowing upward may be used with less chance of clogging.

---

[14] Manka, J.; Rubhun, M.; Mandelbaum, A.; and Bortinger, A. 1974. Characterization of organics in secondary effluents. *Environmental Science and Technology* 8: 1017–20.

The removal rate of organics from wastewater is often much higher than that indicated by laboratory tests [15]. This effect is attributed to bacterial growths on the carbon, which oxidize organics. There is some evidence that biological action is promoted by activated carbon.

Economics require regeneration of the carbon, which is accomplished by heating it to 950 °C in a steam-air atmosphere. This process oxidizes adsorbed organics and regenerates the carbon surface, with an approximately 10% loss of carbon.

Removal of organics may also be accomplished by adsorbent synthetic polymers. Such polymers as Amberlite XAD-4 have hydrophobic surfaces and strongly attract relatively insoluble organic compounds, such as chlorinated pesticides. The porosity of these polymers is up to 50% by volume, and the surface area may be as high as 850 m^2/g. They are readily regenerated by solvents such as isopropanol and acetone. Under appropriate operating conditions, these polymers remove virtually all nonionic organic solutes; for example, phenol at 250 mg/L is reduced to less than 0.1 mg/L by appropriate treatment with Amberlite XAD-4.

Oxidation of dissolved organics holds some promise for their removal. Ozone, hydrogen peroxide, molecular oxygen (with or without catalysts), chlorine and its derivatives, permanganate, or ferrate can be used. Electrochemical oxidation may be possible in some cases.

8.14 REMOVAL OF DISSOLVED INORGANICS

In order for complete water recycling to be feasible, inorganic-solute removal is essential. The effluent from secondary waste treatment generally contains 300–400 mg/L more dissolved inorganic material than does the municipal water supply. It is obvious, therefore, that 100% water recycle without removal of inorganics would cause the accumulation of an intolerable level of dissolved material. Even when water is not destined for immediate reuse, the removal of the inorganic nutrients phosphorus and nitrogen is highly desirable to reduce eutrophication downstream. In some cases, the removal of toxic trace metals is needed.

One of the most obvious methods for removing inorganics from water is distillation. Unfortunately, the energy required for distillation is generally too high for the process to be economically feasible. Furthermore, volatile materials such as ammonia and odorous compounds are carried over to a large extent in the distillation process, unless special preventative measures are taken. Freezing produces a very pure water, but is considered uneconomical with present technology. The three processes considered most promising for bulk removal of inorganics from water are *electrodialysis, ion exchange,* and *reverse osmosis.*

---

[15] Hassler, J.W. 1974. *Purification with activated carbon.* New York: Chemical Publishing Co., Inc.

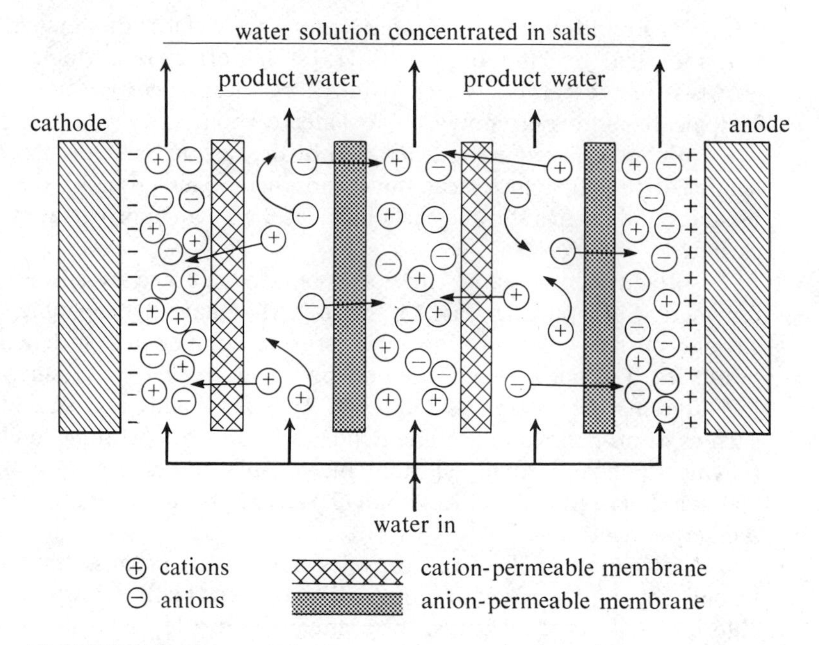

FIGURE 8.8 Electrodialysis apparatus for the removal of ionic material from water.

8.15 ELECTRODIALYSIS

Electrodialysis consists of applying a direct current across a body of water separated into vertical layers by membranes alternately permeable to cations and anions. Cations migrate toward the cathode and anions toward the anode. Cations and anions both *enter* one layer of water, and both *leave* the adjacent layer. Thus, layers of water enriched in salts alternate with those from which salts have been removed. The water in the brine-enriched layers is recirculated to a certain extent to prevent excessive accumulation of brine. The principles involved in electrodialysis treatment are shown in Figure 8.8.

Although the relatively small ions constituting the salts dissolved in wastewater readily pass through the membranes, large organic ions (proteins, for example) and charged colloids migrate to the membrane surfaces, often fouling or plugging the membranes and reducing efficiency. In addition, growth of microorganisms on the membranes can cause fouling.

Experience with pilot plants indicates that electrodialysis is a practical and economical method of removing up to 50% of the dissolved inorganics from secondary sewage effluent, once the effluent has been carefully pretreated to eliminate fouling substances. Such a level of efficiency would permit repeated recycle of water without dissolved inorganic materials reaching unacceptably high levels.

8.16 ION EXCHANGE

The **ion-exchange** method for softening water is described in detail in Section 8.4. The ion-exchange process used for removal of inorganics consists of passing the water successively over a solid cation exchanger and a solid anion exchanger, which replace cations and anions by hydrogen ion and hydroxide ion, respectively. The net result is that each equivalent of salt is replaced by a mole of water. For the hypothetical ionic salt MX, the reactions are:

$$H^{+\,-}Cat(solid) + M^+ + X^- \rightleftharpoons M^{+\,-}Cat(solid) + H^+ + X^- \qquad (8.16.1)$$

$$OH^{-\,+}An(solid) + H^+ + X^- \rightleftharpoons X^{-\,+}An(solid) + H_2O \qquad (8.16.2)$$

where $^-$Cat(solid) represents the solid cation exchanger and $^+$An(solid) represents the solid anion exchanger. The cation exchanger is regenerated with strong acid and the anion exchanger with strong base.

Demineralization by ion exchange generally produces water of a very high quality. Unfortunately, some organic compounds in wastewater foul ion exchangers, and microbial growth on the exchangers can diminish their efficiency. In addition, regeneration of the resins is expensive, and the concentrated wastes from regeneration require disposal in a manner that will not damage the environment. Despite these problems, ion exchange remains a leading contender for the tertiary treatment of wastewater, and further technological advances in the process are to be expected.

Weak base ion-exchange resins have been shown to be useful for the removal of protein and viruses from wastewater on an experimental basis [16]. The resin used in the experiment consisted of a phenol-formaldehyde polymer with secondary amine functional groups:

```
            H
            |
  {resin}—N—H⁺
            |
            R
```

Use of such resins to remove residual proteins, and particularly viruses, could be quite effective in advanced wastewater treatment.

8.17 REVERSE OSMOSIS

Reverse osmosis (also known as **hyperfiltration**) is a very useful technique for the purification of water. Basically, reverse osmosis consists of forcing pure water through a semipermeable membrane that allows the passage of water but not of other material. This process depends on the preferential sorption of water on the surface of the membrane, which is generally composed of porous cellulose acetate. Pure water from the sorbed layer is forced through pores in the membrane under pressure. If the thickness of the sorbed water layer is *d*, the pore diameter for optimum separation should be 2*d*. The optimum pore diameter depends upon the thickness of the sorbed pure water layer and may be several times the diameters of the solute and solvent molecules. Therefore, reverse osmosis is not a simple sieve separation or ultrafiltration process. The principle of reverse osmosis is illustrated in Figure 8.9 (see page 208).

---

[16] Foster, D.H.; Engelbrecht, R.S.; and Snoeyink, V.L. 1977. Application of weak base ion-exchange resins for removal of proteins. *Environmental Science and Technology* 11: 55–61.

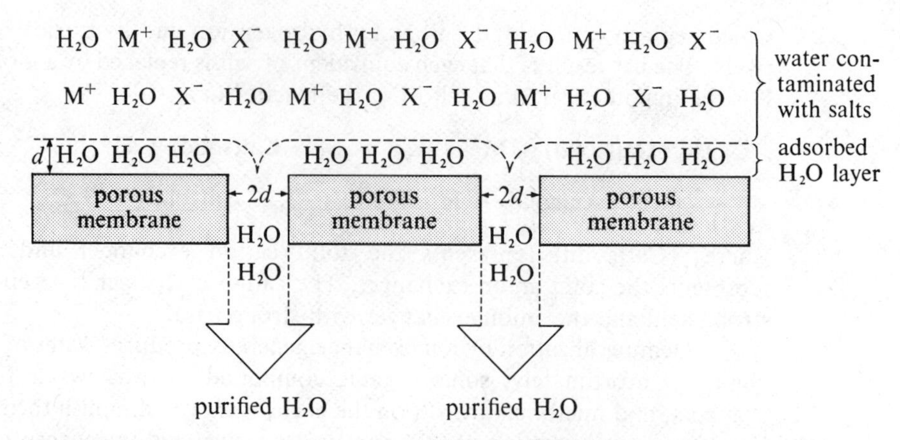

FIGURE 8.9 Solute removal from water by reverse osmosis.

Typical performance for reverse osmosis applied to water desalination has been described [17]. A plant processing 1,500,000 liters per day (L/d) of seawater containing 43,000 mg/L of dissolved solids can produce 300,000 L/d of fresh water containing only 200–500 mg/L dissolved solids, and 1,200,000 L/d of brine containing 52,000 mg/L solids. In another typical application for the treatment of industrial wastewater containing copper, zinc, and chromium metals, as well as other pollutants, 95% of the water was converted to a product pure enough for discharge, and the pollutants were concentrated in the remaining 5% collected as a brine waste.

8.18 PHOSPHORUS REMOVAL

Advanced waste treatment normally requires removal of phosphorus to reduce algal growth. Algae may grow at PO_4^{3-} levels as low as 0.05 mg/L. Growth inhibition requires levels well below 0.5 mg/L. Since municipal wastes typically contain approximately 25 mg/L of phosphate (as orthophosphates, polyphosphates, and insoluble phosphates), the efficiency of phosphate removal must be quite high to prevent algal growth. This removal may occur in the sewage treatment process (1) in the primary settler; (2) in the aeration chamber of the activated sludge unit; or (3) after secondary waste treatment.

Normally, the activated sludge process removes about 20% of the phosphorus from sewage. Thus, an appreciable fraction of largely biological phosphorus is removed with the sludge. Domestic sewage contains such a high concentration of phosphorus from detergents and other sources that considerable phosphate ion remains in the effluent. However, some wastes, such as carbohydrate wastes from sugar refineries, are so deficient in phosphorus that supplementation of the waste with inorganic phosphorus is required for proper growth of the microorganisms degrading the wastes.

Under some sewage-plant operating conditions, much greater than normal phosphorus removal has been observed. In such plants, characterized by high dissolved oxygen and pH levels in the aeration tank, removal of 60–90% of the phosphorus has been attained, yielding two or three times the normal level of phosphorus in the sludge. In a conventionally operated aeration tank of an

[17] 1977. Prognosis on RO. *Environmental Science and Technology* 11: 1052–3.

TABLE 8.1 Chemical Precipitants for Phosphate and Their Products

Precipitant(s)	*Product*
$Ca(OH)_2$	$Ca_5OH(PO_4)_3$ (hydroxyapatite)
$Ca(OH)_2$ + NaF	$Ca_5F(PO_4)_3$ (fluorapatite)
$Al_2(SO_4)_3$	$AlPO_4$
$FeCl_3$	$FePO_4$
$MgSO_4$	$MgNH_4PO_4$

activated sludge plant, the CO_2 level is relatively high because of release of the gas by the degradation of organic material. A high CO_2 level results in a relatively low pH, due to the presence of carbonic acid. The aeration rate generally is not very high because oxygen transfer to water is more efficient at lower dissolved oxygen concentrations. Therefore, the aeration rate normally is not high enough to sweep out sufficient dissolved carbon dioxide to bring its concentration down to low levels. Thus, the pH generally is low enough that phosphate is maintained primarily in the form of the $H_2PO_4^-$ ion. However, at a higher rate of aeration in a relatively hard water, the CO_2 is swept out, the pH rises, and reactions such as the following occur:

$$5\ Ca^{2+} + 3\ HPO_4^{2-} + H_2O \rightarrow Ca_5OH(PO_4)_3(s) + 4\ H^+ \qquad (8.18.1)$$

The precipitated hydroxyapatite or other form of calcium phosphate is incorporated in the sludge floc. Reaction 8.18.1 is strongly hydrogen-ion dependent, and an increase in the hydrogen-ion concentration drives the equilibrium back to the left. Thus, under anaerobic conditions when the sludge medium becomes more acidic due to higher CO_2 levels, the calcium returns to solution.

Chemically speaking, phosphate is most commonly removed by precipitation. Some common precipitants and their products are shown in Table 8.1. Precipitation processes are capable of at least 90–95% phosphorus removal at reasonable cost.

Lime, $Ca(OH)_2$, is the chemical most commonly used for phosphorus removal (Reaction 8.18.1). It has the advantages of low cost and ease of regeneration. The efficiency with which phosphorus is removed by lime is not as high as would be predicted by the low solubility of hydroxyapatite. Some of the possible reasons for this are slow precipitation of $Ca_5OH(PO_4)_3$; formation of nonsettling colloids; precipitation of calcium as $CaCO_3$ in certain pH ranges; and the fact that phosphate may be present as condensed phosphates (polyphosphates) which form soluble complexes with calcium ion.

Phosphate can be removed from solution by adsorption on some solids, particularly activated alumina, Al_2O_3. Removals of up to 99.9% of orthophosphate have been achieved with this method.

8.19 NITROGEN REMOVAL

Next to phosphorus, nitrogen is the algal nutrient most commonly removed as part of advanced wastewater treatment. The techniques most often used for nitrogen removal are summarized in Table 8.2 (see page 210). Nitrogen in municipal wastewater generally is present as organic nitrogen or ammonia. Ammonia is the primary nitrogen product produced by most biological waste-

TABLE 8.2 Common Processes for the Removal of Nitrogen from Wastewater[1]

Process	*Principles and conditions*
Air stripping ammonia	Ammonium ion is the initial product of biodegradation of nitrogenous waste. It is removed by raising the pH to approximately 11 with lime and stripping ammonia gas from the water by air in a stripping tower. Scaling, icing, and air pollution are major disadvantages.
Ammonium ion exchange	This is an attractive alternative to air stripping, made possible by the development of clinoptilolite, a natural zeolite selective for ammonia: Na^+(clinoptilolite) + $NH_4^+ \rightarrow Na^+$ + NH_4^+(clinoptilolite). Regenerated with sodium or calcium salts.
Biosynthesis	The production of biomass in the sewage treatment system and its subsequent removal from the sewage effluent result in a net loss of nitrogen from the system.
Nitrification-denitrification	Several schemes are based on the conversion of ammonium nitrogen to nitrate under aerobic conditions, $2\ NH_4^+ + 3\ O_2 \xrightarrow{Nitrosomonas} 4\ H^+ + 2\ NO_2^- + 2\ H_2O$ $2\ NO_2^- + O_2 \xrightarrow{Nitrobacter} 2\ NO_3^-$ followed by production of elemental nitrogen (denitrification): $4\ NO_3^- + 5\ \{CH_2O\} + 4\ H^+ \xrightarrow[\text{bacteria}]{\text{denitrifying}} 2\ N_2(g) + 5\ CO_2(g) + 7\ H_2O$ Denitrification may be accomplished in an anaerobic activated sludge system or in an anaerobic column. Sometimes additional organic matter (methanol) is added.
Chlorination	Reaction of ammonium ion and hypochlorite (from chlorine) results in denitrification by chemical reactions: $NH_4^+ + HOCl \rightarrow NH_2Cl + H_2O + H^+$ $2\ NH_2Cl + HOCl \rightarrow N_2(g) + 3\ H^+ + 3\ Cl^- + H_2O$

[1]For details, see C. E. Adams, Jr., 1974, Removing nitrogen from waste water, *Environmental Science and Technology* 8: 696–701.

treatment processes. This is because it is expensive to aerate sewage sufficiently to oxidize the ammonia to nitrate through the action of nitrifying bacteria. If the activated sludge process is operated under conditions such that the nitrogen is maintained in the form of ammonia, the latter may be stripped in the form of NH_3 gas from the water by air. For ammonia stripping to work, the ammoniacal nitrogen must be converted to volatile NH_3, and this requires a pH substantially higher than the pK_a of the NH_4^+ ion. In practice, the pH is raised to approximately 11.5 by the addition of lime (which also serves to remove phosphate). The ammonia is stripped from the water by air in a forced-draft, countercurrent, air-stripping tower. Although ammonia removals exceeding 90% have been achieved, a number of problems remain. In climates where subfreezing temperatures occur, freezing of the water in the tower is a major problem. Furthermore, lime precipitates from the water, forming bothersome deposits on the tower surfaces. Questions may be raised about the advisability of venting ammonia directly to the atmosphere, since the substance is a potential air pollutant. The ammonia stripped from the water may well fall back on the watershed and be washed directly back into the body of water the advanced treatment plant was designed to protect!

Nitrification followed by denitrification is a promising technique for the removal of nitrogen from wastewater. The first step is an essentially complete conversion of ammonia and organic nitrogen to nitrate under strongly aerobic conditions, achieved by more extensive than normal aeration of the sewage:

$$NH_4^+ + 2\,O_2 \xrightarrow[\text{(nitrifying bacteria)}]{} NO_3^- + 2\,H^+ + H_2O \qquad (8.19.1)$$

The second step is the reduction of nitrate to nitrogen gas. This reaction is also bacterially catalyzed and requires a carbon source and a reducing agent such as methanol, CH_3OH.

$$6\,NO_3^- + 5\,CH_3OH + 6\,H^+ \xrightarrow[\text{bacteria}]{\text{denitrifying}} 3\,N_2(g) + 5\,CO_2 + 13\,H_2O \qquad (8.19.2)$$

The denitrification process may be carried out either in a tank or on a carbon column. In pilot plant operation, conversions of 95% of the ammonia to nitrate and 86% of the nitrate to nitrogen have been achieved.

8.20 REMOVAL OF HEAVY METALS FROM WASTEWATER

Heavy metals such as copper, cadmium, mercury, and lead are found in wastewaters from a number of industrial processes. Because of the toxicity of many heavy metals, their concentrations must be reduced to very low levels prior to release of the wastewater. A number of approaches are used in heavy-metals removal.

Lime treatment (Sections 8.4 and 8.18) removes heavy metals as insoluble hydroxides, basic salts, or coprecipitated with calcium carbonate or ferric hydroxide. This process does not completely remove mercury, cadmium, or lead, so their removal is aided by addition of sulfide (recall from Chapter 7 that most heavy metals are sulfide-seekers):

$$Cd^{2+} + S^{2-} \rightarrow CdS(s) \qquad (8.20.1)$$

Heavy chlorination is frequently necessary to break down metal-solubilizing ligands (see Chapter 4). Lime precipitation does not normally permit recovery of metals and is sometimes undesirable from the economic viewpoint.

Electrodeposition (reduction of metal ions to metal by electrons at an electrode), reverse osmosis (see Section 8.17), and ion exchange (see Section 8.16) are frequently employed for metal removal. Solvent extraction using organic-soluble chelating substances is also effective in removing many metals. **Cementation,** a process by which a metal deposits by reaction of its ion with a more readily oxidized metal, may be employed:

$$Cu^{2+} + Fe\,\text{(iron scrap)} \rightarrow Fe^{2+} + Cu \qquad (8.20.2)$$

Activated carbon adsorption effectively removes some metals from water at the part per million level. Sometimes a chelating agent is sorbed to the charcoal to increase metal removal.

The removal of trace metals from wastewater has been summarized ([18], Table 8.3). The second column in Table 8.3 shows the ranges of trace-metal levels found in a survey of wastewater from industrial plants in Michigan. These data provide an idea of values to be encountered in industrial wastewaters in other areas.

Even when not specifically designed for the removal of heavy metals, most waste-treatment processes remove appreciable quantities of the more troublesome heavy metals encountered in wastewater. The heavy-metal removal resulting from biological waste treatment is shown in the third column of Table 8.3. These metals accumulate in the sludge from biological treatment, so sludge disposal must be given careful consideration. Average metal contents of biological waste-treatment sludges from 33 biological treatment plants are given in the fourth column of Table 8.3.

Various physical-chemical treatment processes effectively remove heavy metals from wastewaters. One such treatment is lime precipitation followed by activated-carbon filtration. Activated-carbon filtration may also be preceded by treatment with ferric chloride to form a ferric hydroxide floc, which is an effective heavy-metals scavenger. Similarly, alum, which forms aluminum hydroxide, may be added prior to activated-carbon filtration.

It should be noted that the form of the heavy metal has a strong effect upon the efficiency of metal removal. For instance, chromium(VI) is normally more difficult to remove than chromium(III). Chelation may prevent metal removal by solubilizing metals (see Chapter 4).

In the past, removal of heavy metals has been largely a fringe benefit of wastewater-treatment processes. Currently, however, more consideration is being given to design and operating parameters that specifically enhance heavy-metals removal as part of wastewater treatment.

8.21 TOTAL WASTE-TREATMENT SYSTEMS

Figures 8.10 and 8.11 (p. 214) summarize some existing waste-treatment systems combining conventional activated-sludge waste treatment with tertiary treatment. Figure 8.10 is a schematic diagram of a **conventional tertiary treatment system**, so named because of the "conventional" conversion of nitrogen to NH_3, and its subsequent removal by air-stripping rather than by nitrification followed by denitrification [19]. Steps A, B, and C represent conventional sewage treatment by way of primary settling and activated-slude biodegradation. Lime, $Ca(OH)_2$, is added in step D to precipitate phosphate and, after the addition of CO_2, $CaCO_3$ and additional phosphate are precipitated in settler E. The calcium carbonate that

---

[18] Cohen, J.M. 1977. Trace metal removal by wastewater treatment. In *Technology Transfer*, January 1977, pp. 2–7. Washington, D.C.: U.S. Environmetal Protection Agency.

[19] 1970. "FWQA steps up teriary treatment study. *Environmental Science and Technology* 4: 550–1. Figures 8.10 and 8.11 are taken from this paper and are reprinted by permission of the American Chemical Society.

TABLE 8.3 Removal of Trace Metals by Wastewater Treatment Processes[1]

Metal	*Concentration range,[2] mg/L*	*% Removal by biological treatment*	*Concentration in digested sludge,[3] mg/kg*	*% Removal by lime precipitation–activated carbon*	*% Removal by ferric chloride–activated carbon*	*% Removal by alum–activated carbon*
cadmium	<0.008–0.142	20–45	31	99.6	98.6	55.2
chromium(III)	<0.020–0.700	40–80	1100	98.2	99.3	99.3
copper	<0.020–3.36	0–70	1230	90	96	98.3
mercury	<0.0002–0.044	20–75	7	91	99	98.3
nickel	<0.0020–8.80	15–40	410	99.5	37	37
lead	<0.050–1.27	50–90	830	99.4	99.1	96.6
zinc	<0.030–8.31	35–80	2780	76	94	28

[1]Data from Reference [18]

[2]In wastewater from plants in Michigan

[3]Average of sludge from 33 biological waste treatment plants, mg metal/kg dry sludge

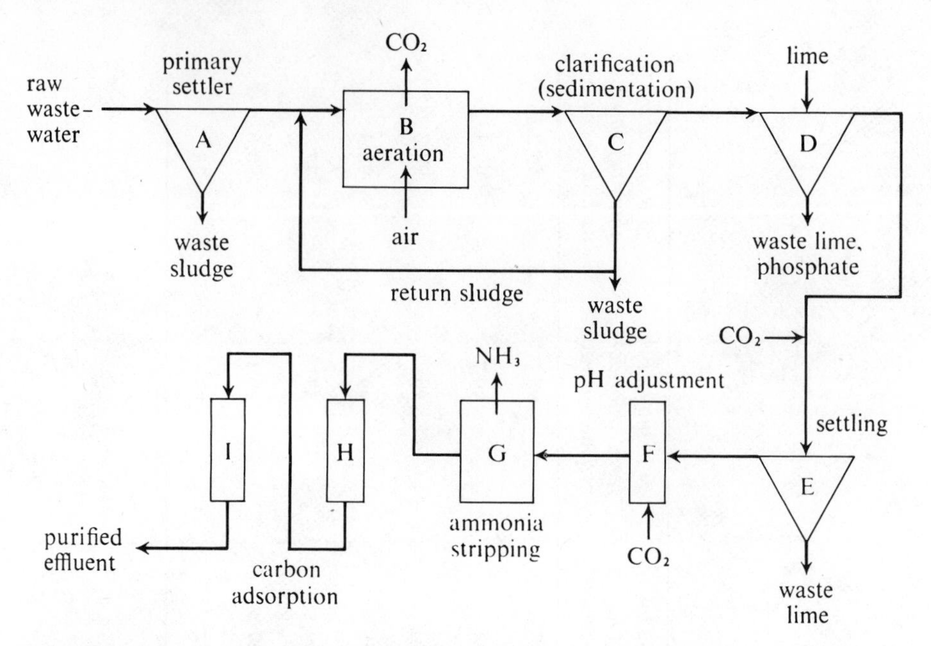

FIGURE 8.10 A schematic diagram showing conventional tertiary treatment of sewage.

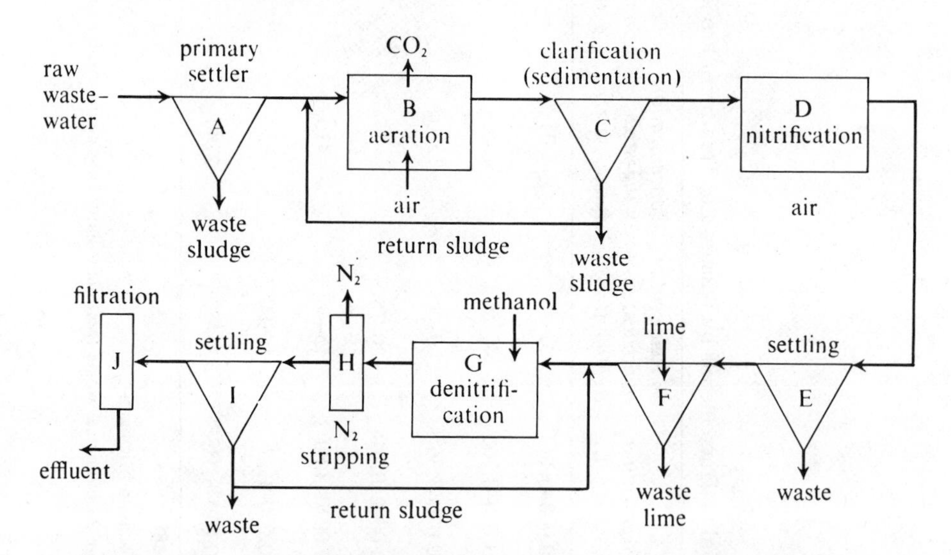

FIGURE 8.11 A schematic diagram showing a conventional treatment-nitrification-denitrification system for the treatment of sewage and wastewater.

is precipitated is recalcined and converted to $Ca(OH)_2$,

$$CaCO_3 \rightarrow CaO + CO_2(g) \tag{8.21.1}$$

$$CaO + H_2O \rightarrow Ca(OH)_2 \tag{8.21.2}$$

and both the $Ca(OH)_2$ and the CO_2 are reused in the treatment process. The pH is adjusted in step F and ammonia is stripped from the water by air in step G. The water is finally "polished" to remove residual organic material on activated carbon in columns I and H.

A slightly different approach, called **conventional treatment–nitrification–denitrification**, is illustrated in Fig. 8.11. Steps A–C are identical to those described in Figure 8.10. In step D, the secondary effluent is saturated with air until essentially all of the nitrogen is converted to nitrate ion through biological oxidation. Additional removal of suspended material occurs in settler E, and lime is added to remove phosphate in step F. A low level of methanol is then added to the water to serve as a carbon and energy source for the denitrifying bacteria growing anaerobically in carbon column G (see Reaction 8.19.2). The nitrogen gas produced by denitrification is stripped off in step H and additional particulate material and biomass are settled out in step I. A final filtration step, J, removes residual particulate matter and produces highly purified water for reuse or release to the environment.

8.22 PHYSICAL-CHEMICAL TREATMENT OF WASTEWATER

Complete physical-chemical wastewater-treatment systems offer both advantages and disadvatages relative to biological treatment systems. A number of physical-chemical treatment systems are currently coming into operation. The capital costs of these facilities are generally less than that of biological treatment facilities, and they usually require less land. They are better able to cope with toxic materials and overloads. However, they require careful operator control.

Basically, a physical-chemical treatment process involves the following steps [20]:

1. Removal of scum and solid objects.
2. Clarification, generally with addition of a coagulant, and frequently with the addition of other chemicals (such as lime for phosphorus removal).
3. Filtration to remove filterable solids.
4. Activated carbon adsorption.
5. Disinfection.

During the early 1970s, it appeared likely that physical-chemical treatment would largely replace biological treatment. However, dramatically higher chemical and energy costs since then have changed the picture appreciably.

The basic steps of a complete physical-chemical wastewater-treatment facility are shown in Figure 8.12 (p. 216). This design is based on a new 50-million-gallon-per-day Cleveland, Ohio waste-treatment plant.

The first physical-chemical wastewater-treatment plant for a municipality in the United States was the Rosemount, Minnesota plant. Its general features are similar to those shown in Figure 8.12, except that a cation exchanger composed of an ammonia-selective zeolite is used to remove ammonium ion at the end of the treatment process. Approximate influent and effluent water parameters for this plant are given in Table 8.4 (see page 217).

---

[20] 1972. Debugging physical-chemical treatment. *Environmental Science and Technology* 6: 984–5.

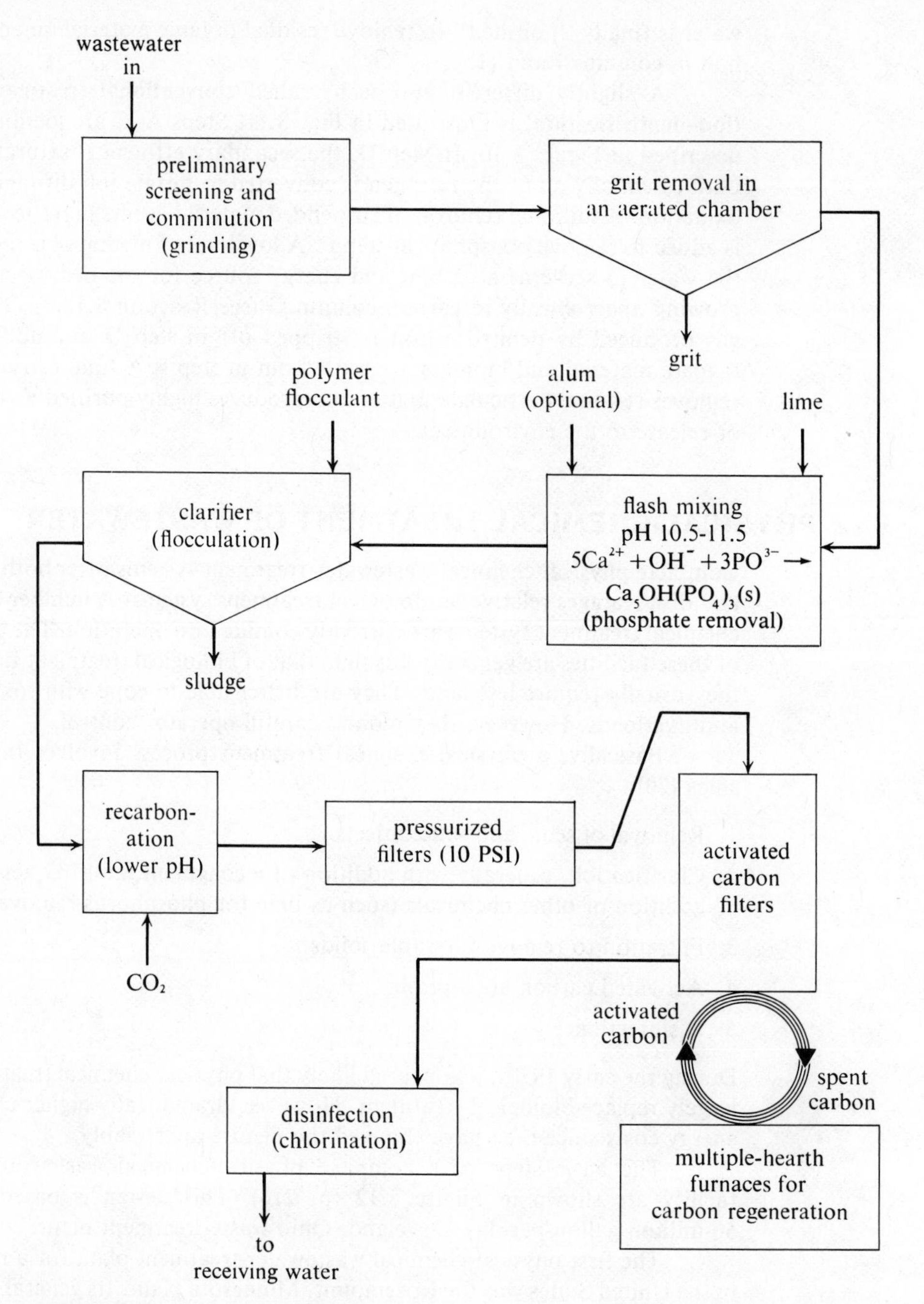

FIGURE 8.12 Major components of a complete physical-chemical treatment facility for wastewater.

TABLE 8.4 Approximate Values of Wastewater Parameters from the Rosemount, Minnesota Physical-Chemical Treatment Plant[1]

Parameter	Concentration of value influent	effluent
suspended solids	300 mg/L	10 mg/L
pH	6.5–8.5	8.5
biological oxygen demand	300 mg/L	10 mg/L
chemical oxygen demand	200–400 mg/L	10 mg/L
N (ammonia)	15–35 mg/L	1 mg/L
P (total)	5–15 mg/L	1 mg/L

[1]Adapted from 1973, Regional agency starts up physical-chemical treatment plant, *Environmental Science and Technology,* **7:** 804–5 (1973).

8.23 SLUDGE

Perhaps the most pressing water treatment problem at this time has to do with sludge collected or produced during water treatment. Finding a safe place to put the sludge or a use for it has proven troublesome, and the problem is aggravated by the growing numbers of water treatment systems.

Some sludge is present in wastewater and may be collected from it. Such sludge includes human wastes, garbage grindings, organic wastes and inorganic silt and grit from storm water runoff, and organic and inorganic wastes from commercial and industrial sources. There are two major kinds of sludge generated in a waste treatment plant. The first of these is organic sludge from activated sludge, trickling filter, or rotating biological reactors. The second is inorganic sludge from the addition of chemicals, such as in phosphorus removal (see Section 8.18).

Most commonly, sewage sludge is subjected to anaerobic digestion in a digester designed to allow bacterial action to occur in the absence of air. This reduces the mass and volume of sludge and ideally results in the formation of a stabilized humus. Disease agents are also destroyed in the process.

Following digestion, sludge is generally conditioned and thickened to concentrate and stabilize it and make it more dewaterable. Relatively inexpensive processes, such as gravity thickening, may be employed to get the moisture content down to about 95% [21]. Sludge may be further conditioned chemically by the addition of iron or aluminum salts, lime, or polymers.

Sludge dewatering is employed to convert the sludge from an essentially liquid material to a damp solid containing not more than about 85% water. This may be accomplished on sludge drying beds consisting of layers of sand and gravel. Mechanical devices may also be employed, including vacuum filtration, centrifugation, and filter presses. Heat may be used to aid the drying process.

Some of the alternatives for the ultimate disposal of sludge include land spreading, ocean dumping, and incineration. Each of these choices has disadvan-

[21] Fronk-Leist, C.A.; Villiers, R.V.; and Farrell, J.B. 1980. Optimization of Blue Plains dewatering technology. *Environmental Research Brief.* Cincinnati, Ohio: U.S. Environmental Protection Agency.

TABLE 8.5 Undesirable Components Commonly Found in Sewage Sludge from a Primary Settler[1]

Component	*Level, ppm by dry weight unless otherwise stated*
Organics	
PCB	0–105
DDT	0–1 (found much less frequently now)
DDD	0–0.5 (found much less frequently now)
dieldrin	0–2
aldrin	0–16
phenol	sometimes encountered
Heavy metals	
cadmium	0–100
lead	up to 400
mercury	3–15
chromium	up to 700
copper	80–1000
nickel	25–400
zinc	300–2000 (common deterrent to use of sludge as a soil conditioner due to its toxicity to plants)
Pathogenic microorganisms	
human viruses	generally present
salmonella (in raw sludge)	500 viable cells/100mL
salmonela (in digested sludge)	30 viable cells/100mL
fecal coliforms (raw sludge)	1×10^7 viable cells/100 mL
fecal coliforms (digested sludge)	4×10^5 viable cells/100mL

[1]Estimates based in part on 1974, Containing the flow of sewage sludge, *Environmental Science and Technology* 8: 702–3.

tages, such as the presence of toxic substances in sludge spread on land, or the high fuel cost of incineration.

Some of the undesirable components found in sewage sludge are shown in Table 8.5. This table refers to sludge from a primary settler, although many of the same components are found in secondary settler sludges.

Rich in nutrients, waste sewage sludge contains around 5% N, 3% P, and 0.5% K on a dry-weight basis and can be used to fertilize and condition soil. The humic material in the sludge improves the physical properties and cation-exchange capacity of the soil. Among the factors limiting this application of sludge are excess nitrogen pollution of runoff water and ground water, survival of pathogens, and the presence of heavy metals in the sludge.

Possible accumulation of heavy metals is of the greatest concern insofar as the use of sludge on cropland is concerned. Sewage sludge is an efficient heavy metals scavenger. On a dry basis, sludge samples from industrial cities have shown levels of up to 9,000 ppm zinc, 6,000 ppm copper, 600 ppm nickel, and up to 800 ppm cadmium! These and other metals tend to remain immobilized in soil by chelation with organic matter, adsorption on clay minerals, and precipitation as insoluble compounds, such as oxides or carbonates. However, increased application of sludge on cropland has caused distinctly elevated levels of zinc and cadmium in both leaves and grain of corn. Therefore, caution has been advised in heavy or prolonged application of sewage sludge to soil. The problem of heavy metals in sewage sludge is one of the many reasons for not allowing mixture of wastes to occur prior to treatment. Sludge does, however, contain nutrients which should not be wasted, given the possibility of eventual fertilizer shortages [22]. Prior control of heavy metal contamination from industrial sources should greatly reduce the heavy metal content of sludge and enable it to be used more extensively on soil.

As increasing problem in sewage treatment arises from **sludge sidestreams**. These consist of water removed from sludge by various treatment processes [23]. Sewage treatment processes can be divided into **mainsteam treatment processes** (primary clarification, trickling filter, activated sludge, and rotating biological reactor) and **sidestream processes**. During sidestream treatment sludge is dewatered, degraded, and disinfected by a variety of processes, including gravity thickening, dissolved air flotation, anaerobic digestion, aerobic digestion, vacuum filtration, centrifugation, belt-filter press filtration, sand-drying-bed treatment, sludge-lagoon settling, wet air oxidation, pressure filtration, and Purifax treatment. Each of these produces a liquid by-product sidestream which is circulated back to the mainstream. These add to the biochemical oxygen demand and suspended solids of the mainstream.

A variety of chemical sludges are produced by various water treatment and industrial processes [24]. Among the most abundant of such sludges is alum sludge produced by the hydrolysis of Al(III) salts used in the treatment of water, which creates gelatinous aluminum hydroxide:

$$Al^{3+}(aq) + 3\ OH^-(aq) \rightarrow Al(OH)_3(s) \qquad (8.23.1)$$

Alum sludges normally are 98% or more water and are very difficult to dewater.

Both iron(II) and iron(III) compounds are used for the precipitation of impurities from wastewater via the precipitation of $Fe(OH)_3$. The sludge contains $Fe(OH)_3$ in the form of soft, fluffy precipitates that are difficult to dewater beyond 10 or 12% solids.

The addition of either lime, $Ca(OH)_2$, or quicklime, CaO, to water is used to raise the pH to about 11.5 and cause the precipitation of $CaCO_3$, along with metal hydroxides and phosphates. Calcium carbonate is readily recovered from lime sludges and can be recalcined to produce CaO, which can be recycled through the system.

Metal hydroxide sludges are produced in the removal of metals such as lead, chromium, nickel, and zinc from wastewater by raising the pH to such a

---

[22] Garcia, W.J.; Blessin, C.W.; Inglett, G.E.; Kwolek, W.F.; Carlisle, J.N.; Hughes, L.N.; and Meister, J.F. 1981. Metal accumulation and crop yield for a variety of edible crops grown in diverse soil media amended with sewage sludge. *Environmental Science and Technology* 15: 793–804.

[23] Ball, R.; Harris, M.; and Deeny, K. 1982. *Evaluation and control of sidestreams generated in publicly owned treatment works.* EPA-600/S2-82-016. Cincinnati, Ohio: Municipal Environmental Research Laboratory, U.S. Environmental Protection Agency.

[24] Vesilind, P.A. 1979. Chemical sludges. In *Treatment and disposal of wastewater sludges,* pp. 235–49. Ann Arbor, Mich.: Ann Arbor Science Publishers, Inc.

level that the corresponding hydroxides or hydrated metal oxides are precipitated. The disposal of these sludges is a substantial problem because of their toxic heavy metal content. Reclamation of the metals is an attractive alternative for these sludges.

Pathogenic (disease-causing) microorganisms may persist in the sludge left from the treatment of sewage. Many of these organisms present potential health hazards, and there is risk of public exposure when the sludge is applied to soil [25]. Therefore, it is necessary both to be aware of pathogenic microorganisms in municipal wastewater treatment sludge and to find a means of reducing the hazards caused by their presence.

The most significant organisms in municipal sewage sludge include the following: (1) **indicators**, including fecal and total coliform; (2) **pathogenic bacteria**, including *Salmonellae* and *Shigellae*; (3) **enteric (intestinal) viruses**, including *Enterovirus* and *Poliovirus*, and (4) **parasites**, such as *Entamoeba histolytica* and *Ascaris lumbricoides*.

Several ways are recommended to significantly reduce levels of pathogens in sewage sludge. **Aerobic digestion** involves aerobic agitation of the sludge for periods of 40 to 60 days (longer times are employed with low sludge temperatures). **Air drying** involves draining and/or drying of the liquid sludge for at least three months in a layer 20–25 cm thick. This operation may be performed on underdrained sand beds or in basins. **Anaerobic digestion** involves maintenance of the sludge in an anaerobic state for periods of time ranging from 60 days at 20 °C to 15 days at temperatures exceeding 35 °C. **Composting** involves mixing dewatered sludge cake with bulking agents subject to decay, such as wood chips or shredded municipal refuse, and allowing the action of bacteria to promote decay at temperatures ranging up to 45–65 °C. The higher temperatures tend to kill pathogenic bacteria. Finally, pathogenic organisms may be destroyed by **lime stabilization** in which sufficient lime is added to raise the pH of the sludge 12 or more.

8.24 WATER RENOVATION BY SOIL

Virtually all of the materials that waste-treatment processes are designed to eliminate may be absorbed by soil or degraded in soil. In fact, most of these materials are essential for soil fertility. Wastewater may provide the water that is essential to plant growth, in addition to the nutrients—phosphorus, nitrogen and potassium—usually provided by fertilizers. Wastewater also contains essential trace elements and vitamins. Stretching the point a bit, the degradation of organic wastes provides the CO_2 essential for photosynthetic production of plant biomass. Most organic matter is readily degraded in soil; in addition, soil is a good cation exchanger and is very effective in removing such ions as ammonium ion from water. Thus, in principle, soil constitutes an excellent treatment system —primary, secondary and tertiary—for water.

Early civilizations, such as the Chinese, used human organic wastes to increase soil fertility, and the practice continues today. The ability of soil to purify

[25] 1981. *Density levels of pathogenic organisms in municipal wastewater sludge—a literature review.* EPA-600/S2-81-170. Cincinnati, Ohio: Municipal Environmental Research Laboratory, U.S. Environmental Protection Agency.

water was noted well over a century ago. In 1850 and 1852, J. Thomas Way, a consulting chemist to the Royal Agricultural Society in England, presented two papers to the Society entitled "Power of Soils to Absorb Manure." Mr. Way's experiments showed that soil is an ion exchanger. Much practical and theoretical information on the ion-exchange process resulted from this work.

Experiments involving the direct application of wastewater to soil have shown appreciable increases in soil productivity. A process called **overland flow** has been described [26] in which wastewater containing pulverized solids is allowed to trickle over sloping soil. Suspended solids, BOD, and nutrients are largely removed. This method is most applicable to rural communities in relatively warm climates. Approximately one acre of land is required to handle the sewage from 200 persons.

SUPPLEMENTARY REFERENCES

Antonie, R.V. 1976. *Fixed biological surfaces—treatment of wastes.* West Palm Beach, Fla.: CRC Press, Inc.

Baumann, D.D., and Dworkin, D.M. 1978. *Planning for water reuse.* Chicago: Maaroufa Press.

Benjes, H.H., Jr. 1979. *Handbook of biological wastewater treatment.* New York: Garland STPM Press (Garland Publishing, Inc.).

Callely, A.G.; Forster, C.F.; and Stafford, D.A., eds. 1977. *Treatment of industrial effluents.* New York: Halsted (John Wiley and Sons, Inc.).

Cherry, K.F. 1982. *Plating waste treatment.* Ann Arbor, Mich.: Ann Arbor Science Publishers, Inc.

Clark, J.W.; Viessman, W., Jr.; and Hammer, M.J. 1977. *Water supply and pollution control.* New York: Harper and Row, Publishers, Inc.

1978. *Cleaning our environment—a chemical perspective.* Washington, D.C.: American Chemical Society.

Cooper, W.J., ed. 1981. *Chemistry in water reuse.* Vol. 1. Ann Arbor, Mich.: Ann Arbor Science Publishers, Inc.

Culp, R.L.; Wesner, G.M.; and Culp, G.L. 1978. *Handbook of advanced wastewater treatment.* 2nd ed. New York: Van Nostrand Reinhold Co.

DeRenzo, D.J., ed. 1980. *Biodegradation techniques for industrial organic wastes.* Park Ridge, N.J.: Noyes Data Corp.

Devik, O., ed. 1976. *Harvesting polluted water.* New York: Plenum Publishing Corp.

D'Itri, F.M., ed. 1982. *Land treatment of municipal wastewater.* Ann Arbor, Mich.: Ann Arbor Science Publishers, Inc.

D'Itri, F.M.; Martinez, J.A.; and Lãmbarri, M.A. 1981. *Municipal wastewater in agriculture.* New York: Academic Press, Inc.

Eikum, A.S., and Seabloom, R.W., eds. 1982. *Alternative wastewater treatment.* Hingham, Ma.: D. Reidel Publishing Co., Inc.

Gillies, M.T., ed. 1981. *Potable water from wastewater.* Park Ridge, N.J.: Noyes Data Corp.

---

[26] Thomas, R.E.; Jackson, K.; and Penrod, L. 1974. *Feasibility of overland flow for treatment of raw domestic wastewater.* EPA-660/2-74-087. Washington, D.C.: Environmental Protection Agency.

Hammer, M.J. 1975. *Water and waste-water technology.* New York: John Wiley and Sons, Inc.

1978. *Health aspects of water recharge.* Huntington, N.Y.: Water Information Center.

Hudson, H.E., Jr. 1981. *Water clarification processes.* New York: Van Nostrand Reinhold Co.

Johnson, J.D., ed. 1975. *Disinfection: water and wastewater.* Ann Arbor, Mich.: Ann Arbor Science Publishers, Inc.

Jolley, R.L.; Gorchey, H.; and Hamilton, D.H., Jr., eds. 1978. *Water chlorination: environmental impact and health effects.* Vol. 2. Ann Arbor, Mich.: Ann Arbor Science Publishers, Inc.

Kitagishi, K., and Yamane, I., eds. 1981. *Heavy metal pollution in soils in Japan.* Tokyo: Japan Scientific Societies Press.

Lehr, J.H. 1980. *Domestic water treatment.* New York: McGraw-Hill Book Co.

Loehr, R.C.; Jewell, W.J.; Novak, J.D.; Clarkson, W.W.; and Friedman, G.S. 1979. *Land application of wastes.* New York: Van Nostrand Reinhold Co.

Mattock, G., ed. 1978. *New processes of waste water treatment and recovery.* New York: Halsted (John Wiley and Sons, Inc.).

Middlebrooks, E.J. 1982. *Water reuse.* Ann Arbor, Mich.: Ann Arbor Science Publishers, Inc.

National Academy of Sciences. 1982. *Drinking water and health.* Washington, D.C.: National Academy Press.

Owen, W.F. 1982. *Energy in wastewater treatment.* Englewood Cliffs, N.J.: Prentice-Hall, Inc.

1979. *Proceedings of the Industrial Waste Conference.* (Conference held at Purdue University.) Ann Arbor, Mich.: Ann Arbor Science Publishers, Inc.

Ramalho, R.S. 1977. *Introduction to wastewater treatment processes.* New York: Academic Press, Inc.

Sandbach, F. 1982. *Principles of pollution control.* New York: Longman, Inc.

Scott, J.S., and Smith, P.G. 1981. *Dictionary of waste and water treatment.* Woburn, Mass.: Butterworth, Inc.

Stuckey, D., and Hamza, A., eds. 1982. *Management of industrial wastewater in developing nations.* Elmsford, N.Y.: Pergamon Press, Inc.

Stumm, W., and Morgan, J.J. 1981. *Aquatic chemistry.* 2nd ed. New York: Wiley-Interscience.

Suffet, I.H., and McGuire, M.J., eds. 1980. *Activated carbon adsorption of organics from the aqueous phase.* Ann Arbor, Mich.: Ann Arbor Science Publishers, Inc.

Vernick, A.S., and Walker, E.C., eds. 1981. *Handbook of wastewater treatment processes.* New York: Marcel Dekker, Inc.

1977. *Water purification in the EEC.* Elmsford, N.Y.: Pergamon Press, Inc.

1971. *Water quality and treatment.* 3d ed. New York: McGraw-Hill Book Co.

Watson, M.R. 1973. *Pollution control in metal finishing.* Park Ridge, N.J.: Noyes Data Corp.

White, G.C. 1978. *Disinfection of wastewater and water for reuse.* New York: Van Nostrand Reinhold Co.

Winkler, M.A. 1981. *Biological treatment of wastewater.* New York: John Wiley and Sons, Inc.

QUESTIONS AND PROBLEMS

1. How many moles of NTA should be added to 1000 liters of water having a pH of 9 and containing CO_3^{2-} at $1.00 \times 10^{-4}M$ to prevent precipitation of $CaCO_3$? Assume a total calcium level of 40 mg/L.
2. What is the purpose of the return sludge step in the activated-sludge process?
3. What are the two processes by which the activated-sludge process removes soluble carbonaceous material from sewage?
4. Why might hard water be desirable as a medium if phosphorus is to be removed by an activated-sludge plant operated under conditions of high aeration?
5. How does reverse osmosis differ from a simple sieve separation or ultrafiltration process?
6. How many liters of methanol would be required daily to remove the nitrogen from a 200,000-L/day waste-treatment plant producing an effluent containing 50 mg/L of nitrogen? Assume that the nitrogen has been converted to NO_3^- in the plant. The denitrifying reaction is Reaction 8.19.2.
7. Discuss some of the advantages of physical-chemical treatment of sewage as opposed to biological wastewater treatment. What are some disadvantages?
8. Why is recarbonation necessary when water is softened by the lime-soda process?
9. Assume that a waste contains 300 mg/L of biodegradable $\{CH_2O\}$ and is processed through a 200,000-L/day sewage-treatment plant which converts 40% of the waste to CO_2 and H_2O. Calculate the volume of air (at 25°, 1 atm) required for this conversion. Assume that the O_2 is transferred to the water with 20% efficiency.
10. If all of the $\{CH_2O\}$ in the plant described in Question 9 could be converted to methane by anaerobic digestion, how many liters of methane (STP) could be produced daily?
11. Assuming that aeration of water does not result in the precipitation of calcium carbonate, which of the following would not be removed by aeration: hydrogen sulfide; carbon dioxide; volatile odorous bacterial metabolites; alkalinity; iron.
12. In which of the following water supplies would moderately high water hardness be most detrimental: municipal water; irrigation water; boiler feed-water; drinking water (in regard to potential toxicity).
13. An *increase* in which of the following *decreases* the rate of oxidation of iron(II) to iron(III) in water? [Fe(II)]; pH; $[H^+]$; $[O_2]$; $[OH^-]$.
14. A wastewater containing dissolved Cu^{2+} ion is to be treated to remove copper. Which of the following processes would *not* remove copper in an insoluble form: lime precipitation; cementation; treatment with NTA; ion exchange; reaction with metallic Fe.
15. Match each water contaminant in the left column with its preferred method of removal in the right column.

(a) Mn^{2+}	(1) Activated carbon
(b) Ca^{2+} and HCO_3^-	(2) Raise pH by addition of Na_2CO_3
(c) Trihalomethane compounds	(3) Addition of lime
(d) Mg^{2+}	(4) Oxidation

16. A cementation reaction employs iron to remove Cd^{2+} present at a level of 350mg/L from a wastewater stream. Given that the atomic weight of Cd is 112.4 and that of Fe is 55.8, how may kg of Fe are consumed in removing all the Cd from 4.50×10^6 liters of water?

CHAPTER 9
ENVIRONMENTAL CHEMICAL ANALYSIS

9.1 THE ROLE AND IMPORTANCE OF ENVIRONMENTAL CHEMICAL ANALYSIS

Our understanding of the environment can only be as good as our knowledge of the identities and quantities of pollutants and other chemical species in water, air, soil, and biological systems. Therefore, proven, state-of-the-art techniques of chemical analysis, properly employed, are essential to environmental chemistry. We are witnessing a very exciting period in the evolution of analytical chemistry, characterized by the development of new and improved analysis techniques that enable detection of much lower levels of chemical species and a vastly increased output of data per unit time. These developments have caused some problems. Because of the lower detection limits of some instruments, it is now possible to detect quantities of pollutants that would have escaped detection previously, resulting in difficult questions regarding the setting of maximum allowable limits of various pollutants. The increased output of data from automated instruments has in many cases overwhelmed our capacity to assimilate and understand it.

Challenging problems still remain in developing and utilizing techniques of environmental chemical analysis. Not the least of these problems is knowing which species should be measured, or even whether or not an analysis should be performed at all. The quality and choice of analyses is much more important than the number of analyses performed. Indeed, it is possible that too many analyses of environmental samples are performed, whereas fewer, more carefully planned analyses would yield more useful information.

This chapter covers the major aspects of environmental chemical analysis. Many techniques are common to water, air, soil, and biological sample analyses, and the most important of these techniques are covered in this chapter. Water analysis is discussed specifically. Later chapters refer back to this material as background for the analysis of nonaquatic samples; for example, Chapter 16 on atmospheric monitoring relies heavily on this chapter.

9.2 TITRATION METHODS

A number of analytical procedures for water analysis, as well as some for air analysis, are based upon **titration**, either manual or automated. Some of the titration procedures used are discussed in this section.

Acidity (see Section 2.13) is determined simply by titration of hydrogen ion with base. Titration to the methyl orange endpoint (pH 4.5) yields the "free acidity" due to strong acids (HCl, H_2SO_4). Carbon dioxide does not, of course, appear in this category. Titration to the phenolphthalein endpoint, pH 8.3, yields total acidity and accounts for all acids except those weaker than HCO_3^-. Where metal ions may be present, the sample is first oxidized with hydrogen peroxide to convert the metal to a more acidic form, such as may occur through atmospheric oxidation. The importance of this step is illustrated by ferrous ion, which is relatively nonacidic but is oxidizable through natural processes to acidic iron(III). The preferred titration practice now is to titrate with 0.02*N* NaOH to the desired pH as indicated by a pH meter.

Alkalinity (Section 2.12) is generally determined by titration with 0.02*N* H_2SO_4 to pH 8.3 (phenolphthalein endpoint) or pH 4.5 (methyl orange endpoint). Titration to pH 8.3 neutralizes bases as strong as, or stronger than, carbonate ion,

$$CO_3^{2-} + H^+ \rightarrow HCO_3^- \quad (9.2.1)$$

whereas titration to pH 4.5 protonates bases weaker than CO_3^{2-} but as strong as, or stronger than, HCO_3^-:

$$HCO_3^- + H^+ \rightarrow H_2O + CO_2(g) \quad (9.2.2)$$

Titration to the lower pH yields total alkalinity.

Free carbon dioxide in water is determined by titration with standard sodium hydroxide or sodium carbonate from the methyl orange to the phenolphthalein endpoint, pH 8.3. This corresponds to conversion of the carbon dioxide to bicarbonate ion. Sorption of carbon dioxide by a base followed by titration of the bicarbonate ion product with acid is a feasible, though no longer the best, method for the determination of CO_2 in air. This method has been used to measure atmospheric carbon dioxide levels in Paris since 1891.

Chloride ion in various environmental samples is determined by titration with silver nitrate,

$$Ag^+ + Cl^- \rightarrow AgCl(s) \quad (9.2.3)$$

and the endpoint may be indicated by several methods. Among these is the classic **Mohr method**, dating back to 1856, in which the endpoint is shown by formation of a red solid, Ag_2CrO_4. Potentiometry is also quite useful in indicating this endpoint.

The ions involved in water hardness, a measure of the total concentration of calcium and magnesium in water, are readily titrated with a solution of EDTA, a chelating agent discussed in Section 4.7. The solution is buffered at pH 10.0 to maintain the EDTA in a relatively nonprotonated form. A higher pH might result in the precipitation of $CaCO_3$ or $Mg(OH)_2$. The titration reaction is

$$Ca^{2+} \text{ (or } Mg^{2+}) + Y^{4-} \rightarrow CaY^{2-} \text{ (or } MgY^{2-}) \quad (9.2.4)$$

Since calcium forms more stable complexes with EDTA than does magnesium, the calcium complex forms preferentially. Eriochrome Black T is used as an indicator, and it requires the presence of magnesium, with which it forms a wine-red complex. When liberated in the free form by complexation of the magnesium with EDTA, Eriochrome Black T forms a blue solution. In order to ensure that sufficient magnesium is present in the solution for observation of the endpoint reaction, a small quantity of magnesium-EDTA complex is generally added. Because of the presence of equal quantities of magnesium and complexing agent in the magnesium-EDTA complex, the volume of EDTA required for the titration is not affected. Hardness also may be calculated as the total of magnesium and calcium concentrations determined by *atomic absorption analysis* (see Section 9.4).

Oxygen is determined in water by the **Winkler test**, a titration method. There are many variations of the basic Winkler test which help to avoid interference from species such as nitrite ion or ferrous ion. Only the main features of the analysis are discussed here.

The first reaction in the Winkler test is the oxidation of manganese(II) to manganese(IV) by oxygen in a basic medium:

$$Mn^{2+} + 2\ OH^- + \frac{1}{2}O_2 \rightarrow MnO_2(s) + H_2O \qquad (9.2.5)$$

Acidification of the brown hydrated MnO_2 in the presence of I^- releases free I_2:

$$MnO_2(s) + 2\ I^- + 4\ H^+ \rightarrow Mn^{2+} + I_2 + 2\ H_2O \qquad (9.2.6)$$

The liberated iodine (actually present as the I_3^- complex) is titrated with standard thiosulfate, using starch as an endpoint indicator:

$$I_2 + 2\ S_2O_3^{2-} \rightarrow S_4O_6^{2-} + 2\ I^- \qquad (9.2.7)$$

A back calculation from the amount of thiosulfate required yields the original quantity of dissolved oxygen (DO) present.

Biochemical oxygen demand, BOD (see Section 7.15), is determined by adding a microbial "seed" to the diluted sample, saturating with air, incubating for five days, and determining the oxygen remaining. The results are calculated to show BOD as mg/L of O_2. A BOD of 80 mg/L, for example, means that biodegradation of the organic matter in a liter of the sample would consume 80 mg of oxygen. The determination of a related parameter, chemical oxygen demand (COD), is discussed in Section 7.15.

Sulfide concentration may be determined by titration. In sewage and in most natural waters it is necessary to free the sulfide from insoluble salts as H_2S and collect it prior to analysis. The solution to be analyzed is acidified with H_2SO_4, which converts HS^- and most sulfides to volatile H_2S:

$$HS^- + H^+ \rightarrow H_2S(g) \qquad (9.2.8)$$

The volatile product is collected in a zinc acetate solution:

$$H_2S + Zn^{2+} + 2\ C_2H_3O_2^- \rightarrow ZnS(s) + 2\ HC_2H_3O_2 \qquad (9.2.9)$$

Following the collection of hydrogen sulfide as zinc sulfide, the collecting solution is acidified with HCl, and a standardized solution of iodine in KI is added. Some of the iodine is reduced to iodide by the sulfide, and the excess is titrated with standard sodium thiosulfate, $Na_2S_2O_3$.

Sulfite concentration is determined by titration with iodide-iodate titrant. This reagent is prepared by reacting a standard quantity of potassium iodate, KIO_3, with excess potassium iodide to produce a solution of triiodide ion, I_3^-.

The sulfite is oxidized by iodine(0) in an acidic medium:

$$I_3^- + SO_3^{2-} + H_2O \rightarrow 3\,I^- + SO_4^{2-} + 2\,H^+ \quad (9.2.10)$$

Starch indicator is added, and the equivalence point is indicated by the appearance of the starch-iodine color.

9.3 ABSORPTION SPECTROPHOTOMETRY

Absorption spectrophotometry of light-absorbing species in solution, called **colorimetry** if visible light is absorbed, remains an important method of analysis for many water and some air pollutants. Basically, absorption spectrophotometry consists of measuring the amount of monochromatic light passing through a light-absorbing solution as compared to the amount passing through a blank solution containing everything in the medium but the sought-for constituent. The transmittance (I) of the blank is set at 100% and the $\%T$ of the unknown solution may range from 0 to 100, depending upon sample concentration. The absorbance (A) is defined as

$$A = \log \frac{100}{\%T} \quad (9.3.1)$$

and the relationship between A and the concentration (C) of the absorbing substance is given by Beer's law:

$$A = abC \quad (9.3.2)$$

In the Beer's law relationship, a is the **absorptivity**, a wavelength-dependent parameter characteristic of the absorbing substance; b is the path length of the light through the absorbing solution; and C is the concentration of the absorbing substance. Normally, b is given in centimeters. If C is expressed as mole/liter, a is in units of liter $\times$ mole^{-1} $\times$ cm^{-1}. A linear relationship between A and C at constant path length indicates adherence to Beer's law. In many cases, analyses may be performed even when Beer's law is not obeyed, if a suitable calibration curve is prepared.

Only a few substances (such as MnO_4^-) absorb visible light strongly enough to be analyzed directly. Therefore, a color-developing step usually is required in which the sought-for substance reacts to form a colored species. Often a colored species is extracted into a nonaqueous solvent to provide a more intense color and a more concentrated solution. It is beyond the scope of this book to describe the principles of absorption spectrophotometry in detail. The reader should refer to any instrumental analysis or advanced quantitative analysis textbook for more information.

Oxygen in water may be determined colorimetrically by the oxidation of several dyes, including indigo carmine, safranine, and methylene blue, to dark-colored forms. The changes in structure of indigo carmine upon oxidation are shown in Reaction 9.3.3. This reaction has been used as the basis of a colorimetric method for the analysis of oxygen in water [1]. Special precautions must be taken to prevent the oxidation of the highly reactive leuco form of the indigo carmine dye prior to analysis. After mixing with water, the degree of formation

---

[1] Gilbert, T.W.; Behymer, T.D.; and Castaneda, H.B. 1982. Determination of dissolved oxygen in natural and wastewaters. *American Laboratory* March 1982.

of the oxidized dark-blue form of the dye may be used as a measure of the oxygen content of the water.

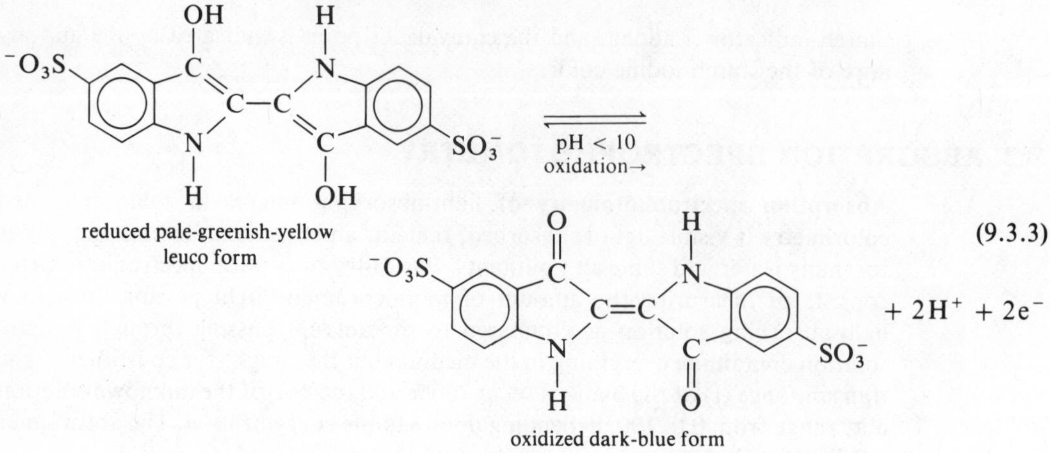

$$\text{reduced pale-greenish-yellow leuco form} \underset{\text{oxidation}\rightarrow}{\overset{\text{pH} < 10}{\rightleftharpoons}} \text{oxidized dark-blue form} + 2H^+ + 2e^- \qquad (9.3.3)$$

(Structures: leuco form, ^-O_3S–benzene ring fused to C(OH) and N(H), linked C–C to C(OH) and N(H) of a second ring bearing SO_3^-; oxidized form, ^-O_3S–ring with C(=O) and N(H), linked C=C to N(H) and C(=O) of a second ring bearing SO_3^-.)

The details of colorimetric analyses will not be given for all pollutants normally analyzed by this method; instead, they are summarized in Table 9.1. In addition, it should be noted that a vast number of procedures have been developed for the colorimetric analysis of metals. Many of these methods are excellent and quite sensitive. However, the convenience, specificity, and sensitivity of atomic absorption have made that instrumental technique the method of choice for most metals commonly analyzed in natural waters.

Ammonia may be determined by the measurement of light absorbed after treatment of the ammonia-containing solution with **Nessler's reagent**. Nessler's reagent, an alkaline solution of mercuric iodide in potassium iodide, reacts with ammonia to form an orange-brown product, which remains for a time in colloidal suspension (Reaction 9.3.4). The wavelength at which the absorbance of the resulting suspension is measured varies from 400–500 nm (1 nanometer = 1×10^{-9} meter) and depends upon the concentration range of the ammonia. A calibration curve must be prepared from standard solutions treated in exactly the same manner as the unknown. To avoid interference by particles in the water or by ions that form precipitates with base, ammonia frequently must be distilled from a basic solution of the sample prior to the analysis.

$$2HgI_4^{2-} + NH_3 + 3OH^- \longrightarrow \mathrm{I{-}Hg{-}O{-}Hg{-}NH_2}(s) + 7I^- + 2H_2O \qquad (9.3.4)$$

Arsenic in amounts down to 1 microgram is determined with silver diethyldithiocarbamate reagent, $(C_2H_5)_2NC(S)S^-Ag^+$. The arsenic is first reduced to arsine, AsH_3, by zinc in acidic solution. The volatile AsH_3 is collected in silver diethyldithiocarbamate reagent, and the absorbance of the resulting colored solution is measured at 535 nm. The reagent blank is used as a reference solution. Alternatively, the AsH_3 can be flushed with N_2 into a hydrogen flame and determined by atomic absorption (see Section 9.4).

The two major forms of chlorine in water, *free chlorine* (primarily HOCl and OCl^-) and *combined chlorine* (chloramines such as NH_2Cl), are determined

TABLE 9.1 Selected Colorimetric Analyses of Chemical Pollutants[1]

Pollutant	*Reagent and method*
Ammonia	Alkaline mercuric iodide reacts with ammonia, producing colloidal orange-brown $NH_2Hg_2I_3$, which absorbs light between 400 and 500 nm
Arsenic	Reaction of arsine, AsH_3, with silver diethylthiocarbamate in pyridine, forming a red complex
Boron	Reaction with curcumin, forming red rosocyanine
Bromide	Reaction of hypobromite with phenol red to form bromphenol blue-type indicator
Chlorine	Development of color with orthotolidine
Cyanide	Formation of a blue dye from reaction of cyanogen chloride, CNCl, with pyridine-pyrazolone reagent, measured at 620 nm
Fluoride	Decolorization of a zirconium-dye colloidal precipitate ("lake") by formation of colorless zirconium fluoride and free dye
Nitrate and nitrite	Nitrate is reduced to nitrite, which is diazotized with sulfanilamide and coupled with *N*-(1-naphthyl)-ethylenediamine dihydrochloride to produce a highly colored azo dye measured at 540 nm
Nitrogen, Kjeldahl-phenate method	Digestion in sulfuric acid to NH_4^+ followed by treatment with alkaline phenol reagent and sodium hypochlorite to form blue indophenol measured at 630 nm
Phenols	Reaction with 4-aminoantipyrine at pH 10 in the presence of potassium ferricyanide, forming an antipyrine dye which is extracted into pyridine and measured at 460 nm
Phosphate	Reaction with molybdate ion to form a phosphomolybdate which is selectively reduced to intensely colored molybdenum blue
Selenium	Reaction with diaminobenzidine, forming colored species absorbing at 420 nm
Silica	Formation of molybdosilicic acid with molybdate, followed by reduction to a heteropoly blue measured at 650 nm or 815 nm
Sulfide	Formation of methylene blue
Sulfur dioxide	Collection of SO_2 gas in tetrachloromercurate solution, followed by reaction with formaldehyde and pararosaniline hydrochloride, to form a red-violet dye measured at 548 nm
Surfactants	Reaction with methylene blue to form blue salt
Tannin and lignin	Blue color from tungstophosphoric and molybdophosphoric acids

[1] Excludes metals normally determined by atomic absorption.

spectrophotometrically by reaction with orthotolidine

H_2N — NH_2

H_3C CH_3

to form a yellow substance whose absorbance is measured at 435 nm or 490 nm. Absorbance or color comparison is made at the time of maximum color development. Free chlorine develops color more rapidly than combined chlorine.

Free cyanide (HCN or CN^-) is a very toxic water pollutant arising from metal refining and cleaning, electroplating, coke ovens, and various industrial processes. It is determined spectrophotometrically. The free cyanide is first reacted with chloramine-T to convert it to cyanogen chloride, CNCl. This compound forms a blue dye absorbing at 620 nm when reacted with pyridine-pyrazolone reagent.

Nitrite ion may be determined spectrophotometrically at low levels with comparative ease and high sensitivity by a diazotization method in which color is developed by the reaction of sulfanilic acid, nitrous acid, and *N*-(1-naphthyl)-ethylenediamine dihydrochloride. The absorbance of the highly colored azo dye product is measured at 540 nm. The same basic method is applied to the analysis of nitrate ion after the reduction of that ion to nitrite with cadmium turnings "copperized" by reaction with $CuSO_4$ solution. If nitrite is present in the original sample, it may be determined prior to the reduction of nitrate, and the nitrate then calculated by difference.

Detergents, specifically anionic surfactants, are determined spectrophotometrically as *methylene blue active substances* (MBAS). Methylene blue forms blue-colored salts with anionic surfactants, including alkyl benzene sulfonate (ABS), linear alkyl sulfonate (LAS), and alkyl sulfates. The salt is extracted into chloroform, and the absorbance of the chloroform solution is measured at 652 nm. The method works over the concentration range 0.025–100 mg/L of ABS. A number of materials interfere with the analysis, however.

Phenol analysis is one of the more important analyses employed in water pollution studies because of the objectionable tastes from chlorinated phenols in chlorinated drinking waters containing as little as 1 μg/L of phenol. Phenol and other phenolic compounds are separated from wastewater by distilling. For example, 450 mL of a 500-mL sample of water may be added to 50 mL of phenol-free water and distilled to a total collected volume of 500 mL. With this procedure, the phenols are distilled at an almost constant rate, leaving the non-volatile impurities behind. Color is developed by reaction of the phenols at pH 10.0 with 4-aminoantipyrine in the presence of potassium ferricyanide. The resulting antipyrine dye is extracted into chloroform, and its absorbance is measured at 460 nm. The method is quite sensitive, with a detection limit of around 1 μg/L of phenol.

Orthophosphate ion, PO_4^{3-}, is determined by a very sensitive spectrophotometric method involving reaction with molybdate ion to form phosphomolybdate, which is selectively reduced (for example, by hydrazine sulfate) to molybdenum blue, an intensely colored blue species of uncertain composition. The method can be used for the analysis of as little as 0.01 mg/L of PO_4. Insoluble orthophosphates do not respond to this method but can be dissolved in boiling acid and subsequently analyzed:

$$M_3(PO_4)_2(s) + 6\ H^+ \rightarrow 3\ M^{2+} + 2\ H_3PO_4 \qquad (9.3.5)$$

Similarly, condensed phosphates, such as those used as detergent builders, do not respond directly to the molybdenum blue method but may be hydrolyzed by strong acid to produce orthophosphoric acid:

$$P_3O_{10}^{5-} + 2\ H_2O + 5\ H^+ \rightarrow 3\ H_3PO_4 \qquad (9.3.6)$$

Organic phosphorus, too, generally is converted to orthophosphate by acid digestion. Thus, boiling with acid yields a solution which may be analyzed for total phosphate, but it does not provide information on specific phosphate species.

The most important colorimetric air-pollutant analysis is the rosaniline procedure for sulfur dioxide first described by West and Gaeke [2] and later perfected [3]. It is applied to the analysis of 0.005–5 ppm by volume SO_2 in air. Chemical reaction with tetrachloromercurate ion,

$$HgCl_4^{2-} + SO_2 + H_2O \rightarrow HgCl_2SO_3^{2-} + 2\,H^+ + 2\,Cl^- \qquad (9.3.7)$$

enables collection of sulfur dioxide from ambient air. Typically, this involves scrubbing 30 liters of air through 10 mL of scrubbing solution, with an excellent collection efficiency of around 95%. Sulfur dioxide in the collecting medium is reacted with formaldehyde,

$$HCHO + SO_2 + H_2O \rightarrow HOCH_2{-}SO_3H \qquad (9.3.8)$$

to form a product which reacts with pararosaniline hydrochloride:

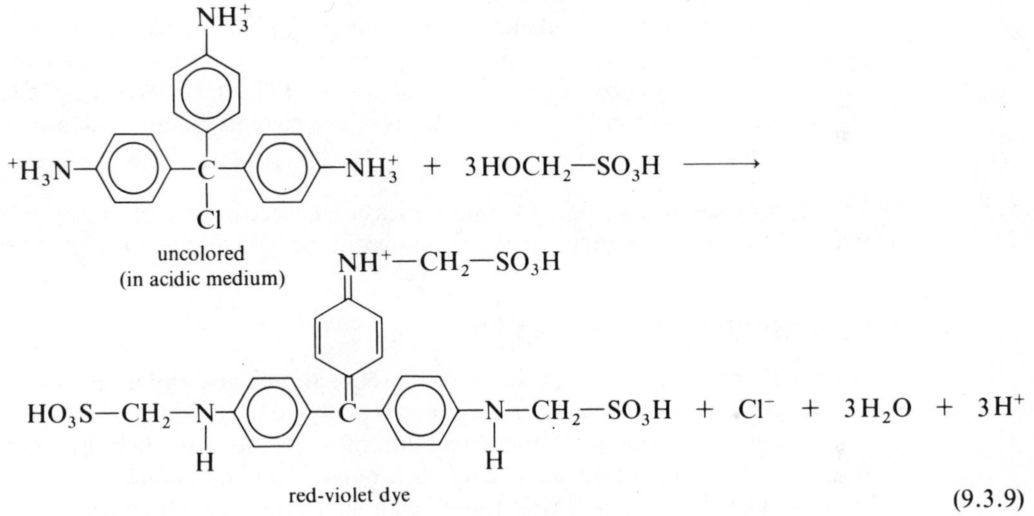

NH_3^+

$^+H_3N-C_6H_4-C(Cl)(C_6H_4NH_3^+)-C_6H_4-NH_3^+ + 3\,HOCH_2{-}SO_3H \longrightarrow$

uncolored
(in acidic medium)

$NH^+{-}CH_2{-}SO_3H$

$HO_3S{-}CH_2{-}N(H){-}C_6H_4{-}C({=}C_6H_4{=}NH^+{-}CH_2{-}SO_3H){-}C_6H_4{-}N(H){-}CH_2{-}SO_3H + Cl^- + 3\,H_2O + 3\,H^+$

red-violet dye

(9.3.9)

The absorbance of the product red-violet dye is measured at 548 nm.

Nitrogen dioxide at levels exceeding 2 ppm constitutes a major interference with the original West-Gaeke method. The interference may be eliminated by the addition of sulfamic acid, H_2NSO_3H, which acts as a reducing agent by converting some of the oxygenated nitrogen species to nitrogen gas.

Although this method of sulfur dioxide analysis is cumbersome and complicated, it has been refined to the point where it is used in continuous-monitoring equipment, as exemplified by the Technicon Air Monitor IV instrument. A block diagram of a continuous sulfur dioxide monitor based upon the West-Gaeke method is shown in Figure 9.1 (see page 232).

---

[2] West, P.W., and Gaeke, G.C. 1956. Fixation of sulfur dioxide as disulfitomercurate(II) and subsequent colorimetric estimation. *Analytical Chemistry* 28: 1816–9.

[3] 1971. Reference method for the determination of sulfur dioxide in the atmosphere (pararosaniline method). *Federal Register* 36: 8168.

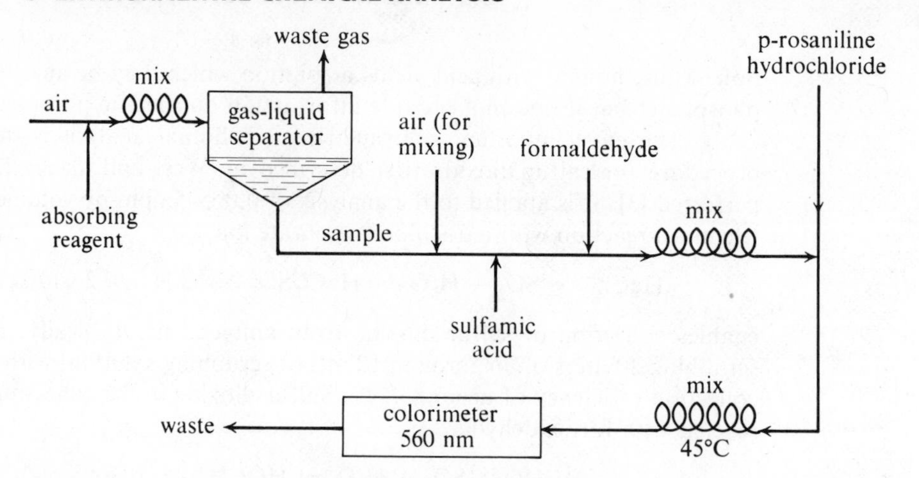

FIGURE 9.1 Schematic diagram of a continuous sulfur dioxide monitor using a modified West-Gaeke method.

Oxidizing agents (oxidants) in the atmosphere, such as ozone, hydrogen peroxide, organic peroxides, and chlorine, may be determined colorimetrically. This is accomplished by collecting the sample in 1% KI buffered at pH 6.8. Oxidants oxidize I^- ion by reactions that produce colored I_3^- ion, such as

$$O_3 + 2\,H^+ + 3\,I^- \rightarrow I_3^- + O_2 + 2\,H_2O \qquad (9.3.10)$$

The absorbance of the colored I_3^- ion is measured spectrophotometrically at 352 nm. If the color is too intense, the solution is diluted with additional I^- solution.

9.4 ATOMIC ABSORPTION ANALYSIS

A revolutionary development during the 1960s, **atomic absorption analysis** has become the method of choice for most metals analyzed in environmental samples. This technique is based upon the absorption of monochromatic light by a cloud of atoms of the analyte metal. The monochromatic light is produced by a source composed of the same atoms as those being analyzed. The source produces intense electromagnetic radiation with a wavelength exactly the same as that absorbed by the atoms, resulting in extremely high selectivity.

The basic components of an atomic absorption instrument are shown in Figure 9.2. The key element is the **hollow cathode lamp**. This device consists of a glass tube containing noble gases, primarily argon, at several mm pressure. It contains an anode and a cathode with a high potential applied between them. An electrical current of several milliamperes flows through the lamp when it is in operation. The cathode is a hollow cylinder, the inside of which is coated with the metal to be analyzed. In operation, the high potential between the electrodes results in the production of positively charged noble gas ions (Ar^+) which impinge upon the negatively charged hollow cathode with a very high energy. Bombardment of the hollow cathode with energetic positive ions causes "sputtering" of metal atoms from the surface of the hollow cathode. These energized metal atoms emit radiation with a very narrow wavelength band characteristic of the metal. This radiation is guided by the appropriate optics through a flame into which the sample is aspirated. In the flame, most metallic compounds are decomposed, and the metal is reduced to the elemental state, forming a cloud of atoms. These atoms absorb a fraction of radiation in the flame. The fraction of radiation

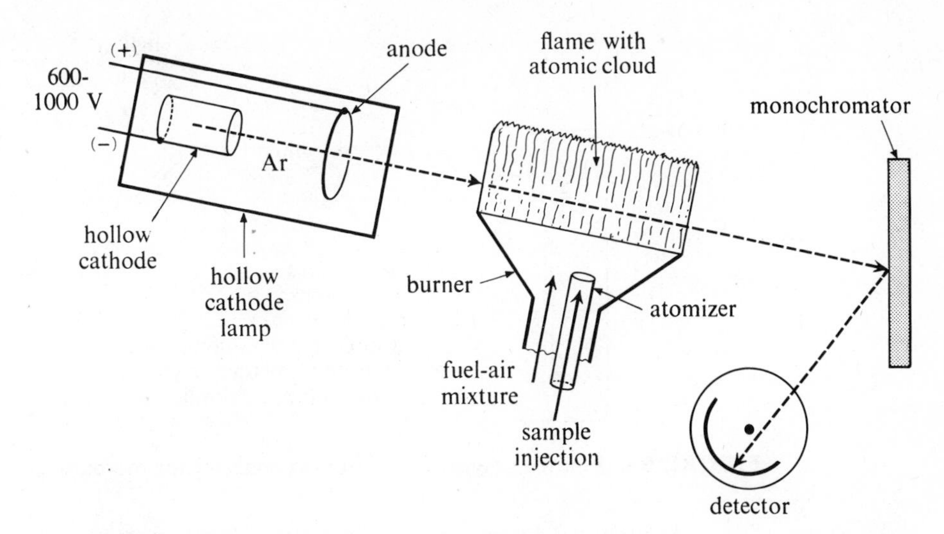

FIGURE 9.2 The basic components of an atomic absorption spectrophotometer.

absorbed increases with the concentration of the sought-for element in the sample according to the Beer's law relationship (Eq. 9.3.2). The attenuated light beam next goes to a monochromator to eliminate extraneous light resulting from the flame, finally passing to a detector and readout system.

Recent advances in atomic absorption technology have eliminated the flame as an atomizer and "sample holder" for some elements. The most commonly used of these new devices is the graphite furnace, which consists of a hollow graphite cylinder placed so that the light beam passes through it. A small sample (up to 100-μL) is inserted in the tube through a hole in the top. An electric current is passed through the tube to heat it—gently at first to dry the sample, then rapidly to incandescence. Metals (particularly the more volatile ones) are vaporized in the hollow portion of the tube, and the absorption of the metal atoms is recorded as a spike-shaped signal. A diagram of a simple graphite furnace and a typical output signal is shown in Figure 9.3.

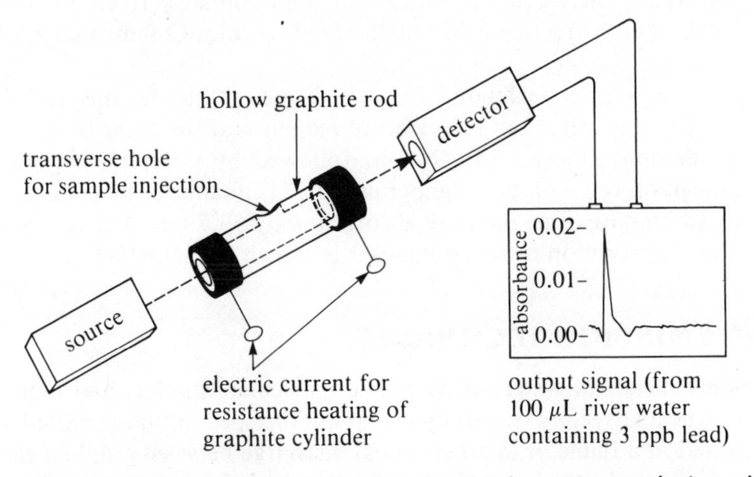

FIGURE 9.3 Graphite furnace for atomic absorption analysis and typical output signal.

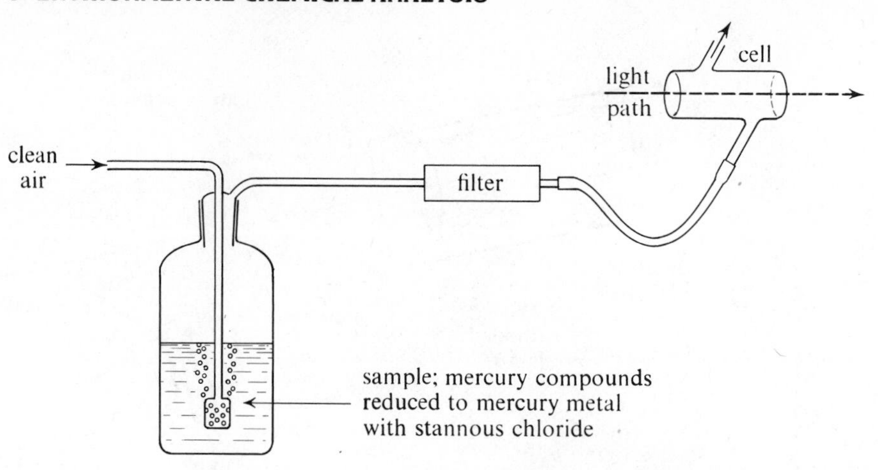

FIGURE 9.4 Flameless atomic absorption analyzer for mercury.

The major advantage of the graphite furnace atomizer over conventional flame techniques is a lower detection limit for many metals. The detection limit for atomic absorption is defined as the concentration of metal (mg/L) producing absorption equivalent to twice the noise level (background fluctuation) at zero absorption. (Another term frequently used to describe the characteristics of a spectrophotometric analysis is **sensitivity**, the minimum concentration of metal that will result in absorption of 1% of the incident light.) The detection limits of graphite furnaces are up to 1000 times lower than those of conventional flame devices. Some of these limits for environmentally important metals found in water are given in Table 9.2, at the end of Section 9.5.

Direct atomization of biological samples in a flameless atomic absorption device would greatly simplify flameless atomic absorption, but it has generally been impractical because of interference from smoke. **Zeeman-effect atomic absorption (ZAA)**, originally developed for the direct determination of mercury in fish by instruments on fishing boats, has enabled direct sample atomization in some cases. When this technique is employed, either the source light or the sample is placed in a strong magnetic field, splitting and polarizing the spectral line used for analysis. The amount of background interference that can be tolerated is increased by a factor of approximately five, and the precision of background correction is increased about tenfold. Commercial ZAA instruments are now available.

A special technique for the flameless atomic absorption analysis of mercury involves room-temperature reduction of mercury to the elemental state by stannous chloride in solution, followed by sweeping the mercury into an absorption cell with air. Nanogram (10^{-9}g) quantities of mercury can be determined by measuring mercury absorption at 253.7 nm. A diagram of a flameless atomic absorption mercury analyzer is shown in Figure 9.4.

9.5 ATOMIC EMISSION TECHNIQUES

Many metals in water, atmospheric particulate matter, and biological samples may be analyzed very well by observing the spectral lines emitted when they are heated in a flame or in an electrical discharge between graphite electrodes, as in conventional **emission spectroscopy**. With the latter technique, the intensities and wavelengths of lines in the ultraviolet and visible regions are recorded on strips of photographic film after passage through a monochromator. These lines are compared to nearby lines from a standard (generally iron). The wavelength pattern of the lines provides a qualitative analysis, and the intensities of selected lines provide a rough quantitative analysis. Among the elements commonly analyzed by emission spectroscopy are beryllium, lead, iron, cadmium, nickel, vanadium, manganese, chromium, copper, zinc, barium, and tin.

An especially promising emission technique that has become a useful routine analytical tool since about 1976 is **inductively coupled plasma atomic emission spectroscopy** [4]. The "flame" for this technique consists of an incandescent plasma (ionized gas) of argon heated inductively by radiofrequency energy at 4–50 MHz and 2–5 kW (Fig. 9.5). The energy is transferred to a stream of argon through an induction coil, producing temperatures up to 10,000 K. The sample atoms are subjected to temperatures around 7000 K, twice those of the hottest conventional flames (for example, nitrous oxide–acetylene at 3200 K). Since emission of light increases exponentially with temperature, lower detection limits are obtained. Furthermore, the technique enables emission analysis of some of the environmentally important metalloids such as arsenic, boron, and selenium. Interfering chemical reactions and interactions in the plasma are minimized as compared to flames. Of greatest significance, however, is the capability of analyzing as many as 30 elements simultaneously, enabling a true multielement analysis technique. Its application to environmental samples has been summarized [5].

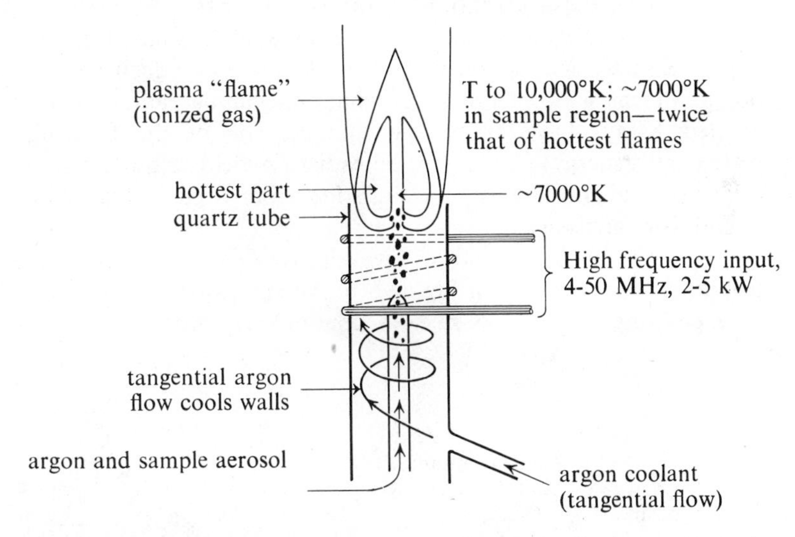

FIGURE 9.5 Schematic diagram showing inductively coupled plasma, used for optical emission spectroscopy.

---

[4] Fassel, V.A. 1978. Quantitative elemental analyses by plasma emission spectroscopy. *Science* 202: 183–91.

[5] Ward, A.F. 1978. Inductively coupled argon plasma spectroscopy. *American Laboratory* November, 1978, pp. 79–87.

TABLE 9.2 Comparison of Detection Limits among Conventional Flame Atomic Absorption, Graphite-Furnace Atomic Absorption, and Inductively Coupled Plasma Optical Emission Spectroscopy

	Detection limit, mg/L		
	Atomic absorption		Plasma emission
Element	*Graphite furnace*	*Conventional flame*	
As	0.003	0.1	0.006
Sb	0.001	0.1	0.004
Hg	0.002	0.5	0.03
Cd	0.000003	0.001	0.0001
Pb	0.0002	0.001	0.001
Be	0.00002	0.002	0.00002
Mn	0.00002	0.002	0.00003
Fe	0.0001	0.005	0.0004
Sn	0.025	0.02	0.025
B	———	0.7	0.04
Cu	0.00002	0.002	0.0004
Cr	0.0001	0.003	0.0002

9.6 X-RAY FLUORESCENCE

X-Ray fluorescence is another multielement analysis technique that is especially useful for the characterization of atmospheric particulate matter, but it can be applied to some water and soil samples as well. It is based upon observation of X-rays emitted when electrons fall back into inner shell vacancies created by bombardment with energetic X-radiation, gamma radiation, or protons. The emitted X-rays have an energy characteristic of the particular atom. The wavelength (energy) of the emitted radiation yields a qualitative analysis of the elements, and the intensity of radiation from a particular element provides a quantitative analysis.

A schematic diagram of a wavelength-dispersive X-ray fluorescence spectrophotometer is shown in Figure 9.6. An excitation source, normally an X-ray tube emitting "white" X-rays (a continuum), produces a primary beam of

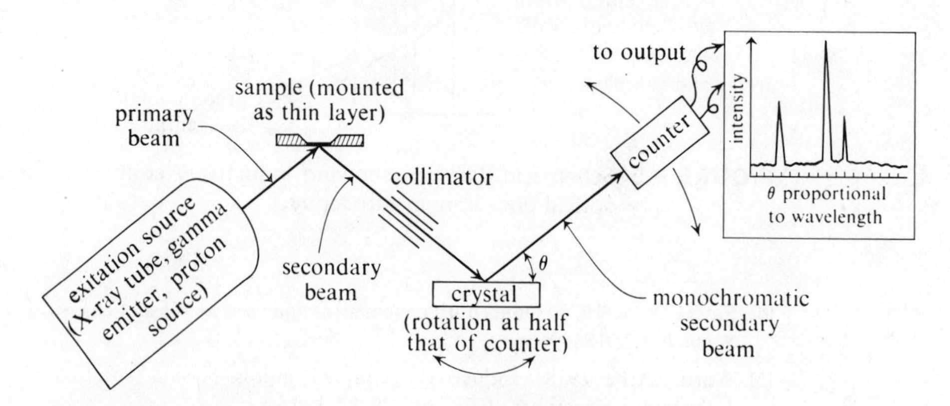

FIGURE 9.6 Wavelength-dispersive X-ray fluorescence spectrophotometer.

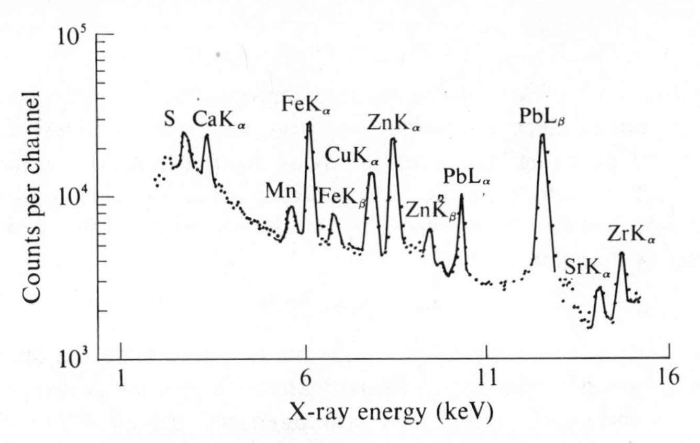

FIGURE 9.7 Energy-dispersive X-ray fluorescence spectrum from an atmospheric particle sample.

energetic radiation which excites fluorescent X-rays in the sample. A radioactive source emitting gamma rays or protons from an accelerator may also be used for excitation. For best results, the sample should be mounted as a thin layer, which means that segments of air filters containing fine particulate matter make ideal samples. The fluorescent X-rays are passed through a collimator to select a parallel secondary beam, which is dispersed according to wavelength by diffraction with a crystal monochromator. The monochromatic X-rays in the secondary beam are counted by a detector which rotates at a degree twice that of the crystal to scan the spectrum of emitted radiation.

Development of energy-selective detectors of the Si(Li) semiconductor type [6] has enabled measurement of fluorescent X-rays of different energies without the need for wavelength dispersion. Instead, the energies of a number of lines falling on a detector simultaneously are distinguished electronically. An energy-dispersive X-ray fluorescence spectrum from an atmospheric particulate sample is shown in Figure 9.7.

Given the current state of the art, wavelength-dispersive X-ray fluorescence is superior to the energy-dispersive technique except in speed and convenience. Even so, the wavelength-dispersive technique can permit the analysis of over 20 preselected elements within approximately two minutes. An additional advantage of X-ray fluorescence multielement analysis is that sensitivities and detection limits do not vary greatly across the periodic table as they do with methods such as neutron activation analysis or atomic absorption. Proton-excited X-ray emission is particularly sensitive.

9.7 NEUTRON ACTIVATION ANALYSIS

Radioactive nuclei decay at a rate ($-dN/dt$) that is proportional to the number of nuclei present, N (see Section 7.12). The rate expression for decay is shown at the top of the following page (Eq. 9.7.1).

---

[6] Rhodes, J.R. 1973. Energy-dispersive X-ray spectrometry for multi-element pollution analysis. *American Laboratory* 5 (7): 57.

$$-\frac{dN}{dt} = \frac{0.693}{t_{1/2}}N \tag{9.7.1}$$

where $t_{1/2}$ is the half-life of the nuclide, or the time in which one-half of the radioactive nuclei decay. Radioactive isotopes (radionuclides) are produced artificially by exposing nuclei to neutrons from a neutron source, usually a nuclear reactor. For example, manganese, which occurs naturally as the manganese-55 isotope, is converted by irradiation with neutrons, designated ${}_0^1n$, to radioactive manganese-56:

$${}_{25}^{55}\text{Mn} + {}_0^1n = {}_{25}^{56}\text{Mn} \tag{9.7.2}$$

In this equation, the superscript is the mass number of the species and the subscript is the atomic number. The manganese-56 product decays with a half-life of 2.6 hours and emits gamma rays with an energy of 0.85 million electron-volts (MeV). If N_i is the number of nonradioactive manganese nuclei present in the sample (manganese-55 nuclei), the rate of production of radioactive nuclei P is given by the formula

$$P = f\sigma N_i \tag{9.7.3}$$

where f is the neutron flux (particles/cm^2/sec) and σ is the **cross section** of the isotope being activated. The cross section expresses the tendency of the inactive nuclei to absorb neutrons and become activated. A high cross section increases the sensitivity of the activation analysis of a particular nucleus.

Basically, **neutron activation analysis** consists of three steps: (1) irradiating a sample with neutrons to produce radionuclides; (2) determining the activities of the radionuclide products; (3) relating the activities to the quantities of the elements originally present in the sample. The activity of a radionuclide depends upon the number of nuclei of the parent isotope and not on its chemical form. For that reason, neutron activation analysis is an elemental analysis and cannot be used to determine individual chemical species.

Assuming access to a reactor, neutron activation analysis can be very useful to the environmental chemist [7]. Its primary advantage for some elements is high sensitivity. As little as 1×10^{-7} g of the elements most amenable to this technique can be analyzed readily by neutron activation. Factors that contribute to high sensitivity are high cross section and high isotopic abundance of the parent isotope; low atomic weight (more nuclei per unit weight); high neutron flux; and high detection efficiency. Cross section is the most important of these characteristics and varies by a factor of over 10^6 for various isotopes. Using detection limit as a criterion, some of the elements commonly found in water that are best determined by neutron activation analysis, as compared to atomic absorption or flame emission, are chlorine, titanium, vanadium, arsenic, selenium, bromine, molybdenum, antimony, iodine, and tungsten.

Although many elements can be determined by direct analysis of the gamma-ray spectrum of the sample after irradiation, a chemical (radiochemical) separation is frequently necessary prior to "counting" the desired radionuclide. This leads to another substantial advantage of neutron activation analysis—the

[7] Kay, M.A.; McKown, D.M.; Gray, D.H.; Eichor, M.E.; and Vogt, J.R. 1973. Neutron activation analysis in environmental chemistry. *American Laboratory* July, 1973, pp. 39–48.

use of **carriers**. After the irradiation is completed, the total number of radionuclides of an element remaining at any given time is determined only by their half-lives and not by any chemical reactions. Therefore, large quantities of the sought-for element may be added as a carrier after activation to facilitate chemical separations of the sought-for element on a macro scale. A carrier can even be used with separations that are less than 100% efficient. If the separation efficiency can be determined, a correction can be applied to calculate the amount of the element originally present. Furthermore, use of a carrier after irradiation prevents the problem of contamination of reagents by the element being analyzed.

Specificity is a big advantage with neutron activation analysis. Radiochemical properties, half-life, gamma energy, and beta energy are all properties that are used to distinguish elements analyzed by neutron activation analysis. Although two different radioisotopes may have one property in common (the same half-lives, for example), they are not at all likely to have two or three identical properties. Thus, it is generally relatively easy to determine the element, and even the particular parent isotope of that element, giving rise to the radionuclide produced by activation.

The calculation of the quantity of an element in a sample from all the factors contributing to the measured radioactivity (neutron flux, isotope cross section, counter efficiency) is possible in principle, but prohibitively difficult in fact. Therefore, a standard consisting of a known quantity of the element being determined is almost always used in activation analysis.

9.8 ELECTROCHEMICAL METHODS OF ANALYSIS

Several useful techniques for water analysis utilize electrochemical sensors. Generally these techniques fall under the classification of either potentiometry or voltammetry.

Potentiometry is based upon the general principle that the relationship between the potential of a measuring electrode and that of a reference electrode is a function of the log of the activity of an ion in solution, according to the Nernst equation applied to ion-selective electrodes,

$$E = E_a + \frac{2.303RT}{z\,F} \log (a_z) \tag{9.8.1}$$

where E is the measured potential; E_a is a constant for the particular electrode system; R is the gas constant; T is the absolute temperature; z is the signed charge (+ for cations, − for anions); F is the Faraday constant; and a_z is the activity of the ion being measured. At a given temperature, the quantity $2.303RT/F$ has a constant value, 0.0592 volt at 25 °C. At constant ionic strength, the activity coefficient in a_z remains constant, and the Nernst equation may be written in terms of concentration; for example,

$$E = E_a + \frac{0.0592}{2} \log [Cd^{2+}] \tag{9.8.2}$$

for cadmium ion at a cadmium electrode, or

$$E = E_a - 0.0592 \log [F^-] \tag{9.8.3}$$

for fluoride ion at a fluoride electrode.

Electrodes that respond more or less selectively to various ions are constructed in a number of ways. Generally, the potential-developing component is a membrane of some kind that allows for selective exchange of the sought-for ion. The glass electrode used for the measurement of hydrogen-ion activity and pH is the oldest and most widely used ion-selective electrode. The basic components of the glass electrode are shown in Figure 9.8. The potential-developing device is a glass membrane that selectively exchanges hydrogen ion in preference to other cations, giving a Nernstian response to hydrogen-ion activity, a_{H^+}:

$$E = E_a + \frac{2.303RT}{F} \log (a_{H^+}) \tag{9.8.4}$$

Glass electrodes have been developed that are selective for monovalent cations other than hydrogen, especially Na^+, NH_4^+, and Ag^+.

Potentiometry as an analytical tool has received much impetus since 1966 with the development of several new types of ion-selective electrodes, sometimes called *specific-ion electrodes*. These are of two general types: the solid membrane type (Fig. 9.8), and the liquid ion-exchanger type (Fig. 9.8). Solid membrane ion-selective electrodes develop a potential through exchange of ions between the solution and the membrane surface. Liquid membrane electrodes involve a similar exchange of ions between solution and a liquid ion-exchanger held in a thin filter disc. The ion exchanger is chosen to be selective for the ion being analyzed. Among the ions for which solid membrane electrodes are available are fluoride, chloride, bromide, iodide, cyanide, thiocyanate, cadmium, cupric ion, lead, silver, and sulfide. Liquid membrane electrodes are available for calcium, nitrate, potassium, and water hardness (divalent cations). Enzyme electrodes

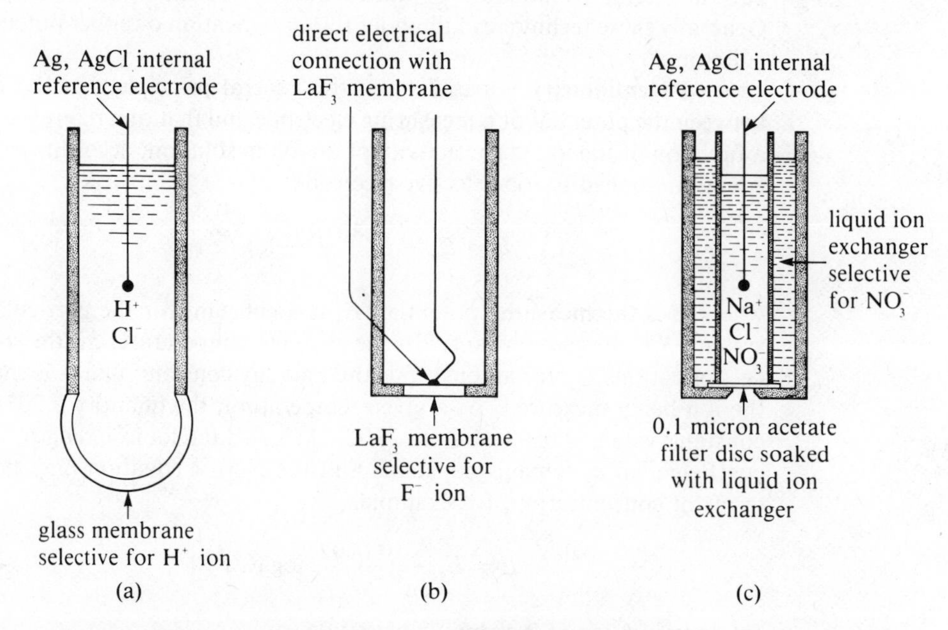

FIGURE 9.8 Ion-selective electrodes: (a) glass electrode selective for hydrogen ion; (b) lanthanum fluoride solid membrane electrode selective for fluoride ion; (c) liquid ion-exchanger electrode selective for nitrate ion.

have been developed containing an immobilized enzyme on the electrode surface, which catalyzes the production of ions from biological materials such as amino acids. The ions are then sensed by the electrode. Thus, even organic compounds may be analyzed potentiometrically. Dissolved gases in water and even some gases in the atmosphere can be analyzed by electrodes having membranes selectively permeable to the gas being analyzed, which causes a pH change in the electrode.

Of the ion-selective electrodes, the fluoride electrode is the most successful. It is well-behaved, relatively free of interference, and has an adequately low detection limit and a long range of linear response. Like all ion-selective electrodes, it suffers the following disadvantage: its electrical output is in the form of a potential signal that is proportional to *log* of concentration, so that relatively small potential errors give high concentration errors.

Voltammetric techniques, the measurement of currents resulting from potential applied to a microelectrode, have found some applications in water analysis. The oldest of these is **classical direct current polarography**. New techniques have been developed that are much more sensitive than classical polarography. One such technique is **differential-pulse polarography**, in which the potential is applied to the microelectrode in the form of small pulses superimposed on a linearly increasing potential. The current is read near the end of the voltage pulse and compared to the current just before the pulse was applied. It has the advantage of minimizing the capacitive current, which sometimes obscures the current due to the reduction or oxidation of the species being analyzed. **Fast linear-sweep voltammetry** utilizes a rapidly changing potential applied to the electrode, displaying the resulting current on an oscilloscope. **Anodic-stripping voltammetry** involves deposition of metals on an electrode surface over a period of several minutes followed by stripping them off very rapidly using a linear anodic sweep [8]. The electrodeposition concentrates the metals on the electrode surface, and increased sensitivity results. An even better technique is to strip the metals off using a differential pulse signal. A differential-pulse anodic-stripping voltammogram of lead, cadmium, and zinc in tap water is shown in Figure 9.9.

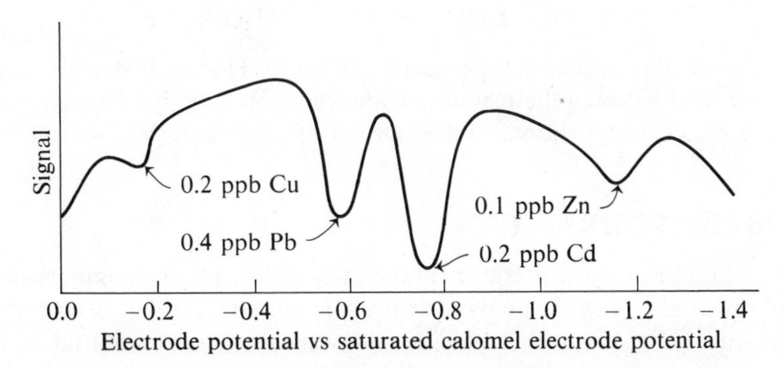

FIGURE 9.9 Differential-pulse anodic-stripping voltammogram of tap water at a mercury-plated, wax-impregnated graphite electrode.

[8] Wang, J. 1982. Anodic stripping as an analytical tool. *Environmental Science and Technology* 16: 104A–109A.

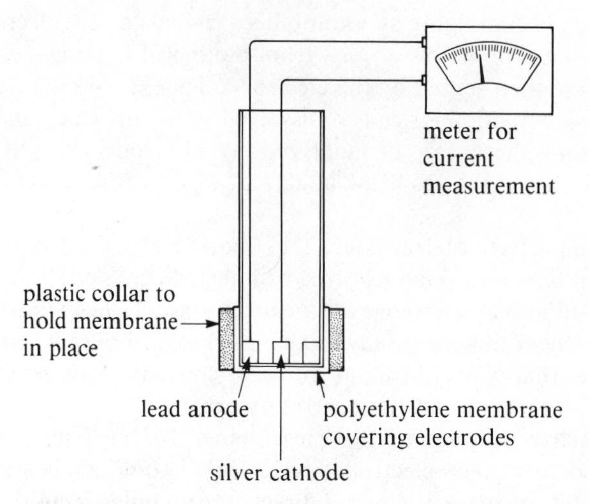

FIGURE 9.10 Membrane-covered electrode for sensing dissolved oxygen.

Voltammetric techniques can also be used to differentiate among the chemical states (speciation) of metals in water. Complexation or other chemical binding of the metal shows up as a shift in the potential at which the metal is reduced or oxidized. Decreasing the pH releases bound metal ions, making them more available for deposition at relatively less negative potentials.

Membrane-covered voltammetric electrodes are now the most common means for the measurement of dissolved oxygen [9]. Such an electrode is shown in Figure 9.10. An end-on view of the electrode shows a silver cathode surrounded by an annular lead anode, both covered by a membrane selectively permeable to oxygen. As oxygen diffuses through the membrane, it undergoes the following half-reaction at the silver cathode:

$$\frac{1}{2}O_2 + H_2O + 2e \rightarrow 2\,OH^- \qquad (9.8.5)$$

The half-reaction at the lead anode is

$$Pb + 2\,OH^- \rightarrow PbO + H_2O + 2\,e \qquad (9.8.6)$$

These reactions occur in the presence of 4M KOH absorbed by a small disc of lens paper held between the membrane and the electrode tip. The amount of current generated is proportional to the amount of oxygen in the solution.

9.9 GAS CHROMATOGRAPHY

First described in the literature in the early 1950s, **gas chromatography** has played an essential role in the analysis of orgnaic materials. It has been particularly successful for low-level pesticide analysis in environmental samples. Gas chromotography is both a qualitative and quantitative technique; for some analytical applications of environmental importance, it is remarkably sensitive and selective.

---

[9] Hitchman, M.L. 1978. *Measurement of dissolved oxygen*. New York: John Wiley and Sons, Inc., pp. 71–129.

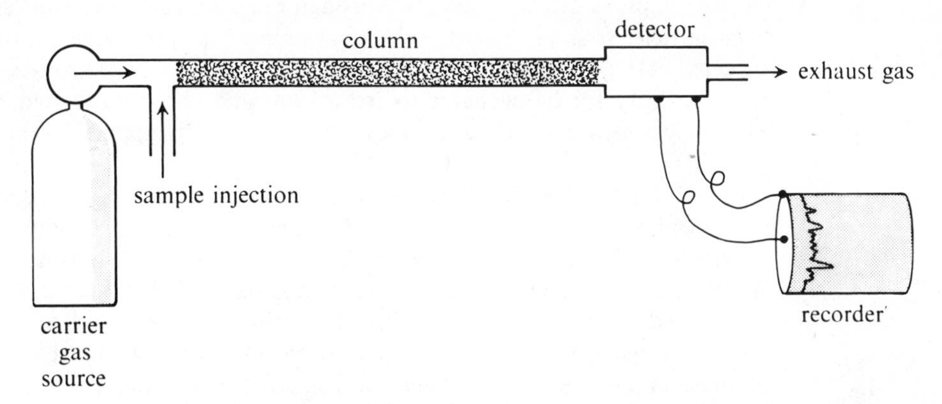

FIGURE 9.11 Schematic diagram of the essential features of a gas chromatograph.

Gas chromatography is based upon the principle that when a mixture of volatile materials transported by a carrier gas is passed through a column containing an adsorbent solid phase, or, more commonly, an absorbing liquid phase coated on a solid material, each volatile component will be partitioned between the carrier gas and the solid or liquid. The length of time required for the volatile component to traverse the column is proportional to the degree to which it is retained by the nongaseous phase. Since different components may be retained to different degrees, they will emerge from the end of the column at different times. If a suitable detector is available, the time at which the component emerges from the column and the quantity of the component are both measured. A recorder trace of the detector response appears as peaks of different sizes, depending upon the quantity of material producing the detector response. Both quantitative and (within limits) qualitative analyses of the sought-for substances are obtained.

The essential features of a gas chromatograph are shown schematically in Figure 9.11. The carrier gas generally is argon, helium, hydrogen, or nitrogen. Precise control of the flow rate of the carrier gas is essential. The sample is injected as a single compact plug into the carrier gas stream immediately ahead of the column entrance. If the sample is liquid, it is essential that the injection chamber be heated to vaporize the liquid rapidly and avoid "tailing" or spreading of the peaks. However, if the sample is heated excessively, pyrolysis (thermal decomposition) may occur and produce erroneous results. Most commonly, the separation column consists of a metal or glass tube packed with an inert solid of high surface area covered with a liquid phase. An active solid itself may be used as a separation medium in some cases **(gas-solid chromatography, GSC)**. When a column containing a liquid phase is used, the technique is known as **gas-liquid chromatography, GLC**. A very long, small-diameter capillary column may be employed in which the liquid phase is coated on the inside of the column. Many different kinds of column packings are used for various purposes. A column is generally chosen for its ability to separate compounds in a given group of compounds—that is, for its selectivity.

The component that primarily determines the sensitivity of gas chromatographic analysis, and, for some classes of compounds, the selectivity as well, is the detector. One such device is the thermal conductivity detector, which responds to changes in the thermal conductivity of gases passing over it. The electron-capture detector operates through the capture of electrons emitted by a beta-particle source. Compounds containing halogens and phosphorus are especially effective in capturing electrons. Therefore, the electron-capture detector is selective for halogenated hydrocarbons and phosphorus compounds. It is extremely sensitive for these compounds and is widely applied to the analysis of pesticides.

The **flame-ionization gas chromatographic detector** is very sensitive for the detection of organic compounds. It is based upon the phenomenon by which organic compounds form highly conducting fragments, such as C^+, in a flame. Application of a potential gradient across the flame results in a small current that may be readily measured. The response of a flame-ionization detector is approximately proportional to the number of carbon atoms in the compound. The presence of atoms such as halogens, nitrogen, or oxygen decreases the response of the detector. The flame-ionization detector has a very wide range of linear response and is sensitive to extremely low levels of hydrocarbons, ranging down to 1×10^{-3} μg. Because of the extremely high sensitivity of the detector, it is sometimes possible to analyze hydrocarbons without a previous concentration step. A schematic diagram of a flame-ionization detector is shown in Figure 9.12.

The *mass spectrometer*, described in Section 9.12, may be used as a detector for a gas chromatograph. The combination of these two instruments is an especially powerful analytical tool for organic compounds.

Gas chromatographic analysis requires that a compound exhibit at least a few mm of vapor pressure at the highest temperature at which it is stable. In many cases, organic compounds that cannot be chromatographed directly may be converted to derivatives that are amenable to gas chromatographic analysis.

It is seldom possible to analyze organic compounds in water by direct injection of the water into the gas chromatograph; higher concentration is usually required (see Section 9.16). Two techniques commonly employed are extraction with solvents such as hexane and purging volatile compounds with helium or N_2.

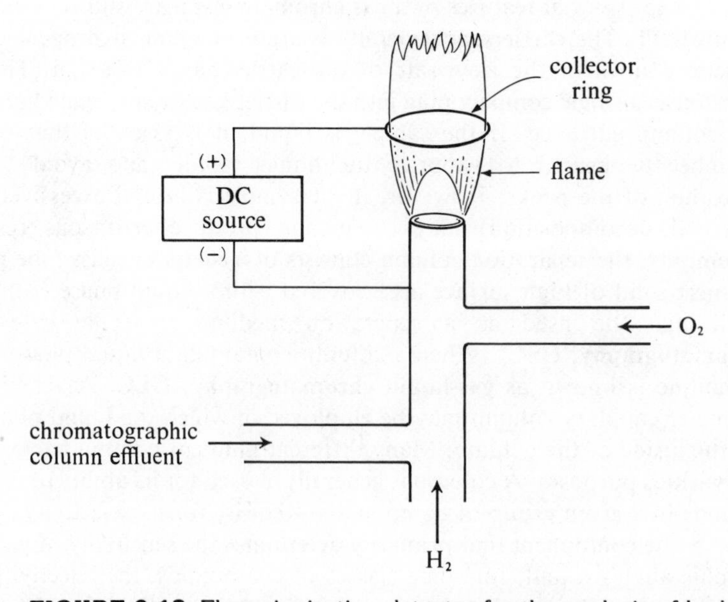

FIGURE 9.12 Flame-ionization detector for the analysis of hydrocarbons.

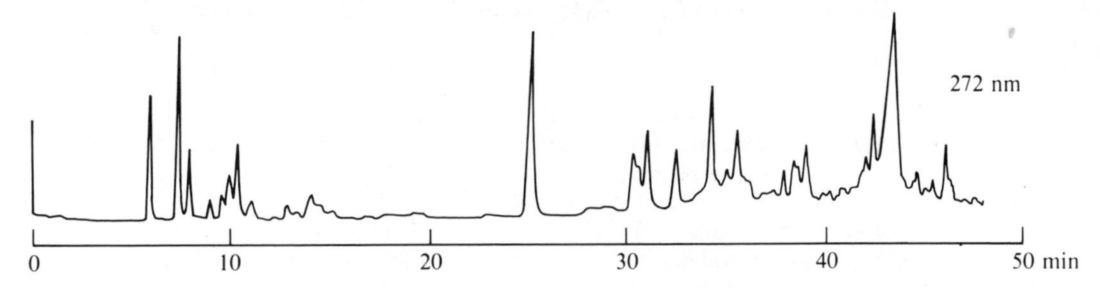

FIGURE 9.13 HPLC chromatogram showing separation of some compounds extracted from water contaminated by coal gasification; UV detection at 272 nm.

9.10 HIGH-PERFORMANCE LIQUID CHROMATOGRAPHY

Recent developments in hardware and materials have enabled the use of a liquid mobile phase and very small column-packing particles in chromatographs, resulting in the high-resolution chromatographic separation of materials in the liquid phase. High pressures up to several thousand psi are required to get a reasonable flow rate in such systems. Analysis using such devices is called **high-performance liquid chromatography (HPLC)** and offers an enormous advantage, in that the materials analyzed need not be changed to the vapor phase, a step that often requires preparation of a volatile derivative or results in decomposition of the sample. The basic features of a high-performance liquid chromatograph are the same as those of a gas chromatograph, shown in Figure 9.11, except that a solvent reservoir and high-pressure pump are substituted for the carrier gas source and regulator. An HPLC chromatogram of some water pollutants is shown in Figure 9.13.

Given present technology, refractive index and ultraviolet detectors are most generally used for the detection of peaks coming from the liquid chromatograph column. Fluorescence detection can be especially sensitive for some classes of compounds. Available detectors are not as sensitive as the best of those used for gas chromatography, although improvements may be anticipated. High-performance liquid chromatography has emerged as a very useful technique for the analysis of a number of water pollutants.

9.11 CHROMATOGRAPHIC ANALYSIS OF WATER POLLUTANTS

The Environmental Protection Agency has developed a number of chromatography-based standard methods for determining water pollutants. These consist of the **600 series** of analyses for various classes of compounds [10]. Some of these methods use the purge-and-trap technique, whereas others use solvent extraction to isolate the analytes from water. The **purge-and-trap technique** involves bubbling gas through a column of water to purge volatile organics from the water and

[10] 1979. Guidelines establishing test procedures for the analysis of pollutants; proposed regulations. *Federal Register* 44(233): 69464–69574.

TABLE 9.3 Methods 601–612 for the Analysis of Water Pollutants

Method number	*Class of compounds*	*Example*
601	Purgeable halocarbons	Carbon tetrachloride
602	Purgeable aromatics	Toluene
603	Acrolein and acrylonitrile	—
604	Phenols	Phenol and chlorophenols
605	Benzidines	Benzidine and 3,3′-dichlorobenzidine
606	Phthalate esters	Bis(2-ethylhexylphthalate)
607	Nitrosamines	*N*-nitroso-*N*-dimethylamine
608	Organochlorine pesticides and PCB's	Heptachlor, PCB 1016
609	Nitroaromatics and isophorone	Nitrobenzene
610	Polynuclear aromatic hydrocarbons	Benzo(a)pyrene
611	Haloethers	Bis(2-chloroethyl) ether
612	Chlorinated hydrocarbons	1,3-Dichlorobenzene

trapping the organics in a column containing sorbents such as Tenax, silica gel, and activated charcoal. The trapping column is then heated from room temperature to 180 °C, and the volatile organics are swept into a chromatograph. **Solvent extraction procedures** commonly make use of dichloromethane (methylene chloride) as an extracting solvent. The first 12 of the 600 series of methods are listed in Table 9.3.

9.12 MASS SPECTROMETRY

Mass spectrometry is particularly useful for the identification of specific organic pollutants. It depends upon the production of ions by an electrical discharge or chemical process, followed by separation based on the charge-to-mass ratio and measurement of the ions produced. The output of a mass spectrometer is a **mass spectrum**, such as the one shown in Figure 9.14.

A mass spectrum is characteristic of a compound and serves to identify it. Computerized data banks for mass spectra have been established and may be stored in computers interfaced with mass spectrometers. Identification of a mass spectrum depends upon the purity of the compound from which the spectrum was taken. Prior separation by gas chromatography with continual sampling of the column effluent by a mass spectrometer, commonly called *GC-mass spec,* is particularly effective in the analysis of organic pollutants.

Elemental analysis at very low levels can be accomplished by **spark-source mass spectrometry**. In this method, the sample is held on an electrode in an evacuated chamber and ionized by an electrical spark discharge. This technique, along with emission spectrometry, X-ray fluorescence, and neutron activation analysis, is useful for multielement analysis.

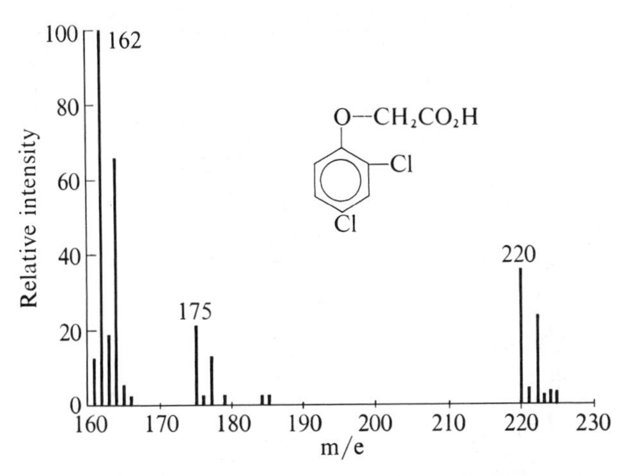

FIGURE 9.14 Partial mass spectrum of the herbicide 2, 4-dichlorophenoxyacetic acid (2, 4-D), a common water pollutant.

9.13 WATER ANALYSIS BY AUTOMATED PROCEDURES

In many cases, large numbers of water samples must be analyzed in order to obtain meaningful results. Automated techniques are now widely used for the analysis of large numbers of samples. Among the species in water for which automated procedures are now used are alkalinity, chloride, fluoride, hardness, ammonia, Kjeldahl nitrogen, nitrate-nitrite, NTA, and sulfate.

One instrument used for automated analysis is the Technicon AutoAnalyzer. It automatically withdraws samples from sample vials and carries them through the procedures required for analysis. Generally, colorimetric procedures are employed, and a filter photometer is used to measure the absorbance of the colored solution. The AutoAnalyzer may be set up to filter the sample, add reagent, dilute, mix, heat, digest, or even extract [11].

A schematic diagram of an AutoAnalyzer set-up for the analysis of alkalinity is shown in Figure 9.15 (see page 248). The liquids are transported through the AutoAnalyzer system by a peristaltic pump, consisting basically of rollers moving over flexible tubing. By using different sizes of tubing, the flow rates of the reagents are proportioned. Air bubbles are introduced into the liquid stream to aid mixing and to separate one sample from another. Mixing of the sample and various reagents is accomplished in mixing coils. Since many color-developing reactions are not rapid, a delay coil is provided that allows the color to develop before reaching the colorimeter. Bubbles are removed from the liquid stream by a debubbler prior to introduction into the flowcell for colorimetric analysis.

Nothing would be more helpful in understanding natural water systems than the capability for continuous monitoring of all physical properties and

---

[11] Allen, H.E. 1971. Automated chemical analysis. In *Instrumental analysis for water pollution control,* K.H. Mancy, ed., Ch. 7. Ann Arbor, Mich.: Ann Arbor Science Publishers, Inc.

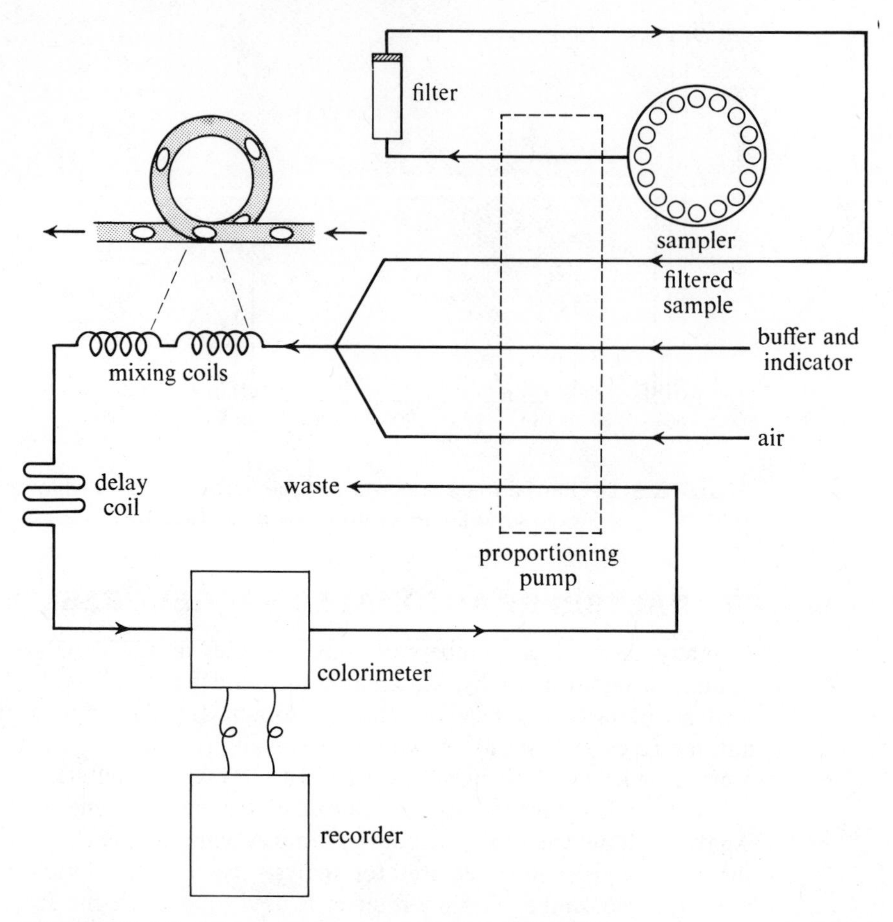

FIGURE 9.15 Schematic diagram of an AutoAnalyzer set up for analysis of total alkalinity. Addition of alkalinity to a methyl orange solution buffered at pH 3.1 causes a loss in color in proportion to the alkalinity.

chemical species of interest. Although that ideal has not been realized, a number of properties of water can be monitored continuously. These properties are primarily ones for which direct sensors are available, such as chloride ion, conductivity, dissolved oxygen, pH, solar radiation, temperature, and turbidity. Redox potential may also be measured, although the reading obtained is generally of dubious value.

In principle, it is possible to measure many chemical species in water on a continuous basis using the appropriate reagents and automatic analysis equipment. In practice, so many things can go wrong in remote continuous monitoring stations, and the expense is so great, that it is not practical to monitor large numbers of chemical species in natural waters on a continuous basis. The ideal of having a trouble-free specific-ion electrode for the measurement of every ion of interest is a long way from realization.

9.14 WATER TEST KITS

Many of the analytical methodologies discussed in this chapter are very sophisticated. For that reason, their use is confined to well-equipped, well-financed laboratories. In many cases, equipment, personnel, and funds are not available for the use of sophisticated tools for water analysis. Frequently the capability of doing an immediate on-site analysis is much more important than high accuracy or ultrasensitivity. For such applications, ingenious water test kits have been devised [12] for simple water analysis by untrained personnel. Over 40 water constituents and physical properties may be measured with such kits, especially where concentrations are at a medium-to-high level. Some utilize simple color comparators for colorimetric analyses, and a simple filter photometer is available for more accurate analyses. These tests, used with full knowledge of their limitations, are a necessary and important tool for water analysis.

9.15 PHYSICAL PROPERTIES MEASURED IN WATER

The commonly determined physical properties of water are color, residue (solids), odor, temperature, specific conductance, and turbidity. Most of these terms are self-explanatory and will not be discussed in detail. All of these properties either influence or reflect the chemistry of the water. Solids, for example, arise from chemical substances either suspended or dissolved in the water and are classified physically as total, filterable, nonfiltrable, or volatile. **Specific conductance** is a measure of the degree to which water conducts alternating current and reflects, therefore, the total concentration of dissolved ionic material. Details on standard methods for determining these parameters have been summarized [13–16].

9.16 WATER SAMPLING

It is beyond the scope of this text to describe water sampling procedures in detail. It must be emphasized, however, that the acquisition of meaningful data demands that correct sampling and storage procedures be used. These procedures may be quite different for different species in water. Details on the acquisition and storage of samples are found in References 13–16.

In general, separate samples must be collected for chemical and biological analysis because the sampling and preservation techniques are quite different. Usually, the shorter the time interval between sample collection and analysis, the more accurate the analysis will be. Indeed, some analyses must be performed in the field within minutes of sample collection. Others, such as the determination of temperature, must be done on the body of water itself. Within a few minutes after collection, water pH may change, dissolved gases (oxygen, carbon dioxide, hydrogen sulfide, chlorine) may be lost, or other gases (oxygen, carbon dioxide)

---

[12] 1976. *Products for water and wastewater analysis.* Loveland, Co.: Hach Chemical Company.

[13] 1974. *Methods for chemical analysis of water and wastes.* Cincinnati, Ohio: Environmental Protection Agency.

[14] 1980. *Standard methods for the examination of water and wastewater.* 15th ed. Washington, D.C.: American Public Health Association.

[15] *Annual book of ASTM standards, water and atmospheric analysis.* Philadelphia: American Society for Testing and Materials. Published annually.

[16] 1973. *Instrumentation for environmental monitoring—water.* Berkeley, Calif.: University of California, Lawrence Berkeley Laboratory. Periodic updates.

may be absorbed from the atmosphere. Therefore, analyses of temperature, pH, and dissolved gases should always be performed in the field. Furthermore, precipitation of calcium carbonate accompanies changes in the pH–alkalinity–calcium carbonate relationship following sample collection. Analysis of a sample after standing may thus give erroneously low values for calcium and total hardness.

Redox reactions may cause substantial errors in analysis. For example, soluble iron(II) and manganese(II) are oxidized to insoluble iron(III) and manganese(IV) compounds as an anaerobic water sample absorbs atmospheric oxygen (see Sections 3.9 and 3.10). Microbial activity may decrease phenol or biological oxygen demand (BOD) values, change the nitrate-nitrite-ammonia balance, or alter the relative proportions of sulfate and sulfide. Iodide and cyanide frequently are oxidized. Chromium(VI) may be reduced to chromium(III), which precipitates readily. Sodium, silica, and boron are leached from glass container walls. Color, odor, and turbidity change as the sample ages. Plastic containers often are sources of trace-metal contamination. Many other things can go wrong, but these are some of the more common problems resulting from sample storage.

Acquiring a truly representative sample is as important as sample preservation. A representative single sample of a body of water must be a composite of many samples taken from a number of different locations over a long period of time.

Generally, it is much more meaningful to analyze a large number of separate samples taken at different times and different locations than it is to compile and analyze a single representative sample. For example, it will be little consolation to fish that the *average* dissolved oxygen level in a eutrophic lake is sufficient to support them, if the dissolved oxygen level falls to near zero before dawn each day!

Very accurate records of the location, time, and conditions of sample collection are essential. Although we can point out many pitfalls to be avoided in sample collection and storage, ultimate success depends upon the judgment of the individual analyst. Written directions for sampling cannot cover every possible case.

Because of the extreme dilution of many water pollutants, the material to be analyzed frequently must be concentrated prior to analysis. An example of a commonly used concentration technique is **carbon adsorption** for the collection of organic matter [17]. With this technique, quantities of water as large as a thousand gallons or more are passed over activated carbon. The organic material is subsequently extracted from the dried carbon, first with chloroform, then with alcohol. The solvents are evaporated and the weights of the extracts are expressed in units of $\mu g/L$ as carbon-chloroform extract (CCE) and carbon-alcohol extract (CAE).

A large number of organic compounds are extracted from the activated carbon. The chloroform extract is particularly significant because it contains compounds responsible for tastes and odors in water. Many of the organics

[17] Andelman, J.B., and Christian, R.F. 1971. Concentration and separation techniques. In *Instrumental analysis for water pollution control.* K.H. Mancy, ed., Ch. 4. Ann Arbor, Mich.: Ann Arbor Science Publishers, Inc.

extracted by chloroform, such as phenols and oils, are relatively insoluble in water. The alcohol extracts, though often exceeding the CCE in quantity, generally are not so odorous.

A number of separation procedures are used to further characterize the organic extracts. Among the organics identified by these procedures are carboxylic acids, phenols, sulfonic acids, fulvic acids, pesticides, and polycyclic aromatic hydrocarbons.

The carbon adsorption method leaves much to be desired as a tool for characterizing organics in water. Chemical reactions can occur on the carbon surface. Some organics are volatilized during the drying process, while others are not completely extracted. Further volatilization may occur when the solvents are evaporated. Nevertheless, the simplicity, speed, and efficiency of carbon adsorption make it a useful tool in determining organics in water.

Several concentration techniques are based upon a phase change in water. The most common such technique is **freeze concentration.** The water in the sample is frozen, yielding pure crystals of ice and leaving water-soluble impurities in a liquid phase of much reduced volume. The low temperature required for freeze concentration avoids loss of volatile constituents. Obviously, it is not feasible to subject the enormous volumes of water sampled by carbon adsorption to freeze concentration.

Solvent extraction (liquid-liquid extraction) is a useful method for isolating organic-soluble materials from water. When water is shaken with a water-immiscible organic solvent, organic impurities in the water become distributed between the water and solvent phases. If the impurity is much more soluble in the organic phase, it is possible to concentrate it in a relatively small volume of organic solvent. A particularly useful application of solvent extraction is the isolation of chlorinated hydrocarbon pesticides; frequently, the solvent is added directly to the sample bottle at the time of collection. Another useful application of solvent extraction involves the concentration of metal ions after reaction with a chelating agent to form nonionic complexes.

Ion exchange (see Section 8.4) is a separation procedure finding wide application in water analysis. It is used to concentrate ions from large volumes of water. Ion exchange is particularly useful for the selective removal of ions that interfere with the analysis of other ions. It may also be used for determining the total equivalents of ionic material present by titration of the hydrogen ion liberated by reactions of the type

$$n\mathrm{H}^{+}\,{}^{-}\mathrm{Cat(solid)} + \mathrm{M}^{n+} = \mathrm{M}^{n+}[{}^{-}\mathrm{Cat(solid)}]_n + n\mathrm{H}^{+} \qquad (9.16.1)$$

where $^{-}$Cat(solid) represents a cation exchanger. Special chelating cation-exchange resins are employed to remove selected heavy metals from water for analysis.

Sample treatment immediately following collection is used to define several different states of metal in water. A sample to be analyzed for *dissolved metals* is filtered through a 0.45-micron membrane filter as soon as possible after collection. The first portion of the filtrate is used to rinse the container, and a sufficient amount is then collected for analysis. The solution collected for analysis is acidified to pH 2 or 3, thus enabling storage for some time prior to analysis. The addition of acid prevents hydrolysis and precipitation.

Suspended material to be analyzed for metals is collected on a 0.45-micron membrane filter. Such a filter retains virtually all the mass of the particulate matter. Filters with a smaller pore diameter are not generally practical because of extremely slow filtration rates. A specific volume, generally 100 mL or more, of a well-agitated sample is filtered, using a vacuum or pressure technique to speed the filtration. After filtration, the membrane filter and the material on it are transferred to a beaker and digested in hot, concentrated HNO_3. The sample is diluted to a definite volume, depending upon the expected level of metal ion, then analyzed by atomic absorption.

Extractable metal is treated in the following way. When collected, the entire sample is acidified with concentrated HNO_3, 5 mL/L. At the time of analysis, the sample is agitated and a representative 100-mL aliquot is placed in a beaker or flask. After the addition of 5 mL of redistilled hydrochloric acid, the sample is heated to near boiling for 15 minutes, then filtered. The volume of the filterable sample is adjusted to 100 mL with distilled water, and the sample is analyzed. This procedure measures something less than the total amount of metal in the sample. The metal found will be higher than that found in the soluble fraction, substantially so in the presence of a large quantity of suspended material. The procedure gives a relatively good estimate of total available metal.

9.17 WATER SAMPLE PRESERVATION

It is not possible to protect a water sample from changes in composition. However, various additives and treatment techniques can be employed to minimize sample deterioration. These methods are summarized in Table 9.4.

TABLE 9.4 Preservatives and Preservation Methods Used with Water Samples[1]

Preservative or technique used	*Effect on sample*	*Type of samples for which the method is employed*
Nitric acid	Keeps metals in solution	Metal-containing samples
Sulfuric acid	Bactericide	Biodegradable samples containing organic carbon, COD, oil, or grease
	Formation of sulfates with volatile bases	Samples containing amines or ammonia
Sodium hydroxide	Formation of sodium salts with volatile acids	Samples containing volatile organic acids or cyanides
Mercuric chloride	Bactericide	Samples containing various forms of nitrogen or phosphorus, or some biodegradable organics
Cooling (4 °C)	Inhibition of bacteria, retention of volatile material	Samples containing microorganisms, acidity, alkalinity, BOD, organic C, P, and N, color, odor
Chemical reaction	Fix a particular constituent	Samples to be analyzed for dissolved oxygen using the Winkler method

[1] From information in 1974, *Methods for Chemical Analysis of Water and Wastes,* Cincinnati, Ohio: Environmental Protection Agency.

The most general method of sample preservation is refrigeration to 4 °C. Freezing normally should be avoided because of physical changes—formation of precipitates and loss of gas—which may adversely affect sample composition.

Sample holding times vary, from zero for parameters such as temperature or dissolved oxygen measured by a probe, to 6 months for metals. Many different kinds of samples, including those to be analyzed for acidity, alkalinity, and various forms of nitrogen or phosphorus, should not be held for more than 24 hours.

Details on water sample preservation are to be found in standard references on water analysis [13, 14]. Instructions should be followed for each kind of sample in order to ensure meaningful results.

9.18 TOTAL ORGANIC CARBON IN WATER

The importance of dissolved organic compounds in water was discussed in Section 7.15. Dissolved organic carbon exerts an oxygen demand in water; often is in the form of toxic substances; and is a general indicator of water pollution. Therefore, its measurement is quite important.

The measurement of **total organic carbon, TOC**, is now recognized as the best means of assessing the organic content of a water sample. This measurement has gained favor recently with the development of methods [18] which, for the most part, totally oxidize the dissolved organic material to produce carbon dioxide. The amount of carbon dioxide evolved is taken as a measure of TOC.

Older TOC instruments used a pyrolysis technique to oxidize organics. This approach is handicapped by difficulties in handling the relatively large quantities of water involved in the analysis of a low TOC-content sample. A newer approach uses a dissolved oxidizing agent promoted by ultraviolet light. Potassium peroxidisulfate, $K_2S_2O_8$, is usually chosen as an oxidizing agent to be added to the sample. Phosphoric acid is also added to the sample, which is sparged with air or nitrogen to drive off CO_2 formed from HCO_3^- and CO_3^{2-} in solution. After sparging, the sample is pumped to a chamber containing a lamp emitting ultraviolet radiation of 184.9 nm. This radiation produces reactive free radical species, such as the hydroxyl radical, $HO\cdot$, discussed as a photochemical reaction intermediate in Section 11.11. These active species bring about the rapid oxidation of dissolved organic compounds as shown in the following general reaction:

$$\text{organics} + HO\cdot \xrightarrow[K_2S_2O_8]{} CO_2(g) + H_2O \qquad (9.18.1)$$

After oxidation is complete, the CO_2 is sparged from the system and measured with a gas chromatographic detector or by absorption in ultrapure water followed by a conductivity measurement. Figure 9.16 (p. 254) is a schematic of a modern TOC analyzer.

9.19 MEASUREMENT OF RADIOACTIVITY IN WATER

There are several sources of naturally occurring radioactive materials that may contaminate water (see Section 7.12). These include particularly uranium, thorium, and radium, as well as the radioactive products of the breakdown of

[18] Poirer, S.J. and Wood, J.H. 1978. A new approach to the measurement of organic carbon. *American Laboratory* December, 1978, pp. 78–89.

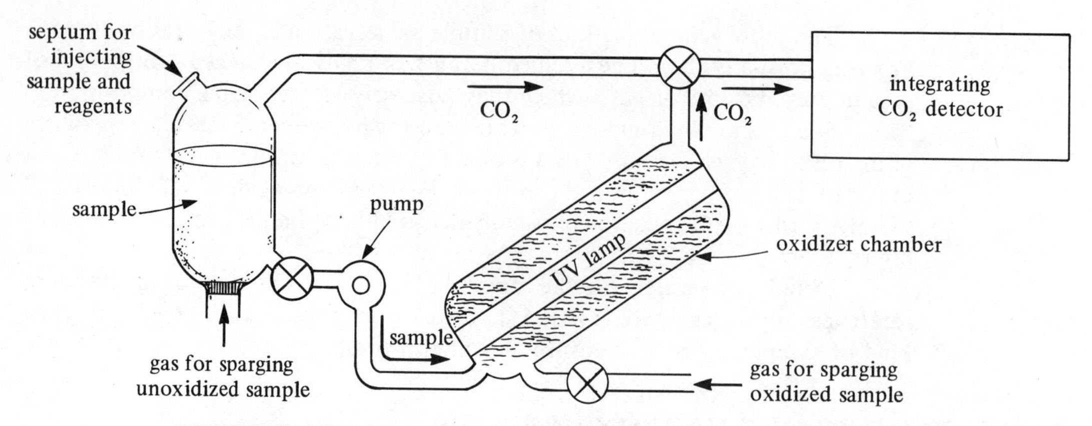

FIGURE 9.16 TOC analyzer employing UV-promoted sample oxidation.

these elements, all of which emit alpha, beta, or gamma radiation in the process of decay to a stable end product. In addition, there are a number of artificial sources of radioactivity that hold the potential for water contamination.

Radioactive contamination of water is normally detected by measurements of gross beta and gross alpha activity, a procedure that is simpler than detecting individual isotopes. The determination is made in the following way. The sample is evaporated to form a very thin layer on a small pan, which is then inserted inside an **internal proportional counter**. This set up is necessary because beta particles can penetrate only very thin detector windows, and alpha particles have essentially no penetrating power. Without *back-scattering* (bouncing of particles back into the detector), over 50% of the beta radiation and slightly more than 50% of the alpha radiation enter the detector and are recorded. With complete back-scattering, 100% of each type of radiation is recorded.

More detailed information can be obtained for radionuclides that emit gamma rays by the use of **gamma spectrum analysis**. This technique employs solid state detectors, such as germanium doped with lithium, Ge(Li), to resolve rather closely spaced gamma peaks in the sample's spectra. In conjunction with multichannel spectrometric data analysis, it is possible to analyze a number of radionuclides in the same sample without chemical separation. The method requires minimal sample preparation. Its biggest disadvantage is that it does not work for isotopes that do not emit gamma radiation.

9.20 SUMMARY OF WATER ANALYSIS PROCEDURES

The main chemical parameters commonly determined in water are summarized in Table 9.5. In addition to these, a number of other solutes, especially specific organic pollutants, may be determined in connection with specific health hazards or incidents of pollution.

TABLE 9.5 Chemical Parameters Commonly Determined in Natural Waters and Water Supplies

Chemical species	*Significance in water*	*Methods of analysis commonly used*
Acidity	Indicative of industrial pollution or acid mine drainage	Titration
Alkalinity	Water treatment, buffering, algal productivity	Titration
Aluminum	Water treatment, buffering	Atomic absorption, colorimetry
Ammonia	Algal productivity, pollutant	Colorimetry
Arsenic	Toxic pollutant	Colorimetry, atomic absorption
Barium	Toxic pollutant	Atomic absorption
Beryllium	Toxic pollutant	Atomic absorption, fluorimetry
Boron	Toxic to plants	Colorimetry, plasma emission
Bromide	Seawater intrusion, industrial waste	Colorimetry, potentiometry
Cadmium	Toxic pollutant	Atomic absorption
Calcium	Hardness, productivity, treatment	Atomic absorption
Carbon dioxide	Bacterial action, corrosion	Titration, calculation
Chloride	Saline water contamination	Titration, potentiometry
Chlorine	Water treatment	Colorimetry
Chromium	Toxic pollutant (hexavalent Cr)	Atomic absorption, colorimetry
Copper	Plant growth	Atomic absorption
Cyanide	Toxic pollutant	Colorimetry, potentiometry
Fluoride	Water treatment, toxic at high levels	Colorimetry, potentiometry
Hardness	Water quality, water treatment	Atomic absorption, titration
Iodide	Seawater intrusion, industrial waste	Catalytic effect, potentiometry
Iron	Water quality, water treatment	Atomic absorption, colorimetry
Lead	Toxic pollutant	Atomic absorption, voltammetry
Lithium	May indicate some pollution	Atomic absorption, flame photometry
Magnesium	Hardness	Atomic absorption
Manganese	Water quality (staining)	Atomic absorption
Mercury	Toxic pollutant	Flameless atomic absorption
Methane	Anaerobic bacterial action	Combustible-gas indicator
Nitrate	Algal productivity, toxicity	Colorimetry, potentiometry
Nitrite	Toxic pollutant	Colorimetry
Nitrogen (albuminoid)	Proteinaceous material	Colorimetry
Nitrogen (organic)	Organic pollution indicator	Colorimetry
Oil and grease	Industrial pollution	Gravimetry

TABLE 9.5 Chemical Parameters (*continued*)

Chemical species	*Significance in water*	*Methods of analysis commonly used*
Oil and grease	Industrial pollution	Gravimetry
Organic carbon	Organic pollution indicator	Oxidation-CO_2 measurement
Organic contaminants	Organic pollution indicator	Activated carbon adsorption
Oxygen	Water quality	Titration, electrochemical
Oxygen demand (biochemical)	Water quality and pollution	Microbiological-titration
Oxygen demand (chemical)	Water quality and pollution	Chemical oxidation-titration
Ozone	Water treatment	Titration
Pesticides	Water pollution	Gas chromatography
pH	Water quality and pollution	Potentiometry
Phenols	Water pollution	Distillation-colorimetry
Phosphate	Productivity, pollution	Colorimetry
Phosphorus (hydro-lyzable)	Water quality and pollution	Colorimetry
Potassium	Productivity, pollution	Atomic absorption, flame photometry
Selenium	Toxic pollutant	Colorimetry, neutron activation
Silica	Water quality	Colorimetry, plasma emission
Silver	Water pollution	Atomic absorption
Sodium	Water quality, saltwater intrusion	Atomic absorption, flame photometry
Strontium	Water quality	Atomic absorption, flame photometry
Sulfate	Water quality, water pollution	Gravimetry, turbidimetry
Sulfide	Water quality, water pollution	Colorimetry, potentiometry, titration
Sulfite	Water pollution, oxygen scavenger	Titration
Surfactants	Water pollution	Colorimetry
Tannin and Lignin	Water quality, water pollution	Colorimetry
Vanadium	Water quality, water pollution	Catalytic effect
Zinc	Water quality, water pollution	Atomic absorption

SUPPLEMENTARY REFERENCES

Albanges, J., ed. 1980, 1982. *Analytical techniques in environmental chemistry*. 2 Vols. Elmsford, N.Y.: Pergamon Press, Inc.

Baiulescu, G.E., and Cosofret, V.V. 1977. *Applications of ion-selective electrodes in organic analysis*. New York: Halsted Press.

Dollberg, D.D., and Verstuyft, A.W. 1980. *Analytical techniques in occupational health chemistry*. Washington, D.C.: American Chemical Society.

Epstein, S. 1977. *Chemical analysis by emission spectroscopy.* Palisades, N.J.: Franklin Publishing Co.

Ewing, G.W. 1977. *Environmental analysis.* New York: Academic Press, Inc.

Frei, R.W., ed. 1979. *Recent advances in environmental analysis.* New York: Gordon and Breach, Science Publishers, Inc.

Fritschen, L.J., and Gay, L.W. 1979. *Environmental instrumentation.* New York: Springer-Verlag, New York, Inc.

Gibb, T.R.P., Jr. 1975. *Analytical methods in oceanography.* ACS Advances in Chemistry Series 147. Washington, D.C.: American Chemical Society.

Hemphill, D.D. *Proceedings of the University of Missouri's Annual Conference on Trace Substances in Environmental Health.* Columbia, Mo.: University of Missouri. Published annually since 1967.

Hitchman, M.L. 1978. *Measurement of dissolved oxygen.* New York: Wiley-Interscience.

1975. *Instrumentation for environmental monitoring. Vol. 3: Radiation.* Berkeley, Calif.: Lawrence Berkeley Laboratory.

1975. *Instrumentation for environmental monitoring. Vol. 4: Biomedical.* Berkeley, Calif.: Lawrence Berkeley Laboratory.

Lawrence, J.F., ed. 1981. *Trace analysis.* Vol. 1. New York: Academic Press, Inc.

1974. *Methods for chemical analysis of water and wastes.* Cincinnati, Ohio: U.S. Environmental Protection Agency.

Middleditch, B.S.; Missler, S.R.; and Hines, H.B. 1981. *Mass spectrometry of priority pollutants.* New York: Plenum Publishing Corp.

Moye, H.A., ed. 1981. *Analysis of pesticide residues.* New York: Wiley-Interscience.

Pinta, M. 1978. *Modern methods for trace element analysis.* Ann Arbor, Mich.: Ann Arbor Science Publishers, Inc.

1976. *Quality criteria for water.* EPA-440/9-76-023. Washington, D.C.: U.S. Environmental Protection Agency.

Risby, T.H., ed. 1979. *Unlratrace metal analysis in biological sciences and environment.* Washington, D.C.: American Chemical Society.

Sawicki, E.; Mulik, J.D.; and Wittgenstein, E., eds. 1978. *Ion chromatographic analysis of environmental pollutants.* Ann Arbor, Mich.: Ann Arbor Science Publishers, Inc.

Schuetzle, D., ed. 1979. *Monitoring toxic substances.* Washington, D.C.: American Chemical Society.

Sittig, M. 1974. *Pollution detection and monitoring.* Park Ridge, N.J.: Noyes Data Corp.

Toribara, T.Y.; Coleman, J.R.; Dahneke, B.E.; and Feldman, I. 1978. *Environmental pollutants detection and measurement.* New York: Plenum Publishing Corp.

Winefordner, J.D. 1976. *Trace analysis spectroscopic methods for elements.* New York: Wiley-Interscience.

Zweig, G., ed. 1980. *Analytical methods for pesticides and plant growth regulators.* New York: Academic Press, Inc.

QUESTIONS AND PROBLEMS

1. A soluble water pollutant forms ions in solution and absorbs light at 535 nm. What are two physical properties of water influenced by the presence of this pollutant?

2. A sample was taken from the bottom of a deep, stagnant lake. Upon standing, bubbles were evolved from the sample; the pH went up; and a white precipitate formed. From these observations, what may be said about the dissolved CO_2 and hardness in the water?

3. For which of the following analytes may nitric acid be used as a water sample preservative? H_2S; CO_2; metals; coliform bacteria; cyanide.

4. In the form of what compound is oxygen fixed in the Winkler analysis of O_2?

5. Of the following analytical techniques, the one that would best distinguish between the hydrated $Ag(H_2O)_x^+$ ion and the complex $Ag(NH_3)_2^+$ ion in a wastewater sample is:
   (a) neutron-activation analysis
   (b) atomic absorption
   (c) inductively coupled plasma atomic emission spectroscopy
   (d) potentiometry
   (e) flame emission.

6. A water sample was run through the colorimetric procedure for the analysis of nitrate, giving 55.0% transmittance. A sample containing 1.00 ppm nitrate run through the exactly identical procedure gave 24.6% transmittance. What was the concentration of nitrate in the first sample?

7. What is the molar concentration of HCl in a water sample containing HCl as the only contaminant and having a pH of 3.80?

8. A 200-mL sample of water required 25.12 mL of 0.0200*N* standard H_2SO_4 for titration to the methyl orange endpoint, pH 4.5. What was the total alkalinity of the original sample?

9. Sulfur dioxide was analyzed in air by the West-Gaeke method. A standard air sample containing 5.00 micrograms of sulfur dioxide gave an absorbance of 0.600. Exactly 10 liters of atmospheric air passed into the analyzer gave an absorbance of 0.480. What was the concentration of sulfur dioxide in air in units of micrograms per cubic meter?

10. Analysis of a lead-containing sample by graphite-furnace atomic absorption analysis gave a peak of 0.075 absorbance units when 50 microliters of pure sample was injected. Lead was added to the sample such that the *added* concentration of lead was 6.0 micrograms per liter. Injection of 50 microliters of "spiked" sample gave an absorbance of 0.115 absorbance units. What was the concentration of lead in the original sample?

11. A fluoride electrode read −0.100 volts *versus* a reference electrode in $2.63 \times 10^{-4}M$ standard fluoride solution and −0.118 volts in an appropriately processed fluoride sample. What was the concentration of fluoride in the sample?

12. The activity of iodine-131 ($t_{1/2}$ = 8 days) in a water sample 24 days after collection was 520 picoCuries/liter. What was the activity on the day of collection?

13. Neutron irradiation of exactly 2.00 mL of a standard solution containing 1.00 mg/L of unknown heavy metal "X" for exactly 30 seconds gave an activity of 1,257 counts per minute, when counted exactly 33.5 minutes after the irradiation, measured for a radionuclide product of "X" having a half-life of 33.5 minutes. Irradiation of an unknown water sample under identical conditions (2.00 mL, 30.0 seconds, same neutron flux) gave 1,813 counts per minute when counted 67.0 minutes after irradiation. What was the concentration of "X" in the unknown sample?

14. Why is magnesium-EDTA chelate added to a magnesium-free water sample before it is to be titrated with EDTA for Ca^{2+}?

15. For what type of sample is the flame-ionization detector most useful?

16. Manganese from a standard solution was oxidized to MnO_4^- and diluted such that the final solution contained 1.00 mg/L of Mn. This solution had an absorbance of 0.316. A 10.00 mL wastewater sample was treated to develop the MnO_4^- color and diluted to 250.0 mL. The diluted sample had an absorbance of 0.296. What was the concentration of Mn in the original wastewater sample?

CHAPTER 10
ENVIRONMENTAL CHEMISTRY OF THE GEOSPHERE AND SOIL

10.1 THE GEOSPHERE

The **geosphere**, or solid Earth, is that part of the Earth upon which humans live and from which they extract most of their food, minerals, and fuels. Once thought to have an almost unlimited buffering capacity against the perturbations of humankind, the geosphere is now known to be rather fragile and may, indeed, be harmed by human activities [1]. For example, some billions of tons of Earth material are mined or otherwise disturbed each year in the extraction of minerals, including coal. Two atmospheric pollutant phenomena—excess carbon dioxide (see Section 11.12) and acid rain (see Section 12.15)—have the potential to cause major changes in the geosphere. The excess carbon dioxide may cause global heating ("greenhouse effect"), which could significantly alter rainfall patterns and turn currently productive areas of the Earth into desert regions. The low pH characteristic of acid rain can bring about drastic changes in the solubilities and oxidation-reduction rates of minerals. Erosion caused by intensive cultivation of land is washing away vast quantities of topsoil from fertile farmlands each year. In some areas of industrialized countries, the geosphere has been the dumping ground for toxic chemicals (see the discussion of hazardous wastes in Chapters 19 and 20). Ultimately, the geosphere must provide disposal sites for the nuclear wastes of the approximately 300 nuclear reactors now operating worldwide, as well as those yet to be completed. It may be readily seen that the preservation of the geosphere in a form suitable for human habitation is one of the greatest challenges facing humankind.

Human activities on the Earth's surface may have an effect upon climate. The most direct such effect is through the change of surface **albedo**, defined as the precentage of incident solar radiation reflected by a land or water surface. For example, if the sun radiates 100 units of energy per minute to the outer limits of the atmosphere, and the Earth's surface receives 60 units per minute of the total and then reflects 30 units upward, the albedo is 50 percent. Some typical albedo

---

[1] Fyfe, W.S. 1981. The environmental crisis: quantifying geosphere interactions. *Science* 213: 105–110.

values for different areas on the Earth's surface are [2]: evergreen forests, 7-15%, dry, plowed fields, 10-15%; deserts, 25-30%; fresh snow, 85-90%; asphalt, 8%. In some heavily developed areas, anthropogenic heat release is comparable to the solar input. The anthropogenic (human-produced) energy release over the 60 square kilometers of Manhattan Island averages about 4 times the solar energy falling on the area; over the 3,500 km^2 of Los Angeles the anthropogenic energy release is about 13% of the solar flux.

One of the greater impacts of humans upon the geosphere is the creation of desert areas through abuse of land with marginal amounts of rainfall. The process of **desertification** is manifested by declining ground water tables, salinization of topsoil and water, reduction of surface waters, unnaturally high soil erosion, and desolation of native vegetation [3]. Large, arid areas of the western U.S. are experiencing at least some of these symptoms as the result of human activities. As the populations of the western states increase, one of the greatest challenges facing the residents is to prevent additional desertification. The problem is severe in other parts of the world, particularly Africa's Sahel (southern rim of the Sahara), where the Sahara advanced southward at a particularly rapid rate during the period 1968-73.

To humans the most important part of the geosphere is the **soil**. Though only a tissue-thin layer compared to the Earth's total diameter, soil is the medium for the production of most food, upon which most living things must depend. Good soil—and a climate conducive to its productivity—is the most valuable asset a nation can have.

In addition to being the site of most food production, soil is the receptor of large quantities of pollutants, such as particulate matter from power plant smokestacks. Fertilizers and some other materials applied to soil often contribute to water and air pollution. Therefore, soil is a key component of environmental chemical cycles.

With increasing population and industrialization, one of the more important aspects of our use of the geosphere has to do with the protection of water sources [4]. Mining, agricultural, chemical, and radioactive wastes all have the potential for contaminating both surface water and ground water. Sewage sludge spread on land may contaminate water by release of nitrate and heavy metals. Landfills may likewise be sources of contamination. Leachates from unlined pits and lagoons containing hazardous liquids or sludges may contaminate drinking water.

It should be noted, however, that many soils have the ability to assimilate and neutralize pollutants. Various chemical and biochemical phenomena in soils operate to reduce the harmful nature of pollutants. These phenomena include oxidation-reduction processes, acid-base reactions, precipitation, sorption, and biochemical degradation. Some hazardous organic chemicals may be degraded to harmless products, and heavy metals may be sorbed by the soil. In general,

---

[2] Barney, G.O. 1980. *The global 2000 report to the President.* 2 vols. Washington, D.C.: U.S. Government Printing Office.

[3] Sheridan, D. 1981. *Desertification of the United States.* Washington, D.C.: Council on Environmental Quality, U.S. Government Printing Office.

[4] Pojasek, R.B. 1977. *Drinking water quality enhancement through source protection.* Ann Arbor, Mich.: Ann Arbor Science Publishers, Inc.

however, extreme care should be exercised in disposing of chemicals, sludges, and other potentially hazardous materials on soil where the possibility of water contamination exists.

10.2 THE NATURE OF SOIL

Soil is a variable mixture of minerals, organic matter, and water, capable of supporting plant life on the Earth's surface. It contains air spaces and generally has a loose texture (Fig. 10.1). The mineral portion of soil is formed from parent rocks by the weathering action of physical, chemical, and biological processes. The organic portion consists of plant biomass in various stages of decay. High populations of bacteria, fungi, and animals such as earthworms may be found in soil.

The solid fraction of a typical productive soil is approximately 5% organic matter and 95% inorganic matter. Some soils, such as peat soils, may contain as much as 95% organic material. Other soils contain as little as 1% organic matter.

Typical soils exhibit distinctive layers with increasing depth (Fig. 10.2). These layers are called **horizons**. The top layer, typically several inches in thickness, is known as the **A horizon**, or **topsoil**. This is the layer of maximum biological activity in the soil and contains most of the soil organic matter. Metal ions and clay particles in the A horizon are subject to considerable leaching. The next layer is the **B horizon**, or **subsoil**. It receives material such as organic matter, salts, and clay particles leached from the topsoil. The **C horizon** is composed of weathered parent rocks from which the soil originated.

10.3 WATER AND AIR IN SOIL

Large quantities of water are required for the production of most plant materials. For example, several hundred pounds of water are required to produce one pound of dry hay. Water is part of the three-phase, solid-liquid-gas system mak-

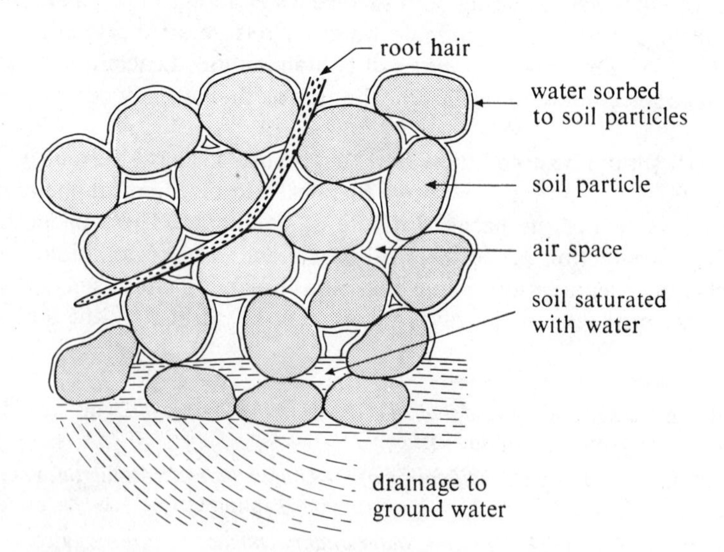

FIGURE 10.1 Fine structure of soil, showing solid, water, and air phases.

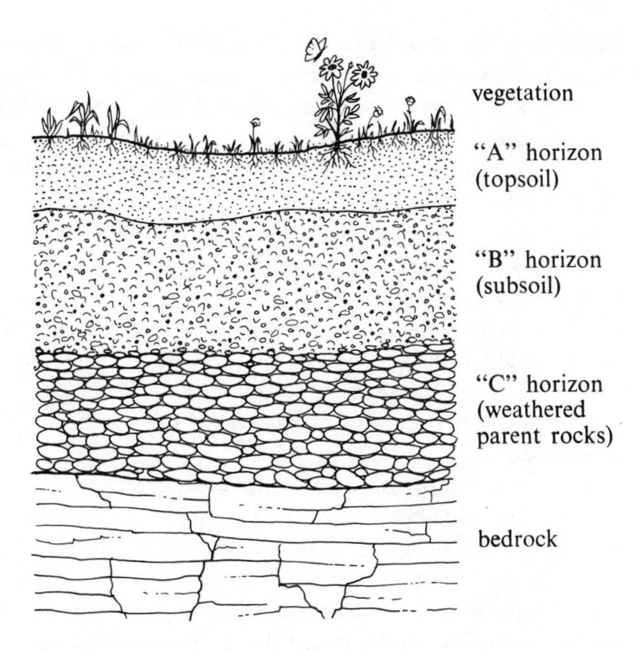

FIGURE 10.2 Soil profile showing soil horizons.

ing up soil. It is the basic transport medium for carrying essential plant nutrients from solid soil particles into plant roots and to the farthest reaches of the plant's leaf structure (Fig. 10.3). The water enters the atmosphere from the plant's leaves, a process called **transpiration.**

Normally, because of the small size of soil particles and the presence of small capillaries and pores in the soil, the water phase is not totally independent

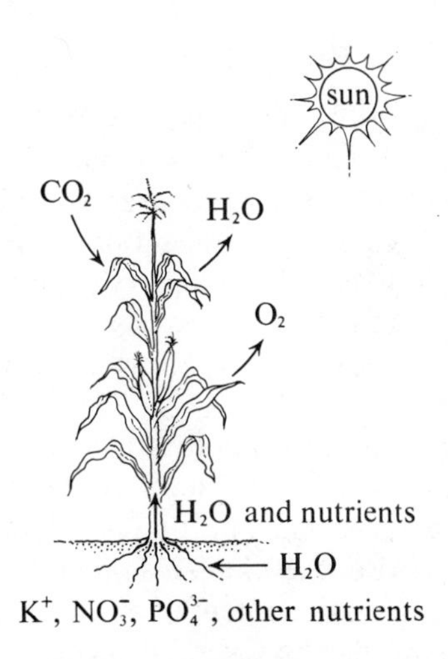

FIGURE 10.3 Plants transport water from the soil to the atmosphere by transpiration. Nutrients are also carried from the soil to the plant extremities by this process. Plants remove CO_2 from the atmosphere and add O_2 by photosynthesis. The reverse occurs during plant respiration.

of soil solid matter [5]. The availability of water to plants is governed by gradients arising from capillary and gravitational forces. The availability of nutrient solutes in water depends upon concentration gradients and electrical potential gradients. Water present in larger spaces in soil is relatively more available to plants and readily drains away. Water held in smaller pores, or between the unit layers of clay particles (see Section 6.5) is held much more strongly. Soils high in organic matter may hold a relatively large amount of water compared to other soils, but it is relatively less available to plants because of physical and chemical sorption of the water by the organic matter.

There is a very strong interaction between clays and water in soil. Water is absorbed on the surfaces of clay particles. Because of the high surface/volume ratio of colloidal clay particles, a great deal of water may be bound in this manner. Water is also held between the unit layers of the expanding clays, such as the montmorillonite clays.

A waterlogged (water-saturated) soil undergoes drastic changes in physical, chemical, and biological properties. Oxygen in such soil is rapidly used up by the respiration of microorganisms degrading organic matter. In such soils, the bonds holding soil colloidal particles together are broken, which causes disruption of soil structure. Thus, the excess water in such soils is detrimental to plant growth, and the soil does not contain the air required by most plant roots. Most useful crops, with the notable exception of rice, cannot grow on waterlogged soils.

One of the most marked chemical effects of waterlogging is a reduction of pE by the action of organic reducing agents acting through bacterial catalysts. Thus, the redox condition of the soil becomes much more reducing, and the soil pE may drop from that of water in equilibrium with air (+13.6 at pH 7) to 1 or less. One of the more significant results of this change is the mobilization of iron and manganese as soluble iron(II) and manganese(II) through reduction of their insoluble higher oxides:

$$MnO_2 + 4\,H^+ + 2\,e \rightarrow Mn^{2+} + 2\,H_2O \qquad (10.3.1)$$

$$Fe_2O_3 + 6\,H^+ + 2\,e \rightarrow 2\,Fe^{2+} + 3\,H_2O \qquad (10.3.2)$$

Although soluble manganese generally is found in soil as Mn^{2+} ion, soluble iron(II) frequently occurs as negatively charged iron-organic chelates. Strong chelation of iron(II) by soil fulvic acids (Chapter 4) apparently enables reduction of iron(III) oxides at more positive pE values than would otherwise be possible. This causes an upward shift in the $Fe(II)$-$Fe(OH)_3$ boundary shown in Figure 3.3.

Soluble metal ions such as Fe^{2+} and Mn^{2+} are toxic to some plants at high levels. Their oxidation to insoluble oxides may cause formation of deposits of Fe_3O_3 and MnO_2, which clog tile drains in fields.

Roughly 35% of the volume of a typical soil is composed of air-filled pores. Whereas the normal dry atmosphere at sea level contains 21% O_2 and 0.03% CO_2 by volume, these percentages may be quite different in soil air because of the decay of organic matter (Eq. 5.10.2). This process consumes oxygen and produces CO_2. As a result, the oxygen content of air in soil may be as low as 15%, and the carbon dioxide content may be several percent. Thus, the decay

---

[5] van Olphen, H. 1975. Water in soils. In *Soil components, Vol. 2, inorganic components,* J.E. Gieseking, ed., Ch. 15, pp. 497–528. New York: Springer-Verlag New York, Inc.

of organic matter in soil increases the equilibrium level of dissolved CO_2 in ground water. This lowers the pH and contributes to weathering of carbonate minerals, particularly calcium carbonate (see Eq. 2.14.1). As discussed later in this chapter, CO_2 also shifts the equilibrium of the process by which roots absorb metal ions from soil (see Eq. 10.6.1).

10.4 THE INORGANIC COMPONENT OF SOIL

The weathering of parent rocks and minerals to form the inorganic soil component results ultimately in the formation of inorganic colloids. These colloids are repositories of water and plant nutrients, which may be made available to plants as needed. Inorganic soil colloids often absorb toxic substances in soil, thus playing a role in detoxification of substances that otherwise would harm plants. The abundance and nature of inorganic colloidal material in soil are obviously important factors in determining soil productivity.

The uptake of plant nutrients by roots often involves complex interactions with the water and inorganic phases. For example, a nutrient held by inorganic colloidal material has to traverse the mineral/water interface and then the water/root interface. This process is often strongly influenced by the ionic structure of soil inorganic matter, a subject that has occupied the attention of many soil scientists [6].

Among the common elements in the Earth's crust are oxygen, 46.6%; silicon, 27.7%; aluminum, 8.1%; iron, 5.0%; calcium, 3.6%; sodium, 2.8%; potassium, 2.6%; and magnesium, 2.1%. Minerals composed of these elements constitute most of the mineral fraction of the soil. Since 74.3% of the Earth's crust consists of silicon and oxygen, minerals containing these two elements are especially prevalent in soil. For instance, finely divided quartz, SiO_2, is a common soil component. Among the silicates, orthoclase ($KAlSi_3O_8$), albite ($NaAlSi_3O_8$), and epidote ($4\ CaO \cdot 3\ (AlFe)_2O_3 \cdot 6\ SiO_2 \cdot H_2O$) are very common components of soil. Iron oxides, particularly geothite, FeO(OH), and magnetite, Fe_3O_4, make up much of the mineral fraction of many soils; some iron oxides give soil a red color. Manganese oxides and titanium oxides are encountered in abundance in some soils. Of the carbonates, calcium carbonate is a common soil constituent.

The clay minerals (see Section 6.5) occur widely and are chemically significant components of soils. These secondary minerals, consisting basically of hydrated aluminum and iron silicates, serve to bind cations such as Ca^{2+}, Mg^{2+}, K^+, Na^+, and NH_4^+. Thus, these cations are protected from leaching by water but are available as plant nutrients. In water, many clays are readily suspended as colloidal particles. As such they may be leached from the soil or carried to lower soil layers.

10.5 ORGANIC MATTER IN SOIL

Though typically comprising less than 5% of a productive soil, organic matter plays a very important role in determining soil productivity. It serves as a source of food for microorganisms; undergoes chemical reactions such as ion exchange;

---

[6] Gieseking, J.E., 1975. *Soil components, Vol. 2, inorganic components.* New York: Springer-Verlag New York, Inc.

and influences the physical properties of soil. Some organic compounds even contribute to the weathering of mineral matter, the process by which soil is formed. For example, it has been shown [7] that oxalate ion, $C_2O_4^{2-}$, produced as a soil fungi metabolite, occurs in soil as the calcium salts whewellite and weddelite. Oxalate in soil water dissolves minerals, thus speeding the weathering process and increasing the availability of nutrient ion species. This weathering process involves oxalate complexation of iron or aluminum in minerals, represented by the reaction

$$3\ H^+ + M(OH)_3(s) + 2\ CaC_2O_4(s) \rightarrow M(C_2O_4)_2^-(aq) + 2\ Ca^{2+}(aq) + 3\ H_2O \qquad (10.5.1)$$

in which M is Al or Fe. Some soil fungi produce citric acid, and other chelating organic acids, which react with silicate minerals and release potassium and other nutrient metal ions held by these minerals.

```
              H
              |
     O   H    O      H   O
     ||  |    |      |   ||
HO — C — C — C ———— C — C — OH        citric acid
         |    |      |
         H   C=O     H
              |
              O
              |
              H
```

The strong chelating agent 2-ketogluconic acid is produced by some soil bacteria. By solubilizing metal ions, it may contribute to the weathering of minerals. It may also be involved in the release of phosphate from insoluble phosphate compounds.

Biologically active components of the organic soil fraction include polysaccharides, amino sugars, nucleotides, and organic sulfur and phosphorus compounds. Humus, a water-insoluble material that biodegrades very slowly, makes up the bulk of soil organic matter. The organic compounds in soil have been described in detail [8]. They are summarized in Table 10.1.

The accumulation of organic matter in soil is strongly influenced by temperature and by the availability of oxygen. Since the rate of biodegradation decreases with decreasing temperature, organic matter does not degrade rapidly in colder climates and tends to build up in soil. In water and in waterlogged soils, decaying vegetation does not have ready access to oxygen, and organic matter accumulates. The organic content may reach 90% in areas where plants grow and decay in soil saturated with water.

Of the organic components listed in Table 10.1, soil humus is by far the most significant. Humas is composed of base-soluble fractions called *humic* and *fulvic acids* (described in Section 4.19) and an insoluble fraction called *humin*. Humus is the residue left when bacteria and fungi biodegrade plant material. The

[7] Graustein, W.C.; Cromack, K., Jr., and Sollins, P. 1977. Calcium oxalate: Occurrence in soils and effect on nutrient and geochemical cycles. *Science* 198: 1252–4.

[8] Gieseking, J.E., ed. 1975. *Soil components, Vol. I, organic components.* New York: Springer-Verlag New York, Inc.

TABLE 10.1 Major Classes of Organic Compounds in Soil

Compound type	*Composition*	*Significance*
Humus	Degradation-resistant residue from plant decay, largely C, H, and O	Most abundant organic component, improves soil physical properties, exchanges nutrients, reservoir of fixed N
Fats, resins, and waxes	Lipids extractable by organic solvents	Generally only several percent of soil organic matter, may adversely affect soil physical properties by repelling water, perhaps phytotoxic (poisonous to plants)
Saccharides	Cellulose, starches, hemicellulose, gums	Major food source for soil microorganisms, help stabilize soil aggregates
Nitrogen-containing organics	N bound to humus, amino acids, amino sugars, uncharacterized compounds	Provide nitrogen for soil fertility
Phosphorus compounds	Phosphate esters, inositol phosphates (phytic acid), phospholipids	Source of plant phosphate

bulk of plant biomass consists of relatively degradable cellulose and degradation-resistant **lignin**, which is a polymeric substance with a higher carbon content than cellulose. Among its prominent chemical components are aromatic rings connected by alkyl chains, methoxyl groups, and hydroxyl groups. Lignin is the precursor of most soil humus.

An increase in nitrogen/carbon ratio is a significant feature of the transformation of plant biomass to humus. This ratio starts at approximately 1/100 in fresh plant biomass. During the humification process, microorganisms convert organic carbon to CO_2 to obtain energy. Simultaneously, the bacterial action incorporates bound nitrogen with the compounds produced by the decay processes. The result is a nitrogen/carbon ratio of about 1/10 upon completion of humification. As a general rule, therefore, humus is relatively rich in organically bound nitrogen.

Humic materials in soil strongly sorb many solutes in soil water and have a particular affinity for heavy polyvalent cations. Soil humic substances may contain levels of uranium more than 10^4 times that of the water with which they are in equilibrium. Thus, water becomes depleted of its cations (or purified) in passing through humic-rich soils. Humic substances in soils also have a strong affinity for organic compounds with low water-solubility such as DDT or Atrazine, a herbicide widely used to kill weeds in corn fields.

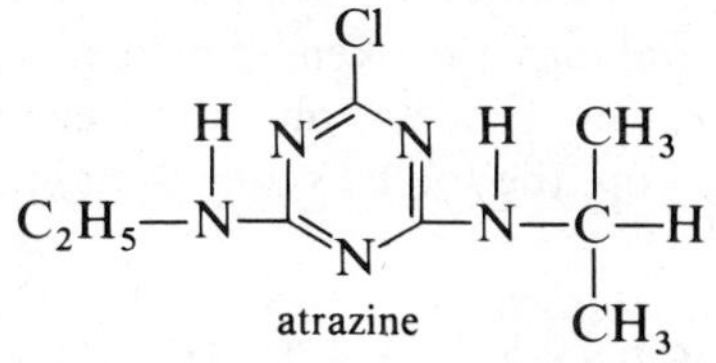

Cl
H N N H CH_3
C_2H_5—N—⟨triazine ring, N⟩—N—C—H
CH_3

atrazine

In some cases, there is a strong interaction between the organic and inorganic portions of soil. This is especially true of the strong complexes formed between clays and humic (fulvic) acid compounds. In many soils, 50–100% of soil carbon is complexed with clay. These complexes play a role in determining the physical properties of soil, soil fertility, and stabilization of soil organic matter. Although the exact type of chemical binding between clay colloidal particles and humic organic particles is not known, one of the mechanisms is probably of the flocculation type (see Chapter 6), in which anionic organic molecules with carboxylic acid functional groups serve as bridges in combination with cations to bind clay colloidal particles together as a floc. Support is given to this hypothesis by the known ability of the cations NH_4^+, Al^{3+}, Ca^{2+}, and Fe^{3+} to stimulate clay-organic complex formation. The synthesis, chemical reactions, and biodegradation of humic materials are affected by interaction with clays. The lower-molecular-weight fulvic acids may be bound in the spaces in layers in clay particles.

The presence of naturally occurring **polynuclear aromatic (PNA) compounds** is an interesting feature of soil organic matter. These compounds, some of which are carcinogenic, are discussed as air pollutants in Sections 13.16 and 15.4. PNA compounds found in soil include fluoranthene, pyrene, chrysene, and carcinogenic benzo(a)pyrene. Organic acids, long-chain alcohols, sterols, and terpenes also occur in soil organic matter. Extraction of soil with ether and alcohol yields the pigments β-carotene, chlorophyll, and xanthophyll. The origin of PNA compounds in soil is unknown.

10.6 ACID-BASE AND ION-EXCHANGE REACTIONS IN SOILS

One of the more important chemical functions of soils is the exchange of cations. As discussed in Section 6.12, the ability of a sediment or soil to exchange cations is expressed as the *cation-exchange capacity (CEC)*, the number of milliequivalents (meq) of monovalent cations that can be exchanged per 100 g of soil (on a dry-weight basis). The CEC should be looked upon as a conditional constant, since it may vary with soil conditions such as pE and pH. Both the mineral and organic portions of soils exchange cations. Clay minerals exchange cations because of the presence of negatively charged sites on the mineral, resulting from the substitution of an atom of lower oxidation number for one of higher number; for example, magnesium for aluminum. Organic materials exchange cations because of the presence of the carboxylate group and other basic functional groups. Humus typically has a very high cation-exchange capacity. The cation-exchange capacity of peat may range from 300–400 meq/100 g. Values of cation-exchange capacity for soils with more typical levels of organic matter are around 10–30 meq/100 g.

Cation exchange in soil is the mechanism by which potassium, calcium, magnesium, and essential trace-level metals are made available to plants. When nutrient metal ions are taken up by plant roots, hydrogen ion is exchanged for the metal ions. This process, plus the leaching of calcium, magnesium, and other metal ions from the soil by water containing carbonic acid, tends to make the soil acidic:

$$\text{soil}\}Ca^{2+} + 2\,CO_2 + 2\,H_2O \rightarrow \text{soil}\}^{H^+}_{H^+} + Ca^{2+}\,(\text{root}) + 2\,HCO_3^- \qquad (10.6.1)$$

Soil acts as a buffer and resists changes in pH. The buffering capacity depends upon the type of soil.

The oxidation of pyrite in soil causes formation of acid-sulfate soils sometimes called "cat clays":

$$FeS_2 + \frac{7}{2} O_2 + H_2O \rightarrow Fe^{2+} + 2\ H^+ + 2\ SO_4^{2-} \qquad (10.6.2)$$

Layers of acid-sulfate soils with pH values as low as 3.0 have been encountered [9]. These soils, which are commonly found in Delaware, Florida, New Jersey, and North Carolina, are formed when neutral or basic marine sediments containing FeS_2 become acidic upon oxidation of pyrite when exposed to air. For example, soil reclaimed from marshlands and used for citrus groves has developed high acidity detrimental to plant growth. In addition, H_2S released by increased acidity is very toxic to citrus roots.

Soils are tested for potential acid-sulfate formation using a peroxide test. This test consists of oxidizing FeS_2 in the soil with 30% H_2O_2,

$$FeS_2 + \frac{15}{2} H_2O_2 \rightarrow Fe^{3+} + H^+ + 2\ SO_4^{2-} + 7\ H_2O \qquad (10.6.3)$$

then testing for acidity and sulfate. Appreciable levels of sulfate and a pH below 3.0 indicate potential to form acid-sulfate soils. If the pH is above 3.0, either little FeS_2 is present or sufficient $CaCO_3$ is in the soil to neutralize the sulfuric acid and acidic Fe^{3+}.

Pyrite-containing mine spoils (residue left over from mining) also form soils similar to acid-sulfate soils of marine origin. In addition to high acidity and toxic H_2S, a major chemical species limiting plant growth on such soils is Al(III). Aluminum ion liberated in acidic soils is very toxic to plants.

Most common plants grow best in a soil with a pH near neutrality. If the soil becomes too acidic for optimum plant growth, it may be restored to productivity by liming, ordinarily through the addition of calcium carbonate:

$$\text{soil}\}\begin{matrix}H^+\\H^+\end{matrix} + CaCO_3 \rightarrow \text{soil}\}\ Ca^{2+} + CO_2 + H_2O \qquad (10.6.4)$$

In areas of low rainfall, soils may become too basic (alkaline) due to the presence of basic salts such as Na_2CO_3. Alkaline soils may be treated with aluminum or iron sulfate, which release acid on hydrolysis:

$$2\ Fe^{3+} + 3\ SO_4^{2-} + 6\ H_2O \rightarrow 2\ Fe(OH)_3(s) + 6\ H^+ + 3\ SO_4^{2-} \qquad (10.6.5)$$

Sulfur added to soils is oxidized by bacterially mediated reactions to sulfuric acid:

$$S + \frac{3}{2}O_2 + H_2O \rightarrow 2\ H^+ + SO_4^{2-} \qquad (10.6.6)$$

and sulfur is used, therefore, to acidify alkaline soils. The huge quantities of sulfur now being removed from fossil fuels to prevent air pollution by sulfur dioxide may make the treatment of alkaline soils by sulfur much more attractive economically.

Competition of different cations for cation exchange sites on soil cation exchangers may be described semiquantitatively by exchange constants. For ex-

---

[9] Calvert, D.V., and Ford, H.W. 1973. Chemical properties of acid-sulfate soils recently reclaimed from Florida marshland. *Soil Sci. Soc. Amer. Proc.* 37: 367–71.

ample, soil reclaimed from an area flooded with seawater will have most of its cation exchange sites occupied by Na^+, and restoration of fertility requires binding of nutrient cations such as K^+:

$$\text{soil}\}Na^+ + K^+ \rightleftharpoons \text{soil}\}K^+ + Na^+ \tag{10.6.7}$$

The exchange constant K_c, where

$$K_c = \frac{N_K[Na^+]}{N_{Na}[K^+]} \tag{10.6.8}$$

expresses the relative tendency of soil to retain K^+ and Na^+. In this equation, N_K and N_{Na} are the equivalent ionic fractions of potassium and sodium, respectively, bound to soil, and $[Na^+]$ and $[K^+]$ are the concentrations of these ions in the surrounding soil water. For example, a soil with all cation exchange sites occupied by Na^+ would have a value of 1.00 for N_{Na}; with one-half of the cation exchange sites occupied by Na^+, N_{Na} is 0.5; etc. The exchange of anions by soil is not nearly so clearly defined as is the exchange of cations. In many cases, the exchange of anions does not involve a simple ion-exchange process. This is true of the strong retention of orthophosphate species by soil. At the other end of the scale, nitrate ion is very weakly retained by soil.

Ion exchange of anions may be visualized as occurring at the surfaces of oxides in the mineral portion of soil. A mechanism for the acquisition of surface charge by metal oxides is shown in Figure 6.6, using MnO_2 as an example. At low pH, the oxide surface may have a net positive charge, enabling it to hold anions such as chloride by simple electrostatic attraction:

$$\begin{array}{l} \text{O—H}^+ \;\; \text{Cl}^- \\ | \\ \text{M—OH}_2 \end{array}$$

At higher pH values, the metal-oxide surface has a net negative charge due to the formation of OH^- ion on the surface, caused by loss of H^+ from the water molecules bound to the surface:

$$\begin{array}{l} \text{O} \\ | \\ \text{M—OH}^- \end{array}$$

In such cases, it is possible for anions such as HPO_4^{2-} to displace hydroxide ion and bond directly to the oxide surface:

$$\begin{array}{l} \text{O} \\ | \\ \text{M—OH}^- \end{array} + HPO_4^{2-} \longrightarrow \begin{array}{l} \text{O} \\ | \\ \text{M—OPO}_3\text{H}^{2-} \end{array} + OH^-$$

10.7 MACRONUTRIENTS IN SOIL

Plant nutrients may be divided into macronutrients and micronutrients. **Macronutrients** are those elements that occur in substantial levels in plant materials or in fluids in the plant. **Micronutrients** (Section 10.11) are elements that are essential only at very low levels and generally are required for the function of essential enzymes.

The elements generally recognized as essential macronutrients for plants are carbon, hydrogen, oxygen, nitrogen, phosphorus, potassium, calcium, magnesium, and sulfur. Of these, carbon, hydrogen, and oxygen are obtained from the atmosphere and from water. In addition, through the action of nitrogen-fixing bacteria, nitrogen may be obtained by some plants directly from the atmosphere. The other essential macronutrients must be obtained from soil. Of these, nitrogen, phosphorus, and potassium are the most likely to be lacking are commonly added to soil as fertilizers. Because of their importance, these elements will be discussed in Sections 10.8, 10.9, and 10.10, respectively.

Calcium-deficient soils are relatively uncommon. Liming, a process used to treat acid soils (see Section 10.6), provides a more than adequate calcium supply for plants. However, calcium uptake by plants and leaching by carbonic acid (Reaction 10.6.1) may produce a calcium deficiency in soil. Acid soils may still contain an appreciable level of calcium which, because of competition by hydrogen ion, is not available to plants. Treatment of acid soil to restore the pH to near-neutrality generally remedies the calcium deficiency. In alkaline soils, the presence of high levels of sodium, magnesium, and potassium sometimes produces calcium deficiency because these ions compete with calcium for availability to plants.

Although magnesium makes up 2.1% of the Earth's crust, most of it is rather strongly bound in minerals. Generally, exchangeable magnesium is considered available to plants and is held by ion-exchanging organic matter or clays. The availability of magnesium to plants depends upon the calcium/magnesium ratio. If this ratio is too high, magnesium may not be available to plants and magnesium deficiency results. Similarly, excessive levels of potassium or sodium may cause magnesium deficiency.

Sulfur is assimilated by plants as the sulfate ion, SO_4^{2-}. In addition, in areas where the atmosphere is contaminated with SO_2, sulfur may be absorbed as sulfur dioxide by plant leaves. Atmospheric sulfur dioxide levels have been high enough to kill vegetation in some areas (see Chapter 12). However, some experiments designed to show SO_2 toxicity to plants have resulted in increased plant growth where there was an unexpected sulfur deficiency in the soil used for the experiment.

Soils deficient in sulfur do not support plant growth well, largely because sulfur is a component of some essential amino acids and of thiamin and biotin. Sulfate ion is generally present in the soil as immobilized insoluble sulfate minerals or as soluble salts, which are readily leached from the soil and lost as soil water runoff. Unlike the case of nutrient cations such as K^+, little sulfate is adsorbed to the soil (that is, bound by ion-exchange binding) where it is resistant to leaching while still available for assimilation by plant roots.

Soil sulfur deficiencies have been found in a number of regions of the world. Whereas most fertilizers used to contain sulfur, its use in commercial fertilizers is declining. If this trend continues, it is possible that sulfur will become a limiting nutrient in more cases.

As noted in Section 10.6, the reaction of FeS_2 with acid in acid-sulfate soils may release H_2S, which is very toxic to plants and which also kills many beneficial microorganisms. Toxic hydrogen sulfide can also be produced by reduction of sulfate ion through microorganism-mediated reactions with organic matter.

$$SO_4^{2-} + 2\,\{CH_2O\} + 2\,H^+ \rightarrow H_2S + 2\,CO_2 + 2\,H_2O \qquad (10.7.1)$$

Production of hydrogen sulfide in flooded soils may be inhibited by treatment with oxidizing compounds, one of the most effective of which is KNO_3.

10.8 NITROGEN IN SOIL

Figure 10.4 summarizes the primary sinks and pathways of nitrogen in soil. In most soils, over 90% of the nitrogen content is organic. This organic nitrogen is primarily the product of the biodegradation of dead plants and animals. It is eventually hydrolyzed to NH_4^+, which can be oxidized to NO_3^- by the action of bacteria in the soil.

Nitrogen bound to soil humus (see Section 10.5) is especially important in maintaining soil fertility. Unlike potassium or phosphate, nitrogen is not a significant product of mineral weathering. Nitrogen-fixing organisms ordinarily cannot supply sufficient nitrogen to meet peak demand. Inorganic nitrogen from fertilizers and rainwater is often largely lost by leaching. Soil humus, however, serves as a reservoir of nitrogen required by plants. It has the additional advantage that its rate of decay, hence its rate of nitrogen release to plants, roughly parallels plant growth—rapid during the warm growing season, slow during the winter months.

Nitrogen is an essential component of proteins and other constituents of living matter. Plants and cereals grown on nitrogen-rich soils not only provide higher yields, but are often substantially richer in protein and, therefore, more nutritious. Nitrogen is most generally available to plants as nitrate ion, NO_3^-. Some plants such as rice may utilize ammonium nitrogen; however, other plants find this form of nitrogen toxic. When nitrogen is applied to soils in the ammonium form, nitrifying bacteria perform an essential function in converting it to available nitrate ion.

Plants may absorb excessive amounts of nitrate nitrogen from soil. This phenomenon occurs particularly in heavily fertilized soils under drought conditions. Forage crops containing excessive amounts of nitrate can poison ruminant animals such as cattle or sheep. Plants containing excessive levels of nitrate can endanger people when used for *ensilage*, an animal food consisting of finely

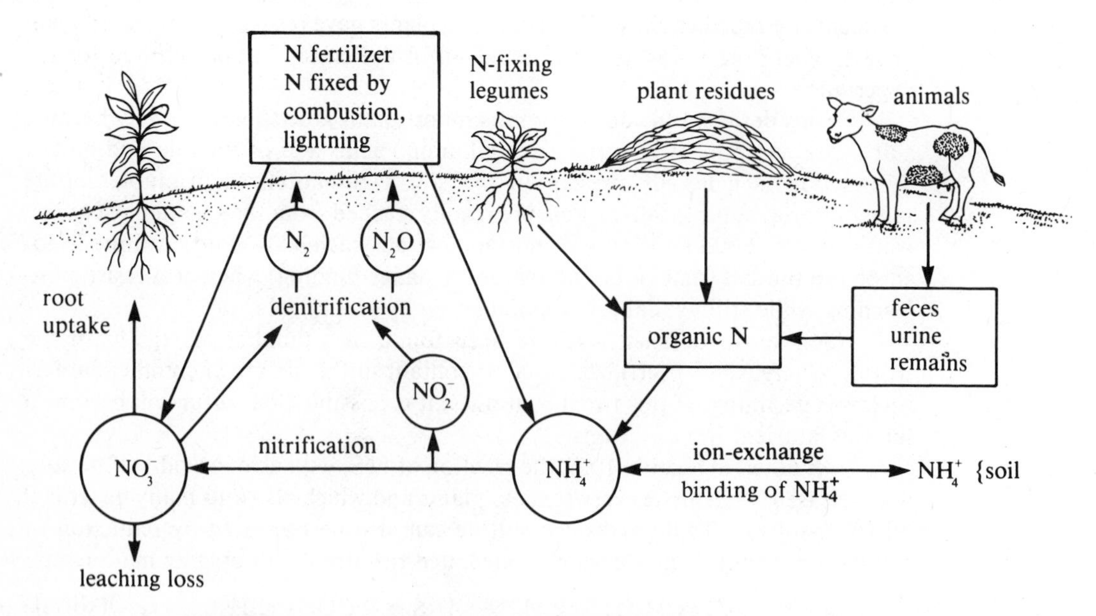

FIGURE 10.4 Nitrogen sinks and pathways in soil.

chopped plant material such as partially matured whole corn plants, fermented in a structure called a *silo*. Under the reducing conditions of fermentation, nitrate in ensilage may be reduced to toxic NO_2 gas, which can accumulate to high levels in enclosed silos. There have been many cases reported of persons being killed by accumulated NO_2 in silos.

Nitrogen fixation is the process by which atmospheric N_2 is converted to nitrogen compounds available to plants (see Section 5.15). As discussed in Section 5.15, human activities are resulting in the fixation of a great deal more nitrogen than would otherwise be the case. Artificial sources now account for 30–40% of all nitrogen fixed. These include chemical fertilizer manufacture; nitrogen fixed during fuel combustion (see Section 12.11); combustion of nitrogen-containing fuels; and the increased cultivation of nitrogen-fixing legumes (see the following paragraph). A major concern with this increased fixation of nitrogen is the possible effect upon the atmospheric ozone layer by N_2O released during denitrification of fixed nitrogen, as discussed in Sections 11.9 and 12.11.

Prior to the widespread introduction of nitrogen fertilizers, soil nitrogen was provided primarily by **legumes**. These are plants such as soybeans, alfalfa, and clover, which contain on their root structures bacteria capable of fixing atmospheric nitrogen. Leguminous plants have a *symbiotic* (mutually advantageous) relationship with the bacteria that provide their nitrogen. Legumes may add significant quantities of nitrogen to soil, up to 100 pounds per acre per year, which is comparable to amounts commonly added as synthetic fertilizers. Soil fertility with respect to nitrogen may be maintained by rotating plantings of nitrogen-consuming plants with plantings of legumes, a fact recognized by agriculturists as far back as the Roman era.

The nitrogen-fixing bacteria in legumes exist in special structures on the roots called **root nodules** (see Fig. 10.5). The rod-shaped bacteria that fix nitrogen

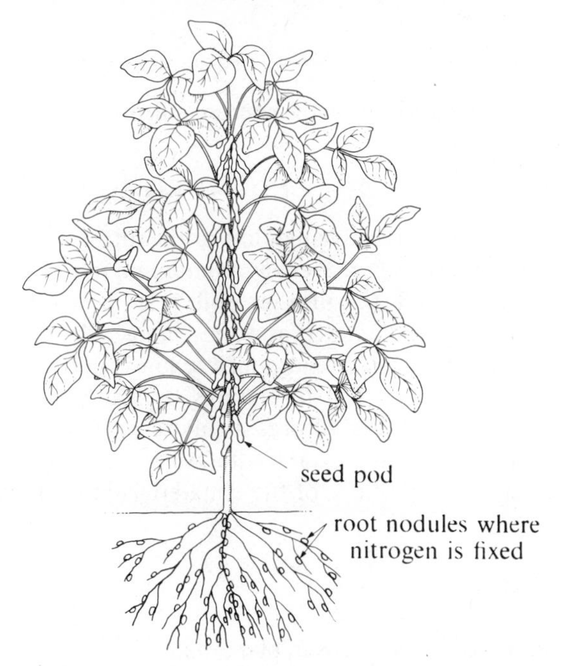

FIGURE 10.5 A soybean plant, showing root nodules where nitrogen is fixed.

are members of a special genus called *Rhizobium*. These bacteria may exist independently, but cannot fix nitrogen except in symbiotic combination with plants. Although all species of *Rhizobium* appear to be very similar, they exhibit a great deal of specificity in their choice of host plants. Curiously, legume root nodules also contain a form of hemoglobin, which must somehow be involved in the nitrogen-fixation process.

Nitrate pollution of some surface waters and ground water has become a major problem in some agricultural areas (see Chapter 7). Although fertilizers have been implicated in such pollution, there is evidence that feedlots are a major source of nitrate pollution. The growth of livestock populations and the concentration of livestock in feedlots have aggravated the problem. Such concentrations of cattle, coupled with the fact that a steer produces approximately 18 times as much waste material as a human, have resulted in high levels of water pollution in areas with small human populations. Streams and reservoirs in such areas frequently are just as polluted as those in densely populated and highly industrialized areas.

Nitrate in farm wells is a common and especially damaging manifestation of nitrogen pollution from feedlots because of the susceptibility of ruminant animals to nitrate poisoning. The stomach contents of ruminant animals such as cattle and sheep constitute a reducing medium (low pE) and contain bacteria capable of reducing nitrate ion to toxic nitrite ion:

$$NO_3^- + 2\,H^+ + 2\,e \rightarrow NO_2^- + H_2O \qquad (10.8.1)$$

The origin of most nitrate produced from feedlot wastes is amino nitrogen present in nitrogen-containing waste products. Approximately one-half of the nitrogen excreted by cattle is contained in the urine. Part of this nitrogen is proteinaceous and the other part is in the form of urea, NH_2CONH_2. As a first step in the degradation process, the amino nitrogen is probably hydrolyzed to ammonia, or ammonium ion:

$$R{-}NH_2 + H_2O \rightarrow R{-}OH + NH_3 \quad (\text{or } NH_4^+) \qquad (10.8.2)$$

This product is then oxidized through microorganism-catalyzed reactions to nitrate ion:

$$NH_3 + 2\,O_2 \rightarrow H^+ + NO_3^- + H_2O \qquad (10.8.3)$$

Under some conditions, an appreciable amount of the nitrogen originating from the degradation of feedlot wastes is present as ammonium ion. Ammonium ion is rather strongly bound to soil (recall that soil is a generally good cation exchanger), and a small fraction is fixed as nonexchangeable ammonium ion in the crystal lattice of clay minerals. Because nitrate ion is not strongly bound to soil, it is readily carried through soil formations by water. Many factors, including soil type, moisture, and level of organic matter, affect the production of ammonia and nitrate ion originating from feedlot wastes, and a marked variation is found in the levels and distributions of these materials in feedlot areas.

10.9 PHOSPHORUS IN SOIL

Although the percentage of phosphorus in plant material is relatively low, it is an essential component of plants. Phosphorus, like nitrogen, must be present in a simple inorganic form before it can be taken up by plants. In the case of phosphorus, the utilizable species is some form of orthophosphate ion. In the pH range that is present in most soils, $H_2PO_4^-$ and HPO_4^{2-} are the predominant orthophosphate species.

Orthophosphate is most available to plants at pH values near neutrality. It is believed that in relatively acidic soils, orthophosphate ions are precipitated or sorbed by species of Al(III) and Fe(III). In alkaline soils, orthophosphate may react with calcium carbonate to form relatively insoluble hydroxyapatite:

$$3\ HPO_4^{2-} + 5\ \underset{\text{calcite}}{CaCO_3(s)} + 2\ H_2O \rightarrow \underset{\text{hydroxyapatite}}{Ca_5(PO_4)_3(OH)(s)} + 5\ HCO_3^- + OH^- \qquad (10.9.1)$$

In general, because of these reactions, little phosphorus applied as fertilizer leaches from the soil. This is important from the standpoint of both water pollution and utilization of phosphate fertilizers. More research remains to be done to establish fully the nature of the chemical interactions determining the availability of orthophosphates in soils.

10.10 POTASSIUM IN SOIL

Relatively high levels of potassium are utilized by growing plants. Potassium activates some enzymes and plays a role in the water balance in plants. It is also essential for some carbohydrate transformations. Crop yields are generally greatly reduced in potassium-deficient soils. The higher the productivity of the crop, the more potassium is removed from soil. When nitrogen fertilizers are added to soils to increase productivity, removal of potassium is enhanced. Therefore, potassium may become a limiting nutrient in soils heavily fertilized with other nutrients.

Potassium is one of the most abundant elements in the Earth's crust, making up 2.6% of the crust, but much of this potassium is not easily available to plants. For example, some silicate minerals such as leucite, $K_2O \cdot Al_2O_3 \cdot 4\ SiO_2$, contain strongly bound potassium. Exchangeable potassium held by clay minerals is relatively more available to plants.

10.11 MICRONUTRIENTS IN SOIL

Boron, chlorine, copper, iron, manganese, molybdenum (for N-fixation), sodium, vanadium, and zinc are considered essential plant **micronutrients**. These elements are needed by plants only at very low levels and frequently are toxic at higher levels. It is entirely possible that other elements will be added to this list as techniques for growing plants in environments free of specific elements improve. Most of these elements function as components of essential enzymes. Manganese, iron, chlorine, zinc, and vanadium may be involved in photosynthesis.

Some plants accumulate extremely high levels of specific trace metals. Those accumulating more than 1.00 mg/g of dry weight are called **hyperaccumulators.** Hyperaccumulation of nickel and copper has been described [10]; for example, *Aeolanthus biformifolius* De Wild growing in copper-rich regions of Shaba Province, Zaire, contains up to 1.3% copper (dry weight) and is known as a "copper flower."

---

[10] Malaisse, F. et al. 1978. *Aeolanthus biformifolius* De Wild.: a hyperaccumulator of copper from Zaire. *Science* 199: 887–8.

10.12 FERTILIZERS

Crop **fertilizers** contain nitrogen, phosphorus, and potassium as major components. Magnesium, sulfate, and micronutrients may also be added. Fertilizers are designated by numbers, such as 6–12–8, showing the respective percentages of nitrogen expressed as N (in this case 6%), phosphorus as P_2O_5 (12%), and potassium as K_2O (8%). Farm manure corresponds to an approximately 0.5–0.24–0.5 fertilizer. The organic fertilizers such as manure must undergo biodegradation to release the simple inorganic species (NO_3^-, $H_xPO_4^{x-3}$, K^+) assimilable by plants.

Most modern nitrogen fertilizers are made by the **Haber process**, in which N_2 and H_2 are combined over a catalyst at temperatures of approximately 500 °C and pressures up to 1000 atm:

$$N_2 + 3\,H_2 \rightarrow 2\,NH_3 \tag{10.12.1}$$

The anhydrous ammonia product has a very high nitrogen content of 82% and may be added directly to the soil. Special equipment is required, however, because of the toxicity of ammonia gas. Aqua ammonia, a 30% solution of NH_3 in water, may be used with much greater safety. It is sometimes added directly to irrigation water.

Ammonium nitrate, NH_4NO_3, is a common solid nitrogen fertilizer. It is made by oxidizing ammonia over a platinum catalyst, converting the nitric oxide product to nitric acid, and reacting the nitric acid with ammonia. The molten ammonium nitrate product is forced through nozzles at the top of a *prilling tower* and solidifies to form small pellets while falling through the tower. The particles are coated with a water repellant. Ammonium nitrate contains 33.5% nitrogen. Although convenient to use, it is explosive and requires considerable care during manufacture and storage.

Urea is easier to manufacture and handle than ammonium nitrate:

$$H_2N-\overset{\overset{\displaystyle O}{\|}}{C}-NH_2$$

Urea is now the favored solid nitrogen-containing fertilizer. The overall reaction for urea synthesis is

$$CO_2 + 2\,NH_3 \rightarrow CO(NH_2)_2 + H_2O \tag{10.12.2}$$

involving a rather complicated process in which ammonium carbamate, $NH_2CO_2NH_4$, is an intermediate.

Other compounds used as nitrogen fertilizers include sodium nitrate (obtained largely from Chilean deposits), calcium nitrate, potassium nitrate, and ammonium phosphates. Ammonium sulfate, a by-product of coke ovens, used to be widely used and may enjoy a resurgence as a coal conversion by-product. The alkali-metal nitrates tend to make soil alkaline, whereas ammonium sulfate leaves an acidic residue.

Phosphate minerals are found in several states, including Idaho, Montana, Utah, Wyoming, North Carolina, South Carolina, Tennessee, and Florida. The principal mineral is fluorapatite, $Ca_5(PO_4)_3F$. The phosphate from fluorapatite is relatively unavailable to plants and is frequently treated with phosphoric or sulfuric acids to produce superphosphates:

$$2\ Ca_5(PO_4)_3F(s) + 14\ H_3PO_4 + 10\ H_2O \rightarrow 2\ HF(g) + 10\ CaH_4(PO_4)_2 \cdot H_2O \quad (10.12.3)$$

$$2\ Ca_5(PO_4)_3F(s) + 7\ H_2SO_4 + 3\ H_2O \rightarrow 2\ HF(g) + 3\ CaH_4(PO_4)_2 \cdot H_2O + 7\ CaSO_4 \quad (10.12.4)$$

The superphosphate products are much more soluble than the parent phosphate minerals. The HF produced as a by-product of superphosphate production can create air pollution problems.

Phosphate minerals are rich in trace elements required for plant growth, such as boron, copper, manganese, molybdenum, and zinc. Ironically, these elements are lost in processing phosphate for fertilizers and are sometimes added to fertilizers later.

Ammonium phosphates are excellent, highly soluble phosphate fertilizers. Liquid ammonium polyphosphate fertilizers consisting of ammonium salts of pyrophosphate, triphosphate, and small quantities of higher polymeric phosphate anions in aqueous solution are becoming very popular as phosphate fertilizers. The polyphosphates are believed to have the additional advantage of chelating iron and other micronutrient metal ions, thus making the metals more available to plants.

Potassium fertilizer components consist of potassium salts, generally KCl. Such salts are found as deposits in the ground or may be obtained from some brines. Very large deposits are found in Saskatchewan, Canada. These salts are all quite soluble in water. One problem encountered with potassium fertilizers is the luxury uptake of potassium by some crops, which absorb more potassium than is really needed for their maximum growth. In a crop where only the grain is harvested, leaving the rest of the plant in the field, luxury uptake does not create much of a problem because most of the potassium is returned to the soil with the dead plant. However, when hay or forage is harvested, potassium contained in the plant as a consequence of luxury uptake is lost from the soil.

10.13 WASTES AND POLLUTANTS IN SOIL

Soil receives large quantities of waste products each year. Much of the sulfur dioxide emitted in the burning of sulfur-containing fuels ends up on soil as sulfates. Atmospheric nitrogen oxides are converted to nitrates in the atmosphere, and the nitrates eventually are deposited on soil. Soil sorbs NO and NO_2 readily, and these gases are oxidized to nitrate in the soil. Carbon monoxide is converted to CO_2 and possibly to biomass by soil bacteria and fungi (see Chapter 12). Particulate lead from automobile exhausts is found at elevated levels in soil along heavily traveled highways. Elevated levels of lead from lead mines and smelters are found on soil near such facilities.

Soil receives enormous quantities of pesticides as an inevitable result of their application to crops. The degradation and eventual fate of these pesticides on soil largely determines the ultimate environmental effects of the pesticides, and detailed knowledge of these effects are now required for licensing of a new pesticide. Among the factors to be considered are the sorption of the pesticide by soil; leaching of the pesticide into water, as related to its potential for water pollution; effects of the pesticide on microorganisms and animal life in the soil; and possible production of relatively more toxic degradation products.

Adsorption of a pesticide by soil is a key step in the degradation of the pesticide. The degree of adsorption and the speed and extent of ultimate degradation are influenced by a number of factors. Some of these, including solubility, volatility, charge, polarity, and molecular structure and size, are properties of the pesticide. Others, including temperature, pH, and pE, are properties of the soil medium. Adsorption of a pesticide by soil components may have several effects. Under some circumstances, it retards degradation by separating the pesticide from the microbial enzymes that degrade it, whereas under other circumstances the reverse is true. Purely chemical degradation reactions may be catalyzed by adsorption. Loss of the pesticide by volatilization or leaching is diminished. The toxicity of a herbicide to plants may be strongly affected by soil sorption.

The forces holding a pesticide to soil particles may be of several types. Physical adsorption involves van der Waals forces arising from dipole-dipole interactions between the pesticide molecule and charged soil particles. Ion exchange is especially effective in holding cationic organic compounds, such as the herbicide paraquat,

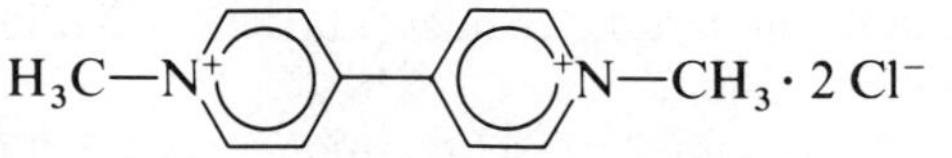

$$H_3C{-}\overset{+}{N}C_5H_4{-}C_5H_4\overset{+}{N}{-}CH_3 \cdot 2\,Cl^-$$

to anionic soil particles. Some neutral pesticides become cationic by protonation and are bound as the protonated positive form. Hydrogen bonding is another mechanism by which some pesticides are held to soil. In some cases, a pesticide may act as a ligand coordinating to metals in soil mineral matter.

The three primary ways in which pesticides are degraded in or on soil are *biodegradation, chemical degradation*, and *photochemical reactions*. Various combinations of these processes may operate in the degradation of a pesticide.

Although insects, earthworms, and plants may play roles in the **biodegradation** of pesticides, microorganisms have the most important role. Several examples of microorganism-mediated degradation of pesticides are given in Section 5.26.

Chemical degradation of pesticides has been observed experimentally in soils and clays sterilized to remove all microbial activity. For example, clays have been shown to catalyze the hydrolysis of *O,O*-Dimethyl *O*-2,4, 5-trichlorophenyl thiophosphate (also called Trolene, ronnel, Etrolene, or trichlorometafos), an effect attributed to $-OH$ groups on the mineral surface:

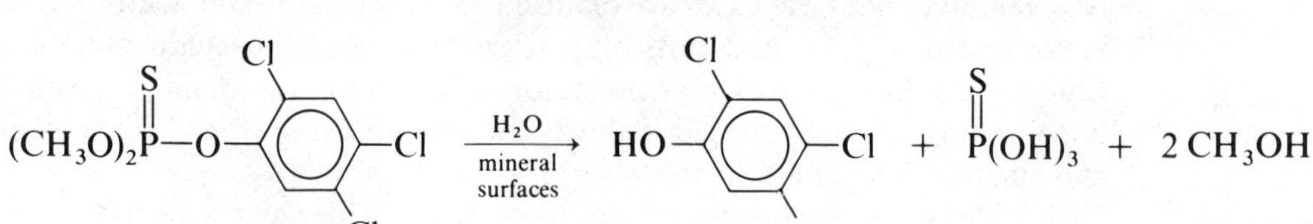

$$(CH_3O)_2P(=S){-}O{-}C_6H_2Cl_3 \xrightarrow[\text{mineral surfaces}]{H_2O} HO{-}C_6H_2Cl_3 + (S{=})P(OH)_3 + 2\,CH_3OH \qquad (10.13.1)$$

Many other purely chemical hydrolytic reactions of pesticides occur in soil.

A number of pesticides have been shown to undergo **photochemical reactions**, that is, chemical reactions brought about by the absorption of light (see Chapter 11). Frequently, isomers of the pesticides are produced as products. Many of the studies reported apply to pesticides in water or on thin films, and the photochemical reactions of pesticides on soil and plant surfaces remain largely a matter of speculation.

Soil is the receptor of many hazardous wastes from landfill leachate, lagoons, and other sources (see Section 20.6). In some cases, **land farming** of degradable hazardous organic wastes is practiced. The degradable material is worked into the soil, and soil microbial processes bring about its degradation. As discussed in Chapter 8, sewage and fertilizer-rich sewage sludge may be applied to soil.

10.14 SOIL EROSION

Soil erosion can occur by the action of both water and wind, although water is the primary source of erosion. To provide an idea of the magnitude of the problem, U.S. Department of Agriculture officials estimate that 15 million tons per minute of topsoil are swept from the mouth of the Mississippi. About one-third of U.S. topsoil has been lost since cultivation began on the continent. At the present time approximately one-third of U.S. cultivated land is eroding at a rate sufficient to reduce soil productivity. It is estimated that 48 million acres of land, somewhat more than 10 percent of that under cultivation, is eroding at unacceptable levels, taken to mean a loss of more than 14 tons of topsoil per acre each year. Specific areas in which the greatest erosion is occurring include northern Missouri, southern Iowa, west Texas, western Tennessee, and the Mississippi Basin. Figure 10.6 shows the pattern of soil erosion in the U.S. in 1977.

Problems involving soil erosion were aggravated in the 1970s and early 1980s when high prices for farmland resulted in the intensive cultivation of high-income crops, particularly corn and soybeans. These crops grow in rows with bare soil in between, which tends to wash away with each rainfall. Furthermore, corn and soybeans began to be planted year after year, without intervening plantings of soil-restoring clover or grass. The problem of decreased productivity due to soil erosion has been masked somewhat by increased use of chemical fertilizers.

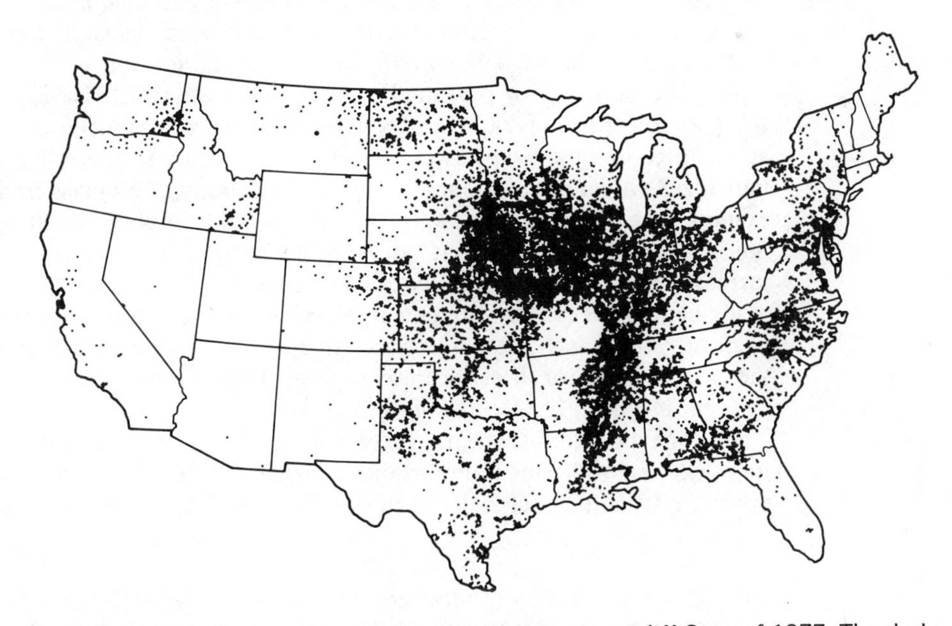

FIGURE 10.6 Pattern of soil erosion in the continental U.S. as of 1977. The dark areas indicate locations where the greatest erosion is occurring. (USDA)

Wind erosion, such as occurs on the generally dry, high plains soils of eastern Colorado, poses another threat. After the Dust Bowl days of the 1930s, much of this land was allowed to revert to grassland, and the topsoil was held in place by the strong root systems of the grass cover. However, in an effort to grow more wheat and improve the sale value of the land, much of it has been cultivated in recent years. For example, from 1979 through 1982, more than 450,000 acres of Colorado grasslands have been plowed. Much of this has been done by speculators who purchase grassland at a low price of $100–$200 per acre, break it up, and sell it as cultivated land at more than double the original purchase price. Although freshly cultivated grassland may yield well for one or two years, the nutrients and soil moisture are rapidly exhausted, and the land becomes very susceptible to wind erosion.

There are a number of solutions to the soil erosion problem. Some are old, well-known agricultural practices, such as terracing, contour plowing, and periodically planting fields with cover crops, such as clover. For some crops **no-till agriculture** greatly reduces erosion. This practice consists of planting a crop among the residue of the previous year's crop, without plowing. Weeds are killed in the newly planted crop row by application of a herbicide prior to planting. The surface residue of plant material left on top of the soil prevents erosion.

Another, more experimental solution to the soil erosion problem is the cultivation of perennial plants, which develop a large root system and come up each spring after being harvested the previous fall. Recently botanists succeeded in developing a perennial corn plant by crossing corn with a distant, wild relative, teosinte, which grows in Central America [11]. Unfortunately, the resulting plant is not outstanding in terms of its grain yields. It should be noted that an annual plant's ability to propagate depends upon producing large quantities of seeds, whereas a perennial plant must develop a strong root system with bulbous growths called *rhizomes*, which store food for the coming year. However, it is possible that the application of genetic engineering (see Section 10.15) may result in the development of perennial crops with good seed yields. The cultivation of such a crop would cut down on a great deal of soil erosion.

The best known perennial plant is the tree, which is very effective in stopping soil erosion [12]. Wood from trees can be used as fuel (see Section 17.4), as a source of raw materials, and as food (see below). There is a tremendous unrealized potential for an increase in the production of biomass from trees. For example, the production of biomass from natural forests of loblolly pine trees in South Carolina has been about 3 dry tons per hectare per year. This has now been increased to 11 tons through selection of superior trees, and 30 tons may eventually be possible [12]. In Brazil, experiments have been conducted with a species of *Eucalyptus*, which has a 7-year growth cycle. With improved selection of trees, the annual yields for three successive cycles of these trees in dry tons per hectare per year has been 23, 33, and 40.

The most important use for wood is, of course, as lumber for construction. This use will remain important as higher energy costs increase the costs of other construction materials, such as steel, aluminum, and cement. Wood is

---

[11] Sullivan, W. 1982. Cross betwen corn and a wild relative yields a perennial crop. *New York Times* 16 February 1982.

[12] Abelson, P.H. 1982. Energy and chemicals from trees. *Science* 215: 1349.

about 50 percent cellulose, which can be hydrolyzed by rapidly improving enzyme processes to yield glucose sugar. The glucose can be used directly as food, fermented to ethyl alcohol for fuel (see *gasohol*, Section 17.17), or employed as a carbon and energy source for protein-producing yeasts. Given these and other potential uses, the future of trees as an environmentally desirable and profitable crop is very bright.

10.15 GENETIC ENGINEERING AND AGRICULTURE

The nuclei of living cells contain the genetic instructions for cell reproduction. These instructions are in the form of a special material called **deoxyribonucleic acid, DNA**. In combination with proteins, DNA makes up the cell chromosomes. During the 1970s the ability to manipulate DNA, called **genetic engineering**, became a reality and is now forming the basis of a major industry. Such manipulation falls into the category of **recombinant DNA technology**. Recombinant DNA gets its name from the fact that it contains DNA from two different organisms, recombined together. This technology promises some exciting developments in agriculture.

The "green revolution" of the mid-1960s used conventional plant-breeding techniques of selective breeding, hybridization, cross-pollination, and back-crossing to develop new strains of rice, wheat, and corn, which, when combined with chemical fertilizers, yielded spectacular results. For example, India's output of grain increased 50 percent. By working at the cellular level, however, it is now possible to greatly accelerate the process of plant breeding. Thus, plants may be developed that resist particular diseases, grow in seawater, or have much higher productivity. The possibility exists for developing entirely new kinds of plants.

One exciting possibility with genetic engineering is the development of plants other than legumes which fix their own nitrogen. For example, if nitrogen-fixing corn could be developed, the savings in fertilizer would be enormous. Furthermore, since the nitrogen is fixed in an organic form in plant root structures, there would be no pollutant runoff of chemical fertilizers.

Another promising possibility with genetic engineering is increased efficiency of photosynthesis. Plants utilize only about 1 percent of the sunlight striking their leaves, so there is appreciable room for improvement in that area.

Cell-culture techniques can be applied in which billions of cells are allowed to grow in a medium and develop mutants which, for example, might be resistant to particular viruses or herbicides or have other desirable qualities. If the cells with the desired qualities can be regenerated into whole plants, results can be obtained that might have taken decades using conventional plant-breeding techniques.

10.16 AGRICULTURE AND HEALTH

Some authorities hold that soil has an appreciable effect upon health. An obvious way in which such an effect might be manifested is the incorporation into food of micronutrient elements essential for human health. One such nutrient (which is toxic at overdose levels) is selenium. It is definitely known that the health of animals is adversely affected in selenium-deficient areas. Human health might be similarly affected.

It is known that there are some striking geographic correlations with the occurrence of cancer. Some of these correlations may be due to soil type. A high incidence of stomach cancer has been shown to occur in areas with certain types of soil in the Netherlands, the United States, France, Wales, and Scandinavia. These soils are high in organic matter content, are acidic, and frequently are waterlogged [13]. A "stomach-cancer-prone life style" has been described [13], which includes consumption of home-grown food, consumption of water from one's own well, and reliance on native and uncommon foodstuffs.

One possible reason for the existence of "stomach-cancer-producing soils" is the production of cancer-causing secondary metabolites by plants and microorganisms. Secondary metabolites are biochemical compounds that are of no apparent use to the organism producing them. It is believed that they are formed from the precursors of primary metabolites when the primary metabolites accumulate to excessive levels.

The role of soil in environmental health is not well known, nor has it been extensively studied. The amount of research on the role of soil in producing foods that are more nutritious and lower in content of naturally occurring toxic substances is quite small compared to research on higher soil productivity. It is to be hoped that the environmental health aspects of soil and its products will receive much greater emphasis in the future.

Sometimes human activities contaminate food grown on soil. Most often this occurs through contamination by pesticides. An interesting example of such contamination occurred in Hawaii in early 1982 [14]. It was found that milk from several sources on Oahu contained very high levels of heptachlor (see Table 7.8). This pesticide causes cancer and liver disorders in mice; therefore, it is a suspected human carcinogen. Remarkably, in this case it was not until 57 days after the initial discovery that the public was informed of the contamination by the Department of Health. The source of heptachlor was traced to contaminated "green chop", chopped-up pineapple leaves fed to cattle. Although heptachlor is banned for most applications, Hawaiian pineapple growers had obtained special Federal permission to use it to control mealybug wilt. Although it was specified that green chop could not be collected within 1 year of the last application of the pesticide, apparently this regulation was violated, and the result was distribution of contaminated milk to consumers.

SUPPLEMENTARY REFERENCES

Ashton, F.D., and Crafts, A.S. 1981. *Mode of action of herbicides.* New York: Wiley-Interscience.

Birch, G.G., ed. 1976. *Food from waste.* Essex, England: Applied Science Publishers, Ltd.

Bolt, G.H., ed. 1979. *Soil chemistry.* New York: Elsevier North-Holland, Inc.

Bolt, G.H., and Bruggenwert, M.G.M., eds. 1976. *Soil chemistry.* New York: Elsevier North-Holland, Inc.

Bowling, D.J.F. 1975. *Uptake of ions by plant roots.* New York: Halsted (John Wiley and Sons, Inc.).

---

[13] Adams, R.S., Jr. 1978. Soil variability and cancer. *Chemical and Engineering News* 12 June 1978, p. 84.

[14] Smith, R.J. 1982. Hawaiian milk contamination creates alarm. *Science* 217: 137–40.

Brownlow, A.H. 1978. *Geochemistry.* Englewood Cliffs, N.J.: Prentice-Hall, Inc.

1979. *Chemistry and agriculture.* London: Chemical Society.

Craig, P.J. 1980. *The natural environment and the biogeochemical cycles.* New York: Springer-Verlag New York, Inc.

Donahue, R.L.; Miller, R.W.; and Shickluna, J.C. 1977. *Soils.* 4th ed. Englewood Cliffs, N.J.: Prentice-Hall, Inc.

Dowdy, R.H. 1981. *Chemistry in the soil environment.* Madison, Wis.: American Society of Agronomy/Soil Science Society of America.

Evans, D.D., and Thames, J.L., eds. 1981. *Water in desert ecosystems.* New York: Academic Press, Inc.

Fahm, L.A. 1980. *The waste of nations: the economic utilization of human waste in agriculture.* Montclair, N.J.: Allanheld, Osmun & Co. Publishers, Inc.

Fairbridge, R.W., and Finkl, C.W., Jr. 1979. *The encyclopedia of soil science.* New York: Academic Press, Inc.

Fitzpatrick, E.A. 1980. *Soils.* New York: Longman, Inc.

Fletcher, W.W., and Little, C.E. 1982. *The American cropland crisis.* Bethesda, Md.: American Land Forum.

1976. *Food and agriculture, a Scientific American book.* San Francisco, Calif.: W.H. Freeman and Company Publishers.

Fortescue, A.C. 1980. *Environmental geochemistry.* New York: Springer-Verlag New York, Inc.

Geissbühler, H., ed. 1979. *World food production—environment—pesticides.* Elmsford, N.Y.: Pergamon Press, Inc.

Greenland, D.J., and Hayes, M.H.B., eds. 1981. *The chemistry of soil processes.* New York: Wiley-Interscience.

Hassall, K.A. 1982. *The chemistry of pesticides: their metabolism, mode of action and uses in crop protection.* Deerfield Beach, Fla.: Verlag Chemie International.

Hobson, P.N., and Robertson, A.M. 1977. *Waste treatment in agriculture.* London: Applied Science Publishers.

Hudson, N. 1981. *Soil conservation.* Ithaca, N.Y.: Cornell University Press.

Krauskopf, K.B. 1979. *Introduction to geochemistry.* 2nd ed. New York: McGraw-Hill Book Company.

Kvenvolden, K.A., ed. 1980. *Geochemistry of organic molecules.* New York: Academic Press, Inc.

Lapedes, D.N. 1977. *McGraw-Hill encyclopedia of food, agriculture, and nutrition.* New York: McGraw-Hill Book Company.

Lerman, A. 1979. *Geochemical processes.* New York: Wiley-Interscience.

Lindsay, W.L. 1979. *Chemical equilibria in soils.* New York: Wiley-Interscience.

Loehr, R.C., ed. 1977. *Food, fertilizer, and agricultural residues.* Ann Arbor, Mich.: Ann Arbor Science Publishers, Inc.

Marshall, C.E. 1977. *The physical chemistry and mineralogy of soils.* New York: Wiley-Interscience.

Middlebrooks, E.J. 1979. *Industrial pollution control, Vol. 1, agro-industries.* New York: Wiley-Interscience.

Morrill, L.G.; Mahilum, B.C.; Mohiuddin, S.H. 1982. *Organic compounds in soils: sorption, degradation, and persistence.* Ann Arbor, Mich.: Ann Arbor Science Publishers, Inc.

Nancollas, G.H., ed. 1982. *Biological mineralization and demineralization.* New York: Springer-Verlag New York, Inc.

Nielsen, D.R., and MacDonald, J.G., eds. 1978. *Nitrogen in the environment, Vol. 1, nitrogen behavior in field soil.* New York: Academic Press, Inc.

Olson, G.W. 1981. *Soils and the environment.* New York: Methuen, Inc.

Page, A.L.; Miller, R.H.; and Keeney, D.R. 1983. *Methods of soil analysis.* Madison, Wisconsin: American Society of Agronomy and Soil Science Society of America.

Paton, T.R. 1979. *The formation of soil material.* Winchester, Mass.: Allen & Unwin, Inc.

Rosenberg, N.J., ed. 1978. *North American droughts.* Boulder, Co.: Westview Press, Inc.

Schnitzer, M., and Khan, S.U., eds. 1978. *Soil organic matter.* New York: Elsevier North-Holland, Inc.

Sears, P.B. 1980. *Deserts on the march.* Norman, Oklahoma: University of Oklahoma Press.

Speidel, G.H., and Agnew, A.F. 1982. *The natural geochemistry of our environment.* Boulder, Co.: Westview Press.

Stevenson, F.J. 1982. *Humus chemistry: genesis, composition, reactions.* New York: Wiley-Interscience.

Trinker, P.B., ed. 1981. *Soils and agriculture.* New York: Halsted (John Wiley and Sons, Inc.).

Walls, J. 1980. *Land, man, and sand.* New York: Macmillan, Inc.

QUESTIONS AND PROBLEMS

1. Give two examples of reactions involving manganese and iron compounds that may occur in waterlogged soil.

2. What temperature and moisture conditions favor the buildup of organic matter in soil?

3. "Cat clays" are soils containing a high level of iron pyrite, FeS_2. Hydrogen peroxide, H_2O_2, is added to such a soil, producing sulfate, as a test for cat clays. Suggest the chemical reaction involved in this test.

4. What effect upon soil acidity would result from heavy fertilization with ammonium nitrate accompanied by good exposure to air and the action of aerobic bacteria?

5. How many moles of H^+ ion are consumed when 200 kilograms of $NaNO_3$ undergo denitrification on soil?

6. What is the primary mechanism by which organic material in soil exchanges cations?

7. Prolonged waterlogging of soil does not:
   (a) increase NO_3^- production,
   (b) increase Mn^{2+} concentration,
   (c) increase Fe^{2+} concentration,
   (d) have harmful effects upon most plants,
   (e) increase production of NH_4^+ from NO_3^-.

8. Of the following phenomena, the one that eventually makes soil more basic is:
   (a) removal of metal cations by roots,
   (b) leaching of soil with CO_2-saturated water,
   (c) oxidation of soil pyrite,
   (d) fertilization with $(NH_4)_2SO_4$,
   (e) fertilization with KNO_3.

9. How many metric tons of farm manure are equivalent to 100 kg of 10–5–10 fertilizer?

10. How are the chelating agents that are produced from microorganisms involved in soil formation?

11. What specific compound is both a particular animal waste product and a major fertilizer?

12. What happens to the nitrogen/carbon ratio as organic matter degrades in soil?

13. To prepare a rich potting soil, a greenhouse operator mixed 75% "normal" soil with 25% peat. Estimate the cation-exchange capacity in milliequivalents/100 g of the product.

14. Explain why plants grown on either excessively acidic or excessively basic soils may suffer from calcium deficiency.

15. What are two mechanisms by which anions may be held by soil mineral matter?

16. What are the three primary ways in which pesticides are degraded in or on soil?

17. Lime from lead mine tailings containing 0.5% lead was applied at a rate of 10 metric tons per acre of soil and worked in to a depth of 20 cm. The soil density was 2.0 g/cm^3. To what extent did this add to the burden of lead in the soil?

18. Match the soil or soil-water constituent in the left column with the soil condition described on the right.

(a) high Mn^{2+} content in soil water	(1) "cat clays" containing initially high levels of pyrite, FeS_2
(b) excess H^+	(2) soil in which biodegradation has not occurred to a great extent
(c) high H^+ and SO_4^{2-} content	(3) waterlogged soil
(d) high organic content	(4) soil whose fertility can be improved by adding limestone

19. What are two major effects that acid rain might have upon minerals in the geosphere?

20. Rank deserts, fresh snow, evergreen forests, and asphalt in increasing order of albedo.

21. What are the processes occurring in soil that operate to reduce the harmful nature of pollutants?

CHAPTER 11
THE NATURE AND COMPOSITION OF THE ATMOSPHERE

11.1 IMPORTANCE OF THE ATMOSPHERE

The atmosphere is a protective blanket which nurtures life on the Earth and protects it from the hostile environment of outer space. The atmosphere is the source of carbon dioxide for plant photosynthesis and of oxygen for respiration. It provides the nitrogen that nitrogen-fixing bacteria and ammonia-manufacturing plants use to produce chemically bound nitrogen, essential for life. As a basic component of the hydrologic cycle, the atmosphere transports water from the oceans to land, thus acting as the condenser in a vast, solar-powered still. Unfortunately, the atmosphere also has been a dumping ground for many pollutant materials—ranging from sulfur dioxide to aerosol-can Freon—a practice which causes damage to vegetation and materials, shortens human life, and possibly alters the characteristics of the atmosphere itself.

The atmosphere serves a vital protective function. It absorbs most of the cosmic rays from outer space and protects living things from their effects. It also absorbs most of the electromagnetic radiation from the sun. Only radiation in the wavelength regions 300–2500 nm and 0.01–40 μm is transmitted to any appreciable extent by the atmosphere. The first region consists of near-ultraviolet, visible, and near-infrared radiation; the second region consists of radio waves. It is particularly fortunate for life on Earth that the atmosphere filters out tissue-damaging ultraviolet radiation below about 300 nm.

The atmosphere is essential in maintaining the heat balance of the Earth. The atmosphere absorbs infrared radiation emitted by the sun and also absorbs energy re-emitted by the Earth in the form of infrared radiation. Therefore, it serves an important heat-stabilizing function and prevents the tremendous temperature extremes that occur on planets and moons lacking substantial atmospheres.

11.2 COMPOSITION OF THE ATMOSPHERE

The two **major components** of dry air at ground level (by volume) are nitrogen (78.08%) and oxygen (20.95%). Argon (0.934%) and carbon dioxide (0.034%) may be classified as **minor components** of the atmosphere. Air may contain from 0.1–5% water by volume, with a normal range of 1–3%. Percentages of noble gases by volume in dry air at ground level are: neon, 1.818×10^{-3}%; helium, 5.24×10^{-4}%; krypton, 1.14×10^{-4}%; and xenon, 8.7×10^{-6}%. A number

TABLE 11.1 Atmospheric Trace Gases in Dry Air Near Ground Level, Expressed as Volume Percent[1]

Gas or species	*Volume percent*[1]	*Major sources*	*Process for removal from the atmosphere*
CH_4	1.6×10^{-4}	Biogenic[2]	Photochemical[3]
CO	$\sim 1.2 \times 10^{-5}$	Photochemical, anthropogenic[4]	Photochemical
N_2O	3×10^{-5}	Biogenic	Photochemical
NO_x[5]	10^{-10}–10^{-6}	Photochemical, lightning, anthropogenic	Photochemical
HNO_3	10^{-9}–10^{-7}	Photochemical	Washed out by precipitation
NH_3	10^{-8}–10^{-7}	Biogenic	Photochemical, washed out by precipitation
H_2	5×10^{-5}	Biogenic, photochemical	Photochemical
H_2O_2	10^{-8}–10^{-6}	Photochemical	Washed out by precipitation
$HO\cdot$[6]	10^{-13}–10^{-10}	Photochemical	Photochemical
$HO_2\cdot$[6]	10^{-11}–10^{-9}	Photochemical	Photochemical
H_2CO	10^{-8}–10^{-7}	Photochemical	Photochemical
CS_2	10^{-9}–10^{-8}	Anthropogenic, biogenic	Photochemical
OCS	10^{-8}	Anthropogenic, biogenic, photochemical	Photochemical
SO_2	$\sim 2 \times 10^{-8}$	Anthropogenic, photochemical, volcanic	Photochemical
I_2	O-trace	—	—
CCl_2F_2[7]	2.8×10^{-5}	Anthropogenic	Photochemical
H_3CCCl_3[8]	$\sim 1 \times 10^{-8}$	Anthropogenic	Photochemical

[1] Levels in the absence of gross pollution.
[2] From biological sources.
[3] Reactions induced by the absorption of light energy as described later in this chapter.
[4] Sources arising from human activities.
[5] Sum of NO and NO_2.
[6] Reactive free radical species with one unpaired electron; described later in the chapter; these are transient species whose concentrations become much lower at night.
[7] A chlorofluorocarbon, Freon F-12.
[8] Methyl chloroform.

of atmospheric **trace gases** are important in atmospheric chemistry. These are listed in Table 11.1.

The density of the atmosphere decreases sharply with increasing altitude as a consequence of the gas laws and gravity. Over 99% of the total mass of the atmosphere is found within approximately 30 km (1 km = 0.62 mi) of the Earth's surface. The total mass of the atmosphere is approximately 5.14×10^{15} metric tons [1]. Large as it is, this mass is only approximately one millionth of the Earth's total mass.

The characteristics of the atmosphere vary greatly, particularly with altitude. Other factors introducing variability are season, latitude, time, and even

[1] Verniani, F. 1966. The total mass of the earth's atmosphere. *J. Geophys. Res.* 71: 385–91.

solar activity. Temperatures in the atmosphere may vary from as low as −138 °C to over 1700 °C. Atmospheric pressure drops from 1.00 atm at sea level to 3.0×10^{-7} atm at 100 km above sea level. Because of these variations, the chemistry of the atmosphere varies greatly with altitude. In addition to temperature differences, the **mean free path** of species in the atmosphere (mean distance traveled before collision with another particle) increases over many orders of magnitude with increasing altitude. A particle having a mean free path of approximately 1×10^{-6} cm at sea level has a mean free path exceeding 2×10^{6} cm at an altitude of 500 km, where the pressure is lower by many orders of magnitude.

11.3 MAJOR REGIONS OF THE ATMOSPHERE

Atmospheric science deals with the movement of air masses in the atmosphere, atmospheric heat balance, and atmospheric chemical composition and reactions. In order to understand atmospheric chemistry, it is necessary first to have some understanding of the physical characteristics of the Earth's atmosphere.

The Earth's atmosphere may be divided into a number of different regions, depending upon the system of classification. The most general classification divides it into the **lower atmosphere** (up to approximately 50 km) and the **upper atmosphere.** Until the development of high-altitude rockets and Earth satellites, only the lower atmosphere was accessible for study. However, use of these devices has resulted in a quantum jump in human understanding of atmospheric phenomena in recent years.

The upper region of the atmosphere contains species appreciably different from those found in the lower region. Since the lower regions are relatively homogeneous in composition, one system of classification divides the atmosphere into regions called the **homosphere** (having little variation in composition) and the **heterosphere** (having a high variation in composition).

The most commonly used system divides the atmosphere according to variations of temperature with altitude. These temperature regions and their characteristics are summarized in Table 11.2 and shown schematically in Figure 11.1. In addition to the regions listed, a fifth region, the **exosphere,** is sometimes included. Some of the species in the exosphere acquire sufficient kinetic energy to break loose from the gravitational field of the Earth and go into outer space.

The boundaries of the **troposphere** are influenced by a number of factors, including temperature and the nature of the underlying terrestrial surface. Even

TABLE 11.2 Major Regions of the Atmosphere and Their Characteristics

Region	*Temperature range, °C*[1]	*Altitude range, km*[1]	*Significant chemical species*
troposphere	15 to −56	0–(10–16)[2]	N_2, O_2, CO_2, H_2O
stratosphere	−56 to −2	(10–16)–50	O_3
mesosphere	−2 to −92	50–85	O_2^+, NO^+
thermosphere	−92 to 1200	85–500	O_2^+, O^+, NO^+

[1]These figures give the boundaries of each region.
[2]The boundary between the troposphere and stratosphere varies from 10–16 km.

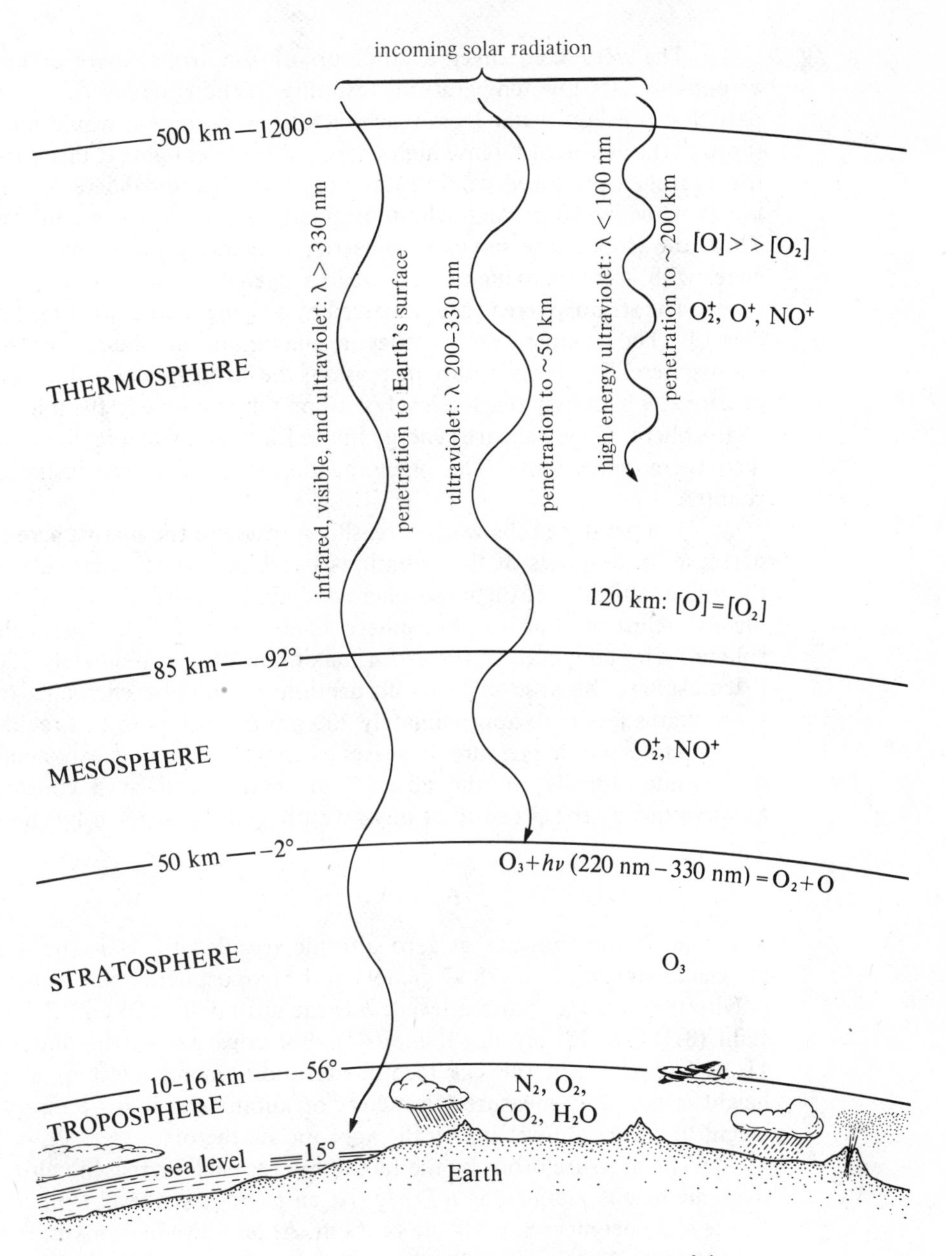

FIGURE 11.1 Major regions of the atmosphere (not to scale).

during a one-day period, the upper limit of the troposphere may vary by a kilometer or more. The troposphere is characterized by falling temperature with increasing altitude, as the distance from the heat-radiating Earth increases. In the absence of air pollution, the composition of the troposphere is quite homogeneous. This homogeneity is due largely to mixing by the constantly circulating air masses in the troposphere. The water content of the troposphere, however, is extremely variable because of cloud formation, precipitation, and evaporation of water from terrestrial water bodies.

The very cold layer at the top of the troposphere is known as the **tropopause.** Its low temperature, resulting in the condensation of water to ice particles, prevents water from reaching altitudes where it would photodissociate through the action of intense high-energy ultraviolet light. If this were to happen, the hydrogen produced would escape the Earth's atmosphere (this happened to much of the hydrogen and helium originally present in the Earth's atmosphere). Thus, the tropopause serves as a barrier preventing most upward transport of water with accompanying net loss of hydrogen from the Earth.

The **stratosphere** is characterized by a rising temperature as the altitude increases. The temperature reaches a maximum at the upper limit of the stratosphere. The temperature increase of the stratosphere is due to the presence of ozone, which may reach a level of 10 ppm by volume in the mid-region of the stratosphere. Ozone absorbs energy in the form of ultraviolet light and causes an increase in temperature. This phenomenon will be discussed further, later in the chapter.

Temperature falls with increasing altitude in the **mesosphere** because of a decrease in the levels of the radiation-absorbing species, particularly ozone. In the higher altitudes of the mesosphere and above, molecules and atoms can completely escape the Earth's atmosphere. (This region is sometimes called the exosphere.) The temperature rises to a maximum of approximately 1200 °C in the **thermosphere** because of the absorption of highly energetic radiation of wavelengths less than approximately 200 nm by species in this region.

Atmospheric pressure decreases as an approximately exponential function of altitude. Ideally, in the absence of mixing and at a constant absolute temperature T, the pressure at any given height P_h is given by the exponential equation

$$P_h = P_0 e^{-Mgh/RT} \qquad (11.3.1)$$

where P_0 is the pressure at zero altitude (sea level); M is the average gram molecular weight of air (28.97 g/mole in the troposphere); g is the acceleration of gravity (981 cm sec^{-2} at sea level); h is the altitude in cm; and R is the gas constant (8.314×10^7 erg deg^{-1} mole^{-1}). For consistency, the units in Equation 11.3.1 are given in the cgs (centimeter-gram-second) system, although since height is normally measured in meters or kilometers, it is necessary to convert height to cm by multiplying by the appropriate factors.

The pressure drops by the factor e^{-1} for each increase in altitude equal to the **scale height,** defined as RT/Mg. At an average sea-level temperature of 288 K, the scale height is 8×10^5 cm or 8 km. At an altitude of 8 km, the pressure is only about 39% of sea-level pressure.

To get a better picture of the variation of pressure with altitude, Equation 11.3.1 may be converted to the logarithmic (base 10) form and h may be expressed in kilometers:

$$\log P_h = \log P_0 - \frac{Mgh \times 10^5}{2.303RT} \qquad (11.3.2)$$

If the pressure at sea level is taken as 1.00 atm, the equation becomes simply

$$\log P_h = -\frac{Mgh \times 10^5}{2.303RT} \qquad (11.3.3)$$

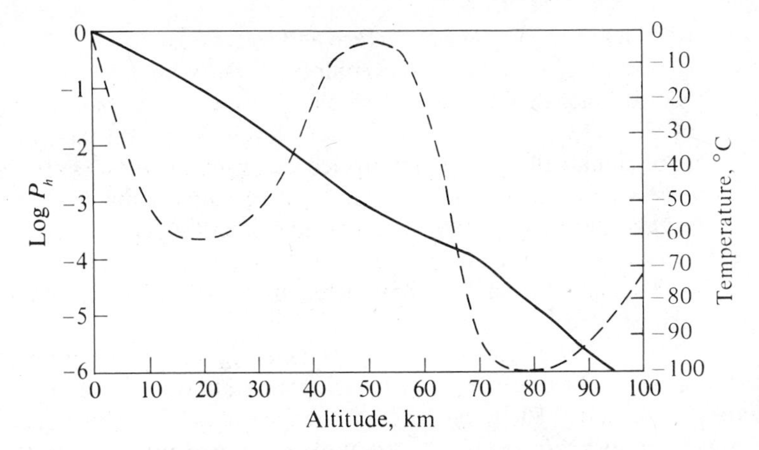

FIGURE 11.2 Variation of pressure (solid line) and temperature (dashed line) with altitude.

Figure 11.2 shows the actual values of log P_h as a function of altitude. The plot is nonlinear because of variations arising from temperature differences and the mixing of air masses. The figure also shows the variation of temperature with altitude.

In terms of bulk, by far the greatest portion of the atmosphere is confined to the region very close to the surface of the Earth. Half of the atmosphere lies within 5.5 km (3.5 miles) of the Earth's surface. Three-fourths lies below 11 km, about the cruising altitude of a high-flying jet passenger plane. Less than 1% of atmospheric gases are more than 40 km from the Earth's surface. Such an altitude is, of course, miniscule compared to the radius of the Earth.

11.4 THE EARTH'S HEAT BALANCE

The energy from the sun that reaches the Earth's upper atmosphere amounts to a very large quantity per unit time, a phenomenon which eventually may form the basis of a large solar-energy industry. At the distance of the Earth from the sun, a square meter of area perpendicular to the line of solar flux receives 19.2 kcal of energy per minute, or 1.34×10^3 watts/m^2 (Figure 11.3). The rate at which the sun's energy reaches the Earth's atmosphere is called the **solar constant**. If all of

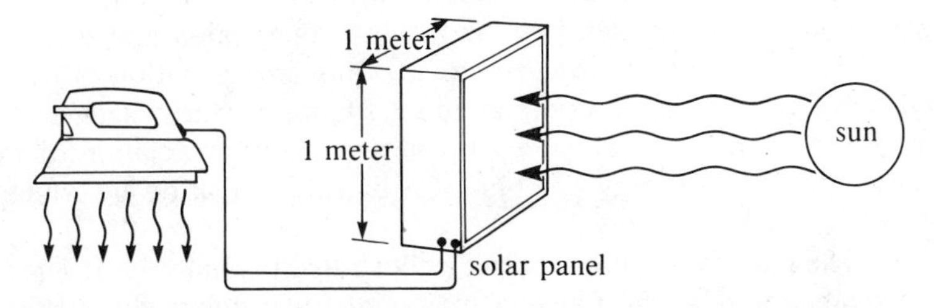

FIGURE 11.3 The solar flux at the distance of the Earth from the sun is 1.34×10^3 watts/m^2.

this energy reached the Earth's surface and were retained, our planet would have vaporized long ago. As it is, the complex factors involved in maintaining the Earth's heat balance within very narrow confines are crucial to retaining conditions of climate that will support present levels of life on the Earth. The great changes of climate that resulted in the ice ages were caused by variations of only a few degrees in average temperature. Marked climate changes within recorded history have been caused by much smaller average temperature changes. The mechanisms by which the Earth's average temperature is retained within this narrow range are complex and not completely understood, but the main features will be explained here.

About 50% of the solar radiation entering the Earth's atmosphere reaches the Earth's surface either directly or after scattering by clouds, atmospheric gases, or particles. The remaining half of the radiation is either reflected directly from the atmosphere or absorbed in the atmosphere and its energy radiated back into space at a later time as infrared radiation. Most of the solar energy reaching the Earth's surface is absorbed, and it must be returned to space in order to maintain the heat balance. In addition, a very small amount of energy (less than 1% of that received from the sun) reaches the Earth's surface by convection and conduction processes (see below) from the Earth's hot mantle, and this, too, must be lost.

Energy transport is crucial to eventual reradiation of energy from the Earth. This is accomplished by three mechanisms. **Radiation** of energy occurs through electromagnetic waves in the infrared region of the spectrum. Radiation can transmit energy through a vacuum and is the only mechanism by which energy can eventually completely escape the planet. **Conduction** of energy occurs through the interaction of adjacent atoms or molecules without the bulk movement of matter. **Convection** involves the movement of whole masses of air, which may be either relatively warm or cold. Convection is the phenomenon that can turn a balmy March morning into a frigid afternoon when a mass of cold arctic air moves into the vicinity. As well as carrying **sensible heat** due to the kinetic energy of molecules, convection carries **latent heat** in the form of water vapor which releases heat as it condenses. An appreciable fraction of the Earth's surface heat is transported to clouds in the atmosphere by conduction and convection before being lost ultimately by radiation.

The radiation that carries energy away from the Earth is of a much longer wavelength than the sunlight that brings energy to the Earth. This is a crucial factor in maintaining the heat balance and one susceptible to upset by human activities. The maximum intensity of incoming radiation occurs at 0.5 micrometers (500 nanometers) in the visible region, with essentially none below 0.2 μm or above 3 μm. This encompasses the whole visible region and small parts of the ultraviolet and infrared adjacent to it. Outgoing radiation is in the infrared region, with maximum intensity at about 10 μm, primarily between 2 μm and 40 μm. Thus the Earth loses energy by electromagnetic radiation of much longer wavelength (lower energy per photon) than the radiation by which it receives energy.

The Earth's radiation budget is illustrated in Figure 11.4. The average surface temperature of the Earth is maintained at a relatively comfortable 15 °C because water vapor and, to a lesser extent carbon dioxide, reabsorb much of the outgoing radiation and reradiate about half of this back to the surface. If this were not the case, the Earth's surface temperature would average around −18 °C.

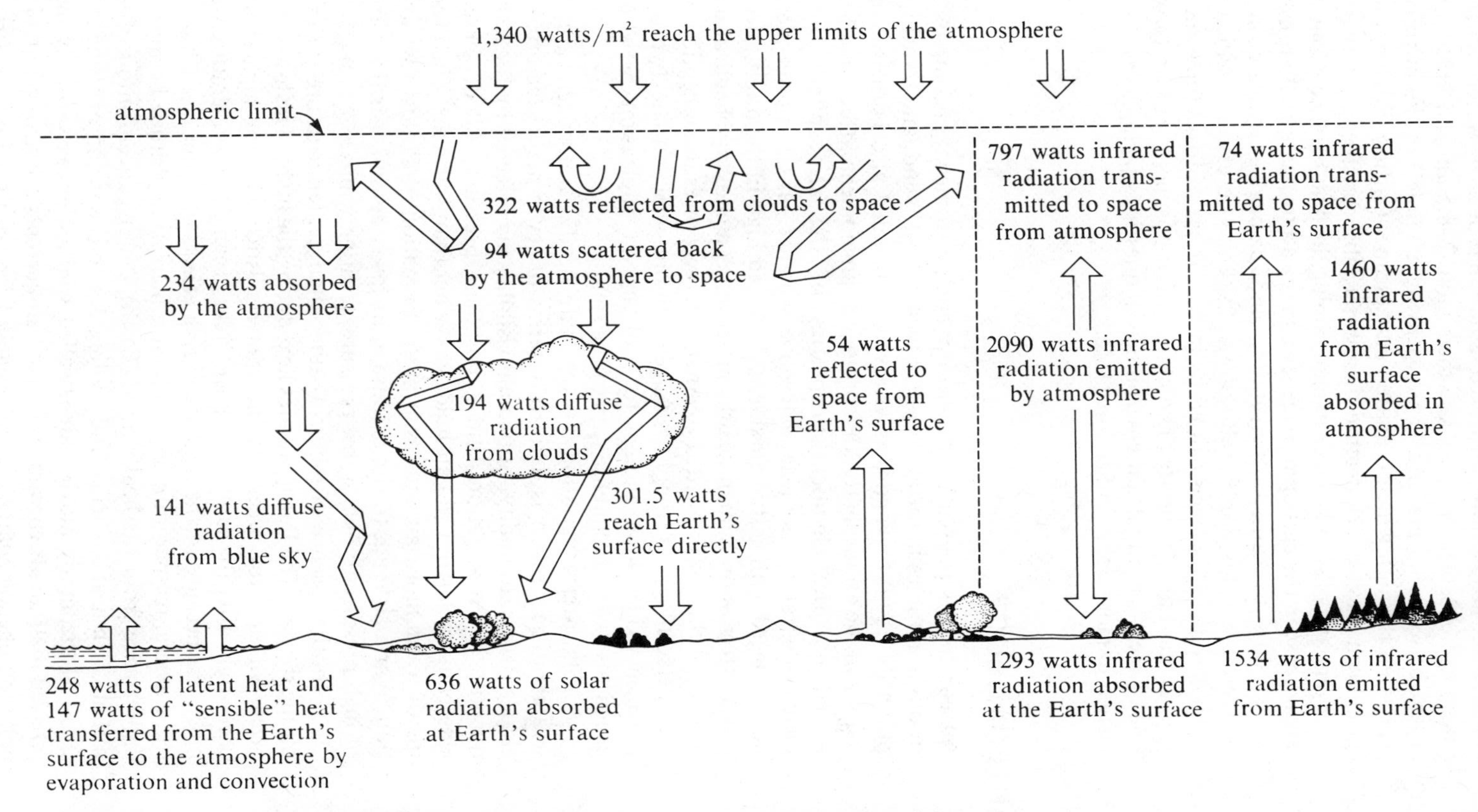

FIGURE 11.4 Earth's radiation "budget" expressed on the basis of portions of the 1340 watts/m² composing the solar flux.

Most of the absorption of infrared radiation is done by water molecules in the atmosphere. Absorption is weak in the regions 7–8.5 μm and 11–14 μm and nonexistent between 8.5 μm and 11 μm, leaving a "hole" in the infrared absorption spectrum through which radiation may escape. Carbon dioxide, though present at a much lower concentration than water, absorbs strongly between 12 μm and 16.3 μm, and plays a key role in maintaining the heat balance. Concern exists that an increase in the carbon dioxide level in the atmosphere could prevent sufficient energy loss to cause a perceptible and damaging increase in the Earth's temperature. This is the well-known "greenhouse effect," which might result from elevated CO_2 levels caused by increased use of fossil fuels and the destruction of the Earth's forests. The matter of atmospheric carbon dioxide content, one of the most important environmental questions facing us at this time, is discussed in more detail in Section 11.12.

11.5 METEOROLOGY

Meteorology is the science of atmospheric phenomena, encompassing the study of physical forces in the atmosphere such as heat, wind, and transitions of water, primarily liquid to vapor, or vice versa. Included also is the movement of air masses in the atmosphere. Meteorological phenomena affect, and in turn are affected by, the chemical properties of the atmosphere. For example, meteorological phenomena determine whether or not power-plant stack gas heavily laced with sulfur dioxide is dispersed high in the atmosphere, with little direct effect upon human health, or settles as a choking chemical blanket in the vicinity of the power plant. Los Angeles largely owes its susceptibility to smog to the meteorology of the Los Angeles basin, which holds hydrocarbons and nitrogen oxides long enough to cook up an unpleasant brew of damaging chemicals under the intense rays of the sun.

Short-term variations in the state of the atmosphere are described as **weather.** The weather is defined in terms of seven major factors: temperature, clouds, winds, humidity, horizontal visibility (as affected by fog, etc.), type and quantity of precipitation, and atmospheric pressure. All of these factors are closely interrelated. Cold air holds less water than warm air. Therefore, the cooling of warm moist air can result in the formation of clouds, fog, and precipitation. Warm air tends to rise because of its lower density. Air flows from a region of high pressure to one of low pressure, creating winds.

In the absence of major weather changes, temperature varies predictably during the day. The Earth radiates heat during the night. The lowest temperatures occur within a few weeks of the winter solstice, and the highest temperatures occur right after the summer solstice. Near oceans, the temperature variations are much less because of the heat-stabilizing effects of water (see Chapter 2). In the troposphere, temperature decreases with increasing altitude because of: increasing distance from the heat source (Earth); lower concentration of water vapor with its high heat capacity; and cooling due to expansion of rising air.

Horizontally moving air is called **wind,** whereas vertically moving air is referred to as an **air current.** Wind and air currents are strongly involved with air pollution phenomena. Wind carries and disperses air pollutants. Prevailing wind direction is an important factor in determining the areas most affected by an air pollution source.

Atmospheric water can be present as vapor, liquid, or ice. The water content of air can be expressed as *specific, absolute,* or *relative humidity*. **Specific humidity** is the number of grams of water vapor per kilogram of moist air. Unless the actual proportion of water in the air changes, specific humidity is a constant, regardless of the temperature and pressure of the air. **Absolute humidity** is a measure of the weight of water vapor in a given volume of air. It varies with air temperature and pressure. **Relative humidity**, expressed as a percentage, describes the amount of water vapor in the air as a ratio of the maximum amount that the air can hold at that temperature. Air with a given relative humidity can undergo any of several processes to reach the saturation point at which water vapor condenses. These processes are: cooling by contact with a cold surface; adiabatic cooling by rising and expansion; mixing of cold and warm air masses; and radiation of heat by the air itself. The temperature below which moisture condensation occurs is called the **dew point**.

Clouds normally form when rising, adiabatically cooling air can no longer hold water in the vapor form, and the water forms very small aerosol droplets. Clouds may be classified in three major forms. **Cirrus clouds** occur at great altitudes and have a thin feathery appearance. **Cumulus clouds** are detached masses with a flat base and frequently a "bumpy" upper structure. **Stratus clouds** occur in large sheets and may cover all of the sky visible from a given point as overcast. Clouds are important absorbers and reflectors of radiation (heat). Their formation is affected by human activities, especially particulate matter pollution and emission of deliquescent gases, such as SO_2 and HCl.

Clouds form by condensation of water vapor. Condensation must occur prior to the formation of precipitation in the form of rain or snow. For this condensation to happen, air must be cooled below the dew point, and nuclei of condensation must be present. These nuclei are hydroscopic substances such as salts, sulfuric acid droplets, and some organic materials. Air pollution is now an important source of condensation nuclei.

Bacteria from the ocean are important organic condensation nuclei [2]. These bacteria become airborne when sea-foam bubbles burst, forming small droplets of water containing entrained bacterial cells. As the water evaporates, bacterial cells and salt crystals remain. These are carried to high altitudes by air currents. Since bacterial cells are hydrophilic colloids (see Section 6.4), they attract water and act as condensation nuclei for very small water droplets and ice particles. (The same general processes also forms salt particles that act as condensation nuclei).

Cloud droplets normally take somewhat longer than a minute to form by condensation. They average about 0.04 mm across and do not exceed 0.2 mm in diameter. Raindrops range from 0.5–4 mm in diameter. Condensation processes do not form particles large enough to fall as precipitation (rain, snow, sleet, or hail). The small condensation droplets must collide and coalesce to form precipitation-size particles. When droplets reach a threshold diameter of about 0.04 mm, they grow more rapidly by coalescence with other particles than by condensation of water vapor. The formation of precipitation is a complicated and important process and is still not completely understood.

---

[2] 1976. Ocean bacteria may trigger precipitation. *Chemical and Engineering News* 9 April 1976.

Distinct air masses are a major feature of the troposphere. These air masses are uniform and are horizontally homogeneous. Their temperature and water-vapor content are particularly uniform. These characteristics are determined by the nature of the surface over which a large air mass forms. Polar continental air masses form over cold land regions; polar maritime air masses form over polar oceans. Air masses originating in the tropics may be similarly classified as tropical continental air masses or tropical maritime air masses. The movement of air masses and the conditions in them may have important effects upon pollutant reactions, effects, and dispersal.

The relatively sharp border areas between air masses are called **fronts.** At the front between a warm air mass and a cold air mass, the cold air forms a wedge beneath the warm air as a consequence of their different densities. Warm, moist air rising along this wedge is cooled, often causing precipitation to occur.

There are several general types of fronts. A **cold front** is one in which warm air is replaced by cold air. Particularly if the cold air wedge along such a front rises steeply, turbulent air conditions, thunderstorms, and brief but intense rainshowers may develop. Replacement of cold air by warm air produces a **warm front**. Warm fronts are normally preceded by cloud cover which may extend several hundred miles. A border between warm and cold air masses that is not in motion is not classified as either a warm front or a cold front, but is called a **stationary front**. When a cold front overtakes a warm front, a mass of warm air is sandwiched between two masses of colder air and is squeezed up between them; this is called an **occluded front**.

Cyclones are particularly important atmospheric phenomena. They are approximately circular, range up to 1000 or more miles in diameter, and are characterized by a low-pressure center surrounded by regions of increasing pressure. The Earth's rotation causes cyclones to spin counterclockwise in the northern hemisphere and clockwise in the southern hemisphere.

A detailed discussion of the general circulation of air masses making up the global climate is beyond the scope of this book. One of the major factors involved is that warm air produced in the tropics around the equator tends to rise, cools by loss of heat through radiation as it migrates northward and southward from the equator, then returns to the tropical regions as cooled air. This circulating air is subject to a **coriolis force**, an inertial force which causes air moving back to the equator to lag behind the Earth's surface in its motion. As a result, wind moving toward the equator acquires a westward component of velocity.

The region of rising air around the equator comprises the **intertropical convergence zone**, an area with a great deal of cloudiness and rain. There is an absence of strong north-south winds across this zone, so that mixing of atmospheric constituents across the hemispheres requires a long time—one to two years. By comparison, mixing within a hemisphere requires one to two months.

The complicated movement of air across the Earth's surface is a crucial factor in the creation and dispersal of air pollution phenomena. When air movement ceases, air stagnation can occur with a resultant build-up of air pollutants in localized regions. Although the temperature of air relatively near the Earth's surface normally decreases with increasing altitude, certain atmospheric conditions can result in the opposite condition—increasing temperature with increasing altitude. Such conditions are characterized by high atmospheric stability and are known as **temperature inversions**. Because they limit the vertical circulation of air, temperature inversions result in air stagnation and the trapping of air pollutants in localized areas.

Temperature inversions may be classified as *frontal, advective,* or *radiational.* A **frontal inversion** occurs by the collision of a warm air mass (warm front) with a cold air mass (cold front). The warm air mass overrides the cold air mass in the frontal area, producing the inversion. **Advective inversion** occurs when a mass of warm air moves over a cold surface, or when warm air is forced up over a mountain range and flows over cool air on the other side of the range. This latter phenomenon frequently occurs east of the Rocky Mountains and is responsible for inversions commonly observed during the winter in Denver, Colorado, for example. Radiation inversions occur on clear nights when lower-level air loses heat to the ground, which, of course, is not receiving solar energy at night and is radiating infrared energy to space. The inversion forms around dawn. Thus, for example, smokestack plumes in the early morning frequently remain close to the ground in the vicinity of the stack, or blow downwind in a narrow horizontal streak.

Air circulation on a relatively small scale is involved in thunderstorms, valley winds, and sea-land breezes. These fall in the category of **mesometeorology** (as compared to **macrometeorology**, which involves the movement of very large masses of air). On an even smaller scale than mesometeorology, very small-scale atmospheric phenomena are sometimes discussed under the heading of **micrometeorology**. For the purposes of this discussion, no distinction need be made between mesometeorology and micrometeorology.

Human activities have succeeded in changing the meteorology of whole cities, a mesometeorological effect. By paving over large areas of a city with nonreflecting asphalt, and by generating heat, humans have created conditions under which the center of a city may be as much as 5 °C warmer than the surrounding area. In such a case, the warmer air rises, bringing in a breeze from the surrounding area. Large cities have been described as "heat islands." Pollutants and carbon dioxide given off from cities may absorb emitted infrared radiation, causing a local greenhouse effect that probably is largely counterbalanced by reflection of incoming solar energy by particulate matter above cities.

Some human activities may even be changing the global climate; these are discussed in Section 11.15.

11.6 EVOLUTION OF THE ATMOSPHERE

It is now widely believed that the Earth's atmosphere originally was very different from its present state and that the changes were brought about by biological activity. Approximately 3.5 billion years ago, when the first primitive life molecule was formed, the atmosphere was chemically reducing, consisting primarily of methane, ammonia, water vapor, and hydrogen. The atmosphere was bombarded by intense, bond-breaking ultraviolet light, which, along with lightning and radiation from radionuclides, provided the energy to bring about those chemical reactions resulting in the production of relatively complicated molecules, including even amino acids and sugars. From the rich chemical mixture in the sea, life molecules evolved. Initially, these very primitive life forms derived their energy from fermentation of organic matter formed by chemical and photochemical processes, but eventually they gained the capability to produce organic matter, "$\{CH_2O\}$," by photosynthesis,

$$CO_2 + H_2O + h\nu \rightarrow \{CH_2O\} + O_2(g) \tag{11.6.1}$$

and the stage was set for the massive biochemical transformation that resulted in the production of almost all the atmosphere's oxygen.

The oxygen initially produced by photosynthesis was probably quite toxic to primitive life forms. However, much of this oxygen was converted to iron oxides by reaction with soluble iron(II):

$$4\ Fe^{2+} + O_2 + 4\ H_2O \rightarrow 2\ Fe_2O_3 + 8\ H^+ \qquad (11.6.2)$$

This resulted in the formation of enormous deposits of iron oxides, which provide major evidence for the liberation of free oxygen in the primitive atmosphere.

Eventually, enzyme systems developed that enabled organisms to mediate the reaction of waste-product oxygen with oxidizable organic matter in the sea. Later, this mode of waste-product disposal was utilized by organisms to produce energy by respiration, which is now the mechanism by which nonphotosynthetic organisms obtain energy.

In time, oxygen accumulated in the atmosphere, of course providing an abundant source of oxygen for respiration. It had an additional benefit in that it enabled the formation of an ozone shield (see Section 11.9). The ozone shield absorbs bond-rupturing ultraviolet light. With the ozone shield protecting tissue from destruction by high-energy ultraviolet radiation, the Earth became a much more hospitable environment for life, and life forms were enabled to move from the sea to land.

11.7 CHEMICAL AND PHOTOCHEMICAL REACTIONS IN THE ATMOSPHERE

Considering the experimental difficulties inherent in the investigation of atmospheric chemical reactions, it is perhaps surprising that so much has been learned about such reactions. One of the primary obstacles encountered in studying atmospheric chemistry is that the chemist generally must deal with incredibly low concentrations, so that the detection and analysis of reaction products is quite difficult. Simulating high-altitude conditions in the laboratory can be extremely hard because of interferences, such as those from species given off from container walls under conditions of very low pressure. Many chemical reactions that require a third body to absorb excess energy occur very slowly in the upper atmosphere, where there is a sparse concentration of third bodies, but occur readily in a container whose walls effectively absorb energy. Container walls may serve as catalysts for some important reactions, or they may absorb important species and react chemically with the more reactive ones.

The absorption of light by chemical species can bring about reactions which do not otherwise occur under the conditions (particularly the temperature) of the medium in the absence of light. Thus, photochemical reactions, even in the absence of a chemical catalyst, occur at temperatures much lower than those which otherwise would be required. Such photochemical reactions, induced by intense solar radiation, play a very important role in determining the nature and ultimate fate of a chemical species in the atmosphere.

Nitrogen dioxide, NO_2, is one of the most photochemically active species found in a polluted atmosphere and is an essential participant in the smog-formation process. A species such as NO_2 may absorb light of energy $h\nu$ in a reaction, producing a molecule in an electronically excited state designated by an asterisk, *:

$$NO_2 + h\nu \rightarrow NO_2^* \qquad (11.7.1)$$

The photochemistry of nitrogen dioxide is discussed in greater detail in Section 12.12 and in Chapter 14.

Electronically excited molecules are one of the three relatively reactive and unstable species that are encountered in the atmosphere and are strongly involved with atmospheric chemical processes. The other two species are atoms or molecular fragments with unshared electrons, called **free radicals**, and ionized atoms or molecular fragments.

Electronically excited molecules are produced when stable molecules absorb energetic electromagnetic radiation in the ultraviolet or visible regions of the spectrum. A molecule may possess several possible excited states, but generally ultraviolet or visible radiation is energetic enough to excite molecules only to several of the lowest energy levels. The nature of the excited state may be understood by considering the disposition of electrons in a molecule. Most molecules have an even number of electrons. The electrons occupy orbitals, with a maximum of two electrons with opposite spin occupying the same orbital. The absorption of light may promote one of these electrons to a vacant orbital of higher energy. In some cases the electron thus promoted retains a spin opposite to that of its former partner, giving rise to a **singlet (excited) state**. In other cases the spin of the promoted electron is reversed, such that it has the same spin as its former partner; this gives rise to a **triplet (excited) state**.

ground state	singlet state	triplet state
	↓	↑
↑ ↓	↑	↑
↑ ↓	↑ ↓	↑ ↓

These excited states are relatively energized compared to the ground state and are chemically reactive species. Their participation in atmospheric chemical reactions, such as those involved in smog formation, will be discussed later in detail.

In order for a photochemical reaction to occur, light must be absorbed by the reacting species. If the absorbed light is in the visible region of the sun's spectrum, the absorbing species is colored. Colored NO_2 is the prime example of such a species in the atmosphere. Normally, the primary photochemical process involves activation of the molecule by the absorption of a single **quantum** of light. The energy of one quantum is equal to the product $h\nu$, where h is Planck's constant, 6.62×10^{-27} erg sec, and ν is the frequency of the absorbed light in sec^{-1}.

A molecule energized by the absorption of light loses energy by a number of processes [3], some of which are described as follows. In this discussion, keep in mind that the symbol * represents an excited state of the species.

---

[3] Wayne, R.P. 1970. *Photochemistry*. New York: Elsevier North-Holland, Inc.

An excited species such as O_2^* may simply lose energy to another molecule or atom, designated M, by a process known as **physical quenching**:

$$O_2^* + M \rightarrow O_2 + M \tag{11.7.2}$$

The eventual result is dissipation of energy to the surroundings as heat, because M acquires translational energy as a result of physical quenching. The excited species may dissociate, a process responsible for the predominance of atomic oxygen in the upper atmosphere:

$$O_2^* \rightarrow O + O \tag{11.7.3}$$

The excited species may undergo a direct reaction:

$$O_2^* + O_3 \rightarrow 2\,O_2 + O \tag{11.7.4}$$

Energy gained by the absorption of electromagnetic radiation may be lost by the emission of electromagnetic radiation, a process known generally as **luminescence:**

$$NO_2^* \rightarrow NO_2 + h\nu \tag{11.7.5}$$

Special cases of luminescence are called **fluorescence** (almost immediate re-emission of light) or **phosphorescence** (delayed emission). If the excited species orginated from a chemical reaction, the emission of light is called **chemiluminescence**. (All of these luminescence phenomena are used in chemical analysis. Chemiluminescence is especially effective for the analysis of some air pollutants such as ozone.) Luminescence and chemiluminescence are responsible for some of the phenomena observed in the sky. For example, the faint atmospheric light emission called *airglow* (see Section 11.9) is caused partially by chemiluminescence from excited hydroxyl radicals (see Section 11.11):

$$O_3 + H \rightarrow HO^{\cdot *} + O_2 \tag{11.7.6}$$

$$HO^{\cdot *} \rightarrow HO\cdot + h\nu \tag{11.7.7}$$

An intermolecular energy transfer may occur in which an excited species transfers energy to another species. A subsequent reaction by the second molecule is called a **photosensitized reaction.** It is believed that part of airglow arises from luminescence by excited sodium atoms, Na*, formed in the atmosphere by an intermolecular energy transfer involving the following reaction:

$$O_2^* + Na \rightarrow O_2 + Na^* \tag{11.7.8}$$

In some cases only energy transfer within the molecule occurs, a process called **intramolecular transfer,**

$$XY^* \rightarrow XY^\dagger \tag{11.7.9}$$

where the dagger, †, denotes an excited state different from the initial one. Finally, the excited species may undergo **spontaneous isomerization**, as in the conversion of *o*-nitrobenzaldehyde to *o*-nitrosobenzoic acid, a reaction used in chemical actinometers to measure exposure to electromagnetic radiation:

$$C_6H_4(NO_2)\text{—C(=O)—H} + h\nu \longrightarrow C_6H_4(NO)\text{—C(=O)—OH} \tag{11.7.10}$$

The absorption of very energetic radiation may result in the detachment of an electron, a process called **photoionization**:

$$N_2 + h\nu \rightarrow N_2^+ + e \quad (11.7.11)$$

Photoionization is often considered a subcategory of photodissociation in which one of the dissociation products is an electron.

Light absorbed in the infrared region is not sufficiently energetic to break chemical bonds. The receptor molecules do gain vibrational and rotational energy. The energy absorbed as infrared radiation ultimately is dissipated as heat and raises the temperature of the atmosphere. As discussed in Section 11.4, the absorption of infrared radiation is very important in the Earth's acquiring heat from the sun and in the retention of energy radiated from the Earth's surface.

11.8 IONS AND RADICALS IN THE ATMOSPHERE

One of the characteristics of the upper atmosphere which cannot be duplicated under laboratory conditions is the presence of significant levels of electrons and positive ions. Because of the rarefied conditions in the upper atmosphere, these ions may exist for long periods before recombining to form neutral species.

At altitudes of approximately 50 km and up, ions are so prevalent that the region is called the **ionosphere.** The presence of the ionosphere has been known since about 1901, when it was discovered that radio waves could be transmitted over long distances, where the curvature of the Earth makes line-of-sight transmission impossible. These radio waves bounce off the ionosphere.

Ultraviolet light is the primary producer of ions in the ionosphere. In darkness, the positive ions slowly recombine with free electrons. The process is especially rapid in the lower regions of the ionosphere, where the concentration of species is relatively high. Thus, the lower limit of the ionosphere lifts at night and makes possible the transmission of radio waves over much greater distances.

The Earth's magnetic field has a strong influence upon the ions in the upper atmosphere. Probably the best-known manifestation of this phenomenon is found in the **Van Allen belts**, discovered in 1958. These regions consist of two belts of ionized particles which circle the Earth. If they are visualized as two doughnuts, then the axis of the Earth's magnetic field extends through the holes in the doughnuts. In the inner belt, the highly energetic ionizing radiation consists of protons. In the outer belt, it consists of electrons. A schematic diagram of the Van Allen belts is shown in Figure 11.5.

Although ions are produced in the upper atmosphere primarily by the action of energetic electromagnetic radiation, they may also be produced in the

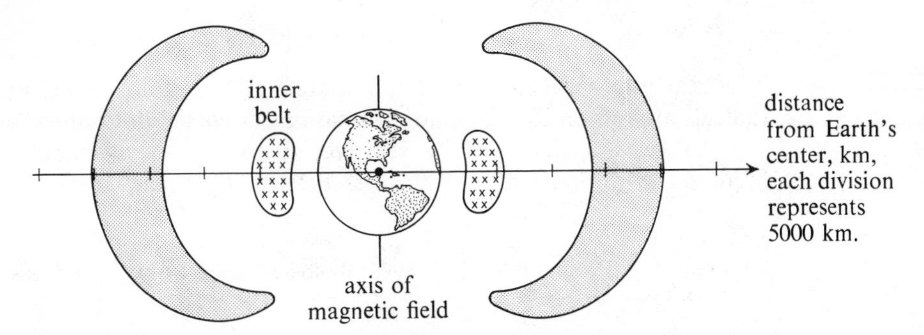

FIGURE 11.5 Cross section of the Van Allen belts encircling the Earth.

troposphere by the shearing of water droplets during precipitation. The shearing may be caused by the compression of descending masses of cold air or by strong winds over hot, dry land masses. The last phenomenon is known as the foehn, sharav (in the Near East), or Santa Ana (in southern California). These hot, dry winds cause severe discomfort. The ions produced by them consist of electrons and positively charged molecular species. The ions produced by the primary ionization process are surrounded by approximately 2.7×10^{19} molecules per cm^3 and undergo an average of approximately 1×10^9 collisions per second. Given sufficient humidity, these ions become surrounded by clusters of one to eight water molecules within approximately 2×10^{-7} sec. These clusters are called **small air ions** and may last for up to several minutes. Small air ions undergo further reactions, among which is incorporation of trace gases from the atmosphere. Eventually, they are neutralized by combination with ions of opposite charge or with uncharged condensation particles. Because of the latter phenomenon, the concentration of small air ions decreases with increased air pollution and rarely exceeds 1×10^4 ions/cm^3 in clean air.

There is some evidence that small air ions adversely affect various life processes [4]. Among these effects are inhibition of bacterial and fungal growth; reduction in viable cells in bacterial aerosols; and alteration of serotonin metabolism in mammals. The latter effect may be particularly significant because of serotonin's role as a powerful neurohormone, and may be partially responsible for the well-documented malaise afflicting persons subjected to the hot, dry, windy conditions which may prevail for one or two days during the sharav weather conditions in the Near East. The possibility should also be considered in microenvironments, such as those created by electrical discharge, where the concentration of small air ions may reach high levels.

In addition to forming ions by photoionization, electromagnetic radiation in the upper atmosphere may produce **free radicals**:

$$H_3C{-}\overset{\overset{O}{\|}}{C}{-}H + h\nu \rightarrow H_3C\cdot + H\dot{C}O \qquad (11.8.1)$$

Free radicals are involved with most significant atmospheric chemical phenomena and are of the utmost importance in the atmosphere. They are atoms, or groups of atoms, with unpaired electrons. Because of the strong pairing tendencies of electrons under most circumstances, free radicals are highly reactive. The upper atmosphere is so rarefied, however, that radicals may have half-lives of several minutes, or even longer. Radicals can take part in chain reactions in which one of the products of each reaction is a radical. Eventually, through processes such as reaction with another radical, one of the radicals in a chain is destroyed and the chain ends:

$$H_3C\cdot + H_3C\cdot \rightarrow C_2H_6 \qquad (11.8.2)$$

This reaction is called a **chain-terminating reaction**. Reactions involving free radicals are responsible for smog formation, discussed in Chapter 14.

Free radicals are quite reactive; therefore, they generally have short lifetimes. It is important to distinguish between high reactivity and instability. A

---

[4] Kreuger, A.P., and Reed, E.J. 1976. Biological impact of small air ions. *Science* 193: 1209–13.

totally isolated free radical or atom would be quite stable. Therefore, free radicals and single atoms from diatomic gases tend to persist under the rarefied conditions of very high altitudes. However, an electronically excited species has a finite, generally very short, lifetime because of energy loss through radiation.

11.9 REACTIONS OF ATMOSPHERIC OXYGEN

Some of the primary features of the exchange of oxygen among the atmosphere, lithosphere, hydrosphere, and biosphere are summarized in Figure 11.6. The oxygen cycle is critically important in atmospheric chemistry, geochemical transformations, and life processes.

Oxygen in the troposphere plays a strong role in processes on the Earth's surface. Atmospheric oxygen takes part in energy-producing reactions, such as the burning of fossil fuels:

$$CH_4 \text{ (in natural gas)} + 2\,O_2 \rightarrow CO_2 + 2\,H_2O \qquad (11.9.1)$$

Atmospheric oxygen is utilized by aerobic organisms in the degradation of organic material. Some oxidative weathering processes consume oxygen, such as

$$4\,FeO + O_2 \rightarrow 2\,Fe_2O_3 \qquad (11.9.2)$$

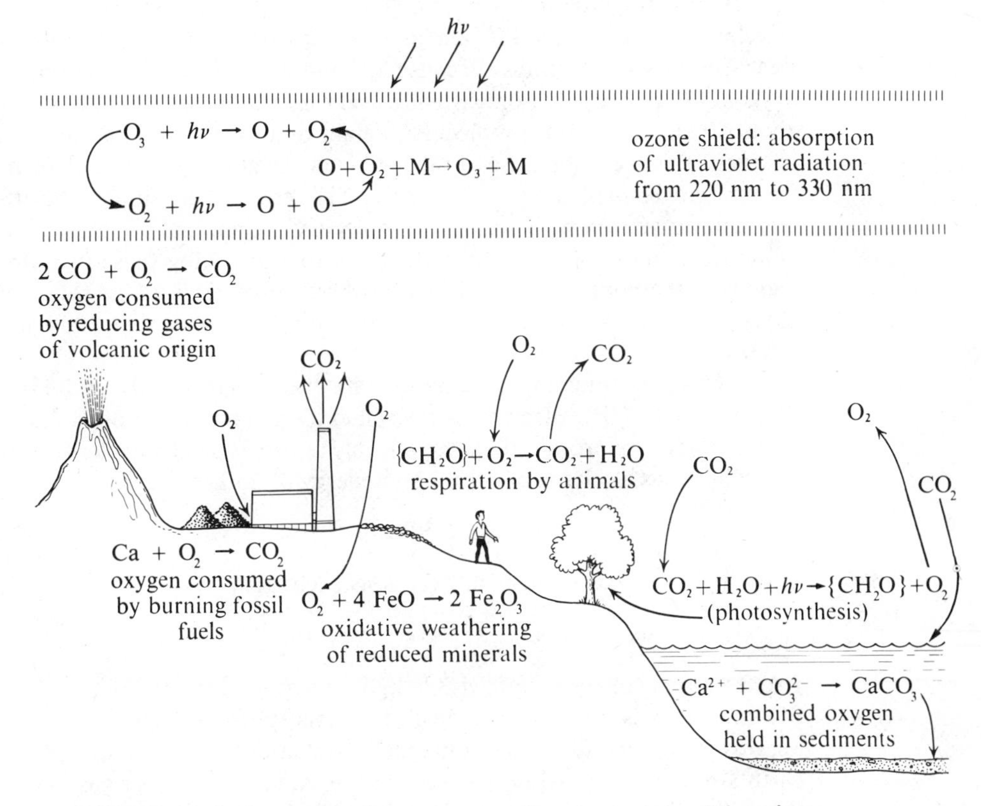

FIGURE 11.6 Oxygen exchange among the atmosphere, lithosphere, hydrosphere, and biosphere.

Oxygen is returned to the atmosphere through plant photosynthesis:

$$CO_2 + H_2O + h\nu \rightarrow \{CH_2O\} + O_2(g) \qquad (11.9.3)$$

All molecular oxygen now in the atmosphere is thought to have originated through the action of photosynthetic organisms, which shows the importance of photosynthetic organisms in the oxygen balance of the atmosphere. It can be shown that most of the carbon fixed by these photosynthetic processes is dispersed in mineral formations as humic material (Section 10.5); only a very small fraction is deposited in fossil fuel beds. Therefore, although combustion of fossil fuels consumes large amounts of O_2, there is no danger of running out of atmospheric oxygen.

Because of the extremely rarefied atmosphere and the effects of ionizing radiation, oxygen in the upper atmosphere exists in some forms that are quite different from those stable at lower levels. In addition to molecular oxygen, O_2, the upper atmosphere contains oxygen atoms, O; excited oxygen molecules, O_2^*; and ozone, O_3.

Atomic oxygen, O, is stable primarily in the thermosphere, where the atmosphere is so rarefied that the three-body collisions necessary for the chemical reaction of atomic oxygen seldom occur. Atomic oxygen is produced by a photochemical reaction:

$$O_2 + h\nu \rightarrow O + O \qquad (11.9.4)$$

The oxygen-oxygen bond is strong (120 kcal/mole). Ultraviolet radiation in the wavelength regions 135–176 nm and 240–260 nm is most effective in causing Reaction 11.9.4 to occur. Because of photochemical dissociation, molecular oxygen, O_2, is virtually nonexistent at very high altitudes. At altitudes exceeding about 80 km, the average molecular weight of air is lower than the 28.97 g/mole observed at sea level because of the high concentration of atomic oxygen resulting from photochemical dissociation of O_2. The resulting division of the atmosphere into a lower section with a uniform molecular weight and a higher region with a nonuniform molecular weight is the basis for classifying these two atmospheric regions as the *homosphere* and *heterosphere*, respectively. Less than 10% of the oxygen in the atmosphere at altitudes exceeding approximately 400 km is present as O_2.

Oxygen atoms in the ground state (not electronically excited) are often designated as $O(^3P)$. Electronically excited oxygen atoms are designated as O* or $O(^1D)$. These are produced by the photolysis of ozone at wavelengths below 308 nm, which has a relatively weak bond energy of 26 kcal/mole,

$$O_3 + h\nu(\lambda < 308\text{ nm}) \rightarrow O^* + O_2 \qquad (11.9.5)$$

or by highly energetic chemical reactions such as

$$O + O + O \rightarrow O_2 + O^* \qquad (11.9.6)$$

Excited atomic oxygen emits visible light at wavelengths of 636 nm, 630 nm, and 558 nm. This is partially responsible for the phenomenon known as **airglow**. Airglow is a very faint electromagnetic radiation continuously emitted by the earth's atmosphere. Although the visible component of airglow is extremely weak, the phenomenon is quite intense in the infrared region of the spectrum.

Oxygen ions, O^+, may be produced by ultraviolet radiation acting upon oxygen atoms (see Eq. 11.9.7).

$$O + h\nu \rightarrow O^+ + e \quad (11.9.7)$$

O^+ is the predominant positive ion in some regions of the ionosphere. It may undergo the following reactions to form other important positive ions:

$$O^+ + O_2 \rightarrow O_2^+ + O \quad (11.9.8)$$

$$O^+ + N_2 \rightarrow NO^+ + N \quad (11.9.9)$$

In intermediate regions of the ionosphere, the species O_2^+ is produced by absorption of ultraviolet radiation at wavelengths of 17–103 nm. The reaction

$$O_2 + h\nu \rightarrow O_2^+ + e \quad (11.9.10)$$

may also be energized by low-energy X-rays. Reaction 11.9.8 and the reaction

$$N_2^+ + O_2 \rightarrow N_2 + O_2^+ \quad (11.9.11)$$

also produce O_2^+ in this region.

Ozone, O_3, is an extremely significant oxygen-containing species found in the stratosphere. Ozone absorbs harmful ultraviolet radiation and serves as a radiation shield, protecting living beings on the Earth from the effects of excessive amounts of such radiation. Ozone is produced by a photochemical reaction [1],

$$O_2 + h\nu \rightarrow O + O \quad (11.9.12)$$

(where the wavelength of the exciting radiation must be less than 242.4 nm), followed by a three-body reaction,

$$O + O_2 + M \rightarrow O_3 + M \quad (11.9.13)$$

in which M is another species, such as a molecule of N_2 or O_2, which absorbs the excess energy given off by the reaction and enables the ozone molecule to stay together. The region of maximum ozone concentration is found within the range of 25–30 km high in the stratosphere. In this region, the ozone concentration may reach 10 ppm.

Ozone absorbs ultraviolet light very strongly in the region 220–330 nm. If this light were not absorbed by ozone, severe damage would result to exposed forms of life on the Earth. Absorption of electromagnetic radiation by ozone converts the radiation's energy to heat and is responsible for the temperature maximum encountered at an altitude of approximately 50 km, the boundary between the stratosphere and the mesosphere. The reason that the temperature maximum occurs at a higher altitude than that of the maximum ozone concentration arises from the fact that ozone is such an effective absorber of ultraviolet light, so that most of this radiation is absorbed in the upper stratosphere, generating heat there, and only a small fraction reaches the lower altitudes, which remain relatively cool.

For the decomposition of ozone,

$$O_3 \rightarrow \frac{3}{2} O_2 \quad (11.9.14)$$

ΔH°_{298} is -34.1 kcal $\times$ mole^{-1}, ΔS°_{298} is $+16.7$ cal $\times$ mole^{-1} $\times$ deg^{-1}, and ΔG°_{298} is -39.1 kcal $\times$ mol^{-1} [5]. From these values it may be seen that ozone is

[5] Johnson, H.S. 1975. Pollution of the stratosphere. In *Advances in physical chemistry*, pp. 315–37. New York: Academic Press, Inc.

thermodynamically unstable. Its decomposition in the stratosphere is catalyzed by a number of natural and pollutant trace constituents, including NO, NO_2, N_2O, H, HO•, HOO•, ClO, Cl, Br, and BrO. The decomposition also occurs on solid surfaces, such as metal oxides and salts produced by rocket exhausts.

Although the mechanisms and rates for the photochemical production of ozone in the stratosphere are reasonably well known, the natural pathways for ozone removal are less well understood. These have been discussed along with a newly proposed removal mechanism [6]. The earliest known reaction for ozone removal is the reaction of ozone with atomic oxygen,

$$O_3 + O \rightarrow O_2 + O_2 \qquad (11.9.15)$$

which obtains some of the required atomic oxygen from another ozone-destroying reaction:

$$O_3 + h\nu \rightarrow O_2 + O \qquad (11.9.16)$$

These reactions can account for only about 20% of the ozone removal. Another approximately 10% removal is accounted for by the reactive hydroxyl radical, HO•, produced by photochemical reactions of H_2, O_2, and H_2O in the stratosphere. A plausible reaction sequence is

$$O_3 + HO\cdot \rightarrow O_2 + HOO\cdot \qquad (11.9.17)$$

$$HOO\cdot + O \rightarrow HO\cdot + O_2 \qquad (11.9.18)$$

It is now believed that the most important agent for stratospheric ozone removal is nitric oxide, which undergoes the following chain reactions:

$$O_3 + NO \rightarrow NO_2 + O_2 \qquad (11.9.19)$$

$$NO_2 + O \rightarrow NO + O_2 \qquad (11.9.20)$$

This removal mechanism is discussed in more detail in Section 12.13 in connection with stratospheric nitrogen oxide pollution by high-flying supersonic airplanes.

Below altitudes of approximately 30 km in the stratosphere, NO is produced largely when with photochemically produced excited oxygen atoms react with nitrous oxide, N_2O:

$$N_2O + O \rightarrow 2\,NO \qquad (11.9.21)$$

Recall that N_2O is a natural component of the atmosphere and is a major product of the denitrification process by which fixed nitrogen is returned to the atmosphere in gaseous form. This is shown in the nitrogen cycle, Figure 5.5.

At altitudes exceeding 30 km, ions and ionizing radiation may play a significant role in the production of ozone-destroying NO [6]. Secondary electrons with an energy range of 10–100 electron volts are produced in the upper regions of the atmosphere by an intense flux of charged particles and cosmic rays. These electrons can reach only the upper regions of the stratosphere before being destroyed, but they produce more highly penetrating bremsstrahlung X-rays which can reach down to 30-km altitude. Both the energetic electrons and X-rays can bring about the dissociation of stratospheric N_2 (see Eq. 11.9.22).

---

[6] Thorne, R.M. 1977. Energetic radiation belt electron precipitation: a natural depletion mechanism for stratospheric ozone. *Science* 195: 287–9.

$$N_2 + h\nu \rightarrow N + N \qquad (11.9.22)$$

to produce nitrogen atoms which react with O_2 to yield NO:

$$O_2 + N \rightarrow NO + O \qquad (11.9.23)$$

The production of secondary electrons needed to initiate this reaction sequence depends upon solar radiation, such as energetic solar protons. On the basis of this mechanism, the ozone layer should decrease during periods of maximum solar activity.

Ozone is toxic, and a mild overdose causes labored breathing, a feeling of chest pressure, cough, and irritated eyes (see Section 14.7). The U.S. Occupational Safety and Health Administration has imposed an 8-hour exposure limit of 0.12 ppm of ozone. In early 1978, a number of cases of alleged ozone sickness were reported among airline passengers and crew members [7]. These were first reported on "jumbo jets" designed with a 13,700-meter ceiling, compared to the 11,600-meter ceiling of most subsonic aircraft. Subsequently, these planes have been equipped with air heating devices using engine heat to destroy ozone in air used to pressurize the cabins. In the summer and fall, the bottom of the ozone layer is normally above 9000 meters, though it may reach as low as 7300 meters in winter and spring, well within the ceiling limits of jet aircraft.

11.10 REACTIONS OF ATMOSPHERIC NITROGEN

The atmosphere contains 78% nitrogen by volume and constitutes an inexhaustible reservoir of that essential element. The nitrogen cycle was discussed in Section 5.14 (see also Figure 5.5), and nitrogen fixation by microorganisms and by humans was discussed in Sections 5.15, 10.8, and 10.12. A small amount of nitrogen is thought to be fixed in the atmosphere by lightning, and some is also fixed by combustion processes, as in the internal combustion engine.

Before the use of synthetic fertilizers reached its current high levels, chemists were concerned that denitrification processes in the soil would lead to nitrogen depletion on the Earth. Now, with millions of tons of synthetically fixed nitrogen being added to the soil each year, we have quite the opposite concern—possible excess accumulation of nitrogen in soil, fresh water, and the oceans.

Unlike oxygen, which is almost completely dissociated to the monatomic form in higher regions of the thermosphere, molecular nitrogen is not readily dissociated by ultraviolet radiation. However, at altitudes exceeding approximately 100 km, atomic nitrogen is produced by photochemical reactions:

$$N_2 + h\nu \rightarrow N + N \qquad (11.10.1)$$

Other reactions which may produce monatomic nitrogen are:

$$N_2^+ + O \rightarrow NO^+ + N \qquad (11.10.2)$$

$$NO^+ + e \rightarrow N + O \qquad (11.10.3)$$

$$O^+ + N_2 \rightarrow NO^+ + N \qquad (11.10.4)$$

In one region of the ionosphere, the so-called E region, NO^+ is one of the predominant ions. A plausible sequence of reactions by which NO^+ is formed is shown at the top of the following page (Eq. 11.10.5).

[7] 1978. Ozone sickness. *Newsweek* 17 April 1978, p. 101.

$$N_2 + h\nu \rightarrow N_2^+ + e \quad (11.10.5)$$

$$N_2^+ + O \rightarrow NO^+ + N \quad (11.10.6)$$

In the lowest region of the ionosphere, the D region, which extends from approximately 50 km in altitude to approximately 85 km, NO^+ is produced directly by ionizing radiation:

$$NO + h\nu \rightarrow NO^+ + e \quad (11.10.7)$$

In the lower part of the D region, the ionic species N_2^+ is formed through the action of galactic cosmic rays:

$$N_2 + h\nu \rightarrow N_2^+ + e \quad (11.10.8)$$

Pollutant oxides of nitrogen, particularly NO_2, are key species involved in air pollution and the formation of photochemical smog. For example, NO_2 is readily dissociated photochemically to NO and reactive atomic oxygen:

$$NO_2 + h\nu \rightarrow NO + O \quad (11.10.9)$$

This reaction is the most important primary photochemical process involved in smog formation. The role played by nitrogen oxides in air pollution will be discussed in Chapters 12 and 14.

11.11 HYDROXYL AND HYDROPEROXYL RADICALS IN THE ATMOSPHERE

During recent years, the importance of the hydroxyl radical, HO•, in atmospheric chemical phenomena has received increasing recognition [8]. This radical may be formed by several processes. At higher altitudes, a common reaction for the formation of hydroxyl radical is photolysis of water, which is also a major contributor to atomic hydrogen in the atmosphere:

$$H_2O + h\nu \rightarrow HO\cdot + H \quad (11.11.1)$$

In the presence of organic matter, hydroxyl radical is produced in abundant quantities as an intermediate in the formation of photochemical smog (see Chapter 14). For purposes of laboratory experimentation, it is convenient to produce hydroxyl radical by the photolysis of nitrous acid vapor, as shown by the following reaction [9, 10]:

$$HONO + h\nu \rightarrow HO\cdot + NO \quad (11.11.2)$$

In the relatively unpolluted troposphere, hydroxyl radical is produced by a sequence of reactions which begins with the photolysis of ozone [11] (Eq. 11.11.3).

---

[8] Wang, C.C.; Davis, L.I.; Wu, C.H.; Japar, S.; Niki, H.; and Weinstock, B. 1975. Hydroxyl radical concentrations measured in ambient air. *Science* 189: 797–800.

[9] Cox, R.A.; Patrick, K.F.; and Chang, S.A. 1981. Mechanism of atmospheric photooxidation of organic compounds. Reactions of alkoxy radicals in oxidation of *n*-butane and simple ketones. *Environmental Science and Technology* 15: 587–592.

[10] Kerr, J.A., and Sheppard, D.W. 1981. Kinetics of the reactions of hydroxyl radicals with aldehydes studied under atmospheric conditions. *Environmental Science and Technology* 15: 960–963.

[11] Davis, D.D., and Chameides, W.L. 1982. Chemistry in the troposphere. *Chemical and Engineering News* 4 October 1982, pp. 39–52.

$$O_3 + h\nu(\lambda < 315\text{ nm}) \rightarrow O(^1D) + O_2 \qquad (11.11.3)$$

to produce an excited oxygen atom, $O(^1D)$. The transition of this atom to the ground state, $O(^3P)$, is *forbidden* (a spectroscopic term meaning that it occurs very slowly), so that the excited oxygen atom has an average lifetime of 110 seconds before decaying to the ground state by photon emission (*luminescence*; see Section 11.7). Most often, the excited oxygen atom loses energy by collision with a nitrogen or oxygen atom:

$$O(^1D) + (N_2 \text{ or } O_2) \rightarrow O(^3P) + (N_2 \text{ or } O_2) \qquad (11.11.4)$$

This simply leads to regeneration of ozone by the reaction

$$O(^3P) + O_2 + M \rightarrow O_3 + M \quad (M = N_2 \text{ or } O_2) \qquad (11.11.5)$$

In some cases, however, the excited oxygen atom collides with a water molecule before going to the ground state, generating two hydroxyl radicals by the reaction

$$O(^1D) + H_2O \rightarrow 2\,HO\cdot \qquad (11.11.6)$$

This reaction, involving oxygen atoms produced by the photolysis of ozone, is the main source of hydroxyl radical in the unpolluted troposphere.

Hydroxyl radical is involved in chemical transformations of a number of trace species in the atmosphere. These are summarized in Figure 11.7 (page 310), and some of these processes are discussed in later chapters. Among the important tropospheric trace species that react with hydroxyl radical are carbon monoxide, sulfur dioxide, hydrogen sulfide, methane, and nitric oxide.

Hydroxyl radical is most frequently removed from the troposphere by reaction with methane or carbon monoxide:

$$CH_4 + HO\cdot \rightarrow H_3C\cdot + H_2O \qquad (11.11.7)$$

$$CO + HO\cdot \rightarrow CO_2 + H \qquad (11.11.8)$$

The highly reactive methyl radical, $H_3C\cdot$, reacts with O_2,

$$H_3C\cdot + O_2 \rightarrow H_3COO\cdot + H_2O \qquad (11.11.9)$$

to form **methylperoxyl radical**, $H_3COO\cdot$. Further reactions of this species are discussed in Section 14.3. The hydrogen atom produced in Reaction 11.11.8 reacts with O_2

$$H + O_2 \rightarrow HOO\cdot \qquad (11.11.10)$$

to produce **hydroperoxyl radical**, $HOO\cdot$. Chain termination may occur by the reaction of this radical:

$$HOO\cdot + HO\cdot \rightarrow H_2O + O_2 \qquad (11.11.11)$$

$$HOO\cdot + HOO\cdot \rightarrow H_2O_2 + O_2 \qquad (11.11.12)$$

The hydrogen peroxide product of the latter reaction is removed from the atmosphere with precipitation. The hydroperoxyl radical can undergo either of the following two reactions, which regenerate $HO\cdot$:

$$HOO\cdot + NO \rightarrow NO_2 + HO\cdot \qquad (11.11.13)$$

$$HOO\cdot + O_3 \rightarrow 2\,O_2 + HO\cdot \qquad (11.11.14)$$

The global concentration of hydroxyl radical, averaged diurnally and seasonally, is estimated to range from 2×10^5 to 1×10^6 radicals per cm^3 in the troposphere. Because of the higher humidity and higher incident sunlight, which

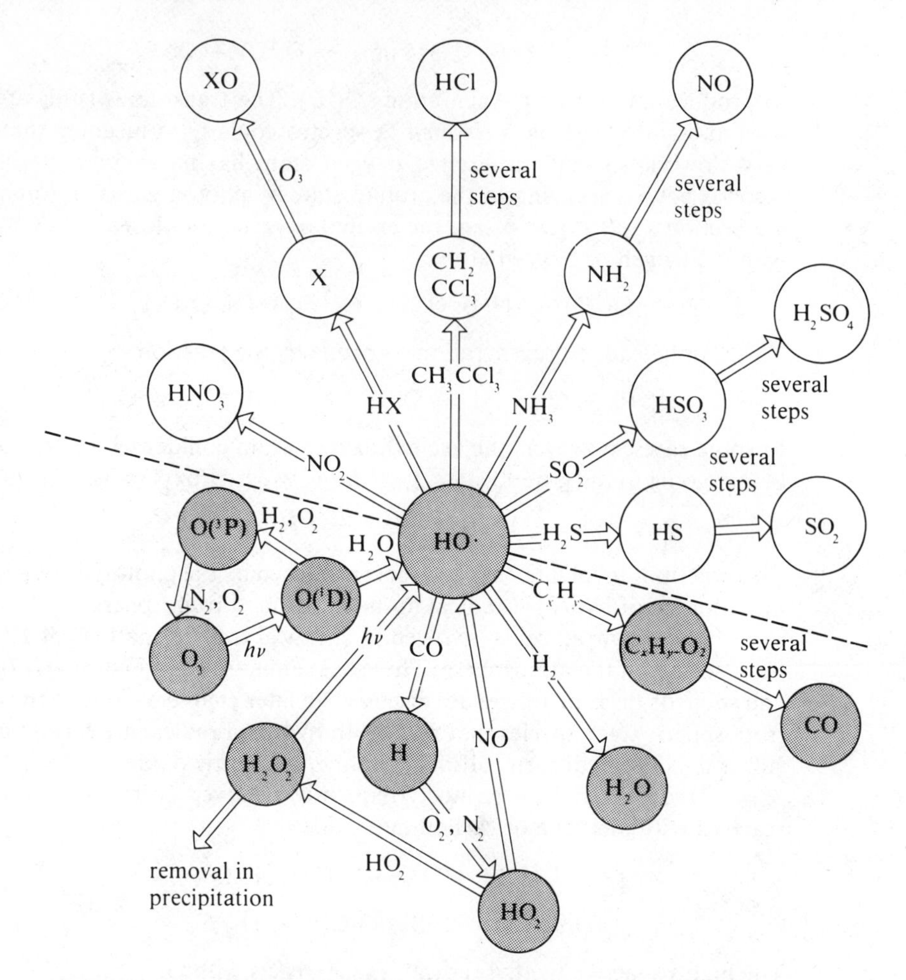

FIGURE 11.7 Control of trace gas concentrations by HO· radical in the troposphere. Processes below the dashed line are those largely involved in controlling the concentration of HO· in the troposphere; those above the line control the concentrations of the associated reactants and products. Reservoirs of atmospheric species are shown in circles, reactions denoting conversion of one species to another are shown by arrows, and the reactants or photons needed to bring about a particular conversion are shown along the arrows. Hydrogen halides are denoted by HX and hydrocarbons by C_xH_y. (Source: D.D. Davis and W.L. Chameides, Chemistry in the troposphere, *Chemical and Engineering News,* 4 October 1982, pp. 39–52. Reprinted by permission of the American Chemical Society.)

result in elevated $O(^1D)$ levels, tropical regions exhibit elevated HO· levels. The southern hemisphere probably has about a 20% higher level of HO· than does the northern hemisphere because of greater production of anthropogenic CO, which consumes HO·, in the northern hemisphere.

The hydroperoxyl radical, HOO·, is an intermediate in some important chemical reactions. There are several ways in which this radical is produced in the atmosphere [12]. One major process is shown by Reactions 11.11.8 and 11.11.10. In polluted atmospheres hydroperoxyl radical is produced by a process starting with the absorption of 260–360 nm ultraviolet light by formaldehyde to give hydrogen atoms and formyl radical, $H\dot{C}O$:

$$HCHO + h\nu \rightarrow H + H\dot{C}O \qquad (11.11.15)$$

The hydrogen atom produces hydroperoxyl radical by Reaction 11.11.10, and the formyl radical produces HOO• by reaction with O_2:

$$H\dot{C}O + O_2 \rightarrow HOO\cdot + CO \qquad (11.11.16)$$

The hydroperoxyl radical reacts more slowly with other species than does the hydroxyl radical. The kinetics and mechanisms of hydroperoxyl radical reactions are incompletely understood. They are difficult to study because it is hard to obtain hydroperoxyl radicals free of the more reactive hydroxyl radicals.

11.12 ATMOSPHERIC CARBON DIOXIDE

Carbon dioxide, composing only 0.034% by volume of the atmosphere, is the nonpollutant species of most concern in the atmosphere, Carbon dioxide, along with water vapor, is primarily responsible for the absorption of infrared energy re-emitted by the Earth and reradiation of this energy back to the Earth's surface. Some scientists fear that changes in the atmospheric carbon dioxide level will cause marked changes in the Earth's climate through the greenhouse effect discussed in Section 11.4 [13].

Valid measurements of overall atmospheric CO_2 levels can only be taken in areas remote from industrial activity. One such point where measurements have been made is near the top of Mauna Loa Mountain in Hawaii. During the 1958–1968 decade, these measurements showed an increase from 316 ppm to 322 ppm, an average annual increase of 0.19%. From 1968 to 1974, the CO_2 level increased from 322 ppm to 331 ppm, a 0.46% annual increase. By 1980 the CO_2 level had reached approximately 338 ppm. Increases of a similar magnitude have been registered in Antarctica. Consideration of both Mauna Loa and Antarctica data, as shown in Figure 11.8 (p. 312), indicates that over the past 22 years has been an increase of atmospheric CO_2 of essentially 1 ppm/year.

Conventional wisdom has maintained that the increase in atmospheric CO_2 has been due almost entirely to increased fossil fuel consumption. A thorough treatment of the carbon dioxide question [14] has raised doubts about this premise, based largely on a seasonal cycle in carbon dioxide levels as shown in the inset in Figure 11.8. These oscillations show a maximum value in the northern hemisphere in April, falling to a minimum in late September or October.

---

[12] Pitts, J.N., Jr., and Finlayson, B.J. 1975. Mechanisms of photochemical air pollution. *Angewandte Chemie* (International Edition in English) 14: 1–15.

[13] Speth, G. 1981. *Global energy futures and the carbon dioxide problem.* Washington, D.C.: Council on Environmental Quality, U.S. Government Printing Office.

[14] Woodwell, G.M. 1978. The carbon dioxide question. *Scientific American* 238: 34–43.

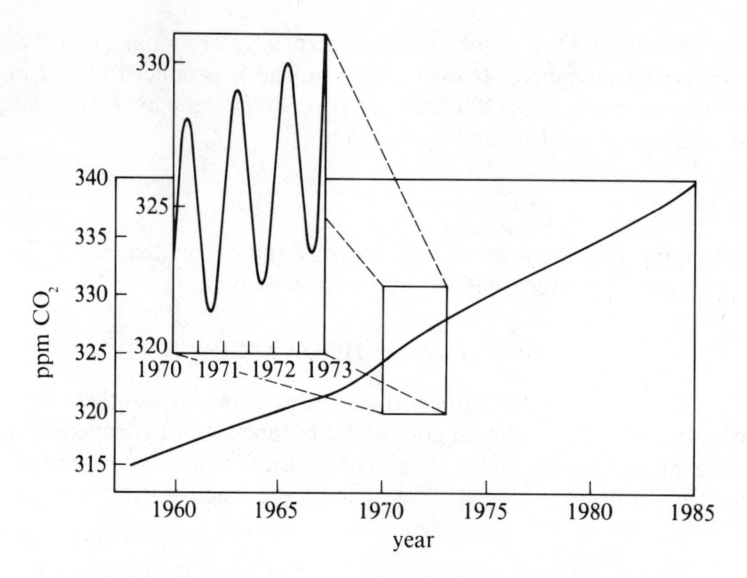

FIGURE 11.8 Increases in atmospheric CO_2 levels in recent years. (Average of data from Mauna Loa Mountain, Hawaii, and from Antarctica. Insert shows seasonal variations of CO_2 level in the northern hemisphere.)

These oscillations are due to the "photosynthetic pulse," influenced most strongly by forests in middle latitudes. Forests have a much greater influence than other vegetation because forests carry out more photosynthesis. Furthermore, forests store enough fixed but readily oxidizable carbon in the form of wood and humus to have a marked influence on atmospheric CO_2 content. Thus, during the summer months, forests carry out enough photosynthesis to reduce the atmospheric carbon dioxide content markedly. During the winter, metabolism of biota, such as bacterial decay of humus, releases a noticeable amount of CO_2. This being the case, destruction of forests results in a net increase in atmospheric carbon dioxide content. Therefore, the current worldwide trend toward converting forest lands to other uses may cause a greater overall increase in atmospheric CO_2 levels than will increased burning of fossil fuels.

In July, 1982, the National Research Council, a division of the National Academy of Sciences, reaffirmed previous warnings about the possible effects of increased CO_2 in the atmosphere. The Council estimated that a doubling of CO_2 in the atmosphere, likely to occur by the middle of the next century, would raise the mean global temperature by 1.5 to 4.5 degrees Celsius. The report tended to discount speculation by some scientists that increased cloud cover resulting from higher temperatures would have a buffering effect upon temperatures.

If the most pessimistic predictions of the effect of increased CO_2 levels on global temperature and climate are accepted, a large increase in atmospheric CO_2 has more potential to cause massive irreversible damage to our environment than any other disaster short of nuclear war. However, some authorities now hold that these predictions have been overestimated by approximately one order of magnitude. One calculation estimates that doubling the atmosphere's CO_2 level would result in an increase in thermal radiation received at the Earth's surface of

2.28 watts per square meter [15]. This compares to a total of 1929 watts/m^2 of solar plus infrared radiation shown in Figure 11.4. The value of 2.28 watts/m^2 must be multiplied by the atmosphere's **surface air temperature response function** of 0.113 °C per watt per square meter to give the change in the mean global surface air temperature. The resulting value is 0.26 °C, about one-tenth that of most previous estimates. It is, however, essentially identical to the value obtained by another group of scientists [16] using an entirely different experimental approach.

The question of the effect of increased global CO_2 levels upon the Earth's temperature is obviously an extremely important one, deserving the most careful research and critical analysis. If the effect is underestimated so that there is little restraint on the increased use of fossil fuels, the result could be a global environmental disaster. If it is grossly overestimated and the use of fossil fuels is restricted too much, modern economic systems and the material well-being of hundreds of millions of people could suffer.

It is of interest to compare quantities of carbon held in various pools of exchangeable carbon. Here we define *exchangeable carbon* as that which can be interchanged with the atmosphere as CO_2 and with other pools of carbon through natural biological and chemical processes. These pools are summarized in Table 11.3. This table shows that the quantities of carbon in atmospheric CO_2, organisms, and soil organic matter are of roughly equal magnitude. Dissolved inorganic carbon in abyssal ocean regions constitutes a huge pool of carbon—it is the ultimate sink for atmospheric CO_2. However, uptake of CO_2 into this pool is too slow for it to appreciably reduce excess atmospheric CO_2 levels within a reasonable time span. It may be seen that the pools of carbon found in organisms (primarily forest trees) and soil organic matter (primarily peat and humus) are large enough compared to atmospheric CO_2 that destruction of forests could well

TABLE 11.3 Major Pools of Exchangeable Carbon

Pool	*Mass of C, metric tons* $\times 10^{-10}$
Atmospheric CO_2	70
Organisms, living or not yet incorporated into soil humus	80
Soil organic matter, primarily humus and peat	150
Dissolved organic carbon in the ocean ("ocean humus")	300
Dissolved inorganic carbon (CO_2–HCO_3^-–CO_3^{2-}) in the 100-meter surface layer of the oceans, from which exchange with the atmosphere may occur readily	60
Dissolved inorganic carbon in the cold abyssal ocean regions below 100 m deep	4000

[15] Idso, S.B. 1980. The climatological significance of a doubling of Earth's atmospheric carbon dioxide concentration. *Science* 207: 1462–3.

[16] Newell, R.E., and Dopplick, T.G. 1979. *Journal of Applied Meteorology* 18: 822.

release enough CO_2 to the atmosphere to affect atmospheric levels of carbon dioxide appreciably.

In terms of strictly chemical reactions in the atmosphere, carbon dioxide is a relatively insignificant species. It is present in low concentrations and is not thought to be highly active in photochemical reactions. However, calculations based on known photochemical reactions, carbon dioxide levels, and ultraviolet radiation intensity indicate that photodissociation of CO_2 by solar ultraviolet light should occur in the upper atmosphere:

$$CO_2 + h\nu \rightarrow CO + O \tag{11.12.1}$$

This reaction could be a major source of CO at higher altitudes. Although CO_2 absorbs infrared radiation strongly, this radiation is not energetic enough to cause chemical reactions to occur.

11.13 WATER IN THE ATMOSPHERE

The water vapor content of the atmosphere varies over a wide range, particularly in the lower atmosphere. The normal range is 1-3% on a volume basis, although air may contain as little as 0.1% or as much as 5%. A global average of 1% water is commonly assumed. The percentage of water in the atmosphere decreases rapidly with increasing altitude. Water circulates through the atmosphere in the hydrologic cycle, as shown in Figure 2.1.

Water absorbs infrared radiation even more strongly than does carbon dioxide, thus greatly influencing the Earth's heat balance. Clouds formed from water vapor reflect light from the sun and have a temperature-lowering effect. On the other hand, water vapor in the atmosphere acts as a kind of "blanket" at night, retaining heat from the Earth's surface by absorption of infrared radiation.

Gaseous water in the upper atmosphere is involved in the formation of hydroxyl and hydroperoxyl radicals, as discussed in Section 11.11. Condensed water vapor in the form of very small droplets is of considerable significance in atmospheric chemistry. The harmful effects of some air pollutants—for instance, the corrosion of metals by acid-forming gases—requires the presence of water which may come from the atmosphere. The presence of water vapor has an important influence upon pollution-induced fog formation under some circumstances. Water vapor interacting with pollutant particulate matter in the atmosphere may reduce visibility to undesirable levels through the formation of aerosol particles (see Section 11.14).

When ice particles in the atmosphere change to liquid droplets, or when these droplets evaporate, heat is absorbed from the surrounding air. Reversal of these processes results in heat release to the air (as latent heat). This may occur many miles from the place where the heat was absorbed and is a major mode of energy transport to the atmosphere. It is the predominant type of energy transition involved in thunderstorms, hurricanes, and tornadoes.

On a global basis, rivers drain only about one-third of the precipitation that falls on the Earth's continents [17]. This means that two-thirds of the

---

[17] Shukla, J., and Mintz, Y. 1982. Influence of land-surface evapotranspiration on the Earth's climate. *Science* 215: 1498-1501.

precipitation is lost as *evapotranspiration* (combined evaporation and transpiration; see Figure 2.1). During the summer, evapotranspiration may exceed precipitation because of the large quantities of water stored in the root zone of the soil. In some cases, evapotranspiration furnishes atmosperic water vapor necessary for cloud formation and precipitation. It is probable, therefore, that large-scale deforestation (or afforestation), soil "destruction" (e.g., plowing up grasslands in semi-arid areas), and irrigation could have an effect on regional climate and rainfall.

As noted in Section 11.3, the cold tropopause serves as a barrier to the movement of water into the stratosphere. The main source of stratospheric water is the photochemical oxidation of methane:

$$CH_4 + O_3 \xrightarrow[\text{reaction steps}]{\text{several photochemical}} CO_2 + 2\,H_2O \qquad (11.13.1)$$

The water thus produced serves as a source of stratospheric hydroxyl radical through Reaction 11.11.1.

$$H_2O + h\nu \rightarrow HO\cdot + H \qquad (11.11.1)$$

11.14 PARTICLES IN THE ATMOSPHERE

Particles are common significant components of the atmosphere, particularly the troposphere. Colloidal-sized particles in the atmosphere are called **aerosols.** Most aerosols from natural sources have a diameter of less than 0.2 micrometers. These particles are called **Aitken particles** and are noted for their ability to induce formation of water droplets in supersaturated atmospheres. Aitken particles originate in nature from sea sprays, smokes, dusts, and the evaporation of organic materials from vegetation. Other typical particles of natural origin in the atmosphere are bacteria, fog, pollen grains, and volcanic ash.

Among the atmospheric phenomena that involve aerosol particles are electrification phenomena in the atmosphere, cloud formation, and fog formation. Particles help determine the heat balance of the Earth's atmosphere by reflecting light. Probably the most important function of particles in the atmosphere is their action as nuclei for the formation of ice crystals and water droplets. Current efforts at rain-making are centered around the addition of condensing particles to atmospheres supersatured with water vapor. Dry ice was used in early attempts; now silver iodide, which forms huge numbers of very small particles, is used.

Particles are involved in many chemical reactions in the atmosphere. Neutralization reactions, which occur most readily in solution, may take place in water droplets suspended in the atmosphere. Small particles of metal oxides and carbon have a catalytic effect on oxidation reactions. Particles may also participate in oxidation reactions induced by light. The effects of particles upon atmospheric environmental chemistry are discussed in greater detail in Chapter 13.

11.15 ARE HUMAN ACTIVITIES CHANGING THE EARTH'S CLIMATE?

There is an old saying regarding the weather: "Everybody talks about the weather, but nobody does anything about it." An unfortunate fact of our time is that we may indeed be doing something about it inadvertently, without knowing what we are doing [18]. Is human activity changing the Earth's climate? A majority of atmospheric scientists would probably answer that question with, "We don't know." The problem is that what we don't know may hurt us.

Certainly the Earth underwent many drastic changes in climate long before human activities could have had any effect. The last ice age is an uncomfortably recent event. The present, relatively warm interval is probably an interglacial period and may terminate within a few centuries [19]. It is not likely that the Earth's present population and current level of civilization could be maintained under the conditions prevalent during an ice age. Indeed, during recorded history, abnormally cold periods lasting several decades caused great suffering.

What have humans done to the Earth's atmosphere that may alter the climate? Some of the possible results of the human activities [12] that may affect the climate are:

1. the shifting of surface water and ground water in massive amounts;
2. the destruction of vegetation and changes in the nature of vegetation;
3. the release of heat from energy-producing sources;
4. the emission of particles and trace gases into the atmosphere;
5. the release of carbon dioxide into the atmosphere from the burning of fossil fuels;
6. the effects of our transportation systems upon land surfaces;
7. the effects of our transportation systems and their emissions upon the upper atmosphere.

These effects are summarized in Figure 11.9.

Of these effects, the most publicity has been given to the greenhouse effect (see Section 11.12) and possible damage to the stratospheric ozone layer by supersonic aircraft emissions or Freon (see Chapter 12). If the greenhouse effect raises the Earth's average temperature sufficiently to melt the polar icecaps, the level of the world's oceans would be raised by perhaps 20–25 feet, and many of the more populated regions of the Earth's surface would be flooded. According to Dr. Hans-Walter Georgii of the University of Frankfurt, West Germany, a rise in the atmospheric carbon dioxide level from approximately 292 ppm in the mid-1800s to an estimated 375 ppm in the year 2000 should be sufficient to increase the average temperature of the Earth by approximately 0.5 °C. However, with an increase in temperature, greater evaporation of water would cause an increase in cloud cover. An increase of only 1% in cloud density should be sufficient to counteract the anticipated 0.5 °C increase in temperature from added carbon dioxide. Perhaps the formation of cloud cover provides an important temperature buffering function.

Some of the quantities of materials added to the atmosphere by human activities are staggering in their magnitude. Dr. Christian Junge of the Max-Planck Institute fur Chemie, Mainz, West Germany, has estimated some of these

---

[18] 1971. *Inadvertent climate modification: report of the study of man's impact on climate.* Cambridge, Mass.: The MIT Press.

[19] Kukla, G.J., and Matthews, R.K. 1972. When will the present interglacial period end? *Science* 178: 190–1.

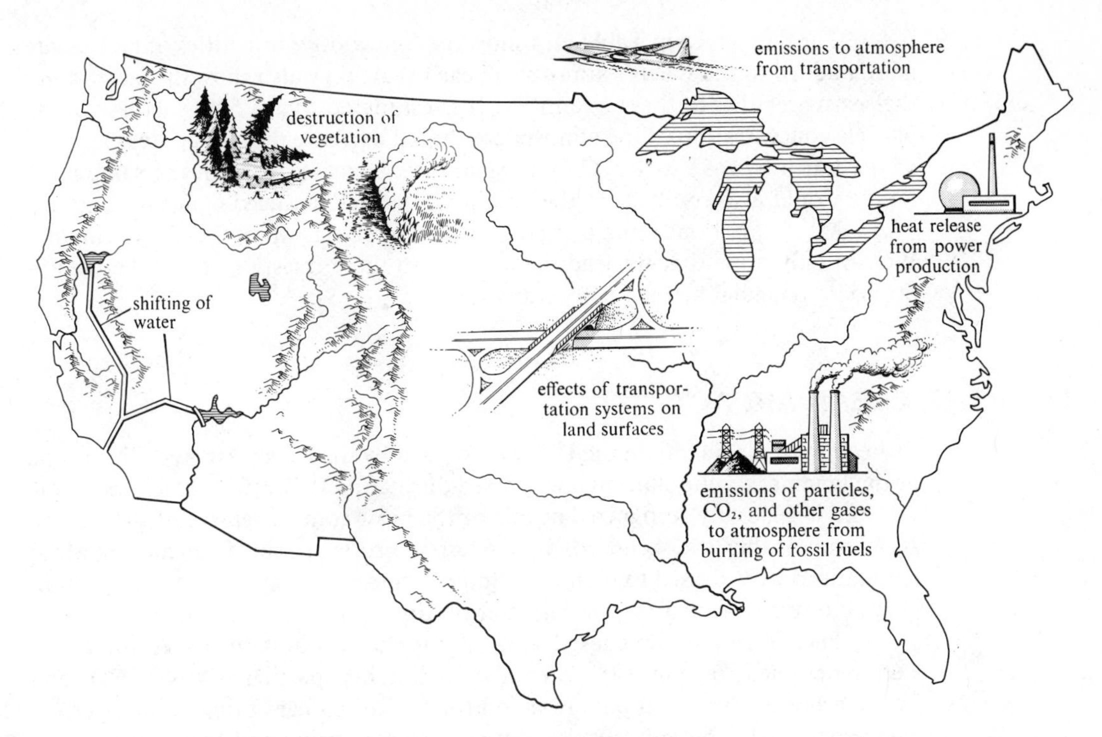

FIGURE 11.9 Human activities that may be changing the global atmosphere.

amounts. Something between approximately 773 million metric tons and 2.2 billion metric tons of particulate matter enter the atmosphere from natural sources each year. The quantities of particles produced by humans that enter the atmosphere each year are between 185 million metric tons and 415 million metric tons. Some of the particulate species found in the atmosphere and their estimated annual production by human and natural sources are given in Table 11.4.

TABLE 11.4 Worldwide Addition of Particulate Matter to the Atmosphere from Human and Natural Sources[1]

	Annual production (metric tons × 10^{-6})	
Material	*Natural sources*	*Human sources*
Total particles	773–2200	185–415
Dust and smoke	—	10–90
Soil, volcanic ash, sea salt, forest fires	438–1100	—
Sulfate particles	130–200	130–200
Nitrates	30–35	140–700
Hydrocarbons	15–20	75–200

[1]From data of C. Junge, *Chemical and Engineering News,* 16 August 1971, p. 40.

The data given in Table 11.4 show the tremendous quantities of particulate matter added to the Earth's atmosphere each year; they also show the uncertainty that exists relative to these quantities. It is estimated that a 50% increase in the particle concentration of the atmosphere should decrease the average temperature of the Earth by 0.5° to 1.0°C. A temperature decrease of about the same magnitude would also result from the additional particle-induced cloud formation.

Many questions remain regarding human effects upon the global climate. We can only hope that we find the answers to these questions and take the appropriate remedial actions before it is too late.

11.16 THE CLEAN AIR ACT

Protection of air quality in the U.S. is provided by the **Clean Air Act**. This is one of the landmark antipollution laws passed during the 1970s (for others, see Table 7.1) and is generally recognized as one of the most comprehensive of these laws. As a result of this law, standards were set up to protect people from air pollution. Regulations were passed to require polluters to meet the standards. An important provision was designed to prevent deterioration of air that was already clean.

The Clean Air Act caused some major changes in industry. Automobiles were redesigned to reduce the output of pollutants, particularly NO, CO, and hydrocarbons. The initial pollution-control devices on cars often limited performance and fuel economy. Smelters, power plants, mills, and factories were required to invest in air pollution control equipment. Costs of these measures and automobile antipollution devices were passed on to consumers, resulting in some opposition to the Act.

Despite its costs, the Clean Air Act has resulted in some marked improvement in air quality. In some cities the air is visibly cleaner with fewer incidents of air pollution alerts each year. The most dramatic changes have come about through increased visibility in industrialized areas. There is even some fragmentary evidence that the public health has improved and mortality rates are lower due to cleaner air.

In 1981 the Clean Air Act came up for revision [20]. In that same year, the National Commission on Air Quality delivered to Congress a 660-page document containing suggestions for revisions of the Clean Air Act. To a large extent, the Commission recommended retaining provisions of the current act. Some changes were recommended, such as setting the automotive carbon monoxide emissions standard at 7.0 grams per mile, rather than the statutory 3.4 grams per mile.

As of 1983, Congress is still debating a new Clean Air Act while the old one remains in effect. Economic and political interests are strongly mixed with scientific concerns in the effort to come up with a satisfactory bill.

SUPPLEMENTARY REFERENCES

Andersen, N.R., and Malahoff, A. 1977. *The fate of fossil fuel CO_2 in the oceans.* New York: Plenum Publishing Corp.

---

[20] 1981. Molding consensus on air law. *Chemical and Engineering News* 9 March 1981, pp. 15–21.

Bach, W.; Pankrath, J.; and Kellogg, W., eds. 1979. *Man's impact on climate.* New York: Elsevier North-Holland, Inc.

Barltrop, J.A., and Coyle, J.D. 1979. *Principles of photochemistry.* New York: John Wiley and Sons, Inc.

Barry, R.G., and Chorley, R.J. 1978. *Atmosphere, weather, and climate.* 3rd ed. New York: Halsted (John Wiley and Sons, Inc.).

Bernard, H.W., Jr. 1980. *The greenhouse effect.* Cambridge, Mass.: Ballinger Publishing Company.

Biswas, M.R., and Biswas, A.K., eds. 1979. *Food, climate and man.* New York: Wiley-Interscience.

Boucher, K. 1976. *Global climate.* New York: Halsted (John Wiley and Sons, Inc.).

Breuer, G. 1980. *Air in danger.* New York: Cambridge University Press.

Broecker, W.S.; Li, Y.-H.; and Peng, T.-H. 1971. *Carbon dioxide—man's unseen artifact.* In *Impingement of man on the oceans,* D.W. Hood, ed., p. 287. New York: Wiley-Interscience.

Budyko, M.I. 1982. *The Earth's climate.* New York: Academic Press, Inc.

Butler, J.D. 1979. *Air pollution chemistry.* New York: Academic Press, Inc.

Chameides, W.L., and Davis, D.D. 1982. The free radical chemistry of cloud droplets and its impact upon the composition of rain. *Journal of Geophysical Research* 87: 4863.

Clark, W.C., ed. 1982. *Carbon dioxide review 1982.* New York: Clarendon.

1981. *Clean air act: summary of GAO reports.* CED-81-84. Washington, D.C.: U.S. Government Printing Office.

1980. GAMETAG papers. Series of 12 papers. *Journal of Geophysical Research* 85: 7285.

Gaskell, T.F., and Morris, M. 1980. *World climate.* New York: Thames and Hudson, Inc.

Georgii, H.W., and Pankrath, J., eds. 1982. *Deposition of atmospheric pollutants.* Hingham, Mass.: D. Reidel Publishing Co., Inc.

Green, F. 1977. *A change in the weather.* New York: W.W. Norton & Co., Inc.

Gribbin, J. 1979. *What's wrong with our weather?* New York: Charles Scribner's Sons.

Hanna, S.R. 1982. *Handbook on atmospheric diffusion* (DE82992945 DOE/TIC-11223). Springfield, Va.: U.S. Department of Commerce, National Technical Information Service.

Heicklen, J. 1976. *Atmospheric chemistry.* New York: Academic Press, Inc.

Hobbs, J. 1980. *Applied climatology.* Boulder, Co.: Westview Press, Inc.

Holland, H.D. 1978. *The chemistry of the atmosphere and oceans.* New York: Wiley-Interscience.

Kellogg, W.W., and Schware, R. 1981. *Climate, change, and society.* Boulder, Co.: Westview Press, Inc.

Lamb, H.H. 1978. *Climate.* New York: Barnes & Noble Books.

Levine, J.S., and Schryer, D.R., eds. 1978. *Man's impact on the troposphere.* NASA Reference Publication 1022. Washington, D.C.: National Aeronautics and Space Adminstration.

1981. *Life on a warmer Earth.* Laxenburg, Austria: International Institute for Applied Systems Analysis.

Lockwood, J.G. 1979. *Causes of climate.* New York: Halsted (John Wiley and Sons, Inc.).

Logan, J.A.; Prather, M.J.; Wofsy, S.C.; and McElroy, M.B. 1981. Tropospheric chemistry: a global perspective. Journal of Geophysical Research 86: 7210.

Lutgens, F.K. 1982. *The atmosphere: an introduction to meteorology.* Englewood Cliffs, N.J.: Prentice-Hall, Inc.

Lutgens, F.K., and Tarbuck, E.J. 1979. *The atmosphere.* Englewood Cliffs, N.J.: Prentice-Hall, Inc.

Macdonald, G.H., ed. 1982. *The long-term impacts of increasing carbon dioxide levels.* Cambridge, Mass.: Ballinger Publishing Co.

McEwan, M.J., and Phillips, L.F. 1975. *Chemistry of the atmosphere.* New York: Halsted (John Wiley and Sons, Inc.).

Meetham, A.R.; Bottom, D.W.; Cayton, S.; Henderson-Sellers, A.; and Chambers, D. 1981. *Atmospheric pollution.* Elmsford, N.Y.: Pergamon Press, Inc.

Mészáros, E. 1981. *Atmospheric chemistry.* New York: Elsevier North-Holland, Inc.

National Academy of Sciences. 1977. *Climate, climatic change, and water supply.* Washington, D.C.

——1978. *The tropospheric transport of pollutants and other substances to the oceans.* Washington, D.C.

Neiburger, M.; Edinger, J.G.; and Bonner, W.D. 1982. *Understanding our atmospheric environment.* 2nd ed. San Francisco: W.H. Freeman and Co. Publishers.

Nieuwstadt, F.T.M., and van Dop, H., eds. 1982. *Atmospheric turbulence and air pollution modelling.* Hingham, Mass.: D. Reidel Publishing Co., Inc.

Parker, H.W. 1977. *Air pollution.* Englewood Cliffs, N.J.: Prentice-Hall, Inc.

1979. *Pathways of pollutants in the atmosphere.* London: Royal Society.

Pearman, G.I., ed. 1980. *Carbon dioxide and climate.* Canberra, Australia: Australian Academy of Science.

Riley, D., and Spolton, L. 1982. *World weather and climate.* 2nd ed. New York: Cambridge University Press.

Schneider, S.H., and Mesirow, L.E. 1977. *The genesis strategy.* New York: Plenum Publishing Corp.

Slater, L.E., and Levin, S.K., eds. 1981. *Climate's impact on food supplies.* Boulder, Co.: Westview Press, Inc.

Smith, C.D., and Parry, M., eds. 1981. *Consequences of climatic change.* Nottingham, England: University of Nottingham, Department of Geography.

Smith, I.M. 1982. *Carbon dioxide—emissions and effects.* London: International Energy Agency.

Stern, A.C. 1976. *Air pollution, Vol. I.* New York: Academic Press, Inc.

Sulman, F.G. 1982. *Short- and long-term changes in climate.* Boca Raton, Fla.: CRC Press, Inc.

Svensson, B.H., and Soderland, R., eds. 1975. Nitrogen, phosphorus, and sulfur—global cycles. SCOPE Report No. 7, *Ecological Bulletin* 22. Orundsboro, Sweden.

Walker, J.C.G. 1977. *Evolution of the atmosphere.* New York: Macmillan, Inc.

Wark, K., and Warner, C.F. 1981. *Air pollution: its origin and control.* 2nd ed. New York: Harper & Row, Publishers, Inc.

Wayne, R.P. 1970. *Photochemistry.* New York: Elsevier North-Holland, Inc.

Wigley, T.M.L.; Ingram, M.J.; and Farmer, G., eds. 1982. *Climate and history: studies in past climates and their impact on man.* New York: Cambridge University Press.

QUESTIONS AND PROBLEMS

1. What phenomenon is responsible for the temperature maximum at the boundary of the stratosphere and the mesosphere?

2. What function does a third body serve in an atmospheric chemical reaction?

3. Why does the lower boundary of the ionosphere lift at night?

4. Why might you expect the reaction of a free radical with NO_2 to be a chain-terminating reaction? (Consider the total number of electrons in NO_2.)

5. The average atmospheric pressure at sea level is 1.012×10^6 dynes/cm^2. The value of g (acceleration of gravity) at sea level is 980 cm/sec^2. What is the mass in kg of the column of air having a cross-sectional area of 1.00 cm^2 at the Earth's surface and extending to the limits of the atmosphere? (Recall that the dyne is a unit of force and force = mass × acceleration of gravity.)

6. Suppose that 22.4 liters of air at STP is used to burn 1.50 g of carbon to form CO_2, and that the resulting gas mixture is adjusted to STP. What is the volume and the average molecular weight of the resulting mixture?

7. If the pressure is 0.01 atm at an altitude of 38 km and is 0.001 atm at 57 km, what is it at 19 km (ignoring temperature variations)?

8. Measured in micrometers, what are the lower wavelength limits of solar radiation reaching the Earth; the wavelength at which maximum solar radiation reaches the Earth; and the wavelength at which maximum energy is radiated back into space?

9. Of the species O, $HO^{\bullet}$, NO_2^*, $H_3C\cdot$, and N^+, which could most readily revert to a nonreactive, "normal" species in total isolation?

10. Of the gases neon, sulfur dioxide, helium, oxygen, and nitrogen, which shows the most variation in concentration in the troposphere?

11. A 12.0-liter sample of air at 25 °C and 1.00-atm pressure was collected and dried. After drying, the volume was exactly 11.50 liters. What was the percentage *by weight* of water in the original air sample?

12. The sunlight incident upon 1 square meter perpendicular to the line of transmission of solar flux just above the Earth's atmosphere provides energy at a rate most closely equivalent to:
    (a) that required to power a pocket electronic calculator;
    (b) that required to provide a moderate level of lighting for a 40-person capacity classroom using fluorescent lights;
    (c) that required to propel a 2500-pound automobile at 55 mph;
    (d) that required to light a 100-watt incandescent light bulb;
    (e) that required to heat a 40-person capacity classroom to 70 °F when the outside temperature is −10 °F.

13. At an altitude of 50 km, the average atmospheric temperature is essentially 0 °C. What is the average number of air molecules per cubic centimeter at this altitude?

14. What two types of condensation nuclei come from bursting sea-foam bubbles?

15. State two factors that make the stratosphere particularly important in terms of its acting as a region where trace atmospheric contaminants are converted to other (chemically less reactive) forms.

16. What is the product when an excited oxygen atom, $O(^1D)$, reacts with a water molecule?

17. What role may N_2 molecules play in the formation of ozone from $O(^1D)$ atoms?

18. What two chemical species are most generally responsible for the removal of hydroxyl radical from the unpolluted troposphere?

19. What is the distinction between the symbols * and · in discussing chemically active species in the atmosphere?
20. What is the distinction between chemiluminescence and luminescence caused when light is absorbed by a molecule or atom?
21. Why does the absorption of infrared light normally not bring about a chemical reaction?

CHAPTER 12

GASEOUS INORGANIC POLLUTANTS AND OXIDES IN THE ATMOSPHERE

12.1 INORGANIC POLLUTANT GASES

A number of gaseous inorganic pollutants are added to the atmosphere by human activities. Those added in the greatest quantities are CO, SO_2, NO, and NO_2. (These quantities are relatively small, compared to the amount of CO_2 in the atmosphere. The possible environmental effects of increased atmospheric CO_2 levels were discussed in Chapter 11.) Other inorganic pollutant gases include NH_3, N_2O, N_2O_5, H_2S, Cl_2, HCl, and HF. Substantial quantities of some of these gases are added to the atmosphere each year by human activities. In the United States alone, it has been estimated [1] that in 1975 automobiles, industry, electricity production, refuse disposal, and space heating produced approximately 96 million metric tons of carbon monoxide, 33 million metric tons of sulfur oxides, and 24 million metric tons of nitrogen oxides as pollutants.

In 1971, the U.S. Environmental Protection Agency set national air quality standards to be met by July 1, 1975, which included carbon monoxide, sulfur dioxide, and nitrogen oxides. These standards are summarized in Table 12.1.

TABLE 12.1 National Air-Quality Standards for CO_2, NO_x, and SO_2

Pollutant	*Limit*[1]
Carbon monoxide	10 mg/m^3 (9 ppm) over an eight-hour period, not to be exceeded more than once per year
	40 mg/m^3 (35 ppm) for a one-hour period, not to be exceeded more than once per year
Nitrogen oxides	100 μg/m^3 (0.05 ppm), annual arithmetic mean
Sulfur dioxide	80 μg/m^3 (0.03 ppm), annual arithmetic mean
	365 μg/m^3 (0.14 ppm) for a 24-hour period, not to be exceeded more than once per year

[1] Primary standards to protect public health. Somewhat more stringent secondary standards to protect public welfare and the general environment are proposed for some pollutants.

[1] 1976. *National air quality and emissions trends report, 1975*. EPA-450/1-76-002. Research Triangle Park, N.C.: Environmental Protection Agency.

In discussing atmospheric pollutants and their global cycles, it is convenient to define and use the following terms and concepts. A **reservoir** is a domain such as the atmosphere or biosphere where a pollutant may "reside" for a time. The amount of a specific pollutant in a reservoir is known as its **burden**. Burdens are commonly expressed in units of 10^{12}g (10^6 metric tons) called **teragrams (Tg)**. The rate of transfer of a pollutant from one sphere or domain to another is called the **flux**. Flux frequently is expressed in units of teragrams per year.

It is useful to determine the **residence time** of a pollutant in a given reservoir. If a reservoir initially contains a pollutant of mass M_i that leaves at a rate R_i governed by a first-order equation,

$$\frac{dM_i}{dt} = \beta_i M_i \qquad (12.1.1)$$

the mean residence time, τ_i, is given by the following equation:

$$\tau_i = \frac{M_i}{R_i} = \frac{1}{\beta_i} \qquad (12.1.2)$$

In this equation, β_i is a constant for a given system and is called the **net transfer coefficient for removal**. It is the sum of removal transfer coefficients involving all major removal processes. If the reservoir is in secular equilibrium, or at a steady state, the pollutant production rate P_i equals the removal rate:

$$\frac{dM_i}{dt} = P_i - R_i = 0 \qquad (12.1.3)$$

12.2 PRODUCTION OF CARBON MONOXIDE

Carbon monoxide, CO, causes problems in cases of locally high concentrations. The overall atmospheric concentration of carbon monoxide has been estimated at 0.1 ppm [2]. Despite intense study, the relative contributions of various sources to atmospheric CO are still uncertain. What is known has been summarized [3]. Much of this information is based upon a study [4] of $^{13}C/^{12}C$ and $^{18}O/^{16}O$ ratios in atmospheric CO. These ratios vary with the source of CO.

The Earth's atmosphere has a burden of approximately 530 million metric tons of CO with an average residence time ranging from 36 to 110 days. Much of this CO is present as an intermediate in the oxidation of methane by hydroxyl radical (see Sections 11.11 and 14.3). From Table 11.1 it may be seen that the methane content of the atmosphere is about 1.6 ppm, more than 10 times the concentration of CO; in fact, methane is the most abundant single hydrocarbon emitted to the atmosphere from any source. Therefore, any oxidation process for methane that produces carbon monoxide as an intermediate is certain to contribute substantially to the overall carbon monoxide burden, probably around two-thirds of the total CO.

---

[2] Strauss, W. 1977. Formation and control of air pollutants. In *Environmental chemistry,* J. O'M. Bockris, ed., pp. 179–212. New York: Plenum Publishing Corp.

[3] Horne, R.A. 1978. *The chemistry of our environment.* New York: John Wiley and Sons, Inc.

[4] Maugh, T.H., II, 1972. Carbon monoxide: natural sources dwarf man's output. *Science* 177: 338–9.

Degradation of chlorophyll during the autumn months releases enormous quantities of CO, amounting to perhaps as much as 20% of the total annual release. Anthropogenic sources account for about 6% of CO emissions. The remainder of atmospheric CO comes from largely unknown sources. These include some plants and marine organisms known as siphonophores, an order of *Hydrozoa*. Carbon monoxide is also produced by decay of plant matter other than chlorophyll.

Because of carbon monoxide emissions from internal combustion engines, maximum levels of this toxic gas tend to occur in congested urban areas at times when the maximum number of people are exposed, such as during rush hours. At such times, carbon monoxide levels in the atmosphere may become as high as 50–100 ppm. Atmospheric levels of carbon monoxide in urban areas show a positive correlation with the density of vehicular traffic, and a negative correlation with wind speed. Whereas urban atmospheres may show average carbon monoxide levels of the order of several ppm, data taken in remote areas show much lower levels.

12.3 EFFECTS OF CARBON MONOXIDE ON HUMAN HEALTH

Carbon monoxide displaces oxygen from hemoglobin to produce carboxyhemoglobin, a reaction with an equilibrium constant of approximately 210.

$$O_2Hb + CO \rightleftharpoons COHb + O_2 \qquad (12.3.1)$$

The result is a reduction in the blood's capacity for carrying oxygen. The level of COHb in the blood increases with the level of exposure to atmospheric CO and with increasing physical activity. The first discernible effect of carbon monoxide poisoning is a reduction in awareness, a probable cause of many automobile accidents. Progressively higher exposures cause impairment of central nervous system functions, changes in cardiac and pulmonary functions, drowsiness, coma, respiratory failure, and finally death. The effects of continuous exposure to various levels of carbon monoxide are summarized in Table 12.2. Industrial workers are allowed an 8-hour time-weighted exposure (see Equation 18.18.1) of 50 ppm. Busy streets frequently show CO levels approaching 100 ppm, and even higher levels are found in automotive tunnels. Average levels of CO in congested areas of large cities have been found to exceed 7 ppm, a level at which some discernible effect on humans has been observed. The question may be raised as to whether exposure to CO has a long-term effect upon human behavior.

TABLE 12.2 Effects of Continuous Exposure to Various Levels of Carbon Monoxide

CO *level, ppm*	*Percent conversion of* O_2Hb *to* COHb	*Effects on humans*
10	2	Impairment of judgment and visual perception
100	15	Headache, dizziness, weariness
250	32	Loss of consciousness
750	60	Death after several hours
1000	66	Rapid death

12.4 FATE OF ATMOSPHERIC CO

It is known that the residence time of carbon monoxide in the atmosphere, τ_{CO}, is not long, perhaps of the order of 4 months. Possible species that might oxidize CO to CO_2 in the atmosphere are HO and HO_2 radicals, atomic oxygen, and ozone. It is now generally agreed that carbon monoxide is removed from the atmosphere primarily by reactions with hydroxyl (HO ·) and hydroperoxyl (HO_2 ·) radicals. These major atmospheric reactants were discussed in Section 11.11. Reactions such as:

$$HO\cdot + CO \rightarrow CO_2 + H \qquad (12.4.1)$$

$$H + O_2 + M \rightarrow HO_2\cdot + M \qquad (12.4.2)$$

$$HO_2\cdot + CO \rightarrow HO\cdot + CO_2 \qquad (12.4.3)$$

$$HO_2\cdot + NO \rightarrow HO\cdot + NO_2 \qquad (12.4.4)$$

$$HO_2\cdot + HO_2 \rightarrow H_2O_2 + O_2 \qquad (12.4.5)$$

$$H_2O_2 + h\nu \rightarrow 2\,HO\cdot \qquad (12.4.6)$$

constitute chains through which CO is consumed [5]. Methane is also involved through the atmospheric CO–HO ·–CH_4 cycle [6]. Of the preceding reactions, it has been shown that Reaction 12.4.3 is too slow to account for much CO removal [7].

It is now well established that soil is a sink for carbon monoxide [8]. Soil microorganisms are the agents that remove CO from the atmosphere. A study of the uptake of CO by various soils performed at 0 °, 10 °, 20 °, and 30 °C has led to an estimate of 4.1×10^{14}g of CO each year taken up by the Earth's total land surface. This is of the same order of magnitude as the amount of CO produced by fossil fuel combustion (6.4×10^{14}g/year) and that produced by CH_4 oxidation by hydroxyl radical (4.0×10^{14}g/year).

12.5 CONTROL OF CARBON MONOXIDE EMISSIONS

Since the internal combustion engine is the primary source of localized carbon monoxide emissions, control measures have been concentrated on the automobile. Carbon monoxide emissions may be lowered by employing a leaner air-fuel mixture, that is, one in which the weight ratio of air to fuel is relatively high. At air-fuel (weight:weight) ratios exceeding approximately 16:1, an internal combustion engine emits virtually no carbon monoxide.

---

[5] Stedman, D.H.; Morris, E.D., Jr.; Daby, E.E.; Niki, H.; and Weinstock, B. 1970. The role of OH radicals in photochemical smog reactions. Paper presented at 160th National Meeting of the American Chemical Society, Chicago, Ill., Sept. 14–18, 1970.

[6] Sze, N.D. 1977. Anthropogenic CO emissions: implications for the atmospheric CO-OH-CH_4 cycle. *Science* 195: 673–5.

[7] Davis, D.D.; Payne, W.A.; and Stiff, L.J. 1973. The hydroperoxyl radical in atmospheric chemical dynamics: reaction with carbon monoxide. *Science* 179: 280–2.

[8] Bartholomew, G.W., and Alexander, M. 1981. Soil as a sink for atmospheric carbon monoxide. *Science* 212: 1389–1391.

Modern automobiles use catalytic exhaust reactors to cut down on carbon monoxide emissions. Excess air is pumped into the exhaust gas, and the mixture is passed through a catalytic converter in the exhaust system, resulting in oxidation of CO to CO_2.

12.6 SULFUR DIOXIDE SOURCES AND THE SULFUR CYCLE

Figure 12.1 shows the main aspects of the global sulfur cycle. This cycle involves primarily H_2S, SO_2, SO_3, and sulfates. There are many uncertainties regarding the sources, reactions, and fates of these atmospheric sulfur species. On a global basis, sulfur compounds enter the atmosphere to a very large extent through human activities. It is estimated that approximately 65 million metric tons of sulfur per year enters the atmosphere through anthropogenic activities, primarily the combustion of fossil fuels. The greatest uncertainties in the cycle have to do

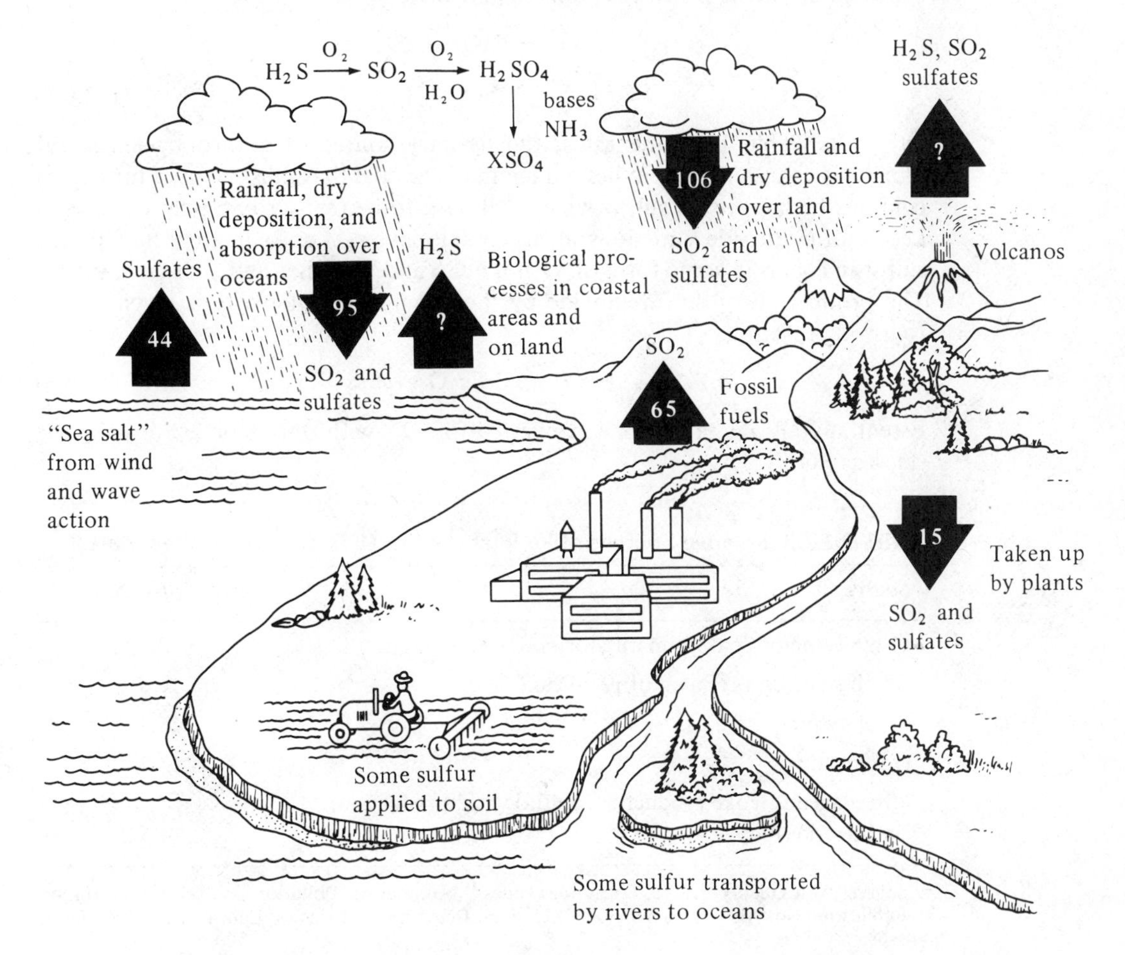

FIGURE 12.1 Atmospheric sulfur cycle. Values are in millions of metric tons (teragrams) of S, as estimated by J.P. Friend, "The Global Sulfur Cycle," in *Chemistry of the Lower Atmosphere*, S.I. Rasool, ed., Ch. 4 (New York: Plenum Publishing Corp.), 1973.

with nonanthropogenic sulfur. Previous estimates indicated that the single greatest source of sulfur dioxide in the atmosphere was H_2S produced by microbial action. However, these estimates were based on differences between the total atmospheric burden of sulfur and the total anthropogenic input of sulfur compounds. As the latter values have been refined, the estimate of H_2S added to the atmosphere by the decay of organic matter and by the biological reduction of sulfate has been drastically reduced [9] and may be as small as 1 teragram per year. Any H_2S that does get into the atmosphere is converted rapidly to SO_2 by the following overall process:

$$H_2S + \frac{3}{2} O_2 \rightarrow SO_2 + H_2O \qquad (12.6.1)$$

The initial reaction is hydrogen ion abstraction by hydroxyl radical,

$$H_2S + HO\cdot \rightarrow HS\cdot + H_2O \qquad (12.6.2)$$

followed by the next two reactions to give SO_2:

$$HS\cdot + O_2 \rightarrow HO\cdot + SO \qquad (12.6.3)$$

$$SO + O_2 \rightarrow SO_2 + O \qquad (12.6.4)$$

As Table 12.3 illustrates, the primary source of anthropogenic sulfur dioxide is coal, and therein lies a dilemma. The coal that must be used for energy contains high levels of sulfur which will have to be removed at great expense to keep sulfur dioxide emissions at acceptable levels. Approximately half of the sulfur in coal is in some form of pyrite, FeS_2, and the other half is organic sulfur. The production of sulfur dioxide by the combustion of pyrite is given by the following reaction:

$$4\ FeS_2 + 11\ O_2 \rightarrow 2\ Fe_2O_3 + 8\ SO_2 \qquad (12.6.5)$$

Essentially all of the sulfur is converted to SO_2, with only 1 or 2% leaving the stack as SO_3.

TABLE 12.3 Relative Sources of Pollutant Sulfur Dioxide in the United States[1]

Source	*Percent of U.S. total*
Combustion of coal (primarily for power generation)	58.2
Combustion and refining of petroleum	19.6
Refinery processes	5.5
Ore smelting	12.2
Miscellaneous (coke production, sulfuric acid production, incineration)	4.5

[1] Source: "Air Quality Criteria for Sulfur Oxides," National Air Pollution Control Administration Publication No. AP-50 (Washington, D.C.: U.S. Department of Health, Education, and Welfare), 1969.

[9] Henry, R.C., and Hidy, G.M. 1980. Potential for atmospheric sulfur from microbial sulfate reduction. *Atmospheric Environment* 14: 1095–1103.

12.7 EFFECTS OF SULFUR DIOXIDE ON HUMAN HEALTH

Sulfur dioxide in the atmosphere has its primary effect upon the respiratory tract, producing irritation and increasing airway resistance. Therefore, exposure to the gas may increase the effort required to breathe. Mucus secretion is also stimulated by exposure to air contaminated by sulfur dioxide. People who already have respiratory weaknesses are particularly susceptible to high levels of the gas in the atmosphere. Although SO_2 causes death in humans at 500 ppm, it has not been found to harm laboratory animals at 5 ppm.

Sulfur dioxide has been at least partially implicated in several acute incidents of air pollution. One of these incidents occurred in December, 1930, in the Meuse River Valley of Belgium when a thermal inversion trapped waste products from a number of industrial sources in the narrow valley. Sulfur dioxide levels reached 38 ppm. Approximately 60 people died in the episode, and some cattle were killed. In October, 1948, a similar incident caused illness in over 40% of the population of Donora, Pennsylvania, and 20 people died. Sulfur dioxide concentrations of 2 ppm were recorded. During a five-day period marked by a temperature inversion and fog in London in December, 1952, approximately 3500–4000 deaths in excess of normal occurred. SO_2 levels reached 1.3 ppm. Autopsies revealed irritation of the respiratory tract, and high levels of sulfur dioxide were suspected of contributing to excess mortality.

12.8 EFFECT OF SULFUR DIOXIDE ON PLANTS

Atmospheric sulfur dioxide is harmful to plants. Acute exposure to high levels of the gas kills leaf tissue (leaf *necrosis*). The edges of the leaves and the areas between the leaf veins are particularly damaged. Chronic exposure of plants to sulfur dioxide causes *chlorosis*, a bleaching or yellowing of the normally green portions of the leaf. Plant injury increases with increasing relative humidity. Plants incur most injury from sulfur dioxide when their stomata (small openings in plant surface tissue that allow interchange of gases with the atmosphere) are open. For most plants, the stomata are open during the daylight hours, and most damage from sulfur dioxide occurs then.

Exposure to a sulfur dioxide concentration of 0.15 ppm for 72 continuous hours has been shown to cut the yields of Durum wheat and barley to 42 and 44 percent of the controls, respectively [10]. Spring wheat yields remained unaffected by a similar treatment. Weekly 3-hour exposures to SO_2 concentrations up to 1.20 ppm had no effect on yields of small grains and alfalfa. The interesting conclusion is that long-term, low-level exposure to sulfur dioxide may be more harmful to crops than short-term, high-level exposure.

As discussed in the following section, sulfur dioxide in the atmosphere is converted to sulfuric acid. In areas with high levels of sulfur dioxide pollution, plants may be damaged by sulfuric acid aerosols. Such damage appears as small spots where sulfuric acid droplets have impinged on leaves.

---

[10] Wilhour, R.G.; Neely, G.E.; Weber, D.E.; and Grothaus, L.C. 1979. Response of selected small grains, range grasses and alfalfa to sulfur dioxide. CERL-050. Corvallis, Oregon: Corvallis Environmental Research Laboratory, U.S. Environmental Protection Agency.

12.9 SULFUR DIOXIDE REACTIONS IN THE ATMOSPHERE

The atmospheric chemical reactions of sulfur dioxide remain subject to some confusion. Many factors, including temperature, humidity, light intensity, atmospheric transport, and surface characteristics of particulate matter, may influence these reactions. Therefore, their study under simulated atmospheric conditions is quite difficult. Like many other gaseous pollutants, sulfur dioxide undergoes chemical reactions resulting in the formation of particulate matter, which then settles or is scavenged from the atmosphere by rainfall or other processes. It is known that high levels of air pollution normally are accompanied by a marked increase in aerosol particles and a consequent reduction in visibility. Reaction products of sulfur dioxide are thought to be responsible for some aerosol formation. Whatever the processes involved, much of the sulfur dioxide in the atmosphere ultimately is oxidized to sulfuric acid and sulfate salts, particularly ammonium sulfate and ammonium hydrogen sulfate. It has been shown that these are the predominant compounds in the light-scattering, submicrometer particulate aerosol encountered around St. Louis [11]. Indeed, it is likely that these sulfates account for the turbid haze that covers much of the eastern part of the U.S. under all atmospheric conditions except those characterized by massive intrusions of Arctic air masses during the winter months. The potential of sulfates to induce climatic change is high and must be taken into account when considering control of sulfur dioxide.

Some of the possible ways in which sulfur dioxide may react in the atmosphere are (1) photochemical reactions; (2) photochemical and chemical reactions in the presence of nitrogen oxides and/or hydrocarbons, particularly alkenes (olefins); (3) chemical processes in water droplets, particularly those containing metal salts and ammonia; and (4) reactions on solid particles in the atmosphere. Keep in mind that the atmosphere is a highly dynamic system with great variations in temperature, composition, humidity, and intensity of sunlight; therefore, different processes may predominate under various atmospheric conditions.

Photochemical reactions are probably involved in some of the processes resulting in the atmospheric oxidation of SO_2. Light with wavelengths above 218.0 nm is not sufficiently energetic to bring about the photodissociation of SO_2 [12]:

$$SO_2 + h\nu \rightarrow SO + O \tag{12.9.1}$$

The lower atmosphere receives only light at wavelengths exceeding 290 nm. In the region of 300–400 nm, solar radiation is absorbed to produce electronically excited states of SO_2. A weak absorption band with a maximum around 384 nm is thought to produce an excited triplet state,

$$SO_2 + h\nu \rightarrow {}^3SO_2 \tag{12.9.2}$$

and a stronger absorption band with a maximum around 294 nm produces an excited singlet state (see Eq. 12.9.3).

---

[11] Charlson, R.J., et. al. 1974. Sulfuric acid-ammonium sulfate aerosol: optical detection in the St. Louis region. *Science* 184: 156–8.

[12] Bufalini, M. 1971. Oxidation of sulfur dioxide in polluted atmospheres—a review. *Environmental Science and Technology* 5: 685–700.

$$SO_2 + h\nu \rightarrow {}^1SO_2 \qquad (12.9.3)$$

In darkness and at SO_2 concentrations of only a few parts per million in pure air, the oxidation of SO_2 to SO_3 and subsequent formation of H_2SO_4 is too slow to measure. However, it has been estimated [12] that, in sunlight at SO_2 concentrations of 5–30 ppm and a relative humidity of 32–91%, the overall reaction

$$SO_2 + \frac{1}{2}O_2 + H_2O \rightarrow H_2SO_4 \qquad (12.9.4)$$

occurs to the extent of approximately 0.1% (of SO_2) per hour. Possibly the photochemical excitation of SO_2 (Eqs. 12.9.2 and/or 12.9.3) is involved in the overall reaction. Several possible reaction paths have been postulated for the overall reaction, but the identities of the intermediate products remain very much in doubt. It is obvious that the oxidation of SO_2 at the parts-per-million level in an otherwise unpolluted atmosphere is a slow process. In general, other pollutant species must be involved in the process in atmospheres receiving SO_2 pollution.

The presence of hydrocarbons and nitrogen oxides greatly increases the oxidation rate of atmospheric SO_2 [13]. As discussed in Chapter 14, hydrocarbons, nitrogen oxides, and ultraviolet light are the ingredients necessary for the formation of photochemical smog. This disagreeable condition is characterized by high levels of various oxidizing species (photochemical oxidants) capable of oxidizing SO_2. In the smog-prone Los Angeles area, the oxidation of SO_2 ranges up to 5–10% per hour. Among the oxidizing species present which could bring about this fast reaction are $HO\cdot$, $HO_2\cdot$, O, O_3, NO_3, N_2O_5, $ROO\cdot$, and $RO\cdot$. As discussed in Chapter 14, the latter two species are reactive, organic free radicals containing oxygen. Although ozone, O_3, is an important product of photochemical smog, it is believed that the oxidation of SO_2 by ozone in the gas phase is too slow to be significant, but it is probably significant in water droplets

One significant pathway for oxidation of SO_2 probably involves the addition of $HO\cdot$ radical to SO_2,

$$HO\cdot + SO_2 \rightarrow HOSO_2\cdot \qquad (12.9.5)$$

forming a very reactive free radical which can add to O_2:

$$HOSO_2\cdot + O_2 \rightarrow HOSO_2O_2\cdot \qquad (12.9.6)$$

The product of this reaction is an oxidizing agent which reacts, for example, with NO to produce NO_2:

$$HOSO_2O_2\cdot + NO \rightarrow HOSO_2O\cdot + NO_2 \qquad (12.9.7)$$

The sulfur-containing radical is eventually converted to a form of sulfate.

In all but relatively dry atmospheres, it is probable that sulfur dioxide is oxidized by reactions occurring inside water aerosol droplets, frequently with H_2O_2 as the oxidizing agent. This oxidation process is much faster in the presence of ammonia, which reacts with sulfur dioxide to produce bisulfite ion and sulfite ion in solution:

$$NH_3 + SO_2 + H_2O \rightarrow NH_4^+ + HSO_3^- \qquad (12.9.8)$$

---

[13] Finlayson, B.J., and Pitts, J.N., 1976. Photochemistry of the polluted troposphere. *Science* 192: 111–9.

$$NH_3 + HSO_3^- \rightarrow NH_4^+ + SO_3^{2-} \quad (12.9.9)$$

One explanation for the catalytic effect of ammonia assumes that the rate of oxidation is given by the equation

$$\frac{d[SO_4^{2-}]}{dt} = k[SO_3^{2-}] \quad (12.9.10)$$

where k has a value of 0.1 min^{-1}. The overall process of sulfur dioxide oxidation in the aqueous phase is rather complicated [14]. It involves the transport of gaseous SO_2 and oxidant to the aqueous phase, diffusion of species in the aqueous droplet, hydrolysis and ionization of SO_2, and oxidation of SO_2 by the following overall process, where {O} represents an oxidizing agent such as H_2O_2, $HO\cdot$, or O_3, and S(IV) is $SO_2(aq)$, HSO_3^-, or SO_3^{2-}:

$$\{O\} + S(IV) \xrightarrow{H_2O} 2\,H^+ + SO_4^{2-} \quad (12.9.11)$$

Laboratory studies have shown that ions of iron(II), iron(III), nickel(II), copper(II), and especially manganese(II) catalyze the oxidation of sulfur dioxide to sulfate in water droplets.

The importance of aqueous-phase oxidation of sulfur dioxide in the atmosphere has been emphasized by a detailed study of available data pertaining to SO_2 oxidation [15]. It was concluded that only about 20% of the SO_2 oxidized in the atmosphere is oxidized by homogeneous gas-phase reaction and the remaining 80% is oxidized in the liquid phase. The same study concluded that about half of the SO_2 in the atmosphere is removed by oxidation to sulfate or sulfuric acid and the other half is removed by dry and wet deposition of SO_2 on land, vegetation, and water.

Heterogeneous reactions on solid particles may also play a role in the removal of sulfur dioxide from the atmosphere. In atmospheric photochemical reactions, such particles may function as **nucleation centers**. Thus, they act as catalysts and grow in size by accumulating reaction products. The final result would be production of an aerosol with a composition unlike that of the original particle. Little research has been done on the role that solid particles play in the oxidation of sulfur dioxide under conditions like those found in the atmosphere. Soot particles, which consist of elemental carbon contaminated with polynuclear aromatic hydrocarbons (see Section 13.16) produced in the incomplete combustion of carbonaceous fuels, have been shown to catalyze the oxidation of sulfur dioxide to sulfate [16]. The presence of sulfate on the soot particles was revealed by photoelectron spectroscopy (ESCA). The sulfate ESCA peak was more intense when the atmosphere was humid than when it was dry. Soot particles are very common in polluted atmospheres, so it is very likely that they are strongly involved in catalyzing the oxidation of sulfur dioxide.

---

[14] Schwartz, S.E., and Freiberg, J.E. 1981. Mass transport limitations to the rate of reaction of gases in liquid droplets: application to oxidation of SO_2 in aqueous solutions. *Atmospheric Environment* 15: 1129–1144.

[15] Möller, D. 1980. Kinetic model of atmospheric SO_2 oxidation based on published data. *Atmospheric Environment* 14: 1067–1076.

[16] Novakov, T.; Chang, S.G.; and Harker, A.B. 1974. Sulfates as pollution particulates: catalytic formation on carbon (soot) particles. *Science* 186: 259–61.

Oxides of metals such as aluminum, calcium, chromium, iron, lead, or vanadium may also be catalysts for the heterogeneous oxidation of sulfur dioxide. These oxides may also adsorb sulfur dioxide. However, considering a normal particulate matter loading of around 200 μg of solids per cubic meter of air, the resulting particulate matter surface area of 1–10 cm^2/m^3 is too low to account for much adsorption.

12.10 SULFUR REMOVAL

A number of processes are being used to remove sulfur and sulfur oxides from fuel before combustion and from stack gas after combustion. Most of these efforts concentrate on coal, since it is the major source of sulfur oxides pollution. Physical separation techniques may be used to remove discrete particles of pyritic sulfur from coal. Chemical methods may also be employed for removal of sulfur from coal. Another technique for lowering local sulfur dioxide pollution levels is the use of extremely tall stacks. The total amount of sulfur dioxide ejected into the atmosphere is not lessened, but it travels a greater distance and is dispersed before reaching ground level.

Fluidized bed combustion of coal promises to eliminate SO_2 emissions at the point of combustion. The process consists of burning granular coal in a bed of finely divided limestone or dolomite maintained in a fluid-like condition by air injection. Heat calcines the limestone,

$$CaCO_3 \rightarrow CaO + CO_2(g) \tag{12.10.1}$$

and the lime produced absorbs SO_2:

$$CaO + SO_2 + \frac{1}{2} O_2 \rightarrow CaSO_4 \tag{12.10.2}$$

Many processes have been proposed or studied for the removal of sulfur dioxide from stack gas. Table 12.4 (see page 334) summarizes major stack gas scrubbing systems. These include throwaway and recovery systems as well as wet and dry systems. A dry throwaway system used with only limited success involves injection of dry limestone or dolomite into the boiler, followed by recovery of dry lime, sulfites, and sulfates. The overall reaction, shown here for dolomite, is

$$CaCO_3 \cdot MgCO_3(s) + SO_2(g) + \frac{1}{2}O_2(g) \rightarrow CaSO_4(s) + MgO(s) + 2\ CO_2(g) \tag{12.10.3}$$

The solid sulfate and oxide products are removed by electrostatic precipitators or cyclone separators. The process has an efficiency of 50% or less for the removal of sulfur oxides.

As may be noted from the chemical reactions shown in Table 12.4, all of these processes, except for catalytic oxidation, depend upon absorption of SO_2 by an acid-base reaction. The first two processes listed are throwaway processes yielding large quantities of wastes; the others provide for some sort of sulfur product recovery.

To date, lime or limestone slurry scrubbing have been used more than other processes for SO_2 removal. These involve acid-base reactions with SO_2. When sulfur dioxide dissolves in water, equilibrium is established between SO_2 gas and dissolved SO_2 (see Eq. 12.10.4, next page).

TABLE 12.4 Major Stack-Gas Scrubbing Systems[1]

Process	*Chemical reactions*	*Major advantages or disadvantages*
Lime slurry scrubbing[2]	$Ca(OH)_2 + SO_2 \rightarrow CaSO_3 + H_2O$	Up to 200 kg of lime are needed per metric ton of coal, producing huge quantities of waste product.
Limestone slurry scrubbing[2]	$CaCO_3 + SO_2 \rightarrow CaSO_3 + CO_2(g)$	Lower pH than lime slurry, and not so efficient.
Magnesium oxide scrubbing	$Mg(OH)_2(slurry) + SO_2 \rightarrow MgSO_3 + 2\,H_2O$	The sorbent can be regenerated, and this need not be done on site.
Sodium-base scrubbing	$Na_2SO_3 + H_2O + SO_2 \rightarrow 2\,NaHSO_3$	There are no major technological limitations. Annual costs are relatively high.
	$2\,NaHSO_3 + heat \rightarrow Na_2SO_3 + H_2O + SO_2$ (regeneration)	
Double alkali[2]	$2\,NaOH + SO_2 \rightarrow Na_2SO_3 + 2\,H_2O$	Allows for regeneration of expensive sodium alkali solution with inexpensive lime.
	$Ca(OH)_2 + Na_2SO_3 \rightarrow CaSO_3(s) + 2\,NaOH$ (regeneration of NaOH)	

[1]For details regarding these and more advanced processes, see the following two books. Satriana, M. 1981. *New developments in flue gas desulfurization technology*. Park Ridge, N.J.: Noyes Data Corp. Hudson, J.L., and Rochelle, G.T., eds. 1982. *Flue gas desulfurization*. Washington, D.C.: American Chemical Society.

[2]These processes have also been adapted to produce a gypsum product by oxidation of $CaSO_3$ in the spent scrubber medium:

$$CaSO_3 + \tfrac{1}{2}\,O_2 + 2\,H_2O \rightarrow CaSO_4 \cdot 2\,H_2O(s)$$

Gypsum has some commercial value, such as in the manufacture of plasterboard, and makes a relatively settleable waste product.

$$SO_2(g) \rightleftharpoons SO_2(aq) \qquad (12.10.4)$$

This equilibrium is described by Henry's Law (Section 2.9),

$$[SO_2(aq)] = K \times P_{SO_2} \qquad (12.10.5)$$

where $[SO_2(aq)]$ is the concentration of dissolved *molecular* sulfur dioxide; K is the Henry's Law constant for SO_2; and P_{SO_2} is the partial pressure of sulfur dioxide gas. In the presence of base, the reactions

$$H_2O + SO_2(aq) \rightleftharpoons H^+ + HSO_3^- \qquad (12.10.6)$$

$$HSO_3^- \rightleftharpoons H^+ + SO_3^{2-} \qquad (12.10.7)$$

shift Reaction 12.10.4 strongly to the right. In the presence of calcium carbonate slurry (as in limestone slurry scrubbing), hydrogen ion is taken up by the reaction

$$CaCO_3 + H^+ \rightleftharpoons Ca^{2+} + HCO_3^- \qquad (12.10.8)$$

The reaction of calcium carbonate with carbon dioxide from stack gas,

$$CaCO_3 + CO_2 + H_2O \rightleftharpoons Ca^{2+} + 2\,HCO_3^- \qquad (12.10.9)$$

results in some sorption of CO_2. The reaction of sulfite and calcium ion to form highly insoluble calcium sulfite hemihydrate

$$Ca^{2+} + SO_3^{2-} + \frac{1}{2}\,H_2O \rightleftharpoons CaSO_3 \cdot \frac{1}{2}\,H_2O(s) \qquad (12.10.10)$$

also shifts Reactions 12.10.6 and 12.10.7 to the right. Gypsum is formed in the scrubbing process by the oxidation of sulfite,

$$SO_3^{2-} + \frac{1}{2}\,O_2 \rightarrow SO_4^{2-} \qquad (12.10.11)$$

followed by reaction of sulfate ion with calcium ion:

$$Ca^{2+} + SO_4^{2-} + 2\,H_2O \rightleftharpoons CaSO_4 \cdot 2\,H_2O(s) \qquad (12.10.12)$$

Formation of gypsum in the scrubber is undesirable because it creates scale in the scrubber equipment. Gypsum is sometimes produced deliberately in the spent scrubber liquid downstream from the scrubber (see Table 12.4, note 2).

When lime, $Ca(OH)_2$, is used in place of limestone (lime slurry scrubbing), a source of hydroxide ions is provided for direct reaction with H^+:

$$H^+ + OH^- \rightarrow H_2O \qquad (12.10.13)$$

The reactions involving sulfur species in a lime slurry scrubber are essentially the same as those just discussed for limestone slurry scrubbing. The pH of a lime slurry is higher than that of a limestone slurry, so that the former has more of a tendency to react with CO_2, resulting in the absorption of that gas:

$$CO_2 + OH^- \rightleftharpoons HCO_3^- \qquad (12.10.14)$$

Current practice with lime and limestone scrubber systems calls for injection of the slurry into the scrubber loop beyond the boilers. A number of power plants are now operating with this kind of system. Experience to date has shown that these scrubbers remove well over 90% of both SO_2 and fly ash when operating properly. (Fly ash is fuel combustion ash normally carried up the stack with flue gas.) Although corrosion and scaling problems are being solved, disposal of lime sludge poses formidable obstacles. The quantity of this sludge may be appreciated by considering that approximately 1 ton of limestone is required for each 5 tons of coal. The sludge is normally disposed of in large ponds. Water seeping through the sludge beds becomes laden with calcium sulfate and other salts. Techniques are badly needed to stabilize this sludge as a structurally stable, nonleachable solid.

Recovery systems in which sulfur dioxide or elemental sulfur are removed from the spent sorbing material, which is recycled, are much more desirable from an environmental viewpoint than are throwaway systems. Many kinds of recovery systems have been investigated, including those that involve scrubbing with magnesium oxide slurry, sodium sulfite solution, ammonia solution, or sodium citrate solution.

Sulfur dioxide trapped in a stack-gas-scrubbing process can be converted to hydrogen sulfide by reaction with synthesis gas (H_2, CO, CH_4),

$$SO_2 + (H_2, CO, CH_4) \rightarrow H_2S + CO_2 \qquad (12.10.15)$$

The Claus reaction is then employed to produce elemental sulfur:

$$2\ H_2S + SO_2 \rightarrow 2\ H_2O + 3\ S \qquad (12.10.16)$$

12.11 NITROGEN OXIDES IN THE ATMOSPHERE

The three oxides of nitrogen normally encountered in the atmosphere are nitrous oxide (N_2O), nitric oxide (NO), and nitrogen dioxide (NO_2). Nitrous oxide, a commonly used anesthetic known as "laughing gas," is produced by microbiological processes and is a component of the unpolluted atmosphere at a level of approximately 0.3 ppm (see Table 11.1). This gas is relatively unreactive and probably does not significantly influence important chemical reactions in the lower atmosphere. Its concentration decreases rapidly with altitude in the stratosphere due to the photochemical reaction

$$N_2O + h\nu \rightarrow N_2 + O \qquad (12.11.1)$$

and some reaction with singlet atomic oxygen:

$$N_2O + O \rightarrow N_2 + O_2 \qquad (12.11.2)$$

$$N_2O + O \rightarrow NO + NO \qquad (12.11.3)$$

These reactions are significant in terms of depletion of the ozone layer. Increased global fixation of nitrogen (Sections 5.15 and 10.8), accompanied by increased microbial production of N_2O, could constitute a threat to the ozone layer.

Colorless, odorless nitric oxide (NO) and pungent red-brown nitrogen dioxide (NO_2) are very important in polluted air. Collectively designated NO_x, these gases enter the atmosphere from natural sources, such as lightning and biological processes, and from pollutant sources. The latter are much more significant because of regionally high NO_x concentrations, which can cause severe air quality deterioration. Practically all anthropogenic NO_x enters the atmosphere as a result of the combustion of fossil fuels in both stationary and mobile sources. Globally, around 86 million metric tons of nitrogen oxides are emitted to the atmosphere from these sources each year, compared to several times that much from widely dispersed natural sources. United States production of nitrogen oxides is around 21 million metric tons per year. In 1965, it was estimated that automobiles supplied approximately half of the anthropogenic nitrogen oxides added to the atmosphere in the United States each year. That proportion is probably somewhat lower now with nitrogen oxide pollutant controls on newer automobiles.

Most NO_x entering the atmosphere does so as NO. At very high temperatures, the following reaction occurs:

$$N_2 + O_2 \rightarrow 2\ NO \qquad (12.11.4)$$

The speed with which this reaction takes place increases steeply with temperature. A mixture of 3% O_2 and 75% N_2, typical of that which occurs in the combustion

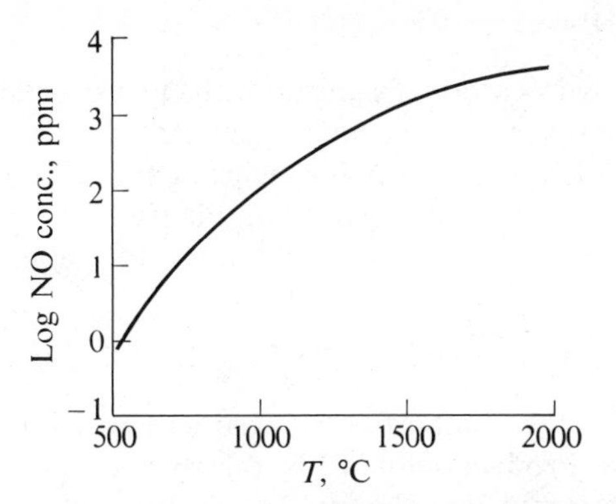

FIGURE 12.2 Log of equilibrium NO concentration as a function of temperature in a mixture containing 75% N_2 and 3% O_2.

chamber of an internal combustion engine, produces 500 ppm of NO in 23 minutes at 1315 °C and in only 0.117 seconds at 1980 °C. The equilibrium concentration of NO in such a mixture is shown as a function of temperature in Figure 12.2. The equilibrium concentration of NO in this mixture at room temperature (27 °C) is only 1.1×10^{-10} ppm. In air containing 80% nitrogen and 20% oxygen, the equilibrium concentrations of NO are approximately three times those shown in the figure.

High temperatures favor both a high equilibrium concentration and a rapid rate of formation of NO. Rapid cooling of the exhaust gas from combustion "freezes" NO at a relatively high concentration because equilibrium is not maintained. Thus, by its very nature, the combustion process both in the internal combustion engine and in furnaces produces high levels of NO in the combustion products.

The mechanism for formation of nitrogen oxides from N_2 and O_2 during combustion is a complicated process. Both oxygen and nitrogen atoms are involved. These are formed at the very high combustion temperatures by the reactions

$$O_2 + M \rightarrow O + O + M \tag{12.11.5}$$

$$N_2 + M \rightarrow N\cdot + N\cdot + M \tag{12.11.6}$$

where M is a highly energetic third body. Breakage of the oxygen bond requires 118 kcal/mole, and breakage of the nitrogen bond requires 225 kcal/mole, so the energies required for these reactions are quite high. Once formed, O and N atoms participate in the following chain reaction for the formation of nitric oxide from nitrogen and oxygen:

$$N_2 + O \rightarrow NO + N \tag{12.11.7}$$

$$N + O_2 \rightarrow NO + O \tag{12.11.8}$$

$$N_2 + O_2 \rightarrow 2\,NO \tag{12.11.9}$$

There are, of course, many other species present in the combustion mixture besides those shown. The oxygen atoms are especially reactive toward hydrocarbon fragments by reactions such as that shown in Equation 12.11.10.

$$RH + O \rightarrow R\cdot + HO\cdot \tag{12.11.10}$$

where RH represents a hydrocarbon fragment with an extractable hydrogen atom. These fragments compete with N_2 for oxygen atoms. It is partly for this reason that the formation of NO is appreciably higher at air/fuel ratios exceeding the stoichiometric ratio (lean mixture), as shown in Figure 14.3.

The hydroxyl radical itself can participate in the formation of NO. The reaction is

$$N + HO\cdot \rightarrow NO + H\cdot \tag{12.11.11}$$

NO is a product of the combustion of coal and petroleum containing chemically bound nitrogen. Production of NO by this route occurs at much lower temperatures than those required for "thermal" NO, discussed previously

Atmospheric chemical reactions convert NO_x to nitric acid, inorganic nitrate salts, organic nitrates, and peroxyacetyl nitrate (see Section 14.4). The lifetime of NO_x has been measured in plumes of polluted air downwind from urban areas [17]. An average pseudo-first-order rate constant for NO_x was found to be 0.18 per hour under daylight conditions, with a range of 0.14 to 0.24 hr^{-1}. (A first-order rate constant is described by the equation for a first-order reaction, although the reaction being discussed does not proceed by a truly first-order mechanism, hence the term *pseudo-first-rate constant.* Rate constants cannot be understood without a thorough coverage of kinetics. Students wishing to delve into this further must study some physical chemistry.) These rates suggest that NO_x lasts only about 5 to 6 hours in the urban atmosphere. Since sulfur dioxide lasts much longer than NO_x in the atmosphere, these findings suggest that acid rain (see Section 12.15) originating with NO_x tends to come much more from local sources than acid rain originating with SO_2.

12.12 ATMOSPHERIC REACTIONS OF NO_x

The principal reactive nitrogen oxide species in the troposphere are NO, NO_2, and HNO_3. These species cycle among each other, as shown in Figure 12.3. Nitric oxide, NO, is the primary form in which NO_x is released to the atmosphere. However, the conversion of NO to NO_2 is relatively rapid in the troposphere.

Nitrogen dioxide is a very reactive and significant species in the atmosphere. It absorbs light throughout the ultraviolet and visible spectrum penetrating the troposphere. At wavelengths below 398 nm, photodissociation occurs,

$$NO_2 + h\nu \rightarrow NO + O \tag{12.12.1}$$

to produce ground state oxygen atoms, sometimes designated $O(^3P)$. Above 430 nm, only excited molecules are formed,

$$NO_2 + h\nu \rightarrow NO_2^* \tag{12.12.2}$$

whereas at wavelengths between 398 nm and 430 nm, either process may occur. Photodissociation at these wavelengths requires input of rotational energy from

[17] Spicer, C.W. 1982. Nitrogen oxide reactions in the urban plume of Boston. *Science* 215: 1095–97.

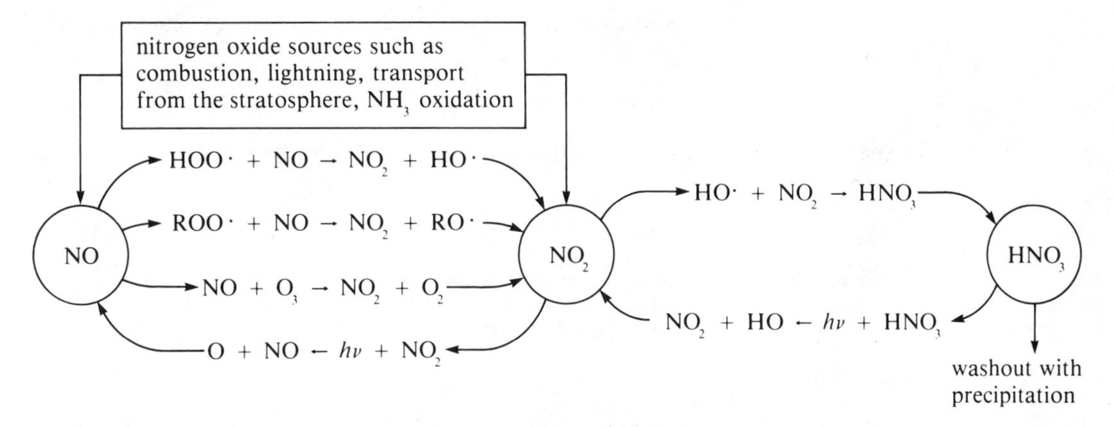

FIGURE 12.3 Principal reactions among NO, NO_2, and HNO_3 in the atmosphere. ROO· represents an organic peroxyl radical, such as the methylperoxyl radical (see Section 11.11).

the NO_2 molecule. (Rotational energy comes from the rotation of the molecule.) The reactivity of NO_2 to photodissociation is shown clearly by the fact that at 40° latitude the half-life of NO_2 is only 85 seconds, much shorter than that of any other atmospheric component.

The photodissociation of nitrogen dioxide can give rise to the following significant inorganic reactions, in addition to a host of atmospheric reactions involving organic species:

$$O + O_2 + M \text{ (a third body)} \rightarrow O_3 + M \quad (12.12.3)$$

$$NO + O_3 \rightarrow NO_2 + O_2 \quad (12.12.4)$$

$$NO_2 + O_3 \rightarrow NO_3 + O_2 \quad (12.12.5)$$

$$O + NO_2 \rightarrow NO + O_2 \quad (12.12.6)$$

$$O + NO_2 + M \rightarrow NO_3 + M \quad (12.12.7)$$

$$NO_2 + NO_3 \rightarrow N_2O_5 \quad (12.12.8)$$

$$NO + NO_3 \rightarrow 2\ NO_2 \quad (12.12.9)$$

$$O + NO + M \rightarrow NO_2 + M \quad (12.12.10)$$

Nitrogen dioxide ultimately is removed from the atmosphere as nitric acid, nitrates, or (in atmospheres where photochemical smog is formed) as organic nitrogen. These are major sources of fixed nitrogen in the biological nitrogen cycle. A minor mechanism for the removal of nitrogen dioxide from the atmosphere is the reaction with water to produce nitric acid and nitric oxide,

$$3\ NO_2 + H_2O \rightarrow 2\ HNO_3 + NO \quad (12.12.11)$$

or to produce nitric acid and nitrous acid:

$$2\ NO_2 + H_2O \rightarrow HNO_3 + HNO_2 \quad (12.12.12)$$

The rates of production of nitric acid by these reactions are too slow to account for significant NO_2 removal. Another path, which may be more rapid, involves formation of N_2O_5 by Reactions 12.12.7 and 12.12.8, followed by reaction of the pentoxide with water:

$$N_2O_5 + H_2O \rightarrow 2\ HNO_3 \qquad (12.12.13)$$

In the stratosphere, nitrogen dioxide reacts with hydroxyl radicals to produce nitric acid:

$$HO\cdot + NO_2 \rightarrow HNO_3 \qquad (12.12.14)$$

In this region, the nitric acid can also be destroyed by hydroxyl radicals,

$$HO\cdot + HNO_3 \rightarrow H_2O + NO_3 \qquad (12.12.15)$$

or by a photochemical reaction,

$$HNO_3 + h\nu \rightarrow HO\cdot + NO_2 \qquad (12.12.16)$$

so that HNO_3 serves as a temporary sink for NO_2 in the stratosphere. Nitric acid produced from NO_2 is removed as precipitation, or reacts with bases (ammonia, particulate lime) to produce particulate nitrates.

12.13 HARMFUL EFFECTS OF NITROGEN OXIDES

Nitric oxide, NO, is biochemically less active and less toxic than NO_2. Like carbon monoxide and nitrite, NO attaches to hemoglobin and reduces oxygen transport efficiency. However, in a polluted atmosphere, the concentration of nitric oxide normally is much lower than that of carbon monoxide so that the effect on hemoglobin is much less.

Acute exposure to NO_2 can be quite harmful to human health, a fact which should be taken into consideration in performing laboratory operations involving this toxic gas. The health effects of nitrogen dioxide vary with the degree of exposure. For exposures ranging from several minutes to one hour, a level of 50–100 ppm of NO_2 causes inflammation of lung tissue for a period of 6–8 weeks, after which time the subject normally recovers. Exposure of the subject to 150–200 ppm of NO_2 causes *bronchiolitis fibrosa obliterans,* a condition fatal within 3–5 weeks after exposure. Death generally results within 2–10 days after exposure to 500 ppm or more of NO_2. "Silo-filler's disease," caused by NO_2 generated by the fermentation of ensilage containing nitrate, is a particularly striking example of nitrogen dioxide poisoning. Deaths have resulted from the inhalation of NO_2-containing gases from burning celluloid and nitrocellulose film. Accidental release of liquid NO_2 being loaded into a Titan II intercontinental missile at Rock, Kansas, on August 24, 1978, resulted in the death of two people and injury to five others. Liquid NO_2 is used in these rockets as an oxidant for hydrazine fuel.

Although extensive damage to plants is observed in areas receiving heavy exposure to NO_x, most of this damage probably comes from secondary products of nitrogen oxides, such as PAN formed in smog (see Chapter 14). Exposure of plants to several parts per million of NO_2 in the laboratory causes leaf spotting and breakdown of plant tissue. Exposure to 10 ppm of NO causes a reversible decrease in the rate of photosynthesis. The effect on plants of long-term exposure to a few tenths of a part per million of NO_2 is not known and should be the subject of some careful research.

Nitrogen oxides are known to cause fading of dyes used in some textiles. This has been observed in gas clothes dryers and is due to NO_x formed in the dryer flame. Much of the damage to materials caused by NO_x comes from secondary nitrates and nitric acid. For example, stress-corrosion cracking of springs used in telephone relays occurs far below the yield strength of the nickel-brass spring metal because of the action of particulate nitrates and aerosol nitric acid formed from NO_x.

Much concern has been expressed about the possibility that NO_x emitted to the atmosphere by supersonic transport planes could catalyze the partial destruction of the stratospheric ozone layer, which absorbs damaging short-wavelength (240–300 nm) ultraviolet radiation (see Chapter 11). Detailed consideration of this effect is quite complicated, and only the main features are considered here.

In the upper stratosphere and in the mesosphere, molecular oxygen is photodissociated by ultraviolet light of less than 242-nm wavelength:

$$O_2 + h\nu \rightarrow O + O \qquad (12.13.1)$$

In the presence of energy-absorbing third bodies, the atomic oxygen reacts with molecular oxygen to produce ozone:

$$O_2 + O + M \rightarrow O_3 + M \qquad (12.13.2)$$

Ozone can be destroyed by reaction with atomic oxygen,

$$O_3 + O \rightarrow O_2 + O_2 \qquad (12.13.3)$$

and its formation can be prevented by recombination of oxygen atoms:

$$O + O + M \rightarrow O_2 + M \qquad (12.13.4)$$

Addition of the reaction of nitric oxide with ozone,

$$NO + O_3 \rightarrow NO_2 + O_2 \qquad (12.13.5)$$

to the reaction of nitrogen dioxide with atomic oxygen,

$$NO_2 + O \rightarrow NO + O_2 \qquad (12.13.6)$$

results in a net reaction for the destruction of ozone:

$$O + O_3 \rightarrow O_2 + O_2 \qquad (12.13.7)$$

Along with NO_x, water vapor is also emitted into the atmosphere by aircraft exhausts, which could accelerate ozone depletion by the following two reactions:

$$O + H_2O \rightarrow HO\cdot + HO\cdot \qquad (12.13.8)$$

$$HO\cdot + O_3 \rightarrow HOO\cdot + O_2 \qquad (12.13.9)$$

However, there are many natural stratospheric buffering reactions which tend to mitigate the potential ozone destruction from those reactions outlined above. Atomic oxygen capable of regenerating ozone is produced by the photochemical reaction,

$$NO_2 + h\nu \rightarrow NO + O \quad \lambda < 420\,\text{nm} \qquad (12.13.10)$$

A competing reaction removing catalytic NO is

$$NO + HOO\cdot \rightarrow NO_2 + HO\cdot \qquad (12.13.11)$$

Currently, atmospheric chemists believe that supersonic aircraft emissions in the stratosphere will not damage the ozone layer and, in fact, may slightly enhance levels of stratospheric ozone [18].

12.14 CONTROL OF NITROGEN OXIDES

The level of NO_x emitted from stationary sources such as power-plant furnaces generally falls within the range of 50–1000 ppm. NO production is favored both kinetically and thermodynamically by high temperatures, as well as by high excess oxygen concentrations. These factors must be considered in reducing NO emissions from stationary sources. Reduction of flame temperature to prevent NO formation is accomplished by adding recirculated exhaust gas, cool air, or inert gases. Unfortunately, this decreases the efficiency of energy conversion as calculated by the Carnot Equation (see Section 17.8).

Low-excess-air firing is known to be effective in reducing NO_x emission when oil and gas are used as fuels and is probably effective when coal is used. As the term implies, low-excess-air firing uses the minimum amount of excess air required for oxidation of the fuel, so that less oxygen is available for the reaction

$$N_2 + O_2 \rightarrow 2\,NO \qquad (12.14.1)$$

in the high-temperature region of the flame. Incomplete fuel burnout, with the emission of hydrocarbons, soot, and CO, is an obvious problem with low-excess-air firing. This may be overcome by a two-stage combustion process consisting of the following steps:

1. A first stage in which the fuel is fired at a relatively high temperature with a substoichiometric amount of air, for example, 90–95% of the stoichiometric requirement. NO formation is limited by the absence of excess oxygen.
2. A second stage in which fuel burnout is completed at a relatively low temperature in excess air. The low temperature prevents formation of NO.

In some power plants fired with gas, the emission of NO has been reduced by as much as 90% by a two-stage combustion process.

Removal of NO_x from stack gas presents some formidable problems. Possible approaches to NO_x removal are catalytic decomposition of nitrogen oxides, catalytic reduction of nitrogen oxides, and sorption of NO_x by liquids or solids.

Approaches have been proposed for the catalytic reduction of NO in stack gas involving methane,

$$CH_4 + 4\,NO \rightarrow 2\,N_2 + CO_2 + 2\,H_2O \qquad (12.14.2)$$

ammonia,

$$4\,NH_3 + 6\,NO \rightarrow 5\,N_2 + 6\,H_2O \qquad (12.14.3)$$

and carbon monoxide:

$$2\,CO + 2\,NO \rightarrow N_2 + 2\,CO_2 \qquad (12.14.4)$$

---

[18] Hudson, R.D., ed. 1979. *The stratosphere: present and future.* Washington, D.C.: National Aeronautics and Space Administration.

Production of undesirable by-products is a major concern in these processes. For example, sulfur dioxide reacts with carbon monoxide to produce toxic carbonyl sulfide, COS:

$$SO_2 + 3\,CO \rightarrow 2\,CO_2 + COS \tag{12.14.5}$$

In general, the catalytic processes do not appear to be promising for coal-fired plants at the present time.

Most sorption processes have been aimed at the simultaneous removal of both nitrogen oxides and sulfur oxides. Sulfuric acid solutions or alkaline scrubbing solutions containing $Ca(OH)_2$ or $Mg(OH)_2$ may be used. The species N_2O_3 produced by the reaction

$$NO_2 + NO \rightarrow N_2O_3 \tag{12.14.6}$$

is most efficiently absorbed. Therefore, since NO is the primary combustion product, the introduction of NO_2 into the flue gas is required to produce the N_2O_3, which is absorbed efficiently. A proposed process involving recycling of NO_2 is the **Tyco modified sulfuric-acid scrubbing process**. This process involves the following steps:

1. *Oxidizer:* The flue gas and NO_2 are introduced into an oxidizer in which the following reaction occurs:

$$NO_2 + SO_2 + H_2O \rightarrow H_2SO_4 + NO$$

2. *Scrubber:* The NO and more NO_2 react to form N_2O_3,

$$NO_2 + NO \rightarrow N_2O_3$$

which contacts sulfuric acid introduced into a scrubber. The nitrogen oxides are removed by the reaction

$$N_2O_3 + 2\,H_2SO_4 \rightarrow 2\,NOHSO_4 + H_2O$$

and the scrubbed flue gas is exhausted to the atmosphere.

3. *Decomposer:* Sulfuric acid is produced in the decomposer and NO_2 is regenerated:

$$2\,NOHSO_4 + \frac{1}{2}\,O_2 + H_2O \rightarrow 2\,H_2SO_4 + 2\,NO_2$$

Some of the sulfuric acid is recycled to the scrubber.

4. HNO_3 *reactor*: Nitric acid is produced in the HNO_3 reactor,

$$3\,NO_2 + H_2O \rightarrow 2\,HNO_3 + NO$$

and excess NO_2 and NO are recirculated through the oxidizer.

The automobile presents the greatest challenge in the area of NO_x removal. Unfortunately, the internal combustion engine produces nitric oxides by over-combustion and undesirable CO and hydrocarbons by under-combustion. Lean fuel mixtures and high temperatures reduce hydrocarbon and CO emissions but increase nitric oxide emissions. Spark retard is one of the few measures that decreases both hydrocarbon emissions and nitrogen oxide emissions. Typically, for the most efficient fuel combustion, advance spark timing of approximately 40 degrees from *top dead center* is used (where top dead center is the point at which the piston begins the power stroke). Retarding the spark to exact top dead center decreases NO_x emissions from approximately 5000 ppm to approximately 1000 ppm. (See Figure 14.2 for a schematic drawing of the combustion process in an automobile engine.)

Since most NO_x in any combustion process results from high peak combustion temperatures, measures that reduce these temperatures generally reduce NO_x emission. One such measure is recirculation of exhaust gas with the fuel-air mixture. The diluting effect of the recirculated exhaust gas decreases the peak combustion temperature. For example, recirculation of approximately 12% of the exhaust gas can reduce NO_x emissions by approximately 60%. Unfortunately, hydrocarbon emission is increased by this measure.

Catalytic exhaust reactors can be used to eliminate nitric oxide from automobile exhausts. Since the removal of nitric oxide requires an excess of reductant in the exhaust gas, and the removal of CO and hydrocarbons requires excess of oxidant, the simultaneous catalytic removal of these pollutants requires a dual catalytic exhaust system. If the fuel mixture supplied to the engine is relatively rich, nitric oxide can be reduced by excess CO and hydrocarbons in the first catalyst bed. After the injection of additional air, the remaining oxidizable impurities are oxidized in the presence of another catalyst.

12.15 ACID RAIN

As discussed in this chapter, much of the sulfur and nitrogen oxides entering the atmosphere are converted to sulfuric and nitric acids, respectively. When combined with hydrochloric acid arising from hydrogen chloride emissions, these acids cause acidic precipitation **(acid rain)** that is now a major pollution problem in some areas.

Headwater streams and high-altitude lakes are especially susceptible to the effects of acid rain and may sustain loss of fish and other aquatic life. Other effects include reductions in forest and crop productivity; leaching of nutrient cations and heavy metals from soils, rocks, and the sediments of lakes and streams; dissolution of metals such as lead and copper from water distribution pipes; corrosion of exposed metal; and dissolution of the surfaces of limestone buildings and monuments.

Soil sensitivity to acid precipitation can be estimated from *cation exchange capacity* (CEC, see Section 6.12) [19]. Soil is generally insensitive if free carbonates are present or if it is flooded frequently. Soils with a cation exchange capacity above 15.4 are also insensitive. Soils with cation exchange capacities between 6.2 and 15.4 are slightly sensitive. Soils with cation exchange capacities below 6.2 normally are sensitive if free carbonates are absent and the soil is not frequently flooded.

As calculated in Section 2.11, pure water in equilibrium with ambient air has a pH of about 5.7, which is slightly acidic, due to the presence of dissolved carbon dioxide. Although measurements of rainwater acidity prior to 1950 are scarce, those that were made, plus measurement of the pH of "fossil precipitation," which fell many years ago and is preserved in the form of glacial ice, in-

---

[19] Glass, N.R., et al. 1982. Effects of acid precipitation. *Environmental Science and Technology* 16: 162A–169A.

dicate that acidic precipitation was not a problem prior to 1930. By 1950, the problem did exist in certain areas, and data from recent years have indicated a worsening trend. The pH values of precipitation in much of the northeastern U.S. now run around 4.0, and values as low as 2.0 have been observed.

The longest-term experimental study of acid precipitation in the U.S. has been conducted at the U.S. Forest Service Hubbard Brook Experimental Forest in New Hampshire's White Mountains. It is downwind from major U.S. urban and industrial centers and is, therefore, a prime candidate to receive acid precipitation. This is reflected by mean annual pH values ranging from 4.0 to 4.2 during the 1964–74 period. During this period, the annual hydrogen ion input ($[H^+] \times$ volume) increased by 36%.

Acid precipitation shows a strong geographic dependence, as illustrated in Figure 12.4, representing the pH of precipitation in the continental U.S. The preponderance of acidic rainfall in the northeastern U.S. is obvious.

Table 12.5 (p. 346) shows typical major cations and anions in pH-4.25 precipitation. Although actual values encountered vary greatly with time and location of collection, this table does show some major features of ionic solutes in precipitation. From the predominance of sulfate anion, it is apparent that sulfuric acid is the major contributor to acid precipitation. Nitric acid makes up a smaller but growing contribution to the acid present. Hydrochloric acid ranks third.

Analyses of the movements of air masses have shown a correlation between acid precipitation and prior movement of an air mass over major sources of anthropogenic sulfur and nitrogen oxides emissions. This is particularly obvious in southern Scandinavia, which receives heavy-burden air pollution from densely populated, heavily industrialized areas of Europe.

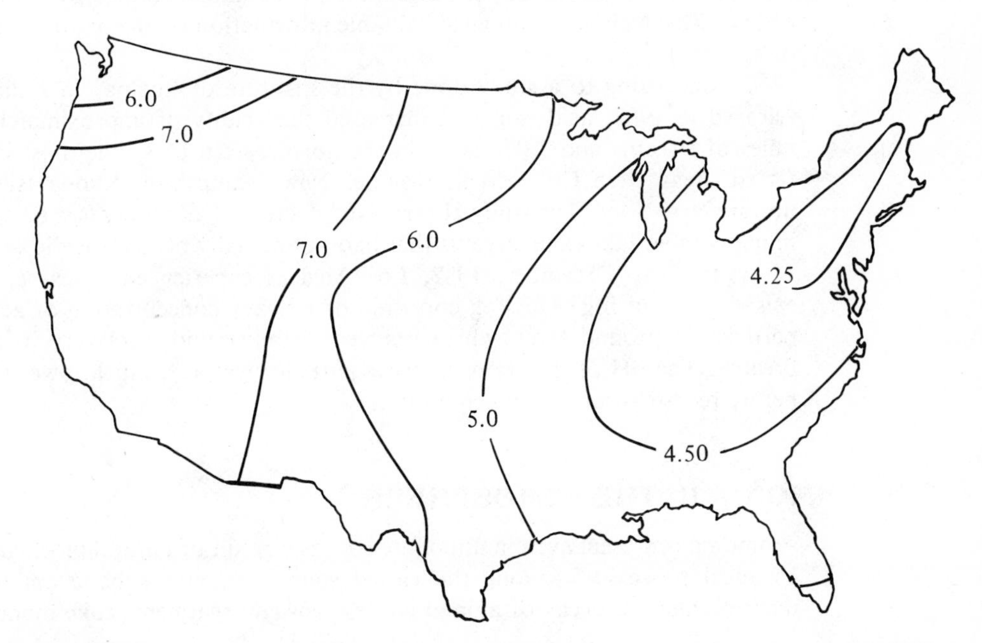

FIGURE 12.4 Typical precipitation-pH pattern in the continental United States. Actual values found vary with the time of year and climatic conditions.

TABLE 12.5 Typical Values of Cation and Anion Concentrations in Acidic Precipitation

Cations		*Anions*	
Ion	*Concentration equivalents/L × 10^6*	*Ion*	*Concentration equivalents/L × 10^6*
H^+	56	SO_4^{2-}	51
NH_4^+	10	NO_3^-	20
Ca^{2+}	7	Cl^-	12
Na^+	5	**Total**	83
Mg^{2+}	3		
K^+	2		
Total	83		

An important factor in the study of acid rain and sulfur pollution involves the comparison of primary sulfate species (those emitted directly by point sources) and secondary sulfate species (those formed from gaseous sulfur compounds, primarily by the atmospheric oxidation of SO_2 [20]). A *low* primary-sulfate content indicates transport of the pollutant from some distance, whereas a *high* primary-sulfate content indicates local emissions. This information can be useful in determining the effectiveness of SO_2 control in reducing atmospheric sulfate, including sulfuric acid. Primary and secondary sulfates can be measured using the oxygen-18 content of the sulfates [20]. This content is higher in sulfate emitted directly from a power plant than it is in sulfate formed by the oxidation of SO_2. This technique can yield valuable information on the origins and control of acid rain.

According to a study done by the Institute of Ecology in Indianapolis, released in 1982, acid rain had increased the acidity of approximately 25,000 miles of streams and 3,000 lakes in the northeastern U.S. The most highly affected areas were Connecticut, Maine, New Hampshire, Rhode Island, and upstate New York. The study also showed decreased alkalinity (see Section 2.12) in many more lakes and streams that had not actually become acidic.

In early December, 1982, Los Angeles experienced a severe, two-day episode of **acid fog**. This fog consisted of a heavy concentration of acidic mist particles at ground level, which reduced visibility and were very irritating to breathe. The pH of the water in these particles was 1.7, much lower than ever before recorded for acid precipitation.

12.16 AMMONIA IN THE ATMOSPHERE

Ammonia is present even in unpolluted air as a result of natural biochemical and chemical processes. Among the varied sources of atmospheric ammonia are microorganisms, decay of animal wastes, sewage treatment, coke manufacture,

---

[20] Holt, B.D.; Kumar, R.; and Cunningham, P.T. 1982. Primary sulfates in atmospheric sulfates: estimation by oxygen isotope ratio measurements. *Science* 217: 51–53.

ammonia manufacture, and leakage from ammonia-based refrigeration systems. High concentrations of ammonia gas in the atmosphere are generally indicative of accidental release of the gas.

The main mechanism for the removal of ammonia from the atmosphere consists of acid-base reactions. Ammonia in polluted atmospheres is a key species in the formation and neutralization of nitrate and sulfate aerosols [21]. Ammonia reacts with these acidic aerosols to form ammonium salts:

$$NH_3 + HNO_3 \rightarrow NH_4NO_3 \quad (12.16.1)$$

$$NH_3 + H_2SO_4 \rightarrow NH_4HSO_4 \quad (12.16.2)$$

Ammonium salts are among the more corrosive salts in atmospheric aerosols.

12.17 CHLORINE, FLUORINE, AND THEIR GASEOUS COMPOUNDS

Chlorine gas, Cl_2, does not occur as an air pollutant on a large scale but can be quite damaging on a local scale. Chlorine was the first poisonous gas used in World War I. It is widely used as a manufacturing chemical, in the plastics industry, for example, as well as for water treatment and as a bleach. Therefore, possibilities for its release exist in a number of locations. Chlorine is quite toxic and is a mucous-membrane irritant. It is very reactive chemically and a powerful oxidizing agent. Chlorine dissolves in atmospheric water droplets, yielding hydrochloric acid and hypochlorous acid, an oxidizing agent:

$$H_2O + Cl_2 \rightarrow H^+ + Cl^- + HOCl \quad (12.17.1)$$

Spills of chlorine gas have caused fatalities among exposed persons. The rupture of a derailed chlorine tank car at Youngstown, Florida, on February 25, 1978, resulted in the deaths of 8 people who inhaled the deadly gas, and a total of 89 people were injured. Many of the victims were exposed as they drove along a highway paralleling the tracks. Evacuation of almost 1000 residents was required.

Hydrogen chloride, HCl, is emitted from a number of sources. Incineration of chlorinated plastics, such as polyvinylchloride, releases HCl as a combustion product.

```
 H  Cl H  Cl H  Cl H  Cl H  Cl
 |  |  |  |  |  |  |  |  |  |
···C—C—C—C—C—C—C—C—C—C···        polyvinylchloride
 |  |  |  |  |  |  |  |  |  |
 H  H  H  H  H  H  H  H  H  H
```

Recently, it has been recognized that much of the chlorine in coal is organically bound and that this chlorine is released as HCl during combustion. Increased use of coal may lead to more HCl contamination of the atmosphere. Some compounds released to the atmosphere as air pollutants hydrolyze to form HCl. One such incident occurred on April 26, 1974, when a storage tank containing 750,000 gallons of liquid silicon tetrachloride, $SiCl_4$, began to leak in South

---

[21] Tuazon, E.C.; Winer, A.M.; and Pitts, J.N., 1981. Trace pollutant concentrations in a multiday smog episode in the California south coast air basin by long path length Fourier transform infrared spectroscopy. *Environmental Science and Technology* 15: 1232–1237.

Chicago, Illinois. This compound reacted with water in the atmosphere to form a choking fog of hydrochloric acid droplets:

$$SiCl_4 + 2\ H_2O \rightarrow SiO_2 + 4\ HCl \quad (12.17.2)$$

Many people became ill from inhaling the vapor.

In February, 1982, in Stroudsburg, Pennsylvania, a wrecked truck dumped 12 tons of powdered aluminum chloride during a rainstorm. This compound produces HCl gas when wet, and more than 1,200 residents had to be evacuated from their homes because of the fumes generated.

Fluorine, hydrogen fluoride, and other volatile fluorides are produced in the manufacture of aluminum, and hydrogen fluoride is a by-product in the conversion of fluorapatite (rock phosphate) to phosphoric acid, superphosphate fertilizers, and other phosphorus products. The wet process for the production of phosphoric acid involves the reaction of fluorapatite, $Ca_5F(PO_4)_3$, with sulfuric acid:

$$Ca_5F(PO_4)_3(s) + 5\ H_2SO_4 + 2\ H_2O \rightarrow 5\ CaSO_4 \cdot 2\ H_2O(s) + HF + 3\ H_3PO_4 \quad (12.17.3)$$

It is necessary to recover most of the by-product fluorine from rock phosphate processing to avoid severe pollution problems. Recovery as fluosilicic acid, H_2SiF_6, has become quite attractive from an economic standpoint. It is estimated that complete recovery of fluosilicic acid from phosphate plants would almost satisfy domestic needs for fluorine compounds.

Hydrogen fluoride gas is irritating to body tissues, and the respiratory tract is very sensitive to it. Brief exposure to HF vapors at the part-per-thousand level may be fatal. The acute toxicity of F_2 is higher than that of HF. Chronic exposure to high levels of fluorides causes *fluorosis*, the symptoms of which include mottled teeth and pathological bone conditions.

Plants are particularly susceptible to the effects of gaseous fluorides. Fluorides from the atmosphere appear to enter the leaf tissue through the stomata. Fluoride is a cumulative poison in plants, and exposure of sensitive plants to even very low levels of fluorides for prolonged periods results in damage. Characteristic symptoms of fluoride poisoning are *chlorosis* (fading of green color due to conditions other than the absence of light), edge burn, and tip burn. Conifers (such as pine trees) afflicted with fluoride poisoning may have reddish-brown necrotic needle tips. The sensitivity of some conifers to fluoride poisoning is illustrated by the fact that fluorine produced by aluminum plants in Norway has destroyed forests of *Pinus sylvestris* up to 8 miles distant. Damage to trees was observed at distances as great as 20 miles from the plant.

Silicon tetrafluoride gas, SiF_4, is another gaseous fluoride pollutant produced during some steel and metal smelting operations that employ CaF_2, fluorspar. Fluorspar reacts with silicon dioxide (sand), releasing SiF_4 gas:

$$2\ CaF_2 + 3\ SiO_2 \rightarrow 2\ CaSiO_3 + SiF_4(g) \quad (12.17.4)$$

Sulfur hexafluoride gas, SF_6, occurs in the atmosphere at levels of about 0.3 parts per trillion. It is extremely unreactive and is used as an atmospheric tracer. It does not absorb ultraviolet light in either the troposphere or stratosphere and is probably destroyed above 50 km by capture of free electrons.

The fluorine-containing air pollutants with the greatest potential for damage to the atmosphere are the **Freons**, or **chlorofluorocarbons (CFC)**, used as fluids in refrigeration mechanisms and, until several years ago, as propellants in spray cans containing deodorants, hair spray, and many other products. The most widely used CFC compounds are dichlorodifluoromethane (CFC-12), Cl_2CF_2 which boils at −28 °C, and trichlorofluoromethane (CFC-11), Cl_3CF, which boils at +24 °C. Other CFC compounds commonly manufactured include CCl_2FCClF_2 (CFC-113) and $CClF_2CClF_2$ (CFC-114). CFC-11 and CFC-12 are essentially inert chemically at lower altitudes. This nonreactivity, combined with worldwide production of approximately one-half million metric tons of CFC compounds per year and deliberate or accidental release to the atmosphere, has resulted in CFC's becoming homogeneous components of the global atmosphere. In 1974 it was convincingly suggested [22] that chlorofluoromethanes could catalyze the destruction of stratospheric ozone, which filters out cancer-causing ultraviolet radiation from the sun. Although quite inert in the lower atmosphere, CFC's undergo photodecomposition by the action of high-energy ultraviolet radiation in the stratosphere through reactions such as

$$Cl_2CF_2 + h\nu \rightarrow Cl\cdot + Cl\dot{C}F_2 \qquad (12.17.5)$$

which release Cl atoms. These atoms react with ozone, destroying it and producing ClO:

$$Cl + O_3 \rightarrow ClO + O_2 \qquad (12.17.6)$$

In this region of the atmosphere, there is an appreciable concentration of atomic oxygen, by virtue of the reaction

$$O_3 + h\nu \rightarrow O_2 + O \qquad (12.17.7)$$

Nitric oxide, NO, is also present. The ClO species may react with either O or NO, regenerating Cl atoms and resulting in chain reactions that cause the net destruction of ozone:

$$ClO + O \rightarrow Cl + O_2 \qquad (12.17.8)$$

$$Cl + O_3 \rightarrow ClO + O_2 \qquad (12.17.6)$$

$$O_3 + O \rightarrow 2\,O_2 \qquad (12.17.9)$$

$$ClO + NO \rightarrow Cl + NO_2 \qquad (12.17.10)$$

$$Cl + O_3 \rightarrow ClO + O_2 \qquad (12.17.6)$$

$$O_3 + NO \rightarrow NO_2 + O_2 \qquad (12.17.11)$$

The potential role of NO shown suggests a synergistic effect between this pollutant, released by supersonic aircraft flying at high altitudes (see Section 12.13), and CFC's.

If the chain reactions just described are in fact significant, it should be possible to detect both ClO and Cl in the 25–45-km region, where local photochemistry controls the concentration of O_3. In fact, both these species have been detected in the 25–45-km region by an instrument measuring resonantly scattered ultraviolet radiation at 118.9 nm from Cl atoms excited by radiation from a plasma discharge lamp [23]. The Cl atoms were measured directly and molecules were measured indirectly by introducing NO and observing Cl atoms

[22] Molina, M.J., and Rowland, F.S. 1974. *Nature* (London) 249: 810.

[23] Anderson, J.G.; Margitan, J.J.; and Stedman, D.H. 1977. Atomic chlorine and chlorine monoxide in the stratosphere: three *in situ* observations. *Science* 198: 501–3.

produced by Reaction 12.17.10. Observations during July and October yielded Cl and ClO concentrations equaling or exceeding those predicted; December data gave lower-than-predicted values. The Cl:ClO ratio was found to be equal to that predicted.

The role of atmospheric CFC's in ozone depletion remains the subject of considerable controversy and uncertainty. Even if production were stopped today, these compounds would persist in the atmosphere for many decades. If CFC's are in fact responsible for atmospheric ozone depletion, it will be many years before the damage can be reversed. In consideration of these uncertainties, CFC propellants are now banned in the U.S. CFC refrigerants are not yet restricted.

On an encouraging note, the current estimate of ozone attrition caused by CFC compounds is about 50% lower than previous estimates [24]. Whereas in 1979 the National Academy of Sciences estimated a reduction of 18.6% in the ozone layer if CFC releases were continued on the 1977 level, the current estimate is 5–9%. Furthermore, increased atmospheric carbon dioxide levels (see Section 11.12) might further decrease the attrition of stratospheric ozone by cooling the stratosphere and slowing down ozone-consuming reactions. The new estimates are based upon more accurately known reaction rates of hydroxyl radical, HO•, particularly with chlorine oxide, ClO. Whereas in 1979 the atmospheric ClO levels below 35 km were one-third of the calculated values, the new estimates of HO• reaction rates yield calculated ClO levels which agree with the observed levels.

Estimates of the incidence of skin cancer resulting from increased ultraviolet-radiation exposure due to an ozone depletion have been revised upward [25]. For each 1% depletion of the ozone layer there is a 2% increase in the amount of ultraviolet (UV) radiation reaching the Earth's surface. With the exception of malignant melanoma, it is likely that about 90% of skin cancers arise from exposure to sunlight, especially in the damaging UV-B region of 290–320 nm. It is estimated that each 1% decrease in ozone will cause a 2–5% increase in highly treatable basal-cell skin cancer (currently 300,000 to 400,000 new cases in the U.S. annually) and a 4–10% increase in more serious squamous-cell skin cancer (currently 100,000 cases per year). Of course, these values are largely speculative.

12.18 HYDROGEN SULFIDE, CARBONYL SULFIDE, AND CARBON DISULFIDE

Hydrogen sulfide is produced by microbial decay of sulfur compounds and microbial reduction of sulfate (see Sections 5.21 and 5.22), from geothermal steam, from wood pulping, and from a number of miscellaneous natural and anthropogenic sources. Most atmospheric hydrogen sulfide is rapidly converted to SO_2 and to sulfates. The organic homologs of hydrogen sulfide, the mercaptans, enter the atmosphere from decaying organic matter and have particularly objectionable odors.

---

[24] 1982. *Causes and effects of stratospheric ozone reduction: an update.* Washington, D.C.: National Academy Press.

[25] Maugh, T.H., II. 1982. New link between ozone and cancer. *Science* 216: 396–7.

Hydrogen sulfide pollution from artificial sources is not as much of a problem as sulfur dioxide pollution. However, there have been several acute incidents of hydrogen sulfide emissions resulting in damage to human health and even fatalities. The most notorious such incident occurred in Poza Rica, Mexico, in 1950. Accidental release of hydrogen sulfide from a plant used for the recovery of sulfur from natural gas caused the deaths of 22 people and the hospitalization of over 300. The symptoms of poisoning included irritation of the respiratory tract and damage to the central nervous system. Unlike sulfur dioxide, which appears to affect older people and those with respiratory weaknesses, there was little evidence of correlation between the observed hydrogen sulfide poisoning and the age or physical condition of the victim.

In a tragic incident which occurred in February 1975, hydrogen sulfide gas leaking from an experimental secondary-recovery oil well near Denver City, Texas, killed nine people trying to flee the lethal fumes. A process was being tried in which carbon dioxide, rather than water, was injected under high pressure to recover petroleum. Leakage from the well released deadly hydrogen sulfide present in the oil-bearing formation. Recent efforts to tap very deep natural gas formations have increased the hazard from hydrogen sulfide. A pocket of H_2S was struck at 15,000 feet while drilling such a well near Athens, Texas, in 1978. Leakage of hydrogen sulfide on May 12, 1978, forced the evacuation of 50 families. As an emergency measure in such cases, the gas may be ignited to form less toxic SO_2.

Hydrogen sulfide at levels well above ambient concentrations destroys immature plant tissue. This type of plant injury is readily distinguished from that due to other phytotoxins. A study of the effects of long-term plant exposure to H_2S [26] has revealed some interesting differences in effects at different exposure levels. The plants studied were lettuce, alfalfa, Thompson seedless grapes, sugar beets, California buckeye Douglas fir, and ponderosa pine. More sensitive species were killed by continuous exposure to 3000 ppb H_2S, whereas other species exhibited reduced growth, leaf lesions, and defoliation. Reduction of the exposure by a factor of 10 resulted in the same effects to a lesser degree. However, exposure to 30 ppb of hydrogen sulfide stimulated the growth of alfalfa, lettuce, and sugar beets. This effect was dependent upon the stage of growth, humidity, and temperature. It was noted that leaves accumulated sulfur at levels proportional to exposure.

Damage to certain kinds of materials is a very expensive effect of hydrogen sulfide pollution. Paints containing lead pigments, $2\ PbCO_3 \cdot Pb(OH)_2$ (no longer widely used), are particularly susceptible to darkening by H_2S. Darkening results from exposure over several hours to as little as 50 ppb H_2S. The lead sulfide originally produced by reaction of the lead pigment with hydrogen sulfide eventually may be converted to white lead sulfate by atmospheric oxygen after removal of the source of H_2S, thus partially reversing the damage.

A black layer of copper sulfide forms on copper metal exposed to H_2S. Eventually, this layer is replaced by a green coating of basic copper sulfate such as $CuSO_4 \cdot 3\ Cu(OH)_2$. The green "patina," as it is called, is very resistant to further corrosion. Such layers of corrosion can seriously impair the function of

---

[26] Thompson, C.R., and Kats, G. 1978. Effects of H_2S fumigation on crop and forest plants. *Environmental Science and Technology* 5: 550–3.

copper contacts on electrical equipment. Hydrogen sulfide also forms a black sulfide coating on silver.

Carbonyl sulfide, COS, is now recognized as a component of the atmosphere at a tropospheric concentration of approximately 500 parts per trillion by volume, with an estimated global burden of about 2.4 teragrams as S [27]. It is, therefore, a significant sulfur species in the atmosphere. It is possible that the HO• radical-initiated oxidation of COS and carbon disulfide (CS_2) would yield 8–12 teragrams as S in atmospheric sulfur dioxide per year. Though this is a small yield compared to pollution sources, this process could account for much of the SO_2 burden in the remote troposphere.

Both COS and CS_2 are oxidized in the atmosphere by reactions initiated by the hydroxyl radical. The initial reactions are

$$HO\cdot + COS \rightarrow CO_2 + HS\cdot \quad (12.18.1)$$

$$HO\cdot + CS_2 \rightarrow COS + HS\cdot \quad (12.18.2)$$

The sulfur-containing products undergo further reactions to sulfur dioxide and, eventually, to sulfate species.

SUPPLEMENTARY REFERENCES

Benarie, M.M., ed. 1982. *Atmospheric pollution 1982.* New York: Elsevier North-Holland, Inc.

Bragg, G.M., and Strauss, W., eds. 1981. *Air pollution control.* Part 4. New York: John Wiley and Sons, Inc.

1979. *Carbon monoxide.* Albany, N.Y.: WHO Publication Center USA.

Chameides, W.L., and Davis, D.D. 1982. The free radical chemistry of cloud droplets and its impact upon the composition of acid rain. *Journal of Geophysical Research* 87:4863.

Clark, W.C.C. 1982. *Carbon dioxide review 1982.* New York: Oxford University Press, Inc.

Delwiche, C.C., ed. 1981. *Denitrification, nitrification, and atmospheric nitrous oxide.* New York: Wiley-Interscience.

D'Itri, F.M., ed. 1982. *Acid precipitation.* Woburn, Mass.: Ann Arbor Science Publishers, Inc./Butterworth.

Drablos, D., and Tollan, A., eds. 1980. *Ecological impact of acid precipitation.* Oslo, Norway: SNSF Project.

Glassman, I. 1977. *Combustion.* New York: Academic Press, Inc.

Guderian, R. 1977. *Air pollution—phytotoxicity of acidic gases and its significance in air pollution control.* New York: Springer-Verlag New York, Inc.

Hutchinson, T.C., and Havas, M., eds. 1980. *Effects of acid precipitation on terrestrial ecosystems.* New York: Plenum Publishing Corp.

Idso, S.B. 1982. *Carbon dioxide: friend or foe?* Tempe, Ariz.: IBR Press.

Johnston, H.S. 1975. Pollution of the stratosphere. In *Advances in physical chemistry,* pp. 315–338. New York: Academic Press, Inc.

---

[27] Sze, N.D., and Ko, M.K.W. 1980. Photochemistry of COS, CS_2, CH_3SCH_3, and H_2S: implications for the atmospheric sulfur cycle. *Atmospheric Environment* 14: 1223–1235.

Kramer, J., and Tessier, A. 1982. Acidification of aquatic systems. *Environmental Science and Technology* 16: 606A–615A.

Lee, S.D., ed. 1980. *Nitrogen oxides and their effects on health.* Ann Arbor, Mich.: Ann Arbor Science Publishers, Inc.

Macdonald, G.H., ed. 1982. *The long-term impacts of increasing atmospheric carbon dioxide levels.* Cambridge, Mass.: Ballinger Publishing Company.

Masschelein, W.J., and Rice, R.G. 1979. *Chlorine dioxide: chemistry and environmental impact of oxychlorine compounds.* Ann Arbor, Mich.: Ann Arbor Science Publishers, Inc.

Miller, S.S. 1979. *Control technologies for air pollution.* Washington, D.C.: American Chemical Society.

National Academy of Sciences. 1977. *Carbon monoxide.* Washington, D.C.: National Academy Press.

1977. *Nitrogen oxides.* Washington, D.C.: National Academy Press.

1982. *Causes and effects of stratospheric ozone reduction.* Washington, D.C.: National Academy Press.

Nriagu, J.O. 1978. *Sulfur in the environment.* New York: Wiley-Interscience.

Oliver, R.C. 1977. *Aircraft emissions—potential effects on ozone and climate.* Springfield, Va.: National Technical Information Service

Record, F.A. 1982. *Acid rain information book.* Park Ridge, N.J.: Noyes Data Corp.

Schryer, D.R., ed. 1982. *Heterogeneous atmospheric chemistry.* Washington, D.C.: AGU Publications.

Shriner, D.S.; Richmond, C.R.; and Lindberg, S.E., eds. 1980. *Atmospheric sulfur deposition: environmental impact and health effects.* Ann Arbor, Mich.: Ann Arbor Science Publishers, Inc.

Sittig, M. 1975. *Environmental sources and emissions handbook.* Park Ridge, N.J.: Noyes Data Corp.

Slack, A.V., and Hollinden, G.A. 1975. *Sulfur dioxide removal from waste gases.* 2nd ed. Park Ridge, N.J.: Noyes Data Corp.

Staff of E.I. du Pont de Nemours and Company. 1982. *World production and release of chlorofluorocarbons 11 and 12 through 1981.* Washington, D.C.: Chemical Manufacturers Association.

Stern, A.C., ed. 1976. *Air pollution, Vol. I, air pollutants, their transformation and transport.* New York: Academic Press, Inc.

Sugden, T.M., and West, T.F., eds. 1980. *Chlorofluorocarbons in the environment.* Chichester, England: Horwood.

Svensson, B.H., and Soderland, R., eds. 1975. Nitrogen, phosphorus, and sulfur—global cycles. SCOPE Report No. 7, *Ecological Bulletin* 22. Orundsboro, Sweden.

Theodore, L., and Buonicore, A.J. 1982. *Air pollution control equipment: selection, design, operation, and maintenance.* Englewood Cliffs, N.J.: Prentice-Hall, Inc.

White, L.J. 1982. *The regulation of air pollutant emissions from motor vehicles.* Washington, D.C.: American Enterprise Institute for Public Policy Research.

QUESTIONS AND PROBLEMS

1. What does it mean if a pollutant reservoir is in secular equilibrium, or steady state?
2. Which unstable, reactive species is responsible for removal of CO from the atmosphere?

3. Which of the following fluxes in the atmospheric sulfur cycle is smallest?
   (a) sulfur species washed out in rainfall over land;
   (b) sulfates entering the atmosphere as "sea salt";
   (c) sulfur species entering the atmosphere from volcanoes;
   (d) sulfur species entering the atmosphere from combustion of fossil fuels;
   (e) hydrogen sulfide entering the atmosphere from biological processes in coastal areas and on land.

4. Of the following agents, the one that would not favor conversion of sulfur dioxide to sulfate species in the atmosphere is:
   (a) ammonia;
   (b) water;
   (c) contaminant reducing agents;
   (d) ions of transition metals such as manganese;
   (e) sunlight.

5. Of the stack-gas scrubber processes discussed in this chapter, which is the least efficient for SO_2 removal?

6. The air inside a garage was found to contain 10 ppm CO by volume at standard temperature and pressure (STP). What is the concentration of CO in mg/L and in ppm by weight?

7. Assume that an incorrectly adjusted lawn mower is operated in a garage such that the combustion reaction in the engine is

$$C_8H_{18} + \frac{17}{2} O_2 \rightarrow 8\ CO + 9\ H_2O$$

   If the dimensions of the garage are 5 × 3 × 3 meters, how many grams of gasoline must be burned to raise the level of CO in the air to 1000 ppm by volume at standard temperature and pressure (STP)?

8. A 12.0-liter sample of waste air from a smelter process was collected at 25 °C and 1.00 atm pressure, and sulfur dioxide was removed. After SO_2 removal, the volume of the air sample was 11.50 liters. What was the percentage by weight of SO_2 in the original sample?

9. What is the oxidizing agent in the Claus Reaction?

10. Carbon monoxide is present at a level of 10 ppm by volume in an air sample taken at 15 °C and 1.00 atm pressure. At what temperature (at 1.00 atm pressure) would the sample also contain 10 mg/m^3 of CO?

11. How many metric tons of 5%-S coal would be needed to yield the H_2SO_4 required to produce a 3.00-cm rainfall of pH 2.00 over a 100 km^2 area?

12. In what major respect is NO_2 a more significant species than SO_2 in terms of participation in atmospheric chemical reactions?

13. How many metric tons of coal containing an average of 2% S are required to produce the SO_2 emitted by fossil fuel combustion shown in Figure 12.1? (Note that the values given in the figure are in terms of elemental sulfur, S.) How many metric tons of SO_2 are emitted?

14. Assume that the wet limestone process requires 1 metric ton of $CaCO_3$ to remove 90% of the sulfur from 4 metric tons of coal containing 2% S. Assume that the sulfur product is $CaSO_4$. Calculate the percentage of the limestone converted to calcium sulfate.

15. Referring to Problems 13 and 14, calculate the number of metric tons of $CaCO_3$ required each year to remove 90% of the sulfur from 600 million metric tons of coal (approximate annual U.S. consumption), assuming an average of 2% sulfur in the coal.

16. If a power plant burning 10,000 metric tons of coal per day with 10% excess air emits stack gas containing 100 ppm by volume of NO, what is the daily output of NO?

17. How many cubic kilometers of air at 25 °C and 1 atm pressure would be contaminated to a level of 0.5 ppm NO_x from the power plant discussed in Question 16?

CHAPTER 13
PARTICULATE MATTER IN THE ATMOSPHERE

13.1 PARTICLES IN THE ATMOSPHERE

Particles in the atmosphere, which range in size from about one-half millimeter (the size of sand or drizzle) down to molecular dimensions, are made up of an amazing variety of materials and discrete objects. They may consist of either solids or liquid droplets. Atmospheric particles are frequently referred to as *particulates,* although *particulate matter* is more correct. Atmospheric **aerosols** are solid or liquid particles smaller than 100 μm in diameter. Particulate matter makes up the most visible and obvious form of air pollution. Pollutant particulate matter in the 0.001 to 10 μm range is commonly suspended in the air near sources of pollution such as urban atmospheres, industrial plants, highways, and power plants.

Very small, solid particles include carbon black, silver iodide, combustion nuclei, and sea-salt nuclei. Larger particles include cement dust, wind-blown soil dust, foundry dust, and pulverized coal. Liquid particulate matter (see Fig. 13.1), generally categorized as **mist**, includes raindrops, fog, and sulfuric-acid mist. Some particulate matter is biological material, such as very small viruses, bacteria, bacterial spores, fungal spores, and pollen. Particulate matter may be organic or inorganic; both types are very important atmospheric contaminants.

Particulate matter originates from a wide variety of processes, ranging from simple grinding of bulk matter to complicated chemical or biochemical syn-

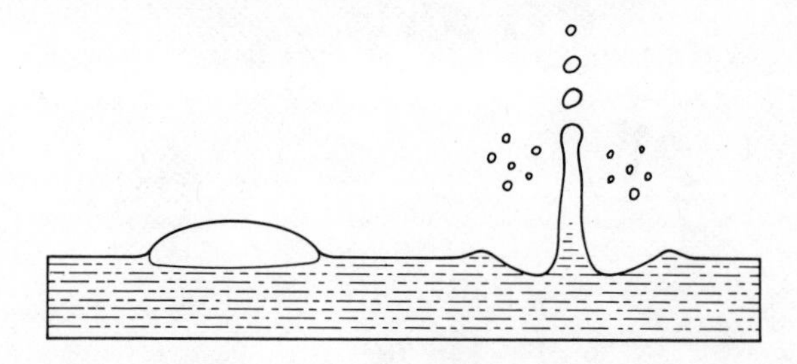

FIGURE 13.1 The bursting of bubbles in seawater forms small liquid aerosol particles. Evaporation of water from the aerosol particles results in the formation of particles (sea-salt nuclei).

theses. The effects of particulate matter are also widely varied. Possible effects on climate were discussed in Chapter 11. Either by itself, or in combination with gaseous pollutants, particulate matter may be detrimental to human health. Atmospheric particles may damage materials, reduce visibility, and cause undesirable esthetic effects.

The composition of atmospheric aerosol particles may be divided into five classes [1]. These are carbonaceous material, metal oxides and glasses, water, dissolved ionic species (electrolytes), and ionic solids. In general, the total aerosol mass can be accounted for by measuring Cl, Br, SO_4, NO_3, Na, K, Ca, Mg, NH_4, Fe, Al, Si, Pb, aerosol water, and carbonaceous material. Of these, carbonaceous material, aerosol water, SO_4, NO_3, NH_4, and Si predominate. Aerosol ionic strengths are very sensitive to relative humidity, which should be measured when an aerosol is sampled.

There is a marked variation of aerosol composition with size. The very small particles generally originate from gases, such as from the conversion of SO_2 to H_2SO_4, and tend to be acidic. Larger particles tend to consist of materials generated mechanically from bulk substances and have a greater tendency to be basic.

As shown in Figure 13.2, atmospheric particles undergo a number of processes in the atmosphere. Small colloidal particles are subject to *diffusion processes.* Smaller particles *coagulate* together to form larger particles. *Sedimentation* is a major processes for the removal of particles from the atmosphere, as is *scavenging* by raindrops and other forms of precipitation. Particles also react chemically with atmospheric gases.

Several terms are used to describe atmospheric particles. Some of the most important are summarized in Table 13.1 (next page).

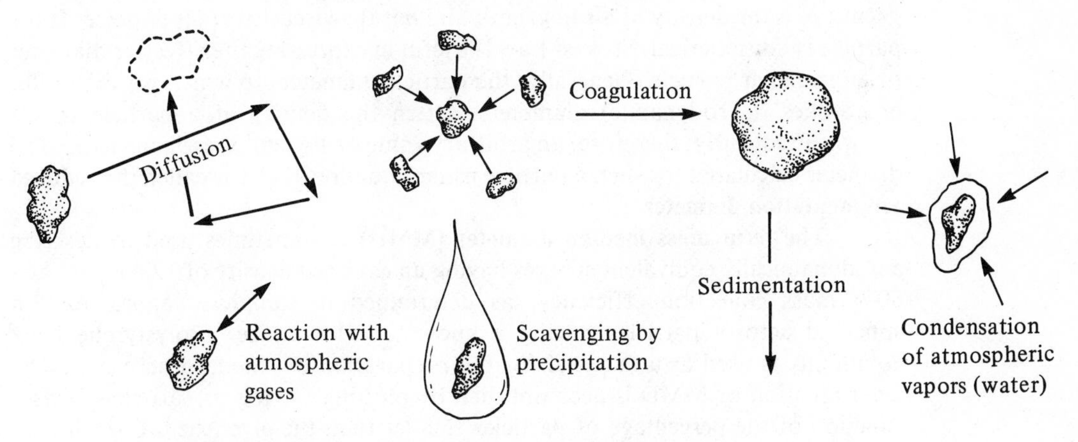

FIGURE 13.2 Processes affecting atmospheric particles.

[1] Stelson, A.W., and Seinfeld, J.H. 1981. Chemical mass accounting of urban aerosol. *Environmental Science and Technology* 15: 671–679.

TABLE 13.1 Important Terms Describing Atmospheric Particles

Term	*Meaning*
Aerosol	Colloidal-sized atmospheric particle
Condensation aerosol	Formed by condensation of vapors or reactions of gases.
Dispersion aerosol	Formed by grinding of solids, atomization of liquids, or dispersion of dusts
Fog	Term denoting high level of water droplets
Haze	Denotes decreased visibility due to presence of particles
Mists	Liquid particles
Smoke	Particles formed by incomplete combustion of fuel

13.2 SIZE AND SETTLING OF ATMOSPHERIC PARTICLES

The term **particle size** is the subject of considerable confusion. In some cases it is used to designate particle *radius*, and in other cases it designates particle *diameter*. The rate at which a particle settles is a function of particle diameter and density. The settling rate is important in determining the effect of the particle in the atmosphere. For spherical particles greater than approximately 1 μm in diameter, Stokes' law applies,

$$\nu = \frac{gd^2(\rho_1 - \rho_2)}{18n} \tag{13.2.1}$$

where ν is the settling velocity in cm/sec; g is the acceleration of gravity in cm/sec^2; d is the diameter of the particle in cm; ρ_1 is the density of the particle in g/cm^3; ρ_2 is the density of air in g/cm^3; and n is the viscosity of air in poise. If the particle is nonspherical, Stokes' Law is useful in expressing the effective diameter of an irregular particle. Generally, the particle diameters to which we shall refer are Stokes' (aerodynamic) diameters. Often the density of a particle is not known. Frequently, therefore, an arbitrary value of 1 g/cm^3 is assigned to ρ_1. The diameter calculated for such a particle using Equation 13.2.1 is called the **reduced sedimentation diameter.**

The term **mass median diameter (MMD)** is sometimes used to describe aerodynamically equivalent spheres having an assigned density of 1.00 g/cm^3 at a 50% mass collection efficiency, as determined in samplers calibrated with spherical aerosol particles having a known, uniform size. (Polystyrene latex commonly is used as a material for the preparation of standard aerosols.) The determination of MMD is accomplished by plotting the log of particle size as a function of the percentage of particles smaller than the given size. Logarithmic probability paper is used, and reasonably linear plots are frequently obtained. Two such plots are shown in Figure 13.3. You can see from the plot that particles of aerosol X have a mass median diameter of 2.0 μm (ordinate corresponding to 50% on the abscissa). In the case of aerosol Y, the sampler was not capable of accurately differentiating the sizes of particles below approximately 0.7 μm, and more than 50% of the particles were smaller than the smallest measurable size. However, by a linear extrapolation to 50%, the MMD is estimated to be 0.5 μm.

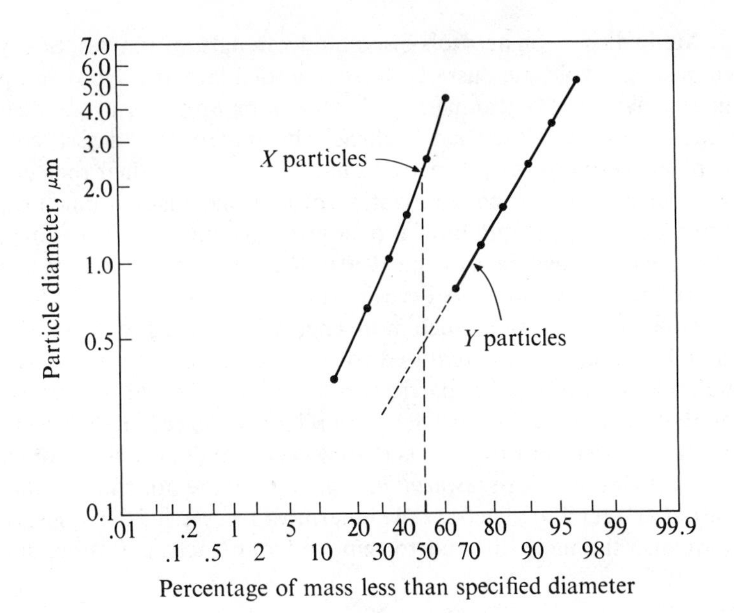

FIGURE 13.3 Particle-size distribution for particles of X (MMD = 2.0 μm) and Y (MMD = 0.5 μm by extrapolation).

Settling particles below approximately 1 μm in diameter slip between air molecules, and the resulting discontinuous resistance of air allows for settling at a rate faster than the rate estimated by Stokes' Law. Extremely small particles are susceptible to **Brownian motion** (random movement resulting from collisions with molecules), and Stokes' Law does not apply. Particles above 10 μm settle rapidly and generally stay in the atmosphere for only very short periods of time. The turbulence generated by very large particles makes Stokes' Law inapplicable to them.

Obviously, the optical properties of particles have an important effect upon air quality. Particles less than approximately 0.1 μm in diameter scatter light much like molecules, causing Rayleigh scattering. Generally, such small particles have an insignificant effect upon visibility in the atmosphere. The light-scattering and intercepting properties of particles larger than 1 μm are approximately proportional to the particle's cross-sectional area. Particles having diameters between 0.1 μm and 1 μm are of sizes having the same order of magnitude as the wavelengths of visible light. Therefore, interference phenomena are important, and the light-scattering properties of these particles are especially significant.

13.3 PHYSICAL PROCESSES FOR PARTICLE FORMATION

As a general rule, particles above approximately 1 μm in size are formed by the disintegration of larger particles. This is a *dispersion* process, and the product is called a **dispersion aerosol**. Dusts are solid dispersion aerosols.

Some of the many ways of forming dispersion aerosols include evolution of dust from coal grinding, formation of spray in cooling towers, and blowing of dirt from dry soil.

Many dispersion aerosols originate from natural sources. Sea spray, wind-blown dust, and volcanic dust of course preceded humans. However, people have invented a myriad of ingenious ways for breaking up material and dispersing it to the atmosphere. "Off-the-road" vehicles churn across the desert, coating fragile desert plants with layers of dispersed sand. Quarries and their rock crushers spew forth a plume of limestone. Cultivation of land has made it much easier for the wind to pick up topsoil and blow it over great distances. This was in part responsible for the dust bowl days of the 1930s, which threaten to return again in some areas (see the discussion of soil erosion in Chapter 10).

Note, however, that much more energy is required to break material down into small particles than is required for or released by the synthesis of particles through adhesion of smaller particles or through chemical processes. This is the reason that most dispersion particles are relatively large. These larger particles in general have fewer harmful effects than do very small particles. For example, the smaller particles are more *respirable*; that is, they are not filtered out in the nose and throat and tend to penetrate the innermost recesses of the lungs. Smaller particles are also the most difficult to remove from stacks and other sources.

13.4 CHEMICAL PROCESSES FOR PARTICLE FORMATION

This section considers only the formation of primarily inorganic particles through chemical processes. The very important formation of organic particulate matter is considered in Section 13.15.

Metal oxides constitute a major class of inorganic particles in the atmosphere. These are formed whenever fuels containing metals are burned. For example, particulate iron oxide is formed during the combustion of pyrite-containing coal:

$$3\,FeS_2 + 8\,O_2 \rightarrow Fe_3O_4 + 6\,SO_2 \qquad (13.4.1)$$

Organic vanadium in residual fuel oil is converted to particulate vanadium oxide. Part of the calcium carbonate in the ash fraction of coal is converted to calcium oxide and is emitted to the atmosphere through the stack:

$$CaCO_3 + \text{heat} \rightarrow CaO + CO_2 \qquad (13.4.2)$$

A special case of particulate matter formed during the combustion of fuel is that of the particulate lead halides produced in the burning of leaded gasoline. Tetraethyl lead in leaded gasoline reacts with oxygen and halogenated scavengers in the form of dichloroethane and dibromoethane, forming lead halides which are volatile enough to exit through the exhaust system but which condense to form particles:

$$Pb(C_2H_5)_4 + O_2 + \begin{pmatrix}\text{halogenated}\\ \text{scavengers}\end{pmatrix} \rightarrow CO_2 + H_2O + PbCl_2 + PbBrCl + PbBr_2 \qquad (13.4.3)$$

By 1975, some 200,000 tons of lead were entering the U.S. atmosphere annually by this pathway. These quantities are currently decreasing as the use of unleaded gasoline becomes more commonplace.

A common process for the formation of aerosol mists involves the oxidation of atmospheric sulfur dioxide to sulfuric acid, a hygroscopic substance which accumulates atmospheric water to form small liquid droplets:

$$2\ SO_2 + O_2 + 2\ H_2O \rightarrow 2\ H_2SO_4 \tag{13.4.4}$$

In the presence of basic air pollutants, such as ammonia or calcium oxide, the sulfuric acid reacts to form salts:

$$H_2SO_4(\text{droplet}) + 2\ NH_3(\text{gas}) \rightarrow (NH_4)_2SO_4(\text{droplet}) \tag{13.4.5}$$

$$H_2SO_4(\text{droplet}) + CaO(\text{particle}) \rightarrow CaSO_4(\text{droplet}) + H_2O \tag{13.4.6}$$

Under conditions of low humidity, water is lost from these droplets, and a solid aerosol is produced.

The preceding examples show several ways in which solid or liquid inorganic aerosols are formed by chemical reactions. Such reactions constitute an important general process for the formation of aerosols, particularly the smaller particles.

13.5 SURFACE PROPERTIES, ELECTRICAL CHARGE, AND REACTIONS OF PARTICLES

The three most important surface properties of particles are *nucleation*, *adhesion*, and *sorption*. **Nucleation** is the phenomenon by which a supersaturated vapor condenses on a particle (nucleus) to form a liquid droplet. Nucleation is involved in raindrop formation. Even in an atmosphere supersaturated with a vapor such as water, strong barriers exist which prevent the molecules from simply aggregating to form particles or droplets. However, if a particle of the condensed material is already present, it is relatively easy for vapor molecules to condense on the existing particle. The effect is largely the same even if the existing particle is composed of material different from that of the vapor but is covered with an adsorbed layer of the vapor material. Thus, particles may act as **nucleation centers**, a property vitally important in atmospheric phenomena such as raindrop formation.

Particles may adhere strongly to each other or to solid surfaces. **Adhesion**, or **coagulation**, is the process by which extremely small particles form larger aggregates and eventually reach a size at which they settle rapidly. Liquid particles of the same composition are very likely to adhere to one another when they collide. The likelihood of solid particles adhering to one another increases with decreasing particle size. The degree of particle adhesion to a surface depends upon particle and surface composition, electrical charge, surface films (moisture or oil), and roughness of the surface. The tendency of particles to adhere to surfaces is responsible for many of their obnoxious properties, for example, soiling of automobiles or house siding. It is used to advantage, however, in some processes for particulate matter removal.

Sorption, as used here, designates the phenomenon by which molecules are taken up by particles. If the vapor or gas is dissolved within the particle, the phenomenon is called **absorption**. **Adsorption** refers to sorption on the surface of the particle. **Chemisorption** refers to sorption involving a specific chemical interaction, such as the reaction of atmospheric CO_2 with particles of $Ca(OH)_2$:

$$Ca(OH)_2(s) + CO_2 \rightarrow CaCO_3 + H_2O \tag{13.5.1}$$

Other examples of chemisorption are the reactions of sulfur dioxide with aluminum oxide or iron oxide aerosols and the reaction of sulfuric acid aerosols and ammonia.

Particles in the atmosphere acquire either a negative or positive **surface electrical charge** by mechanisms such as adhesion of ions to the particle surface or proximity of the particle to a highly charged electrode. The quantity of charge is limited by air's electrical breakdown strength and the particle's surface area. As a result, the maximum charge that a particle can attain in dry air is approximately 8 esu/cm^2. Very small particles may approach this limit, but particles larger than 0.1 μm are always far below it. The charge on an atmospheric particle may be either negative or positive. The sign and degree of charge on a particle is determined by the movement of the particle in an electrical field.

As explained in Section 13.18, the ability of particles to acquire surface charge is used for their removal from stack gases. Particles acquire an electrical charge at one electrode and are collected by electrolytic deposition at an oppositely charged electrode.

13.6 PARTICULATE MATTER CONCENTRATIONS AND SIZE

The levels of particles in the atmosphere may be considered in terms of both numbers and weight per unit volume. The numbers of particles in the atmosphere may vary from several hundred per cubic centimeter in ultraclean air to more than 100,000 per cubic centimeter in highly polluted atmospheres. Most of the particulate mass in the atmosphere is found in the size range of 0.1–10 μm. In remote areas, the levels of atmospheric particulate mass may be as low as 10 μg/m^3. In urban areas, average levels of 60 μg/m^3 to 220 μg/m^3 are commonly encountered, and levels may range up to the vicinity of 2000 μg/m^3.

The total particulate-matter burden of an atmosphere is of relatively less importance than the chemical nature and size of the particles. Particles in the size range of 0.1–1 μm have an especially strong influence on light scattering, cloud nucleation, atmospheric chemical reaction catalysis, and the human respiratory tract [2]. These very small particles are particularly hard to remove from stack gas and other sources by existing control devices such as filters, electrostatic precipitators, and scrubber systems.

The size and composition of particulate matter varies diurnally and with weather conditions. Furthermore, the elemental composition differs markedly with particle size, and this may have a strong influence upon the toxicological effects of particulate matter. For example, calcium tends to occur in large particles produced by wind erosion of soil, rock quarries, cement, and similar sources. Likewise, the common soil components—iron, manganese, potassium, and titanium—are generally found in large particles. Lead and bromine occur to a large extent in smaller particles, those 0.5 μm in diameter. A particular toxicological hazard is presented by particles in the 0.05–1 μm range, because they are most efficiently retained in the respiratory system. Lead and bromine are found together in atmospheric particulate matter and are produced to a large extent by automobile exhausts.

---

[2] Friedlander, S.K. 1973. Small particles in air pose a big control problem. *Environmental Science and Technology* 7: 1115–8.

13.7 THE COMPOSITION OF INORGANIC PARTICULATE MATTER

Figure 13.4 illustrates the basic factors responsible for the composition of inorganic particulate matter. In general, the proportions of elements in atmospheric particulate matter reflect the relative abundances of elements in the parent material. The elemental composition of particulate matter may be used to trace its source [3]. It is, of course, necessary to take into account chemical reactions which may change these proportions. For example, particulate matter largely from ocean-spray origin in a coastal area receiving sulfur dioxide pollution may show anomalously high sulfate and correspondingly low chloride content. The sulfate comes from atmospheric oxidation of sulfur dioxide to form nonvolatile ionic sulfate, whereas some chloride (from NaCl in the seawater) may be lost from the solid aerosol as volatile HCl:

$$2\,SO_2 + O_2 + 2\,H_2O \rightarrow 2\,H_2SO_4 \qquad (13.7.1)$$

$$H_2SO_4 + 2\,NaCl(particulate) \rightarrow Na_2SO_4(particulate) + 2\,HCl(gas) \qquad (13.7.2)$$

Another problem with using particle composition to trace a pollution source is that the particles may undergo changes in composition between source and collection point. Using this approach, estimates have been made of sources of total suspended particulate matter in New York City [3]. The results for 1972–75 showed that these sources were 40% natural or secondary sources, 20–25% from automobile emissions, 10% from fuel oil combustion, and 5% from incineration.

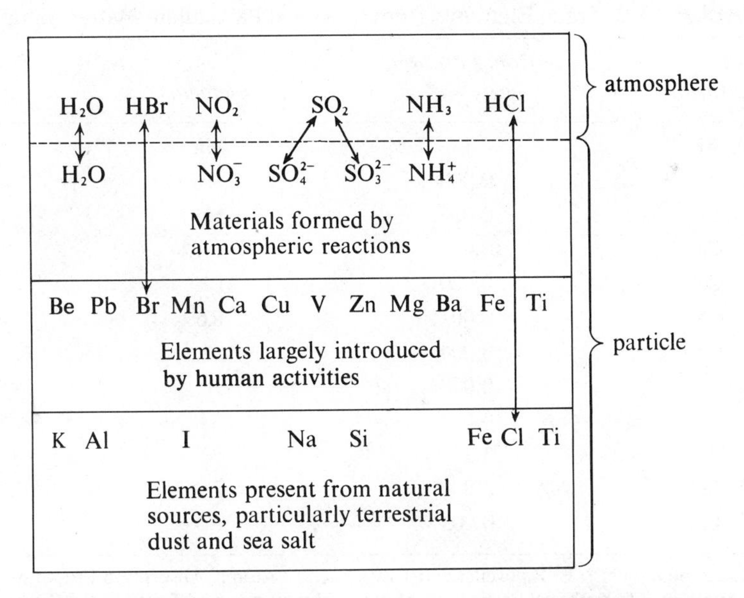

FIGURE 13.4 Some of the components of inorganic particulate matter and their origins.

[3] Kleinman, M.T.; Pasternack, B.S.; Eisenbud, M.; Kneip, T.J.. 1980. Identifying and estimating the relative importance of sources of airborne particulates. *Environmental Science and Technology* 14: 62–65.

Atmospheric particulate matter has an extremely wide diversity of chemical composition. Organic matter, nitrogen compounds, sulfur compounds, a number of metals, and radionuclides are all found in particulate matter in polluted urban atmospheres. An extensive analysis of particulate matter collected from a large number of sampling stations in urban United States locations [4] showed the following composition in units of $\mu g/m^3$ (arithmetic averages of biweekly samples): total suspended particulate matter (105), ammonium (1.3), nitrate (2.6), sulfate (10.6), benzene-soluble organics (6.8), antimony (0.01), arsenic (0.02), cadmium (0.002), chromium (0.015), copper (0.09), iron (1.58), lead (0.79), manganese (0.10), nickel (0.034), tin (0.02), titanium (0.04), vanadium (0.050), and zinc (0.67). Molybdenum was found at an average level of <0.005 $\mu g/m^3$, although a maximum value of 0.78 $\mu g/m^3$ was observed. The average levels of beryllium, bismuth, and cobalt found were all less than 0.0005 $\mu g/m^3$, although maximum values of 0.010, 0.064, and 0.060 $\mu g/m^3$, respectively, were observed. Although not included in this list, sodium is a common element found in particulate matter in coastal areas. Sodium enters the atmosphere as sodium chloride in sea-spray aerosol particles. Levels of approximately 1 $\mu g/m^3$ of sodium are not uncommon in coastal areas. Radionuclides were found in the particulate matter, with a gross beta radioactivity of 0.8 picoCuries/m^3 and a maximum value of 12.4 pCi/m^3.

At various times of the year or day and in different locations, any of the trace elements which are of concern for their toxicological effects are found in atmospheric particles. Table 13.2 shows some elements found in particulate

TABLE 13.2 Trace Elements from a Typical Particulate Matter Sample[1]

Element	*Concentration, $\mu g/m^3$ of air*	*Element*	*Concentration, $\mu g/m^3$ of air*
Al	4.1	Mg	1.3
Be	0.00003	Mn	0.012
Bi	0.003	Na	2.2
Ca	10.4	Ni	0.013
Cd	0.020	Pb	0.8
Co	0.0015	Rb	0.003
Cr	0.004	Si	13.4
Cs	0.0004	Sr	0.022
Cu	0.6	Ti	0.15
Fe	3.3	V	0.009
K	2.8	Zn	0.14
Li	0.003		

[1] Based on data in L.E. Ranweiler and J.L. Moyers, "Atomic Absorption Procedure for Analysis of Metals in Atmospheric Particulate Matter," *Environmental Science and Technology* 8: 152–6 (1974). The cadmium values may be low because of incomplete recovery during ashing; the copper values may be high because of contamination by the sampling pump. Sample collected in Tucson, Arizona.

[4] 1966. *Air Quality Data, 1964–1965*. Cincinnati, Ohio: United States Department of Health, Education, and Welfare, Division of Air Pollution.

TABLE 13.3 Composition as a Function of Size for Boston-Area Atmospheric Particulate Matter[1]

Element (concentration)[2]	*Size*[3]	*Likely source*
Al, 1.475	large	soil erosion, rock dust, coal combustion
Ce, 0.0036	mixed	unknown
Cl, 1.445	large	marine aerosols from evaporation of sea spray
Co, 0.0018	mixed	two or more unknown sources
Fe, 1.425	large	soil erosion, rock dust, coal combustion
Hf, 0.00004	mixed	unknown
Lu, 0.00001	mixed	unknown
Na, 1.910	large	marine aerosols, Na:Cl ratio higher than seawater because of volatile Cl products (such as HCl)
Sb, 0.0097	small	very volatile element, possibly from combustion of oil, coal, or refuse
Se, 0.0029	small	very volatile element, possibly from combustion of oil, coal, or refuse
Sc, 0.00031	large	soil erosion, rock dust, coal combustion
Ta, 0.00006	mixed	unknown
Th, 0.00007	large	soil erosion, rock dust, coal combustion
V, 0.92	small	combustion of residual oil (occurs at very high levels in Venezuelan crude-oil residues)
Zn, 0.30	small	probably from combustion, consistent with occurrence in small particles
Yb, 0.00002	mixed	unknown

[1] Based on data from E.S. Gladney, W.H. Zoller, A.G. Jones, and G.E. Gordon, "Composition and Size Distributions of Atmospheric Particulate Matter in Boston Area," *Environmental Science and Technology 8:* 551–7 (1974).

[2] Average of several runs, $\mu g/m^3$.

[3] Large, above approximately 3 μm; small, below approximately 3 μm.

matter collected in Tucson, Arizona. Many other examples can be cited of elemental analysis of particles, all of which show wide variations in particulate matter composition.

The composition of atmospheric particles as a function of size has been studied in the Boston area. The results of this study, summarized in Table 13.3, show some of the factors contributing to particulate matter composition in an industrial coastal area.

Various forms of particulate elemental carbon, such as graphite, coke, carbon black, and soot, compose one of the more visible and troublesome particulate air pollutants [5]. Because of its good adsorbent qualities, particulate

[5] Pimenta, J.A., and Wood, G.R. 1980. Determination of free and total carbon in suspended air particulate matter collected on glass fiber filters. *Environmental Science and Technology* 14: 556–561.

carbon can be a carrier of gaseous and other particulate pollutants. Carbon may catalyze some heterogeneous atmospheric reactions, including the conversion of SO_2 to sulfate. Sources of particulate carbon include auto and truck exhausts, heating furnaces, incinerators, power plants, and steel and foundry operations.

13.8 TOXIC METALS IN THE ATMOSPHERE

Some of the metals found in polluted atmospheres are known to be hazardous to human health. All except beryllium belong to the classification of heavy metals. Some idea of the relative dangers of these metals may be gained from the values allowed in atmospheres to which healthy industrial workers are exposed during an eight-hour day. The value for each metal is known as a **threshold limit value, TLV**. The Midwest Research Institute has compared TLVs of metals with their concentrations in a number of urban air samples. The results are summarized in Table 13.4, which ranks the metals in increasing order of TLV (from the most toxic to the least toxic) and compares the average ratio of the metal concentrations found in urban air to the TLV concentrations. Lead comes closest to being present at hazardous levels, with mercury second.

Appreciable quantities of lead, zinc, and cadmium are deposited onto forest vegetation in forested areas [6]. It should be noted that the ground/atmosphere interface in forested areas, which cover much of North America, is through the forest canopy. Study of an oak stand in the Tennessee Valley showed that atmospheric deposition on leaves contributed about one-third of the flux of

TABLE 13.4 Toxic Metals in the Atmosphere: a Comparison of TLV's and Average Levels Found in Urban Atmospheres[1]

Metal (listed in descending order of toxicity)	*TLV*, mg/m^3 (*range denotes different toxic forms*)	*Average metal level: TLV ratio*
Be	0.002	less than 0.00025
Hg	0.01–0.05	0.002
Cd	0.1	0.0001
Pb	0.1–0.2	0.004
As	0.2–0.5	0.00004
V	0.5	0.0002
Ni	0.007–1.0	0.00003
Cu	1.0	0.00009
Fe	1.0	0.0003
Zn	1.0	0.0001

[1] Based on data from Midwest Research Institute presented in J. Calving Giddings, *Chemistry, Man and Environmental Change,* San Francisco: Canfield Press, 1973, p. 339.

[6] Lindberg, S.E.; Harriss, R.C.; and Turner, R.R. 1982. Atmospheric deposition of metals to forest vegetation. *Science* 215: 1609–11.

lead, zinc, and cadmium eventually reaching the forest floor, but only added about one-tenth of the manganese [6]. Metals are deposited on the leaves by both wet and dry deposition. The ratio of wet to dry deposition is 0.1 for Mn, 0.8 for Pb, and about 3–4 for cadmium and zinc. The area studied was subject to acid precipitation (average pH of rainfall, 4.1) which can serve to partially dissolve and bind the metals to the leaves, leaving metal concentrations that are much higher than those in the rain, itself. It is thought that the deposition of atmospheric metals onto forest vegetation is an important part of the geochemical cycles of some metals whose cycles are strongly influenced by human activities, such as smelting operations.

13.9 MERCURY IN THE ATMOSPHERE

Mercury in the atmosphere is a cause of great concern because of its toxicity. Elemental mercury is volatile, and much of the mercury entering the atmosphere does so in atomic form. This includes most of the mercury entering the atmosphere from coal combustion and volcanoes.

Some mercury in the atmosphere is associated with particulate matter [7]. Volatile organomercury compounds such as dimethylmercury, $(CH_3)_2Hg$, are also encountered in the atmosphere. Monomethylmercury salts, such as $(CH_3)HgBr$, are also volatile mercury compounds which are sometimes found in the atmosphere. Most coals contain appreciable levels of mercury, and this presents a problem which must be dealt with as the use of coal increases.

13.10 LEAD IN THE ATMOSPHERE

Currently, lead is probably the most serious atmospheric heavy-metal pollutant. Decreasing use of leaded gasoline and increasing use of coal will probably result in its replacement by mercury as the most troublesome atmospheric metal pollutant. As shown in Equation 13.4.3, lead halides from automotive exhausts are the most common form of atmospheric lead. Of these, lead bromochloride, PbBrCl, predominates. Appreciable amounts of $NH_4Cl \cdot 2\,PbBrCl$, an ammonium chloride-lead bromochloride double salt, are also found in auto exhausts.

The atmospheric reactions and ultimate fate of this lead are the subject of conflicting opinions [8]. Determination of Pb:Br ratios in both atmospheric particulate matter and lake sediments tends to show a 1:1 correlation, indicating that the bromochloride tends to remain intact. It is now generally held that loss of halide occurs most readily through photochemical processes. This is viewed as a process in which light provides energy for the transfer of an electron from a halide atom to a lead ion in the lead halide crystalline lattice, leading to the

---

[7] Braman, R.S., and Johnson, D.L. 1974. Selective absorption tubes and emission technique for determination of ambient forms of mercury in air. *Environmental Science and Technology* 8: 996–1003.

[8] Boyer, K.W., and Laitinen, H.A. 1974. Lead halide aerosols: some properties of environmental significance. *Environmental Science and Technology* 8: 1093–6.

formation of elemental halogen and a reduced form of lead:

$$Pb_XCl_XBr_X + h\nu \rightarrow Pb_XCl_XBr_{X-1} + \frac{1}{2}Br_2 \quad (13.10.1)$$

The halogen, of course, reacts quickly in the atmosphere, generally producing a halide salt.

13.11 BERYLLIUM IN THE ATMOSPHERE

Beryllium is used primarily as a component of specialty alloys. Some of the primary uses include electrical equipment, electronic instrumentation, space gear, and nuclear equipment. Only about 350 metric tons of beryllium are used in the United States each year, so that its distribution is by no means comparable to that of other toxic metals such as lead or mercury. Because of its specialty uses, however, consumption of beryllium is increasing.

During the 1940s and 1950s, the toxicity of beryllium and beryllium compounds became widely recognized. One of the main results of these studies was the elimination of this element from phosphors (coatings which produce visible light from ultraviolet light) in fluorescent lamps. Table 13.4 shows that beryllium has the lowest allowable TLV of any metal, and its presence in the atmosphere is very hazardous.

13.12 ASBESTOS IN THE ATMOSPHERE

Asbestos is a naturally occurring fibrous silicate. The fibers are separable, yielding fine filaments. The tensile strength, flexibility, and nonflammability of these fibers, in combination with other desirable properties, have led to many uses of asbestos, ranging from brake linings to pipe insulation.

Small fibers of asbestos irritate lung tissue, causing a condition called "asbestosis" [9]. This condition is characterized by lung fibrosis. It has been estimated [10] that from about 1950 to 1975, approximately 10% of the deaths among New York City insulation workers was caused by asbestosis. Therefore, atmospheric asbestos is a matter of considerable concern.

13.13 MINERAL PARTICULATE MATTER FROM COMBUSTION: FLY ASH

Much of the mineral particulate matter in a polluted atmosphere is in the form of oxides and other compounds produced during the combustion of high-ash fossil fuel. Much of the mineral matter in fossil fuels such as coal or lignite is converted during combustion to a fused, glassy bottom ash which presents no air pollution problems. Smaller ash particles called **fly ash** enter furnace flues and, in a properly equipped stack system, are collected efficiently. However, some of the fly ash particles escape through the stack and enter the atmosphere. Unfortunately, these

---

[9] Gordon, G., and Zoller, W. 1975. *Chemistry in modern perspective.* Reading, Mass.: Addison-Wesley Publishing Co., Inc.

[10] Selikoff, I.J. 1969. Asbestos. *Environment* 11(2): 2.

tend to be the smaller particles which do the most damage to human health, plants, and visibility.

The composition of fly ash varies over a wide range, depending upon the source of fuel. The general ranges of composition to be expected are shown in Table 13.5. Fly ash does contain some trace elements, with some of the more harmful ones concentrated on smaller particles.

TABLE 13.5 Range of Fly-Ash Composition Found in Four Different Investigations[1]

Component	*Expressed as*	*Range of percentages*
Ca	CaO	0.12–14.3
Mg	MgO	0.06–4.77
Fe	Fe_2O_3 or Fe_3O_4	2.0–26.8
Al	Al_2O_3	9.81–58.4
S	SO_3	0.12–24.33
Ti	TiO_2	0.0–2.8
carbonate	CO_3	0.0–2.6
Si	SiO_2	17.3–63.6
P	P	0.03–20.6
K	K_2O	2.8–3.0
Na	Na_2O	0.2–0.9
C	C	0.37–36.2
undetermined		0.08–18.9

[1] W.S. Smith and C.W. Gruber, *Atmospheric Emissions from Coal Combustion. An Inventory Guide,* PHS-Pub-999-AP-24. Cincinnati, Ohio: United States Department of Health, Education and Welfare, Division of Air Pollution, 1966.

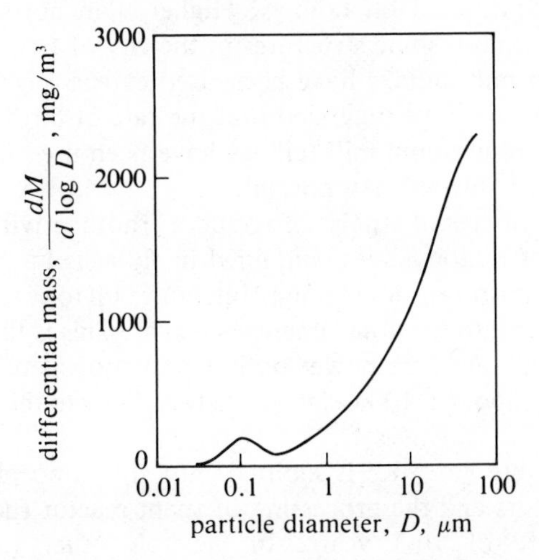

FIGURE 13.5 General appearance of particle-size distribution in coal-fired power-plant fly ash. The data are given on differential mass coordinates, where M is mass, so that the area under the curve in a given size range is the mass of the particles in that size range. Similar plots are shown in reference 11.

The size of fly-ash particles is a very important factor in their removal from stack gas and in their ability to enter the body through the respiratory tract. A study of size distribution of fly ash from coal-fired utility boilers [11] has produced the data represented by Figure 13.5 (see page 369). This plot shows a bimodal (two-peak) distribution of size, with a peak at about 0.1 μm. Although typically this peak represents only about 1–2 percent of the total fly-ash mass, it includes the vast majority of total number of particles and particle surface area. These particles probably result from a volatilization-condensation process during combustion. They are of a size that is most difficult to remove by electrostatic precipitators and bag houses (see Section 13.18). Analysis of the submicrometer particles showed a higher concentration of more volatile elements, such as As, Sb, Hg, and Zn. This finding would support a volatilization-condensation mode of particle formation.

13.14 RADIONUCLIDES IN THE ATMOSPHERE

A significant natural source of radionuclides in the atmosphere is the noble gas **radon**, a product of radium decay [12]. Radon may enter the atmosphere as one of two isotopes, ^{222}Rn or ^{220}Rn. Both are alpha emitters, with half-lives of 3.8 days and 54.5 seconds, respectively. Both are parts of decay schemes ending ultimately in stable isotopes of lead. The initial decay products, ^{218}Po and ^{216}Po, are nongaseous and adhere readily to atmospheric particulate matter. Therefore, some of the radioactivity detected in these particles is of natural origin. Furthermore, cosmic rays act on nuclei in the atmosphere to produce other radionuclides, including ^{7}Be, ^{10}Be, ^{14}C, ^{39}Cl, ^{3}H, ^{22}Na, ^{32}P, and ^{33}P.

One of the more serious problems in connection with radon appears to be that of radioactivity originating from uranium mill tailings. Such tailings have been used in some towns as backfill, as soil conditioner, and as a base for building foundations. Radon produced by the decay of radium exudes from foundations and walls constructed on tailings. Higher than normal levels of radioactivity have been found in some structures in the city of Grand Junction, Colorado, where uranium mill tailings have been used extensively in construction. Some medical authorities have suggested that the rate of birth defects and infant cancer in areas where uranium mill tailings have been used in residential construction are significantly higher than normal.

The combustion of fossil fuels introduces radioactivity into the atmosphere in the form of radionuclides contained in fly ash. Large coal-fired power plants lacking ash-control equipment may introduce up to several hundred milliCuries of radionuclides into the atmosphere each year. This is far more than a nuclear power plant with the same power output. An equivalent oil-burning plant would discharge only about 1/1000 as much radioactivity to the atmosphere annually.

The radioactive noble gas ^{85}Kr is emitted into the atmosphere by the operation of nuclear reactors and the processing of spent reactor fuels. This radionuclide has a half-life of 10.3 years. In general, other radionuclides

---

[11] McElroy, M.W.; Carr, R.C.; Ensor, D.S.; and Markowski, G.R. 1982. Size distribution of fine particles from coal combustion. *Science* 215: 13–19.

[12] Eisenbud, M. 1968. Sources of radioactive pollution. In *Air Pollution,* Vol. I, 2nd ed., A.C. Stern, ed., Ch. 5. New York: Academic Press, Inc.

produced by reactor operation are either chemically reactive and can be removed from the reactor effluent, or have such short half-lives that a short time delay prior to emission prevents their leaving the reactor. Widespread use of fission power will inevitably result in an increased level of ^{85}Kr in the atmosphere. Fortunately, biota will not concentrate this radionuclide because it is unreactive chemically.

The detonation of nuclear weapons above ground contributes a great deal to particulate radioactive material. Such material may be carried through the atmosphere for tremendous distances, eventually to be scavenged from the atmosphere by rain. For example, about a week after a nuclear detonation in China on December 24, 1967, the following radionuclides were detected in rain in Fayetteville, Arkansas [13]: ^{91}Y, ^{141}Ce, ^{144}Ce, ^{147}Nd, ^{147}Pm, ^{149}Pm, ^{151}Sm, ^{153}Sm, ^{155}Eu, ^{156}Eu, ^{89}Sr, ^{90}Sr, ^{115m}Cd, ^{129m}Te, ^{131}I, ^{132}Te, ^{140}Ba. (In these formulas, *m* denotes a metastable state that decays by gamma-ray emission to an isotope of the same element.) The rate of travel of radioactive particles through the atmosphere is a function of particle size. Appreciable fractionation of nuclear debris is observed because of differences in the rates at which various components of nuclear debris move through the atmosphere.

13.15 ORGANIC PARTICULATE MATTER

Organic particulate matter occurs in a wide variety of compounds in the atmosphere. It is interesting to note that a so-called average formula of this organic matter, collected from 200 U.S. areas and isolated as a benzene-extractable fraction, is $C_{32.4}H_{48}O_{3.8}S_{0.083}(halogen)_{0.065}(alkoxy)_{0.12}$ [14]. Much of this organic matter occurs in the respirable 1-μm range.

Organic particles typically are collected for study on a filter designed to remove a minimum of 99.9% of particles having a diameter of 0.3 μm or greater [15]. For analysis, the organic matter is then extracted from the filter with a solvent such as ether or benzene. Next, the extract may be fractionated into groups such as the "weak acid group," the "strong acid group," the "neutral group," and the "basic group." The neutral group, containing largely hydrocarbons, may be further fractionated into an aliphatic fraction, an aromatic fraction, and an oxygenated fraction.

The aliphatic fraction of the neutral group contains a high percentage of long-chain hydrocarbons, predominantly those containing 16–28 carbon atoms. Such hydrocarbons are not thought to be a particularly significant health hazard or to participate strongly in atmospheric chemistry reactions. Many of the hydrocarbons found in the aromatic fraction, however, are polynuclear aromatics (Section 13.16), and some are proven carcinogens (cancer-causing) agents). Therefore, these hydrocarbons are particularly important as air pollutants and will be discussed in some detail. Aldehydes, ketones, epoxides,

---

[13] Thein, M., et al. 1969. Fractionation of bomb-produced rare-earth nuclides in the atmosphere. *Environmental Science and Technology* 3: 667–670.

[14] Sawicki, E. 1967. *Arch. Environ. Health* 14: 46.

[15] Hoffman, D., and Wynder, E.L. 1968. Chemical analysis and carcinogenic bioassays of organic particulate pollutants. In *Air Pollution,* Vol. II, 2nd ed., A.C. Stern, ed. Ch. 20. New York: Academic Press, Inc.

peroxides, esters, quinones, and lactones are found among the oxygenated neutral organic components. A few of these compounds may be mutagenic (mutation-causing) or carcinogenic. The acidic fractions contain long-chain fatty acids and nonvolatile phenols. Among the acids recovered from air-pollutant particulate matter are lauric, myristic, palmitic, stearic, behenic, oleic, and linoleic. The basic fraction consists largely of alkaline *N*-heterocyclic hydrocarbons, such as acridine:

N

acridine

At the present time, comparatively little is known about the chemical constitution and environmental significance of the basic fraction.

A significant portion of the organic particulate matter contained in automotive exhausts has been characterized as "oxidized, polymerized hydrocarbons and nitrogenous azaheterocyclic substances" [16]. The formation of this ominous-sounding material is explained by a series of complicated processes involving chemical reactions of materials in automobile engines and exhausts with nitrogenous compounds. Included among these nitrogenous compounds are some nitrites, methyl nitrate, and alkyl amines. These reactions, involving to a certain extent pyrosynthesis (Section 13.16) in the automobile engine and exhaust system, produce oxidized polymerized hydrocarbons. Lubricating oil and its additives may also contribute to organic particulate matter.

Some specific types of organic compounds and functional groups have been identified in respirable particulate matter along highways [16]. These compounds and groups are summarized in Table 13.6. Many of them are harmful to human health, with carcinogens being a particular concern.

TABLE 13.6 Types of Organic Compounds Found in Respirable Particulate Matter along Highways[1]

Type of compound	*Possible effect*
Polar organics	Extracted by aqueous body fluids, enabling penetration of tissue in a soluble form.
Aldehydes	Suspected of being the primary causal agents in smoke poisoning, postulated denaturation of amino acids and RNA. Inflamation and necrosis of tissue result.
Other oxygenated organics (acids, neutral oxygen compounds)	Some have suspected carcinogenic effects.
Polynuclear azaheterocyclic compounds (benzacridines, dibenzacridines, benzoquinolines)	Some of these, such as dibenz(a,h)acridine and dibenz(a,j)-acridine are known carcinogens.

[1] These compounds are discussed in detail in Reference [16].

[16] Ciacco, L.L.; Rubino, R.L.; and Flores, J. 1974. Composition of organic constituents in breathable airborne particulate matter near a highway. *Environmental Science and Technology* 8: 935–42.

13.16 POLYCYCLIC AROMATIC HYDROCARBONS

Polycyclic (polynuclear) aromatic hydrocarbons (PAH) have received a great deal of attention because of the known carcinogenic effects of some of these compounds. Benzo(a)pyrene is most commonly cited as one of the more carcinogenic PAH compounds found in atmospheric particulate matter. Other actively carcinogenic PAH compounds found in organic particulate matter are benz(a)anthracene, chrysene, benzo(e)pyrene, benz(e)acephenanthrylene, benzo(j)fluoranthene, and indeno(1,2,3-cd)pyrene. Some representative structures of PAH compounds are:

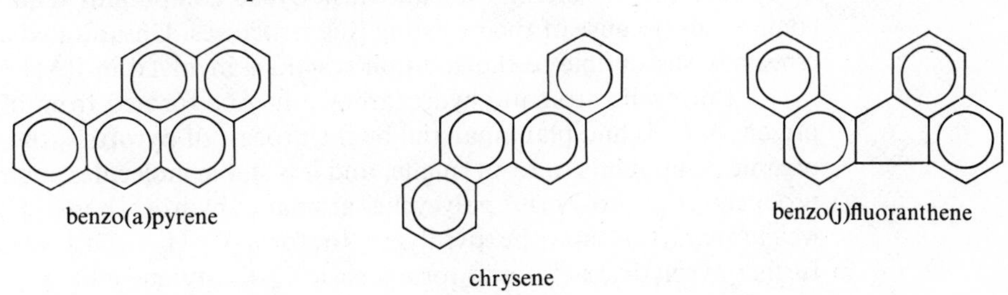

The isolation of the first known cycloalkenyl derivative, cyclopenta(cd)-pyrene, from soots was reported in 1975 [17]:

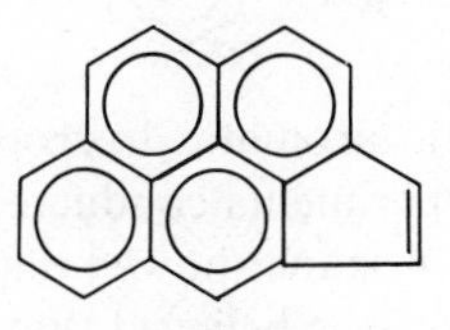
cyclopenta(cd)pyrene

This orange polycyclic aromatic hydrocarbon is a known carcinogen.

PAH compounds are most likely to be found in urban atmospheres. In such atmospheres, the known carcinogenic PAH compounds typically are found at levels ranging up to approximately 20 $\mu g/m^3$. Some specific atmospheres or effluents may contain very high levels of such compounds. Effluent from a coal furnace may contain over 1000 $\mu g/m^3$ of PAH compounds and cigarette smoke almost 100 $\mu g/m^3$ [15].

PAH hydrocarbons may be synthesized from saturated hydrocarbons at high temperatures under oxygen-deficient conditions. Even very-low-molecular-weight hydrocarbons, including methane, may act as precursors for the polycyclic aromatic compounds. The two-carbon hydrocarbons, ethylene and ethane, have yielded more than 10 different polycylic aromatic hydrocarbons. The process by which low-molecular-weight hydrocarbons form PAH compounds is called **pyrosynthesis**. At temperatures exceeding approximately 500 °C, carbon-hydrogen and carbon-carbon bonds are broken to form free radicals. These radicals undergo dehydrogenation and combine chemically to form aromatic ring structures, which are resistant to thermal degradation. The basic process for the formation of aromatic ring structures from pyrosynthesis starting with ethane is shown at the top of page 374. This process may proceed to the formation of

[17] Wallcave, L.; Nagel, D.L.; Smith, J.W.; and Waniska, R.D. 1975. Two pyrene derivatives of widespread environmental distribution: cyclopenta(cd)pyrene and acepyrene. *Environmental Science and Technology* 9: 143–5.

$$H_3C{-}CH_3 \xrightarrow[\Delta]{-H} H_2C{=}CH{-}CH{=}CH_2 + HC{\equiv}CH \xrightarrow[\Delta]{-H} C_6H_6 \longrightarrow \text{polycyclic aromatic hydrocarbons}$$

stable PAH structures. The tendency to form PAH compounds by pyrosynthesis varies with hydrocarbon type. The general order of this tendency is [18] aromatics > cycloolefins > olefins > paraffins. Cyclic compounds tend to form PAH compounds because of their existing ring structures. Unsaturated compounds are especially susceptible to the addition reactions involved in PAH formation.

Polycyclic aromatic hydrocarbons may be formed from higher paraffins present in fuels and plant material by the process of **pyrolysis**, the "cracking" of organic compounds to form smaller and less stable molecules and radicals. These products react to form polycyclic aromatic hydrocarbons. High-molecular-weight paraffins may be pyrolyzed to form $C_{10}H_{22}$. This species undergoes further pyrolytic reactions to form a basic C_6–C_2 styrene unit [19], which leads to formation of PAH compounds:

$$C_{10}H_{22} \xrightarrow[\Delta]{-H} C_6H_5{-}CH{=}CH_2 \longrightarrow \text{polycyclic aromatic hydrocarbons}$$

Polycyclic aromatic hydrocarbons are found almost exclusively in the solid phase. Experiments conducted on benzo(a)pyrene produced by incomplete combustion of propane showed that less than 4% of this PAH compound was in the vapor phase. It is believed that most PAH compounds are sorbed onto soot particles. Soot itself is a highly condensed product of polynuclear aromatic hydrocarbons. X-Ray structure studies have shown that soot particles, like graphite crystals, have a hexagonal symmetry. A soot particle is composed of several thousand interconnected crystallites. The crystallites, in turn, are made up of graphitic platelets, each consisting of approximately 100 condensed aromatic rings. Soot contains 1–3% hydrogen and 5–10% oxygen, the latter due to partial surface oxidation.

Benzo(a)pyrene adsorbed on soot disappears very rapidly in the presence of light. The products are thought to be oxygenated organic compounds. It is believed that most of the polycyclic aromatic hydrocarbon is sorbed on the surface of the soot particle, and that the large surface area of the particle contributes to the high rate of reaction. Although the exact formulas of the photo-oxidation products are not known, oxidation of benzo(a)pyrene with chromic acid or ozone yields anhydrides, carboxylates, dicarboxylates, diacetates, phenols, and quinones.

PAH compounds can react to form other biologically active compounds in smog [20]. Both benzo(a)pyrene and perylene react in simulated smoggy

---

[18] Edwards, J.B. 1974. *Combustion: the formation and emission of trace species.* Ann Arbor, Mich: Ann Arbor Science Publishers, Inc.

[19] Badger, G.M.; Donnelly, J.K.; and Spotswood, T.M. 1965. *Australian Journal of Chemistry* 19: 1023.

[20] Pitts, J.N., Jr., et al. 1978. Atmospheric reactions of polycyclic aromatic hydrocarbons: facile formation of mutagenic nitro derivatives. *Science* 202: 515–9.

atmospheres to form nitro derivatives that are mutagenic (therefore, possibly carcinogenic), according to the Bruce Ames test (see Section 18.19). It is significant that perylene is neither mutagenic nor carcinogenic. Under conditions leading to smog formation, however, perylene forms mutagenic 3-nitroperylene [20]:

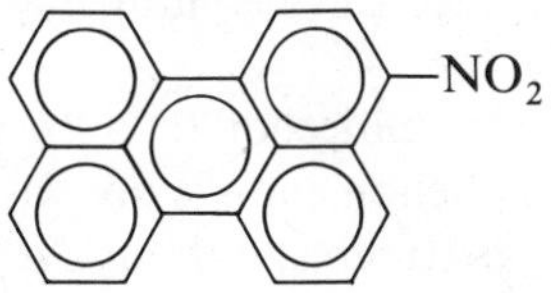

$-NO_2$

13.17 HEALTH EFFECTS OF PARTICULATE MATTER

The only major route by which particles enter the body is through the respiratory tract. Relatively large particles are likely to be retained in the nasal cavity and in the pharynx. Very small particles are likely to reach the lungs and to be retained by them. The respiratory system possesses mechanisms for the removal of particulate matter. In the ciliated region of the respiratory system, particles are carried as far as the entrance to the gastrointestinal tract by a flow of mucus. Macrophages in the nonciliated pulmonary regions carry particles to the ciliated region.

The respiratory system may be damaged directly by particulate material that enters the blood system or lymph system through the lungs. In addition, the particulate material or soluble components of it may be transported to organs some distance from the lungs and have a detrimental effect on these organs. Particles cleared from the respiratory tract largely enter the gastrointestinal tract by swallowing. Material may also enter the blood after being transported to the gastrointestinal tract.

A strong correlation has been found between increases in the daily mortality rate and acute episodes of air pollution. In such cases, high levels of particulate matter are accompanied by high levels of SO_2 and other pollutants, so that any conclusions must be drawn with caution.

13.18 REMOVAL OF PARTICULATE MATTER

The removal of particulate matter from gas streams is the most widely practiced means of air-pollution control. A number of devices have been developed for this purpose, differing widely in effectiveness, complexity, and cost.

The simplest means of particulate matter removal is **sedimentation**, a phenomenon that occurs continuously in nature. Gravitational settling chambers may be employed for the removal of particles from gas streams by simple settling under the influence of gravity. These chambers take up a large amount of space and have a relatively low collection efficiency, particularly for small particles.

Gravitational settling of particles is enhanced by increased particle size, which occurs spontaneously by coagulation. Thus, over time, the size of particles increases and the number of particles decreases in a mass of air containing particles. Brownian motion of particles less than about 0.1 μm in size is primarily responsible for their contact, enabling coagulation. Particles greater than about 0.3 μm in radius do not diffuse appreciably and serve primarily as receptors of smaller particles.

Inertial mechanisms are also employed for particle removal. These depend upon the fact that the radius of the path of a particle in a rapidly moving, curving air stream is larger than the path of the stream as a whole. Therefore, when a gas stream is spun by vanes, a fan, or a tangential gas inlet, the particulate matter may be collected on a separator wall because the particles are forced outward by centrifugal force. Devices utilizing this mode of operation are called **dry centrifugal collectors**.

Particles subjected to a **thermal gradient** move from hotter to cooler regions. This phenomenon is not widely employed as a means of particle removal from waste-gas streams, although it does have some potential for this purpose. It is used to sample aerosol particles from the atmosphere prior to chemical analysis (see Chapter 16).

A number of different kinds of **wet collectors** are employed for the removal of particles, wherein liquid droplets are used to trap the particles and wash them away. Their redispersal is prevented by the liquid.

As noted in Section 13.5, aerosol particles may acquire electrical charges. In an electric field, such particles are subjected to a force, F (dynes), given by the equation

$$F = Eq \tag{13.18.1}$$

where E is the voltage gradient (statvolt/cm) and q is the electrostatic charge on the particle (in esu). This phenomenon has been used in highly efficient **electrostatic precipitators**, as shown in Figure 13.6. The particles acquire a charge when the gas stream is passed through a high-voltage, direct-current corona. Because of their charge, the particles are attracted to a grounded surface, from which they may be later removed. Ozone may be produced by the corona discharge. Similar devices used as household dust collectors may produce toxic ozone if not properly operated.

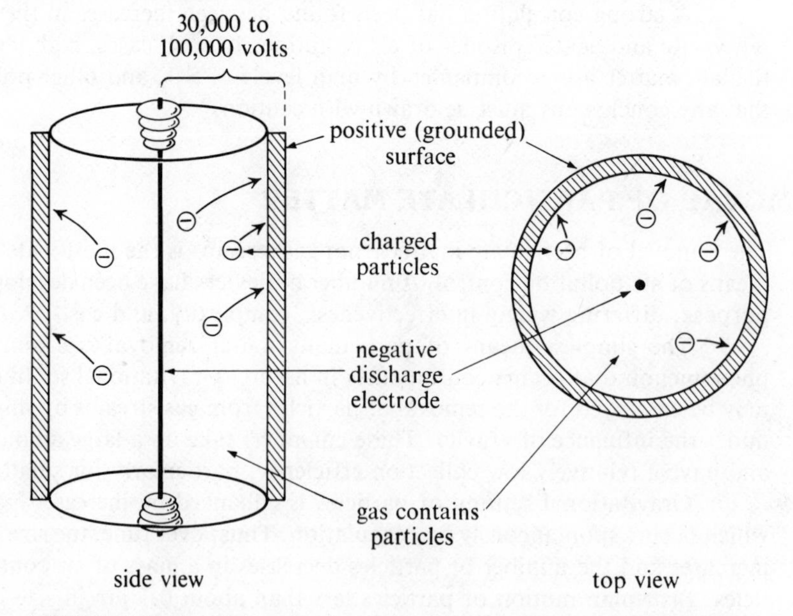

FIGURE 13.6 Schematic diagram of an electrostatic precipitator.

Fabric filters, as their name implies, consist of fabrics that allow the passage of gas but retain particulate matter. These are used to collect dust in bags contained in structures called *baghouses.*

When combustible particles are to be removed, **afterburners** may be used. These devices are also used for the combustion of vapors, fumes, and odorous compounds.

SUPPLEMENTARY REFERENCES

1979. *Airborne particles.* Baltimore: University Park Press.

1969. *Air quality criteria for particulate matter.* National Air Pollution Control Administration Publication No. AP-49. Washington, D.C.

American Chemical Society. 1978. *Cleaning our environment—a chemical perspective.* Washington, D.C.

Bjørseth, A., ed. 1983. *Handbook of Polycyclic aromatic hydrocarbons.* New York: Marcel Dekker, Inc.

Bjørseth, A., and Dennis, A.J., eds. 1980. *Polynuclear aromatic hydrocarbons.* New York: Van Nostrand Reinhold Company.

Bricard, J. 1977. Aerosol production in the atmosphere. In *Environmental chemistry,* J.O'M. Bockris, ed., Ch. 11, pp. 313–30. New York: Plenum Publishing Corp.

Brodeur, P. 1980. *The asbestos hazard.* New York: New York Academy of Sciences.

Cheremisinoff, P.N., ed. 1981. *Air/particulate—instrumentation and analysis.* Woburn, Mass.: Butterworth Publishers.

1969. *Control techniques for particulate air pollutants.* National Air Pollution Control Administration Publication No. AP-51. Washington, D.C.

Crawford, M. 1976. *Air pollution control theory.* New York: McGraw-Hill Book Co.

Dietz, V. 1975. *Removal of trace contaminants from the air.* ACS Symposium No. 17. Washington, D.C.: American Chemical Society.

Fuchs, N.A. 1964. *The mechanics of aerosols.* New York: Macmillan, Inc.

Hinds, W.C. 1982. *Aerosol technology.* New York: Wiley-Interscience.

Horne, R.A. 1978. The chemistry of air pollution. In *The chemistry of our environment,* Ch. 6, pp. 159–234. New York: John Wiley and Sons, Inc.

1971. *Inadvertent climate modification.* Cambridge, MA.: M.I.T. Press.

Kneip, T.J., and Lioy, P.J., eds. 1980. *Aerosols: anthropogenic and natural sources and transport.* New York: New York Academy of Sciences.

Lewtas, J., ed. 1982. *Toxicological effects of emissions from diesel engines.* New York: Elsevier North-Holland, Inc.

Lyon, W.S. 1976. *Trace element measurements at the coal-fired steam plant.* West Palm Beach, Fla.: CRC Press, Inc.

Mercer, T.T. 1973. *Aerosol technology in hazard evaluation.* New York: Academic Press, Inc.

Mercer, T.T.; Morrow, P.E.; and Stöber, W., eds. 1972. *Assessment of airborne particles: fundamentals, applications, and implications to inhalation toxicity.* Springfield, Ill.: Charles C. Thomas, Publisher.

Moore, J.W., and Moore, E.A. 1976. *Environmental chemistry.* New York: Academic Press, Inc.

Sittig, M. 1977. *Particulates and fine dust removal: processes and equipment.* Park Ridge, N.J.: Noyes Data Corp.

1978. *Symposium on environmental effects of sulfur oxides and related particulates.* New York: New York Academy of Medicine.

Wolff, G.T., and Klimisch, R.L., eds. 1982. *Particulate carbon: atmospheric life cycle.* New York: Plenum Publishing Corp.

QUESTIONS AND PROBLEMS

1. As stated in Section 13.5, the maximum electrical charge that an atmospheric particle may attain in dry air is about 8 esu/cm^2. How many electrons per square cm of surface area is this?

2. For charged particles that are 0.1 μm or less in size, an average charge of 4.77×10^{-10} esu, equal to that of the *elementary quantum of electricity* (the charge on 1 electron or proton), is normally assumed for the whole particle. What is the surface charge, in esu/cm^2, for a charged spherical particle with a radius of 0.1 μm?

3. What is the settling velocity of a particle having a Stokes' diameter of 10 μm and a density of 1 g/cm^3 in air at 1.00 atm pressure and 0 °C temperature? (The viscosity of air at 0 °C is 170.8 micropoises. The density of air under these conditions is 1.29 g/L.)

4. A freight train that included a tank car containing anhydrous NH_3 and one containing concentrated HCl was wrecked, and both tank cars leaked, causing a mixing of their contents. What was the chemical formula of the resulting white aerosol?

5. Examination of aerosol fume particles produced by a welding process showed that 2% of the particles were greater than 7 μm in diameter and 2% were less than 0.5 μm. What is the mass median diameter of the particles?

6. What two vapor forms of mercury might be found in the atmosphere?

7. Analysis of particulate matter collected in the atmosphere near a seashore shows considerably more Na than Cl, on a molar basis. What does this indicate?

8. What type of process results in the formation of very small aerosol particles?

9. Which size range encompasses most of the particulate matter mass in the atmosphere?

10. Why are aerosols in the 0.1–1-μm size range especially effective in scattering light?

11. Per unit mass, why are smaller particles more effective catalysts for atmospheric chemical reactions?

12. In terms of origin, what are the three major categories of elements found in atmospheric particles?

13. What are the five major classes of material making up the composition of atmospheric aerosol particles?

14. The size distribution of particles emitted from coal-fired power plants is bimodal. What are some of the properties of the smaller fraction in terms of potential environmental implications?

CHAPTER 14
PHOTOCHEMICAL SMOG

14.1 WHAT IS SMOG?

The atmospheric condition commonly known as **smog** first became a serious nuisance in the Los Angeles area during the 1940s. It is characterized by reduced visibility, eye irritation, and the deterioration of some materials, such as the cracking of rubber. Although smog developed with increasing industrialization and use of automobiles, it should be noted that air pollution has long been a problem in southern California. In 1542, Juan Rodriquez Cabrillo named San Pedro Bay "The Bay of Smokes" because of the heavy haze that covered the area. Complaints of eye irritation from polluted air in Los Angeles were recorded as far back as 1868.

The term *smog* was originally used to describe the unpleasant combination of smoke and fog laced with sulfur dioxide, which was formerly prevalent in London when high-sulfur coal was the primary fuel used in that city. This mixture is actually chemically reducing and is called a **reducing smog** or **sulfurous smog**. The smog that permeates Los Angeles, Denver, and similar areas a number of days each year has a high concentration of oxidants and is an oxidizing mixture. It is therefore called an **oxidizing smog** or **photochemical smog**. Readily oxidized sulfur dioxide has a short lifetime in an atmosphere where oxidizing photochemical smog is present. (Throughout the remainder of the book, all references to *smog* are to photochemical smog only.)

For photochemical smog, air-quality authorities define a smoggy day as one having moderate to severe eye irritation or visibility below 3 miles when the relative humidity is below 60%. The formation of oxidants in the air, particularly ozone, is indicative of smog formation. Serious levels of photochemical smog may be assumed to be present when the oxidant level exceeds 0.15 ppm for more than one hour.

It is now recognized that three ingredients are required for photochemical smog formation: ultraviolet light, hydrocarbons, and nitrogen oxides. Two of these ingredients, hydrocarbons and nitrogen oxides, are produced by the automobile. Therefore, we will discuss pollutant emissions from the automobile next.

14.2 PRODUCTION AND CONTROL OF POLLUTANT HYDROCARBONS FROM AUTOMOBILES

At the high temperatures and pressures encountered in the automobile engine, products of incompletely burned gasoline undergo chemical reactions which produce several hundred kinds of hydrocarbons. Some of these are highly reactive in

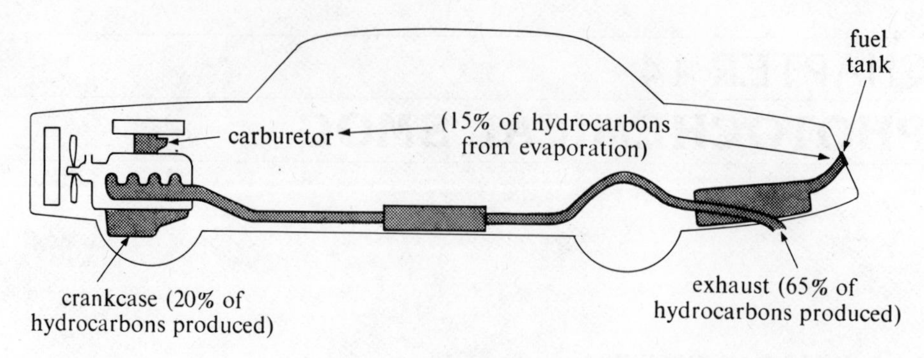

FIGURE 14.1 Potential sources of pollutant hydrocarbons from an automobile without pollution-control devices.

the atmosphere and react to form smog. The production by automobiles of the other major chemical ingredient of smog, nitrogen oxides, was discussed in Section 12.11. Aside from its exhaust, the automobile contains several potential sources of pollutant hydrocarbons (Fig. 14.1). One source is the crankcase, which contains mist composed of lubricating oil and receives "blowby," which consists of exhaust gas and unoxidized carbureted mixture from around the pistons. Blowby is recirculated through the engine intake manifold by way of the PCV (positive crankcase ventilation) valve. These emissions were the first to be controlled.

The fuel system is a second major source of hydrocarbon emissions. Fuel may be given off through fuel tank and carburetor vents. When the engine is shut down and engine heat warms up the fuel system, gasoline may be evaporated and emitted to the atmosphere. In addition, heating during the daytime and cooling at night causes the fuel tank to breathe and emit gasoline fumes.

Several methods have been devised for the control of evaporative emissions from fuel systems. One is a carbon canister that collects evaporated fuel from the fuel tank and fuel system, to be purged and burned when the engine is operating. Reduction of fuel volatility also helps cut down hydrocarbon evaporation from fuel systems; for instance, the diesel automobile has a very low level of emissions from fuel evaporation because of the low vapor pressure of diesel fuel.

As shown in Figure 14.1, the automobile exhaust is the greatest source of hydrocarbons. They are produced by the partial combustion of gasoline in the engine, which for the vast majority of automobiles is a four-cycle internal combustion engine fueled with gasoline. The basic operating characteristics of such an engine are shown in Figure 14.2. The four basic steps in one complete cycle of the engine are:

1. **Intake:** An air-gasoline mixture produced in the carburetor, consisting of about 14–15 parts by weight of air per part gasoline, is drawn into the cylinder through the open intake valve.
2. **Compression:** The combustible mixture is compressed at a ratio of about 7:1. Higher compression ratios favor thermal efficiency and complete combustion of hydrocarbons. However, higher temperatures, premature combustion ("pinging"), and high production of nitrogen oxides also result from higher combustion ratios.

3. **Ignition and power stroke:** As the fuel-air mixture is ignited by the spark plug near top dead center of the combustion stroke, a temperature of about 2500 °C is reached very rapidly at pressures up to 40 atm. The temperature decreases in a few milliseconds. This rapid cooling "freezes" nitric oxide in the form of NO without allowing it time to dissociate to N_2 and O_2, thermodynamically favored at the normal temperatures and pressures of the atmosphere.
4. **Exhaust:** Exhaust gases consisting largely of N_2 and CO_2, with traces of CO, NO, hydrocarbons, and O_2, are pushed out through the open exhaust valve, thus completing the cycle.

The primary cause of unburned hydrocarbons in the engine cylinder is "wall quench." The relatively cool wall in the combustion chamber of the internal combustion engine quenches the flame within several thousandths of an inch of the wall. Part of the remaining hydrocarbon may be retained as residual gas in the cylinder, and part may be oxidized in the exhaust system. The remainder is emitted to the atmosphere as pollutant hydrocarbons. Engine misfire due to improper adjustment and deceleration greatly increases the emission of hydrocarbons. Turbine engines are not subject to the wall quench phenomenon because their surfaces are always hot.

Several engine design characteristics favor lower exhaust hydrocarbon emissions. Since much of the hydrocarbon emission is caused by wall quench, reduction of the surface/volume ratio of the combustion chamber decreases hydrocarbon emissions. Factors which reduce the surface/volume ratio are decreased compression ratio; more nearly spherical combustion chamber shape; increased displacement per engine cylinder; and increased ratio of stroke relative to bore.

Spark retard also reduces exhaust hydrocarbon emissions. Normally, for optimum engine power, the spark is set to fire appreciably before the piston reaches the top of the compression stroke and begins the power stroke. Retarding

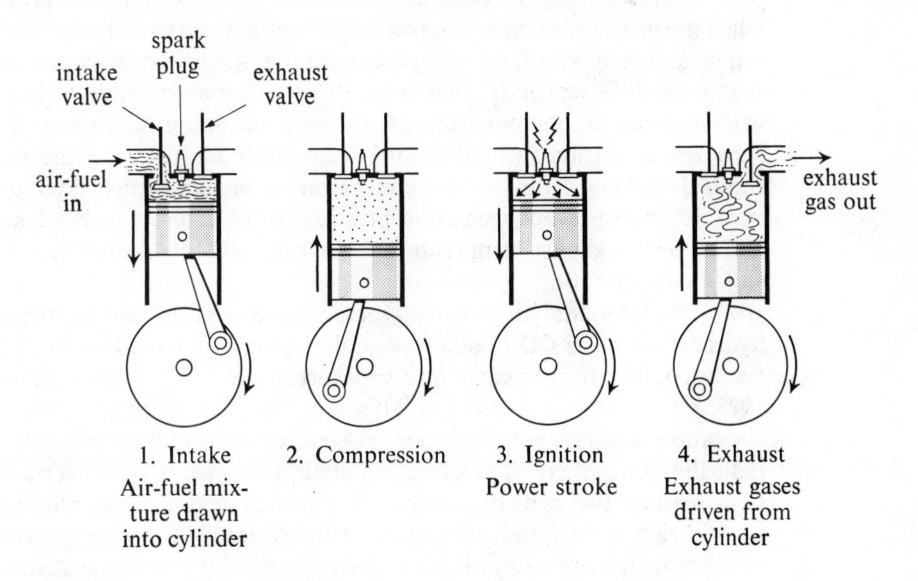

FIGURE 14.2 Steps in one complete cycle of a four-cycle internal combustion engine.

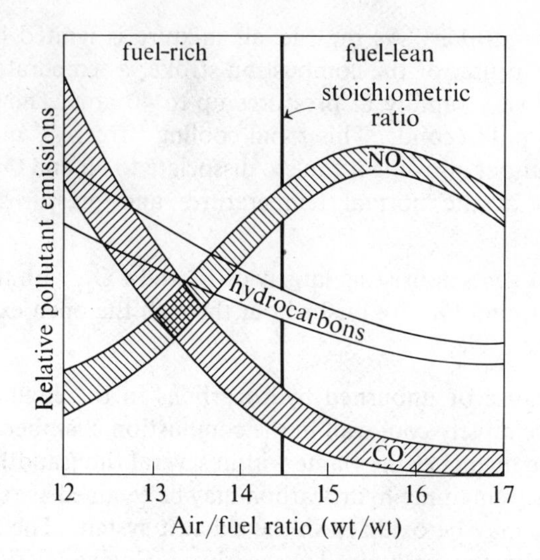

FIGURE 14.3 Effect of air/fuel ratio on pollutant emissions from an internal combustion piston engine.

the spark to a point closer to top dead center reduces the hydrocarbon emissions markedly. One reason for this reduction is that the effective surface/volume ratio of the combustion chamber is reduced, thus decreasing wall quench. Second, when the spark is retarded, the combustion products are purged from the cylinders sooner after combustion. Therefore the exhaust gas is hotter, and reactions consuming hydrocarbons are promoted in the exhaust system.

The air/fuel ratio in the internal combustion engine has a marked effect upon the emission of hydrocarbons, as shown in Figure 14.3. As the air/fuel ratio becomes richer in fuel than the stoichiometric ratio, the emission of hydrocarbons increases sharply. There is a moderate decrease in hydrocarbon emissions when the mixture becomes appreciably leaner in fuel than the stoichiometric ratio requires. The lowest level of hydrocarbon emissions occurs at an air/fuel ratio that is slightly leaner in fuel than the stoichiometric ratio. This behavior is explained by a combination of factors, including minimum quench layer thickness at an air/fuel ratio somewhat richer in fuel than the stoichiometric ratio, decreasing hydrocarbon concentration in the quench layer with a leaner mixture, increasing oxygen concentration in the exhaust with a leaner mixture, and a peak exhaust temperature at a ratio slightly leaner in fuel than the stoichiometric ratio.

Various catalytic exhaust reactors have been devised for the oxidation of hydrocarbons and CO (oxidation catalysts) and the reduction of NO (reduction catalysts). The former came into widespread use in the United States during the 1975 auto model year and the latter were introduced in the 1981 model year. Oxidation catalysts require excess oxygen, whereas reduction catalysts require a reducing atmosphere. Therefore, catalytic removal of both NO and hydrocarbons requires two beds of catalyst. A typical system involves running the engine slightly rich in fuel and passing the exhaust gas first over a catalyst where the nitrogen oxides are reduced. Air is then pumped into the exhaust stream, and the hydrocarbons and carbon monoxide are oxidized on another catalyst bed.

Noble metals (Pd, Pt, Ru) and nonstoichiometric oxides, such as Fe_2O_3 or $CoO \cdot Cr_2O_3$, may be employed as oxidation catalysts [1]. The latter have the property of being able to gain oxygen in their lattice structure, followed by loss of oxygen to the substance being oxidized.

The objective of a reduction catalyst is to reduce NO to harmless N_2. The reduction may be catalyzed by noble metals (Pd, Pt, Ru), base metals (Co, Ni, Cu), or oxides (CuO, $CuCrO_4$). Rhodium is the most widely used reduction catalyst, but tends to be in short supply [2]. With carbon monoxide as the reducing agent, the two major reactions are

$$2\ NO + CO \rightarrow N_2O + CO_2 \tag{14.2.1}$$

$$N_2O + CO \rightarrow N_2 + CO_2 \tag{14.2.2}$$

which add up to the desired net reaction

$$2\ NO + 2\ CO \rightarrow N_2 + 2\ CO_2 \tag{14.2.3}$$

Several side-reactions can occur with a reduction catalyst. One of the least desirable of these is

$$2\ NO + 5\ H_2 \rightarrow 2\ H_2O + 2\ NH_3 \tag{14.2.4}$$

which leads to the formation of ammonia.

Since lead can poison auto exhaust catalysts, automobiles equipped with catalytic exhaust-control devices require lead-free gasoline, and facilities for dispensing this fuel have been installed in all major service stations. Legitimate questions may be raised about the wisdom of using huge quantities of precious platinum metal, most of it imported, to clean up exhausts from engines of basically "dirty" design. Some evidence also suggests that catalytic exhaust reactors result in the production of potentially hazardous levels of sulfuric acid by catalysis of the oxidation of sulfur dioxide from sulfur-containing gasoline:

$$2\ SO_2 + O_2 \rightarrow 2\ SO_3 \tag{14.2.5}$$

$$SO_3 + H_2O \rightarrow H_2SO_4 \tag{14.2.6}$$

The original standards for automotive pollutant emissions were contained in the Clean Air Act of 1970. This act has been amended several times since. The basic standards as of 1982 are given in Table 14.1 (next page). The standards of 1982 are still in effect, as of this writing.

Conditions are favorable for the rapid development of the diesel engine as a major alternative to the gasoline engine in automobiles. The diesel engine employs a very high compression ratio, which gives it an inherently higher efficiency than the gasoline engine. Diesel fuel is less expensive than gasoline and is much less of a fire hazard. Surprisingly, despite the image of a "smelly" diesel truck, a well-adjusted diesel engine emits less hydrocarbon exhaust than does a gasoline engine, although NO_x emissions from diesels are higher. Emissions of particulate matter are also higher.

---

[1] Edwards, J.B. 1974. *Combustion: the formation and emission of trace species* Ann Arbor, Mich: Ann Arbor Science Publishers, Inc.

[2] 1980. Auto emissions control faces new challenges. *Chemical and Engineering News* 17 March 1980.

TABLE 14.1 Levels of Automotive Exhaust Pollutants Permitted under the Clean Air Act of 1970 and Its Amendments

	Allowable emissions, grams per mile[1]							
Pollutant	1975	1976	1977	1978	1979	1980	1981	1982
Hydrocarbons	1.5 (0.9)	1.5 (0.9)	1.5 (0.41)	1.5 (0.41)	1.5 (0.41)	0.41 (0.41)	0.41 (0.41)	0.41 (0.41)
Carbon monoxide	15 (9.0)	15 (9.0)	15 (9.0)	15 (9.0)	15 (9.0)	7.0 (9.0)	3.4 (7.0)	3.4 (7.0)
Nitrogen oxides, NO_x	3.1 (2.0)	3.1 (2.0)	2.0 (1.5)	2.0 (1.5)	2.0 (1.5)	2.0 (1.0)	1.0 (1.0)	1.0 (0.4)

[1] California standards in parentheses.

14.3 REACTIONS OF ORGANIC COMPOUNDS IN THE ATMOSPHERE

Hydrocarbons are eliminated from the atmosphere by a number of chemical and photochemical reactions. These reactions are responsible for the formation of many noxious secondary pollutant products and intermediates from relatively innocuous hydrocarbon precursors. These pollutant products and intermediates are generally classified as photochemical smog.

Hydrocarbons and most other organic compounds in the atmosphere are thermodynamically unstable toward oxidation and tend to be oxidized through a series of steps. The oxidation process terminates with formation of CO_2, solid organic particulate matter which settles from the atmosphere, or water-soluble products (for example, acids, aldehydes) which are removed by rain. As discussed in Section 14.5, inorganic species such as ozone or nitric acid are by-products of these reactions.

Some of the major reactions involved in the oxidation of atmospheric hydrocarbons may be understood by considering the oxidation of methane, the most common and widely dispersed atmospheric hydrocarbon. Like other hydrocarbons, methane reacts with oxygen atoms (generally produced by the photochemical dissociation of NO_2, Reaction 12.12.1, or of O_3, Reaction 11.11.3) to generate the important hydroxyl radical (see Section 11.11) and an organic radical

$$CH_4 + O \rightarrow H_3C\cdot + HO\cdot \qquad (14.3.1)$$

The hydrocarbon (methyl) radical produced reacts rapidly with molecular oxygen to form very reactive peroxyl radicals (in this case, methoxyl) which participate in a number of chain reactions, as discussed in Section 14.5:

$$H_3C\cdot + O_2 + M \rightarrow H_3COO\cdot + M \qquad (14.3.2)$$

The hydroxyl radical reacts rapidly with hydrocarbons to form reactive radicals:

$$CH_4 + HO\cdot \rightarrow H_3C\cdot + H_2O \qquad (14.3.3)$$

Listed at the top of the next page are additional reactions that are involved in the overall oxidation of methane [3].

[3] Sundararaman, B.; Smith, W.; and Rogers, J. 1979. *Federal Aviation Administration High Altitude Pollution Program, Second Biennial Report.* Washington, D.C.: Federal Aviation Administration.

$$H_3COO\cdot + NO \rightarrow H_3CO\cdot + NO_2 \tag{14.3.4}$$

$$H_3CO\cdot + O_3 \rightarrow \text{various products} \tag{14.3.5}$$

$$H_3CO\cdot + O_2 \rightarrow CH_2O + HOO\cdot \tag{14.3.6}$$

$$H_3COO\cdot + NO_2 + M \rightarrow CH_3OONO_2 + M \tag{14.3.7}$$

$$H_2CO + h\nu \rightarrow \text{photodissociation products} \tag{14.3.8}$$

(Reactions of NO with peroxyl radicals, such as Reaction 14.3.4, are a major means of regenerating NO_2 in the atmosphere, after it has been photochemically dissociated to NO.)

As you will see throughout this chapter and Chapter 15, hydroxyl radical, HO ·, and hydroperoxyl radical, HOO ·, are observed throughout photochemical chain-reaction processes. These two species are known collectively as **odd hydrogen radicals**. In urban atmospheres, the major source of odd hydrogen radicals is the following reaction [4]:

$$H\cdot + O_2 + M \rightarrow HOO\cdot + M \tag{14.3.9}$$

A common source of H atoms in the urban atmosphere is the photochemical dissociation of formaldehyde,

$$CH_2O + h\nu \xrightarrow[\lambda\,<\,335\ \mathrm{nm}]{} H\cdot + H\dot{C}O \tag{14.3.10}$$

a process that also yields formyl radical, $H\dot{C}O$. This species, itself, may act as a source of odd hydrogen radical through the reaction,

$$HCO + O_2 + M \rightarrow CO + HOO\cdot + M \tag{14.3.11}$$

(Formaldehyde may also undergo photodissociation at wavelengths below 360 nm to form stable H_2 and CO.)

Reactions such as (14.3.1) and (14.3.3) are **abstraction reactions** involving the removal of an atom, usually hydrogen, by reaction with an active species. **Addition reactions** of organic compounds are also common. Typically, hydroxyl radical reacts with an olefin such as propylene to form another reactive free radical:

$$HO\cdot + H\!-\!\underset{H}{\overset{H}{C}}\!-\!\underset{H}{C}\!=\!C\!\begin{smallmatrix}H\\ \\ H\end{smallmatrix} \longrightarrow H\!-\!\underset{H}{\overset{H}{C}}\!-\!\underset{H}{\dot{C}}\!-\!\underset{H}{\overset{H}{C}}\!-\!OH \tag{14.3.12}$$

Ozone adds to unsaturated compounds to form reactive ozonides:

$$H\!-\!\underset{H}{\overset{H}{C}}\!-\!\underset{H}{C}\!=\!C\!\begin{smallmatrix}H\\ \\ H\end{smallmatrix} + O_3 \longrightarrow H\!-\!\underset{H}{\overset{H}{C}}\!-\!\underset{H}{C}\!\begin{smallmatrix}O\!-\!O\\ \\ O\end{smallmatrix}\!C\!\begin{smallmatrix}H\\ \\ H\end{smallmatrix} \tag{14.3.13}$$

Organic compounds (in the troposphere, almost exclusively carbonyls) can undergo primary photochemical reactions resulting in the direct formation of free

[4] Graedel, T.E.; Farrow, L.A.; and Weber, T.A. 1976. Kinetic studies of the photochemistry of the urban atmosphere. *Atmospheric Environment* 10: 1095–1116.

radicals. By far the most important of these is the photochemical dissociation of aldehydes:

$$CH_3-\overset{\overset{\displaystyle O}{\|}}{C}-H + h\nu \longrightarrow H_3C\cdot + H\dot{C}O \tag{14.3.14}$$

Organic free radicals undergo a number of chemical reactions. Hydroxyl radicals may be generated from organic peroxyl radicals by reactions such as

$$C_3H_7OO\cdot \rightarrow CH_3\overset{\overset{\displaystyle O}{\|}}{C}CH_3 + HO\cdot \tag{14.3.15}$$

leaving an aldehyde or ketone. The hydroxyl radical may react with other organic compounds, maintaining the chain reaction. Chain reactions continue to occur so long as a free radical is the product of the reaction. Gas-phase reaction chains commonly have many steps. Furthermore, chain-branching reactions take place in which a free radical reacts with an excited molecule, causing it to produce two new radicals. Chain termination may occur in several ways, including reaction of two free radicals,

$$2\ HO\cdot \rightarrow H_2 + O_2 \tag{14.3.16}$$

adduct formation with nitric oxide or nitrogen dioxide (which, because of their odd numbers of electrons, are themselves stable free radicals),

$$HO\cdot + NO_2 + M \rightarrow HNO_3 + M \tag{14.3.17}$$

or reaction of the radical with a solid particle surface.

Hydrocarbons may undergo heterogeneous reactions on particles in the atmosphere. Dusts composed of metal oxides or charcoal have a catalytic effect upon the oxidation of organic compounds. Metal oxides may enter into photochemical reactions. For example, zinc oxide photosensitized by exposure to light promotes oxidation of organic compounds.

The general types of reactions just discussed are involved in the formation of photochemical smog in the atmosphere. Having examined the types of reactions that hydrocarbons and other organic compounds undergo in the atmosphere, we may next consider the smog-forming process.

14.4 GENERAL ASPECTS OF PHOTOCHEMICAL SMOG FORMATION

An important characteristic of atmospheres receiving hydrocarbon and NO pollution accompanied by intense sunlight and stagnant air masses is the formation of oxidants. In air-pollution parlance, **gross photochemical oxidant** is a substance in the atmosphere capable of oxidizing iodide ion to iodine(0). Sometimes subtances other than iodide ion are used as reducing agents in the measurement of oxidants. The primary oxidant in the atmosphere is ozone. Other atmospheric oxidants include H_2O_2, organic peroxides (ROOR′), organic hydroperoxides (ROOH), and peroxyacyl nitrates.

Nitrogen dioxide, NO_2, is not considered in the category of gross photochemical oxidants. It is about 15% as efficient as O_3 in oxidizing iodide to iodine(0), and a correction is made in measurements for the positive interference of NO_2. Sulfur dioxide is oxidized by O_3 and produces a negative interference, for which a measurement correction must also be made.

The most notorious organic oxidant is peroxyacetyl nitrate, PAN, which has the following formula:

$$CH_3{-}\overset{\overset{\displaystyle O}{\|}}{C}{-}O{-}O{-}NO_2$$

PAN, as well as other members of a homologous series of compounds, is produced photochemically in atmospheres containing olefins and NO_x. It is damaging to plants, attacking younger leaves and causing "bronzing" and "glazing" of their surfaces. Exposure for several hours to an atmosphere containing PAN at a level of only 0.02–0.05 ppm will damage vegetation. The sulfhydryl group of proteins is susceptible to damage by PAN. In reacting with sulfhydryl groups, PAN may act as both an oxidizing agent and an acetylating agent.

It has been shown that residual PAN from a previous smoggy day enhances the rate of photochemical smog formation [5]. This is due to the following sequence of reactions, which may occur in the absence of light:

$$CH_3{-}\overset{\overset{\displaystyle O}{\|}}{C}{-}O{-}O{-}NO_2 \rightarrow CH_3{-}\overset{\overset{\displaystyle O}{\|}}{C}{-}O{-}O\cdot + NO_2 \qquad (14.4.1)$$

$$CH_3{-}\overset{\overset{\displaystyle O}{\|}}{C}{-}O{-}O\cdot + NO \rightarrow CH_3{-}\overset{\overset{\displaystyle O}{\|}}{C}{-}O\cdot + NO_2 \qquad (14.4.2)$$

$$CH_3\overset{\overset{\displaystyle O}{\|}}{C}{-}O\cdot \rightarrow CH_3\cdot + CO_2 \qquad (14.4.3)$$

$$CH_3\cdot + O_2 + M \rightarrow CH_3OO\cdot + M \qquad (14.4.4)$$

$$CH_3OO\cdot + NO \rightarrow CH_3O\cdot + NO_2 \qquad (14.4.5)$$

$$CH_3O\cdot + O_2 \rightarrow HOO\cdot + H{-}\overset{\overset{\displaystyle O}{\|}}{C}{-}H \qquad (14.4.6)$$

$$HOO\cdot + NO \rightarrow HO\cdot + NO_2 \qquad (14.4.7)$$

$$HO\cdot + NO + M \rightarrow HONO + M \qquad (14.4.8)$$

$$HO\cdot + NO_2 + M \rightarrow HNO_3 + M \qquad (14.4.9)$$

As will be seen in Section 14.5, the decomposition of PAN releases radicals which participate in the photochemical-smog-forming process. Furthermore, NO released at night is converted to photochemically active NO_2 by reactions starting with the decomposition of PAN. In addition, nitrous acid, HONO, is formed, which is a strong photochemical source of HO· radical (see Section 11.11).

---

[5] Carter, W.P.L.; Wincer, A.M.; and Pitts, J.N., Jr. 1981. Effect of peroxyacetyl nitrate on the initiation of photochemical smog. *Environmental Science and Technology* 15: 831–4.

Peroxybenzoyl nitrate, PBN,

$$C_6H_5-\overset{\overset{\displaystyle O}{\|}}{C}-O-O-NO_2$$

is found in polluted atmospheres and is a powerful eye irritant and lachrymator. It may be produced in the laboratory by irradiating mixtures of benzaldehyde, ozone, and nitrogen dioxide in air with ultraviolet light. It is probably produced in the atmosphere by photochemical reactions involving aromatic hydrocarbons, nitrogen oxides, and ozone. In addition to PAN and PBN, some other specific organic oxidants that may be important in polluted atmospheres are peroxypropionyl nitrate (PPN); peracetic acid, $CH_3(CO)OOH$; acetylperoxide, $CH_3(CO)OO(CO)CH_3$; ethyl hydroperoxide, CH_3CH_2OOH; *n*-butylhydroperoxide, $CH_3CH_2CH_2CH_2OOH$; and *tert*-butylhydroperoxide, $(CH_3)_3COOH$.

It has been shown that peroxybenzoylnitrate is a chemically active species even in the dark [6]. It undergoes reactions analogous to those just discussed for PAN and shown in the reaction sequence 14.4.1 through 14.4.9.

Although the two types of smog used to be treated as essentially opposite phenomena, it is now recognized that there are important interactions between photochemical and sulfurous smog. This interaction is discussed in greater detail in Section 14.5.

Smoggy atmospheres show characteristic variations with time of day in levels of NO, NO_2, hydrocarbons, aldehydes, and oxidants. A generalized plot illustrating these variations is shown in Figure 14.4. Examination of the figure shows that, shortly after dawn, the level of NO in the atmosphere decreases markedly, a decrease that is accompanied by a peak in the concentration of NO_2. During midday (significantly, after the concentration of NO has fallen to a very

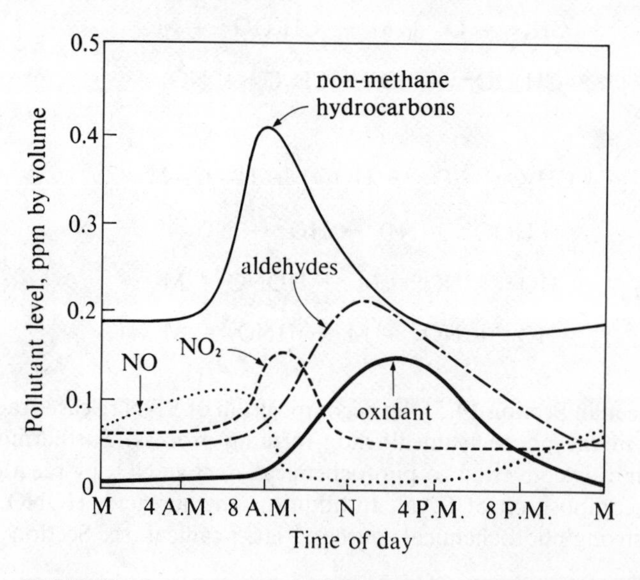

FIGURE 14.4 Generalized plot of atmospheric concentrations of species involved in smog formation as a function of time of day.

[6] Ohta, T., and Mizoguchi, I. 1981. Thermal decomposition rate constant of peroxybenzoyl nitrate in the gas phase. *Environmental Science and Technology* 15: 1229–32.

low level), the levels of aldehydes and oxidants become relatively high. The concentration of total hydrocarbons in the atmosphere peaks sharply in the morning, then decreases during the remaining daylight hours. The chemical basis for this behavior is explained in the following section.

14.5 MECHANISMS OF PHOTOCHEMICAL SMOG FORMATION

In this section, we discuss some of the primary aspects of photochemical smog formation. This discussion is based largely on publications by Graedel [7]; Whitten, Hogo, and Killus [8]; Falls and Seinfeld [9]; and Pitts and Finlayson [10]. The reader is referred to these works and the references in them for more detailed discussion of photochemical smog formation. Since the exact chemistry of photochemical smog formation is not known, many of the reactions given in the following discussion should be viewed as plausible illustrative examples rather than proven mechanisms.

The kind of behavior summarized in Figure 14.4 contains several apparent anomalies which puzzled scientists for many years. The first of these was the rapid *increase* in NO_2 concentration and *decrease* in NO concentration under conditions where it was known that photodissociation of NO_2 to O and NO was occurring (Reaction 14.5.1). Furthermore, it could be shown that the disappearance of olefins and other hydrocarbons was much more rapid than could be explained by the relatively slow reactions of O_3 and O. The anomalies are now explained by chain reactions involving the interconversion of NO and NO_2, the oxidation of hydrocarbons, and the generation of reactive intermediates, particularly hydroxyl radical ($HO\cdot$). These changes are explained by the generalized reaction scheme shown in Figure 14.5 (see page 390). This scheme is based upon the photochemically initiated reactions that occur in an atmosphere containing nitrogen oxides, reactive hydrocarbons, and, of course, oxygen. The time variations in levels of hydrocarbons, ozone, NO, and NO_2 are explained by the following overall reactions:

1. Primary photochemical reaction, which produces ground-state oxygen atoms, $O(^3P)$:

$$NO_2 + h\nu \xrightarrow[\lambda < 420\text{ nm}]{} NO + O \qquad (14.5.1)$$

   (The photochemistry of nitrogen dioxide is discussed in greater detail in Section 12.12.)

2. Reactions involving oxygen species (M is an energy-absorbing third body):

$$O_2 + O + M \rightarrow O_3 + M \qquad (14.5.2)$$

---

[7] Graedel, T.E. 1980. Atmospheric photochemistry. In *The handbook of environmental chemistry,* Volume 2, Part A, O. Hutzinger, ed., pp. 107–143. Berlin: Springer-Verlag.

[8] Whitten, G.Z.; Hogo, H.; and Killus, J.P. 1980. The carbon-bond mechanism: a condensed kinetic mechanism for photochemical smog. *Environmental Science and Technology* 14: 690–700.

[9] Falls, A.H., and Seinfeld, J.H. 1978. Continued development of a kinetic mechanism for photochemical smog. *Environmental Science and Technology* 12: 1398–1406.

[10] Pitts, J.N., Jr., and Finlayson, B.J. 1975. Mechanism of photochemical air pollution. *Angewandte Chemie* (International Edition in English) 14: 1–15.

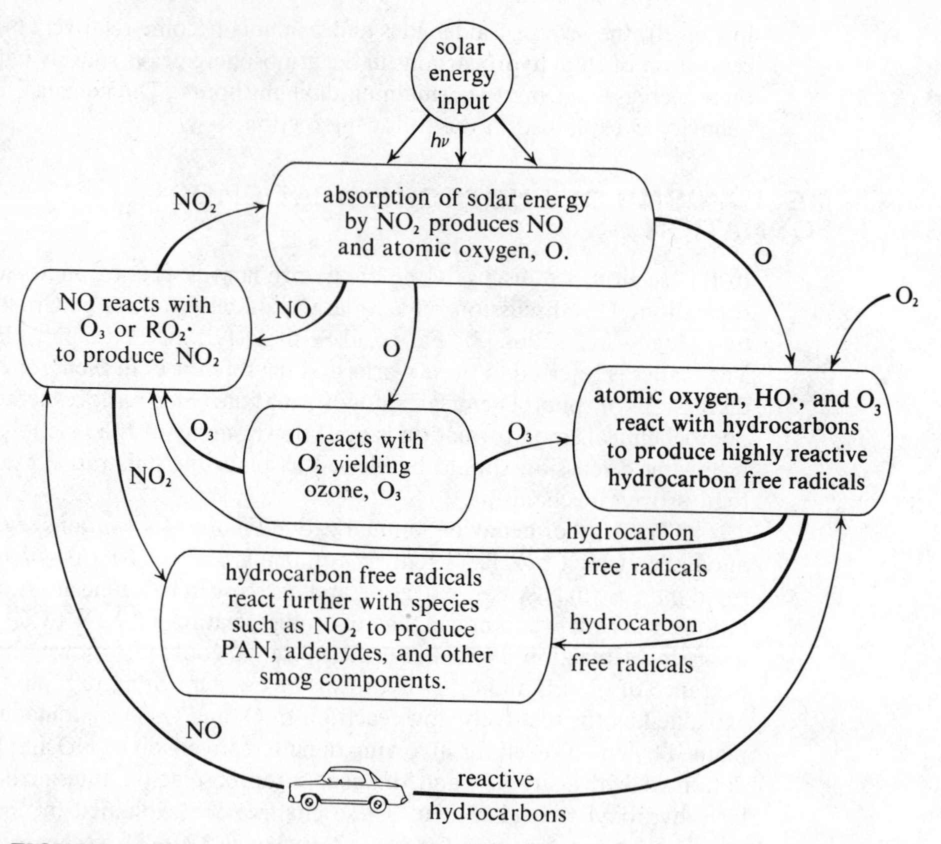

FIGURE 14.5 Generalized scheme for the formation of photochemical smog.

$$O_3 + NO \rightarrow NO_2 + O_2 \quad (14.5.3)$$

The latter reaction is rapid; therefore, the concentration of O_3 remains low until that of NO falls to a low value. Emissions of NO along freeways tend to keep O_3 concentrations low there.

3. Production of organic free radicals from hydrocarbons, RH:

$$O + RH \rightarrow R\cdot + \text{other products} \quad (14.5.4)$$

$$O_3 + RH \rightarrow R\cdot \text{ and/or other products} \quad (14.5.5)$$

(R· is a free radical which may or may not contain oxygen.)

4. Chain propagation, branching, and termination:

$$NO + R\cdot \rightarrow NO_2 + R\cdot \quad (14.5.6)$$

In this case, R contains oxygen and oxidizes NO. This reaction is one of many chain-propagation reactions, some of which involve NO.

$$NO_2 + R\cdot \rightarrow \text{products (e.g., PAN)} \quad (14.5.7)$$

This is the most common chain-terminating reaction, which occurs because NO_2 is a stable free radical present at high concentrations in a smoggy atmosphere. Reaction with NO may also occur. Chains may be terminated by reaction of two R· radicals, although this is uncommon because of their relatively low concentrations. Radical sorption on a particle surface is also possible and may contribute to particle growth.

A number of specific reactions are involved in this overall scheme for the formation of photochemical smog. The formation of atomic oxygen by a primary photochemical reaction (Reaction 14.5.1) leads to several reactions involving oxygen and nitrogen oxide species:

$$O + O_2 + M \rightarrow O_3 + M \quad (14.5.8)$$

$$O + NO + M \rightarrow NO_2 + M \quad (14.5.9)$$

$$O + NO_2 \rightarrow NO + O_2 \quad (14.5.10)$$

$$O_3 + NO \rightarrow NO_2 + O_2 \quad (14.5.11)$$

$$O + NO_2 + M \rightarrow NO_3 + M \quad (14.5.12)$$

$$O_3 + NO_2 \rightarrow NO_3 + O_2 \quad (14.5.13)$$

There are a number of significant atmospheric reactions involving nitrogen oxides, water, nitrous acid, and nitric acid:

$$NO_3 + NO_2 \rightarrow N_2O_5 \quad (14.5.14)$$

$$N_2O_5 \rightarrow NO_3 + NO_2 \quad (14.5.15)$$

$$NO_3 + NO \rightarrow 2\ NO_2 \quad (14.5.16)$$

$$N_2O_5 + H_2O \rightarrow 2\ HNO_3 \quad (14.5.17)$$

(This reaction is slow in the gas phase but may be fast on surfaces.)

$$NO + HNO_3 \rightarrow HNO_2 + NO_2 \quad (14.5.18)$$

$$HNO_2 + HNO_3 \rightarrow H_2O + 2\ NO_2 \quad (14.5.19)$$

$$NO + NO_2 + H_2O \rightarrow 2\ HNO_2 \quad (14.5.20)$$

$$2\ HNO_2 \rightarrow NO + NO_2 + H_2O \quad (14.5.21)$$

Very reactive hydroxyl radicals, $HO\cdot$, can be formed by the reaction of excited atomic oxygen with water,

$$O^* + H_2O \rightarrow 2\ HO\cdot \quad (14.5.22)$$

by photodissociation of hydrogen peroxide,

$$H_2O_2 + h\nu \xrightarrow[\lambda < 350\ nm]{} 2\ HO\cdot \quad (14.5.23)$$

or by the photolysis of nitrous acid,

$$HNO_2 + h\nu \rightarrow HO\cdot + NO \quad (14.5.24)$$

Among the inorganic species with which the hydroxyl radical reacts are oxides of nitrogen,

$$HO\cdot + NO_2 \rightarrow HNO_3 \quad (14.5.25)$$

$$HO\cdot + NO + M \rightarrow HNO_2 + M \quad (14.5.26)$$

and carbon monoxide,

$$CO + HO\cdot + O_2 \rightarrow CO_2 + HOO\cdot \quad (14.5.27)$$

The last reaction is significant in that it is responsible for the disappearance of much atmospheric CO (see Section 12.4) and because it produces the hydroperoxyl radical HOO·. One of the major inorganic reactions of the hydroperoxyl radical is the oxidation of NO:

$$HOO\cdot + NO \rightarrow HO\cdot + NO_2 \qquad (14.5.28)$$

The kinetic rate constants for most of the preceding reactions are reasonably well known, enabling calculations of species concentrations to be expected under various conditions. These calculations, which are supported by laboratory studies of irradiated mixtures, cannot explain the rapid transformation of NO to NO_2 that occurs in an atmosphere undergoing photochemical smog formation. In fact, except at very high CO concentrations (see Reactions 14.5.27 and 14.5.28), the concentration of NO_2 is predicted to remain at very low levels. However, in the presence of reactive hydrocarbons, NO_2 accumulates very rapidly by a reaction process beginning with its photodissociation! It may be concluded, therefore, that the organic compounds form species which react with NO directly, rather than with NO_2.

A number of chain reactions have been shown to result in the general type of species behavior shown in Figure 14.4. Aliphatic hydrocarbons, RH, can react with O or O_3,

$$RH + O + O_2 \rightarrow ROO\cdot + HO\cdot \qquad (14.5.29)$$

or HO· radical,

$$RH + HO\cdot + O_2 \rightarrow ROO\cdot + H_2O \qquad (14.5.30)$$

to produce reactive oxygenated organic radicals, ROO·. Olefins (R = R′) are much more reactive, undergoing overall reactions such as

$$R{=}R' + O + 2\,O_2 \rightarrow ROO\cdot + R''\overset{O}{\overset{\|}{C}}OO\cdot \qquad (14.5.31)$$

$$R{=}CH_2 + O + 2\,O_2 \rightarrow ROO\cdot + HOO\cdot + CO \qquad (14.5.32)$$

$$R{=}R' + HO\cdot + O_2 \rightarrow ROO\cdot + H\overset{O}{\overset{\|}{C}}R'' \text{ (an aldehyde)} \qquad (14.5.33)$$

Aromatic hydrocarbons, ϕH, may also react with O and HO· :

$$\phi H + O + O_2 \rightarrow \phi OO\cdot + HO\cdot \qquad (14.5.34)$$

$$\phi H + HO\cdot + O_2 \rightarrow \phi OO\cdot + H_2O \qquad (14.5.35)$$

Addition reactions of aromatics with HO· are favored [11]. The product is an alcohol (phenol), as shown by the following reaction sequence [12] (see Equations 14.5.36 and 14.5.37).

---

[11] Perry, R.A.; Atkinson, R.; and Pitts, J.N., Jr. 1977. Kinetics and mechanism of the gas phase reaction of OH radicals with aromatic hydrocarbons over the temperature range 296–473 K. *J. Phys. Chem.* 81: 296–304.

[12] Graedel, T.E. 1978. *Chemical compounds in the atmosphere.* New York: Academic Press, Inc.

$$\text{C}_6\text{H}_6 + \text{HO}\cdot \longrightarrow \text{C}_6\text{H}_6(\text{OH})\cdot \tag{14.5.36}$$

$$\text{C}_6\text{H}_6(\text{OH})\cdot + \text{O}_2 \longrightarrow \text{C}_6\text{H}_5\text{OH} + \text{HO}_2 \tag{14.5.37}$$

In the case of alkyl benzenes, such as toluene, the hydroxyl radical attack may occur on the alkyl group, leading to reaction sequences such as those of alkanes.

Aldehydes react with HO ·,

$$\text{R}''\overset{\text{O}}{\overset{\|}{\text{C}}}\text{H} + \text{HO}\cdot + \text{O}_2 \rightarrow \text{R}''\overset{\text{O}}{\overset{\|}{\text{C}}}\text{OO}\cdot + \text{H}_2\text{O} \tag{14.5.38}$$

$$\underset{\text{(formaldehyde)}}{\text{H}_2\text{C}{=}\text{O}} + \text{HO}\cdot + \frac{3}{2}\,\text{O}_2 \rightarrow \text{CO}_2 + \text{HOO}\cdot + \text{H}_2\text{O} \tag{14.5.39}$$

and undergo photochemical reactions:

$$\text{R}''\overset{\text{O}}{\overset{\|}{\text{C}}}\text{H} + h\nu + 2\,\text{O}_2 \rightarrow \text{R}''\text{OO}\cdot + \text{CO} + \text{HOO}\cdot \tag{14.5.40}$$

$$\text{H}_2\text{C}{=}\text{O} + h\nu + 2\,\text{O}_2 \rightarrow \text{CO} + 2\,\text{HOO}\cdot \tag{14.5.41}$$

Earlier theories of hydrocarbon photooxidation assumed that O and O_3 were primarily responsible for the oxidation. However, the concentration of O in a smog-forming atmosphere is so small (about 1×10^{-8} ppm) and the reactions of O_3 with alkanes or aromatics are so slow that it was impossible to account for the rapid disappearance of hydrocarbons in smog. More recent data on reaction rates have shown the crucial roles played by hydroxyl (HO ·) and hydroperoxyl (HOO ·) radicals, especially the former, in hydrocarbon oxidation. Hydroxyl radical, which reacts with some hydrocarbons at rates that are almost diffusion-controlled, is the predominant reactant in early stages of smog formation. Significant contributions are made by hydroperoxyl radical and O_3 after the smog-forming process has been underway for some time (as indicated by significantly lowered values of NO concentrations).

One of the most important reaction sequences in the smog-forming process begins with the abstraction by HO · of a hydrogen atom from a hydrocarbon and leads to the oxidation of NO to NO_2 as follows:

$$\text{RH} + \text{HO}\cdot \rightarrow \text{R}\cdot + \text{H}_2\text{O} \tag{14.5.42}$$

The alkyl radical, R ·, reacts with O_2 to produce a peroxyl radical, ROO · :

$$\text{R}\cdot + \text{O}_2 \rightarrow \text{ROO}\cdot \tag{14.5.43}$$

This strongly oxidizing species oxidizes NO to NO_2, thus explaining what was once a major unknown in the understanding of the smog-forming process:

$$\text{ROO}\cdot + \text{NO} \rightarrow \text{RO}\cdot + \text{NO}_2 \tag{14.5.44}$$

The alkoxyl radical product, RO·, is not so stable as ROO·. In cases where the oxygen atom is attached to a carbon atom that is also bonded to H, a carbonyl compound is likely to be formed by the following type of reaction:

$$H_3CO\cdot + O_2 \rightarrow H{-}\overset{\overset{\displaystyle O}{\|}}{C}{-}H + HOO\cdot \tag{14.5.45}$$

The rapid production of photosensitive carbonyl compounds from alkoxyl radicals is an important stimulant for further atmospheric photochemical reactions. In the absence of extractable hydrogen, cleavage of a radical containing the carbonyl group occurs:

$$H_3\overset{\overset{\displaystyle O}{\|}}{C}O\cdot \rightarrow H_3C\cdot + CO_2 \tag{14.5.46}$$

Another reaction that can lead to the oxidation of NO is of the following type:

$$R\overset{\overset{\displaystyle O}{\|}}{C}OO\cdot + NO + O_2 \rightarrow ROO\cdot + NO_2 + CO_2 \tag{14.5.47}$$

Peroxyacyl nitrates (PAN) are formed by an addition reaction with NO_2:

$$R\overset{\overset{\displaystyle O}{\|}}{C}OO\cdot + NO_2 \rightarrow R\overset{\overset{\displaystyle O}{\|}}{C}OONO_2 \tag{14.5.48}$$

These are highly significant air pollutants, discussed in Section 14.4. Alkyl nitrates and alkyl nitrites may be formed by the reaction of alkoxyl radicals (RO·) with nitrogen dioxide and nitric oxide, respectively:

$$RO\cdot + NO_2 \rightarrow RONO_2 \tag{14.5.49}$$

$$RO\cdot + NO \rightarrow RONO \tag{14.5.50}$$

Addition reactions with NO_2 such as these are important in terminating the reaction chains involved in smog formation. Since NO_2 is involved both in the chain-initiation step (Reaction 14.5.1) and the chain termination step, it is unlikely that moderate reductions in NO_x emissions alone will have much influence on curtailing smog formation.

As shown in Reaction 14.5.45, the reaction of oxygen with alkoxyl radicals produces hydroperoxyl radical. Peroxyl radicals can react with one another to produce reactive hydrogen peroxide, alkoxyl radicals, and hydroxyl radicals:

$$HOO\cdot + HOO\cdot \rightarrow H_2O_2 + O_2 \tag{14.5.51}$$

$$HOO\cdot + ROO\cdot \rightarrow RO\cdot + HO\cdot + O_2 \tag{14.5.52}$$

$$ROO\cdot + ROO\cdot \rightarrow 2\,RO\cdot + O_2 \tag{14.5.53}$$

It may be useful at this time to review the types of compounds capable of undergoing photolysis in the troposphere and thus initiating chain reactions. The most important of these in most tropospheric situations is NO_2:

$$NO_2 + h\nu \xrightarrow[\lambda\, <\, 420\ nm]{} NO + O \tag{14.5.1}$$

In relatively polluted atmospheres, the next most important photodissociation reaction is that of carbonyl compounds, particularly formaldehyde:

$$CH_2O + h\nu \xrightarrow[\lambda\, <\, 335\ nm]{} H\cdot + H\dot{C}O \tag{14.3.10}$$

Hydrogen peroxide photodissociates to produce two hydroxyl radicals:

$$HOOH + h\nu \xrightarrow[\lambda < 350 \text{ nm}]{} 2\ HO\cdot \tag{14.5.54}$$

Finally, organic peroxides may be formed and subsequently dissociate by the following reactions, starting with a peroxyl radical:

$$H_3COO\cdot + HOO\cdot \rightarrow H_3COOH + O_2 \tag{14.5.55}$$

$$H_3COOH + h\nu \xrightarrow[\lambda < 350 \text{ nm}]{} H_3CO\cdot + HO\cdot \tag{14.5.56}$$

It should be noted that the last three photochemical reactions each give rise to two free radical species per photon absorbed. Ozone undergoes photochemical dissociation to produce excited oxygen atoms, $O(^1D)$, at wavelengths less than 315 nm. These atoms may react with H_2O to produce hydroxyl radicals.

It is possible that electronically excited singlet molecular oxygen, designated here as $O_2{}^*(^1\Delta g)$, may be a minor oxidizing agent in the atmosphere. The absorption of light by ground-state oxygen, $O_2(^3\Sigma g)$, occurs in the wavelength regions of 687–692 nm and 759–766 nm. The absorption is too weak to account for any appreciable level of singlet oxygen in the atmosphere. However, oxygen may be excited to the singlet state by energy transfer from another excited species. Ultraviolet irradiation of mixtures of benzaldehyde and oxygen has produced singlet oxygen, as evidenced by its emission of light of 1.27-micron wavelength. Although it has not been established that singlet oxygen participates significantly in atmospheric oxidation processes, it is known to react with olefins to produce peroxidic organic compounds. These compounds could then decompose to form radicals of the type RCO·, thus starting the chain reaction process involved in smog formation. Some authorities have suggested that excited singlet molecular oxygen has adverse health effects because of reactions with tissue.

14.6 REACTIVITY OF HYDROCARBONS IN SMOG FORMATION

The reactivity of hydrocarbons in the smog formation process is an important consideration in understanding the process and in developing control strategies. The most reactive hydrocarbons should be determined, so that their release can be minimized. Less reactive hydrocarbons, of which propane is a good example, may cause smog formation far downwind from the point of release.

Numerous attempts have been made to classify hydrocarbon reactivity. Because of the key role played in hydroxyl radical in smog formation, relative reaction rates with HO· are now widely used to categorize hydrocarbon reactivity [13, 14, 15]. Methane is relatively slow to react and is assigned a reactivity of 1.0.

---

[13] Darnall, K.R.; Lloyd, A.C.; Winer, A.M.; and Pitts, J.N., Jr. 1976. Reactivity scale for atmospheric hydrocarbons based on reaction with hydroxyl radical. *Environmental Science and Technology* 10: 692–6.

[14] Calvert, J.G. 1976. Hydrocarbon involvement in photochemical smog formation in Los Angeles atmosphere. *Environmental Science and Technology* 10: 256–62.

[15] Doyle, G.J.; Lloyd, A.C.; Darnall, K.R.; Winer, A.M.; and Pitts, J.N., Jr. 1975. Gas phase kinetic study of relative rates of reaction of selected aromatic compounds with hydroxyl radicals in an environmental chamber. *Environmental Science and Technology* 9: 237–41.

TABLE 14.2 Relative Reactivities of Hydrocarbons and CO with HO· Radical[1]

Reactivity class	*Reactivity range*[2]	*Approximate half-life in the atmosphere*	*Compounds in increasing order of reactivity*
I	<10	>10 days	methane
II	10–100	24 h–10 d	CO, acetylene, ethane
III	100–1000	2.4–24 h	benzene, propane, *n*-butane, isopentane, methyl ethyl ketone, 2-methylpentane, toluene, *n*-propylbenzene, isopropylbenzene, ethene, *n*-hexane, 3-methylpentane, ethylbenzene
IV	1,000–10,000	15 min–2.4 h	*p*-xylene, *p*-ethyltoluene, *o*-ethyltoluene, *o*-xylene, methyl isobutyl ketone, *m*-ethyltoluene, *m*-xylene, 1,2,3-trimethylbenzene, propene, 1,2,4-trimethylbenzene, 1,3,5-trimethylbenzene, *cis*-2-butene, β-pinene, 1,3-butadiene
V	>10,000	<15 min	2-methyl-2-butene, 2,4-dimethyl-2-butene, *d*-limonene

[1]Based on data from K.R. Darnall, A.C. Lloyd, A.M. Winer, and J.N. Pitts, Jr., "Reactivity Scale for Atmospheric Hydrocarbons Based on Reaction with Hydroxyl Radical," *Environmental Science and Technology* 10: 692–6 (1976).

[2]Based on an assigned reactivity of 1.0 for methane reacting with hydroxyl radical.

(Because of its abundance in the atmosphere, however, methane accounts for a significant fraction of total hydroxyl radical reactions.) Carbon monoxide has a reactivity of 18 compared to methane. One useful classification system categorizes hydrocarbons in classes ranging from I through V on the basis of their reactivity with hydroxyl radicals compared to the reactivity of methane with hydroxyl radical [13]. The specific compounds are listed in Table 14.2. It is interesting to note that β-pinene (produced by conifer trees and other vegetation) and *d*-limonene (which occurs in orange rind) have reactivities of 8750 and 18,800, respectively. The latter compound is the most reactive one listed, showing that even natural products can contribute to pollution.

14.7 EFFECTS OF PHOTOCHEMICAL SMOG

The harmful effects of smog occur mainly in the areas of (1) human health and comfort; (2) effects on material; (3) effects on the atmosphere; and (4) effects on plants. The exact degree to which exposure to smog affects human health is not known, although dire predictions abound. Ozone, a component of smoggy atmospheres, is known to be damaging to health and has a pungent odor. Peroxyacyl nitrates and aldehydes found in smog are eye irritants. Materials are adversely affected by some smog components. Ozone cracks and ages rubber, which has a strong affinity for O_3. Indeed, the cracking of rubber is one test for ozone. (An interesting story is told about two analytical laboratories that differed consistently on the results of ozone analysis until it was pointed out that the one getting lower results was collecting the sample through a rubber tube!)

In 1978, the U.S. Environmental Protection Agency changed the primary ambient air quality standard for ozone, used as a measure of smog, from 0.08

ppm to 0.10 ppm. This move followed laboratory tests showing no adverse effects upon asthmatics, the most sensitive group, after exposure to 0.10 ppm ozone. However, ozone at 0.15 ppm does cause coughing, wheezing, bronchial constriction, and irritation to the respiratory mucous system in healthy, exercising individuals. The current ozone limit is 0.12 ppm.

Ozone attacks natural rubber and similar materials by oxidizing and breaking double bonds in the polymer according to the following reaction:

$$\left[CH_2{-}\overset{\overset{\textstyle H}{|}}{C}{=}\overset{\overset{\textstyle CH_3}{|}}{C}{-}CH_2 \right]_n + O_3 \rightarrow R{-}\overset{\overset{\textstyle H}{|}}{C}\underset{O}{\overset{O{-}O}{\diagdown\!\!\diagup}}\overset{\overset{\textstyle CH_3}{|}}{C}{-}R' \xrightarrow{H_2O} R{-}\overset{\overset{\textstyle O}{\|}}{C}{-}OH + R'{-}\overset{\overset{\textstyle O}{\|}}{C}{-}CH_3$$

(natural rubber polymer)

This oxidative scission type of reaction causes breakage of the bonds that are part of the polymer structure and results in deterioration of the polymer.

Aerosol particles that reduce visibility are formed by the polymerization of the smaller molecules produced in the primary smog-forming reactions. Since these reactions largely involve the oxidation of hydrocarbons, it is not surprising that oxygen-containing organics make up the bulk of the particulate matter produced from smog. Ether-soluble aerosols collected from the Los Angeles atmosphere have shown an empirical formula of approximately CH_2O. Among the specific types of compounds identified in organic smog aerosols are alcohols, aldehydes, ketones, organic acids, esters, and organic nitrates.

It is generally believed that smog aerosol particles form by condensation on existing nuclei rather than by self-nucleation of smog reaction product molecules. Electron micrographs of these aerosols support that view [16], showing that smog aerosol particles in the micron-size region consist of liquid droplets with a core which is opaque to electrons (Fig. 14.6). Thus, particulate matter from a source other than smog may have some influence on the formation and nature of smog aerosols.

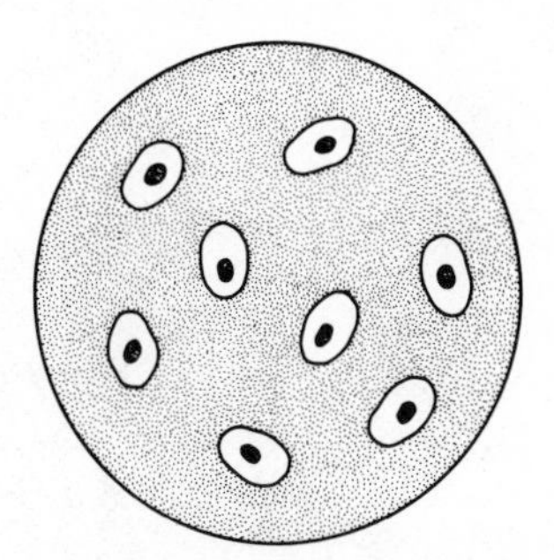

FIGURE 14.6 Representation of an electron micrograph of smog aerosol particles collected by a jet inertial impactor, showing electron-opaque nuclei in the centers of the impacted droplets.

---

[16] Husar, R.B.; White, W.H.; and Blumenthal, D.L. 1976. Direct evidence of heterogeneous aerosol formation in Los Angeles smog. *Environmental Science and Technology* 10: 490–1.

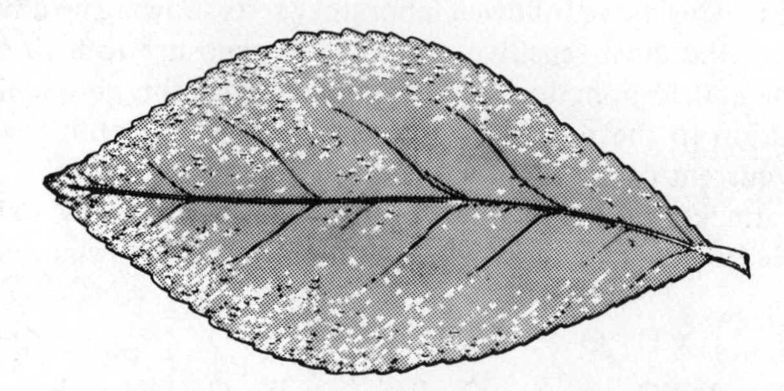

FIGURE 14.7 Ozone damage to a lemon leaf. The spots are yellow chlorotic stippling on the upper surface resulting from ozone exposure. (Color photos of leaves damaged by ozone may be found in "Identification of Air Pollution Damage to Agricultural Crops," E.F. Darley, C.W. Nichols, and J.T. Middleton, *The Bulletin,* 55, Department of Agriculture, State of California, 1966.)

A very frightening aspect of smog, in view of worldwide food scarcities, is the known harmful effect on plants. These effects are largely due to oxidants in the smoggy atmosphere. The three major oxidants involved are ozone, PAN, and nitrogen oxides. PAN has the highest toxicity to plants but is usually present at relatively low levels. Nitrogen oxides occur at relatively high concentrations during smoggy conditions, but their toxicity to plants is relatively low. Therefore, ozone is the greatest problem.

Ozone damage to a lemon leaf is typified by chlorotic stippling (characteristic yellow spots on a green leaf), illustrated in Figure 14.7. Reduction in plant growth may occur without visible lesions on the plant. Brief exposure to approximately 0.06 ppm of ozone may temporarily cut photosynthesis rates in half. Crop damage from ozone and other photochemical air pollutants in California is estimated to cost millions of dollars each year. The geographic distribution of this damage is shown in Figure 14.8.

FIGURE 14.8 Geographic distribution of plant damage from smog in California.

Data gathered by the National Crop Loss Assessment Network has estimated that ozone causes losses of $2 billion to $4 billion per year in the production of corn, wheat, soybeans, and peanuts [17]. These figures point out the very real economic damage caused by air pollution.

The economic costs of measures required to alleviate the effects of severe smog episodes are quite high. For example, on July 14, 1978, the South Coast Air Quality Management District in California required industrial concerns in the Los Angeles area to take severe measures to cut back on emissions. This was the first of what are expected to be a large number of smog alerts. To reduce emissions, coke oven production was cut back, solvent plant operations were stopped, and power companies switched from fuel oil to more scarce natural gas. Affecting 3000 businesses, the alert cost an estimated $5 million in production losses.

14.8 SULFATES AND NITRATES IN PHOTOCHEMICAL SMOG

The formation of sulfate in connection with photochemical smog can aggravate the smog problem. Sulfate may be formed in the atmosphere by the oxidation of either H_2S or SO_2; the latter is the greater pollution problem. Sulfates contribute to acid rainfall, corrosion, reduced visibility, and even adverse health effects.

Although, as noted in Section 12.9, the oxidation of SO_2 to sulfate species is relatively slow in a clean atmosphere, it is much faster under smoggy conditions. During severe photochemical smog conditions, oxidation rates of 5–10% per hour may occur, as compared to only a fraction of a percent per hour under normal atmospheric conditions. Thus, sulfur dioxide pollution in connection with smoggy conditions could result in very high local concentrations of sulfate, which would aggravate an already bad atmospheric condition.

The increased oxidation rate of SO_2 in the presence of photochemical smog is due to one of several oxidizing species. Among the possibilities are oxidizing compounds, including O_3, NO_3, and N_2O_5, as well as reactive radical species, including $HO\cdot$, $HOO\cdot$, O, $RO\cdot$, and $ROO\cdot$. The two major primary reactions are oxygen atom transfer,

$$SO_2 + O\ (\text{from O, RO}\cdot\text{, ROO}\cdot) \rightarrow SO_3 \rightarrow H_2SO_4\text{, sulfates} \qquad (14.8.1)$$

or addition. As an example of addition, $HO\cdot$ adds to SO_2 to form a reactive species which can further react with oxygen, nitrogen oxides, or other species to yield sulfates, other sulfur compounds, or compounds of nitrogen:

$$HO\cdot + SO_2 \rightarrow HOSOO\cdot \qquad (14.8.2)$$

The presence of $HO\cdot$ radical (typically 3×10^6 radicals/cm^3, but appreciably higher in smoggy atmospheres) makes this a likely route. Addition of SO_2 to $RO\cdot$ or $ROO\cdot$ can yield organic sulfur compounds.

It should be noted that the reaction of H_2S with $HO\cdot$ is quite rapid. As a result, the half-life of H_2S in a typical atmosphere is about one-half day. It is much shorter in the presence of photochemical smog.

---

[17] Shabecoff, P. 1982. Losses of 4 crops laid to effect of ozone pollution. *New York Times* 18 February 1982.

Inorganic nitrates or nitric acid are formed by several reactions in a smoggy atmosphere. Among the more important reactions forming nitric acid are the reaction of N_2O_5 with water (Reaction 14.5.17) and the reaction of NO_2 with hydroxyl radical (Reaction 14.5.25). The oxidation of NO or NO_2 to nitrate species may occur after absorption of the gas by an aerosol droplet. Nitric acid formed by these reactions reacts with ammonia in the atmosphere to form ammonium nitrate:

$$NH_3 + HNO_3 \rightarrow NH_4NO_3 \quad (14.8.3)$$

Other nitrate salts may also be formed.

Nitric acid and nitrates are among the more damaging products of photochemical smog formation. In addition to plant damage and possible human health effects, they cause severe corrosion problems. Electrical relay contacts and small springs associated with electrical switches are especially susceptible to damage from nitrate-induced corrosion.

SUPPLEMENTARY REFERENCES

Baker, B.G. 1977. Control of noxious emissions from internal combustion engines. In *Environmental chemistry,* J. O'M. Bockris, ed., Ch. 9, pp. 243–84. New York: Plenum Publishing Corp.

Butler, J.D. 1979. *Air pollution chemistry.* New York: Academic Press, Inc.

Calvert, J.G., and Pitts, J.N., Jr. 1966. *Photochemistry.* New York: John Wiley and Sons, Inc.

Chameides, W.L., and Davis, D.D. 1982. Chemistry in the troposphere. *Chemical and Engineering News* 4 October 1982, pp. 38–52.

Dotto, L., and Schiff, H. 1978. *The ozone war.* Garden City, N.Y.: Doubleday & Co., Inc.

Finlayson, B.J., and Pitts, J.N., Jr. 1976. Photochemistry of the polluted troposphere. *Science* 192: 111–9.

Haagen-Smit, A.J., and Wayne, L.G. 1976. Atmospheric reactions and scavenging processes. In *Air pollution,* 3rd ed., Vol. 1, Ch. 6, pp. 235–89. New York: Academic Press, Inc.

Hampson, R.F., Jr., and Garvin, D., eds. 1978. *Reaction rate and photochemical data for atmospheric chemistry—1977.* Washington, D.C.: National Bureau of Standards.

Heicklen, J. 1976. *Atmospheric chemistry.* New York: Academic Press, Inc.

Hidy, G.M., et al., eds. 1980. *The character and origins of smog particles.* New York: Wiley-Interscience.

Horne, R.A. 1978. *The chemistry of air pollution.* In *The chemistry of our environment,* Ch. 5. New York: John Wiley and Sons, Inc.

Logan, J.A.; Prather, M.J.; Wofsy, S.C.; and McElroy, M.B. 1981. Tropospheric chemistry: a global perspective. *Journal of Geophysical Research* 85: 7210.

McEwan, M.J., and Phillips, L.F. 1975. *Chemistry of the atmosphere.* New York: John Wiley and Sons, Inc.

1977. *Ozone and other photochemical oxidants.* Washington, D.C.: National Academy of Sciences.

1975. *Photochemical oxidant air pollution.* Washington, D.C.: OECD Publications Center.

1979. *Photochemical oxidants.* Albany, N.Y.: WHO Publications Centre USA.

1973. *Proceedings of the Conference on Health Effects of Air Pollutants.* Washington, D.C.: National Academy of Sciences.

Spedding, D.J. 1977. The interaction of gaseous pollutants with materials at the surface of the Earth. In *Environmental chemistry,* J. O'M. Bockris, ed., Ch. 8, pp. 213–42. New York: Plenum Publishing Corp.

Strauss, W. 1977. Formation and control of air pollutants. In *Environmental chemistry,* J. O'M. Bockris, ed., Ch. 7, pp. 179–212. New York: Plenum Publishing Corp.

1976. *Vapor-phase organic pollutants.* Washington, D.C.: National Academy of Sciences.

Wayne, R.P. 1970. *Photochemistry.* New York: Elsevier North-Holland, Inc.

QUESTIONS AND PROBLEMS

1. Of the following species, the one which is the least likely product of the absorption of a photon of light by a molecule of NO_2 is
   (a) O;
   (b) a free radical species;
   (c) NO;
   (d) NO_2^*;
   (e) N atoms.
2. Which of the following statements is true?
   (a) RO· reacts with NO_2 to form alkyl nitrates.
   (b) RO· is not a free radical.
   (c) RO· is not a very reactive species.
   (d) RO· is readily formed by the reaction of stable hydrocarbons and ground state NO_2.
   (e) RO· is not thought to be an intermediate in the smog-forming process.
3. Of the following species, the one most likely to be found in reducing smogs is: ozone; relatively high levels of atomic oxygen; SO_2; PAN; PBN.
4. Why are automotive exhaust hydrocarbons even more damaging to the environment than their quantities would indicate?
5. At what point in the smog-producing chain reaction is PAN formed?
6. What particularly irritating product is formed in the laboratory by irradiation of a mixture of benzaldehyde and nitrogen dioxide with ultraviolet light?
7. Which of the following species reaches its peak value last on a smog-forming day: NO, oxidants, hydrocarbons, or NO_2?
8. What is the main species responsible for the oxidation of NO to NO_2 in a smoggy atmosphere?
9. Give two reasons that a turbine engine should have lower hydrocarbon emissions than an internal combustion engine.
10. What pollution problem does a lean mixture aggravate when employed to control hydrocarbon emissions from an internal combustion engine?
11. Why are two catalytic reactors necessary to control all major automotive exhaust pollutants?

12. What is the distinction between reactivity and instability as applied to some of the chemically active species in a smog-forming atmosphere?

13. Why might carbon monoxide be chosen as a standard against which to compare automotive hydrocarbon emissions in atmospheres where smog is formed? What are some pitfalls created by this choice?

14. What is an undesirable by-product of an oxidation catalyst? Of a reduction catalyst?

15. Some atmospheric chemical reactions are abstraction reactions and others are addition reactions. Which of these two labels applies to the reaction of hydroxyl radical with propane? With propylene?

16. How are oxidants normally detected in the atmosphere?

17. Although NO_x is necessary for smog formation, there are circumstances where it is plausible that moderate reductions of NO_x levels might actually increase the rate of smog formation. Suggest a reason for this belief.

18. Why is ozone especially damaging to rubber?

19. Show how hydroxyl radical, HO•, may react differently with ethylene,

$$\mathrm{H_2C{=}CH_2}\ \text{(H}\backslash\text{C=C/H, H/ }\ \text{\ H),}$$

and methane, CH_4.

20. Name the stable product that results from an initial addition reaction of hydroxyl radical, HO•, with benzene.

CHAPTER 15
ORGANIC POLLUTANTS IN THE ATMOSPHERE

15.1 ORGANIC COMPOUNDS IN THE ATMOSPHERE

Organic contaminants in the atmosphere have been considered in previous chapters. In Chapter 13 we considered some types of particulate organic matter in the atmosphere. Chapter 14 included a discussion of the most important consequence of organic matter in the atmosphere: the formation of noxious photochemical smog. With Chapter 14 as background on chemical processes involving organic compounds in the atmosphere, we may now consider the sources, reactions, and fates of a large variety of organic compounds that enter the atmosphere from various natural and pollutant sources.

The effects of organic pollutants in the atmosphere may be divided into two major categories. The first consists of **direct effects**, such as cancer caused by exposure to vinyl chloride (see Section 18.19). The second, and generally more important effect is the **formation of secondary pollutants**, particularly photochemical smog. In the case of pollutant hydrocarbons in the atmosphere, the latter is the more important effect. In some localized situations, particularly in the workplace, direct effects of organic air pollutants may be equally important.

Natural sources are the most important contributors of organics in the atmosphere. For instance, hydrocarbons contributed by human activities constitute only about 1/7 of the total hydrocarbons in the atmosphere. This ratio is primarily the result of the huge quantities of methane produced by anaerobic bacteria in the decomposition of organic matter in water, sediments, and soil (see Section 5.11):

$$2\,\{CH_2O\} \xrightarrow{\text{bacteria}} CO_2(g) + CH_4(g) \qquad (15.1.1)$$

Flatulent emissions from domesticated animals add about 85 million metric tons of methane to the atmosphere each year. Some authorities now believe that the anaerobic decay of cellulose in the digestive systems of termites is the greatest single source of atmospheric methane, contributing roughly 150 million metric tons of the gas to the atmosphere each year. Methane is considered to be a natural constituent of the atmosphere; the troposphere contains about 1.4 parts per million (ppm) of this gas [1].

Methane in the troposphere contributes to the photochemical production of carbon monoxide and ozone. The photochemical oxidation of methane is a major source of water vapor in the stratosphere.

[1] Enhalt, D.J. 1974. The atmospheric cycle of methane. *Tellus* 26: 58–70.

15.2 ORGANIC COMPOUNDS FROM NATURAL SOURCES

Natural sources produce the greatest quantity and probably the greatest variety of organic compounds found in the atmosphere. Of these sources, vegetation is the most important. A compilation of organic compounds in the atmosphere [2] has shown that a total of 367 different compounds are released to the atmosphere from vegetation sources. Other natural sources include microorganisms, forest fires, animal wastes, and volcanoes.

One of the simplest organic compounds given off by plants is ethylene:

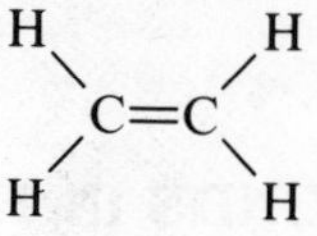

$$\begin{matrix} H & & H \\ & C{=}C & \\ H & & H \end{matrix}$$

This compound is produced by a variety of plants and released to the atmosphere [3]. Because of its double bond, ethylene is highly reactive with hydroxyl radical, HO•, and with oxidizing species in the atmosphere. Ethylene from vegetation sources should be considered an active participant in atmospheric chemical processes.

Most of the hydrocarbons emitted by plants are thought to be of the **terpene** family. Terpenes constitute a large class of organic compounds found in essential oils, obtained when parts of some types of plants are subjected to steam distillation. Most of the plants that produce terpenes belong to the family *Coniferae*, the family *Myrtaceace,* and the genus *Citrus*. One of the most common terpenes emitted by trees is α-pinene, a principal component of turpentine. The terpene limonene, found in citrus fruit and pine needles, is encountered in the atmosphere around these sources. Isoprene (2-methyl-1,3-butadiene), a hemiterpene, has been identified in the emissions from cottonwood, eucalyptus, oak, sweetgum, and white spruce trees. Other terpenes known to be given off by trees include β-pinene, myrcene, ocimene, and α-terpinene.

The structures of α-pinene, isoprene, and limonene are:

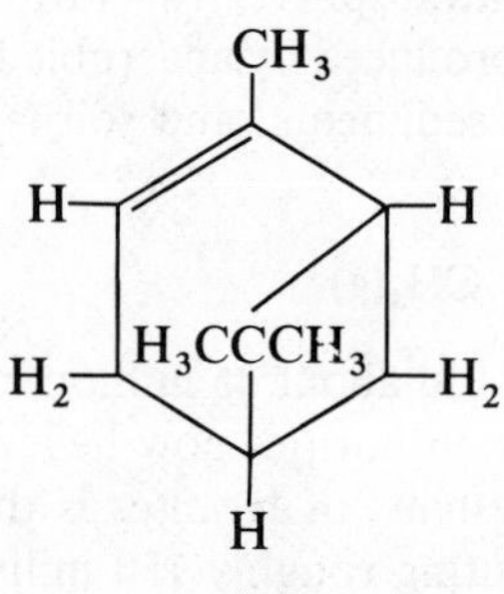

CH_3, H, H, H_3CCH_3, H_2, H_2, H

α-pinene

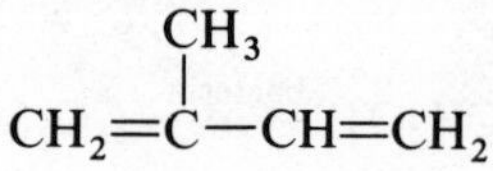

$$CH_2{=}\overset{\displaystyle CH_3}{\overset{|}{C}}{-}CH{=}CH_2$$

isoprene

H_2, H_2, CH_3, H, CH_2, C, CH_3, H, H_2

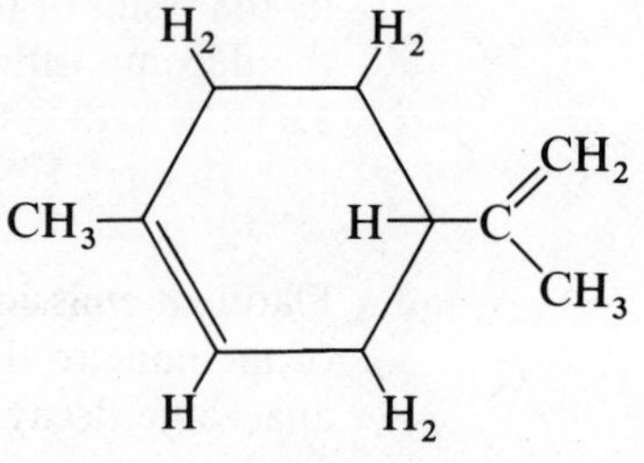

limonene

As exemplified by these structures, terpenes contain olefinic (alkene) bonds,

---

[2] Graedel, T.E. 1978. *Chemical compounds in the atmosphere.* New York: Academic Press, Inc.

[3] Nickell, L.G. 1978. Plant growth regulators. *Chemical and Engineering News* 9 October 1978, pp. 18–34.

usually two or more per molecule. Because of these and other structural features, terpenes are among the most reactive compounds in the atmosphere. The reaction of terpenes with hydroxyl radical is very rapid, and terpenes also react with other oxidizing agents in the atmosphere, particularly ozone, O_3. Turpentine, a common mixture of terpenes, has been widely used in paint because it reacts with atmospheric oxygen to form a peroxide, then a hard resin. It is likely that compounds such as α-pinene and isoprene undergo similar reactions in the atmosphere to form particulate matter. The resulting Aitken nuclei aerosols (see Section 11.14) probably cause the blue haze in the atmosphere above some heavy growths of vegetation.

Smog-chamber experiments have been performed in an effort to determine the fate of atmospheric terpenes. One such study, involving the ultraviolet irradiation of a mixture of α-pinene and NO_x (NO plus NO_2) in air, resulted in the formation of pinonic acid [4]:

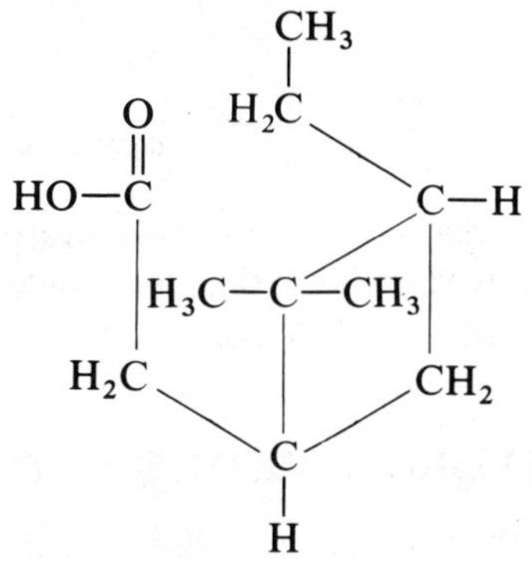

```
                  CH3
                   |
         O      H2C
         ‖          \
     HO—C             C—H
         |           / |
         | H3C—C—CH3   |
         |      |      |
       H2C      |      CH2
           \    |    /
               C
               |
               H
```

This compound occurs in forest aerosol particles; it is almost certainly produced by photochemical processes acting upon α-pinene.

In another smog-chamber experiment, the influence of α-pinene on the smog-forming tendencies of a synthetic mixture of hydrocarbons was investigated [5]. It was found that substitution of the terpene for up to 20% of the synthetic mixture of low-molecular-weight alkanes and alkenes did not appreciably increase the overall reactivity of the mixture as evidenced by the behavior of NO_x and O_3. Therefore, despite the higher reactivity of α-pinene as measured by its reaction with hydroxyl radical, its presence in moderate quantities along with synthetic hydrocarbons does not dramatically increase the smog-forming tendencies of a contaminated atmosphere.

In terms of the variety of compounds emitted from plants, perhaps no class of compounds contains a wider range than **esters.** These compounds are primarily responsible for the fragrances associated with much vegetation. Some typical esters produced as plant volatile matter are shown on page 406.

---

[4] Schwartz, W.; Jones, P.W.; Riggle, C.J.; and Miller, D.F. 1974. *Chemical characterization of model aerosols.* NTIS Document PB-238 557. Columbus, Ohio: Battelle Memorial Institute.

[5] Kamens, R.M., et al. 1981. The impact of α-pinene on urban smog formation: an outdoor smog chamber study. *Atmospheric Environment* 15(6): 969–981.

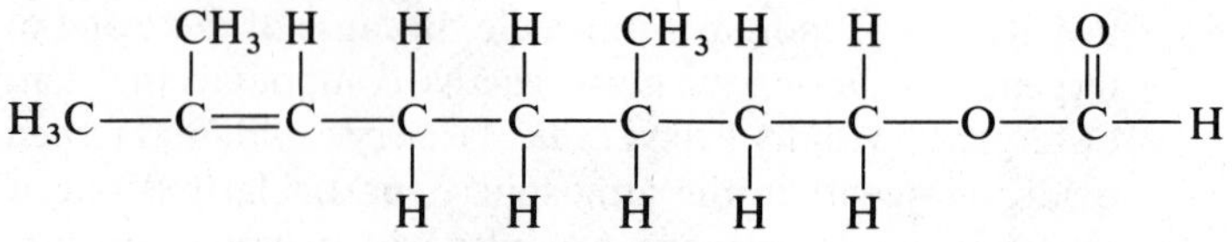

citronellyl formate

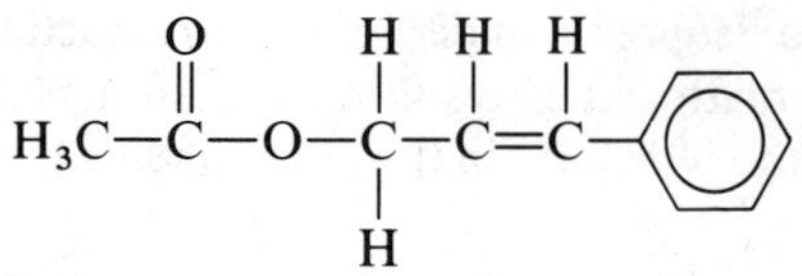

cinnamyl acetate

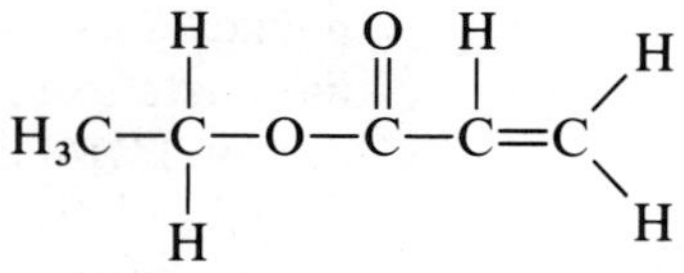

ethyl acrylate

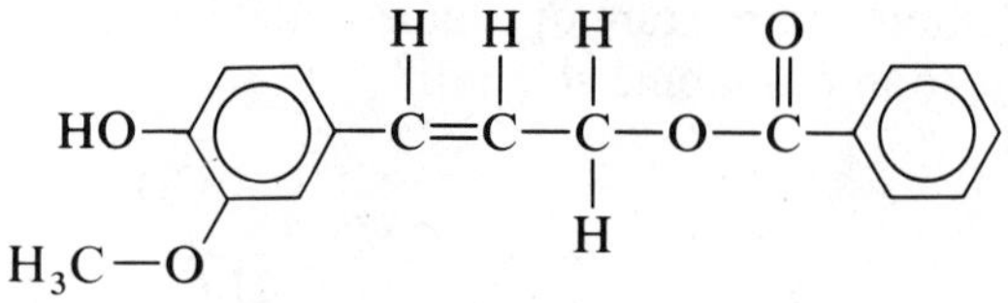

coniferyl benzoate

Despite their great variety, the concentrations of esters released to the atmosphere are small, so it is unlikely that they have a significant effect upon atmospheric chemistry.

15.3 POLLUTANT HYDROCARBONS IN THE ATMOSPHERE

Because of their widespread use in fuels, hydrocarbons predominate among organic atmospheric pollutants. It can be seen from Table 15.1 that petroleum products, primarily gasoline, are the source of most of the anthropogenic pollu-

TABLE 15.1 Global and U.S. Emissions of Pollutant Hydrocarbons

	Emissions, percent of total	
Source	*Global*[1]	*United States*[1]
Petroleum		
Refining	7.2	3.7
Oils and distillates	0.4	4.0
Gasoline	38.5	48.4
Evaporation in transfer and storage	8.8	8.3
Coal		0.6
Heating	2.3	—
Power generation	0.2	—
Industrial uses	0.8	—
Wood (fuel and forest fires)	2.2	8.8
Incinerators and refuse burning	28.3	10.6
Solvent evaporation	11.3	15.6

[1]Based on an estimated annual total of 88 million tons. Adapted from data in E. Robinson and R.C. Robbins, *Sources, Abundance, and Fate of Gaseous Atmospheric Pollutants Supplement,* Menlo Park, Calif.: Stanford Research Institute, 1969.

[2]Based on an estimated annual total of 32 million tons of hydrocarbons. Adapted from data in *National Inventory of Air Pollutant Emissions,* U.S. Department of Health, Education, and Welfare, 1968, p.13.

tant hydrocarbons found in the atmosphere. Hydrocarbons may enter the atmosphere either directly or as by-products of the partial combustion of other hydrocarbons. The latter are particularly important because they tend to be unsaturated and relatively reactive (see Section 14.6 for a discussion of hydrocarbon reactivity in photochemical smog formation). Most hydrocarbon pollutant sources produce about 15% reactive hydrocarbons, whereas those from gasoline are about 45% reactive. It has been shown that the hydrocarbons in uncontrolled automobile exhausts are only about 1/3 alkanes, with the remainder divided approximately equally between relatively reactive alkenes and aromatic hydrocarbons [6], thus accounting for the relatively high reactivity of automotive exhaust hydrocarbons.

Investigators who study smog formation in smog chambers have developed synthetic mixtures of hydrocarbons that mimic the smog-forming behavior of hydrocarbons in a polluted atmosphere. The composition of one such mixture and the structures of the constituent hydrocarbons are given in Table 15.2 (see page 408). The compounds shown in this table provide a simplified idea of the composition of pollutant hydrocarbons likely to lead to smog formation.

We have already discussed one class of hydrocarbons, the terpenes, as atmospheric constituents. We will now discuss **alkanes** (paraffins), **alkenes** (olefins), and **alkynes** (compounds with triple bonds) as atmospheric contaminants. A compilation of the literature documenting organic compounds released to the atmosphere [2] has listed 116 specific alkanes. Predominant among these are the straight-chain alkanes; all of those containing from 1 through 37 carbon atoms have been found in the atmosphere. Most of the various branched-chain isomers of the alkanes with 6 or fewer carbon atoms also occur in the atmosphere. Some typical alkanes found in the atmosphere are listed below:

```
   H  H  H
   |  |  |
H—C—C—C—H
   |  |  |
   H  H  H
```

propane

```
            H
            |
     H  H—C—H      H     H
     |      |       |     |
  H—C———C————C———C—H
     |      |       |     |
     H  H—C—H  H—C—H  H
            |       |
            H       H
```

2,2,3-trimethylbutane

```
   H  H  H  H  H  H  H  H  H  H  H  H  H  H  H  H
   |  |  |  |  |  |  |  |  |  |  |  |  |  |  |  |
H—C—C—C—C—C—C—C—C—C—C—C—C—C—C—C—C—H
   |  |  |  |  |  |  |  |  |  |  |  |  |  |  |  |
   H  H  H  H  H  H  H  H  H  H  H  H  H  H  H  H
```

hexadecane

Because of their high vapor pressures, alkanes with 6 or fewer carbon atoms are normally present in the atmosphere as gases. Alkanes with 20 or more

[6] Morris, W.E., and Dishart, K.T. 1971. Influence of vehicle emission control systems on the relationship between gasoline and vehicle exhaust hydrocarbon composition. In *Effect of automotive emission requirements on gasoline characteristics.* ASTM STP 487. Philadelphia: American Society for Testing and Materials, pp. 63–93.

TABLE 15.2 Hydrocarbons Composing a Typical Synthetic Mixture in Smog Chamber Experiments[1]

Hydrocarbon name	*Hydrocarbon structure*	*Mole percent in hydrocarbon mixture*
Alkanes:		
2-Methylbutane	H | H H—C—H H H | | | | H—C——C——C—C—H | | | | H H H H	0.127
n-Pentane	H H H H H | | | | | H—C—C—C—C—C—H | | | | | H H H H H	0.241
2-Methylpentane	H | H H—C—H H H H | | | | | H—C——C——C—C—C—H | | | | | H H H H H	0.071
2,4-Dimethylpentane	H H | | H H—C—H H H—C—H H | | | | | H—C——C——C——C——C—H | | | | | H H H H H	0.053
2,2,4-Trimethylpentane	H H | | H H—C—H H H—C—H H | | | | | H—C——C——C——C——C—H | | | | | H H—C—H H H H | H	0.064
	Total alkanes:	0.556
Alkenes:		
1-Butene	H H H H | | | \\ C=C—C—C—H / | | H H H	0.027

TABLE 15.2 *(continued)*

Hydrocarbon name	*Hydrocarbon structure*	*Mole percent in hydrocarbon mixture*
cis-2-Butene	H_3C CH_3 C=C H H	0.034
2-Methyl-1-butene	H H H—C—H H H C=C——C—C—H H H H	0.030
2-Methyl-2-butene	H H H—C—H H H H—C——C═C—C—H H H	0.027
Ethylene	H H C=C H H	0.250
Propylene	H H H H C=C—C—C—H H H H	0.075
	Total alkenes:	0.443

[1]Synthetic mixture of hydrocarbons cited in Reference 5.

carbon atoms are present as aerosols or are sorbed to atmospheric particles. Alkanes with 6 to 20 carbon atoms per molecule may be present either as vapor or particles, depending upon conditions.

Alkanes are among the more stable organic species in the atmosphere. As is the case with methane (see Section 14.3), alkanes are attacked primarily by hydroxyl radical, HO•, resulting in the loss of a hydrogen atom and formation of an **alkyl radical,**

$$C_xH_{2x+1}\cdot$$

Subsequent reaction with O_2 causes formation of **alkylperoxyl radical,**

$$C_xH_{2x+1}O_2\cdot$$

These radicals may act as oxidants and form **alkoxyl radicals,**

$$C_xH_{2x+1}O\cdot$$

As a result of these and subsequent reactions, lower-molecular-weight alkanes are eventually oxidized to species that can be precipitated from the atmosphere with particulate matter and eventually undergo biodegradation in soil (see Section 5.12).

Alkenes enter the atmosphere from a variety of processes, including emissions from internal combustion engines and turbines, foundry operations, and petroleum refining. As shown in Table 15.3, several alkenes are among the top fifty chemicals manufactured annually in the U.S. These compounds are used primarily as monomers, which are polymerized to create polymers for plastics (polyethylene, polypropylene, polystyrene), synthetic rubber (styrenebutadiene, polybutadiene) latex paints (styrenebutadiene), and other applications.

All of these compounds, as well as many others manufactured in lesser quantities, are released to the atmosphere. In addition to the direct release of alkenes, alkene compounds are commonly produced by the partial combustion and "cracking" at high temperatures of alkanes, particularly in the internal combustion engine.

Typical alkynes found in the atmosphere are acetylene, used as a fuel for welding torches, and 1-butyne, used in synthetic-rubber manufacture. These compounds are less commonly encountered in the atmosphere than are alkenes.

TABLE 15.3 Alkenes Among the Top 50 Chemicals Manufactured in the U.S.

Compound name	*Structure*	*Rank among top 50 chemicals*[1]	*1980 production, billions of kg*[1]
Ethylene	H H C=C H H	6	12.5
Propylene	H H H C=C—C—H H H	14	6.2
Styrene[2]	H H (benzene ring)—C=C H	21	3.1
Butadiene	H H H H C=C—C=C H H	32	1.3

[1]Data from *Chemical and Engineering News,* 8 June 1981, p. 33.

[2]Also classified as an aromatic hydrocarbon (see next section).

$$\mathrm{H{-}C{\equiv}C{-}H} \qquad\qquad \begin{array}{c} \\ \mathrm{H{-}C{\equiv}C{-}} \\ \\ \end{array}\begin{array}{c} \mathrm{H} \\ | \\ \mathrm{C} \\ | \\ \mathrm{H} \end{array}\begin{array}{c} \\ - \\ \\ \end{array}\begin{array}{c} \mathrm{H} \\ | \\ \mathrm{C} \\ | \\ \mathrm{H} \end{array}\begin{array}{c} \\ \mathrm{{-}H} \\ \\ \end{array}$$

acetylene 1-butyne

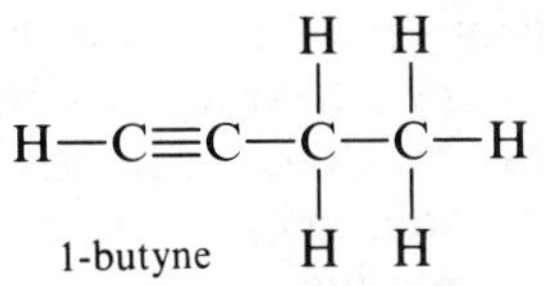

Unlike alkanes, alkenes are highly reactive in the atmosphere, especially in the presence of NO_x and sunlight. Hydroxyl radical can react with alkenes either by abstracting a hydrogen atom or by adding to the double bond. If hydroxyl radical adds to the double bond in propylene, for example, the product is:

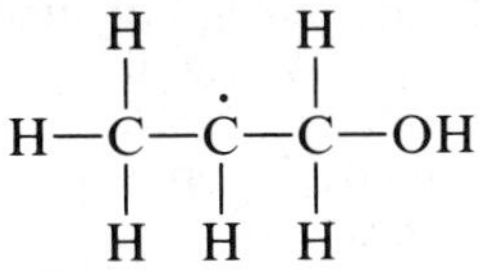

$$\begin{array}{ccccccc} & \mathrm{H} & & \cdot & & \mathrm{H} & \\ & | & & & & | & \\ \mathrm{H}- & \mathrm{C} & - & \mathrm{C} & - & \mathrm{C} & -\mathrm{OH} \\ & | & & | & & | & \\ & \mathrm{H} & & \mathrm{H} & & \mathrm{H} & \end{array}$$

Addition of molecular O_2 to this radical results in the formation of a peroxyl radical,

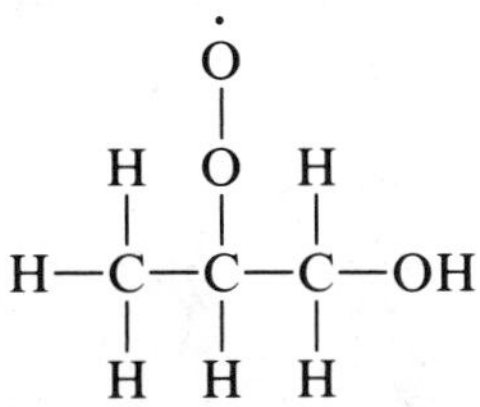

$$\begin{array}{ccccccc} & & & \cdot & & & \\ & & & \mathrm{O} & & & \\ & & & | & & & \\ & \mathrm{H} & & \mathrm{O} & & \mathrm{H} & \\ & | & & | & & | & \\ \mathrm{H}- & \mathrm{C} & - & \mathrm{C} & - & \mathrm{C} & -\mathrm{OH} \\ & | & & | & & | & \\ & \mathrm{H} & & \mathrm{H} & & \mathrm{H} & \end{array}$$

These radicals then take part in reaction chains such as those discussed for the formation of photochemical smog in Chapter 14.

Ozone, O_3, adds across double bonds and is rather reactive with alkenes. The exact chemistry of ozone-alkene reactions in the atmosphere is not well known, but aldehydes are among the products.

15.4 AROMATIC HYDROCARBONS IN THE ATMOSPHERE

Aromatic compounds may be divided into two major classes: those having only one benzene ring and those with multiple rings. The latter are called **polynuclear aromatic hydrocarbons**, commonly abbreviated as PAH or PNA; they were discussed briefly in Section 13.16 as examples of particulate matter. Intermediate between these two major types of aromatic hydrocarbons are the aromatics with two rings, **naphthalene** and its derivatives. Some typical aromatic compounds are:

CH_3 H_3C

benzene 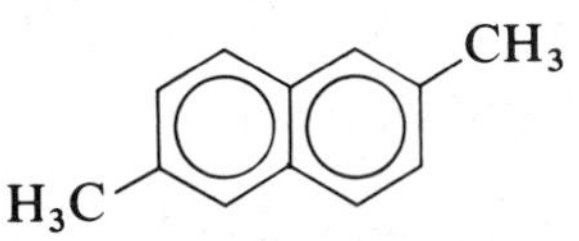 2,6-dimethylnaphthalene 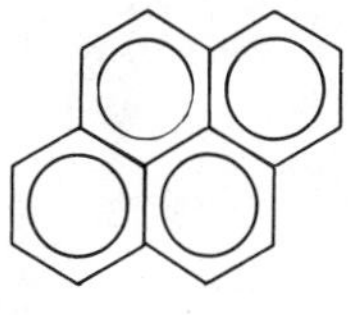 pyrene

In all of these structures, the hexagon with a circle inside it denotes a six-carbon-atom ring with a very stable π(pi)-bonding system. In multi-ring structures, carbon atoms are shared at the "corners" where the rings join. Unless another substituent group is shown, hydrogen atoms are present on corners that are not shared by other rings. Using these criteria, the structure shown for pyrene implies 16 carbon atoms, of which 6 are shared among aromatic rings, and 10 hydrogen atoms attached to "corner" carbon atoms.

As shown in Table 15.4, benzene and several of its derivatives are among the top 50 chemicals manufactured in the U.S. Single-ring aromatic compounds are important constituents of lead-free gasoline, which is steadily replacing leaded gasoline. Aromatic compounds are widely used in industry. In addition to their use as solvents, they are employed to make other chemicals, such as the monomers and plasticizers in polymers. The use of styrene as a plastics monomer

TABLE 15.4 Aromatic Compounds Among the Top 50 Chemicals Manufactured in the U.S.

Compound name	*Structure*	*Rank among top 50 chemicals*[1]	*1980 production, billions of kg*[1]
Toluene	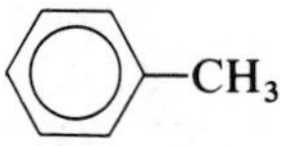	15	5.11
Benzene	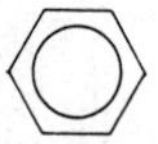	16	4.98
Ethylbenzene	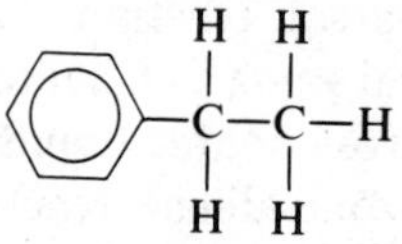	18	3.45
Styrene[2]	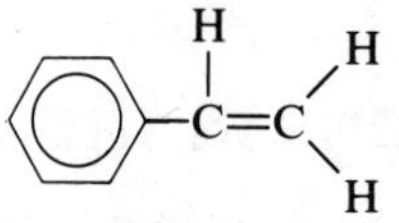	21	3.13
Xylene (all grades)	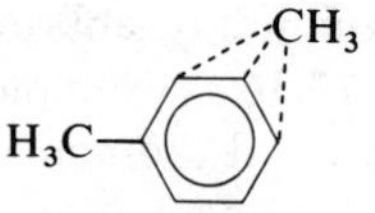	23	2.90
p-Xylene	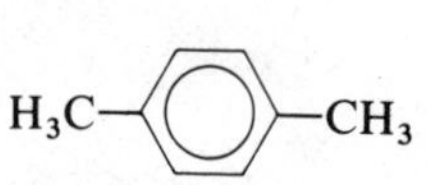	30	1.73
Cumene	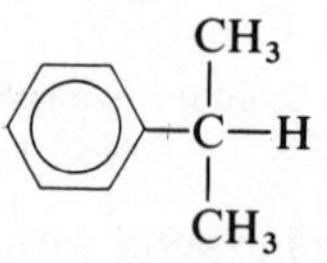	31	1.43

[1]Data from *Chemical and Engineering News*, 8 June 1981, p. 33.

[2]Also listed as an alkene in Table 15.3.

and ingredient of synthetic rubber was mentioned in Section 15.3. Cumene is oxidized to produce phenol and valuable acetone as a by-product. With all these applications, plus production of these compounds as combustion by-products, it is no surprise that aromatic compounds are common atmospheric pollutants.

Approximately 55 hydrocarbons containing a single benzene ring and approximately 30 hydrocarbon derivatives of naphthalene have been found as atmospheric pollutants [2]. In addition, several compounds containing two or more *unconjugated* rings (not sharing the same π electron cloud between rings) have been detected as atmospheric contaminants. One such compounds is biphenyl,

biphenyl

detected in diesel smoke. It should be pointed out that many of these aromatic hydrocarbons have been detected primarily as ingredients of tobacco smoke and are, therefore, of much greater significance in an indoor environment than in an outdoor environment.

As with most atmospheric hydrocarbons, the most likely reaction of benzene and its derivatives is with hydroxyl radical. Addition of $HO\cdot$ to the benzene ring results in the formation of an unstable radical species,

where the dot denotes an unpaired electron in the radical. The electron is said to be **delocalized** and may be represented in the aromatic radical structure by a half-circle with a dot in the middle:

Subsequent abstraction of H from the radical by O_2 results in re-establishment of the aromatic system in the form of stable phenol and reactive $HO_2\cdot$ radical:

$$\text{(OH, H)-radical} + O_2 \longrightarrow \text{phenol (OH)} + HO_2\cdot \quad (15.4.1)$$

Alkyl-substituted aromatics may undergo reactions involving the alkyl group. For example, abstraction of alkyl H by $HO\cdot$ from a compound such as *p*-xylene results in the formation of a radical,

$$\begin{array}{c}\mathrm{H} \\ | \\ \mathrm{H{-}C{-}} \\ | \\ \mathrm{H}\end{array}\!\!\bigcirc\!\!\begin{array}{c}\mathrm{H} \\ | \\ \mathrm{{-}C\cdot} \\ | \\ \mathrm{H}\end{array}$$

which can react further with O_2 to form a peroxyl radical, then enter chain reactions such as those discussed in Chapter 14.

Because of their very low vapor pressures, polynuclear aromatic compounds, PAH's, are present as aerosols in the atmosphere. These compounds are the most stable form of hydrocarbons, having low hydrogen-to-carbons ratios, and are formed by the partial combustion of all hydrocarbons, even those as simple as methane. Coal has a hydrogen-to-carbon ratio of somewhat less than 1 (see Section 17.10) so the partial combustion or pyrolysis of coal is a major source of PAH compounds. Because they exist as particles in the atmosphere, PAH's were discussed with the other particulate atmospheric pollutants in Chapter 13.

15.5 ALDEHYDES AND KETONES IN THE ATMOSPHERE

In discussing organic air pollutants, it is logical to consider **aldehydes** and **ketones** **(carbonyl compounds)** immediately after hydrocarbons because carbonyl compounds are often the first stable species formed by the photochemical oxidation of hydrocarbons. The general formulas of aldehydes and ketones are represented by the following, where R and R ′ represent the hydrocarbon *moieties* (portions), such as a $-CH_3$ group.

$$\underset{\text{aldehyde}}{\mathrm{R{-}\overset{\overset{\displaystyle O}{\|}}{C}{-}H}} \qquad \underset{\text{ketone}}{\mathrm{R{-}\overset{\overset{\displaystyle O}{\|}}{C}{-}R'}} \qquad \underset{\text{carbonyl moiety}}{\mathrm{{-}\overset{\overset{\displaystyle O}{\|}}{C}{-}}}$$

Carbonyl compounds are by-products of the generation of hydroxyl radicals from organic peroxyl radicals by reactions such as the following:

$$\begin{array}{ccccccc} & & & \dot{\mathrm{O}} & & & \\ & & & | & & & \\ & & \mathrm{H} & \mathrm{O} & \mathrm{H} & & \\ & & | & | & | & & \\ \mathrm{H} & - & \mathrm{C} & -\mathrm{C}- & \mathrm{C} & - & \mathrm{H} \\ & & | & | & | & & \\ & & \mathrm{H} & \mathrm{H} & \mathrm{H} & & \end{array} \longrightarrow \underset{\text{ketone}}{\begin{array}{ccccccc} & & \mathrm{H} & \mathrm{O} & \mathrm{H} & & \\ & & | & \| & | & & \\ \mathrm{H} & - & \mathrm{C} & -\mathrm{C}- & \mathrm{C} & - & \mathrm{H} \\ & & | & & | & & \\ & & \mathrm{H} & & \mathrm{H} & & \end{array}} + \mathrm{HO\cdot} \tag{15.5.1}$$

$$\begin{array}{ccccccc} & & & & \dot{\mathrm{O}} & & \\ & & & & | & & \\ & & \mathrm{H} & \mathrm{H} & \mathrm{O} & & \\ & & | & | & | & & \\ \mathrm{H} & - & \mathrm{C} & -\mathrm{C}- & \mathrm{C} & - & \mathrm{H} \\ & & | & | & | & & \\ & & \mathrm{H} & \mathrm{H} & \mathrm{H} & & \end{array} \longrightarrow \begin{array}{ccccccc} & & \mathrm{H} & \mathrm{H} & \mathrm{O} & & \\ & & | & | & \| & & \\ \mathrm{H} & - & \mathrm{C} & -\mathrm{C}- & \mathrm{C} & - & \mathrm{H} \\ & & | & | & & & \\ & & \mathrm{H} & \mathrm{H} & & \text{aldehyde} & \end{array} + \mathrm{HO\cdot} \tag{15.5.2}$$

Aldehydes are particularly likely to oxidize further to organic carboxylic acids, as shown by the reaction below, where {O} is an atmospheric oxidant species:

$$\mathrm{R{-}\overset{\overset{\displaystyle O}{\|}}{C}{-}H} + \{\mathrm{O}\} \rightarrow \mathrm{R{-}\overset{\overset{\displaystyle O}{\|}}{C}{-}OH} \tag{15.5.3}$$

Aldehydes and ketones are widely used industrial chemicals. For example, almost 1 billion kilograms of formaldehyde

$$\begin{array}{ccccc} & & \mathrm{O} & & \\ & & \| & & \\ \mathrm{H} & - & \mathrm{C} & - & \mathrm{H} \end{array}$$

are produced annually in the U.S. to be used in the manufacture of plastics, resins, lacquers, dyes, and explosives. Formaldehyde occurs in the atmosphere primarily in the gas phase [7]. Typical formaldehyde levels associated with aerosols are 40 nanograms (ng) per cubic meter for inland continental atmospheres, 65 ng/m^3 for urban air, and less than 2 ng/m^3 for clean maritime air masses [7]. Although these levels are much lower than those of gas-phase formaldehyde in the atmosphere, they are approximately 1,000 times higher than those predicted from the aqueous solubility of formaldehyde and its concentration in the gas phase. This discrepancy may be attributed to the formation of addition compounds between gaseous formaldehyde and species such as bisulfite, alcohols, phenols, and amines contained in aerosol particles [7].

Among the other important carbonyl compounds, acetaldehyde,

$$\begin{array}{ccccccc} & & \mathrm{H} & & \mathrm{O} & & \\ & & | & & \| & & \\ \mathrm{H} & - & \mathrm{C} & - & \mathrm{C} & - & \mathrm{H} \\ & & | & & & & \\ & & \mathrm{H} & & & & \end{array}$$

is a widely produced organic chemical used in the manufacture of acetic acid, plastics, and other materials. Almost 1 billion kg/year of acetone,

$$\begin{array}{ccccccccc} & & \mathrm{H} & & \mathrm{O} & & \mathrm{H} & & \\ & & | & & \| & & | & & \\ \mathrm{H} & - & \mathrm{C} & - & \mathrm{C} & - & \mathrm{C} & - & \mathrm{H} \\ & & | & & & & | & & \\ & & \mathrm{H} & & & & \mathrm{H} & & \end{array}$$

are produced for use as a solvent and for applications in the rubber, leather, and plastics industries. Another commercially important ketone is methyl ethyl ketone,

$$\begin{array}{ccccccccccc} & & \mathrm{H} & & \mathrm{O} & & \mathrm{H} & & \mathrm{H} & & \\ & & | & & \| & & | & & | & & \\ \mathrm{H} & - & \mathrm{C} & - & \mathrm{C} & - & \mathrm{C} & - & \mathrm{C} & - & \mathrm{H} \\ & & | & & & & | & & | & & \\ & & \mathrm{H} & & & & \mathrm{H} & & \mathrm{H} & & \end{array}$$

employed as a low-boiling solvent for coatings and adhesives and for the synthesis of other chemicals.

In addition to their production from hydrocarbons by photochemical oxidation, carbonyl compounds enter the atmosphere from a large number of sources and processes. These include direct emissions from internal combustion engine exhausts, incinerator emissions, spray painting, polymer manufacture,

---

[7] Klippel, W., and Warneck, P. 1980. The formaldehyde content of the atmospheric aerosol. *Atmospheric Environment* 14: 809–18.

printing, petrochemicals manufacture, and lacquer manufacture. Formaldehyde and acetaldehyde are produced by microorganisms and acetaldehyde is emitted by some vegetation [8].

In addition to their production as intermediates in the photochemical oxidation of hydrocarbons, aldehydes are second only to NO_2 as atmospheric sources of free radicals produced by the absorption of light. This is because the carbonyl group is a **chromophore**, a molecular group that readily absorbs light in the near-ultraviolet region of the spectrum. Commonly, the activated compound produced when a photon is absorbed by an aldehyde dissociates into a formyl radical, $H\dot{C}O$, and an alkyl radical. The photodissociation of acetaldehyde illustrates this two-step process:

$$H_3C{-}C(=O){-}H + h\nu \longrightarrow H_3C{-}C(=O){-}H^* \longrightarrow H_3C\cdot + H\dot{C}O \quad (15.5.4)$$

photochemically excited compound

Photolytically excited formaldehyde, CH_2O may dissociate in two ways. The first of these produces an H atom and an $H\dot{C}O$ radical; the second produces two chemically stable species, H_2 and CO.

Because of the presence of both double bonds and carbonyl groups, olefinic aldehydes are especially reactive in the atmosphere. The most common of these found in the atmosphere is acrolein,

$$H_2C{=}C(H){-}C(=O){-}H$$

a powerful lachrymator (tear-producer) which is used as an industrial chemical and is produced as a combustion by-product.

Ketones commonly undergo photochemical dissociation in the atmosphere at one of the bonds joining the carbonyl group to the hydrocarbon moieties [9]:

$$R{-}C(=O){-}R' + h\nu \rightarrow R{-}\dot{C}(=O) + R'\cdot \quad (15.5.5)$$

The radicals produced then undergo subsequent reactions with O_2 and other chemical species in the atmosphere.

15.6 MISCELLANEOUS OXYGEN-CONTAINING COMPOUNDS IN THE ATMOSPHERE

The most important type of oxygenated organic compound found in the atmosphere, carbonyls, was discussed in the preceding section. Another type, esters, was discussed in Section 15.2, which deals with organic compounds from

[8] Nicholas, H.J. 1973. Miscellaneous volatile plant products. In *Phytochemistry,* Vol. 2, L.P. Miller, ed., pp. 381–399. New York: John Wiley and Sons, Inc.

[9] Vaish, S.P.; McAlpine, R.D.; and Cocivera, M. 1974. Photo-CIDNP studies of 2-butanone and 3-pentanone in solution. *Canadian Journal of Chemistry* 52: 2978–84.

natural sources. This section briefly discusses several other types of oxygen-containing organic compounds in the atmosphere, including **aliphatic alcohols**, in which the hydrocarbon moiety is neither olefinic nor aromatic; **phenols**; **ethers**; and **carboxylic acids**. The general formulas of these compounds are given below, where R and R′ represent the non-aromatic hydrocarbon moieties, and ϕ represents the aromatic moiety, such as a phenyl group (benzene minus an H atom).

$$\underset{\text{aliphatic alcohols}}{R\text{—}OH} \qquad \underset{\text{phenols}}{\phi\text{—}OH} \qquad \underset{\text{ethers}}{R\text{—}O\text{—}R'} \qquad \underset{\text{carboxylic acids}}{R\text{—}\overset{\overset{\large O}{\|}}{C}\text{—}OH}$$

These classes of compounds include many important organic chemicals. Of the alcohols, methanol ranks twentieth in industrial production in the U.S. each year, with annual production of approximately 3 billion kg; ethylene glycol ranks twenty-eighth (annual production, about 1.9 kg); isopropanol ranks forty-third (annual production, about 800 million kg); and ethanol ranks forty-ninth (annual production, about 550 million kg). These and other alcohols are used for a number of purposes, the most common of which is the manufacture of other chemicals. Methanol is widely employed in the manufacture of formaldehyde (see Section 15.5), as a solvent, and as an antifreeze in mixtures with water. Ethanol is a commonly used solvent and is the starting material for the manufacture of acetaldehyde, acetic acid, ethyl ether, ethyl chloride, ethyl bromide, and several important esters. Both methanol and ethanol can be used as motor vehicle fuels, usually in mixtures with gasoline (see gasohol, Section 17.17). Ethylene glycol is a common antifreeze compound.

$$\begin{array}{ccc} & H & \\ & | & \\ H\text{—} & C & \text{—}OH \\ & | & \\ & H & \end{array} \qquad \begin{array}{cccc} & H & H & \\ & | & | & \\ H\text{—} & C\text{—} & C & \text{—}OH \\ & | & | & \\ & H & H & \end{array} \qquad \begin{array}{cccc} & H & H & \\ & | & | & \\ HO\text{—} & C\text{—} & C & \text{—}OH \\ & | & | & \\ & H & H & \end{array} \qquad \begin{array}{ccccc} & H & OH & H & \\ & | & | & | & \\ H\text{—} & C\text{—} & C\text{—} & C & \text{—}H \\ & | & | & | & \\ & H & H & H & \end{array}$$

methanol ethanol ethylene glycol isopropanol

A number of aliphatic alcohols have been reported in the atmosphere. Because of their volatility, the lower alcohols, particularly methanol and ethanol, tend to predominate as atmospheric pollutants. Among the other alcohols released to the atmosphere are 1-propanol, 2-propanol, propylene glycol, 1-butanol, and even those as complex as octadecanol, $CH_3(CH_2)_{16}CH_2OH$, a volatile material given off by plants [8]. Alcohols can undergo photochemical reactions, beginning with abstraction of hydrogen by hydroxyl radical. Mechanisms for scavenging alcohols from the atmosphere should be relatively efficient; the lower alcohols are quite water soluble and the higher alcohols have low vapor pressures.

Olefinic alcohols are known; a number have been found in the atmosphere, largely by-products of combustion. Typical of the olefinic alcohols is 2-buten-1-ol,

$$\begin{array}{cccccc} & H & H & H & H & \\ & | & | & | & | & \\ H\text{—} & C\text{—} & C\text{=} & C\text{—} & C & \text{—}OH \\ & | & & & | & \\ & H & & & H & \end{array}$$

detected in automobile exhausts [10].

Phenols are aromatic alcohols, more noted as water pollutants than as air pollutants. Some typical phenols that have been reported as atmospheric contaminants are shown below. Phenol, itself, ranks thirty-sixth among synthetic chemicals manufactured in the U.S., with annual production of approximately 1 billion kg. It is most commonly used in the manufacture of resins and polymers, such as Bakelite, a phenol-formaldehyde copolymer. Phenols are produced by the pyrolysis of coal and are major by-products of the coking industry. Thus, in local situations involving coal coking and similar operations phenols can be troublesome air pollutants.

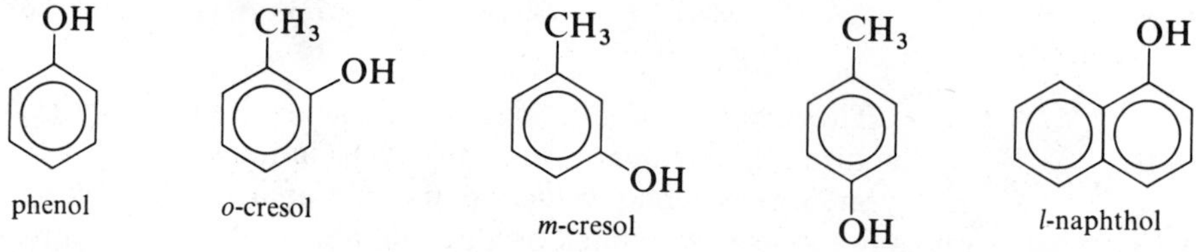

phenol   *o*-cresol   *m*-cresol   *p*-cresol   *l*-naphthol

Ethers are relatively uncommon atmospheric pollutants, although the flammability hazard of diethyl ether vapor in an enclosed work space is well known. In addition to aliphatic ethers, such as dimethyl ether and diethyl ether, several olefinic ethers, including vinyl ethyl ether, have been reported as turbine exhaust products [11]. Two cyclic ethers, ethylene oxide and propylene oxide, are important industrial chemicals, ranking twenty-seventh and forty-fourth, respectively, in annual U.S. production. Ethylene oxide is an intermediate in the production of ethylene glycol, the structure of which was shown previously in this section. Another cyclic ether, tetrahydrofuran, an important industrial solvent, has likewise been reported as an air contaminant.

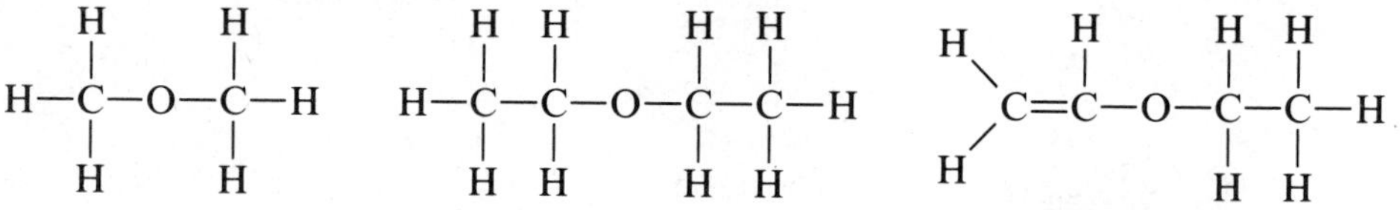

dimethyl ether   diethyl ether   vinyl ethyl ether

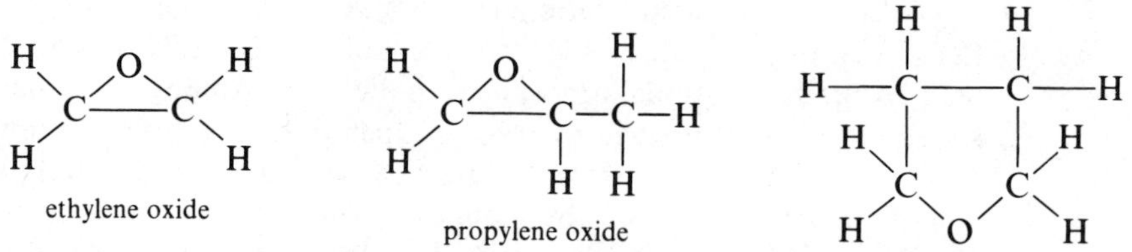

ethylene oxide   propylene oxide   tetrahydrofuran

Carboxylic acids are all characterized by one or more of the functional groups,

$$-\overset{\overset{\displaystyle O}{\|}}{C}-OH$$

---

[10] Seizinger, D.E., and Dimitriades, B. 1972. *Oxygenates in automotive exhaust gas.* Bureau of Mines Report of Investigation 7675. Washington, D.C.: U.S. Department of the Interior.

[11] Conkle, J.P.; Lackey, W.W.; Martin, C.L.; and Miller, R.L. 1976. Organic compounds in turbine combustor exhausts. In *1975 environmental sensing and assessment.* IEEE Publication 75-CH 1004-1 ICESA, pp. 27-2.1 to 27-2.1. New York: Institute of Electrical and Electronic Engineers.

attached to a hydrocarbon moiety. The latter may be aliphatic, olefinic, or aromatic. A large number of carboxylic acids are found in the atmosphere. Most of these are probably the result of the photochemical oxidation of organics, as cited in the example of pinonic acid in Section 15.2. These acids are often at the final result of the photochemical oxidation of other organic compounds in the atmosphere. Because of their low vapor pressures and high water solubilities, they are common constituents of atmospheric aerosols and are one of the most common forms in which organic matter is scavenged from the atmosphere. Organic acids may be formed in the atmosphere either by gas-phase reactions or by reactions of other organics dissolved in aqueous aerosols [2]. The relative importance of these two possible pathways is not known with certainty.

15.7 ORGANOHALIDE COMPOUNDS IN THE ATMOSPHERE

When atoms of fluorine, chlorine, bromine, or iodine are substituted for hydrogen on hydrocarbons, the result is one of the general class of **organohalide** compounds. If the hydrocarbon portion of the molecule is aliphatic, the compound is an **alkyl halide**; substitution onto an aromatic compound yields an **aryl halide**. Organohalide compounds are widely used for solvents, industrial chemicals, raw materials for polymer manufacture, and other applications. Because of their volatility and persistence, they enter the atmosphere readily and are frequently encountered as atmospheric contaminants. Three of the more important low-molecular-weight organochloride compounds are ethylene dichloride (annual production approximately 4.5 billion kg in the U.S., ranking seventeenth in chemical production), vinyl chloride (about 2.8 billion kg/year, ranking twenty-second), and perchloroethylene (annual production about 300 million kg). Vinyl chloride is used to manufacture polyvinyl chloride plastics. As discussed in Section 18.19, long or intense exposure to vinyl chloride is known to cause angiosarcoma, a rare form of live cancer; the possible carcinogenicity of ethylene dichloride remains a matter of controversy.

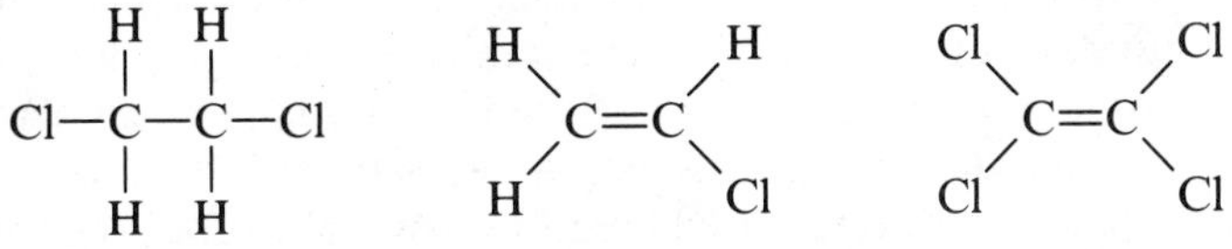

```
   H  H          H       H        Cl       Cl
   |  |           \     /           \     /
Cl—C—C—Cl          C=C               C=C
   |  |           /     \           /     \
   H  H          H       Cl       Cl       Cl
```

ethylene dichloride | vinyl chloride | perchloroethylene

On a global basis, the three most abundant organochlorine compounds in the atmosphere are methyl chloride, CH_3Cl; methyl chloroform, CH_3CCl_3; and carbon tetrachloride, CCl_4 [12]. The average tropospheric concentration of methyl chloride is about 0.61 parts per billion (ppb) by volume, that of methyl chloroform, about 0.1 ppb, and that of carbon tetrachloride, about 0.12 ppb. Carbon tetrachloride and methyl chloroform are of anthropogenic origin. About 7×10^{11} g/year of methyl chloroform are released to the global atmosphere as a result of industrial production and use [13]. This compound is relatively persistent in the atmosphere, with residence times estimated at 1–8 years; therefore, it

---

[12] Singh, H.B.; Salas, L.J.; Shigeishi, H.; and Scribner, E. 1979. Atmospheric halocarbons, hydrocarbons, and sulfur hexafluoride; global distributions, sources, and sinks. *Science* 203: 899–903.

[13] Khalil, M.A.K., and Rasmussen, R.A. 1981. Methyl chloroform: cycles of global emissions. *Environmental Science and Technology* 15: 1506–8.

may pose a threat to the stratospheric ozone layer in the same way that chlorofluorocarbons do (see Section 12.17). Methyl chloride has an atmospheric lifetime of 2–3 years and a relatively high global concentration. It is not one of the most widely produced organic chemicals, so a natural source—probably the sea—is suggested.

Organohalide compounds can enter the atmosphere either in the vapor phase or in association with aerosols. Decreased use of chlorinated hydrocarbon pesticides in recent years has lessened the atmospheric content of these compounds. Until their manufacture was discontinued, polychlorinated biphenyls (PCB's, see Section 7.19) commonly entered the atmosphere from combustion of materials containing them. In one particularly damaging case, a fire in a state office building in Binghamton, New York, in 1981 resulted in the destruction of a PCB-cooled transformer and contamination of the entire building with PCB's via the ventilation system.

A total of 570 million kg of PCB's were produced in North America between 1930 and 1975. When the environmental effects of these compounds became more apparent, their manufacture in the U.S. was banned in 1979. These fire- and degradation-resistant compounds were widely used as coolant-dielectric fluids in capacitors and transformers. Other uses include adhesives, printing and copy papers, plasticizers in coatings, and additives in industrial fluids. The latter were commonly employed in vacuum pumps, hydraulic systems, and turbines. In 1975 about 340 million kg of PCB's were still being used, with some quantities slated to remain in service for approximately 20 years. It has been estimated that 130 million kg of PCB's were placed in landfills, and another 25 million kg had degraded or were incinerated. This leaves 70 million kg circulating in the environment.

A small, but very significant, fraction of PCB's, an estimated 18,000 kg, is thought to be in circulation in the atmosphere over the continental U.S. [14]. Atmospheric transport is probably the primary means by which PCB's are distributed around the globe from their sites of use and disposal. Although atmospheric PCB's are primarily associated with particulate matter in urban regions, these compounds exist almost totally in the vapor phase over rural areas of North America. Vapor-phase PCB's tend to accumulate on plant foliage. Plant species show wide variations in their abilities to sorb PCB's.

As expected from their low vapor pressures, the lighter organohalide compounds are the most likely to be encountered in the atmosphere [2]. Particularly prominent are derivatives of methane, including the compounds with one through four Cl substituents: methyl chloride, methylene chloride, chloroform, and carbon tetrachloride; methyl bromide, CH_3Br; bromoform, $CHBr_3$; and the chlorofluorocarbon compounds (Freons), including CF_2Cl_2, CF_3Cl, $CFCl_3$, and $CHClF_2$. This last group of compounds was discussed in Chapter 12 in regard to their potential threat to the stratospheric ozone layer. A variety of chlorinated ethane compounds may also be encountered in the atmosphere. Several substituted ethylene derivatives have been reported as air contaminants. Trichloroethylene (shown at the top of the following page) a common manufacturing chemical,

---

[14] Buckley, E.H. 1982. Accumulation of airborne polychlorinated biphenyls in foliage. *Science* 216: 520–2.

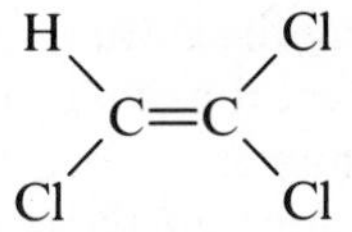

```
H       Cl
 \     /
  C=C
 /     \
Cl      Cl
```

solvent, and dry-cleaning fluid has been reported as an air contaminant, as have vinyl chloride and perchloroethylene, both discussed earlier in this section. A few brominated compounds, such as the solvent ethylene dibromide, CHBr = CHBr, have been reported in air.

Aryl halide PCB air pollutants were mentioned previously in this section. Several halide derivatives of benzene have also been reported in the atmosphere. Prominent among these is chlorobenzene [15], an industrial chemical and solvent.

Alkyl halides in which all hydrogen atoms have been replaced by halogen atoms are very unreactive in the troposphere; the more volatile of these may reach the stratosphere and undergo high-energy photodissociation there. Hydroxyl radical can abstract a hydrogen atom from partially halogenated alkanes, leading to chain reactions similar to those of the alkanes. One end product is HCl, an acidic atmospheric contaminant. The atmospheric reactions of the halogenated alkenes and aryl halides have not been studied in great detail. However, it is known that hydroxyl radical can add across the double bond of halogenated alkenes.

15.8 ATMOSPHERIC ORGANIC SULFUR COMPOUNDS

Although not highly significant as atmospheric contaminants on a large scale, organic sulfur compounds can cause local air pollution problems because of their bad odors. The three major classes of organic sulfur compounds in the atmosphere are **mercaptans**, characterized by the —SH functional group; **organic sulfides**, in which the —S— or —SS— groups are bonded to hydrocarbon moieties; and **heterocyclic sulfur compounds**, which contain sulfur atoms in a ring structure. Examples of these three groups of compounds reported as air pollutants are:

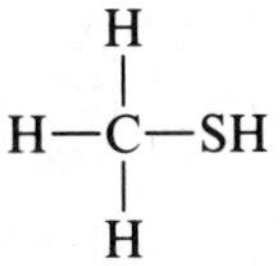

```
                                                                H       H
   H            H     H           H  H       H                  |       |
   |            |     |           |  |       |                H-C-------C-H
 H-C-SH       H-C--S--C-H       H-C--C--SS--C-H                 ||      ||
   |            |     |           |  |       |                  C       C
   H            H     H           H  H       H                /   \   /   \
                                                             H      S      H
methyl mercaptan   dimethyl sulfide   ethylmethyldisulfide        thiophene
```

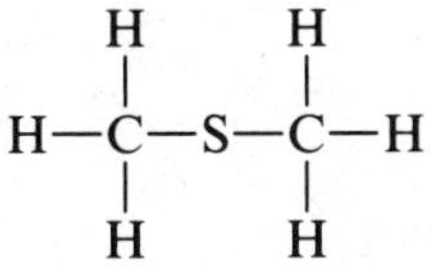

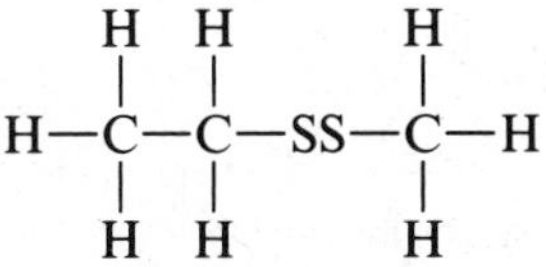

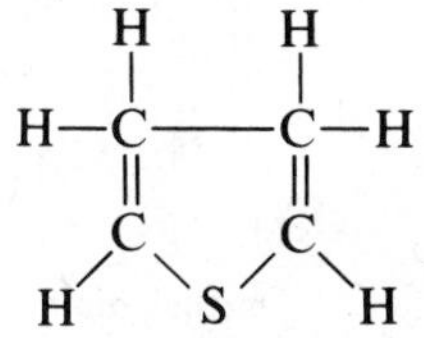

Organic sulfur compounds, including all the structures shown above, occur as impurities in "sour" natural gas. Traces of highly odorous mercaptans are added to natural gas for leak detection. Microbial degradation is an important source of methyl mercaptan and, perhaps, other organic sulfur compounds as well. Wood pulping is one of the more likely sources of atmospheric organosulfur

[15] 1976. *Vapor-phase organic pollutants.* Report of the Committee on Medical and Biologic Effects of Environmental Pollutants. Washington, D.C.: National Academy of Sciences.

compounds. Among the sulfur compounds from this source are methyl mercaptan, ethyl mercaptan, dimethyl sulfide, and ethylmethyl sulfide.

There are numerous other sources of organosulfur compounds in the atmosphere in addition to those discussed above. These include plant volatile matter, animal wastes, packing house and rendering plant wastes, starch manufacture, sewage treatment, and petroleum refining.

Although the impact of organosulfur compounds on atmospheric chemistry is minimal in areas such as aerosol formation or production of acid precipitation components, these compounds can be the worst of all organic compounds in terms of one direct pollution effect: odor. Therefore, it is very important to prevent their release to the atmosphere.

As with all hydrogen-containing organic species in the atmosphere, reaction of organosulfur compounds with hydroxyl radical is a first step in their atmospheric photochemical reactions. The sulfur from both mercaptans and sulfides ends up as SO_2. In both cases there is thought to be a readily oxidized SO intermediate, and HS$\cdot$ radical may be an intermediate in the oxidation of mercaptans. A competing pathway for mercaptan and sulfide degradation is the addition of O atoms to S in the compound [16]. This can result in free radical formation, as shown below for methyl mercaptan:

$$CH_3SH + O \rightarrow H_3C\cdot + HSO\cdot \qquad (15.8.1)$$

The HSO$\cdot$ radical is readily oxidized by atmospheric O_2 to SO_2 by way of an SO intermediate [17].

15.9 ORGANIC NITROGEN COMPOUNDS IN THE ATMOSPHERE

Organic nitrogen compounds that may be found as atmospheric contaminants may be classified as **amines, amides, nitriles, nitro compounds,** or **heterocyclic nitrogen compounds**. Structures of common examples of each of these five classes of compounds reported as atmospheric contaminants are:

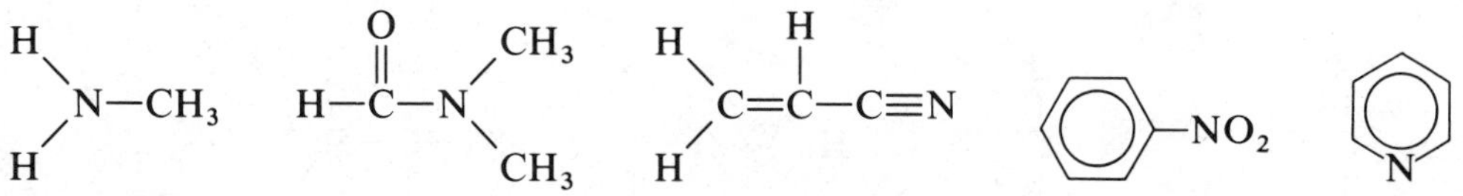

$H_2N{-}CH_3$ (methylamine); $H{-}C(=O){-}N(CH_3)_2$ (dimethyl formamide); $H_2C{=}CH{-}C{\equiv}N$ (acrylonitrile); $C_6H_5{-}NO_2$ (nitrobenzene); C_5H_5N (pyridine)

methylamine dimethyl formamide acrylonitrile nitrobenzene pyridine

Amines consist of compounds in which one or more of the hydrogen atoms in NH_3 has been replaced by a hydrocarbon moiety. Amines with lower molecular weights are volatile. These are prominent among the compounds giving rotten fish their characteristic odor—an obvious reason why air contamination by amines is undesirable. The simplest and most important aromatic amine is

---

[16] Slagle, I.R.; Graham, R.E.; and Gutman, D. 1976. Direct identification of reactive routes and measurement of rate constants in the reactions of oxygen atoms with methanethiol, ethanethiol, and methylsulfide. *International Journal of Chemical Kinetics* 8: 457–458.

[17] Graedel, T.E. 1977. The homogeneous chemistry of atmospheric sulfur. *Rev. Geophys. Space Phys.* 15: 421–428.

aniline, used in the manufacture of dyes, amides, photographic chemicals, and drugs:

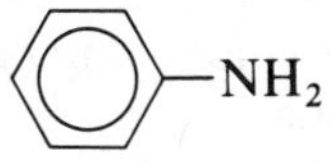

$$C_6H_5-NH_2$$

A number of amines are widely used industrial chemicals and solvents, so that industrial sources have the potential to contaminate the atmosphere with these chemicals. Decaying organic matter produces amines. This is especially true for the decay of protein wastes, so that rendering plants and packing houses are important sources of amines. These compounds are also responsible for some of the odor from sewage treatment plants.

The atmospheric chemistry of amines has not been studied in much detail. As bases (electron-pair donors), their acid-base chemistry in aqueous aerosols may be significant, particularly in a predominantly acidic precipitation (rainfall) environment. As with most organic compounds, the predominant gas-phase reaction is expected to be with hydroxyl radical [18].

Amides are organic compounds with the general formula,

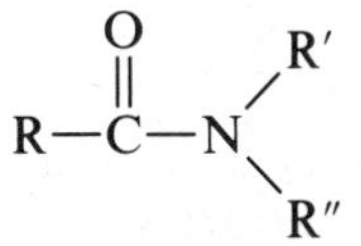

$$R-\overset{\overset{\displaystyle O}{\|}}{C}-N\begin{matrix}\diagup R' \\ \diagdown R''\end{matrix}$$

where the R groups may be hydrogen atoms or hydrocarbon moieties. The amide of greatest commercial importance is dimethylformamide, the structure of which was shown at the beginning of this section. It is used as a solvent for the synthetic polymer, polyacrylonitrile (Orlon, Acrylan), and has been detected occasionally as an atmospheric pollutant from industrial sources. Because of their relatively high boiling temperatures, amides are not commonly found in the atmosphere.

A number of **nitriles**, which are characterized by the $-C\equiv N$ group, have been reported as air contaminants, primarily from industrial sources. Both acrylonitrile (structure shown at the beginning of this section) and acetonitrile, CH_3CN, have been reported in the atmosphere as a result of synthetic rubber manufacture [19]. As expected from their volatilities and levels of industrial production, most of the nitriles reported as atmospheric contaminants are low-molecular-weight aliphatic or olefinic nitriles, or aromatic nitriles with only one benzene ring. Acrylonitrile, used to make polyacrylonitrile polymer, is the only nitrogen-containing organic chemical among the top 50 chemicals, ranking forty-second, with annual production of about 800 million kg.

Nitro compounds have the general formula, RNO_2. Among these reported as air contaminants are nitromethane, nitroethane, and nitrobenzene, where the R portion is, respectively, a methyl, ethyl, and phenyl group. These compounds

---

[18] Atkinson, R.; Perry, R.A.; and Pitts, J.N., Jr. 1978. Rate constants for the reaction of the OH radical with $(CH_3)_2NH$, $(CH_3)_3N$, and $C_2H_5NH_2$ over the temperature range 298–426 °K. *J. Chem. Phys.* 68: 1850–1853.

[19] 1973. *Air pollution emissions factors.* Environmental Protection Agency Report No. AP-42. Research Triangle Park, No. Carolina: U.S. Environmental Protection Agency.

are produced from industrial sources. Highly oxygenated compounds containing the NO_2 group, particularly peroxyacetyl nitrate (PAN) (discussed in Section 14.4), are end products of the photochemical oxidation of hydrocarbons in urban atmospheres but are not commonly classified as nitro compounds in the parlance of organic chemistry.

A large number of **heterocyclic nitrogen compounds** have been reported in tobacco smoke, and it is inferred that many of these compounds can enter the atmosphere from burning vegetation. Coke ovens are another major source of this group of compounds. In addition to the derivatives of pyridine, some of the heterocyclic nitrogen compounds are derivatives of pyrrole. Heterocyclic nitrogen compounds occur almost entirely in association with aerosols in the atmosphere.

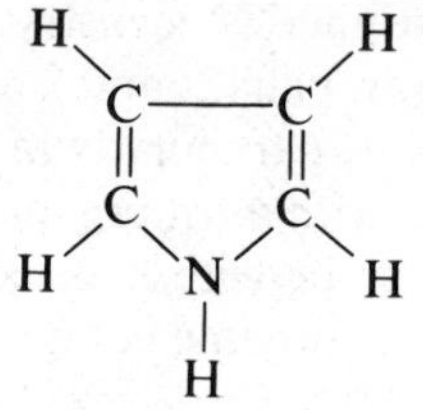

```
H       H
 \     /
  C---C
  ||  ||
  C   C
 / \ / \
H   N   H
    |
    H
```

pyrrole

A class of nitrogen compounds deserving special mention as atmospheric contaminants are **nitrosamines**, which have the following general formula:

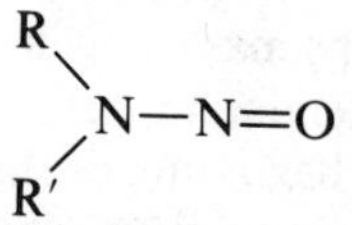

```
R
 \
  N—N=O
 /
R'
```

As discussed in Section 18.19, this group of compounds includes a number of known carcinogens. Both *N*-nitrosodimethylamine (in which both R and R′ are methyl groups) and *N*-nitrosodiethylamine have been detected in the atmosphere [20].

SUPPLEMENTARY REFERENCES

Babich, H., and Stotzky, G. 1974. Air pollution and microbial ecology. *CRC Reviews of Environmental Control* 4(3): 353–421.

Graedel, T.E. 1980. Atmospheric photochemistry. In *The handbook of environmental chemistry,* O. Hutzinger, ed., pp. 107–143. Berlin: Springer-Verlag.

——. 1978. *Chemical compounds in the atmosphere.* New York: Academic Press, Inc.

Hitchcock, D.R., and Wechsler, A.E. 1972. Biological cycling of atmospheric trace gases. Final Report, NASW-2128. Washington, D.C.: U.S. Government Printing Office.

Monteith, J.L., ed. 1977. *Vegetation and the atmosphere.* New York: Academic Press, Inc.

1976. *Vapor-phase organic pollutants.* Washington, D.C.: National Academy of Sciences.

---

[20] Fine, D.H., et al. 1976. *N*-Nitroso compounds in the ambient community air of Baltimore, Maryland. *Analytical Letters* 9: 595–604.

QUESTIONS AND PROBLEMS

1. Match each organic pollutant in the left column with its expected effect in the right column.

   (a) CH_3SH

   (b) $CH_2{=}CH{-}CH_2{-}CH_3$ (structural formula: $H_2C{=}C(H){-}C(H)(H){-}C(H)(H){-}H$)

   (c) $CH_3CH_2CH_2CH_3$

   (1) Most likely to have a secondary effect on the atmosphere.

   (2) Most likely to have a direct effect.

   (3) Should have the least effect of these three.

2. Why are hydrocarbon emissions from uncontrolled automobile exhausts particularly reactive?

3. Assume an accidental release of a mixture of gaseous alkanes and alkenes into an urban atmosphere early in the morning. If the atmosphere at the release site is monitored for these compounds, what can be said about their total and relative concentrations at the end of the day? Explain.

4. Match each radical in the left column with its type in the right column.

   (a) $H_3C\cdot$

   (b) $CH_3CH_2O\cdot$

   (c) $H\dot{C}O$

   (d) $C_xH_{2x+1}O_2\cdot$

   (1) Formyl radical

   (2) Alkylperoxyl radical

   (3) Alkyl radical

   (4) Alkoxyl radical

5. When reacting with hydroxyl radical, alkenes have a reaction mechanism not available to alkanes, which makes the alkenes much more reactive. What is this mechanism?

6. What is the most stable type of hydrocarbon having a very low hydrogen-to-carbon ratio?

7. In the sequence of reactions leading to the oxidation of hydrocarbons in the atmosphere, what is the first stable class of compounds generally produced?

8. Give a sequence of reactions leading to the formation of acetaldehyde from ethane starting with the reaction with hydroxyl radical.

9. What important photochemical property do carbonyl compounds share with NO_2?

CHAPTER 16
ATMOSPHERIC MONITORING

16.1 THE IMPORTANCE OF ATMOSPHERIC MONITORING

A number of chemical analysis techniques that are generally applicable to water, biological, and atmospheric samples were discussed in Chapter 9. This chapter concentrates upon those techniques exclusively devoted to atmospheric monitoring and analysis.

Efforts to control air pollution can be successful only if we are able to determine accurately the nature and levels of pollutants in the atmosphere and from emission sources. Therefore, good analytical methodology, particularly that applicable to automated analysis and continuous monitoring, is essential to the study and alleviation of air pollution. The atmosphere is a particularly difficult analytical system because of a number of factors, including the very low levels of substances to be analyzed; sharp variations in pollutant level with time and location; differences in temperature and humidity; and difficulties encountered in reaching desired sampling points, particularly those substantially above the Earth's surface. Furthermore, although improved techniques for the analysis of air pollutants are continually being developed, a need still exists for new analytical methodology and the improvement of existing methodology.

Much of the earlier data on air pollutant levels (as well as much of the data currently being taken) were unreliable as a result of inadequate analysis and sampling methods. For example, during the mid-1960s, substantial improvements were made in the analytical methods used for nitrogen dioxide and sulfur dioxide. As a consequence, the average values reported by the National Air Surveillance Network for nitrogen dioxide and sulfur dioxide increased approximately 50% and 30%, respectively, in 1966 over values reported in previous years for essentially the same atmospheres. In 1972, however, it was revealed that the reference method for measuring nitrogen dioxide in ambient air consistently overestimated the concentrations of this pollutant at low levels (30–60 $\mu g/m^3$). This was a disturbing development because regulations for the control of nitrogen oxides had been based upon this method, which has since been replaced by an entirely different technique.

An atmospheric pollutant analysis method does not have to give the *actual* value to be useful. One which gives a *relative* value may still be helpful in establishing trends in pollutant levels, determining pollutant effects, and locating pollution sources. Such methods may continue to be used while others are being developed.

16.2 AIR POLLUTANTS MEASURED

The air pollutants generally measured may be placed in five different categories. The first category contains materials for which ambient (surrounding atmosphere) standards have been set by the Environmental Protection Agency [1], as summarized in Table 16.1. The standards are categorized as primary and secondary. **Primary standards** are those defining the level of air quality necessary to protect public health. **Secondary standards** are designed to provide protection against known or expected adverse effects of air pollutants (upon materials, vegetation, animals, and so on).

A second category of atmospheric pollutants which must be analyzed are those known to be specifically hazardous to human health, but for which ambient air quality standards are not applicable. Asbestos, beryllium, and mercury have been placed in this category, and the list will grow as other air pollutants are shown to contribute to serious illness or increases in mortality.

TABLE 16.1 National Air Quality Standards

Pollutant	*Averaging time*	*Primary standard*	*Secondary standard*
sulfur dioxide	annual arithmetic mean	80 $\mu g/m^3$ (0.03 ppm)	—
sulfur dioxide	24 hours	365 $\mu g/m^3$* (0.14 ppm)	—
sulfur dioxide	3 hours	—	1300 $\mu g/m^3$* (0.5 ppm)
nitrogen dioxide	annual arithmetic mean	100 $\mu g/m^3$ (0.05 ppm)	100 $\mu g/m^3$ (0.05 ppm)
carbon monoxide	8 hours	10 mg/m^3 (9 ppm)	10 mg/m^3 (9 ppm)
carbon monoxide	1 hour	40 mg/m^3 (35 ppm)	40 mg/m^3 (35 ppm)
hydrocarbons (exclusive of methane)	3 hours	160 $\mu g/m^3$ (0.24 ppm)	160 $\mu g/m^3$ (0.24 ppm)
particulate matter	annual geometric mean	75 $\mu g/m^3$	60 $\mu g/m^3$**
particulate matter	24 hours	260 $\mu g/m^3$*	150 $\mu g/m^3$*

*Not to be exceeded more than once a year.

**Annual geometric mean, to be used in assessment of plans to achieve the 24-hour standard.

[1] 1977. *Code of Federal Regulations,* 40, Chapter 1, Environmental Protection Agency, Subchapter C—Air Programs, Part 50, National Primary and Secondary Ambient Air Quality Standards, 1 July.

A third category of air pollutants contains those regulated in new installations of selected stationary sources, such as coal-cleaning plants, cotton gins, lime plants, and paper mills. The pollutants in this category are visible emissions, acid (H_2SO_4) mist, particulate matter, nitrogen oxides, and sulfur oxides. These substances often must be monitored in the stack to ensure that emissions standards are being met.

The emissions of mobile sources (motor vehicles) that are now regulated constitute a fourth class of air pollutants. These emissions include hydrocarbons, CO, and NO_x.

Finally, a fifth group of elements and compounds is being studied to determine the degree to which each should be controlled and the best method of control for each. These materials include the following metals: arsenic, cadmium, nickel, chromium, vanadium, manganese, lithium, copper, zinc, barium, and tin. Other species included are fluoride, chlorine, hydrogen chloride, hydrogen sulfide, selenium, boron, phosphorus, polycyclic aromatic hydrocarbons (PAH), polychlorinated biphenyls, odorous compounds, fine particulate matter, pesticides, aero-allergens, reactive organic compounds, and radionuclides.

Much of the remainder of this chapter is devoted to a discussion of the analytical methods for most of the species mentioned above. For some species, an analytical method is well developed and reasonably satisfactory. For others, no really satisfactory method exists. The development of analytical techniques for air pollutants remains a fertile and challenging area for research.

Levels of air pollutants and other air-quality parameters are expressed in several different kinds of units. The recommended units are [2]: for gases and vapors, $\mu g/m^3$ (alternatively, ppm by volume); for weight of particulate matter, $\mu g/m^3$; for particulate matter count, number per cubic meter; for visibility, kilometers; for instantaneous light transmission, percentage of light transmitted; for emission and sampling rates, m^3/min; for pressure, mm Hg; and for temperature, degrees centigrade. Air volumes should be converted to conditions of 10 °C and 760 mm Hg (1 atm), assuming ideal gas behavior.

16.3 SAMPLING

The ideal atmospheric analysis techniques are those that work successfully without sampling, such as long-path laser resonance absorption monitoring. For most analyses, however, various types of sampling are required. In some very sophisticated monitoring systems, samples are collected and analyzed automatically, and the results are transmitted to a central receiving station. Generally, however, a batch sample is collected for later chemical analysis.

The analytical result from a sample can be only as good as the method employed to obtain that sample. A number of factors enter into obtaining a good sample [2]. The size of the sample required (total volume of air sampled) decreases with increasing concentration of pollutant and increasing sensitivity of the analytical method. Often a sample of 10 or more cubic meters is required. The sampling rate is determined by the equipment used and generally ranges from approximately 0.003 m^3/min to 3.0 m^3/min. The duration of sampling time in-

---

[2] Stern, A.C. 1976. *Air pollution.* Vol. III, *Measuring, monitoring, and surveillance of air pollution.* 3rd ed. New York: Academic Press, Inc.

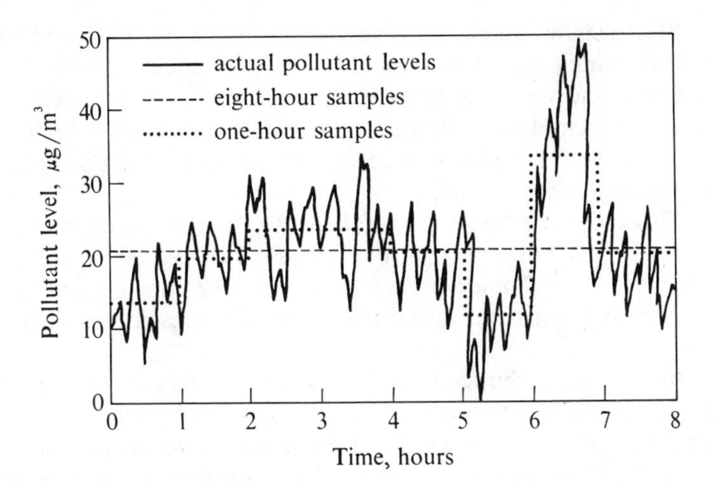

FIGURE 16.1 Effect of duration of sampling upon observed values of air pollutant levels.

fluences the result obtained, as shown in Figure 16.1. The actual concentration of the pollutant is shown by the solid line. A sample collected over an eight-hour period has the concentration shown in the dashed line, whereas samples taken over one-hour intervals exhibit the concentration levels shown by the dotted line.

Filtration is one of the most common techniques for sampling particulate matter. Filters composed of fritted (porous) glass, porous ceramics, paper fibers, cellulose fibers, fiberglass, asbestos, mineral wool, or plastic may be used. A special type of filter is the membrane filter, which yields high flow rates with small, moderately uniform pores. Most of the volume of the filter is composed of pores, and, until clogged by particles, the filters continue to have very high flow rates. The pore diameters are expressed in units of microns, a 0.45-μ filter being a popular representative size. The filter pore diameter does not mean that the filter collects only those particles larger than the stated pore size. Particularly as material collects in the pores, particles smaller than the stated pore diameters may be retained.

Material collected by filters may be analyzed chemically, by weighing, by microscopy, and by particle sizing. The membrane, or controlled pore, filters are particularly useful for particle sizing because they retain particles primarily on their surfaces. In addition, membrane filters made of cellulose esters and similar materials are rendered transparent when treated with clear mineral oils, thus making the particles much more easily visible. The collection of particulate matter by filtration and its subsequent analysis are discussed in more detail in Section 16.12.

The simplest technique for the collection of particles is **sedimentation**. A sedimentation collector may be as simple as a glass jar equipped with a funnel. Liquid is sometimes added to the collector to prevent the solids from being blown out. Sedimentation is generally satisfactory for the collection of particles exceeding 5 μm in diameter. Another device useful for the collection of relatively large particles is the **centrifugal collector**. These devices are small versions of the centrifugal collectors used to remove particles from emission sources. They are designed to impart a spinning motion to a gas stream. As a consequence of their inertia, the dust particles collect on a wall of the collector.

Electrostatic samplers are very efficient for the collection of small particles. These samplers, too, are smaller versions of pollutant control devices. Particles entering the sampler are charged as they pass through an electrical discharge between two electrodes maintained at a potential difference of up to 30,000 volts. The small charged particles in a gas stream are neutralized by, and accumulate on, an electrode downstream.

Thermal precipitators are very effective for the collection of particles over an extremely wide range of sizes. Particles as small as 0.001 μm may be collected with high efficiency. These devices operate upon the principle that suspended particles move to lower-temperature regions when exposed to a high temperature gradient.

Impingers, as the name implies, collect particles from a relatively high-velocity air stream directed at a surface. If the collecting surface is dry, the collection device is called a *dry impinger*. If the surface is wetted to retain the particles, the collection device is called a *wet impinger*. Particle size may be differentiated by dry impingers of the cascade impactor type, as shown in Figure 16.2. The cascade impactor directs the air stream against collection slides through progressively smaller orifices and permits the stepwise collection of smaller particles. Particles may break up into smaller pieces from the impact of impingement; therefore, in some cases impingers yield erroneously high values for small-particle levels. The Andersen sampler and Lundgren impactor are more sophisticated versions of the cascade impactor shown in the figure.

Sampling for vapors and gases may range from methods designed to collect only one specific pollutant to those designed to collect all pollutants. Some of the general methods are described here; other specific sampling methods are found in discussions of the pollutants involved.

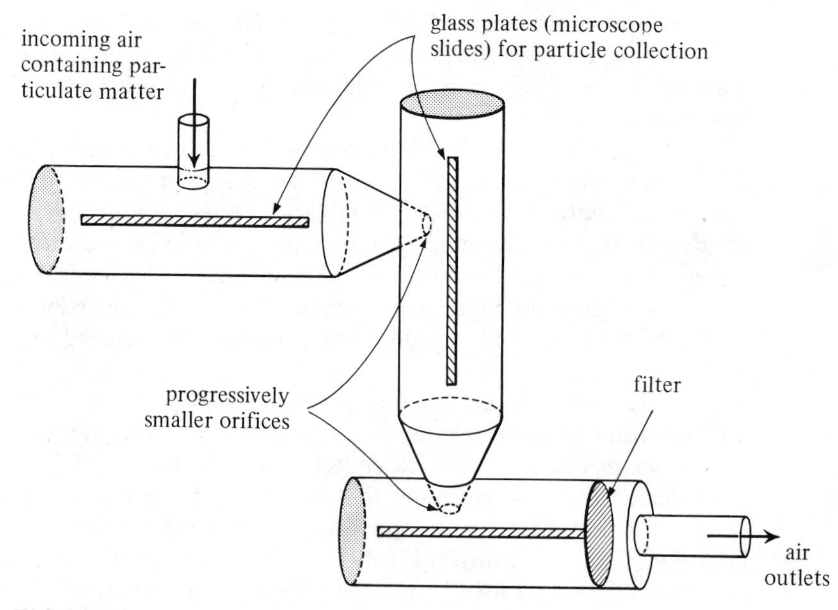

FIGURE 16.2 Schematic diagram of a cascade impactor for the collection of progressively smaller particles.

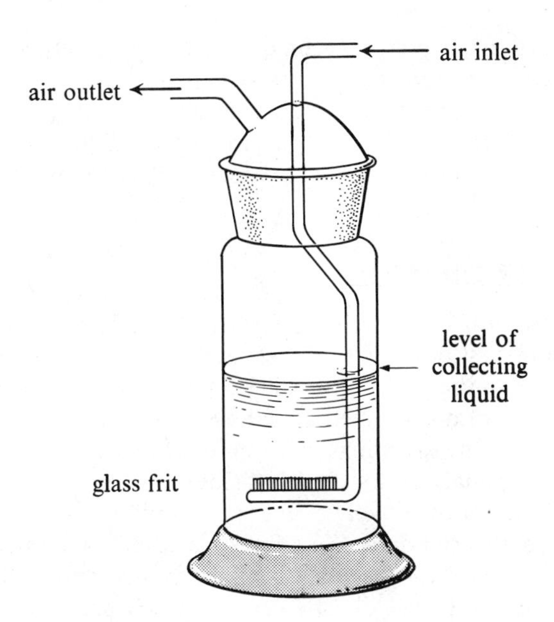

FIGURE 16.3 Scrubber for collecting gaseous air pollutants.

Essentially all pollutants may be removed from an air sample by freezing or by liquefying the air in collectors maintained at a low temperature. Some fractionation may be accomplished by using collectors maintained at progressively lower temperatures. Typically, the first collector is maintained in an ice bath (0 °C), and the last collector is maintained in a liquid-nitrogen bath (– 196 °C).

Absorption in a solvent, such as by bubbling the gas through a liquid, is a very common method for the collection of gaseous pollutants. A number of very pure solvents can be used. Pure water is effective in collecting some gaseous pollutants, such as hydrogen fluoride. Generally, alkaline gases are retained in acidic solutions and acidic gases are collected in basic solutions. Hydrocarbons are collected in some oils. In addition to solubility, diffusivities of the pollutant in both the gas and liquid phases contribute to increased gas collection efficiency. Two characteristics of the absorbing device that increase collection efficiency are small bubble size and increased residence time. Among the types of absorbers used are impingers, countercurrent scrubbers, packed columns, and fritted glass scrubbers, such as the one shown in Figure 16.3.

Adsorption, in which a gas collects on the surface of a solid, is gaining popularity as a means of sample collection. Adsorption is particularly useful for the collection of samples to be analyzed by gas chromatography; in some cases, the sample may be injected directly into the chromatograph from the collecting device. A number of different adsorbent materials are used, including alumina, charcoal, silica gel, and a variety of synthetic substances, such as "molecular sieves." After adsorption, the sample must be removed for analysis by techniques such as heating the column and flushing it with gas, or applying a vacuum.

16.4 METHODS OF ANALYSIS

A very large number of different analytical techniques are used for atmospheric pollutant analysis. Some of these whose uses are not confined to atmospheric analysis were discussed in Chapter 9. Techniques confined largely to atmospheric analysis are discussed in the remainder of this chapter.

A summary of instrumental techniques for air monitoring is presented on the following pages in Tables 16.2, 16.3, and 16.4. A complete, periodically updated summary of instrumentation for environmental monitoring is available [3].

16.5 ANALYSIS OF SULFUR DIOXIDE

The reference method for the analysis of sulfur dioxide is the **colorimetric pararosaniline method** first described by West and Gaeke and subsequently optimized [4]. Applicable to the analysis of 0.005–5 ppm SO_2 in ambient air, it was described as an example of a colorimetric method of analysis in Section 9.3.

Figure 16.4 shows the sampling train employed to sample the atmosphere for sulfur dioxide to be analyzed by the West-Gaeke method. It is shown to illustrate the various components involved in such sampling.

In addition to the standard West-Gaeke method, a number of other techniques have been devised for the analysis of atmospheric sulfur dioxide. One of the oldest of these is the lead dioxide candle, which uses a paste of PbO_2 painted on a gauze-wrapped cylinder or other support. Exposure to sulfur dioxide results in the production of lead sulfate,

$$PbO_2 + SO_2 \rightarrow PbSO_4 \qquad (16.5.1)$$

which is converted to barium sulfate and measured gravimetrically or turbidimetrically.

Conductimetry was used as the basis for a commercial continuous sulfur dioxide analyzer as early as 1929. Generally, the sulfur dioxide is collected in a hydrogen peroxide solution and increased conductance of the sulfuric acid solution is measured:

$$SO_2 + H_2O_2 \rightarrow H_2SO_4 \qquad (16.5.2)$$

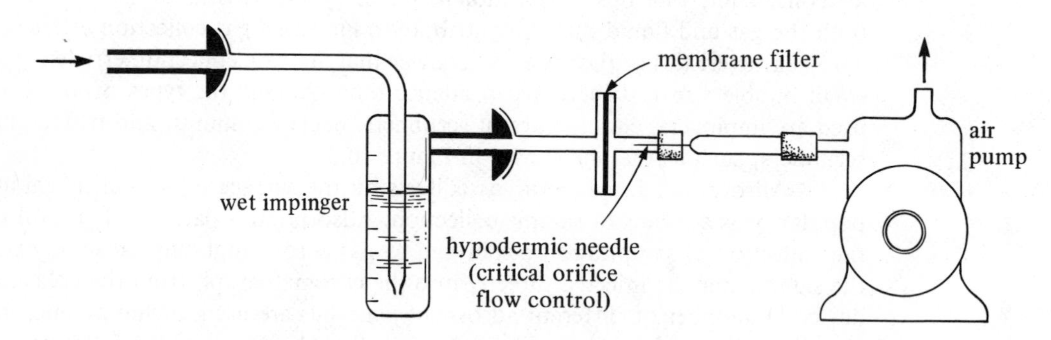

FIGURE 16.4 Sampling train for the collection of sulfur dioxide from the atmosphere.

---

[3] 1972. *Instrumentation for environmental monitoring—air.* Berkeley, Calif.: Lawrence Berkeley Laboratory, University of California. Subsequently updated.

[4] 1971. Reference method for the determination of sulfur dioxide in the atmosphere (pararosaniline method). *Federal Register* 36: 8168.

TABLE 16.2 Instrumental Techniques for Monitoring Gaseous Air Pollutants

	CO	CO_2	SO_2	NO, NO_2	oxidants (O_3)	hydrocarbons	H_2S	mercaptans	HCl	Cl_2	NH_3	gaseous fluorides	gaseous polycyclic aromatic hydrocarbons	gaseous silicates	volatile pesticides
amperometry			X	X	X	X									
chemiluminescence				X	X										
colorimetry			X	X	X	X					X				
conductimetry			X												
catalytic oxidation	X														
chromatography	X	X	X			X	X	X					X	X	X
flame photometry			X												
infrared absorption	X	X	X	X		X									
manual wet chemistry									X	X	X				
potentiometry			X	X							X	X			
tape samplers							X								
ultraviolet absorption			X	X	X	X									
ultraviolet emission			X												
sulfation plates and PbO_2 candle			X												
mass spectrometry						X							X		X

Several types of sulfur dioxide monitors are based on amperometry, in which an electrical current is measured that is proportional to the SO_2 in a collecting solution. Commonly, sulfur dioxide reacts with bromine in solution, causing a change in a measuring electrode potential:

$$SO_2 + Br_2 + 2\,H_2O \rightarrow H_2SO_4 + 2\,HBr \qquad (16.5.3)$$

The potential is restored to its original value by regenerating Br_2 electrochemically and measuring the current required for regeneration.

A method for SO_2 analysis that is now gaining favor consists of bubbling the gas through hydrogen peroxide solution to produce SO_4^{2-}, followed by analysis of the sulfate by **ion chromatography**, a method that separates ions on a chromatography column and detects them very sensitively by conductivity measurement [5]. The ion chromatographic analysis of sulfate can be done in about 10 minutes down to levels of 5 μg/L. It has already been accepted as a method for SO_2 analysis in flue-gas scrubber samples.

Flame photometry, sometimes in combination with gas chromatography, is used for the detection of sulfur dioxide and other gaseous sulfur compounds. The gas is burned in a hydrogen flame, and the sulfur emission line at 394 nm is measured.

---

[5] Wetzel, R. 1979. Ion chromatography: further applications. *Environmental Science and Technology* 13: 1214–1217.

TABLE 16.3 Instrumental Techniques for Monitoring Particulate Matter and Miscellaneous Air Pollutants

	particles, opacity	particles, mass loading	particles, size distribution	particles, velocity	silicates	aeroallergens	pesticides	polycyclic aromatic hydro-carbons (PAH)	fluorides	sulfates
beta-radiation absorption		X								
capacitance		X								
condensation nuclei counters			X							
chromatography					X	X	X	X		
Doppler-shift				X						
dry impingement			X							
electrostatic precipitation			X							
gravimetric		X								
holography			X							
light scattering	X									
light transmission	X		X							
microscopy										
electron spectroscopy										X
nephelometry		X								
paper-tape sampling		X								
piezoelectric microbalance		X								
optical microscopy			X							
electron microscopy			X							
resonant frequency analysis		X								
sedimentation			X							
opacity light meter	X									
Pitot tubes				X						
thermal precipitation		X								
potentiometry									X	

Several direct spectrophotometric methods are used for sulfur dioxide measurement, including nondispersive infrared absorption, ultraviolet absorption, molecular resonance fluorescence, and second-derivative spectrophotometry. The principles of these methods are the same for any gas measured. They are discussed later in this chapter.

16.6 NITROGEN OXIDES

Nitrogen oxides have been among the most difficult atmospheric pollutants to measure accurately. Previous standard methods have depended upon the formation of colored compounds upon reaction of NO_2 with an organic compound. As ''wet chemical'' methods, these have been cumbersome and error-prone; as a consequence, their use has been largely discontinued.

TABLE 16.4 Instrumental Techniques for Monitoring Elements in Atmospheric Pollutants

	arsenic	barium	boron	beryllium	cadmium	chromium	copper	iron	lithium	manganese	lead	nickel	mercury	antimony	phosphorus	tin	zinc	vanadium	selenium	fluorine
atomic absorption	X	X	X	X	X	X	X	X	X	X	X	X	X		X		X			
colorimetry	X			X	X					X	X		X	X				X	X	
emission spectroscopy		X		X	X	X	X	X	X	X	X	X				X	X	X		
neutron activation analysis	X					X	X	X		X			X	X				X	X	
X-ray fluorescence	X	X			X	X	X	X		X	X	X	X	X		X	X	X	X	
potentiometry																				X

The U.S. Environmental Protection Agency now recommends **gas-phase chemiluminescence** as the standard reference method for NO_x analysis. The general phenomenon of chemiluminescence was discussed in Section 11.7. It results from the emission of light from electronically excited species formed by a chemical reaction. In the case of NO, ozone is used to bring about the reaction, producing electronically excited nitrogen dioxide:

$$NO + O_3 \rightarrow NO_2^* + O_2 \qquad (16.6.1)$$

This species loses energy and returns to the ground state through emission of light in the 600–3000 nm range:

$$NO_2^* \rightarrow NO_2 + h\nu \qquad (16.6.2)$$

The emitted light is measured by a photomultiplier; its intensity is proportional to the concentration of NO. A schematic diagram of the device used is shown in Figure 16.5.

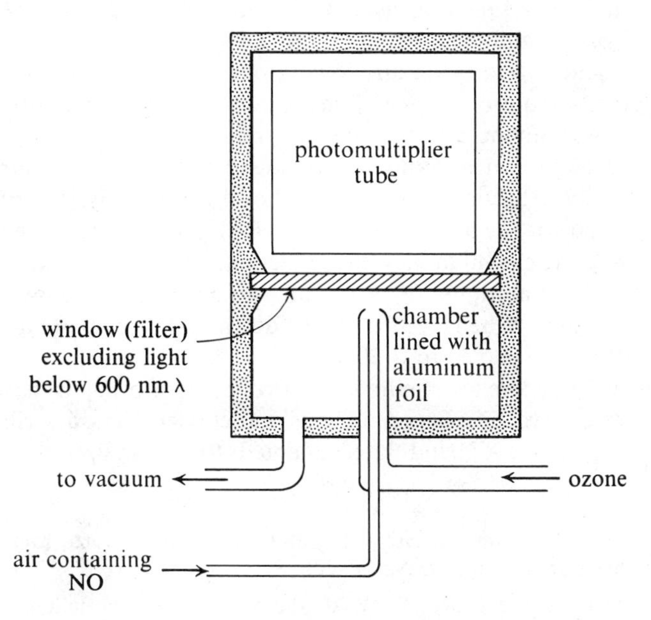

FIGURE 16.5 Chemiluminescence detector for NO.

Since the chemiluminescence detector system depends upon the reaction of O_3 with NO, it is necessary to convert NO_2 to NO in the sample prior to analysis. This is accomplished by passing the air sample over a thermal converter, which brings about the desired conversion. Analysis of such a sample gives NO_x, the sum of NO and NO_2. Chemiluminescence analysis of a sample that has not been passed over the thermal converter gives NO. The difference between these two results is NO_2.

Other nitrogen compounds besides NO and NO_2 undergo chemiluminescence by reacting with O_3, and these may interfere with the analysis if present at an excessive level. Particulate matter also causes interference which may be overcome by employing a membrane filter on the air inlet.

This analysis technique is illustrative of chemiluminescence analysis in general. Chemiluminescence is an inherently desirable technique for the analysis of atmospheric pollutants because it avoids wet chemistry, is basically simple, and lends itself well to continuous monitoring and instrumental methods. Another chemiluminescence method, that employed for the analysis of ozone, is described in the next section.

16.7 ANALYSIS OF OXIDANTS

The atmospheric oxidants that are commonly analyzed include ozone, hydrogen peroxide, organic peroxides, and chlorine. The manual method for the analysis of oxidants, based upon their oxidation of I^- ion, was discussed in Section 9.3 as an example of a spectrophotometric method of analysis. Generally, the level of oxidant is expressed in terms of ozone, although it should be noted that not all oxidants—PAN, for example—react with the same efficiency as O_3. Oxidation of I^- may be employed to determine oxidants in a concentration range of several hundredths of a part per million to approximately 10 ppm. Nitrogen dioxide gives a limited response to the method; reducing substances interfere seriously.

Several instruments have been developed for the analysis of ozone by chemiluminescence. The first chemiluminescent ozone detector, developed by Regener in the early 1960s, measured emission in the 590-nm region from the reaction of ozone and Rhodamine-B on an activated silica-gel surface. The analyzer is standardized using air flowing over an ozone-generating ultraviolet lamp. It does not require a source of gas such as the ethylene or nitric oxide needed by other chemiluminescence ozone analyzers.

The reaction of ozone and nitric oxide to produce excited nitrogen dioxide has been described (Section 16.6) as a method for the analysis of nitric oxide. It can also be used for the analysis of ozone. Reduced pressure is required to prevent quenching the chemiluminescence reaction. Emission is in the 600–3000 nm range, and a photomultiplier detector can only be used in the lower-wavelength region of that range. When levels of ozone down to 0.01 ppm are to be measured, the detector must be cooled to $-20\,°C$.

The EPA reference method for ozone monitoring is based upon the chemiluminescent reaction between ozone and ethylene first described by Nederbragt et al. [6] in 1965 and later adapted to an analytical technique [7].

---

[6] Nederbragt, G.W.; Van Der Horst, A.; and Van Duijn, J. 1965. Rapid ozone determination near an accelerator. *Nature* 206: 87.

[7] Warren, G.J., and Babcock, G. 1970. Portable ethylene chemiluminescence ozone monitor. *Rev. Sci. Instr.* 41: 280.

Chemiluminescence from this reaction occurs over a range of 300–600 nm, with a maximum at 435 nm. The intensity of emitted light is directly proportional to the level of ozone. Ozone concentrations ranging from 0.003–30 ppm may be measured.

16.8 ANALYSIS OF CARBON MONOXIDE

The national Air Quality Standards reference method for the analysis of carbon monoxide is **nondispersive infrared spectrometry**. This technique depends upon the fact that carbon monoxide absorbs infrared radiation strongly at certain wavelengths. Therefore, when such radiation is passed through a long (typically 100 cm) cell containing a trace of carbon monoxide, part of the energy is absorbed by the gas. At higher carbon monoxide levels, more of the infrared radiant energy is absorbed.

A nondispersive infrared spectrometer differs from standard infrared spectrometers in that the infrared radiation from the source is not dispersed according to wavelength by a prism or grating. The nondispersive infrared spectrometer is made very specific for a given compound, or type of compound, by using the sought-for material as part of the detector. A diagram of a nondispersive infrared spectrometer selective for CO is shown in Figure 16.6. Radiation from an infrared source is "chopped" by a rotating device so that it alternately passes through a sample cell and a reference cell. Both beams of light fall on a detector which is filled with CO gas and separated into two compartments by a flexible diaphragm. The relative amounts of infrared radiation absorbed by the CO in the two sections of the detector depend upon the level of CO in the sample. The difference in the amount of infrared radiation absorbed in the two compartments causes slight differences in heating, so that the diaphragm bulges slightly toward one side. Very slight movement of the diaphragm can be detected and

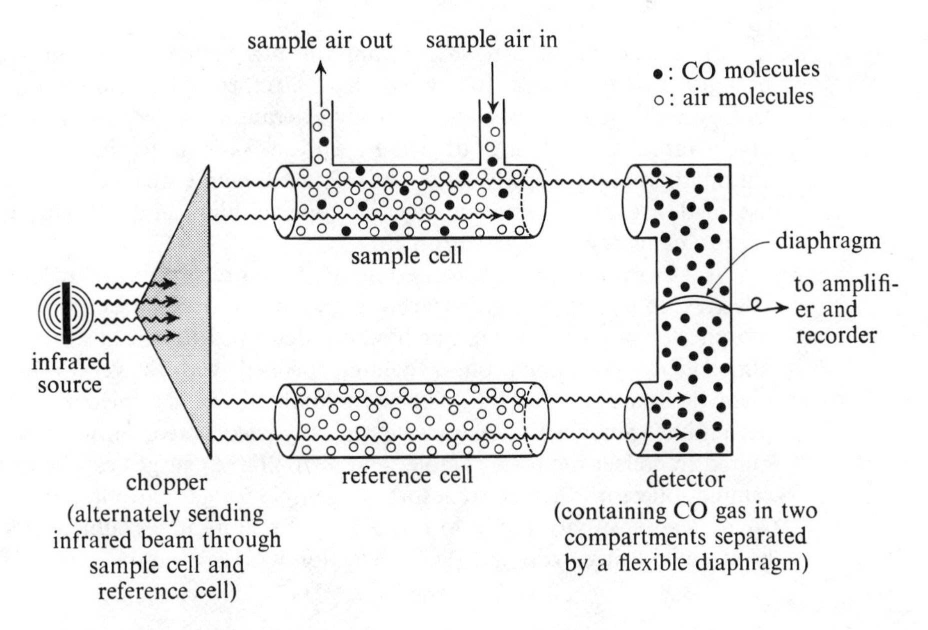

FIGURE 16.6 Nondispersive infrared spectrometer for monitoring carbon monoxide.

recorded. By means of this device, carbon monoxide can be measured from 0–150 ppm with a relative accuracy of ±5% in the optimum concentration range.

Flame-ionization gas chromatography detection can be used for the analysis of carbon monoxide. This detector system was described in detail in Section 9.9. It is selective for hydrocarbons, and conversion of CO to methane in the sample is required. This is accomplished by reaction with hydrogen over a nickel catalyst at 360 °C:

$$CO + 3\,H_2 \rightarrow CH_4 + H_2O \qquad (16.8.1)$$

A major advantage of this approach is that the same basic instrumentation may be used to measure hydrocarbons.

Carbon monoxide may also be analyzed by measuring the heat produced by its catalytic oxidation to CO_2 over a catalyst consisting of a mixture of MnO_2 and CuO. Differences in temperature between a cell in which the oxidation is occurring and a reference cell through which part of the sample is flowing are measured by thermistors. A vanadium oxide catalyst can be used for the oxidation of hydrocarbons, enabling their simultaneous analysis.

16.9 ANALYSIS OF HYDROCARBONS

Sampling of atmospheric hydrocarbons is relatively straightforward. Since methane is a common, naturally occurring, nonpolluting constituent of the atmosphere, it is generally separated from the rest of the sample prior to analysis.

Several techniques are employed for concentration of hydrocarbon samples prior to analysis. "Straw oil," a high-boiling petroleum distillate, is used as a liquid absorber selective for hydrocarbons. Cryogenic (low-temperature) techniques are commonly used to freeze hydrocarbons and other gaseous pollutants. If the analysis of only paraffinic hydrocarbons is desired, oxygenated organics and unsaturated hydrocarbons may be removed by mercuric perchlorate.

The collection of water from the atmosphere constitutes one of the primary disadvantages of cryogenic collectors. Gas chromatographs with hydrogen flame detectors are relatively tolerant of water. Even so, injection of much more than 1 mg of water with a sample to be analyzed by gas chromatography generally cannot be tolerated. Numerous techniques, such as the use of desiccants, freeze-thaw cycles, zone refining, and extraction, have been used to eliminate the water problem.

Adsorption appears to be one of the more promising approaches to the selective collection of hydrocarbons and other organic compounds from atmospheric samples. A technique has been described for the adsorption of organic atmospheric pollutants on a column packed with a very porous styrene-divinylbenzene polymer [8]. With the column, air was sampled at a rate of 4 liters per minute. Since a 10-liter sample typically is collected, only a short time is required to collect the total sample, and air pollution surges can be followed. The sample collector is heated to desorb the sample for gas chromatographic analysis. Analytical sensitivities of as low as 0.1 μg of pollutant per cubic meter of air may be attained using hydrogen-flame detection with this sample collector.

---

[8] Dravnieks, A., et al. 1971. High-speed collection of organic vapors from the atmosphere. *Environmental Science and Technology* 5: 1220–2.

Gas chromatography is the standard reference method for the analysis of non-methane-containing hydrocarbons. This powerful analytical tool has been described in Section 9.9. As discussed in that section, a flame-ionization detector is used for the detection of hydrocarbons that have been separated by gas chromatography.

Two techniques described in Section 16.8 for the analysis of carbon monoxide may also be applied to hydrocarbon analysis. These are nondispersive infrared analysis and catalytic oxidation. Dispersive infrared analysis, in which an analytical band is selected at a particular wavelength with a monochromator, can be used for the quantitative analysis of hydrocarbons. This enables a degree of selectivity among hydrocarbon types.

Chemiluminescence shows some potential for the analysis of reactive hydrocarbons. Reaction of hydrocarbons with atomic oxygen gives intense emission spectra at 700–900 nm. This reaction is carried out in a flow tube in which oxygen atoms are generated with 2450-MHz microwaves. A continuous NO + O spectrum is also generated, but hydrocarbon peaks at 760 nm may be observed. Benzene and toluene at levels as low as 0.3 ppm may be detected.

Mass spectrometry (see Section 9.12) is eminently well suited to the identification of low levels of hydrocarbons, but is normally too expensive and complicated for use in routine monitoring.

16.10 PERMEATION TUBES FOR AIR POLLUTANT STANDARDS

Obtaining accurately known standards is one of the major problems in pollutant analysis. The development of *permeation tubes* [9] has helped greatly to alleviate that problem for sulfur dioxide and other liquefiable gases. Tubes have been developed as standards for ammonia, sulfur dioxide, hydrogen sulfide, hydrogen flouride, nitrogen dioxide, halogenated compounds, a number of hydrocarbons, phosgene, and even organic mercury compounds.

Permeation tubes consist of inert plastic (such as FEP Teflon) tubes containing the liquefied gas to be analyzed, as shown in Figure 16.7. The gas permeates the walls of the tube at a constant rate; this permeation rate is reproducible and depends upon temperature. Permeation tubes may be calibrated

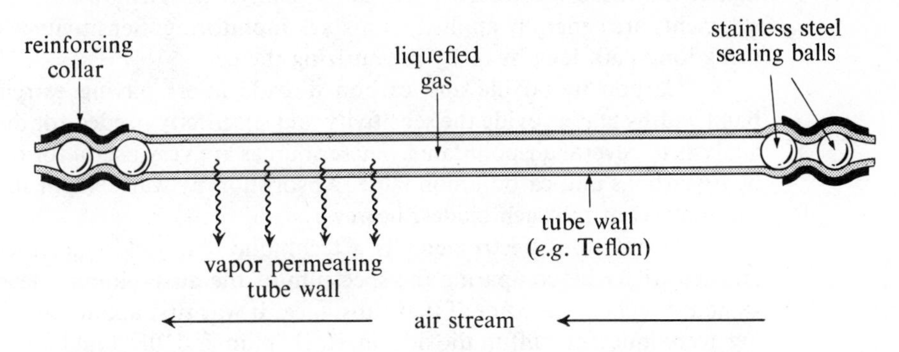

FIGURE 16.7 Permeation tube for atmospheric pollutant standards.

[9] O'Keefe, A.E., and Ortman, G.C. 1966. Primary standards for trace gas analysis. *Analytical Chemistry* 38: 760.

gravimetrically to determine the rate of gas loss, although the tube must be maintained at a carefully controlled temperature for as long as several weeks in order for the weight loss to be accurately measured. Other calibration techniques can be used, including volumetric, colorimetric, and coulometric techniques. A permeation tube may be maintained at a constant temperature by suspending it in a condenser through which thermostatted water is circulated, or by placing the tube inside a holder immersed in a constant-temperature bath.

16.11 DIRECT SPECTROPHOTOMETRIC ANALYSIS OF GASEOUS AIR POLLUTANTS

From the foregoing discussion, it is obvious that measurement techniques that depend upon the use of chemical reagents, particularly liquids, are cumbersome and complicated. It is a tribute to the ingenuity of instrument designers that such techniques are being applied successfully to atmospheric pollutant monitoring. Direct spectrophotometric techniques are much more desirable when they are available and when they are capable of accurate analyses at the low levels required. One such technique, nondispersive infrared spectrophotometry, has been described in some detail for the analysis of carbon monoxide (Section 16.8). Three other spectrophotometric methods are being successfully applied to the monitoring of vapor-phase pollutants in the atmosphere. These are *dispersive absorption spectrometry, correlation spectrometry,* and *second-derivative spectrometry*. These techniques may be used for **point air monitoring**, in which a sample is monitored at a given point, generally by measurement in a long absorption cell. In-stack monitoring may be performed to measure effluents. A final possibility is the collection of long-line data (sometimes using sunlight as a radiation source), an approach which yields concentrations in units of concentration-length (ppm-meters). If the path length is known, the concentration may be calculated. This approach is particularly useful for measuring concentrations in stack plumes.

Dispersive absorption spectrometers are basically standard spectrometers with a monochromator for selection of the wavelength to be measured. They are used to measure air pollutants by determining absorption at a specified part of the spectrum of the sought-for material. Of course, other gases or particulate matter that absorb or scatter light at the chosen wavelength interfere. These instruments are generally applied to in-stack monitoring. Sensitivity is increased by using long path lengths or by pressurizing the cell.

Carbon monoxide and carbon dioxide lasers having extremely narrow band widths may provide the sensitivity and specificity needed for direct infrared analysis of several air pollutants. These sources are very useful for the analysis of hydrocarbons and carbon monoxide. Absorption by water vapor interferes with the analysis of nitrogen oxides, however.

Correlation spectrometry is a technique that yields increased specificity and sensitivity by comparing the spectrum of the air pollutant measured with a standard replica spectrum of that substance. It was first applied as a remote sensing technique for sulfur dioxide in stack plumes [10]. Light that has passed

---

[10] Moffat, A.J., and Millan, M.M. 1969. The applications of optical correlation techniques to the remote sensing of SO_2 plumes using sky light. Paper given at Symposium on Instrumentation for Air Pollution Control Administration of HEW, Cincinnati, Ohio, 26–28 May 1969.

through the sample or through part of the atmosphere is dispersed into a spectrum with a monochromator and the appropriate optics. The resulting spectrum is focused in the plane of a mask upon which is etched a series of slits corresponding to the desired parts of the spectrum of the substance being analyzed. The incoming spectrum and mask are moved relatively in a cyclical fashion, the light passing through the mask is measured with a photomultiplier tube, and the resulting signal is amplified.

Second-derivative spectroscopy [11] is a useful technique for trace gas analysis. Basically, this technique varies the wavelength by a small value around a specified nominal wavelength. The second derivative of light intensity versus wavelength, $d^2I/d\lambda^2$, is obtained. In conventional absorption spectrophotometry, a decrease in light intensity as the light passes through a sample indicates the presence of at least one substance—and possibly many—absorbing at that wavelength. Second-derivative spectroscopy, however, provides information regarding the change in intensity with wavelength, thereby indicating the presence of specific absorption lines or bands which may be superimposed on a relatively high background of absorption. Much higher specificity is obtained. For details on the rather involved theory of second-derivative spectroscopy, the reader is referred to Reference 11.

The spectra obtained by second-derivative spectrometry in the ultraviolet region show a great deal of structure and are quite characteristic of the compounds being observed. Second-derivative spectra of NO, SO_2, and benzene are shown in Figure 16.8.

A schematic diagram of a second-derivative spectrometer for air analysis is shown in Figure 16.9 on page 442. A unique feature is a "wobbler" which varies the position of the absorption slit sinusoidally at a frequency of 45 Hz. This modulates the light signal at a slight amplitude around a wavelength, which in turn is scanned slowly to give the desired spectrum.

Second-derivative spectroscopy is quite sensitive for the analysis of some important air pollutants. Table 16.5 on page 442 gives detection limits for some pollutants that have been analyzed by this technique.

A number of advanced spectroscopic techniques are being developed for the analysis of gas-phase air pollutants. Continued advances in the development

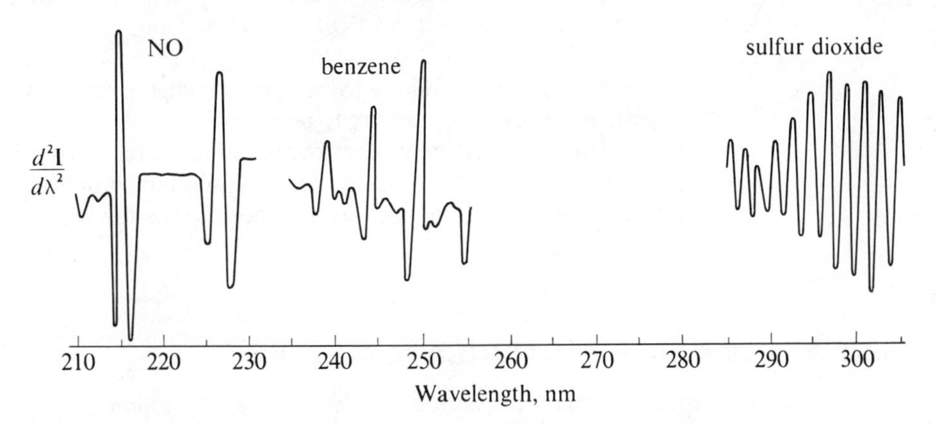

FIGURE 16.8 Second-derivative spectra of NO, benzene, and sulfur dioxide.

[11] Hager, R.N., Jr. 1973. Derivative spectroscopy with emphasis on trace gas analysis. *Anal. Chem.* 45: 1131A–138A.

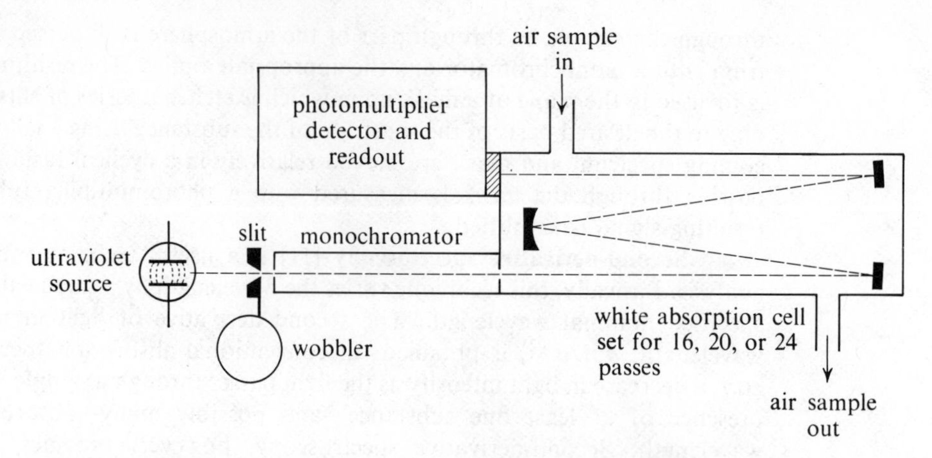

FIGURE 16.9 Second-derivative spectrometer for air analysis.

TABLE 16.5 Detection Limits for the Analysis of Selected Air Pollutants by Second-Derivative Spectroscopy

Pollutant	*Detection limit, ppm*
Ammonia	0.001
Benzene	0.025
Nitric oxide	0.005
Nitrogen dioxide	0.040
Ozone	0.040
Sulfur dioxide	0.001

of lasers, particularly, should make possible more sensitive and selective methods of analysis. For example, laser-induced gas-phase molecular resonance fluorescence techniques are being developed to measure small molecules such as sulfur dioxide, nitrogen oxides, chlorine gas, and formaldehyde. Commercial units are now being used to monitor sulfur dioxide. Back-scattering of laser light is now being used to monitor particulate matter from point sources, such as stacks.

The spectroscopic technique to be applied depends upon the pollutants to be measured, response time required, need for multicomponent analysis, applicability to remote sensing, and other factors. Therefore, it is not likely that any one technique will become a panacea for air pollutant monitoring. Instead, a number of approaches will be used in the applications for which they are most suited.

16.12 ANALYSIS OF PARTICULATE MATTER

The standard reference method for determining the quantity of suspended particles in the atmosphere is the **High Volume Method**, using a Hi-Vol sampler (Figure 16.10). This device is essentially a glorified vacuum cleaner that draws air through a filter. The samplers are usually placed under a shelter which excludes

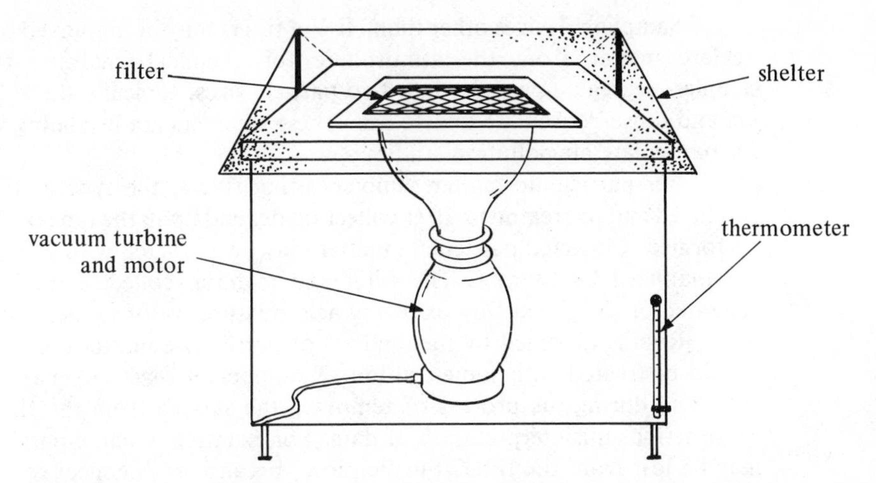

FIGURE 16.10 Hi-Vol sampler for the collection of atmospheric particulate matter.

precipitation and particles larger than about 100 μm in diameter. These devices efficiently collect particles from a large volume of air, typically 2000 m^3.

The filters used in a Hi-Vol sampler are usually composed of glass fibers and have a collection efficiency of at least 99% for particles with 0.8-μm diameter. Particles with diameters exceeding 100 μm remain on the filter surface, whereas particles with diameters down to approximately 0.1 μm are collected on the glass fibers in the filters. Efficient collection is achieved by using very small diameter fibers (less than 1 μm) for the filter material.

The Hi-Vol sampler is most useful for determining total levels of particulate matter. Prior to taking the sample, the filter is maintained at 15–35 °C at 50% relative humidity for 24 hours, then weighed to the nearest milligram. After preparation, the filter is fitted to the Hi-Vol sampler, which draws air through a 406.5-cm^2 portion of the filter at a rate of at least 1.70 m^3/min as measured by a rotameter. After sampling for 24 hours, the filter is removed and equilibrated for 24 hours under the same conditions used prior to its installation on the sampler. The filter is then weighed and the quantity of particulate matter per unit volume of air is calculated. Levels as low as 1 $\mu g/m^3$ can be measured. The filter may be saved for subsequent chemical examination.

An alternative to the somewhat cumbersome weighing of filters is the β **gauge** for aerosol mass measurement. The sample is collected on thin teflon filters, which are then placed between a β-ray source and a detector [12]. Where x is the areal density (mass per unit area), and the count rate with the filter in place and removed are I and I_0, respectively, the following relationship holds:

$$I = I_0 e^{-\mu x} \tag{16.12.1}$$

The parameter μ is the mass absorption coefficient for β radiation in the sample.

---

[12] Courtney, W.J.; Shaw, R.W.; and Dzubay, T.G. 1982. Precision and accuracy of a β gauge for aerosol mass determinations. *Environmental Science and Technology* 16: 236–39.

Sampling devices other than Hi-Vol filters may be employed to collect particulate matter from the atmosphere for chemical analysis. **Dichotomous samplers** segregate particles into two particle sizes, typically those less than 2.5 μm and those betwen 2.5 and 15 μm. These samplers are becoming very popular for particulate air pollution studies.

For particulate matter samplers using filters, the type of filter material employed and its treatment after collection depend upon the type of analysis to be performed. Collected particulate matter may be extracted with a suitable solvent or volatilized by heating. The filter and material collected on it may be incinerated or "digested" by oxidizing acid mixtures prior to elemental analysis.

Results obtained by the analysis of particulate matter collected by filters should be treated with some caution. A number of reactions may occur on the filter and during the process of removing the sample from the filter. This can cause serious misinterpretation of data. For example, volatile particulate matter may be lost from the filter. Furthermore, because of chemical reactions on the filter, the material analyzed may not be the material that was collected.

One of the major difficulties in particle analysis is the lack of suitable filter material. Different filter materials serve very well for specific applications, but none is satisfactory for all applications. Fiber filters composed of polystyrene are very good for elemental analysis because of low background levels of inorganic materials. However, they are not useful for organic analysis. Glass-fiber filters have good weighing qualities and are therefore very useful for determining total particle concentration. However, metals, silicates, sulfates, and other species are readily leached from the fine glass fibers, introducing error into analysis for inorganic pollutant analysis.

Most of the chemical analysis techniques used to characterize atmospheric pollutants were discussed in Chapter 9. These include atomic absorption, optical emission spectroscopy, X-ray fluorescence, neutron activation analysis, and ion-selective electrodes (primarily for fluoride analysis).

Chemical microscopy is an extremely useful technique for the characterization of atmospheric particles [13]. Either visible or electron microscopy may be employed. Particle morphology and shape tell the experienced microscopist a great deal about the material being examined. Reflection, refraction, microchemical tests, and other techniques may be employed to further characterize the materials being examined. Microscopy may be used for determining levels of specific kinds of particles and for determining particle size.

The important class of polycyclic aromatic hydrocarbons (see Section 13.16), which contains several known carcinogens, is extracted from the collecting filter with an organic solvent. Analysis generally is accomplished by fluorescence or ultraviolet absorption spectroscopy. Analysis of these compounds, like analysis of many particulate species, is complicated by losses resulting from incomplete recovery, volatilization, and photooxidation. Among the polycyclic aromatic hydrocarbons found in particulate matter in urban atmospheres are [14] benz(a)anthracene, chrysene, pyrene, fluoranthene,

---

[13] McCrone, W.C., and Delly, J.G. 1973. *The particle atlas* 2nd ed. Vols. 1–4. Ann Arbor, Mich.: Ann Arbor Science Publishers, Inc.

[14] Hoffman, D., and Wynder, E.L. 1968. Chemical analysis and carcinogenic bioassays of organic particulate pollutants. In *Air pollution,* 2nd ed., Vol. II, A.C. Stern, ed., Ch. 20. New York: Academic Press, Inc.

benzo(a)pyrene, benzo(e)pyrene, perylene, benz(e)acephenanthrylene, benzo(j)fluoranthene, benzo(k)fluoranthene, benzo(ghi)perylene, dibenzo(cd,jk)-pyrene, ideno(1,2,3-cd)pyrene, and coronene. Of these polycyclic aromatic hydrocarbons, at least seven are known carcinogens.

In some cases it is necessary to remove gases or vapors from aerosol streams during sampling of aerosols. One example is the removal of water vapor, which might condense and change the mass or qualities of the collected aerosol. In cases where the aerosol is to be analyzed for sulfate, it is necessary to remove gaseous SO_2. This would be the case, for example, in the analysis of stack gas from a coal-fired power plant. Sorption of the SO_2 by the collected aerosol and subsequent oxidation to sulfate can give erroneous values for sulfate. A stripping device has been developed for the removal of SO_2 from an aerosol stream during sampling [15]. This device consists of a fine-mesh inner tube mounted inside a larger diameter solid-walled tube, both 60 cm long. The inner tube has an inside diameter of 1.27 cm and is made of a stainless steel screen with holes approximately 40 μm in diameter. The annular space between the outer and inner tubes is filled with PbO_2. a good sorbent for SO_2 (see Eq. 16.5.1). As the aerosol sample traverses the long inner tube, the SO_2 diffuses through the mesh walls and is sorbed by the PbO_2.

16.13 FINGERPRINTING SOURCES OF POLLUTION

One of the major keys to controlling atmospheric pollution is the ability to identify the pollutants and trace them to their sources. In some cases this can be very hard to do. So far, much emphasis has been placed on monitoring aerosols with "mass-loading" instruments, which measure total quantities of aerosols without identification of the constituents. This approach is generally not useful in identifying sources of pollutants.

The constituents of organic aerosol pollutants can be identified specifically by gas chromatography/mass spectrometry (see Section 9.12). However, this technique is very expensive, requires highly skilled operators, and requires computer-assisted analysis of the huge quantity of data collected from a complex sample.

One relatively straightforward approach to tracing organic aerosol sources is that of generating chromatographic "fingerprints" of the aerosol [16]. This is possible because many polluting chemical processes generate by-products that are characteristic of the process and or the feedstock (raw materials) used. The pattern of by-product production can be used to identify trace discharges from pollutant sources.

Chromatography is the best way to obtain a "fingerprint" of a pollutant source. The organic aerosols can be collected in a sorption column (Tenax), then later desorbed for chromatographic analysis by heating. Detectors selective for elements such as sulfur, nitrogen, or chlorine may add selectivity. Chemical

[15] Kaplan, D.J.; Himmelblau, D.M.; and Kanaoka, C. 1981. A cylindrical PbO_2 diffusion tube for separating SO_2 from an airstream. *Environmental Science and Technology* 15: 558–562.

[16] Bombaugh, K.J., and Lee, K.W. 1981. Fingerprinting pollutant discharges from synfuels plants. *Environmental Science and Technology* 15: 1142–49.

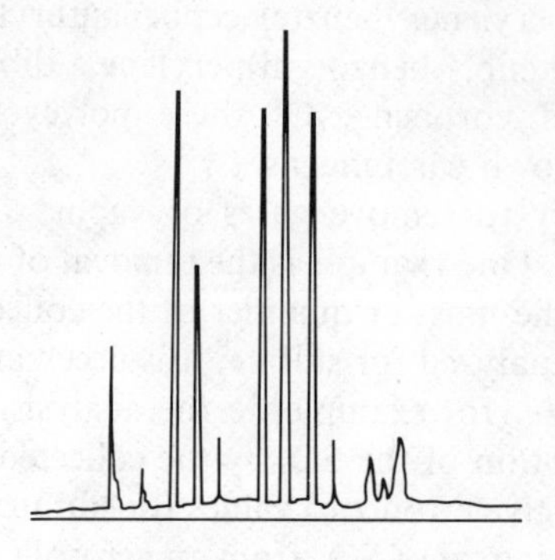

FIGURE 16.11 Capillary gas chromatogram of phenols from a synthetic fuels manufacturing operation.

separations of classes of compounds, such as extraction of phenols into basic solution, may be employed for additional selectivity. Figure 16.11 shows a typical phenol "fingerprint."

SUPPLEMENTARY REFERENCES

Cadle, R.D. 1975. *The measurement of airborne particles.* New York: John Wiley and Sons, Inc.

Coloff, S.G.; Cooke, M.; Drago, R.J.; and Sleva, S.F. 1973. Ambient air monitoring of gaseous pollutants. *American Laboratory* 5: 10.

Hollowell, C.D., and McLaughlin, R.D. 1973. Instrumentation for the monitoring of gaseous air pollutants. *Environmental Science and Technology* 7: 1011.

1972. *Intrumentation for environmental monitoring—air.* Berkeley, Calif.: Lawrence Berkeley Laboratory, University of California. Periodically updated.

Malissa, H. 1977. *Analysis of airborne particles by physical methods.* West Palm Beach, Fla.: CRC Press, Inc.

Schneider, T.; de Doning, H.W.; and Brasser, L.J., eds. 1978. *Air pollution reference measurement methods and systems.* New York: Elsevier North-Holland, Inc.

1976. *Selected methods of measuring air pollutants.* Albany, N.Y.: Q Corporation.

Stern, A.C. 1976. *Air pollution,* Vol. III, *Measuring, monitoring, and surveillance of air pollution.* 3rd ed. New York: Academic Press, Inc.

Warner, P.O. 1976. *Analysis of air pollutants.* New York: Wiley-Interscience.

QUESTIONS AND PROBLEMS

1. What device is employed to make a nondispersive infrared analyzer selective for the compound being determined?

2. After reviewing the discussion of mass spectrometry in Chapter 9, suggest how this technique would be most useful in air pollutant analysis.

3. What is a required characteristic of the absorption cell used for the direct spectrophotometric measurement of gaseous pollutants in the atmosphere, and how may this characteristic be achieved in a cell of manageable dimensions?

4. If 0.250 g of particulate matter is the minimum quantity required for accurate weighing on a Hi-Vol sampler filter, how long must such a sampler be operated at a flow rate of 2.00 m^3/min to collect a sufficiently large sample in an atmosphere containing 5 $\mu g/m^3$ of particulate matter?

5. The atmosphere around a chemical plant is suspected of containing a number of heavy metals in the form of particulate matter. After a review of Chapter 9, suggest several methods that would be useful for a qualitative and roughly quantitative analysis of the metals in the particulate matter.

6. Assume that the signal from a chemiluminescence analyzer for NO is proportional to NO concentration. For the same rate of air flow, an instrument gave a signal of 135 μamp for an air sample that had been passed over a thermal converter and 49 μamp with the converter out of the stream. A standard sample containing 0.233 ppm NO gave a signal of 27 μamp. What was the level of NO_2 in the atmospheric sample?

7. A permeation tube containing NO_2 lost 208 mg of the gas in 124 min at 20 °C. What flow rate of air at 20 °C should be used with the tube to prepare a standard atmospheric sample containing exactly 1.00 ppm by volume of NO_2?

8. What solution should be used in the "wet impinger" shown in Figure 16.4?

9. An atmosphere contains 0.10 ppm by volume of SO_2 at 25 °C and 1.00 atm pressure. What volume of air would have to be sampled to collect 1.00 mg of SO_2 in tetrachloromercurate solution?

10. Assume that 20% of the surface of a membrane filter used to collect particulate matter consists of circular openings with a uniform diameter of 0.45 μm. How many openings are on the surface of a filter with a diameter of 5.0 cm?

11. Some atmospheric pollutant analysis methods have been used in the past that later have been shown not to give the "true" value. In what respects may such methods still be useful?

12. For what purpose is a "critical orifice" used in atmospheric sampling equipment?

13. What is the distinction between *primary* and *secondary* air quality standards?

14. How may ion chromatography be used for the analysis of nonionic gases?

15. What are some of the ways that selectivity for elements or classes of compounds may be introduced into the gas chromatographic "fingerprinting" of organic air pollutants?

CHAPTER 17
NATURAL RESOURCES AND ENERGY

17.1 THE NATURAL RESOURCES-ENERGY-ENVIRONMENT TRIANGLE

Natural resources, energy, and the environment are intimately related (Fig. 17.1). Perturbations in one usually cause perturbations in the other two. For example, reductions in automotive exhaust pollutant levels with the use of catalytic devices, discussed in Chapter 14, have resulted in increased demand for platinum metal, a scarce natural resource, and greater gasoline consumption than would be the case if exhaust emissions were not controlled at all. The availability of many metals depends upon the quantity of energy used and the amount of environmental damage tolerated in the extraction of low-grade ores. Many other such examples could be cited. Because of these intimate interrelationships, resources and energy must be discussed along with environmental chemistry.

Problems of energy, natural resources, and environment came to a head in the mid-1970s, particularly during the 1973–74 "energy crisis." This crisis was accompanied by worldwide shortages of some foods and minerals, followed in

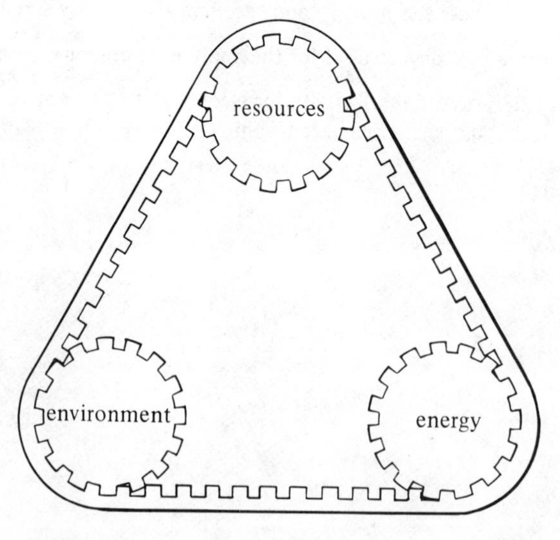

FIGURE 17.1 Strong connections exist among natural resources, energy, and environment.

some cases by surpluses, such as the surplus wheat resulting from increased planting and a copper surplus resulting from the efforts of copper-producing nations to acquire foreign currency by copper export. These surpluses should be regarded as only temporary, so long as growing world population and increasing affluence continue to place demands upon limited resources.

It is beyond the scope of one chapter to discuss natural resources and energy in anything like a comprehensive manner. Instead, we will touch upon the major aspects of natural resources and energy resources as they relate to environmental chemistry.

In recognition of materials and minerals shortages, the National Materials and Minerals Policy Research and Development Act of 1980 was passed by Congress. The overall objective of this law is to assure an adequate and stable supply of minerals. To achieve that goal, the law seeks to encourage a vigorous, comprehensive, coordinated research program in minerals. It furthermore is designed to encourage individual and corporate efforts in the production of minerals. It also encourages federal agencies to make domestic sources more readily available.

In discussing minerals and fossil fuels, two terms related to available quantities are used which should be defined. The first of these is **resources,** which refers to quantities that are estimated to be *ultimately* available. The second term is **reserves,** which refers to well-identified resources that can be profitably utilized with existing technology.

17.2 METALS

With an adequate supply of all of the important elements and energy, almost any needed material can be manufactured. Most of the elements, including practically all of those likely to be in short supply, are metals. Some of these are virtually unavailable in the U.S., which imports almost all its aluminum, chromium, cobalt, manganese, palladium, platinum, and titanium, and the majority of its bismuth, cadmium, mercury, nickel, tungsten and zinc [1].

Some metals are classified as "hot" because of their importance to industrialized societies, uncertain sources of supply, and price volatility in world markets. One of these is antimony, used in auto batteries, fire-resistant fabrics, and rubber. Chromium, another "hot" metal, is used to manufacture stainless steel (especially for parts exposed to high temperatures and corrosive gases), jet aircraft, automobiles, hospital equipment, and mining equipment. The U.S. imports 91% of its chromium, largely from the Republic of South Africa and the U.S.S.R., and obtains the remainder by recycling [2]. Some limited U.S. resources of chromium exist in the Stillwater Complex of Montana and in Oregon beach sands. Cadmium, nickel, and zinc may be substituted for chromium for

---

[1] Huddle, F.H. 1976. The evolving national policy for materials. In *Materials: renewable and nonrenewable resources,* P.H. Abelson and A.L. Hammond, eds., pp. 18–23. Washington, D.C.: American Association for the Advancement of Science.

[2] Ember, L.R. 1981. Many forces shaping strategic minerals policy. *Chemical and Engineering News* 11 May 1981, pp. 20–25.

corrosion protection, whereas substitutes for iron alloys of chromium may be made with nickel, cobalt, molybdenum, and vanadium. The U.S. imports 93% of its cobalt, the bulk of which comes from Zaire, and obtains 7% from recycling. U.S. domestic production stopped in 1979. Cobalt is employed in high-temperature alloys in jet engines, magnets, catalysts, and other uses (see Table 17.1, pp. 452–4). Nickel can be substituted for cobalt in many applications, although the product is generally inferior. Manganese is employed in steel making as ferromanganese. U.S. imports of manganese come largely from Gabon, Brazil, and the Republic of South Africa. More than 80% of identified world resources of this metal are found in the Republic of South Africa and the U.S.S.R. The platinum-group metals (platinum, palladium, iridium, rhodium) are used as catalysts in the chemical industry, in petroluem refining, and in automobile exhaust antipollution devices. The U.S. imports 87% of these metals and receives the remainder from recycling. The Republic of South Africa and the U.S.S.R. are the major import sources. Exploration for domestic sources is underway in the Stillwater Complex in Montana. Substitutes in the electrical and electronic industries include gold, silver, and tungsten. Nickel, vanadium, titanium, and rare earths can substitute in catalytic applications. Titanium is used to manufacture jet engines, steel and other alloys, spacecraft and missile components, and equipment in the chemical processing industry. Titanium dioxide is widely employed as a nontoxic white paint pigment. The U.S. imports 100% of its titanium metal, largely from Australia, but produces most TiO_2 domestically. Adequate domestic supplies of the titanium ore, ilmenite, are available. Vanadium is used to make iron and steel alloys, in the production of titanium alloys, and as a catalyst in sulfuric acid production. Domestic sources can supply U.S. needs although imported vanadium is cheaper. Another critical metal is germanium, used in the electronics industry.

Mining and processing of metal ores involve major environmental concerns, including disturbance of land, air pollution from dust and smelter emissions, and water pollution from disrupted aquifers. This problem is aggravated by the fact that the general trend in mining involves utilization of less rich ores. This is illustrated in Figure 17.2, showing the average percentage of copper in copper ore mined since 1900. The average percentage of copper in ore mined in 1900 was about 4%, but by 1982 it was about 0.6% in domestic ores and 1.4% in richer foreign ores. When the current temporary copper surplus comes to an end, ores as low as 0.1% copper may eventually be processed. Increased demand for a particular metal, coupled with the necessity to utilize lower grade ores, has a vicious multiplying effect upon the amount of ore that must be mined and processed, and accompanying environmental consequences.

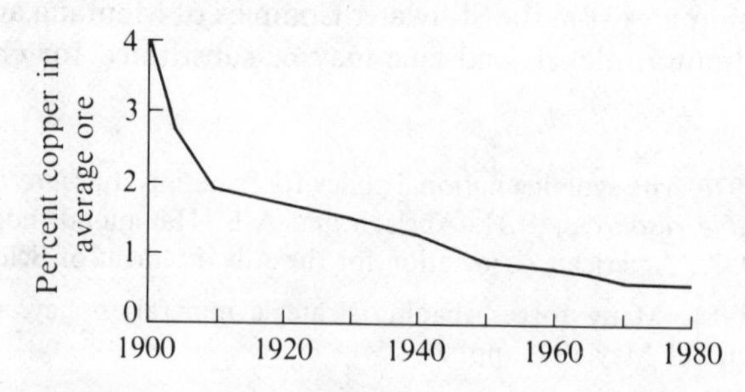

FIGURE 17.2 Average percentage of copper in ore mined, 1900–80.

Discussion of all the major environmental manifestations of the mining of each metal would require a chapter much longer than this one. Instead, these are summarized in Table 17.1. For additional details, the reader is referred to works dealing specifically with mineral resources [3, 4].

It is apparent that we must look to the sea bottom for new sources of minerals. One such source consists of manganese nodules. Located primarily on the Pacific Ocean floor at depths ranging from 2,500 to 6,000 meters, these bodies contain iron, copper, cobalt, and nickel, as well as manganese. Oceanic sulfide deposits may turn out to be sources of about 20 minerals that are strategically short in the U.S. [5]. These deposits are associated with areas where the seafloor is spreading, many of which occur within U.S. territorial waters in the Pacific. The deposits are formed by the upwelling of the Earth's molten silicate magma core from which metals are leached by superheated ocean water infiltrating fractures in the ocean floor. As this mineral-laden water rises through the Earth's crust to the ocean's bottom, it deposits metal sulfide ores, as well as ferric hydroxide.

17.3 NONMETAL MINERAL RESOURCES

A number of minerals other than those used to produce metals are important resources. There are so many of these that it is impossible to discuss them all in this chapter; however, mention will be made of the major ones. As with metals, the environmental aspects of mining many of these minerals are quite important. Typically, even the extraction of ordinary rock and gravel can have important environmental effects.

Asbestos is a term applied to a group of naturally occurring silicate minerals which form relatively strong, flexible fibers. The most common of these is chrysotile, $3\ MgO \cdot 2\ SiO_2 \cdot 2\ H_2O$. Asbestos is used in over 3000 industrial products, including electrical insulation, heat insulation, filters, microbe-resistant paper, and binder in rubber and plastic products. The greatest use is in asbestos-cement products such as roofing and pipe used in sewage and water systems. World production of asbestos has tripled since 1950 and now exceeds 4 million metric tons annually. U.S. asbestos reserves of 4 million metric tons are small compared to annual U.S. consumption of 750,000 metric tons. Global resources, particularly in Quebec, are huge.

About 1.2 million metric tons of barite, composed of finely ground $BaSO_4$, are employed annually in the U.S. to make drilling "mud," which is used in rotary drilling rigs to seal the walls, cool the drill bit, and lubricate the drill stem. World resources of barite exceed 310 million metric tons with about 100 million metric tons in the U.S.

---

[3] Brobst, D.A., and Pratt, W.P., eds. 1973. *United States mineral resources.* Geological Survey Professional Paper 820. Washington, D.C.: United States Government Printing Office.

[4] *Preprints from Mineral Facts and Problems Bulletin,* published monthly for various minerals. Washington, D.C.: Bureau of Mines, U.S. Department of the Interior.

[5] Haggin, J. 1982. Sulfide ores may ease minerals shortage. *Chemical Engineering News* 4 January 1982, pp. 21–23.

TABLE 17.1 Worldwide and Domestic Metal Resources

Metal	*Properties*[1]	*Major uses*	*Ores and aspects of resources*[2]
Aluminum	mp 660 °C, bp 2467 °C, sg 2.70, malleable, ductile	Metal products, including autos, aircraft, electrical equipment. Conducts electricity better than copper per unit weight and is used in electrical transmission lines.	From bauxite ore containing 35–55% Al_2O_3. About 60 million metric tons of bauxite produced worldwide annually, about 27% used in the U.S., which produces about 1.5 million metric tons of bauxite per year. U.S. resources of bauxite are 40 million metric tons, world resources are 15 billion metric tons.
Chromium	mp 1903 °C, bp 2642 °C, sg 7.14, hard, silvery color	Metal plating, stainless steel, wear-resistant and cutting-tool alloys, chromium chemicals, including chromate used as an anticorrosive and cooling-water additive.	From chromite having the general formula [Mg(II), Fe(II)][Cr(III), Al(III), Fe(III)]$_2$O$_4$. Resources of 1 billion metric tons in South Africa and Rhodesia, large deposits in Russia, virtually none in the U.S.
Cobalt	mp 1495 °C, bp 2880 °C, sg 8.71, bright, silvery	Manufacture of hard, heat-resistant alloys such as stellite, permanent magnet alloys, driers, pigments, glazes, catalysts, animal-feed additive.	From a variety of minerals, such as linnaeite, Co_3S_4, and as a by-product of other metals. World consumption is 25,000 metric tons Co per year, 30% used in the U.S. U.S. imports 80% of its cobalt. Abundant global and U.S. resources.
Copper	mp 1083 °C, bp 2582 °C, sg 8.96, dense, ductile, malleable	Electrical conductors, alloys, chemicals. Many uses.	Occurs in low percentages (see Fig. 17.2) as sulfides, oxides, and carbonates in other minerals. U.S. consumption is 1.5 million metric tons per year. World resources of 344 million metric tons, including 78 million in U.S.
Gold	mp 1063 °C, bp 2660 °C, sg 19.3	Jewelry, basis of currency, electronics, increasing industrial uses.	In various minerals at a very low 10 ppm for ores currently being processed in the U.S.; by-product of copper refining. World resources of 1 billion oz, 80 million oz in U.S.

Iron	mp 1535 °C, bp 2885 °C, sg 7.86, silvery metal in (rare) pure form	By far the most widely produced metal, usually as steel, a high-tensile-strength material containing 0.3–1.7% C. Made into many alloys for special purposes.	Occurs as hematite (Fe_2O_3), goethite ($Fe_2O_4 \cdot H_2O$), and magnetite (Fe_3O_4) in abundant supply globally and in the U.S.
Lead	mp 327 °C, bp 1750 °C, sg 11.35, silvery color	Fifth most widely used metal. Storage batteries, gasoline additives, pigments, ammunition.	Major source is galena, PbS. Worldwide consumption of lead is 3.5 million metric tons, about 1/3 in U.S. (not including recycled scrap constituting 40% of use). Global reserves of 140 million metric tons, 39 million metric tons in U.S.
Manganese	mp 1244 °C, bp 2040 °C, sg 7.3, hard, brittle, gray-white	Sulfur and oxygen scavenger in steel, manufacture of alloys, dry cells, chemicals.	Found in a variety of minerals, primarily oxides, in manganese nodules on the ocean floor. About 20 million metric tons of manganese ore produced globally each year, 2 million tons consumed in the U.S., no domestic production. World reserves of manganese are 6.5 billion metric tons.
Mercury	mp −38 °C, bp 357 °C, sg 13.6, shiny liquid metal	Instruments, electronic apparatus, electrodes, chemical compounds (such as fungicides and slimicides).	From cinnabar, HgS. Annual world production of 11,500 metric tons, 1/3 of which is used in the U.S. World resources of 275,000 metric tons, only 6600 metric tons in the U.S.
Molyb-denum	mp 2620 °C, bp 4825 °C, sg 9.01, ductile, silvery-gray	Alloys, pigments, catalysts, chemicals, and lubricants.	Molybdenite (MoS_2) and wulfenite ($PbMoO_4$) are major ores. About two-thirds of global molybdenum production is in the U.S.; global resources are quite large.
Nickel	mp 1455 °C, bp 2835 °C, sg 8.90, silvery color	Alloys, coins, storage batteries, catalysts (e.g., for hydrogenation of vegetable oil).	Found in ores associated with iron. U.S. consumption of nickel is 150,000 metric tons per year, of which less than 10% is produced domestically. Large domestic reserves of low-grade ore are available.

TABLE 17.1 Worldwide and Domestic Metal Resources (*continued*)

Metal	*Properties*[1]	*Major uses*	*Ores and aspects of resources*[2]
Silver	mp 961 °C, bp 2193 °C, sg 10.5, shiny metal	Photographic materials, electronics, sterling ware, jewelry, bearings, dentistry.	Found with sulfide minerals, by-product of copper, lead, and zinc smelting. Annual U.S. consumption of 150 million troy ounces will soon exhaust known resources.
Tin	mp 232 °C, bp 2687 °C, sg 7.31	Coatings, solders, bearing alloys, bronze, chemicals.	Found in many compounds associated with granitic rocks and chrysolites. World consumption of 190,000 metric tons/year, U.S. consumption of 60,000 metric tons/year, world resources of 10 million metric tons.
Titanium	mp 1677 °C, bp 3277 °C, sg 4.5, silvery color	Strong, corrosion-resistant, used in aircraft and their engines, valves, pumps, paint pigments.	Ranks ninth in elemental abundance, commonly as TiO_2; no shortages of titanium are likely.
Tungsten	mp 3380 °C, bp 5530 °C, sg 19.3, gray	Very strong, high boiling point, used in alloys, drill bits, turbines, nuclear reactors, tungsten carbide.	Found as tungstates, such as scheelite ($CaWO_4$); U.S. has 7% of world reserves, China 60%. Could be in short supply in U.S. by year 2000.
Vanadium	mp 1917 °C, bp 3375 °C, sg 5.87, gray	Used to make strong steel alloys.	Occurs in igneous rocks, primarily as V(III), primarily a by-product of other metals, U.S. consumption of 5,000 metric tons/year equals production.
Zinc	mp 420 °C, bp 907 °C, sg 7.14, bluish-white	Widely used in alloys (brass), galvanization, paint pigments, chemicals. Fourth in metal production worldwide.	Found in many ore minerals, including sulfides, oxides, and silicates. World production is 5 million metric tons/year (10% from U.S.) and annual U.S. consumption is 1.5 million metric tons. World resources are 235 million metric tons, 20% in the U.S.

[1] Mp, melting point; bp, boiling point; sg, specific gravity.

[2] All figures are approximate; quantities of minerals considered available depend upon price, technology, recent discoveries, and other factors, so that these quantities are subject to fluctuation.

TABLE 17.2 Major Types of Clays and Their Uses in the United States

Type of clay	*Percent use*	*Composition*	*Uses*
Miscellaneous	72	variable	filler, brick, tile, portland cement, many others
Fireclay	12	variable; can be fired at high temperatures without warping	refractories, pottery, sewer pipe, tile, brick
Kaolin	8	$Al_2(OH)_4Si_2O_5$; is white and can be fired without losing shape or color	paper filler, refractories, pottery, dinnerware, petroleum-cracking catalyst
Bentonite and fuller's earth	7	variable	drilling muds, petroleum catalyst, carriers for pesticides, sealers, clarifying oils
Ball clay	1	variable, very plastic	refractories, tile, whiteware

Clays have been discussed as suspended and sedimentary matter in water (Chapter 6) and as soil components (Chapter 10). Various clays are also used for clarifying oils, as catalysts in petroleum processing, as fillers and coatings for paper, and in the manufacture of firebrick, pottery, sewer pipe, and floor tile. Various types of clays are shown in Table 17.2. Domestic production of clay is about 60 million metric tons per year, and global and domestic resources are abundant.

Fluorine compounds are widely used in industry, with fluorspar, CaF_2, being used in large quantities as a flux in steel manufacture. Synthetic and natural cryolite, Na_3AlF_6, is used as a solvent for aluminum oxide in the electrolytic preparation of aluminum metal. Freon-12, difluorodichloromethane, is widely used as a refrigeration and air-conditioning fluid and is considered to be a major stratospheric pollutant (see Section 12.17). Sodium fluoride is used for water fluoridation. World reserves of high-grade fluorspar are around 190 million metric tons, about 13% of which is in the United States. This is sufficient for several decades at projected rates of use. A great deal of by-product fluorine is recovered from the processing of fluorapatite, $Ca_5(PO_4)_3F$, used as a source of phosphorus (see Section 10.12).

Micas are complex aluminum silicate minerals which are transparent, tough, flexible, and elastic. Muscovite, $K_2O \cdot 3\ Al_2O_3 \cdot 6\ SiO_2 \cdot 2\ H_2O$, is a major type of mica. Better grades of mica are cut into sheets and used in electronic apparatus, capacitors, generators, transformers, and motors. Finely divided mica is widely used in roofing, paint, welding rods, and many other applications. Sheet mica is imported into the United States largely because of high labor costs in the U.S. Domestic production of finely divided "scrap" mica is in excess of demand, so that shortages of this mineral are unlikely.

Phosphorus, along with nitrogen and potassium, is one of the major fertilizer elements (see Chapter 10). Many soils are deficient in phosphate, and its supply is crucial to feeding the world's people. Phosphorus has many other applications, including supplementation of animal feeds, synthesis of detergent builders, and preparation of chemicals such as pesticides and medicines.

The major phosphate minerals are the apatites, having the general formula $Ca_5(PO_4, CO_3)_3(F, OH, Cl)$. The most common of these are fluorapatite, $Ca_5(PO_4)_3F$, and hydroxyapatite, $Ca_5(PO_4)_3(OH)$. Ions of Na, Sr, Th, and U are found substituted for calcium in apatite minerals. Small amounts of PO_4^{3-} may be replaced by SO_4^{2-}, VO_4^{3-}, or AsO_4^{3-}. Because of the latter ion, traces of arsenic are sometimes found in detergent builders and other phosphate products. Vanadium in phosphate minerals may have economic value as a by-product.

Approximately 17% of world phosphate production is from igneous minerals, primarily fluorapatites. About three-fourths of world phosphate production is from sedimentary deposits, generally of marine origin. Vast deposits of phosphate, accounting for approximately 5% of world phosphate production, are derived from guano—droppings of seabirds and bats.

Current U.S. production of phosphate rock is around 40 million metric tons per year, most of it from Florida. Tennessee and several of the western states are also major producers of phosphate. Reserves of phosphate minerals in the United States amount to 10.5 billion metric tons, containing approximately 1.4 billion metric tons of phosphorus. Identified world reserves of phosphate rock are approximately 6 billion metric tons.

Warnings of an impending domestic shortage of phosphate were sounded in a 1979 publication from the General Accounting Office (GAO) in Washington [6]. This report stated that more than half of the phosphate ever mined in the U.S. was mined in the 12-year period prior to 1979, and that the richest deposits of ore were used up. It is estimated that at the rates of mining then taking place, economically recoverable reserves would be gone in 20 to 30 years. However, economic conditions have had a large effect upon the consumption of phosphate and other resources. In 1982, because of poor economic conditions and an inventory of 17 million metric tons of unsold phosphate, much of the Florida phosphate industry shut down. It remains to be seen if former high levels of production will be reached again.

Pigments and fillers of various kinds are used in large quantities. The only naturally occurring pigments still in wide use are those containing iron. These minerals are colored by limonite, an amorphous brown-yellow compound with the formula $2\ Fe_2O_3 \cdot 3\ H_2O$, and hematite, composed of gray-black Fe_2O_3. Along with varying quantities of clay and manganese oxides, these compounds are found in ocher, sienna, and umber. Manufactured pigments include carbon black, titanium dioxide, and zinc pigments. About 1.5 million metric tons of carbon black, manufactured by the partial combustion of natural gas, are used in the U.S. each year, primarily as a reinforcing agent in tire rubber.

Over 7 million metric tons of minerals are used in the U.S. each year as fillers for paper, rubber, roofing, battery boxes, and many other products.

---

[6] 1979. Phosphate—a case study of a valuable depleting mineral in America. Washington, D.C.: General Accounting Office.

Among the minerals used as fillers are asbestos, carbon black, diatomite, barite, fuller's earth, kaolin, mica, limestone, pyrophyllite, and wollastonite ($CaSiO_3$).

Although sand and gravel are the cheapest of mineral commodities per ton, the average annual dollar value of these materials is greater than all but a few mineral products because of the huge quantities involved. In tonnage, sand and gravel production is exceeded only by that of fossil fuels.

At present, old river channels and glacial deposits are used for sand and gravel. Many valuable deposits of sand and gravel are covered by construction and lost to development. Transportation and distance from source to use are especially crucial for this resource. Environmental problems involved with defacing land can be severe, although bodies of water used for fishing and other recreational activities frequently are formed by removal of sand and gravel.

The biggest single use for sulfur is in the manufacture of sulfuric acid. However, the element is employed in a wide variety of other industrial and agricultural products. Current consumption of sulfur amounts to approximately 10 million metric tons per year in the United States.

Sulfur can exist in many forms and undergoes a complicated geobiochemical cycle (Figure 17.3). Sulfur may be expelled from volcanoes as sulfur dioxide or as hydrogen sulfide, which is oxidized to sulfur dioxide and sulfates in the atmosphere. Highly insoluble metal sulfides are oxidized upon exposure to

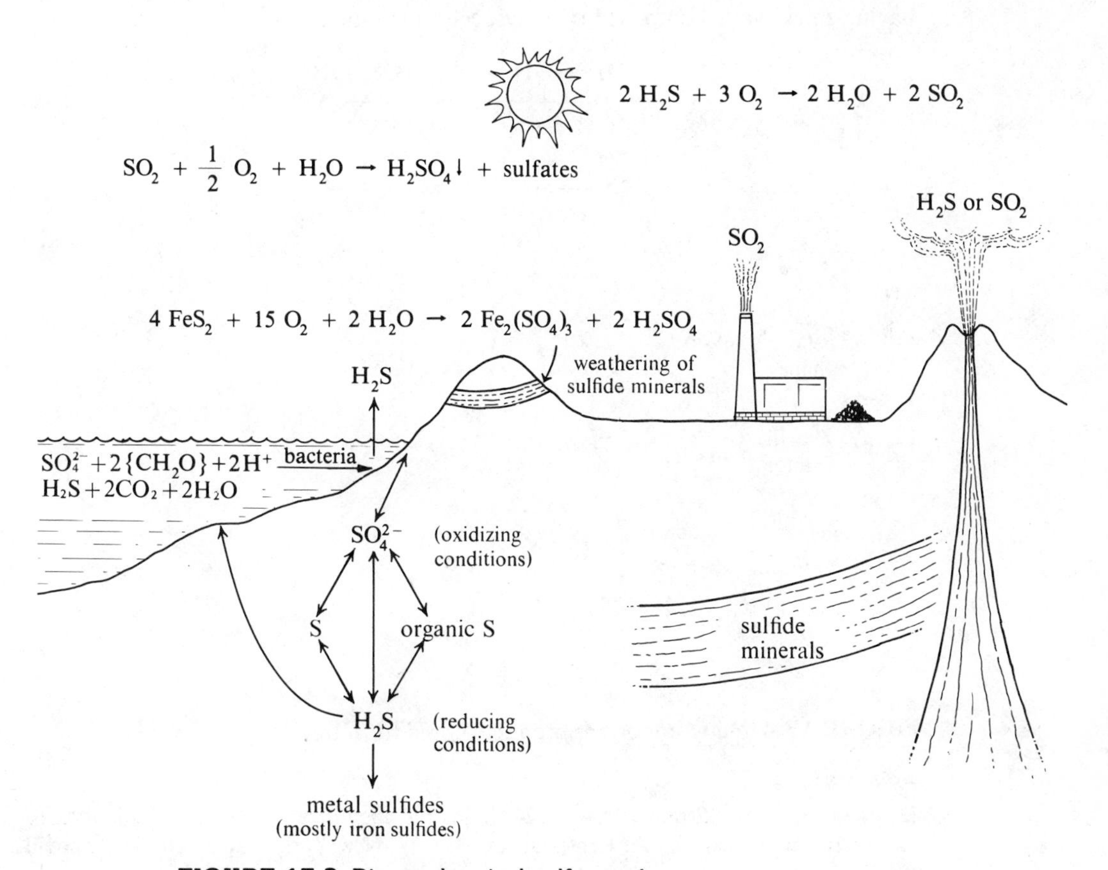

FIGURE 17.3 Biogeochemical sulfur cycle.

atmospheric oxygen to relatively soluble metal sulfates. In the ground and in water, sulfate is converted to organic sulfur by plants and bacteria. Bacteria mediate transitions among sulfate, elemental sulfur, organic sulfur, and hydrogen sulfide. Sulfur(–II) may either escape to the atmosphere as H_2S or precipitate as sulfides of metals, primarily iron.

The four most important sources of sulfur are (in decreasing order): deposits of elemental sulfur; H_2S recovered from sour natural gas; organic sulfur recovered from petroleum; and pyrite (FeS_2). Supply of sulfur is no problem, either in the United States or worldwide. The United States has abundant deposits of elemental sulfur, and sulfur recovery from fossil fuels as a pollution control measure may result in surpluses of this element.

17.4 WOOD—A MAJOR RENEWABLE RESOURCE

Fortunately, one of the major natural resources in the U.S., wood, is a renewable resource. Production of wood and wood products is the fifth largest industry in the United States [7]. Wood ranks first worldwide as a raw material for the manufacture of other products (Fig. 17.4). Forests cover one third of the United States surface area.

Chemically, wood is a complicated substance consisting of long cells having thick walls composed of polysaccharides such as cellulose,

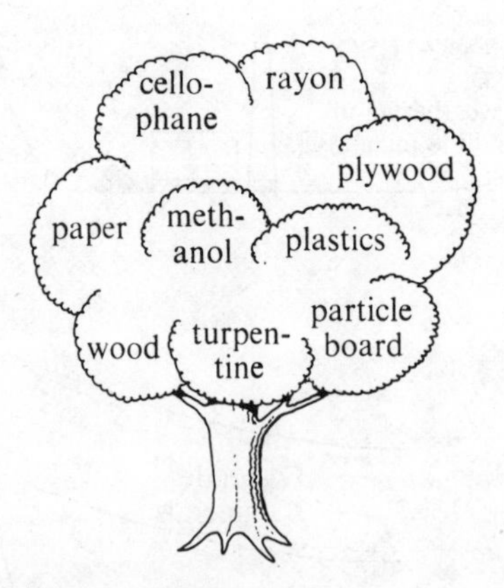

FIGURE 17.4 Many important products come from trees.

[7] Jahn, E.C., and Strauss, R.W. 1974. Industrial chemistry of wood. In *Handbook of industrial chemistry,* J.A. Kent, ed., pp. 435–87. New York: Van Nostrand Reinhold Co.

and lignin, a complex binding material containing groups such as:

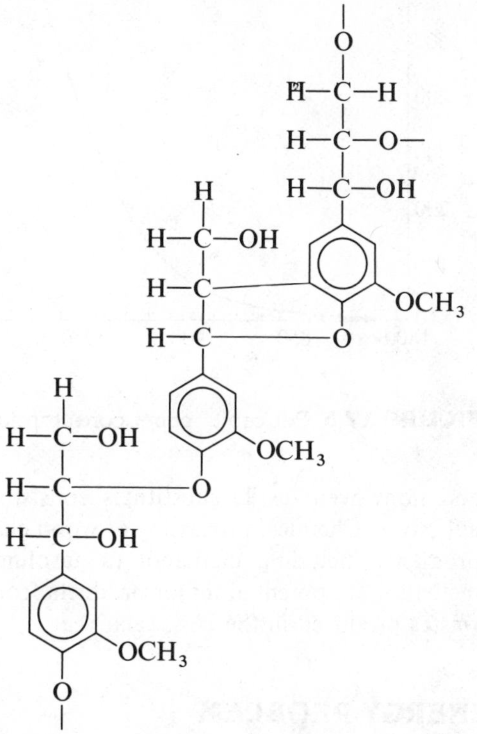

O
H—C—H
H—C—O—
H H—C—OH
H—C—OH
H—C— OCH_3
H—C— O
H
H—C—OH OCH_3
H—C— O
H—C—OH
OCH_3
O

The polysaccharides in cell walls account for approximately three-fourths of *solid wood*, wood from which extractable materials have been removed by an alcohol-benzene mixture. Wood typically contains a few tenths of a percent ash.

A wide variety of organic compounds can be extracted from wood by water, alcohol-benzene, ether, and steam distillation. These compounds include tannins, pigments, sugars, starch, cyclitols, gums, mucilages, pectins, galactans, terpenes, hydrocarbons, acids, esters, fats, fatty acids, aldehydes, resins, sterols, and waxes. Substantial amounts of methanol (sometimes called *wood alcohol*) are extracted from wood. Methanol, once a major source of liquid fuel, is again being considered for that use.

A major use of wood is in paper manufacture. The widespread use of paper is a mark of an industrialized society. Per capita consumption of paper has increased throughout United States history, as shown in Figure 17.5 (on page 460). The manufacture of paper is a highly advanced science and art. Paper consists essentially of cellulosic fibers tightly pressed together. The lignin fraction must first be removed from the wood, leaving the cellulosic fraction. Both the sulfite and alkaline processes for accomplishing this separation have resulted in severe water and air pollution problems, although substantial progress has been made in alleviating these.

Wood fibers and particles can be used for making fiberboard, paper-base laminates (layers of paper held together by a resin and formed into the desired structures at high temperatures and pressures), particle board (consisting of wood particles bonded together by a phenol-formaldehyde or urea-formaldehyde resin)

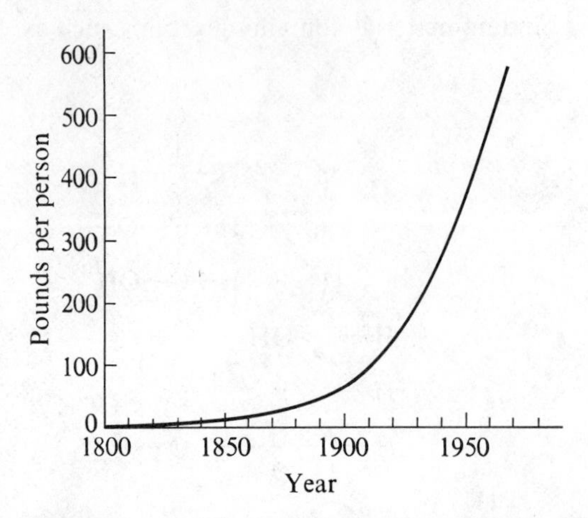

FIGURE 17.5 Per capita paper consumption in the United States.

and nonwoven textile substitutes consisting of wood fibers held together by adhesives. Chemical processing of wood enables the manufacture of many useful products, including methanol (a gasoline substitute) and sugar. Sugar and methanol are potential major products from the 60 million metric tons of wood wastes produced in the U.S. each year.

17.5 THE ENERGY PROBLEM

Since the first "energy crisis" of 1973–74, much has been said and written, many learned predictions have gone awry, and some concrete action has even taken place. Prophecies of catastrophic economic disruption, people "freezing in the dark", and freeways given over to bicycles have not been fulfilled. However, the several-fold increase in crude oil prices since 1973 has exacted a toll. In the U.S. and other industrialized nations, the economy has been plagued by inflation, recession, unemployment, and obsolescence of industrial equipment. The cost to the U.S., alone, may have been as much as $500 billion [8].

The stage was set for the 1973–74 energy crisis by events during the two decades, 1950–70. Abundant oil, which even became somewhat cheaper during the period, led to a worldwide four-fold increase in the use of this resource. In the meantime, the percentage of energy produced by coal—of which tremendous resources are available—declined from 40 to 17 percent of total energy production. Energy conservation was given little attention, and commuting long distances from an energy-inefficient house in the suburbs in a large, gas-guzzling, eight-cylinder automobile became an American way of life. Around 1970, U.S. oil production fell below consumption levels. All that remained to set off the energy crisis were the Arab-Israeli war of 1973 and the resulting oil embargoes by oil-rich Arab countries. As a result, oil prices quadrupled. Further increases resulted from actions to control prices on the part of OPEC (Organization of Petroleum-Exporting Countries) and from a cutback in Iranian production

[8] Martin, D. 1982. Energy shortage eases materially; basic shifts in consumption cited. *New York Times* 8 March 1982.

following a revolution in that country in the late 1970s. In 1982, world oil prices dropped due to decreased demand resulting from depressed economic conditions and successful conservation efforts. Prices were stable in 1983, although price increases have been forecast for the later 1980s.

Rising oil prices have resulted in some action, particularly in the area of conservation, which may prevent another energy crisis of the magnitude of those in the past. For example, the U.S. now uses about 16 million barrels of oil per day; without conservation efforts it would probably be at least 21 million barrels daily. Prior to 1973, total U.S. energy consumption, including all sources, was forecast to be the equivalent of approximately 40 million barrels of oil daily by 1982; instead, it is about 35 million barrels daily. The Exxon Corporation used to forecast a non-Communist world oil demand of 95 million barrels/day by 1985; instead, the forecast is now 60 million barrels/day for the year 2000, compared to current use of 48 million barrels per day.

The current non-Communist world use of *total energy* is the equivalent of 96 million barrels of oil per day. Exxon used to forecast a demand the equivalent of 195 million barrels of oil per day by 1985 but has since reduced the estimate to 160 million barrels by the turn of the century.

Substantial strides have been made in energy efficiency. The average gasoline mileage of 22 miles per gallon for 1982 automobiles (expected to be 30 mpg in 1985) represents a 36 percent increase over 1975. Since the energy crisis of 1973–74, the efficiency of diswashers has increased 58%, that of refrigerator-freezers, 45%, and that of room air conditioners, 17%. The energy required to produce one unit of gross national product declined in the U.S. by 25% from 1973 to 1983 and even greater savings have occurred in some other countries, such as Japan. These savings are now built into the system, making it more immune to disruptions of petroleum supplies.

The solutions to energy problems are strongly tied to environmental considerations. For example, a massive shift of the energy base from petroleum to coal in the U.S. would involve much more strip mining, potential production of acid mine water, use of scrubbers, and other things that are potentially damaging to the environment. Similar examples could be cited for most other energy alternatives.

Clearly, chemists must be involved in developing alternative energy sources. Chemical processes are used in the conversion of coal to gaseous and liquid fuels. New materials developed through the applications of chemistry will be employed to capture solar energy. The environmental chemist has a key role to play in making alternative energy sources environmentally acceptable. The energy problem poses both questions and opportunities for students entering a career in chemistry. It is important, therefore, that these students know the basics of energy resources and alternative energy resources as well as their environmental aspects.

17.6 WORLD ENERGY RESOURCES

At present, most of the energy consumed by humans is produced from fossil fuels. Estimates of the amounts of fossil fuels available differ. Because of undiscovered deposits of petroleum and natural gas, there may be considerably more fossil fuel available than is commonly realized (or less than the more optimistic estimates).

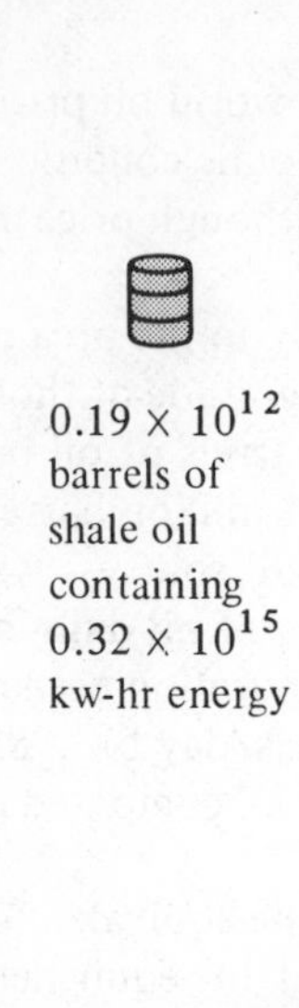

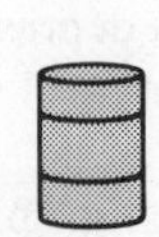

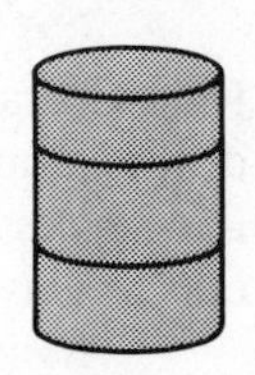

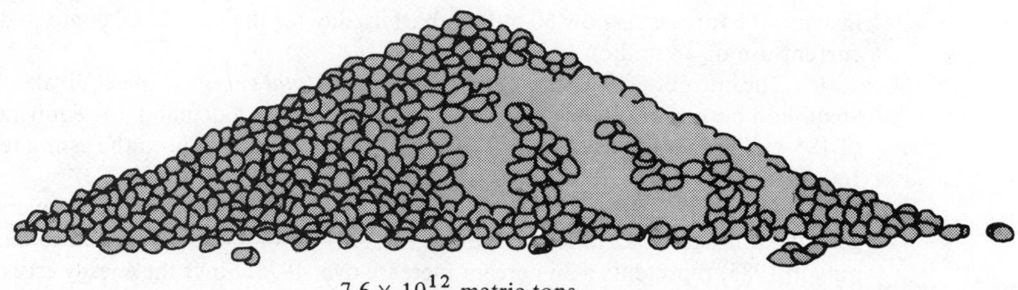

FIGURE 17.6 Initial amounts (before 1800) of the world's recoverable fossil fuels, shown in thermal kilowatt hours of energy. Data taken from M.K. Hubbert, "The Energy Resources of the Earth," in *Energy and Power,* San Francisco, Calif.: W.H. Freeman and Co., 1971.

Estimates of the quantities of recoverable fossil fuels in the world before 1800 are given in Figure 17.6. By far the greatest recoverable fossil fuel is in the form of coal and lignite. Furthermore, only a small percentage of this energy source has been utilized to date, whereas between 10 and 15% of all recoverable petroleum and natural gas has already been consumed. Projected use of these latter resources indicates rapid depletion.

Although it is not the purpose of this chapter to give a detailed discussion of energy resources and projected rates of use, some of the statistics in these areas are interesting as well as sobering. For oil and natural gas in the 48 coterminous United States, it is likely that the peak of production has been reached. The bulk of total production (roughly 80%) will have occurred by the year 2000. Alaskan oil, which may supply up to ten years of U.S. petroleum needs, will help the petroleum supply somewhat, though not by much. Worldwide petroleum resources are approximately 10 times those of the United States, and peak world production will be reached around the year 2000. Appreciable amounts of oil remain in tar sand deposits (about 300 billion barrels in northern Alberta) and in oil shales (about 3000 billion barrels worldwide, of which at least 6% could be recovered economically with existing technology).

The world situation in regard to coal is considerably more favorable. Estimates of recoverable coal (about half of the total present) are approximately

7.6×10^{12} metric tons. Peak production should be reached shortly after the year 2100. If this is the case, about 80% of production will occur over a three-century period ending around the year 2300. It is difficult to make similar estimates for the United States alone because of uncertainty regarding the future fraction of U.S. energy needs to be supplied by coal. However, around 20% of the world's coal resources are found in the United States, providing a substantial reserve of domestic fossil fuel.

Estimates of the total amount of energy that can be obtained from nuclear fission vary widely with the price of reactor fuel and are many orders of magnitude higher if the use of breeder reactors is assumed (these convert abundant nonfissionable uranium-238 and thorium-232 into fissionable plutonium-239 and uranium-233, respectively). If we assume only uranium-235 to be used as a fission fuel source, total recoverable reserves of nuclear fuel are roughly about the same as fossil fuel reserves.

Controlled nuclear fusion offers the prospect of abundant energy. Energy production based on the more feasible deuterium-tritium reaction is limited by relatively small resources of lithium-6, which is required for the production of tritium. Given the amount of lithium-6 available, this fusion reaction would provide energy approximately equal to that from remaining fossil fuel reserves. If the relatively more difficult deuterium-deuterium fusion reaction can be utilized for energy production, the amount of energy available is potentially unlimited. Extraction of only 2% of the deuterium present in the Earth's oceans would yield about a billion times as much energy as was originally present in fossil fuels!

Geothermal power can be a significant source of energy in some areas of the world and is being utilized in northern California, Italy, and New Zealand. Geothermal energy is depleted as subterranean heat is consumed and probably does not have the potential for providing a high percentage of energy worldwide. The same limited potential is characteristic of several renewable energy resources, including hydroelectric energy, tidal energy, and wind power. All of these will continue to contribute significant, but relatively small, amounts of energy.

Solar energy comes as close to being an ideal energy source as any available. It is renewable and nonpolluting. It does not add excess heat to that which must be radiated from the Earth. On a global basis, utilization of only a small fraction of solar energy reaching the Earth could provide for all energy needs. In the United States, for example, with conversion efficiencies ranging from 10–30%, it would only require collectors ranging in area from one-tenth down to one-thirtieth that of the state of Arizona to satisfy present U.S. energy needs. This is still an enormous amount of land, and there are economic and environmental problems related to the use of even a fraction of this amount of land for solar energy collection.

Insofar as U.S. energy resources are concerned, therefore, the picture is both hopeful and grim. Liquid petroleum and natural gas are on the way out as major producers of energy. Utilization of shale oil deposits will give some respite. Nuclear power may provide an increasing share of energy production, though difficult political, safety, technical, and environmental problems will continue to delay its rate of development. Miscellaneous sources such as wind, hydroelectric, tidal, and geothermal power will make useful but limited contributions to total energy production. Looking to the year 2000, we can predict that coal, an abundant and versatile fossil fuel, will certainly fill the energy gap until more exotic sources can be developed. Shale oil and tar sands may also play a role [9]. By the year 2000, technology should be available for the utilization of nuclear fusion or solar power, or both, each of which promises to be an abundant, relatively non-polluting source of energy.

17.7 ENERGY CONSERVATION

From about 1960 until 1973, energy production increased in the United States slightly over 4% per year, a growth rate with a doubling time of approximately 16 years. Such a growth rate can be used or misused to support many different theories. It can be used to show, for example, that within a few decades facilities for the production of power must be greatly expanded to provide for energy needs. It can also be used to show that devastating environmental damage will occur as a result of the exploitation of energy resources. It can even be used to calculate the exact date upon which the whole nation will glow in the dark from exponentially increasing waste heat radiated out into space. What is wrong with such projections is that they need not, should not, and indeed cannot come true.

Any consideration of energy needs and production must take energy conservation into consideration. We refer here not to cold classrooms at 60°F in mid-winter, nor to swelteringly hot homes with no air-conditioning, nor to total reliance on the bicycle for transportation. These are extreme measures which will be unnecessary if our energy resources are managed wisely. But the fact remains that the United States has wasted energy at a deplorable rate. For example, U.S. energy consumption is 50% higher per capita than that of Sweden, a country with the same material standard of living as the U.S. Obviously, a great deal of potential exists for energy conservation that will ease the energy problem.

Table 17.3 shows the percentage of energy consumed and the percentage of waste for the major categories of energy use in the United States. The table is based on 1971 U.S. energy consumption of 63.2×10^{15} Btu and shows the high

TABLE 17.3 Percentage of Energy Consumed and Wasted in the U.S. in Various Categories (Data from Earl Cook, Texas A&M University; adapted from Allen L. Hammond, "Conservation of Energy: The Potential for More Efficient Use," *Science,* 178: 1079–81 (1972).)

Energy use	*Percent of total*	*Percent utilized*	*Percent waste*
Electricity[1]	24.7	35.3	64.7
Industry	30.7	60.2	39.8
Transportation	32.4	15.1	84.9
Household and commercial	36.9	69.9	30.1
Total energy	100.0	40.8	59.2

[1] Utilizable electricity is consumed almost equally in industry, household, and commercial applications. Approximately 2/3 of the energy consumed by electrical power plants is wasted in generation and transmission.

---

[9] Häfele, W. 1980. A global and long range picture of energy developments. *Science* 209: 174–82.

percentage of energy wasted. Despite this waste, there have been some marked improvements in the efficiency of energy utilization. Improvements in the efficiencies of power plants, locomotives, furnaces, and other devices have roughly quadrupled the overall efficiency of energy utilization since 1900.

There is a high potential for energy conservation in industry. Since 1972 the U.S. chemical industry has lowered its energy use by 18% per unit product. From 1972 through 1979, the industry increased output by 28% while increasing energy use by only 3.4% [10]. Fundamental changes in the production of goods and services could result in tremendous energy savings without harming the overall economy. For example, recycling of containers has decreased energy consumption. The manufacture of appliances, automobiles, and other consumer items with emphasis on long life, durability, and ease of repair would be helpful. Policies that encourage employment of people in low-energy-use industries (health care, education, and labor-intensive agriculture) rather than high-energy-use industries would ease the economic impact of a lowered output of disposable material goods.

Transportation is another area where vastly increased efficiencies can be realized. The private auto and airplane are only about one-third as efficient as buses or trains for transportation. Transportation of freight by truck is terribly inefficient compared to rail transport. Truck transport requires about 3800 Btu/ton-mile, compared to only 670 Btu/ton-mile for a train. Major shifts in our current modes of transportation will not come without anguish, but energy conservation dictates that they be made.

Household and commercial uses of energy are relatively efficient. Here again, appreciable savings can be made. The all-electric home requires much more energy (considering the percentage wasted in generating electricity) than a home heated with fossil fuels. The sprawling ranch-house style home uses much more energy per person than does an apartment unit or row house. Improved insulation, sealing around windows, and other measures can conserve a great deal of energy. Electric generating plants centrally located in cities can provide waste heat for commercial and residential heating and cooling.

As scientists and engineers undertake the crucial task of developing alternative energy sources to replace dwindling petroleum and natural gas supplies, energy conservation must receive proper emphasis. In fact, zero energy-use growth, at least on a per capita basis, is a worthwhile and achievable goal. Such a policy would go a long way toward solving many environmental problems. With ingenuity, planning, and proper management, it could be achieved without a reduction in the standard of living or quality of life.

Despite the importance of energy conservation, it should be pointed out that there are pitfalls involved in a total reliance on it as a solution to the energy problem. David Lilienthal, the first chairman of the Atomic Energy Commission, stated this very eloquently in his book on atomic energy [11]: "Proponents of energy conservation who have put it forward as *the answer* to the 'energy crisis' invite the charge that this is an 'elitist' doctrine and a form of the 'no growth'

---

[10] 1981. Industry continues to cut energy demand. *Chemical and Engineering News* 12 January 1981, p. 7.

[11] Lilienthal, D.E. 1980. *Atmoic energy: a new start.* New York: Harper and Row, Publishers, Inc.

concept. . . .To offer it as a basic social doctrine of 'less is better' or as a slogan to rally opposition to nuclear energy expansion and the excesses of industrialization and consumerism will antagonize rather than persuade the general American public and a world in which poverty closely associated with energy shortages is endemic."

17.8 ENERGY CONVERSION PROCESSES

As shown in Figure 17.7, energy occurs in several forms and must be converted to other forms. The efficiencies of conversion vary over a wide range. Conversion of electrical energy to radiant energy by incandescent light bulbs is very inefficient—less than 5% of the energy is converted to visible light and the remainder is wasted as heat. At the other end of the scale, a large electric generator is around 80% efficient in converting mechanical energy to electrical energy. The once much-publicized Wankel rotary engine converts chemical to mechanical energy with an efficiency of about 18%, compared to 25% for a gasoline-powered piston engine and about 37% for a diesel engine. A modern coal-fired steam-generating power plant converts chemical energy to electrical energy with an overall efficiency of about 40%.

One of the most significant energy conversion processes is the conversion of thermal energy to mechanical energy in a heat engine such as a steam turbine.

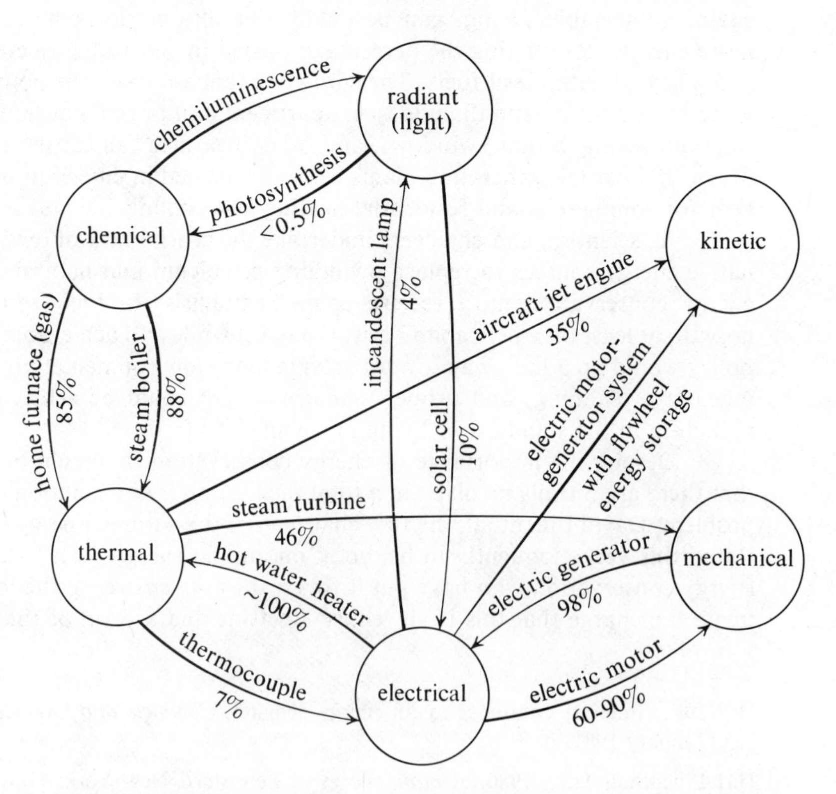

FIGURE 17.7 Kinds of energy and examples of conversion between them, with conversion efficiency percentages.

The Carnot equation,

$$\text{percent efficiency} = \frac{T_1 - T_2}{T_1} \times 100 \qquad (17.8.1)$$

states that the percent efficiency is given by a fraction involving the inlet temperature (for example, of steam), T_1, and the outlet temperature, T_2. These temperatures are expressed in degrees Kelvin (Centigrade temperature plus 273). Typically, a steam turbine engine operates with approximately 810 K inlet temperature and 330 K outlet temperature. These temperatures substituted into Equation 17.8.1 give a maximum theoretical efficiency of 59%. The impossibility of constantly maintaining the incoming steam at the maximum temperature, combined with mechanical energy losses, reduce the overall efficiency of conversion of thermal energy to mechanical energy in a modern steam power plant to approximately 47%. Taking into account losses from conversion of chemical to thermal energy in the boiler, the total efficiency is only about 40%.

Some of the greatest efficiency advances in the conversion of chemical to mechanical or electrical energy have been made by increasing the peak inlet temperature in heat engines. In the early 1900s, T_1 in a steam power plant typically was 550 K; the use of superheated steam has raised that figure to around 830 K in modern power plants. Improved materials and engineering design, therefore, have resulted in large energy savings.

The efficiency of nuclear power plants is limited by the maximum temperatures attainable. Reactor cores would be damaged by the high temperatures used in fossil-fuel-fired boilers and have a maximum temperature of approximately 620 K. Because of this limitation, the overall efficiency of conversion of nuclear energy to electricity is about 30%.

What is the fate of the 60% of energy from fossil-fuel-fired power plants and 70% of energy from nuclear power plants that is not converted to electricity? It is dissipated as heat, either to the atmosphere or to bodies of water and streams. The latter is thermal pollution, which may either harm aquatic life or, in some cases, actually increase bioactivity in the water to the benefit of some species. This waste heat is potentially very useful in applications like home heating, water desalination, and aquaculture (growth of plants in water). Increased production of waste heat by human activity could eventually result in marked changes in climate.

Some devices for the conversion of energy are shown in Figure 17.8 (p. 468). Substantial advances have been made in energy conversion technology over many decades and great advances can be projected for the future. Through the use of higher temperatures and larger generating units, the overall efficiency of fossil-fueled electrical power generation has increased approximately ten-fold since 1900, from less than 4% to a maximum of around 40%. An approximately four-fold increase in the energy-use efficiency of rail transport occurred during the 1940s and 1950s with the replacement of steam locomotives with diesel locomotives. During the coming decades, increased efficiency can be anticipated from such techniques as combined power cycles in which hot exhaust gas from a turbine engine is used to generate steam for a steam turbine. The magnetohydrodynamic electrical generator (see Section 17.10 and Figure 17.11) probably will be developed as a very efficient energy source used in combination with conventional steam generation. Entirely new devices such as thermonuclear reactors for the direct conversion of nuclear fusion energy to electricity will very likely be developed.

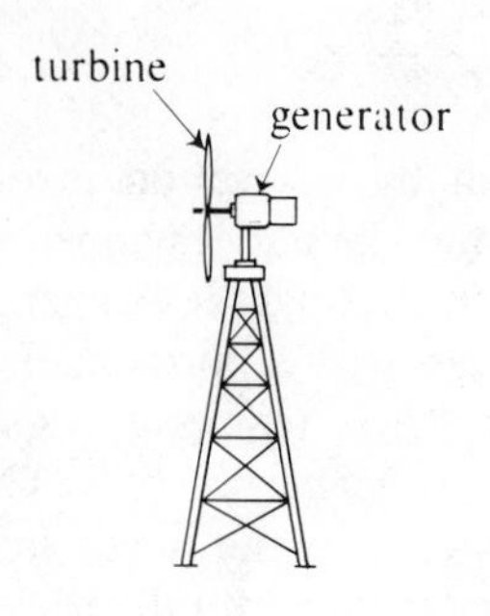

(1) Turbine for conversion of kinetic or potential energy of a fluid to mechanical and electrical energy.

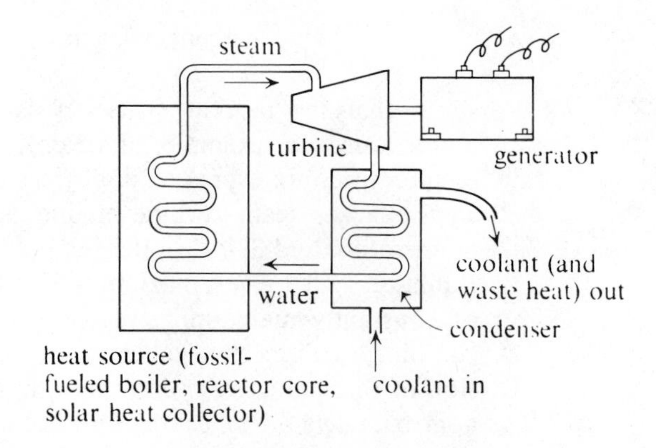

(2) Steam power plant in which high-energy fluid is produced by vaporizing water.

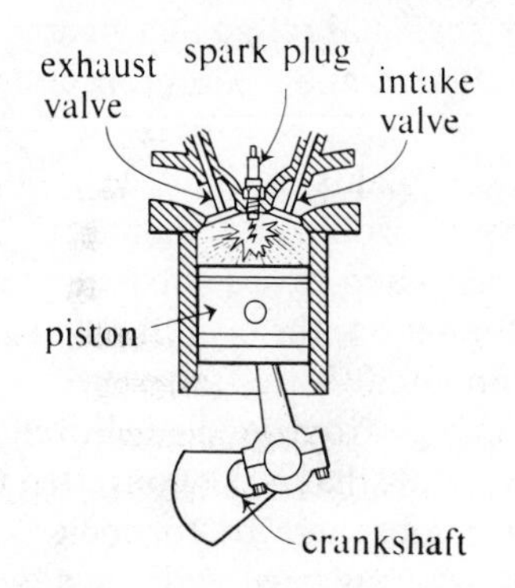

(3) Reciprocating internal combustion engine.

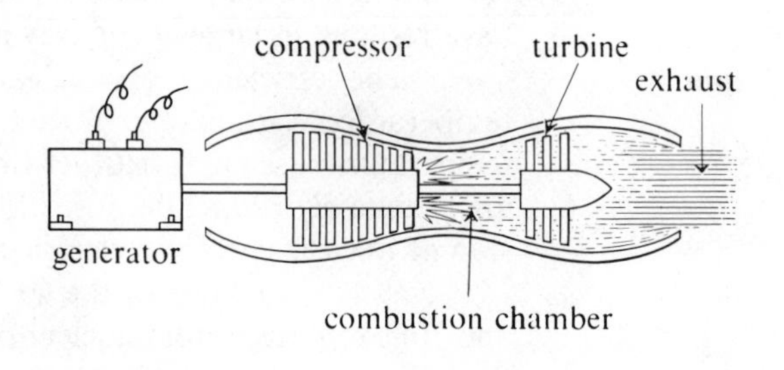

(4) Gas turbine engine. Kinetic energy of hot exhaust gases may be used to propel aircraft.

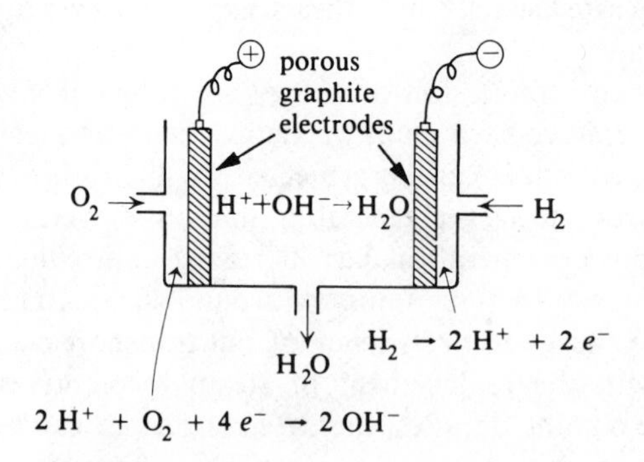

(5) Fuel cell for the direct conversion of chemical energy to electrical energy.

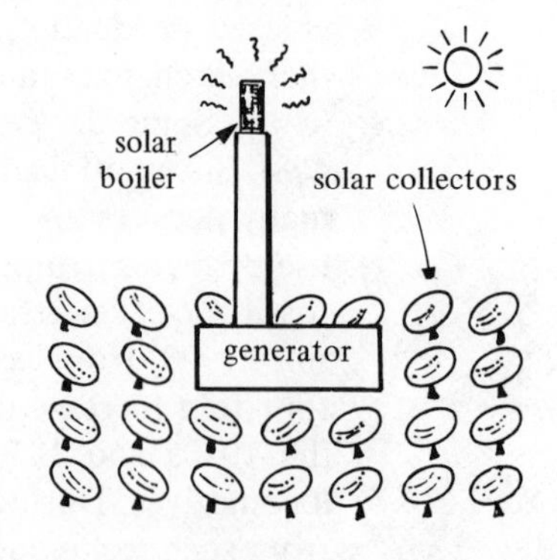

(6) Solar thermal electric conversion.

FIGURE 17.8 Some energy conversion devices.

17.9 PETROLEUM AND NATURAL GAS

The United States petroleum industry began in 1859 with the first commercial oil well in Pennsylvania. Since that time, a total of 100 billion barrels of oil have been produced domestically, most of it in recent years. Reserves of U.S. petroleum were estimated at 30 billion barrels in 1983, about a five-year supply (based on the 1983 consumption rate of 16 million barrels per day). On an encouraging note, Phillips Petroleum Company announced a major oil discovery in the Santa Maria Basin off Santa Barbara, California, in 1982. Production is scheduled to begin in 1987. This field is believed to be the largest domestic oil discovery since the 10-billion-barrel Prudhoe Bay field in Alaska, first drilled in 1968. Estimates of the amount of recoverable petroleum yet to be discovered in the U.S. range from a low of 62 billion barrels [12] to a high of 430 billion barrels [13]. These estimates include Alaskan and offshore oil. Despite the wide variation in estimated oil reserves, it is clear that the peak of U.S. liquid petroleum production has been reached.

As of 1982, U.S. petroleum consumption was on a downward trend, with a drop of 14.3% from 1979. The figures for the year 1981 showed a drop of 4.2% in gasoline consumption. During that same year, consumption of distillates used for diesel fuel and home heating dropped 4.6% and consumption of residual fuel used by power plants and in industry fell 20%.

World production of crude oil in 1981 was 55.6 million barrels per day, a 6.2% decline from the 59.2 million barrels per day produced in 1980. Of this amount, OPEC countries produced 22.6 million barrels. (By mid-1982 these countries were producing only 18.3 million barrels per day.) OPEC's share of the non-Communist world's oil production had fallen to 54% in 1981, compared to 68% in 1974.

In 1981 the Soviet Union produced 11.9 million barrels of oil per day, making it the world's largest petroleum producer. Saudi Arabia was second, with 9.6 million barrels, the U.S. was third, with 8.6 million barrels, and Mexico ranked fourth, with 2.3 million barrels.

Liquid petroleum is found in rock formations ranging in porosity from 10 to 30%. Up to half of the pore space is occupied by water. The oil in these formations must flow over long distances to an approximately 6-inch diameter well from which it is pumped. The rate of flow depends on the permeability of the rock formation, the viscosity of the oil, the driving pressure behind the oil, and other factors. In some cases, oil-bearing formations do not yield oil because of limitations in these factors. This **primary recovery** of oil yields an average of about 30% of the oil in the formation, although sometimes only as little as 15%. More oil can be obtained using **secondary recovery** techniques, which involve forcing water under pressure into the oil-bearing formation to drive the oil out. Primary and secondary recovery together typically extract somewhat less than 50% of the oil from a formation. Finally, **tertiary recovery** techniques can be used to extract even more oil. This normally involves the injection of pressurized carbon dioxide, which forms a mobile solution with the oil and allows it to flow

---

[12] Hubbert, M.K. 1967. *Am. Assoc. Petrol. Geol. Bull.* 51: 2207.

[13] Theobald, P.K.; Schweinfurth, S.P.; and Duncan, D.C. 1972. *U.S. Geol. Surv. Circ. No. 650* July 1972.

more easily to the well. Other chemicals, such as detergents, may be used to aid in tertiary recovery. Currently, some 68% of U.S. oil, 300 billion barrels, is not available through primary recovery, alone. A recovery efficiency of 60% through secondary or tertiary techniques could double the amount of available petroleum. Much of this would come from fields which have already been abandoned or essentially exhausted using primary recovery techniques. Advanced recovery techniques are now being more widely practiced as increasing petroleum prices have made them more feasible.

A more radical approach to the recovery of petroleum from depleted oil fields is mining [14]. A technique known as **drip drainage** can be employed, in which tunnels are drilled beneath the oil-bearing formation and holes are drilled up into the formation at approximately 30-meter intervals. The oil drips from these holes and is collected. This technique can be used at depths down to 1500 meters; beyond this point, heat from the Earth's interior makes mining impractical.

Shale oil is a possible substitute for liquid petroleum. Shale oil is a pyrolysis product of oil shale, a rock containing organic carbon in a complex structure called kerogen. It is believed that approximately 1.8 trillion barrels of shale oil could be recovered from deposits of oil shale in Colorado, Wyoming, and Utah. In the Colorado Piceance Creek basin alone, more than 100 billion barrels of oil could be recovered from prime shale deposits.

Shale oil may be recovered from the parent mineral by retorting the mined shale in a surface retort or by burning the shale underground with an *in situ* process. Both processes present environmental problems. Surface retorting requires the mining of enormous quantities of mineral and disposal of the spent shale, which has a volume greater than the original mineral. *In situ* retorting limits the control available over infiltration of underground water and resulting water pollution. Water passing through spent shale becomes quite saline, so there is major potential for saltwater pollution.

During the late 1970s and early 1980s, several corporations began building facilities for shale oil extraction in northwestern Colorado. The most ambitious of these was the Colony project, a joint construction project of Exxon and Tosco. This plant, situated about 15 miles north of Parachute, Colorado, was to produce 50,000 barrels of synthetic crude oil per day starting around 1987. The amounts of money involved were staggering. Exxon spent $300 million to purchase a 60% share of the project from Tosco, and the two companies spent an additional $400 million on the project. By 1982 the total cost of the project was estimated to be $6 billion or more. The Synthetic Fuels Corporation, an agency set up by the U.S. government, had lent $80 million on the project. In the face of escalating construction costs and a "soft" crude oil market, Exxon withdrew from the project in May, 1982, stopping construction on the plant. That left only the Union Oil Company of California actively involved in shale oil development; they are constructing a plant in western Colorado designed to produce 10,000 barrels of oil per day.

An even larger synthetic fuels project that failed was the $13.1 billion (Canadian) Alsands oil sands project located in Alberta. This project would have

---

[14] Maugh, T.H., II. 1980. Mining could increase petroleum reserves. *Science* 207: 1334–5.

produced 137,000 barrels of oil per day from the mining of tar sand and extraction of heavy oil from it. In April, 1982, both Shell Canada Ltd. and Gulf Canada Ltd. withdrew from the project, the cost of which had escalated from an initial estimate of $5 billion.

The failure of the Colony operation left the whole U.S. synthetic fuels industry in a state of uncertainty. OPEC policy has been to price crude oil just below levels required to make synthetics viable. Conversely, production of synthetics at a significant level could help limit crude oil prices. Such production would require a massive government effort along with price guarantees, a development which does not appear likely at this time.

Because of budgetary and environmental restraints, the development of synthetic fuels such as shale oil in the U.S. is likely to remain slow [15, 16]. It is not likely that any significant synthetic fuels production will be on line before at least 1990.

Natural gas, consisting almost entirely of methane, may ultimately turn out to be the "wild card" in the U.S. energy game. This is because of uncertainties regarding natural gas availability, coupled with the potential for the discovery and development of truly enormous new sources of this premium fuel. In 1968, discoveries of natural gas deposits in the U.S. fell below annual consumption for the first time. This trend continued during the following decade, except for 1970, when gas was discovered in Alaska. During the record-cold winter of 1976–77, severe natural gas shortages occurred in parts of the U.S. Price incentives resulting from the passage of the 1978 Gas Policy Act are expected to reverse the adverse trend of natural gas discoveries versus consumption. As of 1982, there was a natural gas reserve in the U.S. of 5.5 trillion cubic meters (m^3). However, potential reserves of gas from conventional sources may amount to 14 trillion m^3. Another 6 trillion m^3 may be present in deposits trapped in deep basins 5000–10,000 m deep. The cost of drilling a gas well into such a deep deposit runs between $3.5 million and $15 million, and the product is often "soured" by a high hydrogen sulfide content.

In the early 1980s, one of the great natural gas discoveries in U.S. history was developed in the deep gas wells of the Anadarko Basin, covering parts of western Oklahoma and the Texas panhandle. In 1981, alone, this field added 85 billion cubic meters to U.S. natural gas reserves. Three of the best wells in this field produced as much energy as two commercial nuclear power plants, and the 13 best wells matched the energy output of the Tennessee Valley Authority's hydroelectric system. Another new, large gas discovery is in Louisiana's Tuscaloosa Trend.

One of the most promising areas in the U.S. for new natural gas, as well as petroleum, discoveries is the overthrust belt running along a jagged north-south line through parts of Montana, Idaho, Utah, Nevada, and Arizona. This is a geological area in which sandstone 180 and 190 million years old has been thrust over younger geological formations 80 to 110 million years old. The latter formations contain deposits of natural gas and petroleum. These tend to be scattered

---

[15] Hanson, D.J. 1981. Synfuels progress likely to remain slow. *Chemical and Engineering News* 24 August 1981, pp. 13–17.

[16] 1981. Budget cuts threaten synthetic fuels projects. *Chemical and Engineering News* 16 March 1981, pp. 13–14.

throughout the overthrust belt, which is kind of a "geological jumble", so that exploration in the area is of a relatively "hit or miss" nature. Geologists believe that large amounts of natural gas and petroleum will eventually be recovered from the overthrust belt; the natural gas to be found there might even double present U.S. reserves. Much less optimism is expressed concerning the hydrocarbon resources of the eastern overthrust belt, located in the Appalachian Mountains region [17].

There is a tremendous potential for production of natural gas from "unconventional" sources. For an interesting discussion of these sources, the reader is referred to Reference 18. There are five major unconventional sources of natural gas. One of these that is being developed is western tight sands in the Rocky Mountain region, containing an estimated 17 trillion m^3 of natural gas. Attempts during the early 1970s to fracture these concrete-like sandstone formations with nuclear explosions only succeeded in melting the rock and sealing the gas even tighter in the formation. Hydraulic fracturing using sand, glass pellets, or bauxite mixed with fluid under high pressure has been more successful in opening up these formations to allow the gas to escape. **Devonian shales** found in shallow deposits in the northeastern U.S. are similar to western tight sands in that they release gas only slowly [19]. These deposits,holding an estimated 14 trillion m^3 of natural gas, likewise may be tapped by hydraulic fracturing. Coal seams in the U.S. hold an estimated 8.5 trillion m^3 of recoverable methane, the cause of many tragic coal mine explosions and fires. Large quantities of natural gas may be present in methane hydrates, ice-like combinations of methane and water found beneath permafrost and below deep-ocean floors. However, the greatest potential source of natural gas, perhaps amounting to 100 trillion m^3, is found in geopressurized zones in the U.S. Gulf Coast region. These deposits consist of natural gas mixed with hot salt water in underground deposits.

In addition to its use as a fuel, natural gas can be converted to many other hydrocarbon materials. It can be used as a raw material for the Fischer-Tropsch synthesis of gasoline (see Section 17.11). The discovery and development of truly massive sources of natural gas, such as may exist in geopressurized zones, could provide abundant energy reserves for the U.S., though at substantially increased prices.

17.10 COAL

From Civil War times until World War II, coal was the dominant energy source behind U.S. industrial expansion. However, the greater convenience of lower-cost petroleum resulted in a decrease in the use of coal for U.S. energy requirements after World War II. Annual coal production fell by about one-third, reaching a low of approximately 400 million tons in 1958. Production in 1981 was 805 million tons. With diminishing petroleum supplies, increasing uncertainty

---

[17] Hatcher, R.D., Jr. 1982. Hydrocarbon resources of the eastern overthrust belt: a realistic evaluation. *Science* 216: 980–2.

[18] Hodgson, B. 1978. Natural gas: the search goes on. *National Geographic* 154: 632–51.

[19] 1981. Devonian shale deposit resources assessed. *Chemical and Engineering News* 24 August 1981, p. 20.

about the overall safety of nuclear power, and the unavailability of more exotic sources of energy, coal is now emerging as the answer to the energy problem for the next several decades. Coal's potential is shown by the fact that it comprises 87% of U.S. fossil fuel resources while currently supplying only 20% of our energy. The average price of coal burned in electric power plants is about half that of natural gas and one-third that of oil. Coal at $35 per metric ton is equivalent to oil at $7.00 per barrel [20].

The general term *coal* describes a large range of solid fossil fuels derived from partial degradation of plants. Table 17.4 shows the characteristics of the major classes of coal found in the U.S., differentiated largely by percentage of fixed carbon, percentage of volatile matter, and heating value (*coal rank*). Chemically, coal is a very complex material and is by no means pure carbon. For example, a chemical formula for Illinois No. 6 coal, a type of bituminous coal, would be something like $C_{100}H_{85}S_{2.1}N_{1.5}O_{9.5}$.

Anthracite, a hard, clean-burning, low-sulfur coal, is the most desirable of all coals. Approximately half of the anthracite originally present in the United States has been mined. Bituminous coal found in the Appalachian and north central coal fields is the most widely used. It is an excellent fuel with a high heating value. Unfortunately, most bituminous coals have a high percentage of sulfur (an average of 2–3%), so the use of this fuel presents environmental problems. Huge reserves of virtually untouched subbituminous and lignite coals are found in the Rocky Mountain states and in the northern plains of the Dakotas, Montana, and Wyoming. Subbituminous and lignite coals have a relatively high oxygen content, largely in the form of carboxyl, phenolic hydroxyl, and carbonyl groups in a humic acid fraction of the fuel. Some of the lower-grade lignites are almost pure humic acid (see Chapter 4). These fuels have the advantage of being low in sulfur content and are finding increasing use in power plants meeting SO_2 emission standards. They have a number of disadvantages, including low heat content and

TABLE 17.4 Major Types of Coal Found in the United States

	Proximate analysis, percent[1]				
Type of coal	*Fixed carbon*	*Volatile matter*	*Moisture*	*Ash*	*Range of heating value (Btu/pound)*
Anthracite	82	5	4	9	13,000–16,000
Bituminous					
Low-volatile	66	20	2	12	11,000–15,000
Medium-volatile	64	23	3	10	11,000–15,000
High-volatile	46	44	6	4	11,000–15,000
Subbituminous	40	32	19	9	8,000–12,000
Lignite	30	28	37	5	5,500–8,000

[1] These values may vary considerably with the source of coal.

[20] 1980. Report touts expanded use of coal. *Chemical and Engineering News* 19 May 1980, p. 6.

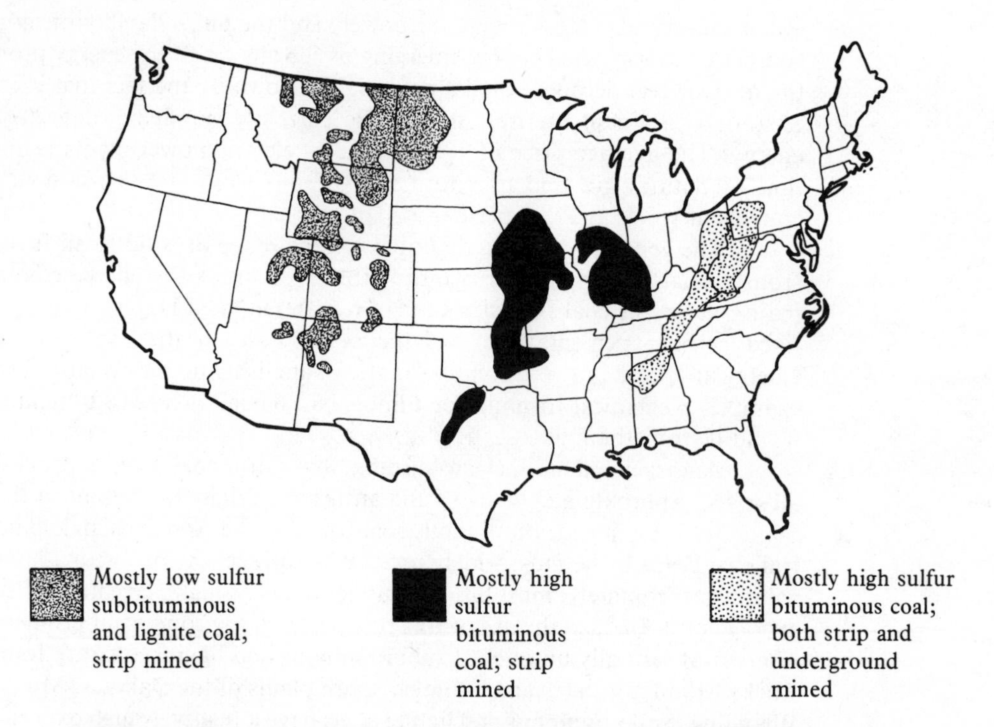

FIGURE 17.9 Areas with major coal reserves in the United States.

high moisture and ash contents. Also, they are generally found great distances from areas having the greatest need for fossil fuels. Lignite in particular tends to lose moisture and crumble when transported, forming what some miners call "bug dust." Despite these disadvantages, the low sulfur content and ease of mining these low-grade fuels is resulting in a rapid increase in their use.

Some geographical areas have very large coal resources. Huge reserves of subbituminous coal are found in the Rocky Mountain states, particularly Montana and Wyoming. An especially striking example is the approximately 350 billion tons of virtually untouched lignite in North Dakota, of which 16–20 billion tons may be mined at present prices. Figure 17.9 shows areas in the U.S. with major coal resources. The Department of Energy estimates U.S. coal reserves at 475 billion tons [21]. This includes 242 billion tons of bituminous coal, 182 billion tons of subbituminous, 43 billion tons of lignite, and 7 billion tons of anthracite.

The extent to which coal can be used as a fuel depends upon solutions to several problems, including (1) minimizing the environmental impact of coal mining; (2) removing ash and sulfur from coal prior to combustion; (3) removing ash and sulfur dioxide from stack gas after combustion; and (4) conversion of coal to liquid and gaseous fuels free of ash and sulfur (see Section 17.11). Progress is being made on minimizing the environmental impact of mining. As more is

---

[21] 1981. DOE estimates of coal reserves up eight percent. *Energy Insider*. Washington, D.C.: U.S. Department of Energy, 25 May 1981.

learned about the processes by which acid mine water is formed, measures can be taken to minimize the production of this water pollutant. Particularly on flatter lands, strip-mined areas can be reclaimed with relative success. Inevitably, some environmental damage will result from increased coal mining, but the environmental impact can be minimized by sensible control measures. Washing, flotation, and chemical processes can be used to remove some of the ash and sulfur from coal prior to burning. Approximately half of the sulfur in the average coal occurs as pyrite, FeS_2, and half as organic sulfur. Although little can be done to remove the latter, much of the pyrite can be separated from most coals by physical and chemical processes.

If even liberalized air pollution emission standards are to be maintained, the burning of high-sulfur coal requires the removal of sulfur dioxide from stack gas. Stack gas desulfurization faces some economic and technological uncertainties; the major processes available for it are summarized in Section 12.10.

The major use for coal is as fuel for electric power plants. The dominance of coal in this application is shown in Figure 17.10, which illustrates the contribution of various energy sources to electric power. The percentage of electric power generated by coal is certain to increase in the near future, for several reasons. The high cost of petroleum will inhibit the construction of any more oil-fired plants, and the Fuel Use Act of 1980 prohibits use of natural gas in power plants after 1990. Most suitable locations for hydroelectric plants have already been developed, and new plants face stiff environmentalist opposition. Orders for new nuclear power plants have essentially ceased. This leaves coal as the major alternative for the generation of large quantities of electricity.

Magnetohydrodynamic power combined with conventional steam generating units promises a major breakthrough in the efficiency of coal utilization. A schematic diagram of a magnetohydrodynamic (MHD) generator is shown in

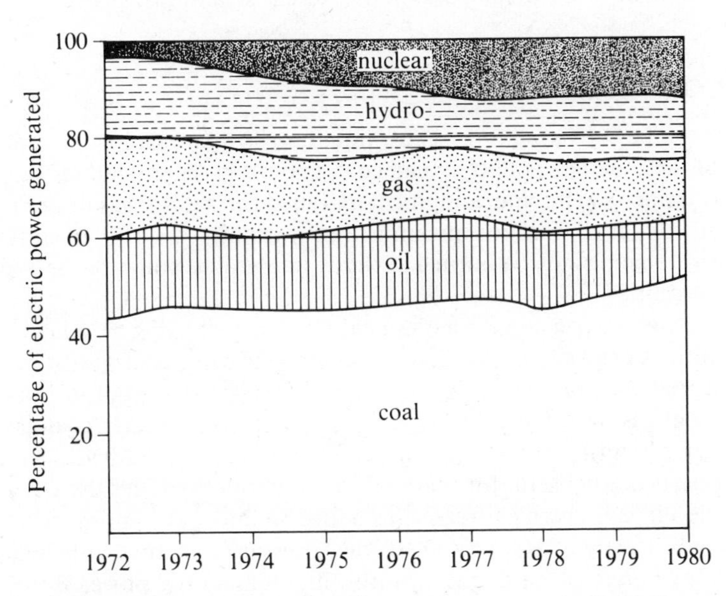

FIGURE 17.10 Major sources of electric power in the U.S. (from Department of Energy data).

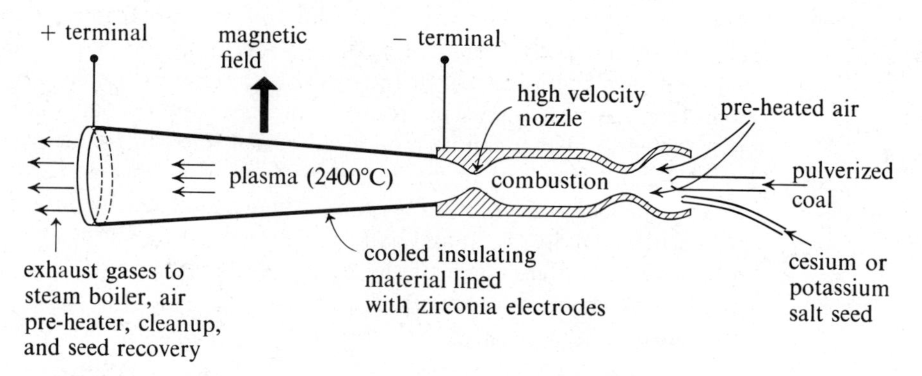

FIGURE 17.11 A magnetohydrodynamic power generator.

Figure 17.11. This device generates direct current by way of a plasma of ionized gas at around 2400 °C blasting through a very strong magnetic field of at least 50,000 gauss. The ionization of the gas is accomplished by injecting a "seed" of cesium or potassium salts. In a coal-fired MHD generator, the ultra-high-temperature gas issuing through a supersonic nozzle contains ash, sulfur dioxide, and nitrogen oxides which severely erode and corrode the materials used. This hot gas is used to generate steam for a conventional steam power plant, thus increasing the overall efficiency of the process. The seed salts combine with sulfur dioxide and are recovered along with ash in the exhaust. Pollutant emissions are low. Despite some severe technological difficulties, it appears that MHD power is feasible on a large scale. In fact, an experimental MHD generator has been tied to a working power grid in the U.S.S.R. for several years. Most important, the overall efficiency of combined MHD-steam power plants should reach 60%, one and one-half times the maximum of present steam-only plants.

17.11 COAL CONVERSION

Coal can be converted to gaseous, liquid, or low-sulfur, low-ash solid fuels [11]. All of these are less polluting than coal, and the gases and liquids can be used with distribution systems and equipment designed for natural gas or petroleum. A major advantage of coal conversion is that it enables use of high-sulfur coal which otherwise could not be burned without intolerable pollution or expensive stack gas cleanup.

Coal conversion is an old idea; a house belonging to William Murdock at Redruth, Cornwall, England, was illuminated with coal gas in 1792. The first municipal coal gas system was employed to light Pall Mall in London in 1807. The coal-gas industry began in the U.S. in 1816. The early coal-gas plants used *coal pyrolysis* (heating in the absence of air) to produce a hydrocarbon-rich product particularly useful for illumination. Later in the 1800s the *water-gas process* was developed, in which steam was added to hot coal to produce a mixture consisting primarily of H_2 and CO. It was necessary to add volatile hydrocarbons to this "carbureted" water-gas to bring its illuminating power up to that of gas prepared by coal pyrolysis. The U.S. had 11,000 coal gasifiers operating in the

1920s [22]. At the peak of its use in 1947, the water-gas method accounted for 57% of U.S.-manufactured gas [23]. The gas was made in low-pressure, low-capacity gasifiers which by today's standards would be inefficient and environmentally unacceptable. During World War II, Germany developed a major synthetic petroleum industry based on coal, which reached a peak capacity of 100,000 barrels per day in 1944. A plant now operating in Sasol, South Africa, converts about 30,000 tons of coal per day to synthetic petroleum.

The major routes for coal conversion are shown in Figure 17.12. The coal-derived fuels that are most likely to be produced commercially in the United States in the future are the following:

1. Solvent-refined coal (SRC), a solid high-Btu product with low sulfur, ash, and water content.
2. Low-sulfur boiler fuels that are liquid at elevated temperatures.
3. Liquid hydrocarbon fuels, including gasoline, diesel fuel, and naphtha.
4. Synthetic natural gas (SNG), essentially pure methane.
5. Low-sulfur, low-Btu gas for industrial use.

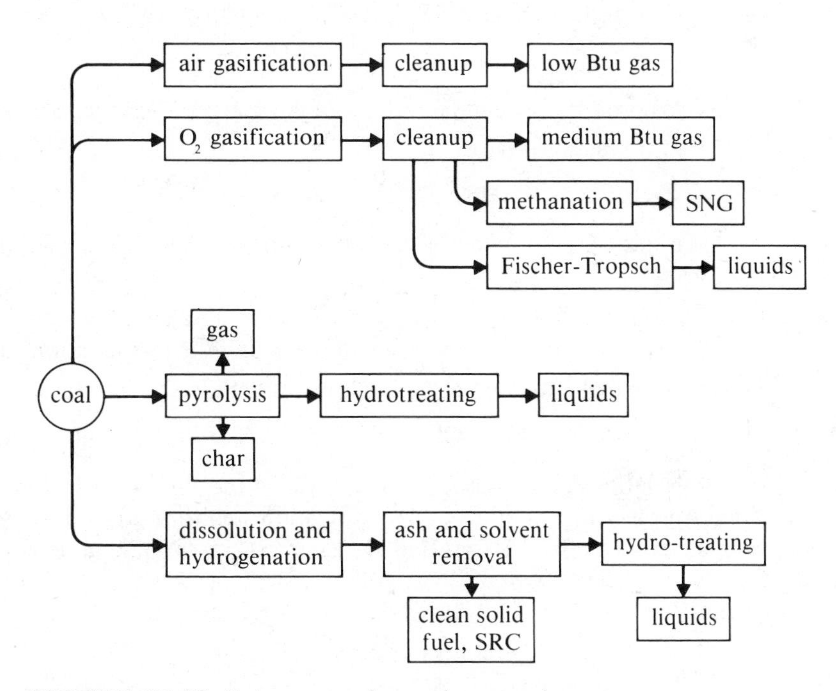

FIGURE 17.12 Major routes for coal conversion.

---

[22] Spencer, D.F.; Gluckman, M.J.; and Alpert, S.B. 1982. Coal gasification for electric power generation. *Science* 215: 1571–6.

[23] Hirsch, R.L.; Gallagher, J.E., Jr.; Lessard, R.R.; and Wesselhoft, R.D. 1981. Catalytic coal gasification: an emerging technology. *Science* 215: 121–127.

Of these processes, the production of gaseous fuels probably is closest to widespread commercial development.

A low-Btu gas consisting of over 60% by volume noncombustible nitrogen and carbon dioxide may be prepared by reacting steam and air with hot coal. This gas would not be suitable for home use, and transporting it for any distance is too expensive because of the noncombustible gas content. However, it is readily desulfurized and serves well as a fuel for gas turbines at the mine site. The hot turbine exhausts can be used to raise steam for steam turbines, thus increasing the system's overall efficiency.

High-Btu synthetic natural gas, SNG, can be produced by any of several processes. The steps in a typical process for SNG production are:

1. Steam and oxygen react with coal in a gasifier, producing a low-Btu gas consisting of carbon monoxide, carbon dioxide, hydrogen, and some methane. The major reactions, simplifying coal as C, are:

$$2\,C + O_2 \rightarrow 2\,CO \qquad \Delta H = -94.1 \text{ kcal (per mole of carbon at 25 °C)} \tag{17.11.1}$$

$$2\,CO + O_2 \rightarrow 2\,CO_2 \tag{17.11.2}$$

$$C + H_2O \rightarrow CO + H_2 \qquad \Delta H = +31.4 \text{ kcal} \tag{17.11.3}$$

$$C + 2\,H_2 \rightarrow CH_4 \qquad \Delta H = -17.9 \text{ kcal} \tag{17.11.4}$$

2. The gas product is freed from tar and dust by scrubbing with water.
3. The shift reaction,

$$H_2O + CO \rightarrow H_2 + CO_2 \tag{17.11.5}$$

   over a shift catalyst increases the molar ratio of H_2 to CO to an optimum value of approximately 3.5:1.
4. Acid gases (H_2S, COS, CO_2) are removed in an alkaline scrubber. Complete sulfur removal is necessary to avoid poisoning the methanation catalyst.
5. The catalytic hydrogenation of CO (methanation step) produces methane gas:

$$CO + 3\,H_2 \rightarrow CH_4 + H_2O \qquad \Delta H = -19.3 \text{ kcal} \tag{17.11.6}$$

Variations on this process include direct reaction of hydrogen from an external source with coal during the gasification step to produce an initially higher methane content (IGT Hygas and Hydrane processes); and addition of calcined lime, which reacts with CO_2 to produce heat needed to sustain endothermic gasification reactions (CO_2-acceptor process):

$$CaO + CO_2 \rightarrow CaCO_3 + \text{heat} \tag{17.11.7}$$

Gasification generally is carried out at temperatures of about 1000 °C and pressures of approximately 1000 psia.

A number of relatively advanced "second generation" coal conversion processes have reached the large pilot-plant stage. One of these is the Synthane gasification process developed by the U.S. Bureau of Mines. A 72-ton/day pilot plant using this process was operated from late 1975 into 1976. A diagram of a Synthane gasification reactor is shown in Figure 17.13. The Synthane gasification process follows a free-fall carbonization step with fluidized bed gasification.

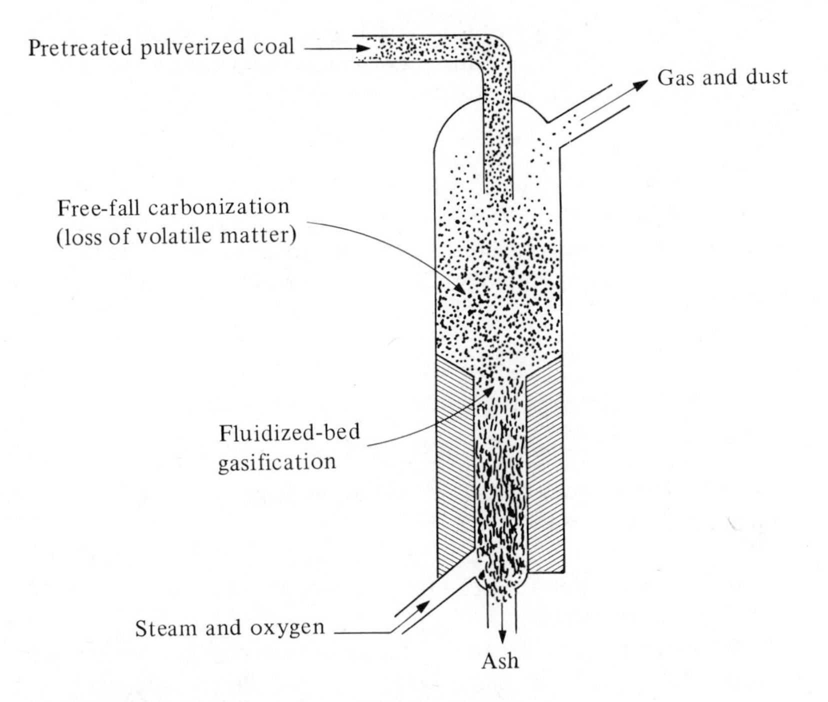

FIGURE 17.13 Synthane gasification reactor.

Caking coal is heat treated to make it non-caking prior to introduction into the reactor. In the second stage of the reactor, steam reacts with char to produce synthesis gas:

$$C + H_2O \rightarrow CO + H_2 \tag{17.11.3}$$

Typical composition of gas from a Synthane gasifier operated with bituminous coal is 10.5% CO, 17.5% H_2, 18.2% CO_2, 37.1% H_2O, 15.4% CH_4, and 0.3% H_2S (all values in mole %). The heating value of this gas after it is dried is 405 Btu/ft^3, as compared to about 1000 Btu/ft^3 for methane. In addition to these gases, tar, oil, water, ash, and sulfur are produced by the gasification process and must be removed. Other gasifiers produce gas with a somewhat different composition. Those having an external source of hydrogen produce gas containing up to 70% CH_4.

The gas coming from the gasifier contains some tar and dust which must be removed by water scrubbing. In some plants, the used scrubbing water is treated (including a biological treatment step) and used in cooling towers. Generally, the gas produced in these plants is too rich in CO to allow for stoichiometric conversion of CO and H_2 to methane. Therefore, it is necessary to react the gas over a shift catalyst,

$$CO + H_2O \rightarrow CO_2 + H_2 \tag{17.11.5}$$

to achieve the necessary H_2/CO ratio (exceeding 3:1).

After the catalytic-shift conversion step, the gas still contains a high percentage of noncombustible CO_2, as well as H_2S and COS which poison the methanation catalyst. These are removed by an alkaline scrubber.

The final step in producing high-Btu gas is catalytic methanation over a nickel catalyst, which produces a product that is mostly methane, a small amount of hydrogen, and a trace of carbon monoxide:

$$CO + 3\,H_2 \rightarrow H_2O + CH_4 \qquad (17.11.6)$$

Because of the unreacted hydrogen gas in the product, it has a heating value of approximately 50 Btu/ft^3 less than that of commercial natural gas.

Higher hydrocarbons equivalent to those in crude oil can be produced from an H_2-CO mixture [24]. The overall reaction for the production of alkanes is

$$n\,CO + (2n + 1)\,H_2 \rightarrow C_nH_{2n+2} + n\,H_2O \qquad (17.11.8)$$

The reaction for the production of alkenes (olefins) is

$$n\,CO + 2n\,H_2 \rightarrow C_nH_{2n} + n\,H_2O \qquad (17.11.9)$$

The Fischer-Tropsch process is used at the large synthetic fuels and petrochemical plant in Sasol, South Africa.

Probably the most promising route for coal gasification is the **Texaco process**, which gasifies a water-slurry of coal at temperatures of 1250 °C to 1500 °C and pressures of 350 to 1200 pounds per square inch. Under these severe conditions no tar by-products are produced. Texaco coal gasification is the subject of a $300 million privately financed project in southern California [25]. It is very efficient, in part because the gas coming from the gasifier is cooled by raising steam in a heat exchanger, and the steam is employed to drive a steam turbine. After particles, NH_3, and H_2S are removed from the gas, it is used to fire a gas turbine engine. The heat from the engine exhaust is used to raise steam to drive another steam turbine. The southern California plant is designed to produce 100 megawatts of power, consuming 1000 tons of coal per day.

As of 1983 the only large-scale commercial coal-conversion plant under construction in the U.S. was the Great Plains Coal Gasification Plant near Beulah, North Dakota. This plant is designed to produce synthetic natural gas with an energy output equivalent to 20,000 barrels of petroleum per day. The plant has been subsidized by the U.S. Synthetic Fuels Corporation, a federally financed agency.

Chemical addition of hydrogen to coal can liquify it and produce a synthetic petroleum product. This can be done with a hydrogen donor solvent, which is recycled and itself hydrogenated with H_2 during part of the cycle. Such a process forms the basis of the successful **Exxon Donor Solvent process**, which has been used in a 250 ton/day pilot plant, operational for periods of up to 3,900 hours [26].

A number of environmental implications are involved in the widespread use of coal gasification. If projected needs for pipeline gas (gas distributed by pipeline for commercial and home use) were to be met by goal gasification, it is

---

[24] Haggin, J. 1981. Fischer-Tropsch: new life for old technology. *Chemical and Engineering News* 26 October 1981, pp. 22–32.

[25] Abelson, P. 1982. Clean fuels from coal. *Science* 215: 351.

[26] Vick, G.K., and Epperly, W.R. 1982. Status of the development of EDS coal liquefaction. *Science* 217: 311–316.

estimated that annual production of coal would have to more than double. Strip mining would have to increase and large quantities of water would be required. Scarcity of water is a particularly severe constraint on gasification processes in the western U.S. For example, the fabulously coal-rich Fort Union formation in the Powder River basin of Wyoming and Montana is located in a region where the average rainfall is only 15–30 cm per year. Indigenous water supplies could not support massive coal gasification operations. Another consideration involves the thermal efficiency of the conversion of coal to gas, which has been estimated at approximately 65%. Therefore, huge quantities of heat would be produced by widespread gasification, causing thermal pollution. Since coal gasification is carried out in closed systems, however, there is little potential for air pollution.

Underground gasification of coal is a potential method for utilizing many of the coal reserves that cannot be mined [27]. For example, many coal deposits are found in "lenses," individual deposits that are either too small or too deep for strip mining and are unsuitable for underground mining. Basically, underground gasification consists of injecting steam and air (or oxygen) into a coal deposit that has been ignited at high temperature and withdrawing product gas through another well. An experimental project using air as an oxidizing agent at Hanna, Wyoming, has yielded a product gas containing 4.5% CH_4, 21.40% H_2, 9.0% CO, 17.1% CO_2, 46.4% N_2, and 0.11% H_2S. Such a fuel, once cleansed of hydrogen sulfide, could be used for on-site electric power generation. A much higher-Btu fuel would be obtained using oxygen as the oxidizing agent.

Hydrogenation removes most sulfur and ash from coal and converts it to a liquid or semisolid fuel. In general, upgrading the quality of coal as a fuel means adding hydrogen. The most desirable products of hydrogenation are those in which the carbon in coal has been reduced to the -4 oxidation state, as in CH_4 and other alkanes. The lowest degree of hydrogenation produces solvent refined coal, SRC, a solid product much like high-grade ash-free coal. To produce SRC, pulverized coal suspended in a solvent is treated with 2% of its weight of hydrogen at 1000 psig and 450 °C to yield a semisolid product which melts at around 170 °C. The product has a heating value of 16,000 Btu/lb, comparable to the best anthracite coals. Sulfur removal is high—approximately 100% of pyritic sulfur and 60% of organic sulfur. The product, much like coal in appearance, is a desirable fuel for industrial and power-plant use. Other processes hydrogenate coal to a higher degree over catalysts and produce a low-sulfur product resembling crude oil. This product can be used directly as fuel or refined to yield gasoline, kerosene, and other petroleum products.

Methanol, CH_3OH, is a convenient liquid fuel which can be produced from coal. Methanol is usually produced on a small scale by the destructive distillation of wood (see Section 17.4). It was widely used for heating and lighting in mid-19th century France and for automobile fuel in gasoline-short countries during World War II. On a large scale, methanol is produced by the reaction of carbon monoxide and hydrogen at a lower H_2/CO ratio than that required for the production of methane:

$$CO + 2\,H_2 \rightarrow CH_3OH \qquad (17.11.10)$$

[27] Lamb, G.H. 1977. *Underground coal gasification.* Park Ridge, N.J.: Noyes Data Corp.

Earlier processes utilized a zinc-chromium oxide catalyst at 300 atm pressure and 200 °C temperature. A later process utilizes a copper-based catalyst at 50 atm and 250 °C. The carbon monoxide and hydrogen used for methanol production are both produced from coal, oxygen, and steam. At levels up to 15%, methanol makes an excellent additive for gasoline. It has a high octane number (106) and improves fuel economy and acceleration time in automobiles. It also reduces the emissions of practically all automotive pollutants.

The most important advantage of methanol is that it can be substituted for present liquid fuels with minimum disruption in existing processing, transportation, and end-use facilities. For example, conversion of a farm tractor for use with methanol fuel is a relatively simple process compared to conversion for use with methane, hydrogen, or electricity.

17.12 NUCLEAR FISSION POWER

The awesome power of the atom revealed at the end of World War II held out enormous promise for the production of abundant, cheap energy. This promise has never really come to fruition, although nuclear energy currently provides a small but significant percentage of electric energy.

Nuclear power reactors currently in use depend upon the fission of uranium-235 nuclei by reactions such as

$$^{235}_{92}\text{U} + ^{1}_{0}n \rightarrow ^{133}_{51}\text{Sb} + ^{99}_{41}\text{Nb} + 4\,^{1}_{0}n \qquad (17.12.1)$$

to produce two radioactive fission products, an average of 2.5 neutrons, and an average of 200 MeV of energy per fission. The neutrons, initially released as fast-moving, highly energetic particles, are slowed to thermal energies in a moderator medium. For a reactor operating at a steady state, exactly one of the neutron products from each fission is used to induce another fission reaction in a chain reaction:

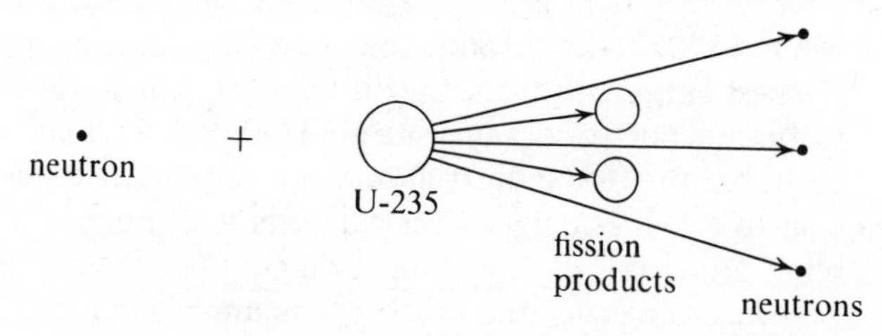

The energy from these nuclear reactions is used to heat water in the reactor core and produce steam to drive a steam turbine, as shown in Figure 17.14.

Because of limitations of structures and materials, nuclear reactors operate at a maximum temperature of around 625 K, compared to approximately 800 K in a fossil-fuel power plant. The thermal efficiency of nuclear power generation is limited by the Carnot relationship and is therefore inherently low. The overall efficiency for production of electricity in a nuclear-fission plant does not exceed 30%. That leaves 70% of the energy to be disposed of in the environment.

The course of nuclear power development may have been permanently altered by the now-famous reactor accident at Three Mile Island (TMI) in Pennsylvania. The incident began on March 28, 1979 with a partial loss of coolant

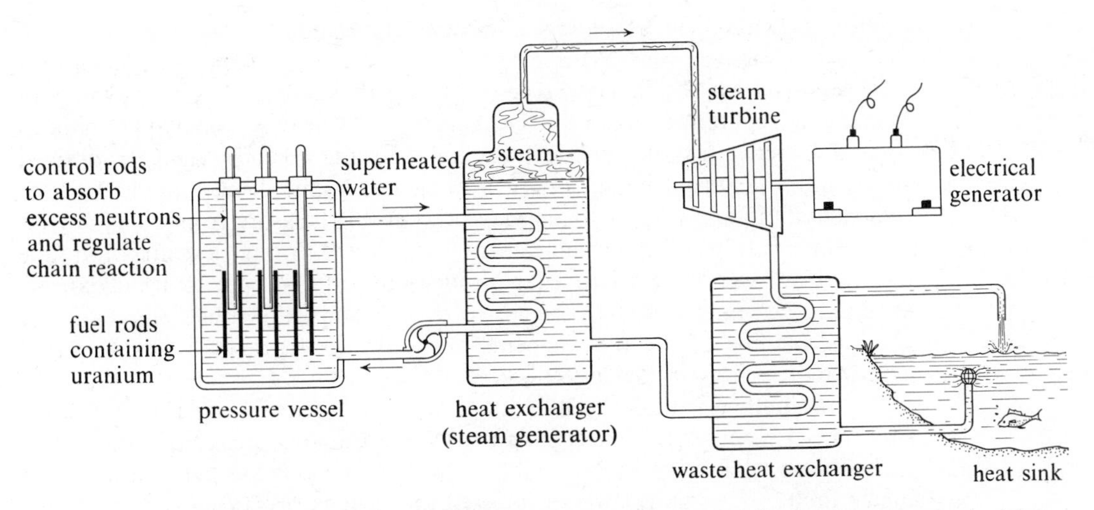

FIGURE 17.14 A typical nuclear power plant, using fission to generate electricity.

water from the Metropolitan Edison Company's nuclear reactor located on Three Mile Island in the Susquehanna River, 28 miles outside of Harrisburg, Pennsylvania. The loss resulted in overheating and partial disintegration of the reactor core [28]. The core temperatures were so high that hydrogen gas was produced by the reaction of water with zirconium in the core:

$$Zr + 2\,H_2O \rightarrow ZrO_2 + 2\,H_2 \tag{17.12.2}$$

Approximately 450 kg of hydrogen were produced, creating a 30-cubic meter hydrogen "bubble" in the reactor's pressure vessel, which prevented complete cooling of the core with water and threatened a possible chemical explosion. The hydrogen was released to the containment structure (the building surrounding the pressure vessel) and removed by catalytic oxidation [28].

For about 14 hours after the incident began, the reactor was essentially out of control and the core was partially exposed. At one point, evacuation was suggested for pregnant women and young children in the vicinity. From time to time over a period of several days, "puffs" of radioactive xenon and krypton gas were released and some radioactive water was dumped into the Susquehanna River. Release of radioactive iodine prompted some officials to recommend that exposed persons take KI antidote to reduce the fraction of total iodine uptake by the body from radioactive iodine, which damages the sensitive thyroid gland. During the period, worried citizens throughout the nation anxiously watched news reports on what might have been a major disaster requiring the evacuation of as many as 500,000 people and costs running into the billions of dollars. Although the incident ended without such a catastrophe, cleanup efforts costing millions of dollars continue. As of mid-1982, the cleanup effort at TMI had consumed 270 million dollars and the total cost was estimated at $1 billion.

One of the major problems involved in the cleanup was the 2,400,000 liters of highly radioactive water standing at a depth of slightly less than 3 meters in the

[28] Mynatt, F.R. 1982. Nuclear reactor safety research since Three Mile Island. *Science* 216: 131–5.

reactor's containment building. This water contained 560,000 curies of strontium-90 and cesium-137 (see Section 7.12), major fission products which were released from the damaged reactor core. Removal of these isotopes with ion-exchanging zeolite was completed in August, 1983, leaving water that contains some radioactive tritium. Although the treated water could probably be drained slowly into the Susquehanna with little danger to human health, public fears of radioactivity in areas downstream may prevent this measure from being taken.

In late July, 1982, a small camera was inserted through a control rod tube of the TMI reactor vessel to enable videotapes of the core to be taken. As expected, the center section of the core proved to consist of rubble where the zirconium cladding of the fuel elements had disintegrated. There was no visual evidence of melting in the core.

In another setback for nuclear power, the Nuclear Regulatory Commission (NRC) suspended the license of the Diablo Canyon nuclear plant near San Luis Obispo, California, in November, 1981. The plant is located near an earthquake fault, and concerns were expressed that it is not sufficiently earthquake-proof.

Corrosion (see Sections 3.13 and 5.27) is a severe problem in steam generators in nuclear power plants (see Figure 17.14). Originally expected to last the 40-year life expectancy of a typical reactor, some steam generators may have to be replaced after fewer than 10 years of operation. As of 1981, at least 11 nuclear power reactors had had, or were having, problems with their steam generators. Each steam generator contains thousands of tubes with walls about 1.25 mm thick. The internal parts of the steam generator are quite radioactive, so repair personnel can only work on one for a few days out of the year before receiving the maximum allowable dose of radiation. Temporary laborers called "sponges" or "jumpers" are hired for this purpose, given minimal training, then sent to the "hot" (highly radioactive) areas of the steam generator to plug leaking pipes or put sleeves on them [29].

Neutron embrittlement of the 20-cm-thick walls of the pressure vessels holding reactor cores is also turning out to be a problem with some reactors. This could lead to cracks in the welds holding the pressure vessel castings together, especially in the event of a rapid cooling of the pressure vessel [30]. In the summer of 1983, the NRC ordered the closing of five reactors for inspection for possible cooling water pipe weld cracks.

Another limitation of fission reactors is the fact that only 0.71% of natural uranium is fissionable uranium-235. This situation could be improved by the development of **breeder reactors**, which convert uranium-238 (natural abundance 99.28%) to fissionable plutonium-239.

A major consideration in the widespread use of nuclear fission power is the production of large quantities of highly radioactive waste products. These remain lethal for thousands of years. They must either be stored in a safe place or disposed of permanently in a safe manner. At the present time, spent fuel elements are being stored under water at the reactor sites. This gives some of the highly radioactive, short-lived fission products a chance to decay so that the fuel

[29] Hileman, B. 1982. Trends in nuclear power. *Environmental Science and Technology* 16: 373A–378A.

[30] Marshall, E. 1982. NRC reviews brittle reactor hazard. *Science* 215: 1596–7.

may be handled more safely. Currently, there are approximately 7,500 metric tons of commercial reactor wastes and 300 million liters of military wastes awaiting disposal. Several alternatives for permanent disposal, such as burial in the ocean bed or salt formations, are being considered. One of the most promising is deep burial in basalt [31], a hard, stable rock which occurs in massive formations in Washington State.

Another problem to be faced with nuclear fission reactors is their eventual decommissioning. There are three possible solutions [32]. One is dismantling soon after shutdown, in which the fuel elements are removed, various components are flushed with cleaning fluids, and the reactor is cut up by remote control and buried. "Safe storage" involves letting the reactor stand 30–100 years to allow for radioactive decay, followed by dismantling. The third alternative is entombment, encasing the reactor in a concrete structure. It is estimated that it would cost $50 to $100 million in 1982 dollars to dismantle a large, 1200-megawatt reactor.

Because of all the problems outlined above, as well as unfavorable economics, the prospects for growth of nuclear fission as an energy source have dimmed appreciably. A total of 139 reactors have been ordered since 1971, but none since the Three Mile Island incident. Since 1971, 79 reactors orders have been cancelled. Although 64 plants are still being built, the Nuclear Regulatory Commission estimates that construction on at least 10 of these will eventually be called off.

17.13 NUCLEAR FUSION POWER

The two main reactions by which energy can be produced from the fusion of two light nuclei into a heavier nucleus are the deuterium–deuterium reaction,

$$^{2}_{1}\mathrm{H} + ^{2}_{1}\mathrm{H} \rightarrow ^{3}_{2}\mathrm{He} + ^{1}_{0}n + 1\ \mathrm{MeV} \qquad (17.13.1)$$

and the deuterium–tritium reaction:

$$^{2}_{1}\mathrm{H} + ^{3}_{1}\mathrm{H} \rightarrow ^{4}_{2}\mathrm{He} + ^{1}_{0}n + 17.6\ \mathrm{MeV} \qquad (17.13.2)$$

The second reaction is more feasible because less energy is required to fuse the two nuclei than is required by the first reaction. The total energy from this type of fusion is limited by the availability of tritium, which is made from nuclear reactions of lithium-6 (natural abundance, 7.4%). The supply of deuterium, however, is essentially unlimited; one out of every 6700 atoms of hydrogen is the deuterium isotope. The $^{3}_{2}\mathrm{He}$ by-product of Reaction 17.13.1 reacts with neutrons, which are abundant in a nuclear fusion reactor, to produce tritium required for Reaction 17.13.2.

The power of nuclear fusion has not yet been harnessed in a sustained, controlled reaction of appreciable duration. Most approaches have emphasized "squeezing" a plasma (ionized gas) of fusionable nuclei in a strong magnetic

---

[31] Coons, W.E., et al. 1980. The functions of an engineered barrier system for a nuclear waste repository in basalt. RHO-BWI-LD-23. Richland, Washington: Rockwell Hanford Operations.

[32] Norman, C. 1982. A long-term problem for the nuclear industry: worn-out reactors may remain radioactive too long to entomb. *Science* 215: 376–379.

field. A plasma of ionized hydrogen gas was generated for 1/50,000 second in a successful test of the $314 million Princeton Tokamak Fusion Test Reactor in December, 1982. Another approach to obtaining the high temperatures required for fusion involves the use of a number of high-powered lasers focused on a pellet of fusionable material.

Controlled nuclear fusion processes would produce almost no radioactive waste products. However, tritium is very difficult to contain, and some release of this isotope would occur. The deuterium–deuterium reaction promises an unlimited source of energy. Either of these reactions would be preferable to fission in terms of environmental considerations. Therefore, despite the possibility of insurmountable technical problems involved in harnessing fusion energy, the promise of this abundant, relatively nonpolluting energy source makes its pursuit well worth a massive effort.

17.14 GEOTHERMAL ENERGY

Underground heat in the form of steam, hot water, or hot rock used to produce steam is already being used as an energy resource. This energy was first harnessed for the generation of electricity at Larderello, Italy, in 1904, and has since been developed in Japan, Russia, New Zealand, the Phillipines, and at the Geysers in northern California.

Underground dry steam is relatively rare but is the most desirable from the standpoint of power generation. More commonly, energy reaches the surface as superheated water or a mixture of water and steam. In some cases, the water is quite pure and can be used for irrigation and livestock; in other cases, the water is loaded with corrosive, scale-forming salts [33]. Utilization of the heat from contaminated geothermal water generally requires that the water be reinjected into the hot formation after heat removal to prevent contamination of surface water.

The utilization of hot rocks for energy requires fracturing of the hot formation, followed by injection of water and withdrawal of steam. This technology is still in the experimental stage, but promises approximately ten times as much energy production as steam and hot water sources.

Land subsidence and seismic effects are environmental factors that may hinder the development of geothermal power. However, this energy source holds considerable promise, and its development continues.

17.15 THE SUN: AN IDEAL ENERGY SOURCE

The recipe for an ideal energy source calls for one that is unlimited in supply, inexpensive, does not add to the Earth's total heat burden, and does not produce chemical air and water pollutants. Solar energy fulfills all of these criteria. It is so abundant that with a collection efficiency of only 10%, collectors covering less than 5% of the United States' surface area could provide the country's projected energy needs for the year 2000. It can be used directly for heating and air-

---

[33] Sung, R.; Houser, G.; Strehler, D.; and Scheyer, K. 1981. Sampling and analysis of potential geothermal sites. EPA-600/S7-81-138. Cincinnati, Ohio: U.S. Environmental Protection Agency, Industrial Environmental Research Laboratory.

conditioning, or can be used to generate electricity or to synthesize energy-containing chemical fuels.

Solar power cells for the direct conversion of sunlight to electricity have been developed and are widely used for energy in space vehicles. With present technology, however, they remain far too expensive for large-scale generation of electricity. Therefore, most schemes for the utilization of solar power depend upon the collection of thermal energy, followed by conversion to electrical energy. The simplest such approach involves focusing sunlight on a steam-generating boiler. Parabolic reflectors can be used to focus sunlight on pipes containing heat-transporting fluids. Selective coatings on these pipes can be used so that only a small percentage of the incident energy is reradiated from the pipes.

A major disadvantage of solar energy is its intermittent nature. It has been pointed out, however, that this problem has been exaggerated [34]. Flexibility inherent in an electric power grid would enable it to accept up to 15% of its total power input from solar energy units without special provision for energy storage. Existing hydroelectric facilities may be used for pumped-water energy storage in conjunction with solar electricity generation. Heat or cold can be stored in water, in a latent form in water (ice) or eutectic salts, or in beds of rock. Mechanical energy can be stored with compressed air or flywheels.

Hydrogen gas, H_2, is an ideal chemical fuel that may serve as a storage medium for solar energy. Electricity generated by solar means can be used to electrolyze a salt solution containing an anion that is very difficult to oxidize, so that oxygen is released at the anode and hydrogen is produced at the cathode. The net reaction is

$$2\,H_2O + \text{electrical energy} \rightarrow H_2(g) + O_2(g) \qquad (17.15.1)$$

Hydrogen, and even oxygen, can be piped some distance and the hydrogen burned without pollution or used in a fuel cell (Figure 17.8). This option is so promising that some experts foresee a ''hydrogen economy''. Disadvantages include: the fact that hydrogen has a heating value per unit volume of about one-third that of natural gas; and the explosive nature of a wide range of hydrogen-air mixtures.

No really insurmountable barriers exist to block the development of solar energy, such as might be the case with fusion power. In fact, the installation of solar space and water heaters is increasing rapidly. With the installation of more heating devices and the probable development of some cheap, direct solar electrical generating capacity, it is likely that by the year 2000 solar energy will be providing an appreciable percentage of energy needs in areas receiving abundant sunlight.

17.16 ENERGY FROM BIOMASS

All fossil fuels originally came from photosynthetic processes. Photosynthesis does hold some promise of producing combustible chemicals to be used for energy production [35] and could certainly produce all needed organic raw materials. It suffers from the disadvantage of being a very inefficient means of

[34] Metz, W.D. 1978. Energy storage and solar power: an exaggerated problem. *Science* 200: 1471–3.

[35] Calvin, M. 1982. New sources for fuel and materials. *Science* 219: 24–6.

solar energy collection (a collection efficiency of only several hundredths of a percent is typical of most common plants). However, the overall energy conversion efficiency of several plants, such as sugarcane, is around 0.6%. Furthermore, some plants, such as *Euphorbia lathyrus* (gopher plant), a small bush growing wild in California, produce hydrocarbon emulsions directly. The fruit of the Philippine plant, *Pittsosporum resiniferum*, can be burned for illumination due to its high content of hydrocarbon terpenes (see Section 15.2), primarily α-pinene and myrcene [35]. Conversion of agricultural plant residues to energy could be employed to provide some of the energy required for agricultural production. Indeed, until about 70 years ago, virtually all of the energy required in agriculture originated from plant materials produced on the land (hay and oats for horses, home-grown food for laborers, and wood for home heating).

Annual world production of biomass is estimated at 146 billion metric tons, mostly from uncontrolled plant growth [36]. Many farm crops and trees can produce 10–20 metric tons per acre per year of dry biomass. Some algae and grasses can produce as much as 50 metric tons per acre per year. The heating value of this biomass is 5000–8000 Btu/lb for a fuel having virtually no ash or sulfur (compare heating values of various coals in Table 17.4). Current world demand for oil and gas could be met with about 5% of the global production of biomass. Meeting U.S. demands for oil and gas would require that about 6% of the land area of the coterminous 48 states be cultivated intensively for biomass production. Another advantage of this source of energy is that use of biomass for fuel would not add any net carbon dioxide to the atmosphere.

Biomass is already significant as a heating fuel. According to a 1981 study of the Worldwatch Institute, wood now provides more of the total U.S. energy than does nuclear power. More than 7% of U.S. homes have woodburning stoves or furnaces. In Vermont, a state that is 83% forested, wood is now used for about 34% of residential heating, compared to only 27% for fuel oil. Other New England states in which wood has become a major source of fuel for residential heating are Maine (90% forested) and New Hampshire (86% forested). Air pollution from wood burning stoves and furnaces is a growing problem in some areas. Currently wood provides about 8% of world energy needs. This percentage could increase through the development of "energy plantations" consisting of trees grown solely for their energy content.

Seed oils show promise as fuels, particularly for use in diesel engines [37]. The most common plants producing seed oils are sunflowers and peanuts. More exotic species include the buffalo gourd, cucurbits, and Chinese tallow tree. Prior to use of such oils in a diesel engine, gums, waxes, and particles larger than 4 μm must be removed.

Biomass could be used to replace much of the U.S. petroleum and natural gas currently consumed in the manufacture of primary chemicals [38]. These amount to about 54 million metric tons per year. Among the sources of biomass

---

[36] 1976. IGT weighs potential of fuels from biomass. *Chemical and Engineering News* 23 February, pp. 24–6.

[37] Morgan, R.P., and Shultz, E.B., Jr. 1981. Fuels and chemicals from novel seed oils. *Chemical and Engineering News* 7 September 1981, pp. 69–77.

[38] Lipinsky, E.S. 1981. Chemicals from biomass: Petrochemical substitution options. *Science* 212: 1465–71.

that could be used for chemical production are grains and sugar crops (for ethanol manufacture), and oilseeds, animal by-products, and animal manure (for methane generation). The biggest potential source of chemicals is the lignocellulose making up the bulk of most plant material. For example, both phenol and benzene might be produced directly from lignin. Brazil has a major program underway for the production of chemicals from fermentation-produced ethanol [39].

17.17 GASOHOL

A major option for converting photosynthetically-produced biochemical energy to a form suitable for internal combustion engines is the production of either methanol or ethanol. Methanol is created by the destructive distillation of wood (Section 17.4) or from synthesis gas manufactured from coal or natural gas (Section 17.11). Ethanol is most commonly manufactured by fermentation of carbohydrates. Either one of these chemicals can be used by itself as fuel in a suitably designed internal combustion engine. More commonly, it has been proposed to blend these alcohols in proportions of up to 20% with gasoline to give gasohol, a fuel that can be used in existing internal combustion engines with little or no adjustment.

Gasohol offers a number of advantages. A high octane rating is obtained without adding tetraethyl lead. The reduction in exhaust emissions can be substantial, up to 50% for carbon monoxide and NO_x. Most important, because of its photosynthetic origin, alcohol may be considered a renewable resource rather than a depletable fossil fuel. The manufacture of alcohol can be accomplished by the fermentation of sugar obtained from the hydrolysis of cellulose in wood wastes (see Section 17.4) and crop wastes. Fermentation of these waste products offers an excellent opportunity for recycling.

Brazil has been a leader in the manufacture of ethanol for fuel uses [40], with 4 billion liters produced in 1982. Brazil now has over 450,000 automobiles that can run on pure alcohol. The extensive involvement of this country with ethanol production stems from the fact that it has few fossil resources but does have ideal conditions for the growth of large quantities of biomass. To date, most of Brazil's alcohol has come from the fermentation of sugarcane. However, a potentially much greater source of fermentable biomass is casava, or manioc, a root crop growing abundantly throughout the country. The United States is producing gasohol, but sales of fuel designated as gasohol have been phased out. Instead, the main use of alcohol in gasoline is as an octane-ratings booster, rather than as a fuel supplement.

Methanol, another potential ingredient of gasohol, can also be made from biomass. This is normally accomplished by converting biomass, such as wood, to CO and H_2, and synthesizing methanol from these gases [41].

---

[39] 1981. Brazil targets more ethanol for chemicals. *Chemical and Engineering News* 9 November 1981, p. 17.

[40] Hoge, W. 1980. Brazil's shift to alcohol as fuel. *New York Times* 13 October 1980.

[41] Haggin, J. 1982. Methanol from biomass draws closer to market. *Chemical and Engineering News* 12 July 1982, pp. 24–25.

17.18 FUTURE ENERGY SOURCES

As discussed in this chapter, a number of options are available for the supply of energy in the future. The major possibilities are summarized in Table 17.5. For additional details, the reader is referred to Reference 42.

TABLE 17.5 Possible Future Sources of Energy

Source	*Principles*
Coal conversion	Manufacture of gas, hydrocarbon liquids, alcohol, or solvent-refined coal from coal.
Oil shale	Retorting petroleum-like fuel from oil shale.
Geothermal	Utilization of underground heat.
Gas-turbine topping cycle	Utilization of hot combustion gases in a turbine, followed by steam generation.
MHD	Electrical generation by passing a hot gas plasma through a magnetic field.
Thermionics	Electricity generated across a thermal gradient.
Fuel cells	Conversion of chemical to electrical energy.
Solar heating and cooling	Direct use of solar energy for heating and cooling through the application of solar collectors.
Solar cells	Use of silicon semiconductor sheets for the direct generation of electricity from sunlight.
Solar thermal electric	Conversion of solar energy to heat followed by conversion to electricity.
Wind	Conversion of wind energy to electricity.
Ocean thermal electric	Use of ocean thermal gradients to convert heat energy to electricity.
Nuclear fission	Conversion of energy released from fission of heavy nuclei to electricity.
Breeder reactors	Nuclear fission combined with conversion of nonfissionable nuclei to fissionable nuclei.
Nuclear fusion	Conversion of energy released by the fusion of light nuclei to electricity.
Bottoming cycles	Utilization of waste heat from power generation for various purposes.
Solid waste	Combustion of trash to produce heat and electricity.
Photosynthesis	Use of plants for the conversion of solar energy to other forms by a biomass intermediate.
Hydrogen	Generation of H_2 by thermochemical means for use as an energy-transporting medium.

SUPPLEMENTARY REFERENCES

American Association for the Advancement of Science. 1982. *Energy, economics, and the environment.* Boulder, CO.: Westview Press, Inc.

Auer, P. 1979. *Advances in energy systems and technology.* New York: Academic Press, Inc.

---

[42] Herman, S.W., and Cannon, J.S. 1977. *Energy futures.* Cambridge, Mass.: Ballinger Publishing Company.

Benn, F.R.; Edewor, J.O.; and McAuliffe, C.A. 1981. *Production and utilisation of synthetic fuels.* London: Applied Science Publishers and New York: Halsted (John Wiley and Sons, Inc.).

Boyen, J.L. 1980. *Thermal energy recovery.* 2nd ed. New York: Wiley-Interscience.

Bungay, H.R. 1981. *Energy, the biomass options.* New York: Wiley-Interscience.

Calvert, N.G. 1980. *Windpower principles.* New York: Halsted (John Wiley and Sons, Inc.).

Cameron, E.N., ed. 1973. *The mineral position of the United States.* Madison, Wisconsin: Univ. of Wisconsin Press.

Cheremisinoff, N.F. 1980. *Wood for energy production.* Ann Arbor, Mich.: Ann Arbor Science Publishers, Inc.

Cheremisinoff, N.F., and Regino, T.C. 1978. *Principles and applications of solar energy.* Ann Arbor, Mich.: Ann Arbor Science Publishers, Inc.

Chigier, N.A., ed. 1978. *Energy from fossil fuels and geothermal energy.* Elmsford, N.Y.: Pergamon Press, Inc.

1977. *Coal as an energy resource: conflict and consensus.* Washington, D.C.: National Academy of Sciences.

Commission of European Communities. 1981. *Fusion technology 1980.* Elmsford, N.Y.: Pergamon Press, Inc.

Coonley, D.R. 1979. *Wind.* Philadelphia: Franklin Institute Press.

Cowling, T.G., and Hilger, A. 1976. *Magnetohydrodynamics.* New York: Crane, Russak & Co., Inc.

Cowser, K.E., and Richmond, C.R., eds. 1980. *Synthetic fossil fuel technology.* Ann Arbor, Mich.: Ann Arbor Science Publishers, Inc.

Crabbe, D., and McBride, R. 1979. *The world energy book.* Cambridge, Mass.: M.I.T. Press

Deju, R.A. 1975. *Extraction of minerals and energy.* Ann Arbor, Mich.: Ann Arbor Science Publishers, Inc.

Derry, D.R. 1980. *World atlas of geology and mineral deposits.* New York: Halsted (John Wiley and Sons, Inc.).

Eckes, A.E., Jr. 1979. *The United States and the global struggle for minerals.* Austin, Texas: Univ. of Texas Press.

Eldridge, F.R. 1979. *Wind machines.* 2nd ed. New York: Van Nostrand Reinhold.

Elliott, M.A., ed. 1981. *Chemistry of coal utilization.* Second Supplementary Volume. New York: John Wiley and Sons, Inc.

EMCON Associates. 1980. *Methane generation and recovery from landfills.* Ann Arbor, Mich.: Ann Arbor Science Publishers, Inc.

Falbe, J. ed. 1982. *Chemical feedstocks from coal.* New York: Wiley-Interscience.

Fischman, L.L. 1981. *World mineral trends and U.S. supply problems.* Baltimore, Maryland: Johns Hopkins University Press.

Glasby, G.P., ed. 1977. *Marine manganese deposits.* New York: Elsevier North-Holland, Inc.

Goodger, E.M. 1980. *Alternative fuels, chemical energy resources.* New York: John Wiley and Sons, Inc.

Gouleke, C.G. 1977. *Biological reclamation of solid wastes.* Emmaus, Pa.: Rodale Press Books.

Gray, M., and Rosen, I. 1982. *The warning: accident at Three Mile Island.* New York: W.W. Norton & Co., Inc.

Harder, E.L. 1982. *Fundamentals of energy production.* New York: Wiley-Interscience.

Hautala, R.R.; King, R.B.; and Kutal, C., eds. 1979. *Solar energy: chemical conversion and storage.* Clifton, N.J.: Humana Press.

Hedley, D. 1981. *World energy: the facts and the future.* New York: Facts on File, Inc.

Herman, S.W., and Cannon, J.S. 1977. *Energy futures.* Cambridge, Mass.: Ballinger Publishing Co.

Hess, L.Y. 1979. *Reprocessing and disposal of waste petroleum oils.* Park Ridge, N.J.: Noyes Data Corp.

Hinckley, A.D. 1981. *Renewable resources in our future.* Elmsford, N.Y.: Pergamon Press, Inc.

Hoffman, E.J. 1982. *Synfuels: the problems and the promise.* Laramie, Wyoming: Energon Co.

Hollander, J.M.; Simmons, M.K.; and Wood, D.O. 1981. *Annual review of energy.* Vol. 6. Palo Alto, Calif.: Annual Reviews.

Hoyle, F. 1977. *Energy or extinction? The case for nuclear energy.* London: Heinemann.

Hunt, V.D. 1979. *Energy dictionary.* New York: Van Nostrand Reinhold.

——. 1981. *The gasohol handbook.* New York: Industrial Press, Inc.

——. 1982. *Handbook of energy technology.* New York: Van Nostrand Reinhold.

Inglis, D.R. 1978. *Wind power and other energy options.* Ann Arbor, Mich.: Univ. of Michigan Press.

Jackson, F.R. 1975. *Recycling and reclaiming municipal solid wastes.* New York: John Wiley and Sons, Inc.

Jensen, M.L., and Bateman, A.M. 1979. *Economic mineral deposits.* New York: John Wiley and Sons, Inc.

Johansen, R.T., and Berg, R.L., eds. 1979. *Chemistry of oil recovery.* ACS Symposium Series 91. Washington, D.C.: American Chemical Society.

Johansson, T.B., and Steen, P. 1981. *Radioactive waste from nuclear power plants.* Berkeley, Calif.: Univ. of California Press.

Jones, J.L., and Radding, S.B., eds. 1979. *Thermal conversion of solid wastes and biomass.* ACS Symposium Series 130. Washington, D.C.: American Chemical Society.

Karr, C., Jr. 1978. *Analytical methods for coal and coal products.* New York: Academic Press, Inc.

Kesler, S.E. 1976. *Our finite mineral resources.* New York: McGraw-Hill Book Co.

Kestin, J.; DiPippo, R.; Khalifa, H.E.; and Ryley, D.J., eds. 1980. *Sourcebook on the production of electricity from geothermal energy.* DOE/RA/4051-1. Washington, D.C.: U.S. Department of Energy.

Kiang, Y.-H. 1981. *Waste energy utilization technology.* New York: Marcel Dekker, Inc.

Kircher, H.B., and Wallace, D.L. 1982. *Our natural resources.* 5th ed. Danville, Ill.: Interstate Printers and Publishers, Inc.

Klass, D.L., and Emert, G.H., eds. 1981. *Fuels from biomass and wastes.* Woburn, Mass.: Butterworth Publishers, Inc.

Larsen, J.W., ed. 1978. *Organic chemistry of coal.* ACS Symposium Series No. 1. Washington, D.C.: American Chemical Society.

Lee, S.S., and Sengupta, S., eds. 1979. *Waste heat management and utilization.* Washington, D.C.: Hemisphere Publishing Corp.

Lenihan, J., and Fletcher, W.W., eds. 1976. *Energy resources and the environment.* New York: Academic Press, Inc.

Lindblom, U., and Gnirk, P. 1982. *Nuclear waste disposal: can we rely on bedrock?* Elmsford, N.Y.: Pergamon Press, Inc.

Little, D.L.; Dils, R.E.; and Gray, J., eds. 1982. *Renewable natural resources.* Boulder, Co.: Westview Press, Inc.

Lyons, T.P., ed. 1982. *Gasohol: a step to energy independence.* New York: Methuen, Inc.

1979. *Manganese nodules.* Hingham, Mass.: D. Reidel Publishing Co., Inc.

Martin, A.E., ed. 1982. *Small scale resource recovery systems.* Park Ridge, N.J.: Noyes Data Corp.

Metz, W.D., and Hammond, A.L. 1978. *Solar energy in America.* Washington, D.C.: American Association for the Advancement of Science.

1975. *Mining in the outer Continental Shelf and in the deep ocean.* Washington, D.C.: National Academy of Sciences.

Moghissi, A.A.; Ozker, M.S.; and Carter, M.W., eds. 1978. *Nuclear power waste technology.* New York: American Society of Mechanical Engineers.

Murray, R.L. 1980. *Nuclear energy.* Elsmford, N.Y.: Pergamon Press, Inc.

——. 1982. *Understanding radioactive waste.* Columbus, Ohio: Battelle Press.

1981. *New coal chemistry.* London: The Royal Society.

Office of Technology Assessment. 1979. *The direct use of coal.* Cambridge, Mass.: Ballinger Publishing Co.

1978. *Oil and gas resources.* Guildford, Surrey, England: Science and Technology Press.

Oppenheimer, E.J. 1980. *Natural gas.* New York: Pan and Podium Productions.

Paul, J.K., ed. 1978. *Ethyl alcohol production and uses as a motor fuel.* Park Ridge, N.J.: Noyes Data Corp.

——, ed. 1980. *Large and small scale ethyl alcohol manufacturing processes from agricultural raw materials.* Park Ridge, N.J.: Noyes Data Corp.

——, ed. 1978. *Methanol technology and application in motor fuels.* Park Ridge, N.J.: Noyes Data Corp.

Priest, J. 1979. *Energy for a technological society.* 2nd ed. Reading, Mass.: Addison-Wesley Publishing Co., Inc.

Probstein, R.F., and Hicks, R.E. 1981. *Synthetic fuels.* New York: McGraw-Hill Book Co.

Rapp, D. 1981. *Solar energy.* Englewood Cliffs, N.J.: Prentice-Hall, Inc.

Reed, T.B., ed. 1981. *Biomass gasification—principles and technology.* Park Ridge, N.J.: Noyes Data Corp.

Reynolds, J.P.; McCarthy, W.N., Jr.; and Theodore, L., eds. 1981. *Environmental and economic considerations in energy utilization.* Woburn, Mass.: Butterworth Publishers, Inc.

Rider, D.K. 1981. *Energy: hydrocarbon fuels and chemical resources.* New York: Wiley-Interscience.

Rogers, P., ed. 1976. *Future resources and world development.* New York: Plenum Publishing Corp.

St.-Pierre, L.E., ed. 1979. *Resources of organic matter for the future.* Montreal: Multiscience Publishers.

St.-Pierre, L.E., and Brown, G.R., eds. 1980. *Future sources of organic raw materials.* Elmsford, N.Y.: Pergamon Press, Inc.

Sarkanen, K.V., and Tillman, D.A., eds. 1979. *Progress in biomass conversion.* Vol. 1. New York: Academic Press, Inc.

Schmidt, F.W., and Willmott, A.J. 1981. *Thermal energy storage and regeneration.* New York: McGraw-Hill Book Co.

Schmidt, R.A. 1980. *Coal in America.* New York: McGraw-Hill Book Co.

Schumacher, M.M., ed. 1980. *Enhanced recovery of residual and heavy oils.* 2nd ed. Park Ridge, N.J.: Noyes Data Corp.

——, ed. 1982. *Heavy oil and tar sands recovery and upgrading.* Park Ridge, N.J.: Noyes Data Corp.

Scott, C.D., ed. 1979. *Biotechnology in energy production and conservation.* New York: John Wiley and Sons, Inc.

Sills, D.L.; Wolf, C.P.; and Shelanski, V.B. 1982. *Accident at Three Mile Island.* Boulder, Co.: Westview Press, Inc.

Singer, S.F. 1979. *Energy: readings from* Scientific American. San Francisco: W.H. Freeman & Co., Publishers.

Sittig, M. 1975. Resources recovery and recycling handbook. Park Ridge, N.J.: Noyes Data Corp.

Sjöström, E. 1981. *Wood chemistry.* New York: Academic Press, Inc.

Sofer, S.S., and Zaborsky, O.R. 1981. *Biomass conversion processes for energy and fuels.* New York: Plenum Publishing Corp.

Sørensen, B. 1979. *Renewable energy.* New York: Academic Press, Inc.

Stephens, M. 1981. *Three Mile Island.* New York: Random House, Inc.

Stephenson, R.M. 1981. *Living with tomorrow.* New York: Wiley-Interscience.

1981. *Strategic minerals: a resource crisis.* Washington, D.C.: Council on Economics and National Security.

Sullivan, R.F., ed. 1981. *Upgrading coal liquids.* ACS Symposium Series 156. Washington, D.C.: American Chemical Society.

Taher, A.H. 1982. *Energy: a global outlook.* Elmsford Park: Pergamon Press, Inc.

1982. *Three Mile Island: thirty minutes to meltdown.* New York: Penguin Books.

Tillman, D.A. 1978. *Wood as an energy resource.* New York: Academic Press, Inc.

Tillman, D.A.; Rossi, A.J.; and Kitto, W.D. 1981. *Wood combustion.* New York: Academic Press, Inc.

Tomlinson, M. 1979. *Chemistry for energy.* ACS Symposium Series 90. Washington, D.C.: American Chemical Society.

Tybach, L., and Muffler, L.J.P. 1981. *Geothermal systems.* New York: Wiley-Interscience.

Wen, C.Y., and Lee, E.S., eds. 1979. *Coal conversion technology.* Reading, Mass.: Addison-Wesley Publishing Co., Inc.

White, L.P., and Plaskett, L.G. 1981. *Biomass as fuel.* New York: Academic Press, Inc.

Whitehurst, D.D., ed. 1980. *Coal liquefaction fundamentals.* ACS Symposium Series 139. Washington, D.C.: American Chemical Society.

Whitehurst, D.D.; Mitchell, T.O.; and Farcasiu, M. 1980. *Coal liquefaction: the chemistry and technology of thermal processes.* New York: Academic Press, Inc.

Wilson, C.L. 1980. *Coal—bridge to the future.* Cambridge, Mass.: Ballinger Publishing Co.

QUESTIONS AND PROBLEMS

1. Considering the average percentage of copper in copper ore mined and the annual U.S. consumption of copper, how many metric tons of copper ore must be mined per year to satisfy U.S. demand?

2. If U.S. demand for copper were to increase by 2% per year and the percentage of copper in the ore were to decrease, on a relative basis, by 2% per year, what would be the quantity of copper ore processed for U.S. consumption after 10 years (see Question 1)?
3. What do clays and mica have in common chemically?
4. Using values for the tonnage of wood wastes produced in the U.S. each year (from Section 17.4) and the heating values of biomass and bituminous coal (Table 17.4), estimate the number of metric tons of coal that could be replaced in the U.S. by wood wastes each year.
5. Using the efficiency figures given in Figure 17.7, what would be the effect of complete conversion of gasoline auto engines to diesel engines insofar as fuel consumption is concerned?
6. Considering the Carnot equation, what would be the maximum theoretical efficiency if the inlet temperature to a steam turbine could be raised to 1000 K at an outlet temperature of 300 K?
7. Why does a diesel engine have a higher theoretical efficiency than a gasoline engine?
8. What would be the effect of a 60% recovery of petroleum from oil-bearing formations, as compared to current recovery levels?
9. Chemically, coal is largely a hydrocarbon material. What does the low atomic ratio of hydrogen to carbon, as shown by the "chemical formula" of Illinois No. 6 coal, imply about the chemical structure of coal?
10. What are the major forms of sulfur in coal?
11. When air is used as the oxidizing agent in the gasification of coal, why is it possible to obtain only a low-Btu gas product?
12. Why is complete sulfur removal essential in the manufacture of SNG?
13. What is the purpose of heat pretreatment for coal to be used in a Synthane gasifier?
14. Why do nuclear power reactors produce more waste heat per unit of electrical energy than do fossil-fueled power plants?
15. What is the distinction between *reserves* and *resources*?
16. What are the major high-efficiency features of the Texaco coal gasification process?

CHAPTER 18
ENVIRONMENTAL BIOCHEMISTRY AND CHEMICAL TOXICOLOGY

18.1 POLLUTION AND THE BIOSPHERE

Most of this book has been devoted to environmental chemical phenomena in water, air, and soil. However, the ultimate concern must be for the biosphere, made up of the living organisms that inhabit the physical environment. These organisms interact in a complex way with the chemical species around them, so that chemical and living species have a strong influence on each other. Life forms vary greatly in their response to environmental chemicals. For example, oxygen, necessary for most organisms, is very poisonous to some anaerobic bacteria. Many chemicals commonly considered as pollutants, such as carbon monoxide, sulfur dioxide, nitrogen dioxide, cyanides, nitrate ion, nitrite ion, mercury, and cadmium are toxic above certain levels, but all appear in the environment naturally and, when present at low levels, can be detoxified by a variety of organisms. It is reasonable to assume that evolutionary processes have led to mechanisms by which living organisms can cope with low levels of heavy metals, sulfur dioxide, and nitrogen oxides. These pollutants cause problems only when human activities lead to their presence in excessively high levels. However, it is unlikely that natural processes ever resulted in the synthesis of a polybrominated biphenyl, a chlorinated hydrocarbon pesticide, or any one of thousands of "exotic" chemicals synthesized and introduced into the environment anthropogenically in recent decades. The evolutionary process has not had time to develop mechanisms for coping with many of these exotic chemical species. Even where such mechanisms do exist, exotic chemicals have been introduced into the environment in such large quantities that the natural biological pathways for their elimination and detoxification have been overloaded.

Therefore, in order to understand how pollutants actually affect the biosphere, it is important to know about environmental biochemistry and chemical toxicology. Recently, increased awareness of the toxic effects of chemicals, along with legislation regulating their use and human exposure to them, have made toxic chemicals the concern of all chemists. In this chapter we will discuss the biochemical and toxicological effects of selected chemical pollutants.

18.2 FUNDAMENTALS OF BIOCHEMISTRY

The chemistry of life processes is called **biochemistry**. Since is is not possible to cover this subject in a few pages, the reader without any biochemical background may want to consult any one of a number of texts on this subject. This section discusses the most basic fundamentals of biochemistry and introduces some major biochemical terms in order to help in understanding the remainder of the chapter.

Cells approximately one micrometer in size are the basic building blocks of living organisms. Although bacteria, yeasts, and some algae consist of single cells, higher organisms are multicellular and contain cells specializing in different functions. Thus, a human has liver, muscle, brain, and skin cells which have quite different functions and different susceptibilities to environmental damage.

As shown in Figure 18.1, an animal cell has many distinct parts. The cell is enclosed within a **cell membrane** composed of protein and lipid. The membrane permeability varies for different substances, enabling the membrane to control the chemical species entering and leaving the cell. Thus, the membrane protects the cell from toxic materials. The cell **cytoplasm** contains proteins, nucleic acids, and other biochemical substances involved in cell processes. Proteins are synthesized in cell **ribosomes.** The **nucleus** is the cell "control center," containing deoxyribonucleic acid (DNA), the genetic material that includes the directions for cell reproduction. As discussed later in this chapter, alteration of DNA structure is one of the major detrimental effects of some toxic substances. Energy-yielding cellular metabolic processes are carried out in **mitochondria.** By a complex series of metabolic processes in the mitochondria, carbohydrates, proteins, and fats are broken down to yield carbon dioxide, water, and energy. The best example of this cellular respiration process is the oxidation of glucose sugar:

$$C_6H_{12}O_6 + 6\ O_2 \rightarrow 6\ CO_2 + 6\ H_2O + \text{energy} \tag{18.2.1}$$

Cellular respiration was discussed in Chapter 5 (see aerobic respiration, Table 5.1) and in Section 8.10 as the major pathway for eliminating oxygen-demanding organic materials from wastewater.

As shown in Figure 18.2 (on page 498), plant cells differ from animal cells in several important respects. Plant cells have **cell walls** composed of cellulose (see Section 17.4) which provide stiffness and strength. Large **vacuoles** in the cells contain material dissolved in water. Plant cells involved in photosynthesis (see

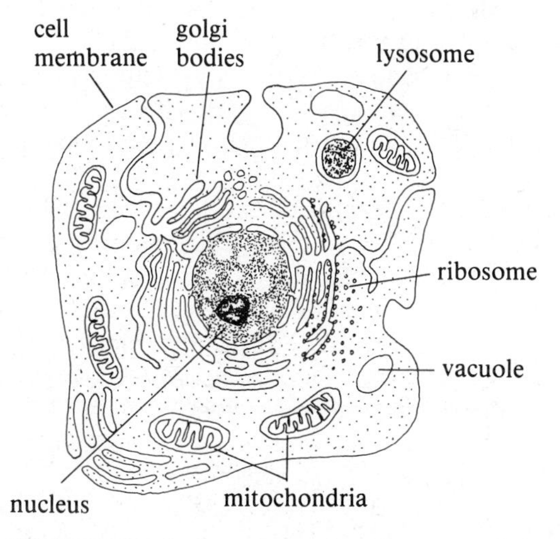

FIGURE 18.1 Structure of a human cell (from S.E. Manahan, *General Applied Chemistry,* 2nd ed. Boston: Willard Grant Press, 1982).

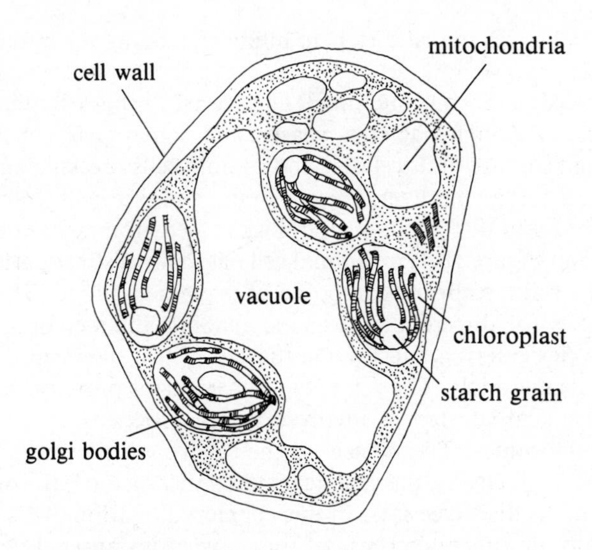

FIGURE 18.2 Structure of a plant cell (from S.E. Manahan, *General Applied Chemistry*, 2nd ed. Boston: Willard Grant Press, 1982).

Section 5.5) also contain **chloroplasts** that absorb light energy and convert it to chemical energy (food). Much of this food is stored as starch in starch bodies.

Plants cannot always get the energy they need from sunlight; during the night they must use stored food. Plant cells, like animal cells, contain mitochondria in which stored food is converted to energy by cellular respiration.

Plant cells, which use sunlight as a source of energy and CO_2 as a source of carbon, are said to be *autotrophic* (see Section 5.6). Basically, they use the reaction

$$6\,CO_2 + 6\,H_2O + \text{sunlight} \rightarrow \underset{\text{sugars}}{C_6H_{12}O_6} + 6\,O_2(g) \qquad (18.2.2)$$

to produce sugars with a high energy content. These sugars can be converted to water-insoluble starch, cell-wall cellulose, and other materials, and serve as the basic energy source for the rest of the plant's life processes.

Animal cells, which depend upon organic materials manufactured by plants for their food, are called *heterotrophic* cells. They use the energy from the chemical reaction between oxygen and food material (Reaction 18.2.1) to carry out their life processes.

The biochemical substances contained in cells are reactants, products, and catalysts in the complex biochemical processes carried out by living organisms. Just as contamination of a chemical reactant or poisoning of a chemical catalyst may ruin a laboratory synthesis or an industrial chemical process, toxic impurities can totally disrupt biochemical reactions. The major biochemical reactants in metabolic processes may be generally classified as *proteins, carbohydrates,* and *lipids*. The biochemical catalysts are called *enzymes*.

Proteins contained in cell cytoplasm are composed of 20 naturally occurring **amino acids.** Each amino acid has the general structure

$$H_2N-\overset{\overset{\displaystyle H}{|}}{\underset{\underset{\displaystyle R}{|}}{C}}-\overset{\overset{\displaystyle O}{\|}}{C}-OH$$

where R is the group determining the identity of the amino acid (for example, $-H$ in glycine, $-CH_3$ in alanine).

Proteins are large polymers of amino acids bonded through **peptide linkages**, the special name given to the amino bonds in proteins:

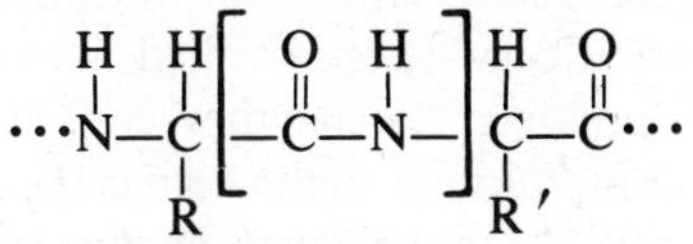

$$\cdots\overset{\displaystyle H}{\overset{|}{N}}-\underset{\underset{\displaystyle R}{|}}{\overset{\overset{\displaystyle H}{|}}{C}}\left[-\overset{\overset{\displaystyle O}{\|}}{C}-\overset{\overset{\displaystyle H}{|}}{N}-\right]\underset{\underset{\displaystyle R'}{|}}{\overset{\overset{\displaystyle H}{|}}{C}}-\overset{\overset{\displaystyle O}{\|}}{C}\cdots$$

The sequence of a large number of amino acids bonded together through amide bonds determines the identity of the protein. In addition, a secondary structure is formed through hydrogen bonds, electrostatic bonds, and disulfide bonds between amino acids, and a tertiary structure is produced by the twisting and folding of the helix making up the secondary structure, held in place principally by hydrogen bonds.

Another major type of biological compound consists of **carbohydrates,** which have an approximate empirical formula of $\{CH_2O\}$. Glucose (see Reactions 18.2.1 and 18.2.2) is one of the simplest carbohydrates. Cellulose, a major component of wood discussed in Section 17.4, is also a carbohydrate. Glucose is a monosaccharide (simple sugar), whereas cellulose is a polysaccharide composed of many monosaccharide units bonded together with a loss of one unit of H_2O per monosaccharide unit.

Lipids constitute a class of compounds that can be extracted from biological materials by organic solvents such as benzene, chloroform, or ether. A typical lipid is the fat, tristearin:

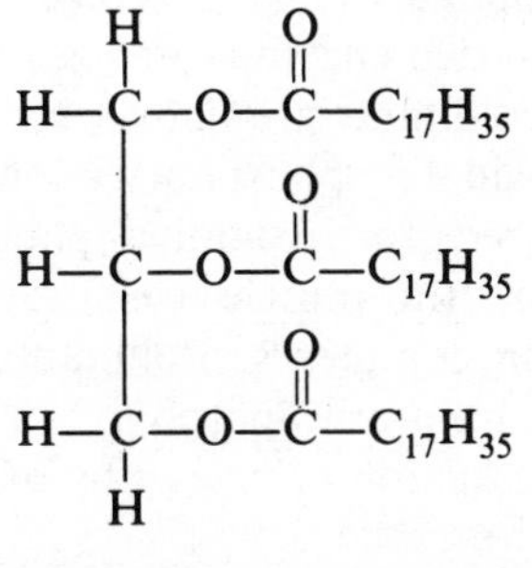

$$\begin{array}{l}
\quad\;\; H \qquad\quad O \\
\quad\;\; | \qquad\quad\;\; \| \\
H-C-O-C-C_{17}H_{35} \\
\quad\;\; | \qquad\quad\; O \\
\quad\;\; | \qquad\quad\;\; \| \\
H-C-O-C-C_{17}H_{35} \\
\quad\;\; | \qquad\quad\; O \\
\quad\;\; | \qquad\quad\;\; \| \\
H-C-O-C-C_{17}H_{35} \\
\quad\;\; | \\
\quad\;\; H
\end{array}$$

The most common lipids are fats (solid) and oils (liquid), but waxes and steroids (such as cholesterol and hormones) are also lipids. Because of their nonpolar, "organic-seeking" nature, lipids tend to accumulate hydrophobic, organic-soluble, water-insoluble pollutants, such as many of the organochlorine pesticides. This contributes to biomagnification of pollutants by organisms, a process by which pollutant concentrations increase as the pollutants move up through higher organisms via the food chain. Lipids are frequently extracted from the flesh of wildlife samples for the analysis of lipid-soluble organic compounds. Degradation-resistant synthetic organic compounds have been found in human fat and in the butterfat of human milk.

Enzymes are biological catalysts that enable chemical reactions to occur at the temperature of a living organism. Each enzyme is assigned a name indicating the reaction that it catalyzes, plus the suffix *-ase*. The substance upon which an enzyme acts is called the **substrate.** Enzymes are proteins, the secondary and tertiary structures of which are essential to their function. Alteration of these structures by environmental insult (heat or toxic chemicals) destroys the effectiveness of the enzyme and may be fatal to the organism. Every enzyme contains an **active site** where the substrate binds to the enzyme. Alteration of this active site, such as by pesticides or heavy metals, **denatures** the enzyme and inhibits its effect. Other substances, usually similar to the natural substrate, may fit into and block the active site, which also inhibits the enzyme's function

An additional species called a **coenzyme** is sometimes required for the enzyme to function. Coenzymes consist of a variety of materials, including vitamins and metal ions such as zinc(II) or iron(III). Environmental damage to coenzymes will likewise prevent enzymes from functioning properly.

Deoxyribonucleic acid (DNA) is a self-replicating material made up of heterocyclic amines, phosphate, and the pentose sugar, deoxyribose. It is the basic template determining heredity and life processes. DNA consists of very complicated molecules arranged in a double-helix structure. The sequence of the four heterocyclic amine bases along the DNA structure provides the genetic code giving the hereditary characteristics of all organisms, including the ability to produce specific enzymes. Alteration of this code by a change in the base sequence changes the order of amino acids made using the DNA as a template. Thus, altered DNA can produce the wrong template for a specific enzyme synthesis, yielding an incorrect amino acid sequence and, therefore, an inactive enzyme. The detrimental effects of DNA alteration are discussed further in Section 18.17.

Energy required for synthesis and other processes in the cell is provided by adenosine triphosphate (ATP), a molecule containing two high-energy phosphate bonds. Hydrolysis of one mole of ATP releases approximately 12 kilocalories of energy and produces one mole of adenosine diphosphate (ADP) plus one mole of hydrogen phosphate ion. ATP is regenerated from ADP through the oxidation of glucose via the two enzyme-catalyzed pathways of glycolysis and the citric acid cycle. Many pollutants are known to inhibit enzymes involved in the citric acid cycle and in oxidative phosphorylation, a step in the overall respiration process. Oxygen required for oxidative phosphorylation is carried to the cells by hemoglobin or (in muscle cells) by myoglobin, proteins containing iron-protoporphyrin complexes. Pollutants that interfere with either the oxidation state of iron within protoporphyrin or with the synthesis of protoporphyrin can inhibit the production of energy by oxidative phosphorylation.

18.3 TOXIC SUBSTANCES

Toxic substances damage biological systems, disturbing the functioning of biochemical processes and resulting in detrimental, and even fatal, effects. The study of toxic substances and their modes of action compose the science of **toxicology**. The mechanisms by which toxic substances affect biological systems and the overall results of these interactions are many and varied. This chapter covers some of the more important examples of the actions of toxic substances. For a more detailed discussion of this topic, the reader is referred to an extensive work on the subject [1].

---

[1] Doull, J.; Klaasen, C.D.; and Amdur, M.O., eds. 1980. *Casarett and Doull's toxicology: the basic science of poisons.* 2nd ed. New York: Macmillan Publishing Co., Inc.

Toxic substances consist of a vast variety of materials. These include organic compounds, inorganic compounds, organometallic compounds, metals, trace elements in various forms, solvents, vapors, and compounds of plant or animal origin. By their very nature, pesticides constitute a large class of toxic substances.

Every part of the human body is susceptible to damage by toxic substances. For instance, the respiratory system may be damaged by inhalation of toxic gases such as chlorine or nitrogen dioxide. Organophosphate insecticides and "nerve gases" disturb the function of the central nervous system, rapidly resulting in death. The liver and kidneys are especially vulnerable to many toxic substances. The sensitive reproductive system can be damaged by toxic substances to the extent that either reproduction becomes impossible or deformed young are born.

In addition to death and disabling illnesses, there are other gross effects caused by exposure to toxic substances. Two of the most important of these are: **mutagenesis**, alteration of parental DNA that results in mutations in offspring; and **teratogenesis**, or birth defects. Of course, the mention of toxic substances immediately brings to mind the large number of chemicals suspected of causing cancer. At the present time, chemical carcinogens head the list of chemicals over which there is concern.

The list of toxic substances is far too long to cover in this chapter. Particularly disturbing is the fact that in many cases it is not known whether a specific chemical compound belongs on this list. Thus, living things are being exposed to harmful chemicals that are not currently recognized as being toxic. On the other hand, important and useful chemicals are being regulated excessively because their harmlessness to humans has not been proven.

Toxic substances may be classified, according to their overall effects, as mutagens, carcinogens, or teratogens. They may also be classified chemically, using such categories as heavy metals, metal carbonyls, or organochlorine compounds. Classification according to function, such as food additive, pesticide, or solvent, is also useful.

18.4 ROUTES OF TOXIC SUBSTANCES THROUGH THE BODY

The effects of toxic substances are acutely dependent upon the **absorption, distribution,** and **excretion** of the toxicants [2]. Another important factor is the rate of metabolism, or biochemical transformation, of the toxic substance in the body.

Except for caustic agents (such as strong acids or bases), a toxicant normally must enter the bloodstream before doing any harm. In decreasing order of importance, the major routes of entry are the gastrointestinal (GI) tract, lungs, and skin. Once in the bloodstream, the toxicant may be transported to a **target organ**, such as the liver or kidney. In some cases, such as hemolysis caused by arsine gas, the harmful effect may occur in the bloodstream, itself.

---

[2] Klaasen, C.D. 1980. Absorption, distribution, and excretion of toxicants. In *Casarett and Doull's toxicology: the basic science of poisons.* 2nd ed., J. Doull, C.D. Klaasen, and M.O. Amdur, eds., Ch. 3. New York: Macmillan Publishing Co., Inc.

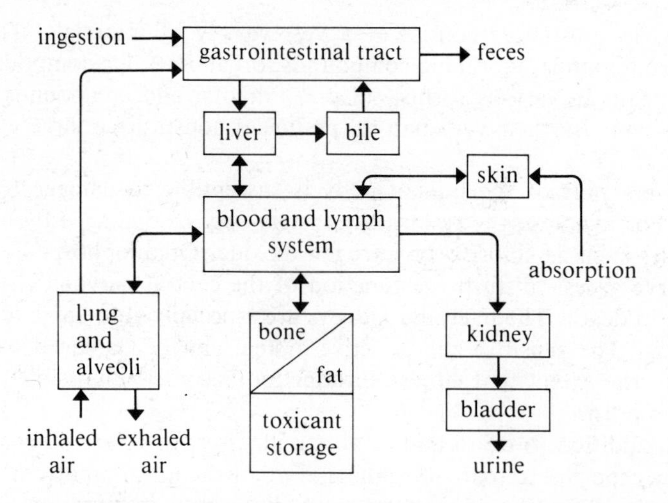

FIGURE 18.3 Major routes by which toxicants are absorbed, distributed, and excreted in the body.

Figure 18.3 shows the main routes by which toxicants are absorbed, distributed, and excreted in the body. As shown, ingestion into the gastrointestinal tract is the main route for toxicant entrance. Inhaled foreign substances released by lung mucus may also enter the gastrointestinal tract. Toxicants in the form of gases, vapors, and extremely small particles may enter the lung and be absorbed by the alveoli, thin-walled pouches that separate the air in the lung from the blood in the bloodstream, through which exchange occurs between inhaled air and blood.

Although human skin is relatively impermeable to most toxic agents, it can absorb enough of some to do systemic (body system) damage. One example is the absorption of carbon tetrachloride through the skin, producing injury to the liver. One of the characteristics of nerve gases, such as sarin, is that they readily pass through skin. The same is true of some insecticides, which have caused death in humans by absorption through the skin.

Distribution of a toxicant throughout the body can occur once it has dissolved in the blood plasma. Distribution depends upon the affinity of the toxicant for various parts of the body and on its ability to pass through cell membranes (see Figure 18.1). These membranes consist of a double layer of lipid molecules penetrated in various locations by protein. A toxicant may pass through a cell membrane by simple diffusion from a region of higher concentration to a region of lower concentration, or by special transport mechanisms in which the cell takes an active part; such mechanisms are capable of moving a toxicant against a concentration gradient.

The **blood-brain barrier** deserves special mention in terms of the distribution of toxicants in the body. This is the barrier preventing access to the central nervous system, and it is considerably less permeable to toxicants than are most other areas of the body. Thus, the central nervous system is partially protected from a number of toxic substances afflicting other parts of the body.

Toxic substances may be concentrated or stored in several places in the body. The liver and kidneys concentrate toxic substances, probably because these organs are involved in eliminating such substances from the body. The liver is particularly effective in metabolizing foreign substances.

Fat tissue has a tendency to collect the many toxic substances that are lipophilic ("fat-loving") and are relatively insoluble in water. Among such chemicals are DDT, PCB's, and PBB's (see Section 7.19), and chlordane.

Bone serves as a storage depot for several inorganic species because it contains the mineral hydroxyapatite, $Ca_5OH(PO_4)_3$. Lead and strontium tend to substitute for calcium in the hydroxyapatite, and F^- ion substitutes for OH^-. Too much accumulation of fluoride in bone can cause a condition known as *skeletal fluorosis*. Accumulation of radioactive strontium in bone can cause osteosarcoma (bone cancer).

Excretion of toxic substances most commonly occurs via the kidneys. In addition, the liver and biliary system excrete some toxic substances, including lead and DDT, and the lungs are important in excreting gaseous and volatile toxic substances. All body secretions have been shown to include toxic substances of one kind or another. These secretions include milk, sweat, tears, and feces.

18.5 METABOLISM OF TOXIC SUBSTANCES

Toxic substances are more readily eliminated from the body if their water solubility is relatively high. Much of the body's metabolic action on these substances has the purpose of increasing the water solubility of lipophilic materials that would otherwise tend to accumulate in the body. The enzyme reactions that accomplish this are of two types [3]. **Phase I reactions** involve oxidation, reduction, and hydrolysis. These reactions convert toxic compounds to derivatives that can undergo **Phase II reactions** of conjugation, or synthesis. Reactions of both phases occur primarily in the liver but may also take place in the kidneys, intestines, lungs, brain, and skin.

Phase I reactions are generally carried out by the cytochrome P-450 monooxygenases, consisting of NADPH-cytochrome c reductase and cytochrome P-450 enzymes. The following are the most important types of reactions brought about by these enzymes. **Epoxidation** involves addition of an O atom across a double bond:

$$R-\underset{H}{\underset{|}{C}}=\underset{H}{\underset{|}{C}}-R' + \{0\} \longrightarrow R-\underset{H}{\underset{|}{C}}\overset{O}{\diagup\diagdown}\underset{H}{\underset{|}{C}}-R' \qquad (18.5.1)$$

Aromatic hydroxylation involves epoxidation of an aromatic ring, followed by rearrangement to a phenol:

$$C_6H_6 + \{0\} \longrightarrow C_6H_5{-}OH \qquad (18.5.2)$$

Aliphatic hydroxylation consists of the addition of an —OH group to an aliphatic compound (see Equation 18.5.3).

---

[3] Neal, R.A. 1980. Metabolism of toxic substances. In *Casarett and Doull's toxicology: the basic science of poisons,* 2nd ed., J. Doull, C.D. Klaasen, and M.O. Amdur, eds., Ch. 4. New York: Macmillan Publishing Co., Inc.

$$\text{R—}\underset{\text{H}}{\overset{\text{H}}{\text{C}}}\text{—}\underset{\text{H}}{\overset{\text{H}}{\text{C}}}\text{—H} + \{\text{O}\} \longrightarrow \text{R—}\underset{\text{H}}{\overset{\text{H—O}}{\text{C}}}\text{—}\underset{\text{H}}{\overset{\text{H}}{\text{C}}}\text{—H} \qquad (18.5.3)$$

N, O, or **S dealkylation** consists of reactions of the following type:

$$\text{R—}\overset{\text{H}}{\text{N}}\text{—CH}_3 + \{\text{O}\} \rightarrow \text{R—NH}_2 + \text{CH}_2\text{O} \qquad (18.5.4)$$

$$\text{R—(O, S)—CH}_3 + \{\text{O}\} \rightarrow \text{R—(OH, SH)} + \text{CH}_2\text{O} \qquad (18.5.5)$$

Desulfuration involves replacement of S by O:

$$\text{R}_2\overset{\text{S}}{\overset{\|}{\text{P}}}\text{—X} + \{\text{O}\} \rightarrow \text{R}_2\overset{\text{O}}{\overset{\|}{\text{P}}}\text{—X} + \{\text{S}\} \qquad (18.5.6)$$

Sulfoxidation is the addition of an oxygen atom to organically bound sulfur:

$$\text{R—S—R}' + \{\text{O}\} \rightarrow \text{R—}\overset{\text{O}}{\overset{\|}{\text{S}}}\text{—R}' \qquad (18.5.7)$$

Phase III reactions generally result in the formation of a compound that is more water soluble and more readily excreted than the original toxic substance. These reactions are conjugation reactions, the most common of which involves synthesis of a glucuronic acid derivative. These glucuronides have the general formula,

[Structure: glucuronide ring bearing C(=O)—OH, ring O, OH, HO, OH, and —X]

where X represents a substituent group, such as one derived from alcohols, —OR; from carboxylic acids, $\text{—O—}\overset{\text{O}}{\overset{\|}{\text{C}}}\text{—R}$; from sulfhydryl compounds, —SR; or from amines, $\text{—}\overset{\text{H}}{\text{N}}\text{—R}$. Glucuronides of toxic substances are normally excreted in bile or urine.

Toxic and other foreign substances in the body undergo a number of other biochemical reactions in addition to those described above. Although these reactions generally serve to detoxify and facilitate the elimination of toxic substances, some metabolic processes can increase toxicity. In particular, the metabolic formation of epoxides of aromatic and olefinic compounds (see Reaction 18.5.1) is thought to be responsible for the mutagenic and carcinogenic activity of some of these compounds.

18.6 DISTURBANCE OF ENZYME ACTION

Since enzymes are the biological catalysts that enable most essential metabolic functions to be carried out in living cells, disturbance of enzyme function is most harmful. The active site where the substrate is bound and the three-dimensional structure of the enzyme determine its specificity for a substrate. An enzyme's active site and structure are particularly susceptible to damage by toxic substances, which can cause enzyme inhibition, preventing normal enzyme function.

Enzymes may require **activators**, which influence the structure of the active site and enable the enzyme to fit the substrate. **Inhibitors** have the reverse effect. Many inhibitors are normally associated with certain enzymes, serving as controls on the reactions that the enzymes catalyze; however, some toxic substances may also act as inhibitors, preventing essential enzyme action. Heavy-metal ions, particularly Hg^{2+}, Pb^{2+}, and Cd^{2+}, are especially effective enzyme inhibitors. These are sulfur-seeking metals that bind to $-SCH_3$ and $-SH$ groups in methionine and cysteine amino acids, which are part of the enzyme structure:

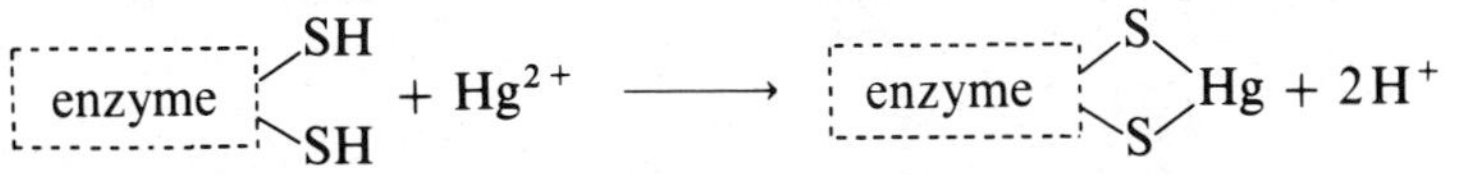

$$\text{enzyme}\begin{matrix}\diagup SH\\ \diagdown SH\end{matrix} + Hg^{2+} \longrightarrow \text{enzyme}\begin{matrix}\diagup S \diagdown\\ \diagdown S \diagup\end{matrix}Hg + 2H^{+} \qquad (18.6.1)$$

The $-SH$ (sulfhydryl) groups are often found on enzyme active sites, where the binding of a heavy-metal ion is especially detrimental.

A number of enzymes (called **metalloenzymes**) contain metals. Substitution of one of these metal ions by another metal ion with the same charge and a similar size inhibits the enzyme. Zinc is a common metalloenzyme component. Cadmium, which is directly below zinc in the periodic table, substitutes for zinc. Despite the chemical similarities between Zn^{2+} and Cd^{2+} ions, the cadmium-containing enzyme does not function properly. This is a common cause of cadmium toxicity. Among the enzymes inhibited by Cd^{2+} ion are adenosine triphosphatase, alcohol dehydrogenase, amylase, carbonic anhydrase, peptidase activity in carboxypeptidase, and glutamic-oxaloacetic transaminase. Both As^{3+} and As^{5+} inhibit pyruvate dehydrogenase. The Pb^{2+} ion inhibits acetylcholinesterase, alkaline phosphatase, adenosine triphosphatase, carbonic anhydrase, and cytochrome oxidase. Mercuric ion, Hg^{2+}, inhibits alkaline phosphatase, glucose-6-phosphatase, and lactic dehydrogenase.

Interference in the synthesis of heme (a complex between a substituted porphyrin and Fe^{2+}, found in hemoglobin and cytochromes) is one of the most important biochemical effects of lead. Lead inhibits several of the key enzymes involved in the overall process of heme synthesis, resulting in accumulation of metabolic intermediates. One of these intermediates is delta-aminolevulinic-acid. A major step in the overall scheme of heme synthesis is the conversion of delta-aminolevulinic acid to porphobilinogen:

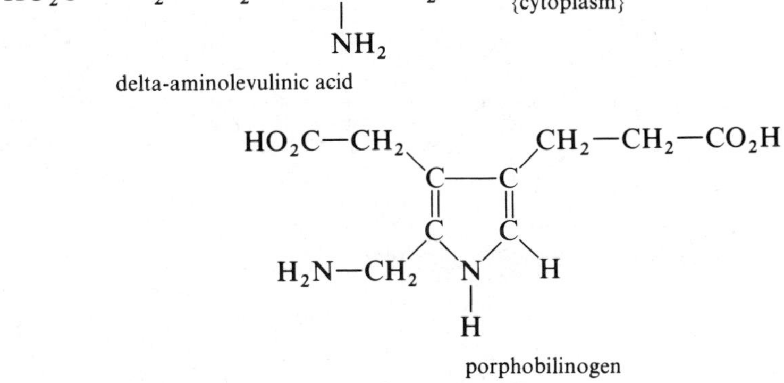

$$HO_2C{-}CH_2{-}CH_2{-}\overset{\overset{\displaystyle O}{\|}}{C}{-}\overset{\overset{\displaystyle H}{|}}{\underset{\underset{\displaystyle NH_2}{|}}{C}}{-}CO_2H \xrightarrow[\{\text{cytoplasm}\}]{\text{ALA dehydrase}}$$

delta-aminolevulinic acid

$$\begin{array}{ccccc} HO_2C{-}CH_2 & & & & CH_2{-}CH_2{-}CO_2H \\ & \diagdown & & \diagup & \\ & C & {-} & C & \\ & \| & & \| & \\ & C & & C & \\ \diagup & & \diagdown \;\; \diagup & & \diagdown \\ H_2N{-}CH_2 & & \underset{\underset{\displaystyle H}{|}}{N} & & H \end{array} \qquad (18.6.2)$$

porphobilinogen

Lead inhibits the ALA-dehydrase enzyme, so that delta-aminolevulinic acid accumulates, rather than forming porphobilinogen. This results in the urinary excretion of large quantities of delta-aminolevulinic acid. Lead also inhibits other steps in the overall synthesis of heme. The net result is the impairment of the synthesis of hemoglobin and other respiratory pigments, such as cytochromes, that require heme. Ultimately, therefore, lead prevents utilization of oxygen and glucose for life-sustaining energy production.

Arsenic provides another example of the interference of a heavy element with enzyme action. Arsenic(III) in the form of arsenite ion reacts with sulfhydryl groups on the enzyme, forming a very stable five-member ring structure:

$$\boxed{\text{enzyme}}\begin{matrix}\text{SH}\\ \text{SH}\end{matrix} + \begin{matrix}{}^-\text{O}\\ {}^-\text{O}\end{matrix}\!>\!\text{As}-\text{O}^- \longrightarrow \boxed{\text{enzyme}}\begin{matrix}\text{S}\\ \text{S}\end{matrix}\!>\!\text{As}-\text{O}^- + 2\text{OH}^- \qquad (18.6.3)$$

Among the many enzymes whose activity is destroyed by binding of the sulfhydryl group, those that produce cellular energy in the citric acid cycle are particularly affected. The first step in this cycle is inhibited by arsenic(III). This step involves the generation of ATP through the oxidation of pyruvic acid (a keto acid) to CO_2 and acetyl CoA (coenzyme A) in the mitochondria, a reaction catalyzed by pyruvate dehydrogenase. Alpha lipoic acid, a sulfhydryl-containing cofactor required for the action of pyruvate dehydrogenase, is inactivated by complexation with arsenic(III):

$$\begin{array}{ccc} & \text{HS}-\text{CH}_2 & \\ \text{O}^- & \text{CH}_2 & \\ -\text{O}-\text{As} \quad + & \text{HS}-\text{CH} & \\ \text{O}^- & \text{CH}_2 & \\ & \text{CH}_2 & \\ & \text{CH}_2 & \\ & \text{CH}_2 & \\ & \text{C}=\text{O} & \\ & \text{(protein)} & \end{array} \longrightarrow \begin{array}{cc} & \text{S}-\text{CH}_2 \\ \text{O}=\text{As} & \text{CH}_2 \\ & \text{S}-\text{CH} \\ & \text{CH}_2 \\ & \text{CH}_2 \\ & \text{CH}_2 \\ & \text{CH}_2 \\ & \text{C}=\text{O} \\ & \text{(protein)} \end{array} \qquad (18.6.4)$$

dihydrolipoic acid-protein — inactivated protein complex with arsenic(III)

Covalent bonding of enzymes with organic compounds may result in enzyme inhibition. This can happen when hydroxyl groups on enzyme active sites become bound. An important example of such binding is provided by the nerve gas diisopropylphosphorfluoriidate, DFP [4]. This compound is thought to bind to hydroxyl groups on the enzyme acetylcholinesterase by reaction 18.6.5.

[4] Moore, J.W., and Moore, E.A. 1976. *Environmental chemistry*. New York: Academic Press, Inc.

$$(C_3H_7O)_2\overset{O}{\overset{\|}{P}}—F + HO—(\text{acetylcholinesterase}) \rightarrow$$

$$(C_3H_7O)_2\overset{O}{\overset{\|}{P}}—O—(\text{acetylcholinesterase}) + HF \qquad (18.6.5)$$

This results in inhibition of the acetylcholinesterase, which is essential to the function of the central nervous system (see Section 18.16). Other enzymes with hydroxyl groups on their active sites may be similarly inhibited.

Coordinate covalent bonding or redox reactions of metals in metalloenzymes is a common mode of toxicity. Fluoride ion, F^-, may complex with magnesium(II) in magnesium-containing enzymes, thus inhibiting enzyme activity. Respiratory oxidation of carbohydrates and other energy-yielding materials is inhibited by binding of CO, H_2S, or CN^- to the complex enzyme, cytochrome a + a_3.

From the preceding discussion, it is obvious that many chemical substances adversely affect enzyme action and that this is a major mode of chemical toxicity. The following sections discuss the biochemical effects of selected toxic substances, most of which impair enzyme function. Among the toxic substances discussed are the heavy elements lead and mercury, and arsenic. Also included are cyanide, nitrite ion, carbon monoxide, sulfur dioxide, and pesticides.

18.7 BIOCHEMICAL EFFECTS OF ARSENIC

Arsenic was cited in an example of enzyme inhibition in Section 18.6. Although it is not a metal, arsenic has many of the properties of a heavy metal, and its toxic effect resembles that of mercury and lead. Arsenic(III), or arsenite, compounds are the most toxic, and discussion will be confined to these compounds.

Because of its chemical similarity to phosphorus, arsenic interferes with some biochemical reactions involving phosphorus. One such reaction occurs in the biochemical generation of the crucial energy-yielding substance adenosine triphosphate (ATP), discussed in Section 18.2. A key step in ATP generation is the enzymatic synthesis of 1,3-diphosphoglycerate from glyceraldehyde 3-phosphate, shown in Figure 18.4 (on page 508). When arsenite is present, it substitutes for phosphate to produce 1-arseno-3-phosphoglycerate instead of 1,3-diphosphoglycerate. Instead of phosphorylation, arsenolysis occurs, consisting of spontaneous hydrolysis to 3-phosphoglycerate and arsenate. No ATP is formed in this reaction.

Inorganic arsenicals at high concentrations will coagulate proteins [5]. This is probably caused by the reaction of arsenic with the sulfur bonds maintaining the secondary and tertiary structures of proteins. It may also be due to reaction of arsenic at the active site.

In summary, the three major known biochemical effects of arsenic are coagulation of proteins, complexation with coenzymes (discussed in Section 18.6), and inhibition of ATP production.

---

[5] O'Brien, R.D. 1967. *Insecticides action and metabolism.* New York: Academic Press, Inc.

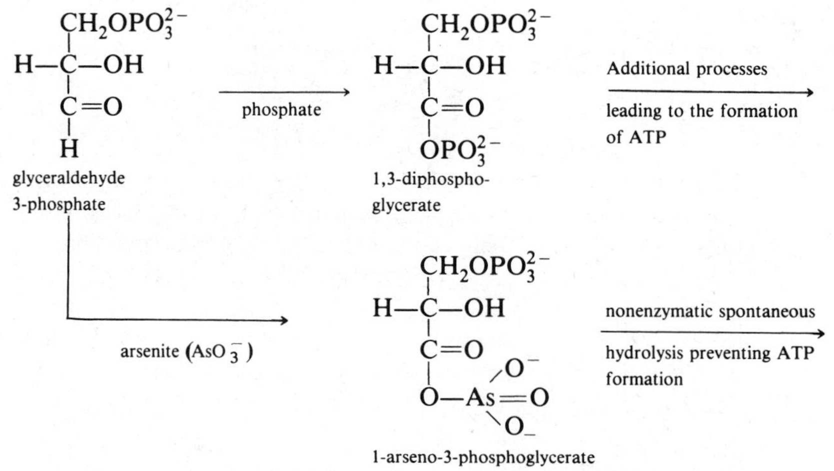

$$\begin{array}{c} CH_2OPO_3^{2-} \\ | \\ H-C-OH \\ | \\ C=O \\ | \\ H \end{array} \xrightarrow{\text{phosphate}} \begin{array}{c} CH_2OPO_3^{2-} \\ | \\ H-C-OH \\ | \\ C=O \\ | \\ OPO_3^{2-} \end{array} \xrightarrow[\text{leading to the formation of ATP}]{\text{Additional processes}}$$

glyceraldehyde 3-phosphate → 1,3-diphosphoglycerate

$$\xrightarrow{\text{arsenite } (AsO_3^-)} \begin{array}{c} CH_2OPO_3^{2-} \\ | \\ H-C-OH \\ | \\ C=O \\ | \\ O-As(=O)(O^-)O^- \end{array} \xrightarrow[\text{hydrolysis preventing ATP formation}]{\text{nonenzymatic spontaneous}}$$

1-arseno-3-phosphoglycerate

FIGURE 18.4 Interference of arsenic in the phosphorylation reaction leading to ATP production.

The common antidotes for arsenic poisoning are compounds that have sulfhydryl groups capable of bonding to arsenite. One such compound is 2,3-dimercaptopropanol (BAL), discussed in Section 18.8 as an antidote for lead poisoning. This compound removes arsenite from protein and restores normal enzyme function.

18.8 BIOCHEMICAL EFFECTS OF LEAD

One of the major biochemical effects of lead, its interference with heme synthesis, was discussed as an example of enzyme inhibition in Section 18.6. This effect is manifested as hematological damage. In addition, ingestion of lead is harmful to the central nervous system, gastrointestinal tract, and kidneys. Lead damages the central nervous system through largely unknown biochemical effects upon cells and nerves in the brain. This can result in symptoms ranging from fatigue and headache through convulsions, cerebral palsy, blindness, and mental retardation. Renal (kidney) function is impaired by lead-induced inhibition of the metabolic processes by which the kidney absorbs glucose, phosphates, and amino acids prior to urinary secretion.

Because of the chemical similarities between Pb^{2+} and Ca^{2+}, bones serve as a repository for lead accumulated by the body (see Section 18.4). Later this lead may be remobilized along with phosphates from the bone. When transported to soft tissues, the lead released from bone has a toxic effect.

Some detrimental effects of lead, such as inhibition of heme synthesis, can be reversed once the lead is removed, as long as exposure has not been too prolonged. Chelating agents can be used for the treatment of lead poisoning. In addition to treatment with calcium EDTA, British anti-Lewisite (BAL) is frequently used. This compound was developed as an antidote for Lewisite, an arsenic-containing poison gas used during World War I. BAL forms a complex with lead

[6], which is then harmlessly excreted through the kidney and liver:

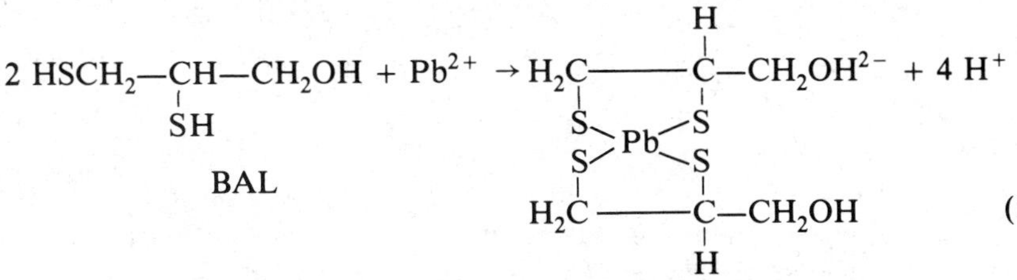

$$
\underset{\text{BAL}}{2\,\mathrm{HSCH_2{-}\underset{\displaystyle SH}{CH}{-}CH_2OH}} + \mathrm{Pb^{2+}} \rightarrow
\begin{array}{ccc}
\mathrm{H_2C} & \text{———} & \mathrm{\overset{\displaystyle H}{C}{-}CH_2OH^{2-}} \\
\mathrm{S} & & \mathrm{S} \\
& \mathrm{Pb} & \\
\mathrm{S} & & \mathrm{S} \\
\mathrm{H_2C} & \text{———} & \mathrm{\underset{\displaystyle H}{C}{-}CH_2OH}
\end{array}
+ 4\,\mathrm{H^+} \qquad (18.8.1)
$$

An interesting theory that has been the subject of recent spirited debate is that the ancient Roman aristocracy suffered from chronic lead poisoning due to the use of lead containers for wine and cooking. Proponents of this theory suggest that chronic lead poisoning was the major cause of the gout prevalent among Roman aristocracy and the "degenerate behavior" of Roman emperors from Augustus in 30 B.C. to Elagabalus in 220 A.D. Some scholars have hotly contested the idea of epidemic lead poisoning among the Roman ruling class, pointing out that a political system that endured for more than 500 years (the longest in world history) could not have been run by gouty, colic-ridden, anemic men suffering from severe lead poisoning.

According to a study released in June, 1983, by Dr. Joseph L. Annest of the National Center for Health Statistics, the amount of lead in the average American's body dropped almost 37% between 1976 and 1980. This change coincides with a 55% decrease in the use of leaded gasoline during the same period. The wide-ranging study of 27,801 people in 64 areas showed that 1980 blood-lead levels averaged 92 μg/L, compared to 146 μg/L in 1976. The Federal Centers for Disease Control consider a blood-lead level exceeding 300 μg/L to be excessive.

18.9 BIOCHEMICAL EFFECTS OF MERCURY

Mercury as a water pollutant was discussed in Section 7.7. Its toxicity varies with chemical species and with kind and length of exposure. If swallowed, elemental mercury is often excreted without severe damage. However, inhalation of mercury vapor results in transfer of mercury to the brain via the bloodstream, causing severe damage to the central nervous system. Whether swallowed or inhaled, methyl mercury is readily absorbed by the bloodstream and carried to the brain, where it is very damaging and may be retained for many months.

Conversion of elemental mercury or inorganic mercury salts to methyl mercury by anaerobic methane-synthesizing bacteria is one of the most distressing aspects of mercury pollution. A cobalt(III)-containing vitamin B_{12} coenzyme is the essential cofactor for this conversion. A methyl group bonded to cobalt on the coenzyme is transferred enzymatically by methyl cobalamin to the mercuric ion, forming methyl mercury ion, CH_3Hg^+, or dimethyl mercury, $(CH_3)_2Hg$, as shown in Reaction 18.9.1.

---

[6] Sunderman, F.W., and Sunderman, F.W., Jr. 1970. *Laboratory diagnosis of diseases caused by toxic agents.* St. Louis, Mo.: Warren H. Green, Inc.

$$\begin{array}{c} CH_3 \\ | \\ -Co(III) \end{array} + Hg^{2+} \xrightarrow[\text{(ATP)}]{\text{methyl cobalamin}} -Co(III) + \begin{array}{c} CH_3Hg^+ \\ \text{or} \\ (CH_3)_2Hg \end{array} \tag{18.9.1}$$

Acidic conditions favor the conversion of dimethyl mercury to methyl mercury, which is soluble in water. Methyl mercury enters the food chain through planktonic life and is concentrated (often several-fold) by fish as it goes through the food chain.

Largely because of their organic portion, alkyl mercurials are very soluble in organic material, particularly the lipids in membranes and brain tissue. The covalent carbon-mercury bond is not readily broken, and the alkyl mercury remains in cells for prolonged periods. Of particular concern is the ability of alkyl mercury compounds to pass through the placental barrier and enter fetal tissues. This phenomenon results from the membrane solubility of alkyl mercury and is not exhibited by other mercury compounds.

Because of its strong affinity for sulfur, mercury bonds strongly to sulfhydryl groups of proteins, including enzymes. It bonds to hemoglobin and serum albumin, both of which contain sulfhydryl groups.

Bonding of mercury to the cell membrane may inhibit active transport of sugars across the membrane and cause increased permeability of the membrane to potassium. The resulting lack of sugar transport to brain cells may result in energy deficiency in the cells. Increased potassium permeability has an effect upon the transmission of brain nerve impulses (see Section 18.16). These physiological effects may explain why babies born to mothers suffering from methyl mercury poisoning frequently suffer irreversible damage to the central nervous system, including cerebral palsy, mental retardation, and convulsions. These same physiological effects may account for the general numbness, hearing impairment, and slurring of speech characteristic of individuals exposed to methyl mercury. Another known effect of methyl mercury poisoning is segregation of chromosomes, chromosome breakage in cells, and inhibited cell division.

The exact nature of the effect of mercury on cell membranes is not known, nor is the identity of the affected enzymes. It will be difficult to determine the exact nature of these effects since there are many possible sulfhydryl-containing enzymes that might be inhibited by mercury.

18.10 BIOCHEMICAL EFFECTS OF CYANIDE

The Romans used cyanide from natural seed sources, such as apple seeds, for executions and suicides. Fruits such as apples, apricots, cherries, peaches, and plums all have seeds containing cyanide. Birth defects in pigs have been traced to ingestion of cyanide-containing cherry seeds by pregnant sows. Other natural sources of cyanide include arrowgrass, sorghum, flax, velvet grass, and white clover.

Cyanide in plants is bonded to a glycoside (sugar) called amygdalin. The cyanide is released by enzymatic or acidic hydrolysis (such as in the stomach). This reaction occurs during the extraction of bitter almond oil (benzaldehyde) from apricot seeds, as shown in Reaction 18.10.1.

$$\underset{\text{amygdalin}}{\text{C}_6\text{H}_5\text{—CH(C}\equiv\text{N)—O—}\underbrace{\text{C}_6\text{H}_{10}\text{O}_4\text{—O—C}_6\text{H}_{11}\text{O}_5}_{\text{glucose units}}} + 2\,H_2O \longrightarrow HCN + \underset{\text{glucose}}{2\,C_6H_{12}O_6} + \underset{\text{benzaldehyde}}{C_6H_5\text{—CH=O}} \qquad (18.10.1)$$

Cyanide, either a water or air pollutant, may enter the environment from many sources. Hydrogen cyanide (HCN) is used as a fumigating agent to kill rodents and other pests in grain bins, buildings, and holds of ships. As a gas, HCN can penetrate to all areas of the structure being fumigated. It can even penetrate pest eggs. Cyanide is used as a material for several chemical syntheses and in various types of research chemicals. Some synthetic fibers, such as the polyacrylics, give off HCN during combustion. This may hasten asphyxiation of persons inhaling smoke from fires. Cyanide salts are used in various mining and metal refining operations, such as gold mining. These salts are also used in metal cleaning and in plating metals, such as silver.

Unlike carbon monoxide or nitrite ion, cyanide does not interfere with processes by which cells receive oxygen. However, it does inhibit oxidative enzymes in mediating the process by which oxygen is utilized to complete the production of ATP in the mitochondria. In doing so, cyanide binds to ferricytochrome oxidase, an iron-containing metalloprotein. This species reacts specifically with oxygen and is the last cytochrome in the oxidative phosphorylation pathway. The sequence of reactions is the following, where Fe(III)-oxid is ferricytochrome oxidase, Fe(II)-oxid is ferrouscytochrome oxidase, and P_i designates inorganic phosphate:

Step I: $$\text{Fe(III)-oxid} + \text{reducing agent} \rightarrow \text{Fe(II)-oxid} + \text{oxidized reducing agent} \qquad (18.10.2)$$

Step II: $$\text{Fe(II)-oxid} + 2\,H^+ + \frac{1}{2}\,O_2 \xrightarrow[\text{ADP} + P_i \;\rightarrow\; \text{ATP}]{} \text{Fe(III)-oxid} + H_2O \qquad (18.10.3)$$

Basically, Fe(III)-oxid is the final acceptor of electrons released from the oxidation of glucose. The Fe(II)-oxid formed transfers the electrons to oxygen. Water is formed upon reaction with hydrogen ion and energetic ATP is produced in the process. Cyanide interferes by bonding to the Fe(III) of ferricytochrome oxidase, inhibiting the electron transfer in Step I. The reaction of oxygen in the overall energy-producing process is thereby prevented. Although the primary inhibitory effect of cyanide occurs with ferricytochrome oxidase, cyanide does complex other hematin compounds.

The metabolic pathway for the detoxification of cyanide involves conversion to the less toxic thiocyanate by a reaction requiring thiosulfate or colloidal sulfur as a substrate:

$$CN^- + S_2O_3^{2-} \xrightarrow{\text{rhodanase}} SCN^- + SO_3^{2-} \qquad (18.10.4)$$

The reaction is catalyzed by the enzyme rhodanase, also called mitochondrial sulfur tranferase. Although not found in the blood, this enzyme does occur abundantly in liver and kidney tissue. Thiosulfate, therefore, can be administered as an antidote for cyanide poisoning.

It is interesting to note that cyanide can also be detoxified by nitrite ion, itself a toxic substance. This process is discussed in the following section.

18.11 BIOCHEMICAL EFFECTS OF NITRITE ION

Nitrite ion affects hemoglobin. To understand this phenomenon, it is necessary to know that hemoglobin contains iron(II). Although the function of hemoglobin is to carry O_2 to cells, iron(II) is not oxidized in the process. However, nitrite ion does oxidize iron(II) in hemoglobin, HbFe(II), to create methemoglobin, HbFe(III), a brown substance that is ineffective in carrying oxygen to tissues:

$$\text{HbFe(II)} \xrightarrow{NO_2^-} \text{HbFe(III)} \qquad (18.11.1)$$

This reaction can result in oxygen deficiency and sometimes death. The mechanism is similar to that involved in carbon monoxide poisoning, discussed in the next section. The condition resulting from excess production of methemoglobin is called methemoglobinemia. It is most common in infants less than six months old, causing the child's skin to turn blue from lack of oxygen. In infants, methemoglobinemia is often caused by ingestion of nitrate with food. Conditions in the stomach of an infant are conducive to conversion of nitrate to nitrite, which is then absorbed into the blood. Fetal hemoglobin, which constitutes 60–80% of the hemoglobin in the bloodstream of an infant shortly after birth, apparently produces methemoglobin more readily than does adult hemoglobin.

The production of methemoglobin is reversible, and two enzymatic systems exist for the reduction of HbFe(III) to HbFe(II). One of these, called methemoglobin diaphorase, is coupled with the coenzyme NADH. It appears that infants are deficient in this coenzyme, increasing their susceptibility to nitrite poisoning.

In the previous section we mentioned that nitrite is effective in detoxifying cyanide. This is because methemoglobin, HbFe(III), bonds to cyanide, reversing the reaction of cyanide with ferricytochrome oxidase, Fe(III)-oxid. This bonding occurs because the iron in Fe(III)-oxid is in the +3 oxidation state. When a patient suffering from cyanide poisoning is treated intravenously with sodium nitrite or by inhalation of amyl nitrite, methemoglobin is formed and the reaction

$$\text{HbFe(III)} + \text{Fe(III)-oxid-CN} \rightarrow \text{HbFe(III)-CN} + \text{Fe(III)-oxid} \qquad (18.11.2)$$

results in the freeing of ferricytochrome oxidase from the cyanide ion. Additional treatment with thiosulfate results in elimination of the cyanide:

$$\text{HbFe(III)-CN} + S_2O_3^{2-} \rightarrow SCN^- + \text{HbFe(III)} + SO_3^{2-} \qquad (18.11.3)$$

In May, 1978, specialists from the Chinese Cancer Institute in Peking, People's Republic of China, announced the finding of correlations between high soil nitrate and nitrite levels and the very high incidence of cancer of the esophagus in China. In addition to high soil nitrate and nitrite levels, other contributing factors were found to be the presence of certain fungi and the absence of molybdenum in the soil. In some areas of China, the incidence of esophageal cancer is 150 per 150,000 persons per year, 45 times the U.S. average.

The synthesis of nitrite and nitrate from organic compounds in the aerobic upper region of the small intestine has been demonstrated [7]. This discovery suggests a greater role for nitrite as a cause of human cancer because of the possibility that nitrite so synthesized will form carcinogenic compounds lower in the intestinal tract, in the relatively acidic cecum and colon.

18.12 BIOCHEMICAL EFFECTS OF CARBON MONOXIDE

Carbon monoxide is toxic to higher animals when inhaled. Its toxicity is due to its reaction with the heme fraction of hemoglobin to form carboxyhemoglobin. Normally, the four iron-containing heme molecules composing a single molecule of hemoglobin bind to four molecules of oxygen. The hemoglobin picks up oxygen in the lungs and carries it, in the form of oxyhemoglobin, to tissue capillaries for release. In the tissues, carbon dioxide is dissolved by blood plasma and is carried to the lungs for release. The carbon monoxide (CO) molecule is very similar in size and shape to oxygen and is readily picked up by hemoglobin. The presence of only one carbon monoxide molecule on any one of the four heme groups in a hemoglobin molecule disturbs oxygen transport. The carboxyhemoglobin does not dissociate nearly so rapidly as oxyhemoglobin, preventing release of oxygen and uptake of carbon dioxide. Deprived of an oxygen supply, the tissues do not function properly and protein, nucleic acid, and lipid synthesis do not occur. This is because glycolysis without oxidative phosphorylation cannot supply the energy required for synthesis of biomolecules.

Other difficulties occur because of the formation of carboxyhemoglobin. Carbon dioxide builds up in the bloodstream, resulting in increased acidity. The cells build up pyruvic acid, lactic acid, and other acids due to anaerobic glycolysis. The resulting condition of acidosis increases the difficulty with which the cells accept oxygen. Furthermore, the acidity denatures or inactivates proteins, resulting in a directly toxic effect to cells.

The formation of carboxyhemoglobin is reversible. When CO exposure ends, and particularly if oxygen is administered, the effect of carbon monoxide poisoning can be reversed. Permanent brain damage results, however, if the brain cells are without adequate oxygen for more than a few minutes.

18.13 BIOCHEMICAL EFFECTS OF SULFUR DIOXIDE

Despite its choking odor, sulfur dioxide is one of the less toxic of the pollutant gases. As discussed in Section 12.9, sulfur dioxide has been implicated in several acute incidents of air pollution resulting in death and is suspected of contributing to emphysema and bronchitis. Combined exposure to sulfur dioxide and very small particles appears to be more toxic and damaging to lung tissue than exposure to either alone.

At very high sulfur dioxide concentrations (of the order of hundreds of parts per million), the gas is largely absorbed by the upper respiratory tract. This is due to the solubility of SO_2 in water contained in the tissues in the upper

---

[7] Tannenbaum, S.R., et al. 1978. Nitrite and nitrate are formed by endogenous synthesis in the human intestine. *Science* 200: 1487–8.

respiratory tract. At concentrations of 1 ppm or less, 95% or more of the sulfur dioxide penetrates the respiratory tract and enters the lungs. A characteristic physiological response to sulfur dioxide inhalation consists of mild bronchial constriction [8].

Inhaled sulfur dioxide is rapidly distributed throughout the body, as shown by animal studies using SO_2 containing radioactive sulfur. Removal of inhaled sulfur dioxide through the respiratory tract is relatively slow, requiring up to a week or more. It is possible that some of the gas is bound to protein. One of the more troublesome possibilities is that sulfuric acid may be formed by oxidation of SO_2 in the lungs. The high acidity and hygroscopic nature of sulfuric acid can result in the hydrolysis and dehydration of cells. Localized low-pH conditions resulting from the formation of sulfuric acid may result in proteins being denatured, as the electrostatic and hydrogen bonding are destroyed when the amino acid carboxyl groups are protonated.

Very little is known about the biochemical effects of sulfur dioxide in plants and animals. Plasma of animals exposed to sulfur dioxide has been found to contain S-sulfonate, which can be produced by the reaction of sulfite with proteins containing disulfide bonds. This is the first biochemical effect of SO_2 observed in animals.

18.14 BIOCHEMICAL EFFECTS OF OZONE AND PAN

As discussed in Chapter 14, ozone is the major oxidant produced during the formation of photochemical smog. Smaller quantities of organic oxidants are also produced, the best-known one being peroxyacetyl nitrate (PAN):

$$CH_3{-}\overset{\overset{\displaystyle O}{\|}}{C}{-}OO{-}NO_2$$

The biochemical effects of ozone and other atmospheric oxidants are not well understood, although these air pollutants are certainly detrimental to both animals and plants. Some of the general effects of photochemical oxidants were discussed in Section 14.7.

Both ozone and PAN irritate the eyes and respiratory passages of humans. Although human mortality data for these substances are scarce, animal experiments indicate that exposure to ozone for several hours at a level of approximately 50 ppm will cause death from the accumulation of fluid in the lungs, a condition known as pulmonary edema. Exposure to much lower levels causes nonlethal accumulation of fluid in the lungs and damage to lung capillaries. Increased temperature and exercise aggravate the effects of oxidants. Young animals, and presumably young humans, are more susceptible to oxidant toxicity than are older subjects. Intermittent exposure with intervening periods when fresh air is inhaled is less damaging than continuous exposure.

---

[8] Amdur, M.O. 1980. Air pollutants. In *Casarett and Doull's toxicology: the basic science of poisons,* 2nd ed., J. Doull, C.D. Klaasen, and M.O. Amdur, eds., Ch. 24, pp. 608–31. New York: Macmillan Publishing Co., Inc.

Ozone and other photochemical oxidants cause a number of biochemical effects [9]. It appears that many of these result from the generation of free radicals, discussed as atmospheric reaction intermediates in Sections 11.8 and 14.4.

Sulfhydryl (–SH) groups on enzymes are particularly susceptible to damage by photochemical oxidants. These groups are oxidized by both ozone and PAN and may be acetylated by the latter. Of the sulfur-containing amino acids, only cysteine, the only amino acid having a sulfhydryl group, is strongly affected by PAN. Under some conditions, methionine is oxidized by PAN to methionine sulfoxide. Other amino acids are not affected by PAN. For example, no detrimental effect is observed upon pancreatic ribonuclease (the enzyme catalyzing the hydrolysis of RNA) from exposure to PAN. Since this enzyme does not contain cysteine residues, it is not susceptible to damage by PAN.

Among the enzymes inactivated by photochemical oxidants are isocitric dehydrogenase, malic dehydrogenase, and glucose-6-phosphate dehydrogenase. These enzymes are involved in the citric acid cycle and in the degradation of glucose to produce cellular energy. It has been further shown that oxidants inhibit the activity of enzymes that synthesize cellulose and lipids in plants.

The double bonds in unsaturated fats are susceptible to attack by ozone. Ozone attack results in bond cleavage and formation of oxidation products. The mechanism is similar to that for the oxidative cleavage of double bonds in natural rubber, discussed in Section 14.7.

18.15 BIOCHEMICAL EFFECTS OF NITROGEN OXIDES

Nitrogen dioxide, NO_2, is less harmful to plants than are other major air pollutants, including sulfur dioxide, photochemical oxidants, and gaseous fluoride. Nitric oxide, NO, is even less phytotoxic. However, it should be stressed that these pollutants lead to the formation of highly phytotoxic photochemical smog components, as discussed in Chapter 14.

As mentioned in Section 12.13, NO_2 is quite toxic to humans. Exposure to 50-100 ppm of NO_2 for up to an hour or so causes lung tissue inflammation for several weeks but normally is not fatal. Exposure to 150–200 ppm of NO_2 causes irreversible lung damage called "bronchiolitis fibrosa obliterans," resulting in death 3–5 weeks after exposure. After exposure to levels exceeding approximately 500 ppm, death normally occurs within 2–10 days. Nitric oxide is less toxic than NO_2.

The biochemical mechanisms of NO_2 toxicity are not well established [10]. Some cellular enzyme systems are susceptible to disruption by NO_2, including catalase and lactic dehydrogenase, whereas cathepsin D does not appear to be damaged by the gas.

Although a weaker oxidant than ozone, NO_2 is probably similar to ozone in its mode of biochemical action. This would include free radical formation and

[9] Committee on Medical and Biologic Effects of Environmental Pollutants. 1977. *Ozone and other photochemical oxidants.* Washington, D.C.: National Academy of Sciences.

[10] Committee on Medical and Biologic Effects of Environmental Pollutants. 1977. *Nitrogen oxides.* Washington, D.C.: National Academy of Sciences.

subsequent reactions. Treatment with antioxidant Vitamin E appears to lessen some of the toxic effects of nitrogen dioxide.

18.16 BIOCHEMICAL EFFECTS OF PESTICIDES

The role of pesticides as water pollutants has been covered in Section 7.18. The reader is referred to Table 7.8 for a list of the names, structures, uses, and characteristics of the major pesticides.

The biochemistry of pesticides is important for several reasons. First, it is through their biochemical effects upon target organisms that pesticides perform their designated functions. Second, biochemical changes in nontarget organisms determine the undesirable side effects of pesticides. Finally, biochemical processes are the main mechanisms by which pesticides are degraded and detoxified.

It is impossible to discuss the biochemistry of pesticides to any appreciable extent in just a few pages. Instead, a few of the major aspects will be mentioned. The reader may refer to works dealing specifically with the subject for a more detailed coverage.

Probably more attention has been given to the biological action of DDT than to that of any other pesticide. Despite all this effort, the exact mode of biochemical action of this widely used pesticide upon target insects and nontarget animals is not completely understood. Like many insecticides, DDT acts upon the central nervous system. Like other chlorinated hydrocarbons, DDT dissolves preferentially in lipid (fat) tissue and may accumulate in the fatty membrane surrounding nerve cells. This may result in interference with the transmission of nerve impulses along the axons, which are projections connecting nerve cells. It is believed that the resulting disruption of the central nervous system kills the target organism.

It is known that interference with the axon transmission of nerve impulses is the cause of the toxicity of organophosphate insecticides. To understand why this is so, it is necessary to know that the axon membrane is more permeable to potassium ion than to sodium ion. As a result, the K^+/Na^+ ratio is much higher inside the axon than outside, resulting in a potential difference of about 60 millivolts across the cell membrane. During transmission of a nerve impulse, a substance called acetylcholine is released, and it attaches to the membrane at one end of the axon. This increases the membrane permeability to Na^+, causing a potential reversal. As the nerve impulses traverse the nerve, allowing the membrane to become permeable to K^+ again, resting potential is restored. This is accomplished by the acetylcholinesterase-catalyzed degradation of acetylcholine,

$$\underset{\text{acetylcholine}}{(CH_3)_3N^{\pm}{-}CH_2{-}CH_2{-}O{-}\overset{\overset{\displaystyle O}{\|}}{C}{-}CH_3} + H_2O \xrightarrow{\text{acetylcholinesterase}}$$

$$\underset{\text{choline}}{(CH_3)_3N^{\pm}{-}CH_2{-}CH_2{-}OH} + \underset{\text{acetic acid}}{CH_3CO_2H} \qquad (18.16.1)$$

Inhibition of the acetylcholinesterase enzyme by organophosphate pesticides, or other substances such as nerve gases, results in accumulation of acetylcholine. This accumulation results in overstimulation of muscles, nerves, and other parts of the organism. Eventually, convulsions, paralysis, and death result. The effects of organophosphate insecticides can be reversed by release of phosphate ester from the enzyme or production of new cholinesterase. Antagonists may be employed which prevent the generation of nerve impulses by blocking acetylcholine, even in the absence of acetylcholinesterase. One such antagonist is atropine, itself a poison.

Insecticidal carbamate esters likewise react with cholinesterase:

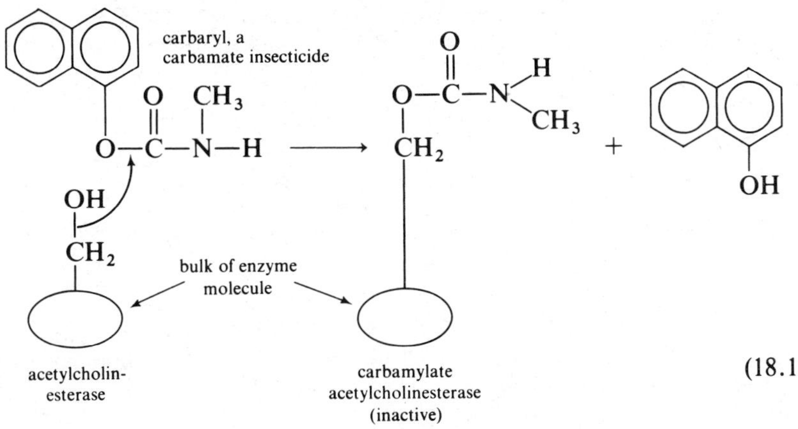

carbaryl, a carbamate insecticide

bulk of enzyme molecule

acetylcholinesterase

carbamylate acetylcholinesterase (inactive)

(18.16.2)

The resulting carbamylated enzyme is much more readily broken down than is the phosphorus ester formed with organophosphates. Therefore, carbamate insecticides are reversible acetylcholinesterase inhibitors.

There are many pathways by which pesticides are metabolized by target and nontarget organisms. For example, in humans, DDT is metabolized very slowly, as shown in Figure 18.5, and the products are excreted from the body in the urine. Some of these metabolic products are themselves insecticides and may persist for a long time in the environment.

Aldrin, dieldrin, and other chlorinated hydrocarbons apparently are metabolized in a manner similar to that used for DDT. Aldrin is oxidized to dieldrin in the metabolic process. Dieldrin is even more toxic to fish than is aldrin. This illustrates an important point—in some cases, pollutants are converted metabolically by organisms in the environment to more toxic chemical species.

Phosphorus and carbamate esters are much more readily broken down to nontoxic metabolites than are chlorinated hydrocarbon insecticides. One path for

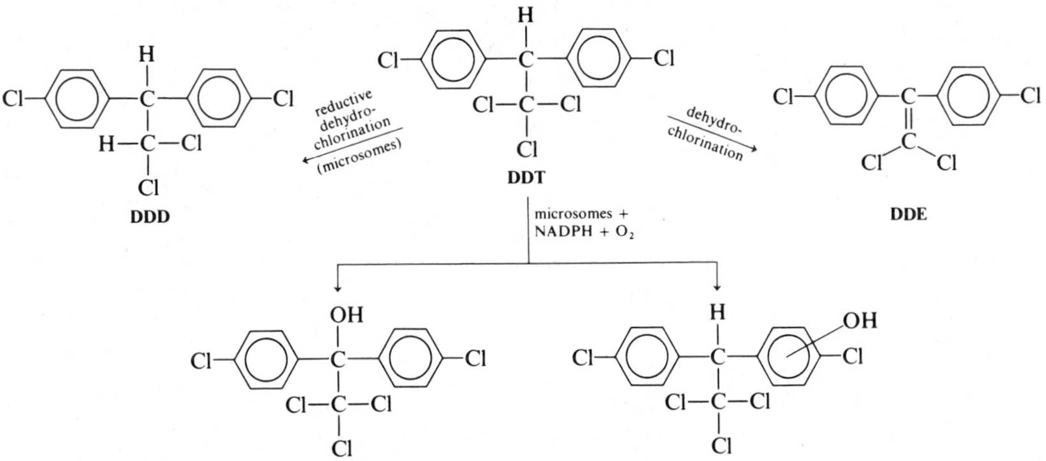

DDD

reductive dehydrochlorination (microsomes)

DDT

dehydrochlorination

DDE

microsomes + $NADPH + O_2$

FIGURE 18.5 Metabolism of DDT in humans.

the degradation of phosphorus esters (phosphorothionates) is hydrolysis, catalyzed by esterase enzymes. For example, the hydrolysis of parathion yields ethanol, thiophosphoric acid, and nitrophenol:

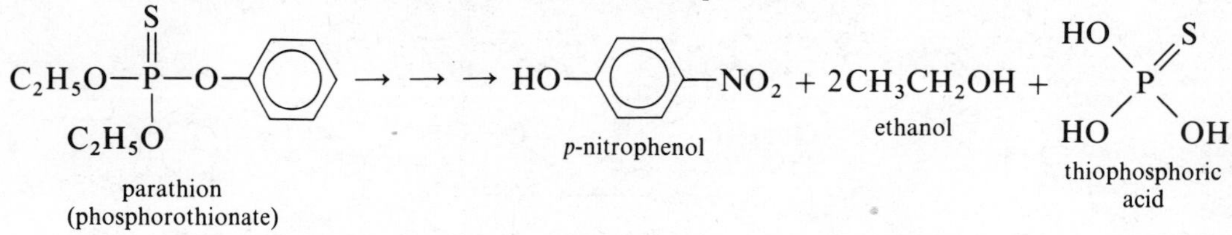

$$(C_2H_5O)_2P(=S)\text{—O—}C_6H_4\text{—}NO_2 \rightarrow \rightarrow \rightarrow HO\text{—}C_6H_4\text{—}NO_2 + 2CH_3CH_2OH + (HO)_2P(=S)OH \quad (18.16.3)$$

parathion (phosphorothionate); p-nitrophenol; ethanol; thiophosphoric acid

Phosphorus esters are also oxidized by oxidases (microsomal enzymes). This process, like hydrolysis, produces *p*-nitrophenol when parathion is degraded. The degradation of carbamates may also be accomplished through the action of esterases and oxygen-requiring microsomal enzymes. Thus, carbaryl degrades by several different pathways to produce 1-naphthol and other products (see Figure 18.6).

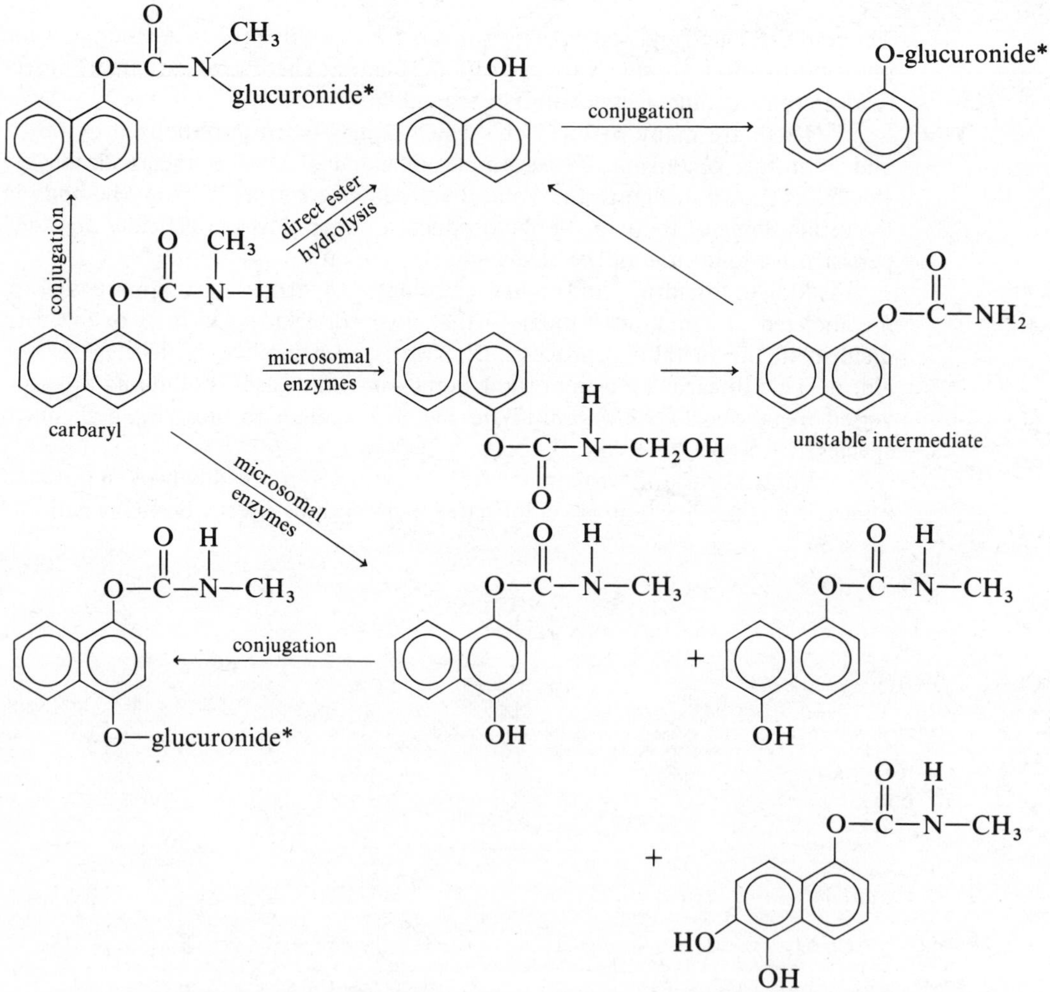

O—C(=O)—N(CH₃)—glucuronide*
conjugation
O—C(=O)—N(CH₃)—H
carbaryl
direct ester hydrolysis
OH
conjugation
O-glucuronide*
microsomal enzymes
O—C(=O)—N(H)—CH₂OH
O—C(=O)—NH₂
unstable intermediate
microsomal enzymes
O—C(=O)—N(H)—CH₃
O-glucuronide*
conjugation
OH
+
HO
OH

FIGURE 18.6 Degradation of carbaryl. *As discussed in Section 18.5, glucuronides are relatively soluble conjugation products that are readily eliminated in urine or bile.

18.17 MUTAGENESIS

Mutagenesis is a fundamental change in the structure of DNA in the paternal or maternal gametocyte. Such mutations may be passed on to progeny. A mutation often results in either the formation of an enzyme having a function other than that intended, or total loss of essential enzyme function. Mutation is the process that gives rise to differences in individual organisms, enabling natural selection to favor individuals, and ultimately new species, most fit for survival. However, most mutations are detrimental. It is for this reason that **mutagens,** chemical substances causing mutations, are of especial concern.

To understand mutagenesis, it is necessary to have some knowledge of deoxyribonucleic acid, DNA. As mentioned in Section 18.2, DNA is the basic genetic material found in cell nuclei. DNA molecules are made up of simple sugars, amines, and derivatives of phosphoric acid, H_3PO_4. The simple sugar is deoxyribose, with the structure

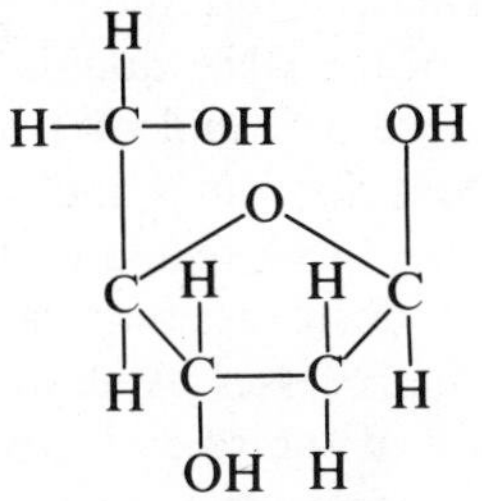

The four amines involved are cyclic amines, or nitrogenous bases. These are adenine, guanine, cytosine, and thymine (Fig. 18.7).

A material related to DNA is obtained if deoxyribose is replaced by the sugar ribose, and if thymine is replaced by the cyclic amine, uracil (Fig. 18.8). This material is **ribonucleic acid, RNA,** which functions in coordination with DNA in the synthesis of proteins.

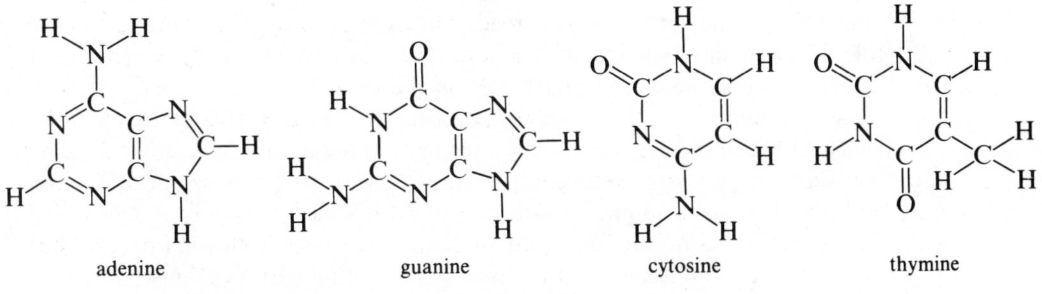

adenine guanine cytosine thymine

FIGURE 18.7 The nitrogenous bases contained in DNA.

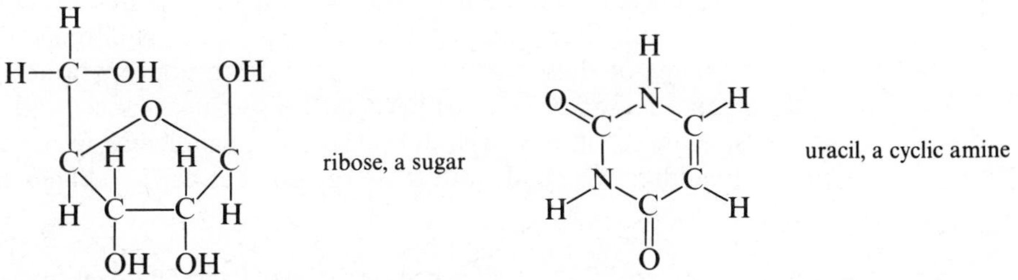

ribose, a sugar

uracil, a cyclic amine

FIGURE 18.8 Molecules that are present in RNA in place of deoxyribose and thymine in DNA.

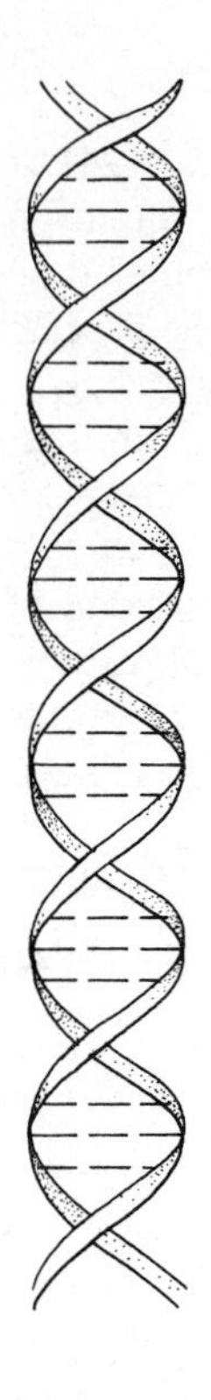

FIGURE 18.9 Representation of the DNA double helix structure. The vertical "ribbons" consist of strings of cyclic amine bases. The dashed lines indicate hydrogen bonding betwen complementary bases.

Molecules of DNA are huge, with molecular weights greater than one billion. Molecules of RNA are also quite large. The structure of DNA is that of the famed "double helix" of two spiral ribbons, as shown in Figure 18.9.

A DNA molecule is like a coded message. The message is written by variations in the order of the cyclic amine bases along the molecule, somewhat like a message sent by telegraph, which consists only of dots, dashes, and spaces in between. The two strands of DNA are complementary, which means that a particular portion of one strand fits like a key in a lock with the corresponding portion of the other strand. If the two strands are pulled apart, each manufactures a new complementary strand, so that two copies of the original double helix result. This occurs during cell reproduction.

Portions of the DNA double helix may unravel, and one of the strands of DNA may produce a strand of RNA. This substance then goes from the cell nucleus out to the ribosomes and regulates the synthesis of new protein or enzymes. In this way, DNA regulates the function of the cell and acts to control life processes. Toxic substances may cause changes in protein structure, resulting in disruption of life processes. Alteration of DNA structure is an even more serious concern. Because of the role of DNA in controlling reproduction and life processes, such a disruption may cause serious damage to an organism. Altered DNA often keeps reproducing itself, frequently causing a visible expression of the mutation.

Mutations may result from exchanges of bases, addition of a base along the DNA chain, deletion of a base, reaction of a base, or disruption of the gross DNA structure. The most common type of exchange of base along the DNA chain is exchange of thymine for cytosine or vice versa, or exchange of adenine and guanine. Addition or deletion of a base is more serious because of the manner in which the genetic code is read. The code consists of triplets of bases along the DNA chain. Addition or deletion of a base may result in the reading of the genetic code being out of phase beyond the point at which the error occurred.

A common example of the reaction of a base on the DNA chain is that caused by nitrous acid, frequently used as a bacterial mutagen. Nitrous acid converts amino groups to keto groups. Thus, it converts the base adenine to hypoxanthine, the base guanine to xanthine, and the base cytosine to uracil, which occurs in RNA but not in DNA (see the structures on the following page). The product bases will no longer pair by hydrogen bonding with the same bases as in the original DNA. Thus, when the DNA strand is duplicated, the base sequence will no longer be the same as in the original.

A mutagen that received a great deal of attention several years ago is "tris," a flame retardant chemical used to treat children's sleepwear. The chemical name of this compound is tris(2,3-dibromopropyl)phosphate. In addition to being mutagenic, it has been shown to cause cancer and sterility in test animals. A metabolite of tris, 2,3-dibromopropanol, has been measured in the urine of children who had worn tris-treated sleepwear [11]. This finding suggests

[11] Blum, A., et al. 1978. Children absorb tris-BP flame retardant from sleepware: urine contains the mutagenic metabolite, 2,3-dibromopropanol. *Science* 201: 1020–23.

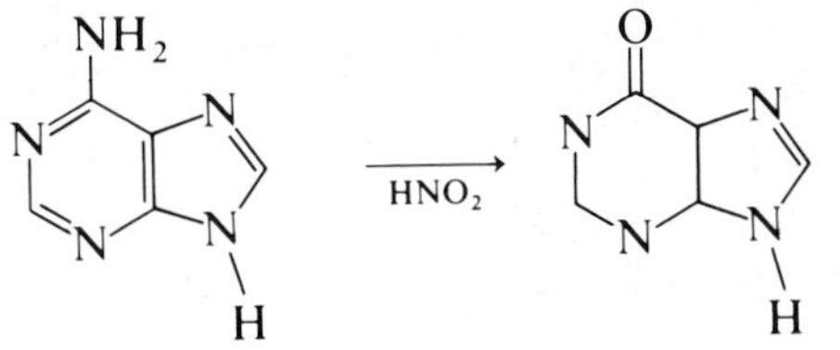

adenine $\xrightarrow{HNO_2}$ hypoxanthine

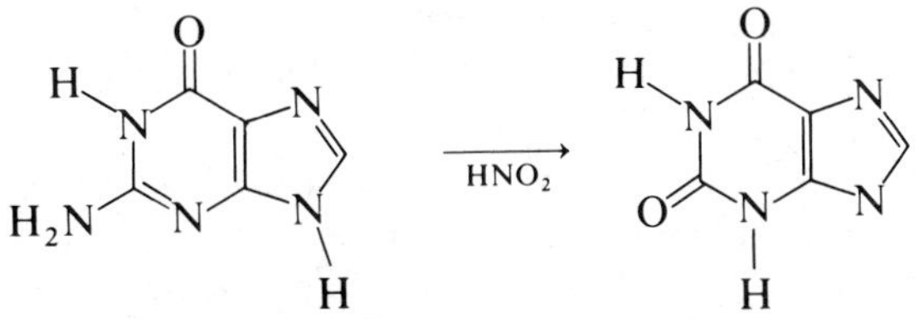

guanine $\xrightarrow{HNO_2}$ xanthine

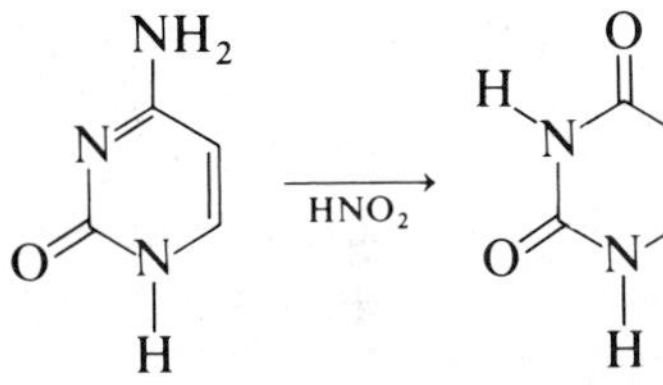

cytosine $\xrightarrow{HNO_2}$ uracil

absorption of tris through the skin. There is probably some oral intake as well. It has not been demonstrated that the levels of tris absorbed from treated sleepwear are sufficient to cause mutations or other harmful effects. However, the possibility is sufficiently high that garments treated with tris have been removed from the market.

18.18 TERATOGENESIS

The study of birth defects caused by foreign agents constitutes the science of **teratology**, a term derived from the French word *terat*, meaning "monster". The four major classes of teratogens are **viruses, radiation, drugs,** and **chemicals** [12]; in this section, environmental chemicals are considered primarily.

Two to three percent of all babies born are afflicted with birth defects. Only 5–10% of birth defects arise solely from the teratogens listed above, whereas about 25% have genetic causes. This leaves 60–65% due to unknown causes, which probably arise from the interaction of genetic and environmental factors. Only about 4–6% of the birth defects caused by environmental agents are due to chemicals. Only about 25 chemicals are known to be teratogenic in humans, whereas 800 animal teratogens are known. Of course, many of the latter are likely human teratogens as well.

The best-known example of a human teratogen is thalidomide, a sedative-hypnotic drug that came into use in Europe and Japan in 1960–61. Taken between

---

[12] Kurzel, R.B., and Cetrulo, C.L. 1981. The effect of environmental pollutants on human reproduction, including birth defects. *Environmental Science and Technology* 15: 626–640.

the 35th and 50th days of pregnancy, it results in the birth of babies with incompletely formed limbs. Approximately 10,000 children with thalidomide birth defects were born in Japan, Europe, and elsewhere.

In addition to thalidomide, other substances known to be teratogenic in humans include excessive doses of vitamin D, androgens, estrogens, and some chemotherapeutic agents used to fight cancer. Of the latter, antimetabolites such as amethopterin and alkylating agents such as busulfan and nitrogen mustards are teratogens.

Tests with animals have shown that a number of environmental pollutants are animal teratogens. These include carbon monoxide (from automobile exhausts and smoking), which may diffuse across the placenta and result in elevated levels of carboxyhemoglobin in the blood of the fetus. Other environmental pollutants that are teratogenic in animals include caffeine (in rats given the equivalent of 12–24 cups of coffee per day), TCDD (see Section 7.18), cadmium, lead, and mercury.

The most dangerous time for fetal exposure to a teratogenic agent is during the period from the 18th through the 55th day of human pregnancy, when organ differentiation is occurring. A single, intense exposure tends to cause defects more readily than the same dose given over several shorter exposures. Exposure of the unborn fetus is through transfer of chemicals across the placenta, whereas exposure of the mother occurs by ingestion, inhalation, and absorption through the skin.

The biochemical mechanisms of teratogenicity are generally unknown. Teratogens come from many classes of chemicals; this is in contrast to mutagens (see Section 18.17), which are generally alkylating agents. It should be noted that, although there is a strong connection between carcinogens and mutagens, teratogens cannot be equated with either of these two classes of substances.

18.19 CARCINOGENESIS

Cancer is uncontrolled cell growth. Chemicals that cause cancer to occur in animals and humans are called **carcinogens**. It is thought that there is a close relationship between mutagenesis and carcinogenesis. Practically all carcinogens are also mutagens, although the reverse has not been shown to be true. Thus, it is believed that carcinogens act upon DNA, perhaps preventing it from giving the necessary directions for the synthesis of substances that control cell growth.

The nature of carcinogenesis is complicated by the fact that, in many cases, cancer results from a single exposure to an **initiator** chemical, followed by continuous exposure to a **promoter chemical**. Table 18.1, published by the Public Health Service of the U.S. Department of Health and Human Services, is a list of substances that are either known to be carcinogens or may reasonably be assumed to be carcinogens and to which a significant number of persons in the U.S. are exposed. Table 18.2 lists those compounds for which good epidemiological evidence exists of their being confirmed human carcinogens. In some cases, the carcinogenicity of these compounds has been established from records of exposed workers who subsequently developed cancer. In other cases, the carcinogenic hazard has been inferred from studies of cancer-susceptible mice or mutagenicity in bacteria, using such tests as the **Bruce Ames test** for screening of *Salmonella typhimurium*.

TABLE 18.1 Known or Strongly Suspected Carcinogens to Which a Significant Number of Persons in the U.S. Are Exposed*

2-Acetylaminofluorene
Acrylonitrile
Aflatoxins
4-Aminobiphenyl
Amitrole
Aramite
Arsenic and certain arsenic compounds
Asbestos
Auramine and the manufacture of auramine
Benz(a)anthracene
Benzene
Benzidine
Benzo(a)pyrene
Benzo(b)fluoranthene and benzo(j)fluoranthene
Beryllium and certain beryllium compounds
N,N-Bis(2-chloroethyl)-2-naphthylamine
Bis(chloromethyl)ether and technical grade chloromethyl methyl ether
Cadmium and certain cadmium compounds
Carbon tetrachloride
Chlorambucil
Chloroform
Chromium and certain chromium compounds
Coke oven emissions
p-Cresidine
Cycasin
Cyclophosphamide
2,4-Diaminotoluene
Dibenz(a,h)acridine
Dibenz(a,j)acridine
Dibenz(a,h)anthracene
7*H*-Dibenzo(c,g)carbazole
Dibenzo(a,h)pyrene
Dibenzo(a,i)pyrene
1,2-Dibromo-3-chloropropane
1,2-Dibromoethane
Dichlorobenzidine
1,2-Dichloroethane
Diethylstilbestrol
4-Dimethylaminoazobenzene
Dimethylcarbamoyl chloride
Dimethyl sulfate
1,4-Dioxane
Formaldehyde
Hematite: underground hematite mining
Hydrazobenzene
Ideno(1,2,3-cd)pyrene
Iron dextran
Isopropyl alcohol manufacturing (strong acid process)
Kepone
Lead acetate and lead phosphate
Lindane and other hexachlorocyclohexane isomers
Melphalan
Mirex
Mustard Gas
2-Naphthylamine
Nickel, certain nickel compounds, and nickel refining
N-Nitrosodi-*N*-butylamine
N-Nitrosodiethanolamine
N-Nitrosodiethylamine
N-Nitrosodimethylamine
N-Nitrosodi-*N*-propylamine
N-Nitroso-*N*-ethylurea
N-Nitroso-*N*-methylurea
N-Nitrosomethylvinylamine
N-Nitrosomorpholine
N-Nitrosonornicotine
N-Nitrosopiperidine
N-Nitrosopyrrolidine
N-Nitrososarcosine
Oxymetholone
Phenacetin
Phenazopyridine hydrochloride
Phenytoxin
Polychlorinated biphenyls
Procarbazine and procarbazine hydrochloride
β-Propiolactone
Reserpine
Saccharin
Safrole
Soots, tars, and mineral oils
Streptozotocin
2,3,7,8-Tetrachlorodibenzo-*p*-dioxin (TCDD)
Thorium dioxide
o-Toluidine hydrochloride
Toxaphene
Tris(1-aziridinyl)phosphine sulfide
Tris(2,3,-dibromopropyl)phosphate
Vinyl chloride

*For a detailed discussion of each of these substances see Public Health Service, U.S. Department of Health and Human Services. 1981. *Second annual report on carcinogens.* Washington, D.C.: U.S. Department of Health and Human Services.

TABLE 18.2 Substances for Which There Is Significant Epidemiological Evidence of Human Carcinogenicity

Substance	*Uses*	*Hazards*
4-Aminobiphenyl	Formerly used as a rubber antioxidant	Exposed workers have shown a high incidence of bladder cancer
Asbestos	Used in almost 5000 different products, such as insulation	Known to cause lung cancer, larynx cancer, and pleural and peritoneal mesotheliomas in exposed individuals
Benzidine	Manufature of dyes, rubber, plastics, printing inks; no longer widely used	Causes bladder cancer
N,N-Bis(2-chloroethyl)-2-naphthylamine	Formerly used as a drug to treat leukemia and related cancers	Causes bladder cancer
Bis(chloromethyl)ether	Chemical intermediate in the manufacture of plastics and ion-exchange resins	Causes lung cancer
Chlorambucil	Used for chemotherapy of some types of cancers	Causes leukemia
Coke oven emissions	By-product of the manufacture of coke from coal	Cause cancer of the lung and urinary tract
Diethylstilbestrol	Growth promoter for cattle and sheep, formerly used as a drug to prevent spontaneous abortions	Causes vaginal cancer in females exposed *in utero*
2-Naphthylamine	Formerly used as an antioxidant and in dye and color film manufacture, now used only in research	Causes bladder cancer
Thorium dioxide	Nuclear reactors, gas lighting mantles, radio-opaque medium for X-ray imaging (no longer used for this purpose)	Specific causative agent for haemangioendothelioma, a form of liver cancer
Vinyl chloride	Polyvinyl chloride manufacture	Causes angiosarcoma, a rare form of liver cancer

The genetics and biochemistry of the Bruce Ames test are rather complicated [13]. Basically, the test depends upon the use of mutant strains of *S. Typhimurium* that do not readily undergo spontaneous reverse mutation, but

[13] Skopeck, T.R.; Liber, H.L.; Krolewski, J.J.; and Thilly, W.G. 1978. Quantitative forward mutation assay in *Salmonella typhimurium* using 8-azaguanine resistance as a genetic marker. *Proceedings of the National Academy of Science (U.S.A.)* 75: 410–14.

which do revert to the wild phenotype upon exposure to wild mutants. The mutation results in the bacteria not being able to synthesize their own histidine, so that they do not grow on a culture lacking histidine. Addition of a chemical mutagen (and, therefore, suspected carcinogen) to the culture will cause some of the bacteria to revert to the wild phenotype, which can synthesize its own histidine, so that viable colonies of bacteria will develop on the histidine-free culture. Extracts of rat liver that contain enzymes capable of metabolizing noncarcinogens to carcinogens are also used in the test.

A specific compound that has been found to cause cancer in exposed workers is vinyl chloride, used to make polyvinyl chloride plastics:

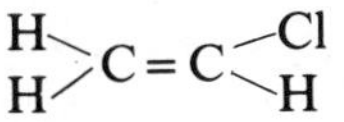

$$\begin{matrix} H \\ H \end{matrix}\!\!>\!C=C\!<\!\!\begin{matrix} Cl \\ H \end{matrix}$$

In the early 1970s, it was found that some workers who had been exposed to vinyl chloride over periods of 14–27 years had developed angiosarcoma, a rare form of liver cancer. Exposure levels of 500 ppm were commonplace in many industrial operations; these levels have been lowered drastically, as required by subsequent regulation.

In September, 1978, the National Cancer Institute released a report showing that ethylene dichloride causes cancer in laboratory animals [14]:

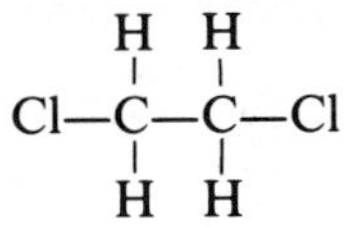

$$\begin{matrix} & & H & & H & & \\ & & | & & | & & \\ Cl & - & C & - & C & - & Cl \\ & & | & & | & & \\ & & H & & H & & \end{matrix}$$

This study was conducted by administering ethylene dichloride immersed in corn oil through a stomach tube directly into the stomachs of rats and mice over an 18-month period. Dose rates ranged from 47–299 mg per kg of body weight per day. This treatment caused stomach, spleen, and subcutaneous cancers in male rats; lung and uterine cancer in mice; and mammary cancers in both female rats and mice. Tests involving inhalation of ethylene dichloride, sponsored by the European chemical industry and the U.S. Manufacturing Chemists Association, have failed to produce cancer in test animals.

Possible carcinogenic activity by ethylene dichloride is of considerable concern because of the widespread use of this chemical as an industrial solvent, dry-cleaning agent, grain fumigant, and gasoline additive for lead scavenging. Only 17% of the approximately 4.75 billion kg of ethylene dichloride manufactured each year ever leaves the manufacturing site; the remainder is used at the site to make other chemicals, primarily vinyl chloride.

A related compound, ethylene dibromide,

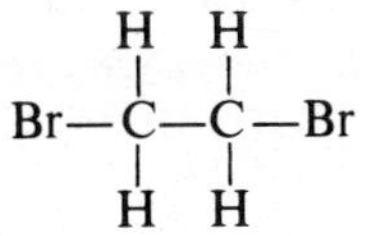

$$\begin{matrix} & & H & & H & & \\ & & | & & | & & \\ Br & - & C & - & C & - & Br \\ & & | & & | & & \\ & & H & & H & & \end{matrix}$$

[14] 1978. Ethylene dichloride cancer issue heats up. *Chemical and Engineering News* 25 September 1978, p. 6.

is suspected of being a carcinogen, mutagen, and teratogen. In late 1981, OSHA proposed lowering the worker exposure limit from 20 ppm to 130 ppb [15]. Ethylene dibromide is used as a fumigant and antiknock compound.

Polycyclic aromatic hydrocarbons constitute another class of carcinogens. These are discussed as air pollutants in Section 13.16. Carcinogenic nitrosamines, such as *N*-ethyl-*N*-nitroso-*N*-butylamine,

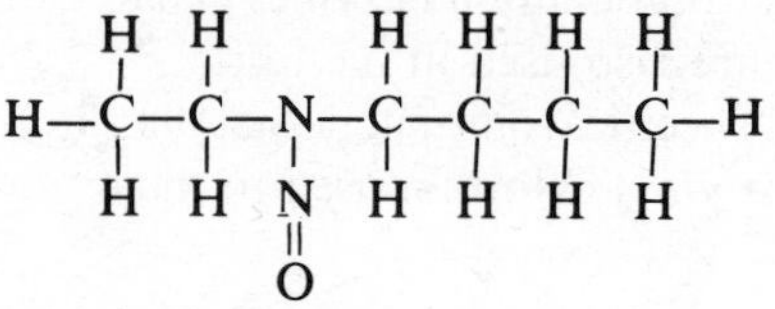

```
  H  H      H  H  H  H
  |  |      |  |  |  |
H—C——C——N——C——C——C——C—H
  |  |  |   |  |  |  |
  H  H  N   H  H  H  H
        ||
        O
```

used in the past as a gasoline and lubricant additive and insecticide, are receiving considerable attention as carcinogens. There is some evidence suggesting that nitrosamines are formed during the cooking of meats containing nitrite additives.

Some carcinogens are produced in nature. The most prominent of these are the aflatoxins, which, for some test organisms, are the most potent known hepatocarcinogens (causing liver cancer). There are ten known aflatoxins. Aflatoxin B_1 has a structure showing the main features of the aflatoxins:

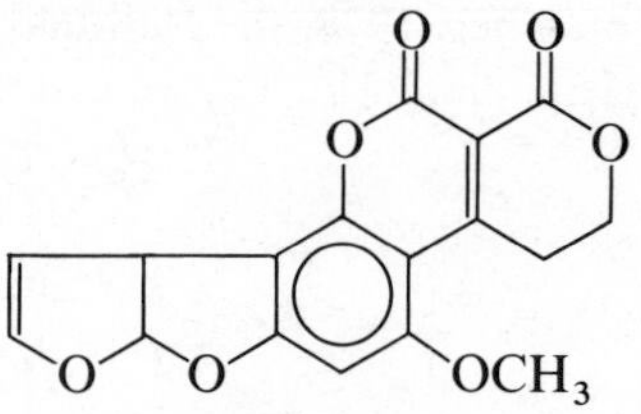

Aflatoxins are produced by some fungi (molds), such as those of the *Aspergillus flavus* subgroup. The biochemical formation of aflatoxins is interesting in that they are secondary metabolites without apparent use to the organism, in contrast to essential primary metabolites such as carbohydrates, proteins, nucleic acids, and lipids. It is believed that secondary metabolites are formed from the precursors of primary metabolites, such as acetate, malonate, pyruvate, and amino acids, when these substances accumulate to excessive levels. Other examples of secondary metabolites are alkaloids, mycotoxins, antibiotics, and pigments.

The tendency of aflatoxin-producing molds to form on food crops is a matter of considerable concern, especially in countries where humid conditions and inadequate grain storage and drying facilities exist. It is known that aflatoxins are produced by fungi growing on corn, peanuts, and coconuts. The aflatoxins persist when these commodities are processed into prepared foods, such as peanut butter.

18.20 HUMAN EXPOSURE TO TOXIC AND HAZARDOUS SUBSTANCES IN THE ATMOSPHERE

A number of chemicals have been designated as hazards to exposed humans, and standards have been set for employee exposure [16]. Table 18.3 is a partial list of

[15] 1982. OSHA to lower ethylene dibromide limit. *Chemical and Engineering News* 4 January 1982, p. 16.

[16] 1977. Selected general industry and health standards. U.S. Department of Labor Occupational Health and Safety Administration. *Federal Register* 42: 62734–62890.

major chemicals for which standards have been designated. Ceiling values have been set for exposure to those chemicals that have a "**C**" preceding the chemical name in the table. Employee exposure may not exceed the ceiling value at any time.

Exposure to chemicals not having a designated ceiling value may not exceed an 8-hour, time-weighted average, assuming a 40-hour work week. In addition to this time-weighted average, compounds listed in Table 18.4 have an acceptable ceiling concentration and an "acceptable maximum peak above the acceptable ceiling concentration for an 8-hour shift." This designates a concentration limit and time limit within which the exposure may exceed the acceptable ceiling concentration during an 8-hour shift. For example, as shown in Table 18.4, the acceptable ceiling concentration of toluene is 300 ppm. However, once during an 8-hour shift, exposure may exceed 300 ppm (but may never be more than 500 ppm) for a maximum exposure time of 10 minutes. This excess exposure must be compensated by a period of exposure to less than 200 ppm so that the 8-hour, time-weighted average will be less than 200 ppm.

The 8-hour, time-weighted average exposure is calculated by the formula

$$E = \frac{C_a T_a + C_b T_b + \cdots + C_n T_n}{8} \qquad (18.20.1)$$

in which the value of the contaminant concentration C_x is constant over a time period T_x (in hours). For example, exposure to 50 ppm acetone for 3 hours, 225 ppm for 4 hours, and 75 ppm for 1 hour gives an E value of 140.6 ppm, well below the 8-hour, time-weighted average limit of 200 ppm.

More detailed regulations have been written for the control of a number of substances that are known or suspected carcinogens [16]. These include asbestos, 4-nitrobiphenyl, alphanaphthylamine, 4,4′-methylene bis(2-chloroaniline), methyl chloromethyl ether, 3,3′-dichlorobenzidine, bis-chloromethyl ether, beta-naphthylamine, 4-aminodiphenyl, ethyleneimine, beta-propiolactone, 2-acetylaminofluorene, 4-dimethylaminoazobenzene, *N*-nitrosodimethylaniline, and vinyl chloride. Most of these compounds are listed in Table 18.1. These regulations specify maximum exposures, special equipment, warning signs, and many other aspects regarding the handling of carcinogens.

Formaldehyde is a very common chemical over which considerable controversy has arise in recent years. Annual U.S. production of this versatile chemical is about 2.5 billion kilograms, with a value of about $300 million. It is used in the manufacture of industrial chemicals, particle board, plywood, polyacetal resins, insulation, hard plastics, fiberboard, paints, lacquers, paper, paperboard, and textiles. Most of the problems associated with formaldehyde have resulted from improperly formulated and installed urea-formaldehyde foam insulation. This can lead to release of formaldehyde into a home [17]. Some people thus exposed become sensitized to the chemical, so that the slightest subsequent exposure from sources such as permanent-press clothing or cigarette smoke can cause the subject to become ill. In an animal study of formaldehyde carcinogenicity it was found that nearly 40% of rats continuously exposed to an atmosphere of 15 parts per million of formaldehyde developed nasal cancer. However, it should be noted that these are levels that humans cannot tolerate for even a short period of time, much less for a lifetime. Studies now underway of

---

[17] Lewin, T. 1982. Insulation lawsuits abound. *New York Times* 25 May, 1982.

TABLE 18.3 Maximum Exposure to Chemicals in the Atmosphere[1]

Substance[2]	*Parts per million by volume of vapor in air at 25 °C, 1.00 atm pressure (ppm)*	*Milligrams per cubic meter of air (mg/M)*[3]
Acetaldehyde	200	360
Acetone	1000	2400
Ammonia	50	35
Arsenic and compounds (as As)	——	0.5
Bromine	0.1	0.7
Calcium oxide	——	5
Carbaryl (Sevin®)	——	5
Carbon black	——	3.5
Carbon dioxide	5000	9000
Carbon monoxide	50	55
Chlorine	0.1	0.3
Coal tar pitch volatiles (benzene-soluble fraction), anthracene, BaP, phenanthrene	——	0.2
Cobalt, metal fume and dust	——	0.1
Cotton dust (raw)	——	1
2,4-D	——	10
DDT-skin	——	1
1,2-Dichloroethylene	200	790
Dinitrobenzene (all isomers)—skin exposure	——	1
Dioxane (diethylene oxide)—skin exposure	100	360
Ethyl acetate	400	1400
Ethyl alcohol (ethanol)	1000	1900
C Ethylene glycol dinitrate and/or nitroglycerin—skin exposure	0.2	1
Fluoride (as F)	——	2.5
Fluorine	0.1	0.2
Heptachlor—skin exposure	——	0.5
Heptane (*n*-heptane)	500	2000
C Hydrogen chloride	5	7
Hydrogen peroxide (90%)	1	1.4
C Iodine	0.1	1
Lead[3]	——	0.050
Methyl alcohol (methanol)	200	260
Nickel carbonyl	0.001	0.007
Osmium tetroxide	——	0.002
Phosphorus (yellow)	——	0.1
Sodium hydroxide	——	2
Sulfur dioxide	5	13
Tetraethyl lead (as Pb)—skin exposure	——	0.07
Turpentine	100	560
Uranium (soluble compounds)	——	0.05

[1]Partial list from Reference 16. See also American Conference of Governmental Industrial Hygienists. 1983. *Threshold limit values for chemical substances and physical agents in the work environment.* Cincinnati, Ohio: ACGIH.

[2]Levels of compounds preceded by a "**C**" (to designate *ceiling*) may not at any time exceed those listed in the table.

[3]As of February 1, 1979, a "medical removal" provision requires removal of a worker from lead exposure when blood-lead levels exceed prescribed levels.

TABLE 18.4 Maximum Exposure to Chemicals in the Atmosphere, Including the Acceptable Maximum Peak Above the Acceptable Ceiling Concentration (from Reference 16)

Substance	*8-Hour, time-weighted average*	*Acceptable ceiling concentration*	*Acceptable maximum peak above acceptable ceiling concentration for an 8-hour shift*	
			Concentration	*Maximum duration*
Benzene	10 ppm	25 ppm	50 ppm	10 min
Beryllium and compounds	2 $\mu g/m^3$	5 $\mu g/m^3$	25 $\mu g/m^3$	30 min
Carbon tetrachloride	10 ppm	25 ppm	200 ppm	5 min in any 4 hrs
Ethylene dibromide	20 ppm	30 ppm	50 ppm	5 min
Formaldehyde	3 ppm	5 ppm	10 ppm	30 min
Methyl chloride	100 ppm	200 ppm	300 ppm	5 min in any 3 hrs
Methylene chloride	500 ppm	1000 ppm	2000 ppm	5 min in any 2 hrs
Styrene	100 ppm	200 ppm	600 ppm	5 min in any 3 hrs
Toluene	200 ppm	300 ppm	500 ppm	10 min

workers who have experienced long-term exposure to formaldehyde should answer the question of possible carcinogenicity in humans.

The now-banned pesticide, 1,2-dibromo-3-chloropropane (DBCP) has been shown to cause male sterility. In April, 1983, a California state court ruled that workers manufacturing this pesticide had not been adequately warned of its dangers.

Sulfites and sulfur dioxide—used for centuries to perserve food and maintain its fresh appearance, as well as to sanitize food containers—have recently been suspected of causing severe allergic reactions, particularly among asthmatics. As a consequence, in March, 1983, the U.S. Food and Drug Administration recommended state regulation of sulfite preservatives. About 3 million kilograms of sulfites are used in the U.S. each year in the preservation of 20 categories of food and beverages, such as wine, baked products, and grain.

In August, 1982, the Consumer Product Safety Commission banned the use of formaldehyde-containing insulation in homes and schools. In April, 1983, a Federal Appeals Court rescinded the ban, ruling that it was based on insufficient evidence.

Another chemical to which workers are commonly exposed is ethylene oxide. This gas is used as a pesticide and in hospitals as a sterilizing agent. In April 1983, the U.S. Occupational Safety and Health Administration proposed lowering the maximum level of exposure to this gas from 50 to 1 ppm in air.

The growing awareness of chemical hazards will have a profound effect upon all professional chemists in the future. It is in the best interest of all chemistry students to acquire a basic knowledge of chemical toxicology. Such an awareness will certainly be required of all chemists, as well as other people working with chemicals during their professional careers.

SUPPLEMENTARY REFERENCES

Aizawa, H. 1982. *Metabolic maps of pesticides.* New York: Academic Press, Inc.

Anderson, K., and Scott, R. 1981. *Fundamentals of industrial toxicology.* Ann Arbor, Mich.: Ann Arbor Science Publishers, Inc.

Arcos, J.C.; Woo, Y.; and Argus, M.F. 1982. *Chemical induction of cancer, structural bases and biological mechanisms.* Vol. 3A: *Aliphatic hydrocarbons.* New York: Academic Press, Inc.

Beroza, M., ed. 1976. *Pest management with insect sex attractants.* ACS Symposium Series No. 23. Washington, D.C.: American Chemical Society.

Boland, B.M., and Lehmann, P.E., eds. 1977. *Cancer and the worker.* New York: New York Academy of Sciences.

Bowman, M.C. 1979. *Cacinogens and related substances.* New York: Marcel Dekker, Inc.

——, ed. 1982. *Handbook of carcinogens and hazardous substances.* New York: Marcel Dekker, Inc.

Burgess, W.A., ed. 1981. *Recognition of health hazards in industry.* New York: Wiley-Interscience.

Choudhary, G., ed. 1981. *Chemical hazards in the workplace.* ACS Symposium Series No. 149. Washington, D.C.: American Chemical Society.

Clayton, G.D., and Clayton, F.E. 1978. *Patty's industrial hygiene and toxicology.* 3rd revised ed. Vol. 1: *General principles.* New York: Wiley-Interscience.

——. 1981. *Patty's industrial hygiene and toxicology.* 3rd ed. Vol. 2A: *Toxicology.* New York: Wiley-Interscience.

——. 1981. *Patty's industrial hygiene and toxicology.* 3rd ed. Vol. 2B: *Toxicology.* New York: Wiley-Interscience.

——. 1982. *Patty's industrial hygiene and toxicology.* Vol. 2C. New York: Wiley-Interscience.

Committee on Medical and Biologic Effects of Environmental Pollutants. 1979. *Ammonia.* National Research Council. Baltimore, Md.: University Park Press.

——. 1977. *Carbon monoxide.* National Research Council. Washington, D.C.: National Academy of Sciences.

Council on Environmental Quality. 1981. *Chemical hazards to human reproduction.* Washington, D.C.

Cralley, L.J., and Cralley, L.V. 1979. *Patty's industrial hygiene and toxicology.* Vol. 3: *Theory and rationale of industrial hygiene practice.* New York: Wiley-Interscience.

De Bruin, A. 1976. *Biochemical toxicology of environmental agents.* Amsterdam, The Netherlands: Elsevier.

de Serres, F.J., and Hollaender, A. 1982. *Chemical mutagens: principles and methods for their detection.* Vols. 6 and 7. New York: Plenum Publishing Corp.

de Serres, F.J., and Shelby, M.D. 1982. *Comparative chemical mutagenesis.* New York: Plenum Publishing Corp.

Fedtke, C. 1982. *Biochemistry and physiology of herbicide action.* New York: Springer-Verlag New York, Inc.

Filov, B.A. 1980. *Quantitative toxicology.* New York: John Wiley and Sons, Inc.

Fishbein, L. 1979. *Potential industrial carcinogens and mutagens.* New York: Elsevier North-Holland, Inc.

Flamm, W.G., and Mehlman, M.A., eds. 1978. *Mutagenesis.* New York: Halsted (John Wiley and Sons, Inc.).

Gelboin, H.V., and Ts'o, P.O.P. 1978. *Polycyclic hydrocarbons and cancer.* New York: Academic Press, Inc.

Gorman, J. 1979.. *Hazards to your health.* New York: New York Academy of Sciences.

Haque, R., ed. 1980. *Dynamics, exposure, and hazard assessment of toxic chemicals.* Ann Arbor, Mich.: Ann Arbor Science Publishers, Inc.

Hedin, P.A., ed. 1977. *Host plant resistance to pests.* ACS Symposium Series No. 62. Washington, D.C.: American Chemical Society.

Hutson, D.H., and Roberts, T.R., eds. 1981. *Progress in pesticide biochemistry.* Vol. 1. New York: John Wiley and Sons, Inc.

Ivie, G.W., and Dorough, H.W., eds. 1977. *Fate of pesticides in large animals.* New York: Academic Press, Inc.

Khan, M.A.Q.; Lech, J.J.; and Menn, J.J., eds. 1979. *Pesticide and xenobiotic metabolism in aquatic organisms.* ACS Symposium Series No. 99. Washington, D.C.: American Chemical Society.

Klass, D.L. 1981. *Biomass as a nonfossil fuel source.* ACS Symposium Series No. 144. Washington, D.C.: American Chemical Society.

Koren, H. 1980. *Handbook of environmental health and safety.* Elmsford, N.Y.: Pergamon Press, Inc.

Lave, L.B., and Seskin, E.P. 1978. *Air pollution and human health.* Baltimore, Md.: Johns Hopkins University Press.

Lee, S.D. 1977. *Biochemical effects of environmental pollutants.* Ann Arbor, Mich.: Ann Arbor Science Publishers, Inc.

——, ed. 1980. *Nitrogen oxides and their effects on health.* Ann Arbor, Mich.: Ann Arbor Science Publishers, Inc.

Lee, S.D., and Mudd, J.B., eds. 1979. *Assessing toxic effects of environmental pollutants.* Ann Arbor, Mich.: Ann Arbor Science Publishers, Inc.

1979. *Long-term hazards from environmental chemicals.* London: The Royal Society.

McRae, A.; Whelchel, L.; and Rowland, H., eds. 1978. *Toxic substances control source-book.* Germantown, Md.: Aspen Systems Corp.

Matsumura, F. 1975. *Toxicology of insecticides.* New York: Plenum Publishing Corp.

Mehlman, M.A.; Shapiro, R.E.; and Blumenthal, H., eds. 1979. *Advances in modern toxicology.* New York: Halsted (John Wiley and Sons, Inc.).

Mennear, J.H., ed. 1979. *Cadmium toxicity.* New York: Marcel Dekker, Inc.

National Academy of Sciences. 1981. *Health effects of exposure to diesel exhaust.* Washington, D.C.: National Academy Press.

National Research Council. 1979. *Zinc.* Baltimore, Md.: University Park Press.

Newberne, P.M., ed. 1982. *Trace substances and health, part 2.* New York: Marcel Dekker, Inc.

Nisbet, C.T., and Karch, N.J. 1983. *Chemical hazards to human reproduction.* Park Ridge, N.J.: Noyes Data Corp.

Nriagu, J.O. 1981. *Cadmium in the environment.* Part 2: *Health effects.* New York: Wiley-Interscience.

——, ed. 1978. *The biogeochemistry of lead in the environment.* New York: Elsevier North-Holland, Inc.

Ochiai, E.-I. 1977. *Bioinorganic chemistry.* Boston: Allyn & Bacon, Inc.

Oehme, F.W., ed. 1978. *Toxicity of heavy metals in the environment.* New York: Marcel Dekker, Inc.

Paulson, G.D.; Frear, D.S.; and Marks, E.P., eds. 1979. *Xenobiotic metabolism: in vitro methods.* ACS Symposium Series No. 97. Washington, D.C.: American Chemical Society.

Plunkett, E.R. 1976. *Handbook of industrial toxicology.* New York: Chemical Publishing Co., Inc.

Preger, L., et al. 1978. *Asbestos-related diseases.* New York: Grune & Stratton, Inc.

1978. *Principles and methods for evaluating the toxicity of chemicals.* Part 1. Albany, N.Y.: WHO Publications Centre USA.

Purdom, P.W., ed. 1980. *Environmental health.* 2nd ed. New York: Academic Press, Inc.

Rao, R.K. 1979. *Pentachlorophenol: chemistry, pharmacology, and environmental toxicology.* New York: Plenum Publishing Corp.

Reeves, A.L., ed. 1981. *Toxicology: principles and practice.* Vol. 1. New York: John Wiley and Sons, Inc.

Scanlan, R.A., and Tannenbaum, S.R., eds. 1982. *N-nitroso compounds.* Washington, D.C.: American Chemical Society.

Schuetzle, D., ed. 1979. *Monitoring toxic substances.* ACS Symposium Series No. 94. Washington, D.C.: American Chemical Society.

Selikoff, I.J., and Lee, D.H.K. 1978. *Asbestos and disease.* New York: Academic Press, Inc.

Shepard, T.H. 1980. *Catalog of teratogenic agents.* 3rd ed. Baltimore, Md.: Johns Hopkins University Press.

Slein, M.W., and Sansone, E.B. 1980. *Degradation of chemical carcinogens: an annotated bibliography.* New York: Van Nostrand Reinhold Co.

Sontag, J.M., ed. 1981. *Carcinogens in industry and the environment.* New York: Marcel Dekker, Inc.

Stutte, C.A., ed. 1977. *Plant growth regulators: chemical activity, plant responses, and economic potential.* ACS Advances in Chemistry Series No. 159. Washington, D.C.: American Chemical Society.

Sugimura, T.; Kondo, S.; Takebe, H., eds. 1982. *Environmental mutagens and carcinogens.* New York: Tokyo and Liss.

Tsuchiya, K.; Sted, S.K.; and Hamagami, C.M., eds. 1978. *Cadmium studies in Japan.* New York: Elsevier North-Holland, Inc.

Verschueren, K. 1977. *Handbook of environmental data on organic chemicals.* New York: Van Nostrand Reinhold Co.

Waldbott, G.L. 1978. *Health effects of environmental pollutants.* 2nd ed. St. Louis, Mo.: C.V. Mosby Co.

Waldbott, G.L.; Burgstahler, A.W.; and McKinney, H.L. 1978. *Fluoridation: the great dilemma.* Lawrence, Kansas: Coronado Press.

Walters, D.B., ed. 1980. *Safe handling of chemical carcinogens, mutagens, teratogens, and highly toxic substances.* Ann Arbor, Mich.: Ann Arbor Science Publishers, Inc.

Weber, L.J., ed. 1982. *Aquatic toxicology.* New York: Raven Press.

Whiteside, T. 1979. *The pendulum and the toxic cloud: the course of dioxin contamination.* New Haven, Conn.: Yale Univ. Press.

Winek, C.L., and Shanor, S.P., eds. 1979. *Toxicology annual.* Published annually since 1977. New York: Marcel Dekker, Inc.

Zajic, J.E., and Himmelman, W.A. 1978. *Highly hazardous materials spills and emergency planning.* New York: Marcel Dekker, Inc.

QUESTIONS AND PROBLEMS

1. What role do you think a cell's membrane might play in determining the toxicity of a substance relative to that cell?
2. What is the distinction between a cell wall and a cell membrane?
3. What are two characteristics of enzymes that determine their function and are susceptible to alteration by toxic substances?
4. What is DNA, and how is it important in the action of some toxic substances?
5. Using a chemical reaction, show how you might expect Pb^{2+} to interact with an enzyme and act as an inhibitor. At what location on the enzyme is this interaction especially harmful?
6. What sort of biochemical effect may be indicated by the abnormal accumulation of a metabolite in the body, or by excessive excretion of it in urine?
7. Recalling material covered in Chapter 4 dealing with chelating agents in natural waters, explain why EDTA might be used as an antidote for certain toxic substances. Why, in such an application, is EDTA normally administered as a calcium salt?
8. Speculate regarding the relative permeabilities to elemental mercury of lung cell membranes and gastrointestinal-tract membranes.
9. Which toxic substances discussed in this chapter is an antidote for another toxic substance?
10. What are two toxic substances that bind to hemoglobin?

CHAPTER 19
THE NATURE AND SOURCES OF HAZARDOUS WASTES

19.1 LOVE CANAL

The name **Love Canal** has come to symbolize the problem of hazardous chemical wastes [1]. The Love Canal story had its beginnings in the late 1800s with a scheme promoted by the flamboyant William T. Love. His idea was to build a seven-mile-long canal from the upper Niagara River at Niagara Falls, New York. The canal would be used for navigation and would terminate in a large hydroelectric generating station, where the water would be emptied back into the Niagara River. A city was to be constructed at this site, with an eventual population of up to a million people, supported by industry attracted by inexpensive hydroelectric power. Showered with accolades, Mr. Love was given authority to take all the water his canal could divert and to condemn all the land needed for the project. Like so many gradiose projects of that era, Love's dream rather quickly came to financial grief, leaving a trench ranging in depth from about 3 to 13 meters, approximately 15 meters wide, and 1.6 kilometers (1 mile) long.

For several decades following its excavation, the abandoned Love Canal remained as a harmless feature of the southeastern Niagara Falls landscape. One of its most popular uses was for skating during the long, cold winters. In the meantime, however, the availability of cheap hydroelectric power was changing the character of the area. First produced at the Niagara Falls site in 1881, hydroelectric power rapidly developed in the area, its growth spurred by the huge quantities of water available and the 57-meter drop of the falls. Inexpensive electricity gave rise to a large chemical industry based upon electrochemical processes. Eventually, large quantities of chlorine, sodium hydroxide, aluminum, and ferroalloys were produced in the area. This provided a solid industrial base for the area's economy, although it created a number of unfortunate environmental side effects.

Starting around 1940, Hooker Chemical Company, a manufacturer of caustic soda, organochlorine pesticides, plasticizers, and other chemicals, began using Love Canal for the disposal of chemical waste by-products. During the 1940s the canal received about 20,000 metric tons of chemical wastes. Prominent among these were *still bottoms*, residues left from the distillation of various

[1] Brown, M.H. 1981. *Laying waste: the poisoning of America by toxic chemicals.* New York: Washington Square Press.

chemicals. One of the major chemicals that were distilled was hexachlorocyclopentadiene,

```
 Cl          Cl
   \        /
    C------C
    ||  Cl  ||
    C   |   C
   / \  |  / \
 Cl    \C/    Cl
        |
        Cl
```

an intermediate in the manufacture of the cyclodiene class of insecticides, including chlordane (see Table 7.8). Both it and impurities left from its distillation were found in Love Canal wastes. Eventually, at least 80 different chemicals were dumped at the site.

In 1953 the Love Canal dump site was deeded to the Niagara Falls Board of Education, and a school was constructed on the property. Housing developments grew around the site, attracted in part by the nearby school. During the 1950s there were some reports of eye and respiratory tract irritation among children playing on the playground atop the waste dump, and even some reports of chemical burns to children from chemicals seeping up through the soil cover atop the dump site.

During the mid-1970s, heavy precipitation caused the Love Canal dump site to become thoroughly saturated with water, so that by the winter of 1977–1978 parts of the area were oozing masses of chemically contaminated sludge. In some locations, old barrels that had originally contained the waste chemicals floated through the softened ground cover above the dump site. Noxious chemical leachate was found in yards and basements of surrounding homes. Various maladies, including chloracne (a skin disorder that can result from exposure to noxious chemicals), birth defects, and an elevated level of miscarriages were reported among nearby residents.

In February, 1978, an Environmental Protection Agency study showed the presence of 26 synthetic organic compounds in the basements of some of the first ring of homes around the Love Canal site (see Figure 19.1). In August of that

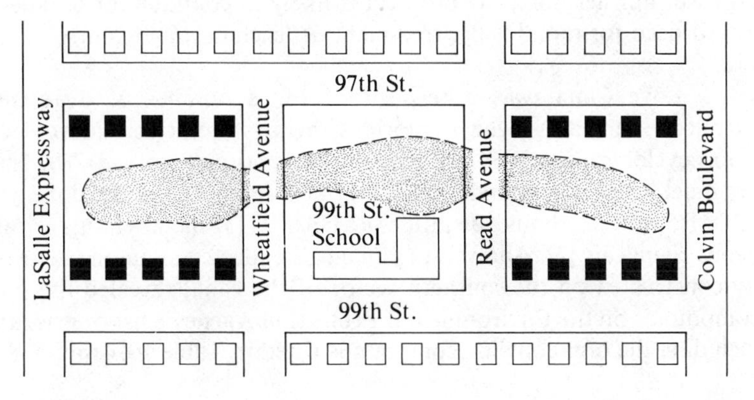

FIGURE 19.1 Map of Love Canal area: --- boundary of canal; ■ first ring of homes; □ second ring of homes.

same year, a more detailed study identified 82 chemical contaminants, 11 of them suspected human carcinogens. At that time, President Carter declared the southern section of the canal area a disaster area, and plans were made to evacuate 239 families. By May, 1980, an additional 710 families had been declared eligible for relocation assistance.

As of late 1981, more than $80 million in federal, state, and local money had been spent on the Love Canal problem. Since 1979, access to the 235 first-and second-ring houses had been largely prohibited, and these homes were enclosed in an area surrounded by a barb-wire-topped, chain-link fence. The 100 homes in the inner ring were scheduled to be razed. Just outside the fence enclosing the canal site and the first two rings of homes there were 558 homes and a housing project in an area where some dwellings were boarded up and others were still occupied. A total of $15 million in state and federal funds were spent to buy dwellings in the third ring, and about 400 families moved out. About 150 dwellings remained in the third ring, and approximately 50 owners had refused to move.

By 1981, efforts to control leachate from Love Canal had resulted in appreciably lower levels of ground water pollutants. The Environmental Protection Agency had committed an additional $4 million of Superfund money to clean up waste chemical pollution in the Niagara Falls area. (The Superfund consists of money specifically designated by federal law for cleaning up waste chemical sites.)

On July 14, 1982, the Environmental Protection Agency announced the results of a study declaring that the area beyond the first two rings of homes surrounding the dump site was safe for humans. This finding would enable the State of New York to go ahead with plans to rehabilitate the region from one and a half to six blocks from the dump site. This area had 400 empty homes which could be resold if safety standards were met.

The federal study released by the EPA further declared that waste chemicals had been controlled by a clay cap placed over the dump and a drainage system constructed around it. The report also indicated that there was no evidence that *swales*—shallow underground channels—had served as routes for the leakage of chemicals from the dump.

From the above discussion it is clear that, in addition to possible health effects, the Love Canal problem has caused great personal anguish and tremendous expense. Furthermore, the problem is likely to continue for decades with a continued need for monitoring, treatment of leachates, and control of access to the site.

Love Canal was a respository for a number of different hazardous chemicals, including benzyl chlorides, benzoyl chloride, chlorobenzenes, hexachlorocyclohexane (BHC, Lindane), trichlorophenol (TCP), miscellaneous organochlorine compounds, acid chlorides, thionyl chloride, dodecyl mercaptans (DDM), sulfides, liquid disulfides (LDS/MCT), metal chlorides, and miscellaneous chemicals [2]. Analysis of chemicals collected at an on-site treatment facility constructed on the southern sector of the canal revealed the presence of 27 compounds on the Environmental Protection Agency's list of priority pollutants, including the organohalide compounds listed in Table 19.1 (pp. 538–9). In addi-

[2] Ember, L.R. 1980. Uncertain science pushes Love Canal solutions to political, legal arenas. *Chemical and Engineering News* 11 August 1980, pp. 22–9.

tion, the analyzed samples revealed appreciable levels of antimony, arsenic, cadmium, copper, chromium, lead, nickel, selenium, silver, and zinc.

A number of organic compounds have been reported near other waste sites in the Niagara Falls area. One investigation [3] dealt with three sampling sites in Niagara Falls, all of which drain into the Niagara River. These sites are the 102nd Street dump site; Bloody Run Creek, described as "little more than a drainage ditch for the Hyde Park landfill" [3]; and Gill creek, which traverses an industrial complex. Among the compounds from these sites identified by gas chromatography—mass spectrometry (GC—MS) were phenols, trichlorophenol, PCB's, polycyclic aromatic hydrocarbon derivatives, chlorobenzenes, and chlorotoluenes. Some benzyl derivatives were also found, which were attributed to by-products of benzyl chloride manufacture. Some "unusual fluorinated compounds" were also found, which were attributed to (trifluoromethyl)chlorobenzene manufacture.

19.2 THE HAZARDOUS WASTE THREAT

Although Love Canal is the most publicized waste disposal site in the U.S., it is not necessarily the biggest or most hazardous one. Some of these sites are of long standing. For example, chemical companies, glue-making factories, and tanneries were in operation in Woburn, Massachusetts, as far back as the 1850s. In those times, waste hides were dumped in back of tanneries, in piles that were sometimes over 13 meters high. Toxic arsenic, chromium, and lead left over from the treatment of hides and other uses were dumped into pools near the factories that used them. This long-term, improper disposal of chemicals has left Woburn with a 60-acre area of contaminated land as a legacy, as well as two contaminated wells which had to be disconnected from the city's water supply. In differing degrees, the same kinds of problems are faced by many communities in the U.S.

On the other side of the nation is another major hazardous waste depository, the Stringfellow Acid Pits, containing the by-products of modern, post-World War II technology. Located in a canyon near Riverside, California, about 50 miles east of Los Angeles, this 22-acre site received about 130 million liters of wastes from 1956 to 1972. This is about six times the quantity of hazardous chemicals dumped at the Love Canal site. Almost one-fifth of the waste chemicals placed in the Stringfellow site came from a company that was once the nation's largest manufacturer of DDT, during the time before that pesticide was banned in the U.S. Approximately 30,000 people live in the ground water basin that could be contaminated by the Stringfellow Pits. During heavy rains in the spring of 1983, the removal of effluent from sumps downslope from the pits had to be increased from 20,000 to 160,000 liters per day. In April, 1983, the U.S. and California governments brought suit against 18 of the 224 industrial concerns and public agencies that have used the pits to discard wastes, for the payment of cleanup expenses, which may amount to as much as $40 million.

As Love Canal faded from the headlines in 1982 as a symbol of hazardous waste threats, the names "dioxin" and "Times Beach" became headline items

[3] Elder, V.A.; Proctor, B.L.; and Hites, R.A. 1981. Organic compounds found near dump sites in Niagara Falls, New York. *Environmental Science and Technology* 15: 1237–43.

TABLE 19.1 Compounds on the EPA List of Priority Pollutants Collected at Love Canal Leachate Treatment Facility

Compound name	*Compound structure*
Hexachloro-1,3-butadiene	$Cl_2C{=}C(Cl){-}C(Cl){=}CCl_2$
1,2,4-Trichlorobenzene	$C_6H_3Cl_3$
Hexachlorobenzene	C_6Cl_6
α-, β-, and γ-benzene hexachloride (hexachlorocyclohexane)	$C_6H_6Cl_6$
Heptachlor	$C_{10}H_5Cl_7$
Phenol	C_6H_5OH
2,4-Dichlorophenol	$C_6H_3Cl_2OH$
Methylene chloride (dichloromethane)	CH_2Cl_2

Compound name	*Compound structure*
1,1-Dichloroethylene	$Cl_2C{=}CH_2$
Chloroform	$Cl{-}CCl(H){-}Cl$ ($CHCl_3$)
Carbon tetrachloride	CCl_4
Trichloroethylene	$Cl_2C{=}CClH$
Dibromochloromethane	$Br{-}C(Cl)(H){-}Br$
Chlorobenzene	$C_6H_5{-}Cl$
1,1,2,2-Tetrachloroethylene	$Cl_2C{=}CCl_2$

symbolizing hazardous wastes in 1983. The term *dioxin* refers to 2,3,7,8-tetrachloro-*p*-dioxin, or TCDD (see Section 7.18), a by-product of the manufacture of 2,4,5-T herbicide and other chlorinated benzene compounds. Times Beach is a flood-prone community southwest of St. Louis, Missouri, in which high levels of soil-TCDD were found during the winter of 1982–83; this discovery compounded the misery caused by two disastrous floods that took place earlier that winter. The TCDD contamination stemmed from the spraying of TCDD-contaminated waste oil on area dirt roads in the early 1970s, in an attempt to control dust. At that time some horses, dogs, and birds died when several local horse arenas were sprayed with the same TCDD-contaminated oil. However, it was not until several years later that the connection was made between the deaths of these animals and their exposure to TCDD. As of mid-1983 a number of TCDD-contaminated sites resulting from the spraying of contaminated oil had been found in Missouri. However, the hazard from TCDD strongly bound to soil has not yet been thoroughly assessed.

By the early 1980s the hazardous waste problem had become the most pressing environmental issue in the U.S. Numerous studies of waste chemical disposal sites have been published. The Environmental Protection Agency has documented 170,000 surface waste impoundments, some of which are hazardous. These impoundments include waste-holding areas from sources such as agricultural runoff and septic-tank cleaning operations, as well as assorted ponds, pits, and lagoons holding wastes [4]. These include approximately 2100 industrial waste sites, 250 of which may pose immediate hazards to public drinking water supplies through contamination of ground water.

The Environmental Protection Agency has assigned a "hazard rating" to various hazardous waste disposal sites [5]. Love Canal ranked only twenty-fifth in the nation, with a 60.96 rating. Typical of the more hazardous sites is an old mining area covering parts of northeastern Oklahoma, Kansas, and Missouri, with a hazard rating of 78.97. This area is the forty-square-mile Tar Creek drainage basin in which the Eagle-Picher Company conducted extensive zinc and lead mining operations from the early 1900s until the mid-1960s. Like most sites with a high hazard rating, it threatens to contaminate drinking water supplies.

The threat of hazardous waste disposal sites is nationwide. This may be seen by examining the geographical distribution of 250 sites believed to pose the greatest threat to drinking water supplies because they are unlined, are within 1 mile of a public water supply well, and are believed to contain hazardous chemical wastes [4]. The state with the most such sites is Florida, with 54. South Carolina, Ohio, Pennsylvania, and Connecticut have 23, 21, 20, and 18, respectively. Other states with appreciable numbers are Mississippi (14), Missouri (13), Michigan (11), Tennessee (10), Arizona (9), and Indiana (8). New York and New Jersey have 6 each, California and Massachusetts have 4 each, and Louisiana has 3. States having two such sites include Colorado, Kansas, Iowa, Maryland, North Carolina, Wisconsin, and West Virginia, whereas those having one each are Minnesota, New Mexico, Vermont, Washington, Delaware, Illinois, Kentucky, Montana, Oregon, Texas, Virginia and Wyoming.

Most waste sites are from souces other than the chemical industry. Oil and gas brine pits make up the single greatest source of waste sites and impoundments. In generally declining order of occurrence are municipal, industrial, coal-mining, agricultural, miscellaneous, and noncoal-mining sources of impoundments and waste sites.

Many case studies have been published on hazardous waste disposal sites and the threats they pose to the environment. From the preceding discussion it should be clear that there are many locations where hazardous materials have been disposed of without proper regard for environmental safety. Despite stringent new laws regulating the disposal of hazardous materials, there remain vast quantities of such materials improperly dumped in a number of places around the nation, some of which may not have been located yet. These sites represent problems that will take decades to solve. As of 1976, when Congress passed the Resource Conservation and Recovery Act (RCRA), EPA estimated that 35 billion kilograms of hazardous waste were being produced annually and estimated that

[4] 1980. Waste sites pose risk to water supply. *Chemical and Engineering News* 6 October 1980, p.6.

[5] 1981. Three-state waste site called nation's worst. *New York Times* 14 November 1981.

only 10% of this amount was being disposed of in an environmentally acceptable manner [6]. In August, 1983, on the basis of data supplied by waste producers and handlers, this figure was raised to 150 billion kilograms.

There are about 14,000 abandoned hazardous waste sites in the U.S., according to the Environmental Protection Agency [7]. As of September, 1983, the EPA had slated 546 of these for priority cleanup using funds raised by taxes on the petroleum and chemical industry levied by the 1980 Superfund law. According to the EPA the 10 most dangerous abandoned hazardous waste dumps are those listed in Table 19.2. These have the highest priority for cleanup.

In a settlement announced on June 8, 1983, the FMC Corporation agreed to spend $6 million to clean up the Fridley, Minnesota site, ranked as the nation's most hazardous. The company agreed to finance the cleanup despite the fact that most of the wastes were placed in the dump before FMC acquired the facility in 1964. The renovation process is to entail digging a 56,000-cubic-meter vault, which will be used to store all soil containing more than 1 ppm trichloroethylene, the predominant contaminant at the site. (Some small plots of soil at the Fridley site contain up to 200,000 ppm (20%) trichloroethylene.) Leakage of pollutants from the vault will be prevented by a thick clay floor, already present at the site; plastic linings; and a plastic cap covered by 2-meter-thick layer of low-permeability clay. The contaminated soil will be treated by sumps, filters, and other purification apparatus installed in and around the vault. Even with this set-up, it may take as long as 15 years before the contaminants in the soil are reduced to acceptable levels.

19.3 WHAT ARE HAZARDOUS WASTES?

Hazardous wastes consist of individual waste materials or combinations of wastes that are presently or potentially dangerous to humans or other living organisms. Such wastes may pose a direct danger, such as explosion, fire, or direct toxicity. They may have a tendency to cause cumulative detrimental effects, such as birth defects. The toxicity of the wastes may increase in passing through various biological systems (biomagnification). Many wastes considered to be hazardous do not degrade, thus persisting in the environment.

The issuance and enforcement of regulations to protect human health and the environment from improper management and disposal of hazardous wastes in the U.S. is the charge of the EPA, through the mandate of Subtitle C of the Solid Waste Disposal Act, as amended by the Resource Recovery and Conservation Act of 1976. The EPA is required to compile a list of hazardous wastes and their characteristics and to see to it that these wastes are controlled from the time of their origin until their destruction or proper disposal. Firms generating and transporting hazardous wastes must keep proper records; issue reports on their activities; maintain a manifest system to ensure proper tracking of hazardous wastes through transportation systems; use approved containers and labels; and deliver wastes to facilities approved for treatment, storage, and disposal. The

[6] Deland, M.R. 1982. Hazardous wastes: the controversy continues. *Environmental Science and Technology* 16: 103A.

[7] American Chemical Society. 1983. Most dangerous U.S. waste dumps identified. *Chemical and Engineering News,* p.8.

TABLE 19.2 Ten Sites Ranked the Most Dangerous Abandoned Hazardous Waste Dumps in the Nation

Site name	*Hazard*	*Years of disposal*	*Types of wastes*	*Contaminants*
FMC Corp. Fridley, Minnesota	Contamination of drinking water source for cities of Fridley and Brooklyn Center; pollution of Mississippi River, the source of potable water for Minneapolis	Early 1950s to early 1970s	Solvents, paint sludges, plating wastes	Trichloroethylene, methyl chloride, benzene, others
Tybouts Corner Landfill New Castle County, Delaware	Surface water and ground water contamination	na	Sanitary, industrial wastes	na
Bruin Lagoon Bruin Borough, Pennsylvania	Ground water and surface water contamination. Lagoon lies adjacent to Bear Creek, which joins the Allegheny River, a water source for Pittsburgh	na	Wastes from coal mines, oil fields, and chemical firms	na
Industri-Plex 128 (Mark Phillip Trust) Woburn, Massachusetts	Surface water and ground water contamination	1853 to 1981	Wastes from manufacture of insecticides, explosives, acids, tanned hides and residues	Arsenic, lead, chromium, others
Lipari Landfill Gibbsboro and Pitman Townships, New Jersey	Surface water, ground water, and air pollution; site is located in area of fruit orchards	1958 to 1971	Domestic and industrial wastes	Benzene, toluene, bis(2-chloroethyl) ether, beryllium, mercury
Sinclair Refinery Wellsville, New York	Surface water and possibly ground water contamination	na	Refinery wastes, including oil sludges and fly ash	Mercury, polychlorinated biphenyls, oil components
Price Landfill Pleasantville, New Jersey	Ground water contamination of potable water source of Pleasantville; plume of contamination threatens Atlantic City	1969 to 1976	Sanitary and industrial wastes	Benzene, chloroform, trichloroethylene
Pollution Abatement Services Oswego, New York	Surface water and ground water pollution; polluted surface water discharges into Lake Ontario	1970 to 1976	Polymer gels, plating wastes, metal sludges, paint wastes, laboratory chemicals	Large quantities of polychlorinated biphenyls, others

Site name	*Hazard*	*Years of disposal*	*Types of wastes*	*Contaminants*
Labounty Site Charles City, Iowa	na	na	na	na
Helen Kramer Landfill Mantua Township, New Jersey	Surface water and ground water contamination	1970 to 1980	Sanitary, construction, and nonchemical industrial wastes	na

Note: na = not available

Data courtesy of American Chemical Society; see Reference 7.

core of EPA regulations pertaining to hazardous wastes has been published [8].

For regulatory purposes, hazardous wastes are defined in terms of *ignitability, corrosivity, reactivity,* and *toxicity.* The general meanings of these terms are self-explanatory for the most part, and each will be discussed in some detail. An **ignitable waste** obviously presents a fire hazard; a **corrosive waste** may attack its container or other materials by destructive chemial action; a **reactive waste** may undergo violent, spontaneous chemical reaction, including explosion; and a **toxic waste** may act as an acute or chronic poison to living organisms.

The EPA's definition of *hazardous waste* [9] may not necessarily be clear in scientific terms. According to current regulations in the U.S., a material must first be a "solid waste" to classify as a "hazardous waste". However, the EPA definition of *solid waste* is very broad, indeed: "garbage, refuse, sludge or other waste materials". The last category includes [8] "any solid, liquid, semisolid, or contained gaseous material resulting from industrial, commercial, mining, or agricultural operations or from community activities, which is discarded or is being accumulated, stored, or physically, chemically, or biologically treated prior to being discarded; or has served its original intended use and sometimes is discarded; or is a manufacturing or mining by-product and sometimes is discarded." However, the EPA definition excludes some materials from the solid waste category. Among these are [8] "domestic sewage, wastes that mix with domestic sewage in a sewer system before entering a publicly owned treatment works (POTW), industrial wastewater discharge from point sources that are subject to regulation under the Clean Water Act, and source, special nuclear, or byproduct materials as defined by the Atomic Energy Act of 1954".

Perhaps the main certainty about definitions such as these is that they will change over time as perceptions of problems, national administrations, and other factors change. From a scientific viewpoint the definition of hazardous materials should be obvious, despite whatever regulations are in vogue at a given time. Therefore, the environmental chemist should seek to apply the principles of environmental chemistry to the control of all hazardous materials in the environment, despite changing regulatory definitions of what these materials are.

---

[8] 1980. *Federal Register* 45(98): Book 2, 19 May 1980, pp. 33063–33285.

[9] 1980. Hazardous wastes: a matter of definition. *Chemical and Engineering News* 16 June 1980, pp. 21–2.

19.4 IGNITABILITY

A waste characterized as hazardous because of *ignitability* falls into at least one of the following categories: (1) liquids having a **flash point** (the lowest temperature at which the vapors above a flammable liquid ignite when tested with a flame) below 140 °F (60 °C); (2) nonliquid liable to cause fires through friction, absorption of moisture, or spontaneous chemical change and liable, when ignited, to burn so vigorously or persistently as to create a hazard; (3) ignitable compressed gases; and (4) oxidizers. In addition to wastes capable of burning during transportation, storage, or disposal, ignitable wastes include those (particularly oxidizers) that are likely to aggravate an already burning fire. In addition to posing a direct danger, wastes that are burning may generate toxic fumes and create convection currents that may transport particulate matter to surrounding areas. The criteria for ignitable liquids encompasses the National Fire Protection Association's classification scheme for Class I and Class II liquids. The other categories of ignitable substances have been defined largely according to U.S. Department of Transportation criteria [10].

Some substances that will ignite or burn are excluded from the ignitable category. These include wine and some latex paints containing less than 24% alcohol; despite their relatively low flash points, these substances do not support combustion for prolonged periods. A number of other substances including tree bark, wood chips, sawdust, and paper boxes do, in fact, "when ignited, burn so vigorously or persistently to create a hazard", but they are excluded from the ignitable category because they are not generally "liable to cause fires through friction, absorption of moisture, or spontaneous chemical change"

As with all such definitions, there is a degree of arbitrariness in the definition of ignitability. This is especially true in the case of solids because there is not a reliable standard test procedure for the ignitability of solids similar to the flash-point test for liquids.

19.5 CORROSIVITY

Wastes defined by the EPA as *corrosive* include: (1) aqueous wastes with a pH less than or equal to 2.0 or greater than or equal to 12.5; and (2) liquid wastes capable of corroding steel at a rate greater than or equal to 0.250 inches (0.635 cm) per year [8]. The rationale for choosing the pH criterion was that liquids with extreme pH's harm tissue and aquatic life, promote the migration of toxic contaminants from other wastes, and may react dangerously with other wastes. Neither total alkalinity (Section 2.12) nor total acidity (Section 2.13) were included as criteria for corrosivity, although some authorities have argued for their inclusion. It should be noted that a very dilute solution of a strong (completely ionized) acid may have a very low pH but can be neutralized by a very small quantity of base, whereas a solution containing a high concentration of a weak acid may have a relatively high pH but require a large quantity of base for neutralization. Similar arguments apply to hazardous waste solutions of strong and weak bases.

Originally the pH limits for defining corrosivity were set at 3.0 and 12.0. Critics argued successfully against these more stringent limits. A pH of 12.0 is less

---

[10] 1979. *Code of Federal Regulations* 49: 100–199, revised as of October 1, 1979.

than that of many lime-stabilized wastes and sludges, and an upper limit of pH 12.0 might prevent valuable use of lime stabilization for rendering some wastes nonhazardous (see Section 20.3). At the acidic extreme, many industrial wastes have pH values between 2.0 and 3.0, so that a pH limit of 3.0 would categorize these wastes as hazardous, despite the fact that they are usually readily neutralized by alkaline material.

As noted above, the rate of metal corrosion was chosen as the other major criterion for corrosivity. This choice obviously was made because wastes capable of corroding metals can destroy metal containers, leading to their spillage and, possibly, mobilization of surrounding wastes. A standardized version of the NACE metal-corrosion test is used in setting this criterion.

19.6 REACTIVITY

Wastes defined as having the characteristic of **reactivity** are those that "(1) readily undergo violent chemical change; (2) react violently or form potentially explosive mixtures with water; (3) generate toxic fumes in a quantity sufficient to present a danger to human health or the environment when mixed with water or, in the case of cyanide or sulfide-bearing wastes, when exposed to mild acidic or basic* conditions; (4) explode when subjected to a strong initiating force; (5) explode at normal temperatures and pressures; or (6) fall within the Department of Transportations's forbidden explosives, Class A explosives, or Class B explosives classifications". Because of their propensity to react violently or explode, these kinds of wastes pose problems at all stages of management. They are defined largely in terms developed by the National Fire Protection Association.

The standard tests that are available for specific kinds of substances are not general enough to define the reactivity of a wide range of potentially hazardous wastes. Despite the lack of such a test, generators of reactive hazardous wastes are almost invariably aware of their reactive nature because of direct dangers to plant and personnel. Major problems can arise in the identification of improperly disposed, abandoned reactive wastes. Two tests that are major candidates for the measurement of aspects of reactivity are the **Explosion Temperature Test** and the **Bureau of Explosives shock instability test**. The latter is used for the Department of Transportation's definition of Class A explosives.

19.7 TOXICITY

For regulatory purposes, *toxicity* is defined in terms of levels of contaminants in a leachate from hazardous waste obtained by means of a specified **extraction procedure**. Basically, a waste is classified as toxic if leachate from it, extracted by an acidic leaching medium with a pH of 5.0 ± 0.2, contains 100 times or more the approved levels of any of the toxic contaminants identified in the National Interim Primary Drinking Water Standards [11]. Since much of the concern over hazardous wastes involves potential contamination of ground water, this test was

---

* Within the pH range 2.0–12.5. Wastes outside this range are automatically designated as hazardous under the pH criterion for corrosivity.

[11] For a listing and discussion of these standards, see: 1978. *Cleaning our environment: a chemical perspective.* Washington, D.C.: American Chemical Society, pp. 197–200.

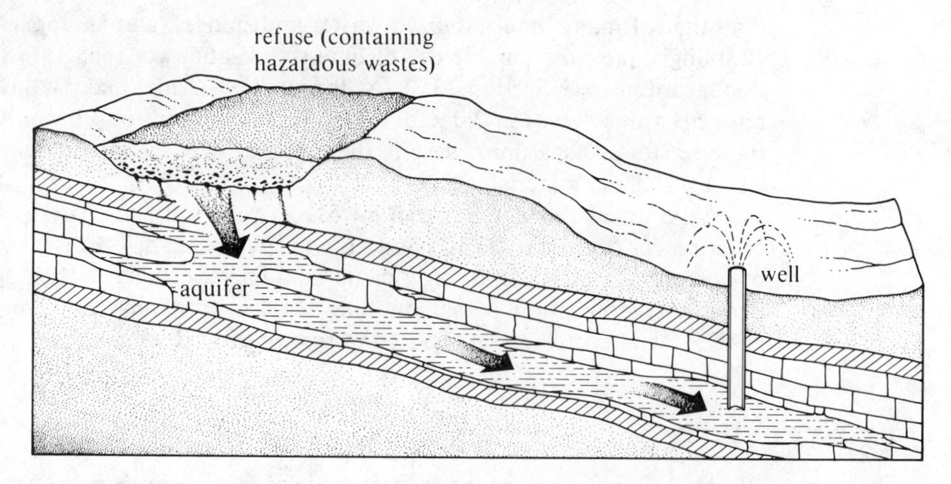

FIGURE 19.2 Codisposal of hazardous wastes and municipal refuse in a landfill can result in contamination of ground water in the underlying aquifers.

designed on the basis of a worst-case "mismanagement scenario" that could result in contamination of ground water from improperly discarded hazardous wastes. This scenario visualizes codisposal of hazardous wastes and municipal refuse in a landfill overlying a ground-water aquifer (Fig. 19.2).

The organic matter in a landfill undergoes aerobic oxidation by bacterially mediated reactions (see Section 5.10 and Table 5.1),

$$O_2 + \{CH_2O\} \rightarrow CO_2 + H_2O \quad (19.7.1)$$

such that any residual oxygen is quickly exhausted. Anaerobic bacteria continue degradation of the biodegradable "garbage" by reactions such as methane fermentation,

$$2\,\{CH_2O\} \rightarrow CO_2 + CH_4 \quad (19.7.2)$$

and fermentation represented by the general reaction,

$$2\,\{CH_2O\} + H_2O \rightarrow HCO_2H + CH_3OH \quad (19.7.3)$$

In a typical fermentation reaction, organic matter is converted to an organic acid, here shown as formic acid, HCO_2H, and a more reduced form, represented here by an alcohol, CH_3OH. The main significance of this reaction is that it yields weak organic acids which ionize to produce hydrogen ion, thus lowering the pH:

$$HCO_2H \rightleftharpoons H^+ + HCO_2^- \quad (19.7.4)$$

Furthermore, carbon dioxide produced by biological processes such as those shown in Reactions 19.7.1 and 19.7.2 reacts with water to lower the pH:

$$CO_2 + H_2O \rightleftharpoons H^+ + HCO_3^- \quad (19.7.5)$$

The weak organic acids produced by biological degradation of organic municipal wastes provide the rationale for using acetic acid as a leaching agent for the standard extraction procedure used to define toxicity. At lower pH values a number of otherwise insoluble hazardous substances codisposed with municipal refuse are mobilized. The most important of these are heavy metal ions. For example, lead in carbonate and basic carbonate salts is solubilized by H^+:

$$PbCO_3(s) + H^+(aq) \rightarrow Pb^{2+}(aq) + HCO_3^-(aq) \tag{19.7.6}$$

There are several other consequences of fermentation in municipal refuse that can result in pollutant mobilization. Consumption of O_2 in covered refuse (Reaction 19.7.1) and the subsequent action of anaerobic bacteria make the medium reducing, as manifested by a low pE (see Chapter 3). Reducing agents, which are electron donors, react with oxidized forms of some metals to form more reduced forms:

$$MnO_2(s) + 4\ H^+(aq) + 2\ e^- \rightleftharpoons Mn^{2+}(aq) + 2\ H_2O \tag{19.7.7}$$

These are the more soluble, and therefore more mobile, forms of the metals. (For the influence of both pE and pH on the forms of a typical metal, see Figure 3.3).

Some of the organic products of the degradation of municipal waste can react directly with some pollutants, resulting in pollutant mobilization. One of the most important examples of this kind of organic product is citric acid,

```
               H
               |
      O   H    O   H   O
      ||  |    |   |   ||
  HO—C—C—C—C—C—OH
          |    |   |
          H    |   H
              C=O
               |
               O
               |
               H
```

Deprotonated citric acid, the citrate anion, functions as a chelating agent (see Chapter 4) and binds to metal ions, forming metal chelates. These chelates generally are soluble. Furthermore, they are normally anionic (negatively charged) species and are not retained by cation-exchange processes that predominate in soil and clay (see Sections 6.5, 6.12, and 10.6).

The action of anaerobic methylating bacteria in active landfills can result in the formation of mobile methylated forms of trace elements (see Section 7.7). Monomethylmercury ion, CH_3Hg^+, and dimethylarsinic acid, $(CH_3)_2AsO(OH)$, are typical of the kinds of methylated species formed.

In setting standards for potential ground-water pollution by toxic substances, one of the most difficult criteria to judge is the **attenuation factor** representing the degree to which solutes may be sorbed and diluted in migrating from a source to a point at which the water is withdrawn from an aquifer. Most of the attenuation of pollutants in water occurs as the water flows downward through geological strata from the disposal site to the aquifer. In addition to dilution by uncontaminated water, pollutants are removed from the leachate by ion exchange, absorption, and adsorption. Colloidal pollutants are also removed by the filtering action of the strata. Attenuation is generally thought to be less for polluted water once it is in the aquifer, where the predominant movement is horizontal, rather than vertical. The contaminated water moves as a **slug** or **plume** through the aquifer, as shown in Figure 19.3 (see page 548). Although some dilution by uncontaminated ground water does occur in the aquifer, as well as some sorption by the ground-water mineral strata, less removal of pollutants usually occurs than does in the vertical movement of the water. Typically, ob-

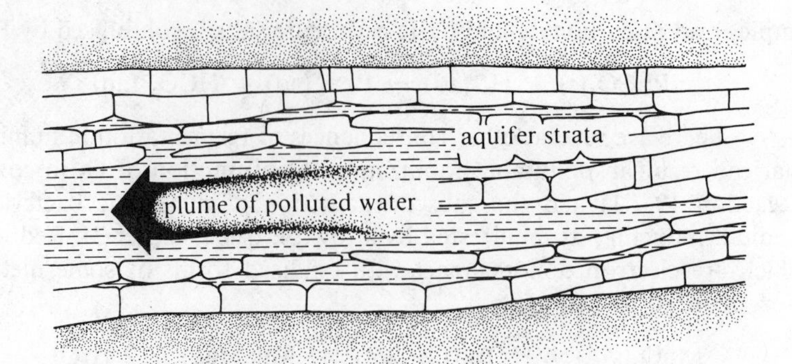

FIGURE 19.3 Polluted water tends to move through an aquifer as an undiluted plume or slug.

served attenuation factors range from 1 to 500. These factors are, of course, quite site-specific. Furthermore, after prolonged exposure to water pollutants, the sorption capacity of some strata may become saturated, lowering the attenuation factor. In extreme cases, circumstances may develop in which sorbed pollutants are liberated from strata, giving attenuation factors of less than 1.

An attenuation factor of 100 was set as a compromise in determining the criterion of toxicity. It is actually a kind of educated guess, due to the lack of good empirical data upon which to base estimations of attenuation in soil and mineral strata. Several factors were considered in arriving at this value. Among these is the reasonable assumption that strata underlying most disposal sites do, in fact, attenuate pollutant solutes and do not simply serve to delay the migration of the solutes. Another factor considered is that under current legislation there is no mechanism for "de-listing" or granting a variance to wastes that fail the extraction procedure. This increases the gravity of the situation that can arise when a waste material is classified as hazardous on the basis of the extraction procedure alone but is subsequently found to be nonhazardous in the environment where it is placed.

19.8 HAZARDOUS COMPOUNDS, FORMULATIONS, AND CLASSES OF SUBSTANCES

Some compounds, formulations, and classes of substances have been specifically designated as hazardous [8]. The Environmental Protection Agency has assigned a **hazard code** to each; these codes are: **ignitable waste, (I); corrosive waste, (C); reactive waste, (R); extraction procedure toxic waste, (E); acute hazardous waste, (H);** and **toxic waste, (T).** In addition, each hazardous waste is assigned an EPA **hazardous waste number**. These material are placed in several classifications, according to source. Examples of materials from each of these classifications are given in Tables 19.3–19.6. Space does not permit listing each substance in each classification here; complete lists may be found in Reference 8. Table 19.3 lists examples of hazardous wastes from **nonspecific sources.** Hazardous wastes from **specific sources** are listed in Table 19.4. Table 19.5 lists some wastes from **discarded commercial chemical products, off-specification materials, containers** of these chemicals, and **spill residues** of such chemicals. Many of the items on this list consist of brand-name commercial formulations, largely of pesticides.

TABLE 19.3 Hazardous Waste from Nonspecific Sources[1]

Hazardous waste number	*Hazardous waste*	*Hazard code*
F001	The spent halogenated solvents used in degreasing: tetrachloroethylene, trichloroethylene, methylene chloride, 1,1,1-trichloroethane, carbon tetrachloride, and the chlorinated fluorocarbons; and sludges from the recovery of these solvents in degreasing operations.	(T)
F004	The spent non-halogenated solvents: cresols, cresylic acid, and nitrobenzene; and the still bottoms from the recovery of these solvents.	(T)
F007	Spent plating-bath solutions from eletroplating operations.	(R, T)
F010	Quenching-bath sludge from oil baths from metal heat-treating operations.	(R,T)
F016	Dewatered air pollution control scrubber sludges from coke ovens and blast furnaces.	(T)

[1]Complete list in [8], p. 33123.

TABLE 19.4 Hazardous Waste from Specific Sources[1]

Hazardous waste number	*Hazardous waste*	*Hazard code*
Wood preservation		
K001	Bottom sediment sludge from the treatment of wastewaters from wood-preserving processes that use creosote and/or pentachlorophenol.	(T)
Inorganic pigments		
K002	Wastewater treatment sludge from the production of chrome yellow and orange pigments.	(T)
K008	Oven residue from the protection of chrome oxide green pigments.	(T)
Organic chemicals		
K009	Still bottoms from the production of acetaldehyde from ethylene.	(T)
K020	Heavy ends (residue) from the distillation of vinyl chloride in vinyl chloride monomer production.	(T)
K027	Centrifuge residue from toluene diisocyanate production.	(R,T)
Pesticides		
K031	By-product salts generated in the production of MSMA and cacodylic acid.	(T)
K037	Wastewater treatment sludges from the production of disulfoton.	(T)
K043	2,6-Dichlorophenol waste from the production of 2,4-D.	(T)
Explosives		
K045	Spent carbon from the treatment of wastewater containing explosives.	(R)
K047	Pink/red water from TNT operations.	(R)

TABLE 19.4 Hazardous Waste from Specific Sources (*continued*)

Hazardous waste number	*Hazardous waste*	*Hazard code*
Petroleum refining		
K048	Dissolved air flotation (DAF) float from the petroleum-refining industry.	(T)
K049	Slop oil emulsion solids from the petroleum-refining industry.	(T)
Leather tanning and finishing		
K053	Chrome (blue) trimmings generated by the following subcategories of the leather tanning and finishing industry: hair pulp/chrome tan/retan/wet finish; hair save/chrome tan/retan/wet finish; retan/wet finish; no beam-house; through-the-blue; and shearling.	(T)
Iron and steel		
K060	Ammonia lime still sludge from coking operations.	(T)
Primary zinc		
K067	Electrolytic anode slimes/sludges from primary zinc production.	(T)

[1]Complete list in [8], pp. 33123–4.

TABLE 19.5 Discarded Commercial Chemical Products, Off-Specification Species, Containers, and Spill Residues[1]

Hazardous waste number	*Hazardous waste*
P003	Acrolein, $H_2C=C(H)-C(=O)-H$
P013	Barium cyanide, $Ba(CN)_2$
P024	*p*-Chloroaniline, $H_2N-C_6H_4-Cl$
P050	Endosulfan, (6,7,8,9,10,10-hexachloro-1,5,5a,6,9,9a-hexahydro-6,9-methano-243-benzodioxathiapen 3-oxide)
P063	Hydrocyanic acid, HCN
P065	Mercury fulminate, $HgC_2N_2O_2$
P081	Nitroglycerine, $H-C(H)(ONO_2)-C(H)(ONO_2)-C(H)(ONO_2)-H$

Hazardous waste number	*Hazardous waste*
P095	Phosgene, $Cl-C(=O)-Cl$
P105	Sodium azide, NaN_3
P110	Tetraethyl lead, $Pb(C_2H_5)_4$
P115	Thallium(I) sulfate, Tl_2SO_4
P122	Zinc phosphide (R,T), Zn_3P_2

[1]Complete list in [8], pp. 33124–5.

TABLE 19.6 Chemical Intermediates from the Manufacture of Commercial Chemicals or Chemical Formulations[1]

Hazardous waste number	*Hazardous waste*
U001	Acetaldehyde, $H-C(H)_2-C(=O)-H$
U021	Benzidine, $H_2N-C_6H_4-C_6H_4-NH_2$
U072	1,4-Dichlorobenzene, $Cl-C_6H_4-Cl$
U115	Ethylene oxide (I,T), H_2C-CH_2 (epoxide ring, bridged by O)
U133	Hydrazine (R,T), N_2H_4
U147	Maleic anhydride, $HC=CH$ with $O=C$ and $C=O$ joined through O (anhydride ring)
U220	Toluene, $C_6H_5-CH_3$

[1]Complete list in [8], p. 33126.

Finally, Table 19.6 lists some **chemical intermediates** from the manufacture of commercial chemicals or chemical formulations. The majority of these are organic compounds or have an organic component. Many are involved in the manufacture of pesticides.

19.9 CHEMICAL CLASSIFICATION OF HAZARDOUS WASTES

As discussed previously, hazardous wastes may be classified largely on the basis of their characteristics: ignitability, reactivity, corrosivity, or toxicity. In addition, a number of specific compounds and formulations have been designated as hazardous. In this section, we shall see that it is also useful to consider hazardous waste compounds on the basis of their chemical classification.

Most of the compounds found in hazardous wastes are organic. Inorganic substances may be considered dangerous either because they are directly hazardous (for example, reactive or toxic) or because they contain elements generally regarded as toxic, such as heavy metals. For example, most forms of cadmium are toxic. Selenium is toxic at higher levels, but essential for nutrition at low levels. Some chemical compounds are very hazardous, even though they are composed of elements generally considered to be nonhazardous in themselves. A typical example is potassium cyanide, KCN. The three elements in this simple compound—potassium, carbon, and nitrogen—are all normal constituents of the body and essential for life processes. Combined in the cyanide form, however, they compose a deadly compound. Another example is perchloric acid, $HClO_4$, composed of three elements that are essential for life processes. In a hot, concentrated form, this compound is a very strong oxidizing agent that reacts vigorously with organic matter. Thus, $HClO_4$, like other perchlorates, pose explosion hazards if used improperly in the laboratory.

Hazardous organic compounds may be classified largely according to **functional groups** consisting of specific arrangements of atoms characteristic of a particular class of organic compounds. Several general types of organic compounds that commonly contain hazardous chemicals are discussed in this section.

Oxygenated compounds comprise the most diverse general class of hazardous organic compounds. These include alcohols, phenols, ethers, cyclic ethers (oxides), aldehydes, ketones, carboxylic acids, and esters. Structures of some representative hazardous waste chemicals containing oxygen are shown below with the functional group characteristic of each class outlined with a dotted line.

```
H     H  H
 \    |  |
  C=C—C—OH
 /       |
H        H
```

allyl alcohol (an olefinic alcohol)

```
   H  H  H  O
   |  |  |  ||
H—C—C=C—C—H
   |
   H
```

crotonaldehyde

```
   H  O  H  H
   |  ||  |  |
H—C—C—C—C—H
   |      |  |
   H      H  H
```

methyl ethyl ketone

```
      O
      ||
Cl3C—C—OH
```

trichloroacetic acid

```
H      H  O       H  H
 \     |  ||      |  |
  C=C—C—O—C—C—H
 /                |  |
H                 H  H
```

ethyl acrylate (ester)

Acrolein, $CH_2=CH—C(O)H$, a highly reactive olefinic aldehyde with a choking odor, caused a serious explosion and fire at Union Chemical's Taft, Louisiana, chemical complex on June 11, 1983. About 17,000 people within a 5-mile radius of the plant were evacuated. The fire in the acrolein tank was allowed to burn out to prevent the spread of toxic fumes to the atmosphere.

Among the oxygen-containing compounds, some **ethers** are regarded as hazardous. These compounds contain an —O— atom bridge between two hydrocarbon *moieties* (major portions of a molecule). The most familiar ether compound is diethyl ether, shown below with the characteristic functional group outlined:

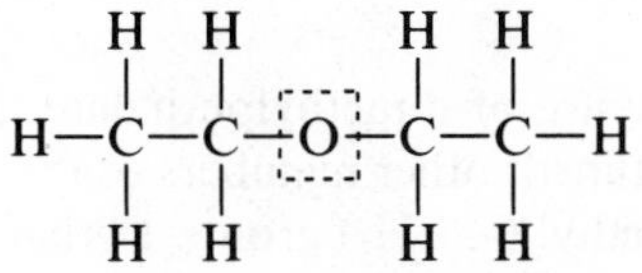

```
    H  H      H  H
    |  |      |  |
H—C—C—[O]—C—C—H
    |  |      |  |
    H  H      H  H
```

Used as an anesthetic, as a solvent, and for many other purposes, diethyl ether is ignitable and toxic. Ethers may occur in a cyclic form. The most common of these is ethylene oxide,

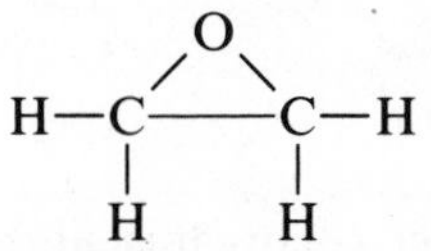

```
        O
       / \
  H—C—————C—H
     |     |
     H     H
```

widely produced as an intermediate in the manufacture of ethylene glycol. This compound is classified as hazardous because of its ignitability and toxicity. Halogenated ethers constitute another class of hazardous compounds, some of which are suspected carcinogens. Typical of these is 2-chloroethyl vinyl ether,

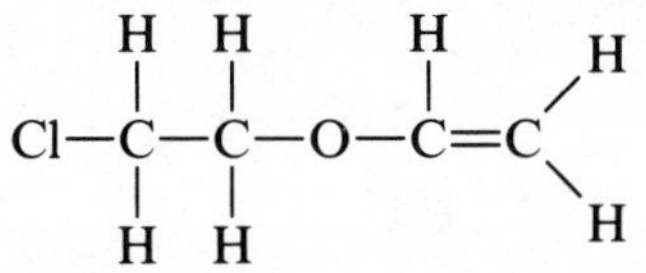

```
     H  H     H    H
     |  |     |   /
Cl—C—C—O—C=C
     |  |         \
     H  H          H
```

A particular hazard with ethers is their tendency to form highly explosive peroxides. These are compounds containing the —O—O— group. Peroxides form in and around the lids of cans of ether that have been stored with some exposure of their contents to air. Di-isopropyl ether has an especially strong tendency to form peroxides:

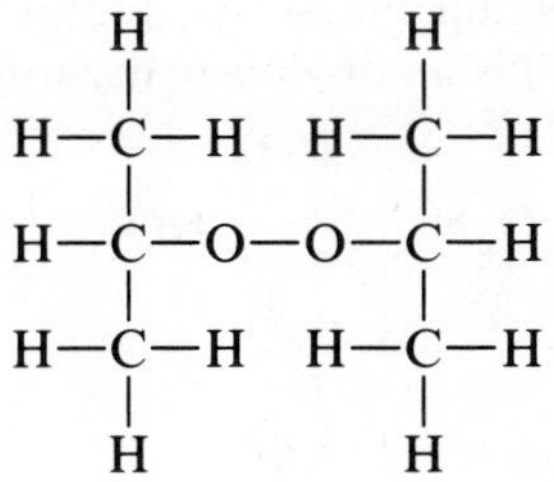

```
    H         H
    |         |
H—C—H   H—C—H
    |         |
H—C—O—O—C—H      peroxide of
    |         |      di-isopropyl ether
H—C—H   H—C—H
    |         |
    H         H
```

The **phthalates** constitute another class of oxygen-containing compounds sometimes listed as hazardous. Dimethylphthalate is the simplest of these ester compounds:

$$C_6H_4(C(=O)\text{—}O\text{—}CH_3)_2$$

In the structure of dimethylphthalate, the moiety characteristic of all phthalate esters is outlined; other members of the class contain hydrocarbon moieties other than the methyl (—CH_3) group. Phthalates are widely used as plasticizers to improve the properties of plastics and are common water pollutants.

Phenols make up another class of hazardous oxygenated compounds. These compounds have an —OH group bonded to an aryl (aromatic) moiety. They are widely used industrial chemicals and are produced as by-products of the pyrolysis (coking) of coal.

C_6H_5—OH phenol

A number of organic nitrogen compounds are classified as hazardous wastes. The most important of these are the *amines, nitro compounds, nitriles,* and *nitrosamines.*

Amines consist of compounds in which one of the three binding sites on the N atom is attached to a hydrocarbon group, and each of the other two groups is attached to either a hydrocarbon moiety or an H atom. Ignitable dimethylamine,

$$H_3C\text{—}N(H)\text{—}CH_3$$

is a typical amine and can be hazardous. A special type of amine consists of **heterocyclic nitrogen compounds,** in which a nitrogen atom is part of an aromatic ring structure. A typical and important example is pyridine:

$$C_5H_5N$$

Nitro compounds contain the —NO_2 group bonded to a hydrocarbon moiety. Trinitrotoluene (TNT) is an obviously hazardous nitro compound:

$$CH_3C_6H_2(NO_2)_3$$ trinitrotoluene, TNT

Nitriles contain the —C≡N group. This class is exemplified by ignitable and toxic acetonitrile:

$$H_3C-C\equiv N$$

Nitrosamines are characterized by the $>N-N=O$ group. An example of these carcinogenic compounds is *N*-ethyl-*N*-nitroso-*N*-butylamine,

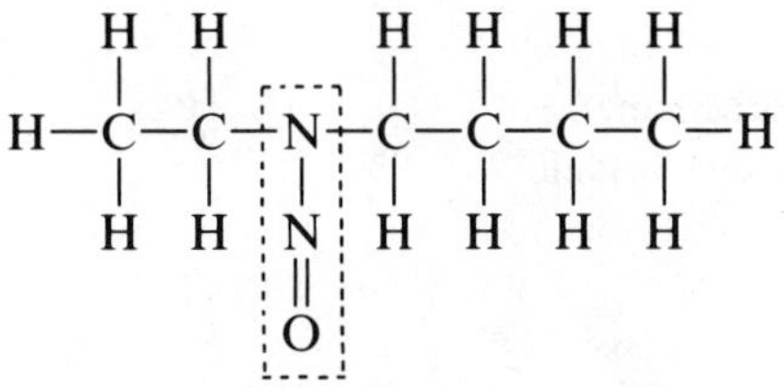

$$CH_3CH_2-N(N{=}O)-CH_2CH_2CH_2CH_3$$

Sulfur compounds make up a separate class of potentially hazardous organic compounds. Many of these are particularly notorious for their strong, unpleasant odors. Sulfur may be bonded to organic compounds in several ways, including the —SH group, as —S— or —SS— bridging between organic moieties, as part of a cyclic structure involving a ring of C atoms (heterocyclic sulfur), and as the C=S group. Examples of these are:

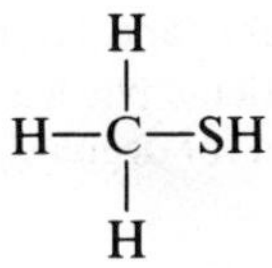

CH_3SH methyl mercaptan

CH_3-S-CH_3 dimethyl mercaptan

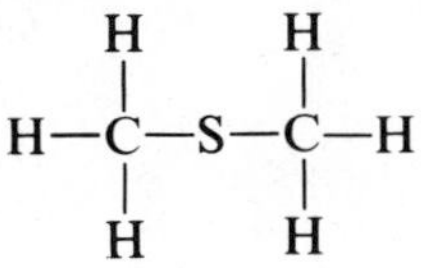

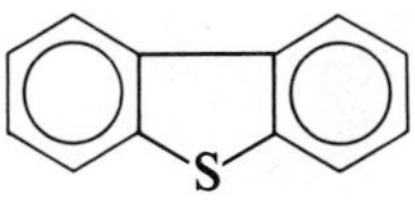

diphenylene sulfide

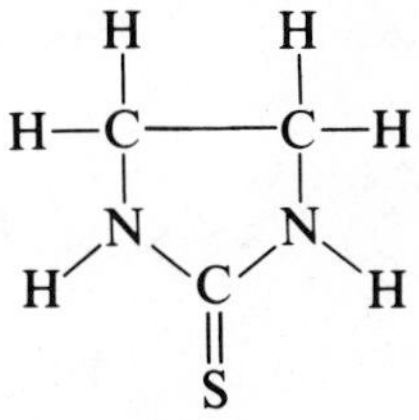

ethylene thiourea

Aromatic compounds containing the π-bonded, six-carbon ring designated in chemical formulas as

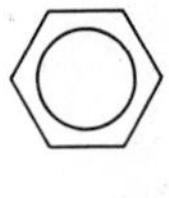

make up a class of hazardous waste compounds; these are usually hazardous by virtue of attached functional groups, as well. **Polynuclear aromatic hydrocarbons (PNA)**, consisting of condensed aromatic rings, are specifically hazardous because many are carcinogenic. The best-known example is benzo(a)pyrene:

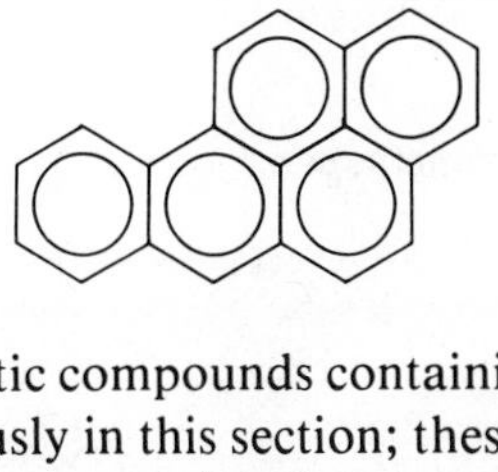

Examples of hazardous aromatic compounds containing atoms other than carbon or hydrogen were cited previously in this section; these include phenols, pyridine, and trinitrotoluene. Included, too, are the phthalates, which contain an aromatic ring as part of the basic phthalate structure.

As noted previously in discussions of their roles as water pollutants, air pollutants, and toxic compounds, **halogenated hydrocarbons** are among the most

TABLE 19.7 Halogenated Hydrocarbons in Hazardous Wastes

Name	*Formula*	*Uses*
Ethyl chloride	CH_3-CH_2-Cl	Tetraethyl lead manufacture
1,2-Dichloroethane (ethylene dichloride)	$ClCH_2-CH_2Cl$	Used to manufacture vinyl chloride (see below)
Vinyl chloride	$H_2C=CHCl$	Used to manufacture polyvinyl chloride, known to cause liver cancer
Tetrachloroethene (tetrachloroethylene, perchloroethylene)	$Cl_2C=CCl_2$	Low flammability and high vapor density make tetrachloroethene, trichloroethene, and 1,1,1-trichloroethane particularly useful for degreasing metal parts.
Trichloroethene (trichloroethylene)	$Cl_2C=CHCl$	Solvent
1,1,1-Trichloroethane (methyl chloroform)	Cl_3C-CH_3	
Carbon tetrachloride (tetrachloromethane)	CCl_4	Solvent, chlorofluorohydrocarbon (Freon) manufacture
Dichlorodifluoromethane	CCl_2F_2	Refrigerant (of the Freon type)
Chloroform	$CHCl_3$	Manufacture of chlorofluorocarbons and fluorocarbon (Teflon) plastics

Name	*Formula*	*Uses*
1,2-Dibromoethane (ethylenedibromide)	H H Br—C—C—Br H H	Lead scavenger in gasoline, fumigant, solvent
Hexachlorocyclo-pentadiene	Cl Cl C—C C C Cl C Cl Cl Cl	Manufacture of cyclodiene "diene" pesticides
Chlordane	Cl Cl Cl Cl —Cl_2 Cl Cl	First of the cyclodiene pesticides, very effective against termites, use now banned in the U.S. because of carcinogenic potential

significant and widespread environmental contaminants. As their name implies, this family of compounds consists of hydrocarbons bonded to one or more halogen atoms (fluorine, chlorine, bromine, or iodine). Most common are the **chlorinated hydrocarbons**, which are particularly prevalent among hazardous waste compounds. Important examples of halogenated hydrocarbons in hazardous wastes are given in Table 19.7.

Halogenated hydrocarbons are widely used as solvents (for example, in decreasing of parts) due to their good solvent properties, nonflammability, and ease of redistillation (enabling solvent reclamation). Among the halogenated solvents used in degreasing are tetrachloroethylene, carbon tetrachloride, and some chlorinated fluorocarbons. These spent solvents and the sludges and still bottoms from their recovery are classified as hazardous [8].

In addition to their uses as solvents, halogenated hydrocarbons are used in the synthesis of a wide variety of pesticides (see Section 7.18 and Table 7.8). They are also used for the synthesis of polymers; for example, tetrafluoroethene is used to synthesize polytetrafluoroethene (Teflon), and vinyl chloride is polymerized to form polyvinylchloride (PVC).

Polychlorinated biphenyls (PCB's) constitute a separate class of chlorinated hydrocarbons; they are particularly persistent in the environment. These compounds were discussed as water pollutants in Section 7.19. They consist of compounds in which from 1 to 10 chlorine atoms are bonded to the two benzene rings making up the biphenyl group, as shown in the generalized formula:

$(Cl)_{1-10}$

TABLE 19.8 Pesticides Considered as Hazardous Constituents of Waste

Pesticide name	*Formula*	*Uses and characteristics*
Endosulfan	Cl, Cl, Cl, Cl, Cl, Cl; $-O-S(=O)-O-$ (cyclic sulfite)	Insecticide, so toxic to fish that it was considered for use as a piscicide
Captan (trichloromethanethiol)	$N-S-CCl_3$ (imide, with two C=O groups, fused to a cyclohexene ring)	Most widely used home foliage protectant fungicide, bird repellent (birds dislike taste on seeds)
Methyl parathion	$(H_3CO)_2P(=S)-O-C_6H_4-NO_2$	Insecticide with a broad range of insect control and relatively low human toxicity
Parathion	Same as methyl parathion, except ethyl groups in place of methyl groups	Insecticide with less range of insect control and higher human toxicity than methyl parathion
Disulfoton	$(CH_3CH_2O)_2P(=S)-S-CH_2CH_2-S-CH_2CH_3$	Plant systemic insecticide (taken up by roots and toxic to sucking insects)
Aldicarb	$H_3C-S-C(CH_3)_2-C(H)=N-O-C(=O)-N(H)-CH_3$	Insecticide and nematicide used as a plant systemic
Sodium fluoroacetate	$F-CH_2-C(=O)-O^-Na^+$	Rodenticide, extremely toxic to humans and domestic animals

19.10 PESTICIDES AS HAZARDOUS WASTES

Because of their widespread manufacture and use, pesticides are commonly encountered as "hazardous constituents" of wastes [12]. Selected examples of such pesticides are given in Table 19.8. In addition to their environmental effects, these and other pesticides are dangerous because of their direct hazard to humans. This hazard is a function of pesticide properties, toxicity, dose, absorption route, and length of exposure. On these bases, Ware [13] defines *hazard* as "an expression of the potential of a pesticide to produce human poisoning".

19.11 INDUSTRIAL PRODUCTION OF HAZARDOUS WASTES

The U.S. chemical industry is huge. Thirty-five of the top U.S.-manufactured chemicals are each produced in quantities exceeding 1 million metric tons per year. Annual chemical sales exceed $100 billion. The production and use of such large quantities of inorganic and organic chemicals in the U.S. has resulted in the output of a great deal of by-product and waste material, much of which is hazardous. Over many years the improper disposal of these by-products has created severe problems, as exemplified by the Love Canal. New legislation regulating the disposal of hazardous chemicals has sharply curtailed the improper disposal of these materials. Furthermore, high costs of raw materials, such as minerals and petroleum, have made it economically attractive to recover resources from waste products that formerly were discarded. However, as mentioned previously, the present waste dumps remain a major problem, even if new dumps are not being created.

Space does not permit a detailed discussion of all the chemically related industries that have the potential to produce hazardous wastes. However, each is discussed briefly in this section. These industries are listed in Table 19.9. Addi-

TABLE 19.9 Major Industries Associated with Chemical Production and Use

Inorganic chemicals manufacture	Battery manufacturing
Organic chemicals manufacture	Electrical and electronic components
Mining and mineral extraction	Leather tanning and finishing
Iron and steel	Plastics manufacture and processing
Nonferrous metals	Adhesives and sealants
Wood products	Paint and ink formulation
Rubber manufacture and processing	Petroleum refining
Power plants	Soap and detergent manufacture
Textile manufacture	Water treatment and publicly owned treatment works
Electroplating	

[12] Reference 8, pp. 33132–3, Appendix VIII.

[13] Ware, G.W. 1978. *The pesticide book.* San Francisco: W.H. Freeman & Co. Publishers.

tional details regarding each may be found in a manual published by the EPA [14].

In terms of quantity, *inorganic chemicals* are by far the most widely manufactured and used. Among the big-volume inorganic chemicals are sulfuric acid, ammonia, sodium hydroxide, chlorine, phosphoric acid, and nitric acid, all of which are potentially hazardous. Several inorganic pigments are also classified as hazardous. Typical inorganic pigment formulations are [15]: chrome yellows and oranges, consisting the $PbCrO_4$ and its modifications, prepared by the reaction of $Na_2Cr_2O_7$, NaOH, and $PbNO_3$; molybdate orange, prepared from $PbCrO_4$ with $PbMoO_4$ additive; and zinc yellow, consisting of a non-stoichiometric formulation precipitated from a mix of zinc oxide, hydrochloric acid, sodium dichromate, and potassium chloride. Some inorganic chemicals, such as H_2S, are ignitable, whereas others, such as N_2O, are oxidizing agents. Mercury fulminate, $Hg(CNO)_2$, is an example of a very reactive inorganic chemical; it is used as a detonator for explosives.

Given the large variety of *organic chemicals* already cited in this chapter, the potential hazards from organic chemicals manufacture should be obvious. Although the quantities of organic chemicals produced are much smaller than those of inorganics, organic chemicals make up a far greater variety of chemicals with a wide range of properties. Some of these properties endow some organic chemicals with a variety of hazardous characteristics.

Plastics manufacture and processing involves a large variety of hazardous chemicals. The greatest quantities of these are monomers used in polymer manufacture, including ethylene, phenol, formaldehyde, propylene, styrene, and vinyl chloride. Another major class of hazardous chemicals used in this industry are phthalate ester plasticizers (see Section 19.9).

Textile manufacture involves hazardous chemicals employed in the manufacture of synthetic fiber polymers. In some cases, the textile polymers are dissolved in potentially hazardous solvents (*spinning solvents*). A number of other chemicals are used for various textile manufacturing processes. These include bleaching, desizing, scouring, mercerizing, and dyeing, as well as the application of lubricants, antistatic agents, and functional finish chemicals. Dye manufacture and use also involves a number of potentially hazardous chemicals. Among these are nitroso compounds, nitrophenols, azo compounds (containing the —N=N— group), triarylmethane compounds, aromatic amines, and sulfur compounds [16].

Rubber manufacture and processing involves a number of ignitable monomers, such as styrene, butadiene, and isoprene, for rubber polymer manufacture. Other potentially hazardous chemicals are used; for example, tire manufacture involves fillers, extenders, reinforcing agents, antioxidants, and

---

[14] 1980. *Treatability manual.* Vol. II: *Industrial descriptions.* EPA-600/8-80-042b. Washington, D.C.: Office of Research and Development, U.S. Environmental Protection Agency.

[15] Martens, C.R. 1974. Pigments, paints, varnishes, lacquers, and printing inks. In *Riegel's handbook of industrial chemistry,* J.A. Kent, ed., Ch. 22, pp. 653–66. New York: Van Nostrand Reinhold Co.

[16] Hampel, C.A., and Hawley, G.G. 1979. *The encyclopedia of chemistry.* 3rd ed., New York: Van Nostrand Reinhold Co. pp. 358–362.

pigments. Rubber is subject to degradation processes through the action of a number of hazardous materials. Prominent among these are oxidants, which attach double bonds in rubber polymers, resulting in bond breakage and deterioration of the rubber.

The *manufacture of adhesives and sealants* involves assorted hazardous chemicals. Some adhesives, for example, make use of ignitable and toxic organic solvents as **vehicles** in which the adhesive is dissolved and spread on the surface to be bonded. The phenol and formaldehyde in urea-formaldehyde and phenol-formaldehyde resins used to bond plywood together are classified as hazardous chemicals. As another example, epoxy resin adhesives are formed in place by the reaction of epichlorohydrin,

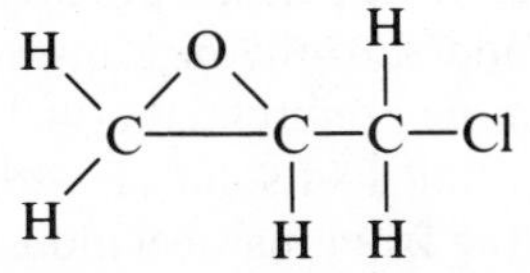

```
H    O     H
 \  / \    |
  C----C---C--Cl
 /     |   |
H      H   H
```

and bisphenol A, or by reactions of their derivatives. Epichlorhydrin is a toxic, ignitable liquid with an *open-cup flash point* of 40.6 °C. (The open-cup flash test involves heating an ignitable liquid in an open container near a flame and noting the temperature at which the vapor from the liquid ignites.)

Paint and ink formulation involve a number of hazardous materials, such as ignitable monomers used to make paint resins, volatile solvents used as paint and ink vehicles, toxic paint and ink pigments, and compounds used as drying agents.

Leather tanning is a general term that describes various operations involved in converting animal hides to leather. Chemicals are employed to dissolve hair, to convert the hide to leather by the infusion of tannin and various metal salts, and for other purposes. Hair pulping may be accomplished by partially dissolving the hair in a concentrated solution of lime and other chemical agents, such as sodium sulfhydrate. Chrome tanning involves treatment of the hide with chromium(III) salts, usually chromic sulfate, along with additives such as sodium formate and soda ash. The potential of hazardous waste chemical production from leather tanning and finishing is relatively high.

The **manufacture of wood products and paper** requires a large variety of potentially hazardous chemicals. Mention has already been made of phenol and formaldehyde used in resins employed to bond plywood together. As another example, paper pulping makes use of toxic and corrosive chemicals, such as NaOH and Na_2S.

Soap and detergent manufacture employ some potentially hazardous chemicals, such as corrosive NaOH used for saponification of fats and sulfuric acid used for sulfonation in surfactant manufacture. If strongly chelating sodium nitrilotriacetate is ever legalized as a detergent builder in the U.S. (see Section 4.15), it could increase the potential of improperly disposed detergents to mobilize heavy metals in disposal sites.

Mining and mineral extraction employ a wide variety of hazardous materials, such as cyanide salts used to extract gold and silver from ores. Among the specific hazardous wastes from this industry listed by the EPA are acid-plant blowdown slurry/sludge resulting from the thickening of blowdown slurry from primary copper production; surface impoundment solids contained in and dredged from surface impoundments at primary lead smelting facilities; sludge from treatment of process wastewater and/or acid-plant blowdown from primary zinc production; electrolytic anode slimes/sludges from primary zinc production; cadmium-plant leach residue (iron oxide) from primary zinc production; and emission control dust/sludge from secondary lead smelting.

The **iron and steel industry** is a huge one with as many as 24 subcategories. It uses large quantities of some chemicals and produces a variety of chemical by-products. This industry has appreciable potential for the production of hazardous wastes. The manufacture of coke required by this industry produces dozens of hazardous chemicals, including phenols, benzene, toluene, and PNA compounds in coal tar. Steel is "pickled" by immersion in sulfuric, hydrochloric, nitric, or hydrofluoric acids, a process that dissolves iron oxide coatings. Organic inhibitors are added to the pickling solution to prevent acid attack on the base metal and promote preferential attack on the oxide; organic wetting agents promote contact of the acid solution with the metal surface. Among the hazardous wastes originating from the steel industry and listed by the EPA are dewatered air pollution control scrubber sludges from coke ovens and blast furnaces; ammonia-still lime sludge from coking operations; emission control dust/sludge from electric furnace production of steel; spent pickling liquor from steel finishing operations; and sludge from lime treatment of spent pickling liquor.

The **nonferrous metals industry** involves the primary and secondary smelting and refining of metals other than iron. Among the wastes from this industry designated as hazardous by the EPA are flotation tailings from selective flotation from mineral metals recovery operations; acid-plant blowdown slurry/sludge resulting from the thickening of blowdown slurry from primary copper production; surface impoundment solids contained in and dredged from surface impoundments at primary lead smelting facilities; sludge from treatment of primary wastewater and/or acid-plant blowdown from primary zinc production; electrolytic anode slimes/sludges from primary zinc production; cadmium-plant leach residue (iron oxide) from primary zinc production; and emission control dust/sludge from secondary lead smelting.

Electroplating is an operation in which a thin layer of a metal is coated onto another metal through the application of an electrical current. There is considerable potential for the production of hazardous wastes from electroplating. Many of the metals pose environmental and toxicological hazards. The plating bath solutions are usually hazardous. For example, highly toxic cyanide is a common plating bath ingredient. Chelating agents, such as NTA and EDTA (see Chapter 4), are often used in plating baths. These may act to transport heavy metals in disposal sites. In addition to their toxicity, some plating bath solution ingredients are hazardous because of their reactivity.

Battery manufacture is an industry that produces lead storage batteries, silver-cadmium storage batteries, dry cells, and other specialty batteries. The greatest potential for hazardous waste production is in lead storage battery manufacture, given the huge size of the industry and the use of corrosive sulfuric acid and toxic lead.

The **electrical and electronic components industry** requires operations, such as electroplating, which make use of hazardous materials. The quantities of materials are generally rather small. In 1983 leaks were found in some underground tanks of trichloroethylene used to wash electronic parts in California's "Silicone Valley."

Steam electric-power generation includes the generation of electricity from steam raised in boilers heated by coal, oil, gas, wood, peat, or nuclear fission. Each of these sources produces some potentially hazardous material. The important question of nuclear wastes was covered in Section 17.12. The largest quantities of solid wastes produced by power generation are from coal combustion. These wastes include fly ash, dry bottom ash, wet bottom ash, and cyclone boiler slag. In 1978, combustion of 480 million metric tons of coal in power plants resulted in the production of 68 million metric tons of ash [17]. This included 49 million metric tons of fly ash, 14 million metric tons of dry bottom ash, and 5 million metric tons of wet bottom ash and cyclone boiler slag. Increasing use of wet scrubber systems to remove sulfur dioxide from stack gas is increasing the amount of solid waste produced by electric power generation. The use of wet lime to scrub stack gas results in the production of large quantities of wet sludge, posing a substantial disposal problem.

Water treatment for municipal and industrial use, as well as wastewater treatment, was discussed in Chapter 8. Some potentially hazardous chemicals, particularly corrosive ones, are employed for water treatment. Of course, the hazards of chlorine (used to disinfect water) are well-known. The biggest waste problem with water treatment is the disposal of sludges resulting from water treatment processes. In some cases where a wastewater treatment plant receives industrial wastewater, these sludges may contain heavy metals and hazardous refractory (difficult to degrade) organic chemicals (see Section 7.17).

Petroleum refining produces a number of waste products. Some of these result from impurities in crude oil, which must be removed as part of the production of refined products. In addition, spent catalyst materials are among the solid waste by-products. Among specific sources of hazardous wastes from petroleum refining are dissolved air flotation (DAF) float; slop-oil emulsion solids; heat-exchanger bundle sludge; API separator sludge; and leaded tank bottoms.

19.12 MUNICIPAL SOLID WASTE

So far, this chapter has covered hazardous wastes, primarily from industrial sources. A less ominous, but still troublesome, problem is the 270 million metric tons of **municipal solid waste** ("garbage") produced in the U.S. each year. The trash that is thrown away from homes and businesses amounts to more than one ton per person per year. Approximately 50% of this waste consists of paper and paper products, 25% consists of metal cans, about 22% is made up of bottles and jars, and the remainder is miscellaneous material such as rubber and plastics. These figures are encouraging because the three major items can be recycled. Paper is already being recycled on a small scale; metal cans are also being recycled, since they can be melted down for a fraction of the energy used to refine the metals from their ores; and returnable bottles are being used more commonly.

Unfortunately, the most common present means of disposing of refuse is burial. Rubbish burial used to occur in the ignominious "city dump", which

---

[17] Murtha, M.J., and Burnet, G. 1979. Processes to increase utilization of power solid wastes. In *Solid waste research and development needs for emerging coal technologies,* D.W. Weeter, ed., pp. 135–152. New York: American Society of Civil Engineers.

usually consisted of an old quarry, some natural depression or ravine, or a swamp. Refuse was hauled to these areas and simply dumped. Sometimes it was burned, either deliberately or accidentally, and an unpleasant cloud of smoke issued from it. In most places, the city dump has given way—in name, at least—to "**sanitary landfill**" operations. Open burning of garbage at such places is prohibited. The solid wastes are hauled to a location where it is dumped and covered daily with a layer of soil. In the best cases, grass is eventually planted on the soil, and a park is established. It is even possible to construct buildings over the landfill, although it is not usually recommended because of seepage of explosive methane gas from the buried refuse.

A much better way of handling municipal refuse is through **resource recovery**. It is possible to recover energy, ferrous metals (steel), non-ferrous metals (particularly aluminum), glass, plastics, and paper from municipal solid waste. Recovery of all these materials requires a rather sophisticated system. However, a lot of benefit could be derived from recovering ferrous metals (by magnetic separation) and energy (by burning combustible materials), leaving a smaller volume and weight of solid for disposal in a landfill. A diagram of such a system is shown in Figure 19.4 The process involves several steps. First, a dryer dries the refuse, and a shredder reduces it in size. Ferrous metals separation is accomplished by passing the refuse between two belts. On the bottom side of the top belt is an electromagnet that attracts iron objects to this belt and holds them there until they leave the magnetic field and are dumped away from the rest of the waste. Air is then used to separate out combustible lighter wastes (mostly paper) to produce **refuse derived fuel (RDF)**. This product may be compacted into

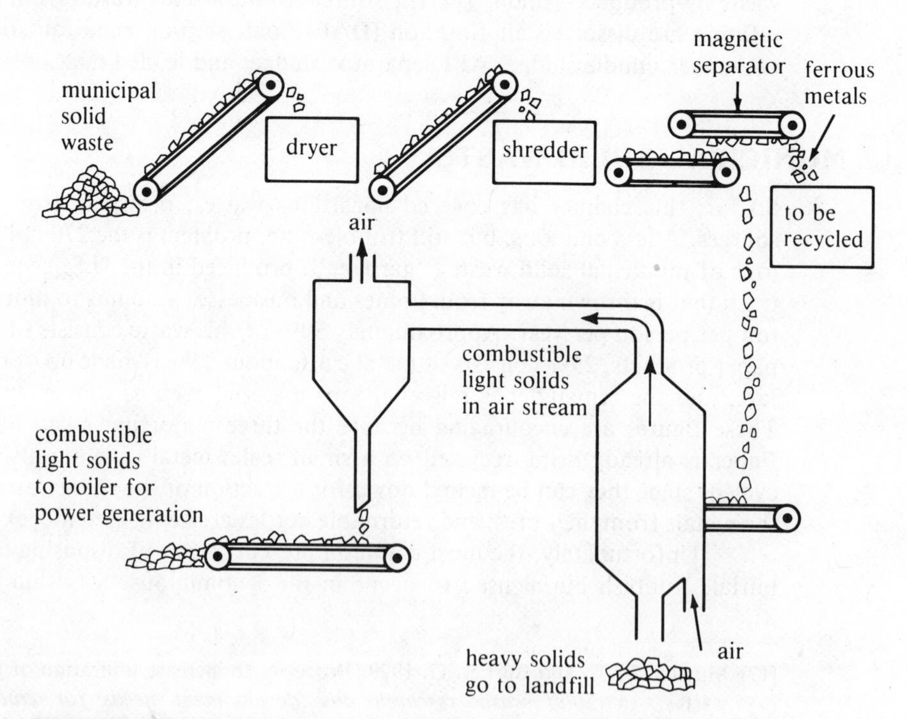

FIGURE 19.4 System for the recovery of fuel and ferrous metals from municipal solid waste.

agglomerated pieces which can substitute for coal in spreader-stroker boilers [18].

Unlike many industrial materials of known, homogeneous composition, municipal waste is somewhat difficult to handle. The refuse consists of many different materials in a variety of shapes and sizes. Garbage tends to become putrid through bacterial decay, and it can be a fire hazard. It does not flow well, may contain abrasive materials, and tends to compact upon standing. These problems present special challenges to the various operations shown in Figure 19.4.

Many areas in Europe have a long-standing tradition of burning refuse to raise steam and generate electricity. For example, Luxembourg, Denmark, and Switzerland each burn about a kilogram of refuse per capita per day. Most European plants are of simple design in which the refuse is simply burned in a furnace and the heat used to generate steam. These plants work well but must be scrupulously maintained. Rufuse-burning technology in the U.S. has tended to concentrate on more exotic systems designed to produce refuse-derived solid fuel or pyrolysis products. Some of these plants have been plagued with operational problems. One plant in Bridgeport, Connecticut was operated for a year and a half in 1979–80, but only converted a three-and-a-half-weeks' supply of garbage to powdery, refuse-derived fuel [19]. Furthermore, one state environmental inspector termed the odor from the plant, "bad enough to gag a maggot". The Hempstead Resources Recovery facility in Garden City, New York, was closed in 1980 after it was allegedly found to be emitting highly toxic TCDD (see Section 7.18).

Along with other densely populated parts of the country, New York City has an increasing need for functional garbage incineration facilities. The city produces about 21,000 tons of solid waste each day but is rapidly running out of places to put it. In late 1981, a large incinerator/power-generating facility was proposed for a location in the former Brooklyn Navy Yard. This plant, designed to burn 3,000 tons of garbage per day, could save a million barrels of oil per year.

Despite technical difficulties and citizen opposition to the location of incineration facilities, refuse burning with energy recovery is an idea whose time has come. Shortages of energy, raw materials, and land for landfill sites dictate that resource recovery from refuse be practiced by any modern society.

Clean air and clean water regulations have resulted in the use of pollution-control technologies that produce solid wastes, the disposal of which can be difficult. A prime example of this is sewage sludge from publicly owned treatment works. Whereas requirements for increased wastewater treatment have increased the amount of sludge produced, regulation of solid waste disposal has restricted the disposal options [20]. Legislation has been passed to encourage the beneficial use and recycling of such waste materials, but there may be some adverse consequences from such uses. In the case of sewage sludge, for example, the high heavy

---

[18] Campbell, J.; Renard, M.L.; and Winter, E.J. 1981. *Densification of refuse-derived fuels: preparation, properties and systems for small communities.* Environmental Protection Agency Report EPA-600-S2-81-188. Springfield, Va.: National Technical Information Service.

[19] Barron, J. 1981. Garbage is garbage, burning it for energy is difficult. *New York Times* 2 August 1981.

[20] Dawson, J.F., et al. 1981. Sewage sludge pathogen model. EPA-600/S1-81-049. Washington, D.C.: Environmental Protection Agency.

metal content and the potential for the release of human pathogens (bacteria and viruses) may pose risks in the reuse of this material in applications such as a land conditioner.

19.13 SOLID WASTES FROM NON-MUNICIPAL SOURCES

Although municipal solid wastes are the most widespread type of solid wastes, there are other major solid waste sources. Some of these sources produce large quantities of wastes in a single location. Much of the solid waste from non-municipal sources consists of material dredged from harbors, streams, and other bodies of water. In the U.S. about 360 million cubic meters of dredged material requires disposal each year.

Analysis of ground water underneath and downgradient from dredge disposal sites tends to show high levels of calcium and magnesium ions, which contribute to degraded water quality due to excess hardness. Dredged material normally contains sufficient organic matter to produce a low pE due to the action of anaerobic bacteria (see Chapters 3 and 5). This tends to result in excessive levels of reduced iron and manganese in the form of Fe^{2+} and Mn^{2+} ions. Levels of alkalinity, potassium, sodium, chloride, and total organic carbon (TOC) are often excessive in dredged material leachate.

Material is often dredged from streams and estuaries draining industrial areas and from harbors that have been exposed for decades to spills from ships and runoff from surrounding industrialized areas. Thus, the dredged material may contain pollutants, such as heavy metals and refractory organic compounds. Dredging tends to mobilize these materials and fine particulate matter from sediments, with a temporary degradation of water quality in the body of water dredged. However, dredging is necessary in many cases to keep navigable waters open and to open harbors to larger ships. An example of the latter is the work that needs to be done to open selected U.S. harbors to the large colliers needed to serve a thriving coal export market.

SUPPLEMENTARY REFERENCES

Bahme, C.W. 1978. *Fire officer's guide to dangerous chemicals.* 2nd ed. Boston: National Fire Protection Association.

Bennett, G.F.; Feates, F.S.; and Wilder, I. 1982. *Hazardous materials spills handbook.* New York: McGraw-Hill Book Company.

Bowman, M.C., ed. 1982. *Handbook of carcinogens and hazardous substances.* New York: Marcel Dekker, Inc.

Bretherick, L. 1979. *Handbook of reactive chemical hazards.* 2nd ed. Woburn, Mass.: Butterworth Publishers, Inc.

Bridgewater, A.V., and Mumford, C.J. 1980. *Waste recycling and pollution control handbook.* New York: Van Nostrand Reinhold.

Chemical Manufacturer's Association. 1982. *A hazardous waste site management plan.* Washington, D.C.

Conway, R.A., ed. 1981. *Environmental risk analysis for chemicals.* New York: Van Nostrand Reinhold.

Cornaby, B.W., ed. 1981. *Management of toxic substances in our ecosystems.* Ann Arbor, Mich.: Ann Arbor Science Publishers, Inc.

Dlouhy, Z. 1982. *Disposal of radioactive wastes.* New York: Elsevier-North Holland, Inc.

Dyer, J.C.; Vernick, A.S.; and Feiler, H.D. 1981. *Handbook of industrial wastes pretreatment.* New York: Garland Publishing, Inc.

Epstein, S.S.; Brown, L.O.; and Pope, C. 1982. *Hazardous waste in America.* San Francisco: Sierra Club Books.

Freney, J.R., and Galbally, I.E., eds. 1982. *Cycling of carbon, nitrogen, sulfur, and phosphorus in terrestrial and aquatic ecosystems.* New York: Springer-Verlag New York, Inc.

Jitendra, S., and Fisher, F. 1981. *Hazardous assessment of chemicals.* Vol. 1. New York: Academic Press, Inc.

Jitendra, S. 1983. *Hazardous assessment of chemicals.* Vol. 2. New York: Academic Press, Inc.

Johansson, T.B., and Steen, P. 1981. *Radioactive waste from nuclear power plants.* Berkeley, Calif.: University of California Press.

Ketchum, B.H.; Kester, D.R.; and Park, P.K. 1981. *Ocean dumping of industrial wastes.* New York: Plenum Publishing Corp.

Klang, Y.-H., and Metry, A.A. 1982. *Hazardous waste processing technology.* Ann Arbor, Mich.: Ann Arbor Science Publishers, Inc.

Leidner, J. 1981. *Plastic waste.* New York: Marcel Dekker, Inc.

Levine, A.G. 1982. *Love Canal.* Lexington, Mass.: Lexington Books (D.C. Heath & Co.).

Long, F.A., and Schweitzer, G.E., eds. 1982. *Risk assessment at hazardous waste sites.* Washington, D.C.: American Chemical Society.

Mallow, A. 1981. *Hazardous waste regulations.* New York: Van Nostrand Reinhold.

Murray, M.Y.; Robinson, C.W.; and Farquhar, G.J., eds. 1981. *Waste treatment and utilization—theory and practice of waste management.* Elmsford, N.Y.: Pergamon Press, Inc.

Nader, R.; Brownstein, R.; and Richard, J. 1981. *Who's poisoning America?* San Francisco: Sierra Club Books.

Overcash, M.R. 1982. *Behavior of organic priority pollutants in the terrestrial system.* Raleigh, N.C.: Water Resources Research Institute of the University of North Carolina.

Pierce, J.J., and Vesilind, P.A., eds. 1981. *Hazardous waste management.* Ann Arbor, Mich.: Ann Arbor Science Publishers, Inc.

Pojasek, R.B., ed. 1979. *Toxic and hazardous waste disposal.* Vol. 1: *Processes for stabilization/solidification.* Ann Arbor, Mich.: Ann Arbor Science Publishers, Inc.

1982. *Proceedings of the 36th Industrial Waste Conference.* Ann Arbor, Mich.: Ann Arbor Science Publishers, Inc.

Santhanam, C.J. 1981. *Flue gas cleaning wastes—disposal and utilization.* Park Ridge, N.J.: Noyes Data Corp.

Sax, N.I. 1979. *Dangerous properties of industrial materials.* 5th ed. New York: Van Nostrand Reinhold.

Shuckrow, A.J.; Pajack, A.P.; and Touhill, C.J. 1982. *Hazardous waste leachate management manual.* Park Ridge, N.J.: Noyes Data Corp.

Sittig, M. 1979. *Incineration of industrial hazardous wastes and sludges.* Park Ridge, N.J.: Noyes Data Corp.

——. 1979. *Landfill disposal of hazardous wastes and sludges.* Park Ridge, N.J.: Noyes Data Corp.

Smith, A.J., Jr. 1981. *Managing hazardous substances accidents.* New York: McGraw-Hill Book Company.

Stevens, B.J. 1980. *Handbook of municipal waste management systems.* New York: Van Nostrand Reinhold.

Student, P.J., ed. 1981. *Emergency handling of hazardous materials in surface transportation.* Washington, D.C.: Association of American Railroads.

1980. *Toxic substances sourcebook.* Series 2. New York: Environmental Information Center.

Verschueren, K. 1977. *Handbook of environmental data on organic chemicals.* New York: Van Nostrand Reinhold.

Weiss, G., ed. 1980. *Hazardous chemicals data book.* Park Ridge, N.J.: Noyes Data Corp.

Wilson, D.C. 1981. *Waste management.* New York: Clarendon.

Zajic, J.E., and Himmelman, W.A. 1978. *Highly hazardous materials spills and emergency planning.* New York: Marcel Dekker, Inc.

QUESTIONS AND PROBLEMS

1. What is *chloracne*?
2. What is the predominant class of organic compounds found in leachate from the Love Canal waste disposal site?
3. What is the greatest single source of waste chemical sites and impoundments?
4. What are the four criteria by which hazardous wastes are defined for regulatory purposes?
5. Would a quantity of discarded "non-safety" matches be likely to be classified as a hazardous waste? If so, on what basis?
6. Why was the upper pH limit for the legal criterion for corrosivity raised from 12.0 to 12.5?
7. What use is made of the National Interim Primary Drinking Water Standards in defining hazardous wastes?
8. What does fermentation have to do with the mobilization of hazardous wastes in landfills?
9. Describe the role of citrate ion in mobilizing heavy metals in landfills.
10. If *attenuation factor* is defined as the ratio,

$$\frac{\text{concentration of pollutant in wastewater stream}}{\text{concentration of pollutant in water sample after contact with geological strata}}$$

explain an attenuation factor of 0.5.

11. Match each waste in the right column with its classification in the left column.

(a) Hazardous wastes from nonspecific sources	(1) Spent carbon from the treatment of wastewater containing explosives
(b) Hazardous waste from specific sources	(2) $H_2C{=}CH{-}C({=}O){-}H$
(c) Discarded commerical chemical products, off-specification species, containers, and spill residues	(3) Spent plating bath solutions from electroplating operations

12. According to what structural characteristic may hazardous organic compounds be classified?

13. Match each of the functional groups in the structures in the right column with its name in the left column. (*R* represents a hydrocarbon moiety and ϕ represents an aromatic moiety.)

(a) Amine	(1) $R{-}C({=}O){-}R$
(b) Aldehyde	(2) $R{-}C({=}O){-}O{-}R$
(c) Phenol	(3) $\phi{-}OH$
(d) Nitro	(4) $R{-}NH_2$
(e) Ketone	(5) $R{-}C({=}O){-}H$
(f) Nitrosamine	(6) $R{-}NO_2$, $\phi{-}NO_2$
(g) Ester	(7) $R{-}N(N{=}O){-}R$

14. In addition to the hazards of ignitability and toxicity, what is an additional hazard presented by ethers?

15. Match each inorganic chemical in the left column with one of its major characteristics or uses listed in the right column.

(a) $Hg(CNO)_2$	(1) Oxidizing agent
(b) $PbCrO_4$	(2) Extremely reactive
(c) H_2S	(3) Highly toxic salt composed of elements generally considered to be innocuous
(d) KCN	(4) Pigment ingredient
(e) N_2O	(5) Gas that is ignitable in air

16. What type of chemical substance would be regarded as equivalent to a corrosive material in terms of its reaction with rubber?

17. What measure adopted to alleviate air pollution from power plants is increasing problems of solid waste disposal?

18. Two major products can be removed from municipal solid waste and put to beneficial use. What are these products, and what are the basic methods for their removal?

CHAPTER 20
ENVIRONMENTAL CHEMISTRY OF HAZARDOUS WASTES

20.1 SEGREGATION AND RECOVERY OF HAZARDOUS WASTE COMPONENTS

Well-managed hazardous materials never enter the environment at all. Therefore, it is very important to manage potentially hazardous substances properly from the moment of production. It is particularly vital to keep the components of waste products segregated to facilitate resource recovery. In a "worst-case scenario", wastes would be contained in a mixture of organic and inorganic material in the form of dilute aqueous sludge [1]. It is extremely difficult either to recover products from such a mixture or to isolate and treat harmful constituents. The ideal situation would involve wastes containing only one major constituent in a well-defined matrix.

Wastes make up a diverse variety of substances which may be classified in numerous ways. Most potentially hazardous wastes may be classified as (1) **organics**, (2) **aqueous wastes**, or (3) **sludges**. These, in turn, may be divided into **levels of segregation**. Let's consider organics first. The lowest level of segregation of organics (that is, none at all) was described above as a mixture of organic and inorganic material in the form of dilute aqueous sludge. At the next higher level of segregation, organics are present as **mixed organics** combined with little or no water. Barring the presence of refractory constituents, such a mixture might be burned as fuel. At the next higher level of segregation, organics may be present as relatively pure materials or as mixtures of compounds with similar properties. Among the examples in ths category is a mixture of hydrocarbons, which could easily be burned as fuel. Chlorinated solvents used in degreasing can be distilled for reuse. Waste lubricating oils can be reclaimed. Other organics at this level of segregation, including some pesticides, may need to be subjected to special detoxification or destruction procedures; if these wastes have been kept from contamination with other wastes, detoxification and destruction procedures are greatly facilitated.

---

[1] Lyman, W.J., and Contos, G. 1980. Inorganic hazardous waste treatment. In *Treatment of hazardous waste.* Environmental Protection Agency Publication EPA-600/9-80-011, pp. 62–71. Springfield, Va.: National Technical Information Service.

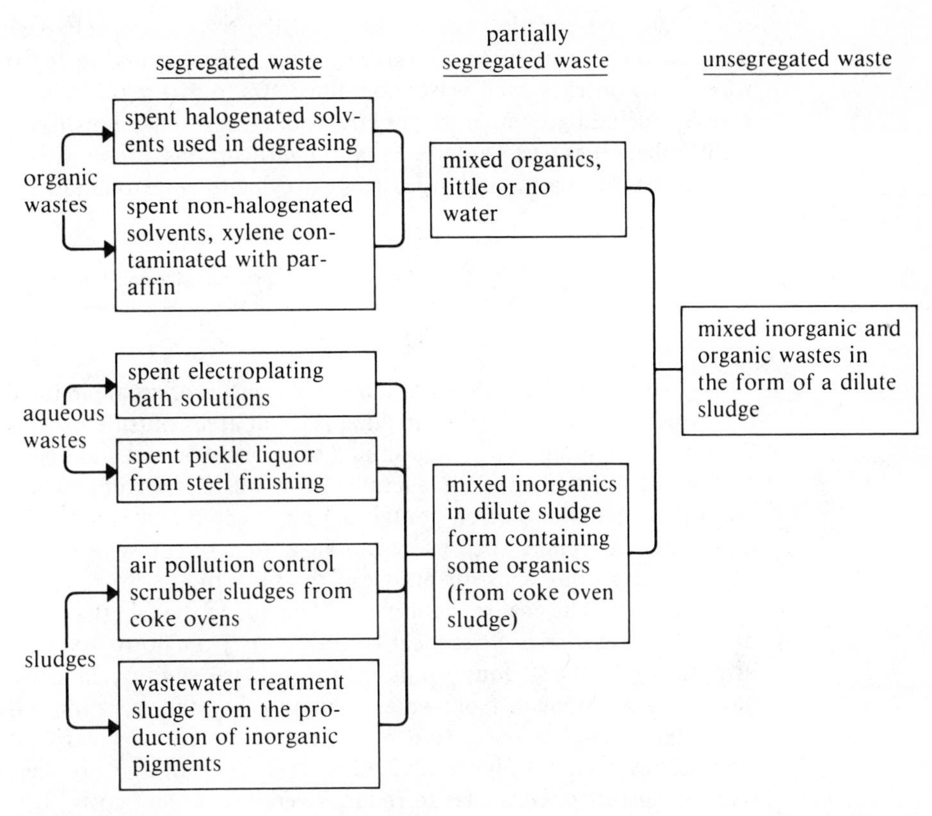

FIGURE 20.1 Sample hazardous wastes showing various degrees of segregation.

As with organic wastes, inorganic constituents in a totally unsegregated waste are hard to reclaim or detoxify. The problem is somewhat simplified at the partially segregated waste stage, due to the removal of most of the organic matter (see Fig. 20.1). At this stage, the inorganic wastes are in the form of a dilute sludge. In some cases, treatment of such an inorganic waste with lime and a coagulant might result in isolation of the hazardous constituents in the form of a more concentrated sludge. At the highest level of inorganic waste segregation are two general classes of segregated wastes—aqueous wastes and sludges. These are, in principle, relatively easy to treat by techniques such as activated carbon sorption of contaminants from aqueous wastes or dewatering and concentration of sludges.

An example of an unsegregated hazardous waste has been described in the literature [1]. The particular waste discussed in Reference 1 was collected by a New England waste hauler from the plating industry and consisted of mixed acidic plating wastes stored in a common tank prior to disposal in a secure landfill. Being a composite of acidic materials, the waste had a very low pH of 0.5–3. The dissolved solids ranged from 8,000 to 10,000 mg/L, and the suspended solids ranged from 2,000 to 10,000 mg/L, so that both dissolved and suspended solids had very high concentration values. Among the metals present at varying levels in both the dissolved and suspended fractions were aluminum, sodium, chromium, iron, nickel, copper, zinc, lead, and tin. Boron was also present, as were high concentrations of the anions Cl^-, NO_3^-, F^-, SO_4^{2-}, and PO_4^{3-}; the more strongly basic of these were, of course, protonated in the strongly acidic waste. The unsegregated waste contained 100 to 500 mg/L of organics (oil and grease), including comparatively large quantities of phenols (see Section 19.9), dichlorobenzenes (see Table 19.6), and hydrocarbons, along with some alkylated polynuclear aromatics and at least one alkylbenzenesulfonamide.

$$R-C_6H_4-S(=O)_2-NH_2$$

generalized formula of an alkylbenzene sulfonamide

Appreciable quantities of hazardous wastes originate in the municipal sector, which as been defined to include [1] "all areas outside the direct control of the generating industry. . . not limited to publicly owned and operated treatment plants and disposal sites". Generally, such waste is a complex mixture of solid and liquid materials, often containing an organic component. Treatment processes that can render such waste nonhazardous or can separate out small quantities of hazardous constituents from the bulk of the waste can greatly reduce disposal costs. The cost reduction would be due to the fact that disposal of hazardous wastes in a secure chemical landfill costs from 10 to 100 times as much as disposal of nonhazardous solids in an ordinary sanitary landfill. Removal of hazardous components from water in an unsegregated waste may allow release of the water through a sewer to a publicly owned treatment works (POTW), at a tremendous savings. Figure 20.2 illustrates some of the options available for treating unsegregated wastes to reduce overall treatment costs.

Although the major objective of any hazardous waste treatment system must be either to render the waste nonhazardous or to reduce the volume of the hazardous component, good environmental practice requires that consideration also be given to resource and/or energy recovery. From a cost viewpoint, transportation expense is often the overwhelming factor. Thus, if the hazardous component of a mixed waste can be reduced to a very small fraction at, or near, the

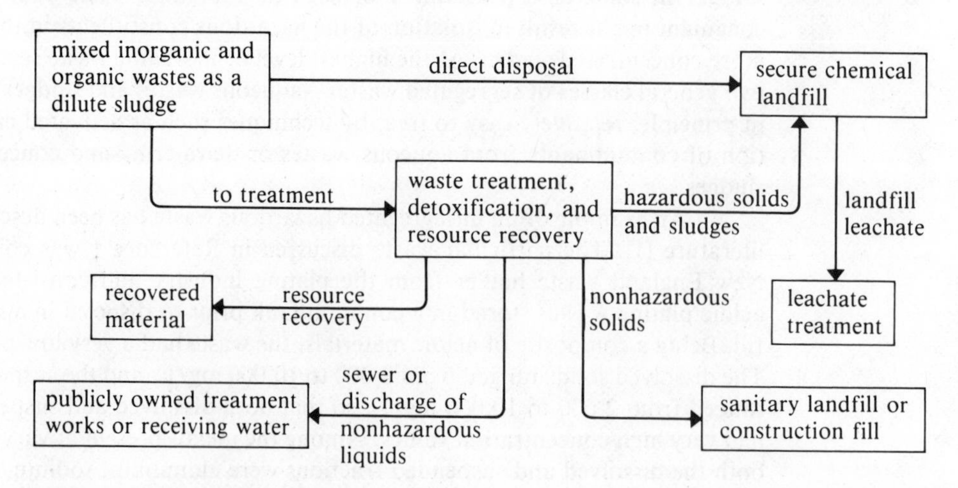

FIGURE 20.2 Waste treatment options for unsegregated wastes from the municipal sector.

site, the extremely high costs of transporting large quantities of hazardous wastes can be avoided.

Incineration is one of the most effective techniques for the destruction of hazardous materials. It may even be used for the destruction of organochlorine compounds, such as trichlorophenol (a raw material for the manufacture of the herbicide 2,4,5-T) and pentachlorophenol, a wood preservative. Destruction of such refractory compounds in an incinerator requires temperatures of 1,200 °C and a residence time at that temperature of 2–4 seconds [2].

20.2 TREATMENT PROCESSES FOR UNSEGREGATED WASTES

A number of processes are employed for the treatment of unsegregated hazardous wastes in order to render all or the major part of the wastes nonhazardous. The nonhazardous portion remaining may then be discharged to a sewer or otherwise disposed of by safe and relatively inexpensive means.

One of the most common such treatment processes is **neutralization** of excess acid or base:

$$\underset{\text{acid waste}}{2\,H^+} + \underset{\text{lime}}{Ca(OH)_2} \rightarrow Ca^{2+} + 2\,H_2O \quad (20.2.1)$$

$$\underset{\text{alkaline waste}}{2\,OH^-} + \underset{\text{sulfuric acid}}{H_2SO_4} \rightarrow 2\,H_2O + SO_4^{2-} \quad (20.2.2)$$

Another common treatment process designed to render specific waste components nonhazardous is **oxidation-reduction**. For example, toxic cyanide ion can be oxidized to less toxic cyanate,

$$\underset{\text{cyanide waste}}{CN^-} + \underset{\text{oxidizing agent}}{\{O\}} \rightarrow \underset{\text{cyanate}}{CNO^-} \quad (20.2.3)$$

or to N_2 and HCO_3^-:

$$2\,CN^- + 5\,Cl_2 + 10\,OH^- \rightarrow N_2 + 2\,HCO_3^- + 10\,Cl^- + 4\,H_2O \quad (20.2.4)$$

Another common example is the reduction of soluble, toxic chromium(VI) to insoluble chromium(III):

$$\underset{\text{chromate}}{8\,CrO_4^{2-}} + \underset{\text{sodium sulfide reductant}}{15\,Na_2S} + 20\,H_2O \rightarrow 4\,Cr_2S_3(s) + 3\,SO_4^{2-} + 40\,OH^- + 30\,Na^+ \quad (20.2.5)$$

or

$$2\,CrO_4^{2-} + 3\,SO_2 + 4\,H_2O \rightarrow 2\,Cr(OH)_3(s) + 3\,SO_4^{2-} + 2\,H^+ \quad (20.2.6)$$

Precipitation is the most common method for removing a soluble material as an insoluble sludge. The most widely used precipitating agent is lime,

---

[2] Miller, S. 1981. Destroying hazardous wastes. *Environmental Science and Technology* 15: 1268–9.

$Ca(OH)_2$, which removes heavy metals and anions by reactions such as the following:

$$3\ Pb^{2+} + 2\ HCO_3^- + 2\ Ca(OH)_2 \rightarrow 2\ PbCO_3 \cdot Pb(OH)_2(s) + 2\ Ca^{2+} + 2\ H_2O \quad (20.2.7)$$

Basic lead carbonate containing 2 formula units $PbCO_3$ per formula unit $Pb(OH)_2$

$$3\ PO_4^{3-} + 5\ Ca(OH)_2 \rightarrow Ca_5OH(PO_4)_3(s) + 9\ OH^- \quad (20.2.8)$$

$$2\ F^- + Ca(OH)_2 \rightarrow CaF_2(s) + 2\ OH^- \quad (20.2.9)$$

Flocculation by means of chemical flocculating agents is employed to remove freshly formed precipitates from suspension, along with colloidal suspended solids originally present in the waste (see Section 6.6).

Adsorption, particularly by activated carbon, may be employed to remove dissolved and very fine suspended materials. If necessary, treated water from unsegregated waste may be further purified by reverse osmosis, ion exchange, or electrodialysis (see Chapter 8).

The most challenging problem involved in treating unsegregated wastes generally is the treatment of sludge isolated from the waste. Usually, the sludge contains a high percentage of water and must be **dewatered**. This is most commonly done by sedimentation, filtration, centrifugation, or evaporation. Once dewatered, the sludge may be further treated to make it suitable for disposal in a secure landfill; techniques for so doing are discussed in the following section. In some cases, the sludge provides a good opportunity for resource recovery because it is a relatively concentrated form of the waste.

To understand some of the treatment processes that can be applied to an unsegregated hazardous waste, consider the case of the mixed acidic plating wastes described in the preceding section [1]. A treatment scheme was developed such that all hazardous components could be removed and/or neutralized, enabling discharge of the remaining water to publicly owned water treatment works. In this scheme, the pH was raised gradually by adding $Ca(OH)_2$, and precipitated heavy metals were continually removed as the pH increased. Thus, each metal was removed at a minimum pH, which enabled selective recovery of some metals. Above pH 8, Na_2S was added to reduce and precipitate Cr(IV) (see Reaction 20.2.5) and to precipitate other heavy metals as sulfides. The addition of lime was found to remove phosphates and fluorides very effectively (see Reactions 20.2.8 and 20.2.9) and to precipitate 90% of the sulfate as $CaSO_4$. Final purification of the water and removal of residual organics was accomplished with powdered activated carbon used at both high and low pH values so that both organic acids and bases would be sorbed. In this case, the activated carbon was not reclaimed.

20.3 FIXATION OF HAZARDOUS SOLID WASTES PRIOR TO DISPOSAL

Despite the best efforts to convert hazardous wastes to a nonhazardous form and to reclaim products from wastes, disposal in a landfill will always be necessary for some wastes. Potentially dangerous solid wastes should be **fixed** before they are placed in landfills. Fixing consists of physical and chemical treatment that prevent leaching or migration of any of the harmful materials in the wastes.

An ideal fixative makes the toxic contaminants immobile and chemically nonreactive. For example, if a toxic heavy metal ion can be incorporated in an insoluble crystal lattice, the metal will be prevented from harming the environment. If the pH of the waste is maintained between 9 and 11, the more harmful metals are generally immobilized as insoluble hydroxides. Anions such as Cl^- and SO_4^{2-} are much more difficult to fix in insoluble forms than are cations. However, with the notable exceptions of some toxic anionic species such as CN^- ion and CrO_4^{2-} ion, anions are generally less dangerous species than heavy metal cations; furthermore, anions are generally more easily destroyed (see Reactions 20.2.3 and 20.2.4).

In addition to limiting solubility of pollutants, fixation should accomplish several other goals. Generally, it should improve the physical handling characteristics of the waste. Furthermore, since the rate at which soluble contaminants are leached from a solid depends upon the surface area of the solid, fixation should reduce the surface area of the solid relative to its mass through aggregation (see Fig. 20.3).

There are a number of specific schemes for fixing wastes. **Cement-based fixation** techniques were developed for safe disposal of low-level nuclear wastes. Basically, the waste is mixed with common cement (Portland cement) such as that used in construction. The mixture hardens to form a concrete material in which the hazardous material is immobilized. Some wastes weaken the cement, whereas others, such as plastics, asbestos, sulfide salts, and latex, may strengthen it.

Self-cementing techniques can be used with some special wastes that can be processed to form a kind of cement. Sludges left from flue-gas desulfurization can be used for this purpose. About 10% of the wet sludge is strongly heated (calcined) and then mixed with the wet sludge, along with fly ash and some special chemical additives. The resulting mixture sets up a concrete-like material.

Lime fixation can be used for some solids that react with $Ca(OH)_2$ and water. The solids that can be fixed this way are generally fine-grained solids derived from materials containing silicon oxides. Such substances are called **pozzolanic material**. Among the most common such wastes are ground blast-furnace

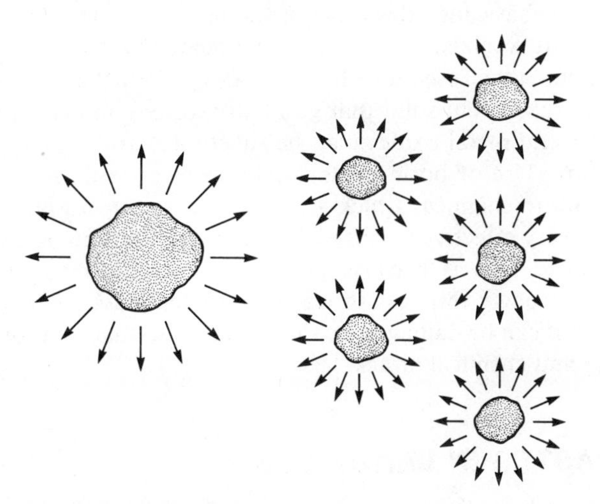

FIGURE 20.3 Soluble substances are much more readily leached from finely divided solid wastes than from those that are aggregated to form larger particles.

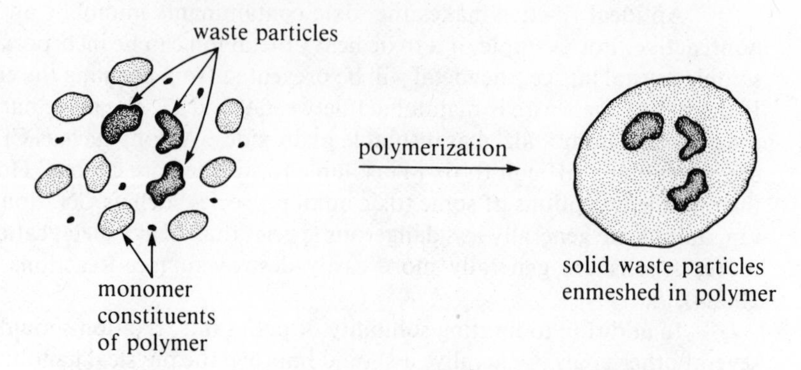

FIGURE 20.4 Wastes can be enmeshed in a polymer, thus immobilizing any hazardous materials in the waste.

slag and fly ash. These substances react with lime and water to produce a form of concrete. Sometimes other wastes can be incorporated with this mixture, enabling their disposal as well.

Thermoplastic techniques can be used for fixing some solid wastes. Largely developed for the disposal of radioactive wastes, these techniques make use of **thermoplastic substances**, which are liquid at temperatures of 120–230 °C and solids at lower temperatures. Such substances include bitumen (the asphalt-like residue of crude oil), paraffin, and polyethylene. Unfortunately, over time these materials tend to deteriorate when exposed to water or oxidizing agents.

Organic polymers are used to fix some wastes. Such polymers are commonly urea-formaldehyde or vinyl ester-styrene polymers. The monomers (small molecules) making up the polymers are mixed with the wastes, along with a catalyst to enable the polymerization to occur. When this takes place, the waste is enmeshed in a mass of polymer, as shown in Figure 20.4.

One of the most effective ways to contain very dangerous materials, such as radioactive wastes, is to incorporate the material into glass. This technique is called **glassification**. To understand how this is done, it is helpful to consider the nature of glass. The basic ingredient of glass is silicon dioxide, SiO_2, which occurs as ordinary sand. It is mixed with sodium carbonate (Na_2CO_3) and calcium carbonate ($CaCO_3$) and then fused together by heating to a high temperature to form glass. Glass is a transparent solid that gradually softens and changes to a liquid when heated. Several metal oxides may be substituted for calcium carbonate in glass manufacture. Use of boric oxide, B_2O_3, gives borosilicate glass, which is especially resistant to chemical attack and changes in temperature.

Glass is a very effective means of fixing wastes that can be incorporated in the glass structure. This is because glass is one of the most inert synthetic materials and is strongly resistant to leaching. When glass contains radioactive wastes, however, it can be damaged by the radiation so that it becomes less resistant to leaching and chemical attack.

20.4 HAZARDOUS WASTES IN LANDFILLS

The EPA defines a *hazardous waste landfill* as "a disposal facility where hazardous waste is placed in or on land and which is not a land treatment facility, a sur-

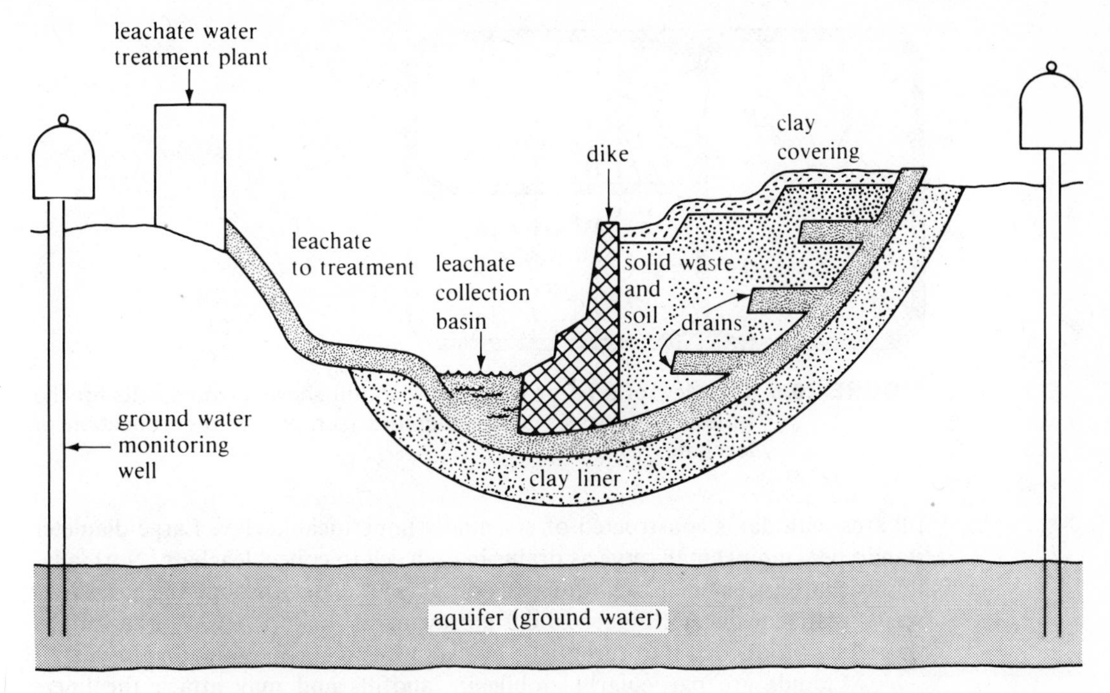

FIGURE 20.5 Properly designed landfill for the disposal of solid hazardous wastes.

face impoundment, or an injection well'' [3]. The same document defines a **landfill cell** as ''a discrete volume of a hazardous waste landfill which uses a liner to provide isolation of wastes from adjacent cells or wastes''. Trenches and pits are examples of landfill cells.

Figure 20.5 shows the main features of a secure landfill designed to contain solid hazardous wastes. Such a landfill has several features not found in a landfill designed for the disposal of ordinary municipal wastes. These include a clay seal that prevents migration of chemicals into ground water; a drainage system that enables collection of leachate, which can be treated to remove contaminants; and monitoring wells for detecting contamination of ground water.

When several different kinds of wastes are placed in the same landfill, it is often necessary to keep them segregated. This is needed because most wastes are pretreated to minimize the solubilities of hazardous solids (for example, making the waste basic to precipitate heavy metals), and the same chemical environment must be maintained to prevent the hazardous materials from redissolving [4]. Isolation of different wastes is accomplished by placing the wastes in different **cells** in the landfill [5]. Usually, these cells are constructed by dividing the land-

[3] 1980. *Federal Register* 45(98): Book 2, May 19, 1980, pp. 33063–33285.

[4] McGahan, J.F. 1980. The secure landfill disposal of hazardous wastes. In *Toxic and hazardous waste disposal,* Vol. 2, *Options for stabilization/solidification,* R.B. Pojasek, ed., Ch. 6. Ann Arbor, Mich.: Ann Arbor Science Publishers, Inc.

[5] Josephson, J. 1981. Hazardous waste landfills. *Environmental Science and Technology* 15: 250–253.

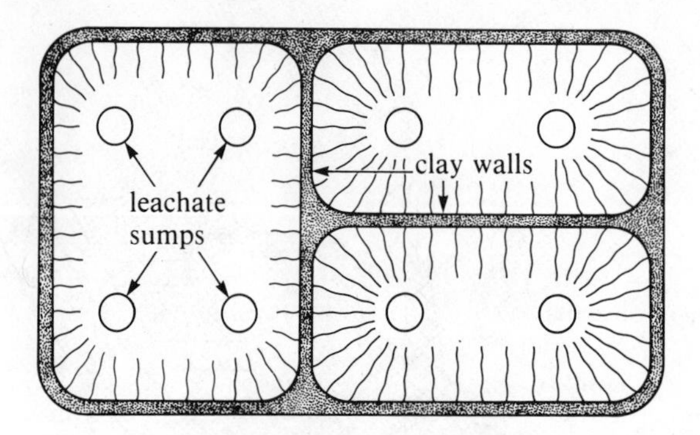

FIGURE 20.6 Top view of a hazardous waste landfill showing three cells for the isolation of wastes and standpipes (sumps) for the collection of leachate.

fill area with dams constructed of essentially impermeable clay. Large-diameter standpipes (sumps) may serve as drains in each cell to collect leachate. Normally, this leachate is treated as hazardous waste and recirculated through the hazardous waste treatment system. Figure 20.6 is a drawing of the top view of a hazardous waste landfill divided into three cells by clay walls (called **segregating berms**).

Liquids are particularly mobile in landfills and may attack the liners designed to prevent the spread of hazardous substances. For that reason, on November 19, 1981, the EPA banned the disposal of liquids in landfills. In late February, 1982, EPA Administrator Anne M. Gorsuch lifted the ban for 90 days, temporarily allowing the disposal of waste containing free liquid and containerized liquid waste in landfills [6]. At the same time, the agency proposed that up to 25% of a landfill's capacity could be used for the disposal of free liquids. A number of groups immediately questioned the wisdom of this decision, especially given growing evidence that the most commonly used lining material, clay, may not be a very effective retainer of liquid organic wastes, and that the lifetime of synthetic liners is limited. As a result of public hearings, the EPA reversed its decision before the 90-day period was up and substituted an interim rule preventing landfill disposal of barrels containing observable toxic liquids. One of the major questions regarding this action was what to do about the inspection of barrels for hazardous contents at the landfill site.

On July 13, 1982, the EPA set forth regulations for the containment of hazardous wastes at all landfill disposal facilities, including landfills and impoundments. These rules govern the more than 2,000 facilities in which hazardous materials are disposed of on land. The new rules were estimated to cost industry at least $1 billion per year. Among the specific regulations are requirements for: liners for new facilities (old facilities are exempt from this requirement); control of water infiltration and runoff; capping the facility when it is full; and monitoring for 30 years after the facility ceases operation. This set of regulations had been delayed since the passage of the RCRA Act of 1976. The Environmental Defense Fund, which had brought legal action to bear in getting the

[6] Deland, M.R. 1982. Disposal of liquid hazardous waste. *Environmental Science and Technology* 16: 225A.

regulations issued, complained about the failure to require installation of liners in existing land dumps and the lack of air-pollution control criteria for landfills.

Finding disposal sites for low-level radioactive wastes, such as spent filters from power reactors and glassware, plastic booties, rubber gloves, and animal carcasses from laboratories, is the center of some controversy. Until recently sites at Hanford, Washington; Barnwell, South Carolina; and Nevada have taken such wastes. According to legislation passed by Congress in 1980, states are responsible for disposal of their own wastes but may form multi-state compacts for that purpose, resulting in disposal in one state, only. Low-level radioactive wastes (in contrast to fission products from reactor cores) are among the least hazardous and most manageable of all hazardous wastes.

According to guidelines approved by the Nuclear Regulatory Commission (NRC) in October, 1982, low-level nuclear wastes can be divided into three categories. Class A wastes, constituting about 60% of all low-level wastes, decay to safe levels within 100 years. These wastes may be buried as liquids in containers in areas to be protected for 100 years. Class B wastes take up to 300 years to decay. Their disposal requires solidification or burial in high-integrity containers. Class C wastes, requiring up to 500 years of decay, must be solidified and buried at least 15 meters deep.

Some authorities are opposed to any landfill disposal of hazardous wastes. In early 1983 the National Governor's Association forwarded a resolution to the Environmental Protection Agency stating that "the agency should be required to develop regulations phasing out the burial of hazardous wastes where alternative treatment technologies are reasonably available." It furthermore stated that "the land disposal of wastes which are highly toxic or persistent should be immediately prohibited."

Alternatives to landfill include the following: (1) reduction of waste generation by changing industrial processes; (2) recovery and recycling of wastes for conversion to useful products; (3) treatment to reduce waste volume and toxicity by biological, chemical, or physical means; (4) land or sea incineration; and (5) disposal of dense, insoluble wastes in deep ocean water. Secure surface storage has been proposed for wastes lacking satisfactory means of treatment until treatment processes can be developed. It has even been proposed that any product or process producing wastes for which safe means of handling (excluding landfill) are unavailable should be banned. Enforcement of such a regulation could be very damaging to the national economy, and an intense program of research for handling such wastes is preferable.

20.5 HAZARDOUS WASTES IN THE HYDROSPHERE

A large variety of hazardous and pollutant chemicals enter the hydrosphere (streams, reservoirs, ground water) through leaching from waste disposal sites, accidental or intentional disposal in wastewaster, volatilization followed by scavenging from the atmosphere by rainwater, and other routes. Once they enter an aquatic system, these species are subject to a number of chemical and biochemical phenomena, including acid-base, oxidation-reduction, precipitation-dissolution, and hydrolysis reactions, as well as biodegradation. Ultimately, the waste or its degradation products end up in a permanent or semipermanent repository, known as a **sink**, or **receptor**.

In considering the processes that hazardous wastes undergo in water, it is important to recall the nature of aquatic systems and the unique properties of water (see Chapter 2). Water encountered outside the laboratory is far from pure. Just as the atmosphere is a constantly changing mass of bodies of moving air with different temperatures, pressures, and humidities, so bodies of water are highly dynamic systems [7]. Rivers, impoundments, and ground-water aquifers are subject to the input and loss of a variety of materials from both natural and anthropogenic sources. These materials may be gases, liquids, or solids. They interact chemically with each other and with living organisms, particularly bacteria, in the water. They are subject to dispersion and transport by stream flow, convection currents, and other physical phenomena. Hazardous substances or their byproducts in water may undergo bioaccumulation through aquatic organism food chains.

Several physical, chemical, and biochemical processes are particularly important in determining the transformations and ultimate fates of hazardous chemical species in the hydrosphere. These include **hydrolysis reactions**, which involve the cleavage of a molecule through the addition of H_2O; **precipitation reactions** (see Section 6.2), generally accompanied by **aggregation** of colloidal particles suspended in water (see Section 6.6); **oxidation-reduction reactions**, generally mediated by microorganisms (see Chapters 3 and 5); **sorption** of hazardous solutes by sediments and by suspended mineral and organic matter; **biochemical processes**, often involving hydrolysis and oxidation-reduction reactions; **photolysis** reactions; and miscellaneous chemical phenomena.

A number of organic compounds encountered as hazardous wastes are formed from their precursor compounds by a synthesis reaction involving the loss of a molecule of water for each molecule of the compound formed. It is not surprising, therefore, that one of the most common modes of degradation of such compounds in water is hydrolysis. This is illustrated for three hazardous waste compounds—an ether, an acid anhydride, and an ester—in the following reactions:

$$\underset{\text{isopropyl ether}}{(CH_3)_2CH\text{—}O\text{—}CH(CH_3)_2} + HOH \underset{\text{synthesis}}{\overset{\text{hydrolysis}}{\rightleftharpoons}} 2\,\underset{\text{isopropyl alcohol}}{HO\text{—}CH(CH_3)_2} \qquad (20.5.1)$$

$$\underset{\text{acetic anhydride}}{CH_3\text{—}C(=O)\text{—}O\text{—}C(=O)\text{—}CH_3} + HOH \underset{\text{synthesis}}{\overset{\text{hydrolysis}}{\rightleftharpoons}} 2\,\underset{\text{acetic acid}}{CH_3\text{—}C(=O)\text{—}OH} \qquad (20.5.2)$$

---

[7] Larson, T.E., ed. 1978. The water environment. In *Cleaning our environment: a chemical perspective,* Ch. 5, pp. 188–274. Washington, D.C.: American Chemical Society.

$$\underset{\text{methyl methacrylate}}{H_2C{=}C(CH_3){-}\overset{O}{\overset{\|}{C}}{-}O{-}CH_3} + HOH \underset{\text{synthesis}}{\overset{\text{hydrolysis}}{\rightleftharpoons}} \underset{\text{methacrylic acid}}{H_2C{=}C(CH_3){-}\overset{O}{\overset{\|}{C}}{-}OH} + \underset{\text{methanol}}{HO{-}CH_3} \quad (20.5.3)$$

The rates at which compounds hydrolyze in water vary widely. Acetic anhydride hydrolyzes very rapidly. In fact, the great affinity of this compound for water (including water in skin) is one of the reasons that it is hazardous. Once in the aquatic environment, though, acetic anhydride is converted very rapidly to essentially harmless acetic acid. Many ethers, esters, and other compounds hydrolyze very slowly, although the rate may be greatly increased by the action of enzymes in microorganisms (biochemical processes). Hydrolysis of some compounds results in the loss of halogen atoms. For example, bis(chloromethyl) ether hydrolyzes rapidly to produce HCl and formaldehyde:

$$\underset{\text{bis(chloromethyl) ether}}{Cl{-}CH_2{-}O{-}CH_2{-}Cl} + H_2O \rightarrow 2\ \underset{\text{formaldehyde}}{H{-}\overset{O}{\overset{\|}{C}}{-}H} + 2\ HCl \quad (20.5.4)$$

Hydrolysis of a large quantity of this chloro ether in the aquatic environment could produce harmful amounts of corrosive HCl and toxic formaldehyde.

As discussed in Section 20.2, the formation of precipitates in the form of sludges is one of the most common means of isolating hazardous components from an unsegregated waste. Although solid inorganic ionic compounds are often discussed in terms of very simple formulas, such as $PbCO_3$ for lead carbonate, much more complicated species (for example, $2\ PbCO_3 \cdot Pb(OH)_2$) generally result when precipitates are formed in the aquatic environment. For example, a hazardous heavy metal ion in the hydrosphere may be precipitated as a relatively complicated compound, coprecipitated as a minor constituent of some other compound, or be sorbed by the surface of another solid.

The major anions present in natural waters and wastewaters are OH^-, CO_3^{2-}, and SO_4^{2-}; these are all capable of forming precipitates with cationic impurities. Therefore, cationic pollutants tend to precipitate as hydroxides, carbonates, and sulfates. Sometimes a distinction can be made between hydroxides and hydrated oxides with similar, or identical, empirical formulas. For example, iron(III) hydroxide, $Fe(OH)_3$, is a relatively uncommon species; iron(III) usually is precipitated from water as hydrated iron(III) oxides, such as β ferric oxide monohydrate, $Fe_2O_3 \cdot H_2O$. Basic salts containing OH^- ion along with some other anion are very common in solids formed by precipitation from water. A typical example is azurite, $2\ CuCO_3 \cdot Cu(OH)_2$. Two or more metal ions may be present in a compound, such as in chalcopyrite, $CuFeS_2$.

Two aspects of the precipitation process are particularly important in determining the fate of hazardous ionic solutes in water. If precipitation occurs too rapidly, with too high a degree of supersaturation, the solid tends to form as a large number of small colloidal particles, which may persist in the colloidal state for a long time. In this form, hazardous substances are much more mobile and accessible to organisms than as precipitates. A second important characteristic of precipitation is that in many cases substances whose own solubilities are not exceeded are coprecipitated with other substances. Many heavy metals are coprecipitated with hydrated iron(III) oxide ($Fe_2O_3 \cdot x\,H_2O$) or manganese(IV) oxide ($MnO_2 \cdot x\,H_2O$).

Sorption processes are particularly common methods for the removal of trace-level hazardous materials from water. As discussed in Section 6.11, freshly precipitated $MnO_2 \cdot x\,H_2O$ very effectively scavenges other metal ions, such as Ba^{2+}, from water.

As shown by the examples in Table 20.1, oxidation-reduction reactions are important means for the transformation of hazardous wastes in water. The degradation of practically all organic wastes in water proceeds via oxidation. Details of oxidation-reduction reactions in natural waters and wastewaters were covered in Chapter 3.

Under many circumstances, biochemical processes largely determine the fates of hazardous chemical species in the hydrosphere. The most important such processes are those mediated by microorganisms, discussed in Chapter 5. In particular, the oxidation of biodegradable hazardous organic wastes in water generally occurs by means of microorganism-mediated biochemical reactions. The role of bacteria in producing organic acids and chelating citrate, which have

TABLE 20.1 Oxidation-Reduction of Wastes in Water

Reaction	*Significance*
Oxidation half-reactions	
$SO_2(aq) + 2\,H_2O \rightarrow 4\,H^- + SO_4^{2-} + 2e^-$	Conversion of dissolved SO_2 gas to sulfuric acid
$CH_3C(=O){-}H + H_2O \rightarrow CH_3C(=O){-}OH + 2\,H^+ + 2\,e^-$	Conversion of acetaldehyde to acetic acid
$\{CH_2O\} + H_2O \rightarrow CO_2 + 4\,H^- + 4\,e^-$	Degradation of biomass
$C_nH_{2n+2} + 2n\,H_2O \rightarrow n\,CO_2 + (6n+2)\,H^+ + (6n+2)\,e^-$	Degradation of hydrocarbons
Reduction half-reactions	
$O_2(aq) + 4\,H^+ + 4\,e^- \rightarrow 2\,H_2O$	Removal of O_2 from water; O_2 is an electron receptor (source of O_2) for oxidation half-reactions above*
$Fe_2O_3 \cdot x\,H_2O(s) + 6\,H^+ + 2\,e^- \rightarrow 2\,Fe^{2+} + (3+x)\,H_2O$	Formation of soluble Fe^{2+}
$MnO_2(s) + 4\,H^+ + 2\,e^- \rightarrow Mn^{2+} + 2\,H_2O$	Production of soluble Mn^{2+}

*For example, the overall reaction for the oxidation of dissolved SO_2 by O_2 is obtained as follows:

$$2\,\{SO_2(aq) + 2\,H_2O \rightarrow 4\,H^+ + SO_4^{2-} + 2\,e^-\}$$
$$O_2(aq) + 4\,H^+ + 4\,e^- \rightarrow 2\,H_2O$$
$$\overline{2\,SO_2(aq) + O_2(aq) + 2\,H_2O \rightarrow 4\,H^+ + 2\,SO_4^{2-}}$$

the effect of solubilizing hazardous heavy metal ions, was discussed in Section 19.7. Some mobile methylated species of hazardous trace elements, such as the methylated forms of arsenic and mercury, are produced by bacterial action.

Photolysis reactions are those initiated by the absorption of light. These were discussed in relation to atmospheric chemistry in Chapters 11–15. The effect of photolytic processes on the production of hazardous wastes in the hydrosphere is minimal, although some photochemical reactions of hazardous waste compounds can occur when the compounds are present as surface films on water exposed to sunlight.

From the standpoint of potential damage by hazardous wastes, the most vulnerable part of the hydrosphere is **ground water** [8]. Although surface water supplies are most subject to direct contamination by hazardous wastes, ground water can become almost irreversibly contaminated by the improper land disposal of hazardous chemicals. In order to understand ground-water contamination by hazardous wastes, it is necessary to understand the geologic considerations involved with hazardous wastes in the geosphere; this is the topic of the following section.

20.6 HAZARDOUS WASTES IN THE GEOSPHERE

The sources, transport, interactions, and fates of contaminant hazardous wastes in the geosphere involve a complex scheme, some aspects of which are illustrated in Figure 20.7. The primary environmental concern regarding hazardous wastes in the geosphere is the possible contamination of ground water aquifers by waste

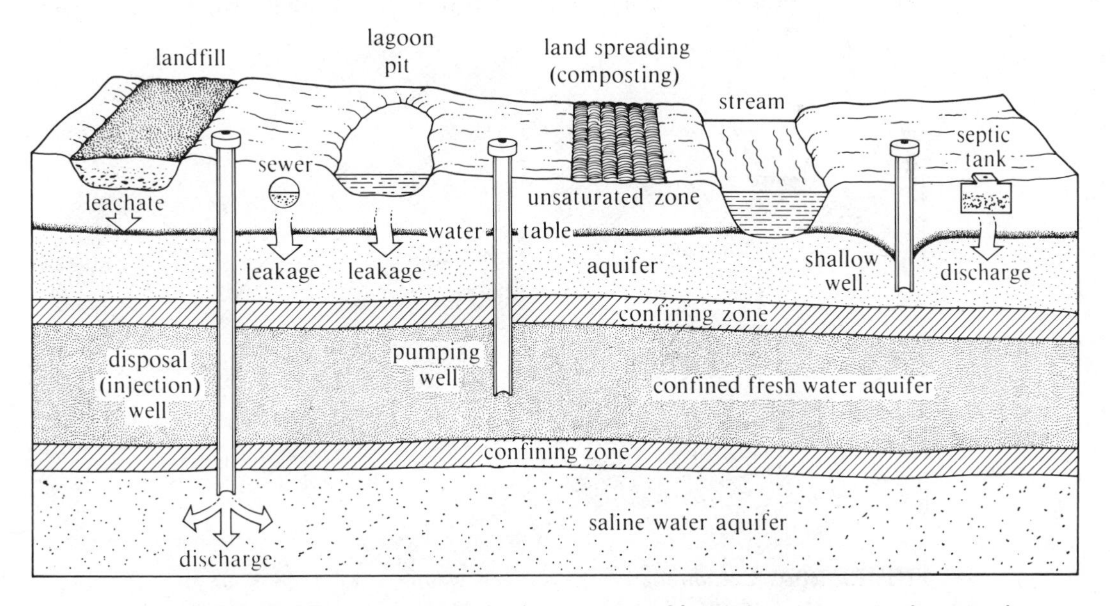

FIGURE 20.7 Sources and other aspects of hazardous wastes in the geosphere.

---

[8] Braids, O.C.; Wilson, G.R.; and Miller, D.W. 1977. Effects of industrial hazardous waste disposal on the ground water resource. In *Drinking water quality enhancement through source protection,* R.B. Pojasek, ed., pp. 179-208. Ann Arbor, Mich.: Ann Arbor Science Publishers, Inc.

leachates and leakage from wastes. As can be seen in the figure, there are a number of possible contamination sources. The most obvious one is leachate from landfills containing hazardous wastes. In some cases, liquid hazardous materials are placed in lagoons, which can leak into aquifers. Leaking sewers can also result in contamination, as can the discharge from septic tanks. Hazardous wastes spread on land can result in aquifer contamination by leachate. Hazardous chemicals are sometimes deliberately disposed of underground in waste disposal wells. There is an opportunity for interchange of contaminated water between surface water and ground water at discharge and recharge points. A detailed discussion of possible ground water contamination was given in an EPA report dealing with that topic [9].

The transport of contaminants in the geosphere depends largely upon the hydrologic factors governing the movement of water underground and the interactions of hazardous waste constituents with geological strata, particularly unconsolidated earth materials. As shown in Figure 19.3, ground water contaminated by hazardous wastes tends to flow as a relatively undiluted plug or plume along with the ground water in an aquifer. The ground water flow rate depends upon the water table gradient and aquifer characteristics, such as permeability and cross-sectional area. The rate of ground water flow is generally quite slow, with a rate of a meter or two per day considered to be very fast.

Figure 20.8 shows how contaminated ground water can result in contamination of a surface water source. This can occur at a discharge area where the ground water flows into a lake or stream.

As discussed in Section 19.7, hazardous waste dissolved in ground water can be attenuated by soil or rock by means of various sorption mechanisms. The degree of attenuation depends upon the surface properties of the solid, particularly surface area. The chemical nature of the attenuating solid is also important,

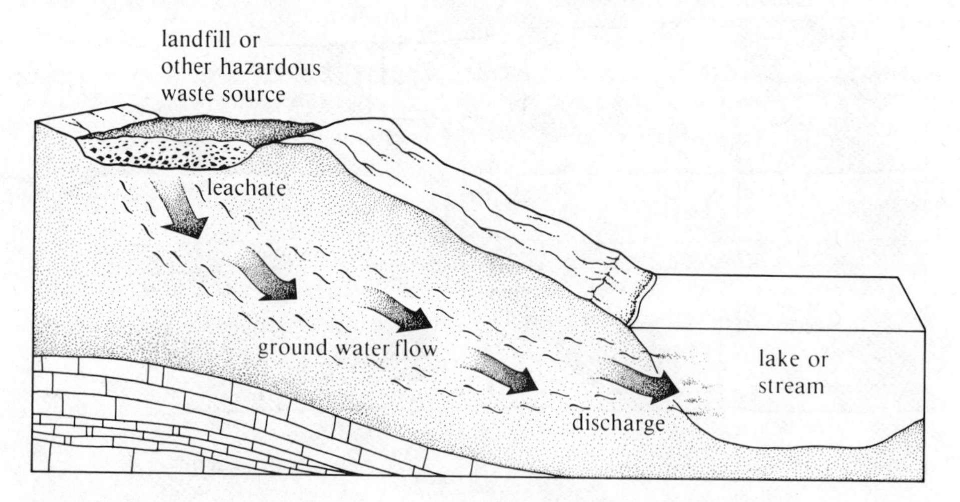

FIGURE 20.8 Contamination of surface water by hazardous wastes via ground water.

[9] 1977. *Report to Congress: Waste disposal practices and their effects on ground water.* Washington, D.C.: U.S. Environmental Protection Agency.

since attenuation is a function of the organic matter (humus) content, presence of hydrous metal oxides, and the content and types of clays present. The chemical nature of the leachate affects attenuation a great deal. For example, attenuation of metals is very poor in acidic leachate. Organic solvents in leachates tend to prevent attenuation of organic hazardous waste constituents.

The degree of attenuation of a pollutant by soil depends upon the water content of the soil. Above the water table there is an unsaturated zone of soil in which attenuation is more highly favored. Normally soil has a greater surface area at liquid-solid interfaces in this zone [10] so that adsorption and ion-exchange processes are favored. Aerobic degradation (see Section 5.10 and Table 5.1) is possible in the unsaturated zone, enabling more rapid and complete degradation of biodegradable hazardous wastes.

Heavy metals are particularly damaging to ground water, and their movement through the geosphere is of considerable concern. Heavy-metal ions may be sorbed by the soil, held by ion-exchange processes, interact with organic matter in soil, undergo oxidation-reduction processes leading to mobilization or immobilization, or even be volatilized as organometallic compounds formed by methylating bacteria. A large number of factors affect heavy-metal mobility and attenuation in soil. These include pH, pE (see Chapter 3), temperature, cation-exchange capacity (see Section 6.12), nature of soil mineral matter, and nature of soil organic matter.

Normally, the mobility of heavy metals in soil and mineral matter is relatively low. One study of relative mobilities in clay mineral columns [11] showed that Pb, Zn, Cd, and Hg were strongly attenuated by the clay, primarily by precipitation and exchange processes. Iron was only moderately attenuated, largely because of the reduction of iron(III) to soluble iron(II). Manganese was actually eluted from the clay, probably because of reduction to soluble Mn^{2+}.

As illustrated by a study of the mobilization of radionuclides [12], codisposal of chelating agents with heavy metals can have a strong effect upon the mobility of metal ions in soil. This was noted in a study of the effects of codisposal of intermediate-level nuclear wastes with chelating agents during the period 1951–1965 at Oak Ridge National Laboratory. The presence of chelating agents resulted from the use of salts of EDTA, ethylenediaminetetraacetic acid (see Section 4.3), in decontaminating facilities exposed to nuclear wastes.

Radionuclide wastes have been buried in shallow trenches on the grounds of Oak Ridge National Laboratory since 1944, so ample time has elapsed to observe the effects of this means of radioactive waste disposal. At these burial sites the predominant geological formation is Conasauga shale. This bedrock material has a very high sorptive capacity for most of the radionuclides produced as by-products of nuclear fission, particularly those that are cationic. Despite

---

[10] Brunner, D.R., and Keller, D.J. 1972. *Sanitary landfill design and operations.* USEPA Solid Waste Management Series. Washington, D.C.: U.S. Environmental Protection Agency.

[11] Griffin, R.A., and Shimp, N.F. 1978. *Attenuation of pollutants in municipal landfill leachate by clay minerals.* Washington, D.C.: U.S. Environmental Protection Agency.

[12] Means, J.L.; Crerar, D.A.; and Duguid, J.O. 1978. Migration of radioactive wastes: radionuclide mobilization by complexing agents. *Science* 200: 1477–81.

this, some migration of radionuclides has been observed from sites used to dispose solid and liquid wastes. Some of this migration has been attributed to the high rainfall in the area, shallow ground-water levels, fractures in the underlying rock that allow for rapid infiltration of dissolved wastes, and other physical factors.

In addition to the factors listed above as contributing to the migration of radionuclides from waste-disposal trenches, it has been found that chelating agents used for decontamination, as well as naturally occurring humic-substance chelators, are responsible for migration in excess of that expected. Most notably, ^{60}Co has been found outside the disposal trenches. Levels of radioactive contamination from this isotope adjacent to the disposal trenches have been observed as high as 1×10^5 disintegrations per minute (dpm) per gram (45,000 picoCuries/g) in soil and as high as 1×10^3 dpm/mL in soil water. In addition, traces of various isotopes of the alpha-emitters, uranium, plutonium, radium, thorium, and californium, have been found outside the disposal area.

Experiments were conducted to determine the **distribution coefficients,** K_d, for ^{60}Co between water sampled from wells at the disposal sites and the shale at the sites. The distribution coefficient is a measure of the affinity of a solute for a solid phase; the higher its value, the greater the tendency of the solute to be sorbed by the solid. For well-water samples ranging in pH from 6.0 to 8.5, values of K_d were measured in the range of 7 to 70, with an averge of about 35. By way of comparison, a standard ^{60}Co solution buffered at pH 6.7, in the absence of chelating agent, had a K_d value of 7.0×10^4, indicating a tremendous affinity for the shale. Similar solutions prepared containing $1 \times 10^{-5} M$ EDTA and cobalt at the same pH gave K_d values of only 2.9. The actual EDTA concentration found in the well samples was $3.4 \times 10^{-7} M$, thus explaining why the distribution coefficient in the well-water samples was somewhat higher than that observed in the experimental samples containing $1 \times 10^{-5} M$ EDTA.

Although in the study cited above it was concluded that EDTA was responsible for the excessive mobilization of ^{60}Co, other species have the potential to mobilize radionuclides or heavy metals. Of these, palmitic acid and phthalic acid were found in leachate from the disposal trenches [13]. Other

$C_6H_4(COOH)_2$ (benzene ring with two adjacent –C(=O)–OH groups)

phthalic acid

$CH_3(CH_2)_{14}COOH$ (H–C(H)(H)–…–C(=O)–OH, fifteen carbons each bearing H atoms, terminal carboxyl)

palmitic acid

[13] Means, J.L.; Crerar, D.A.; and Duguid, J.O. 1976. *Oak Ridge National Laboratory Report* ORNL/TM-5438. Oak Ridge, Tenn.

species that might be codisposed with radionuclides and have the potential to increase their mobility are citrate, fluoride, oxalate, and gluconate salts.

In another study of radioactive waste disposal sites [14], it was observed that organic chelating agents, particularly EDTA, can dramatically increase the migration of radionuclides from the site. According to this study, water samples at the Maxey Flats disposal site in Kentucky, where EDTA-containing plutonium wastes were disposed, showed levels of 300,000 picoCuries/liter, far in excess of those found in the absence of the chelating agent.

The evidence just cited suggests that strong chelating agents would have a tendency to transport heavy metal ions from disposal sites. Landfill disposal of strong chelating agents, such as EDTA, should therefore be avoided.

20.7 HAZARDOUS WASTES IN THE ATMOSPHERE

The fate of hazardous wastes in the atmosphere has been the subject of a detailed study [15]. Two classes of compounds were singled out for study. The first consisted of about 40 chemicals and classes of chemicals under preliminary assessment. The second consisted of five chemicals that are likely to get into the atmosphere from hazardous waste sources and are particularly likely to be harmful. These are acrylonitrile, ethylene dichloride, perchloroethylene, vinylidene chloride and benzo(a)pyrene.

$H_2C{=}CH{-}C{\equiv}N$

acrylonitrile

$H{-}CHCl{-}CHCl{-}H$

ethylene dichloride

$Cl_2C{=}CCl_2$

perchloroethylene

$Cl_2C{=}CH_2$

vinylidene chloride

Benzo(a)pyrene

One of the most important considerations regarding hazardous waste chemicals in the atmosphere is the residence time and the mode of removal. Removal mechanisms may be divided into **physical removal processes** and **chemical removal processes**.

One of the more obvious means by which a vapor-phase hazardous substance may be removed from the atmosphere is by **dissolution** in water in the form of cloud or rain droplets. For vapors of compounds such as ethylene chloride, tetrachloroethylene, and vinylidene chloride, which are not highly soluble in water, Henry's Law (see Section 2.9) combined with estimates of annual

[14] 1981. Organic agents increase plutonium leaching. *Chemical and Engineering News* 8 June 1981, p. 7.

[15] Cupitt, L.T. 1980. *Fate of toxic and hazardous materials in the air environment.* EPA-600/S3-80-08e. Washington, D.C.: U.S. Environmental Protection Agency.

rainfall and mixing in the atmosphere may be used to estimate the atmospheric half-life, $\tau_{1/2}$, of vapors in the atmosphere. Solubility rates may be used to estimate half-lives for substances that are more miscible in water. Using such an approach, the following terms may be defined:

p_{is} = the saturation vapor pressure of the solute in equilibrium with bulk quantities of the solute and a saturated aqueous solution of the solute, measured in torr.

x_{is} = the saturation mole-fraction of the solute in water.

α_i = a dimensionless constant describing the ratio of the concentration of the solute in water to that of its concentration in air.

Using these parameters and assuming that removal occurs only by dissolution in raindrops, estimates may be made of the atmospheric half-lives of the five hazardous chemicals listed above [15]. The half-lives are listed in Table 20.2. As you can see, three of the half-lives are quite long. Therefore, it may be concluded that dissolution of these chemicals into raindrops and their subsequent precipitation from the atmosphere is not a significant mechanism for their removal from the atmosphere.

Another possible mechanism for the removal of vaporized hazardous waste chemicals from the atmosphere is **adsorption by aerosol particles**. With this removal mechanism, the lifetime of sorbed hazardous waste material in the atmosphere is limited to that of the sorbing atmospheric particles, typically about 7 days, plus the time spent in the vapor phase before adsorption. The other parameter involved is the fraction of the compound sorbed by atmospheric particles, designated as ϕ. Therefore, substances sorbed to particles lasting an average of 7 days in the atmosphere have an average atmospheric lifetime of $7/\phi$ days (not including the time before adsorption). The major difficulty arises in trying to evaluate ϕ. Using monolayer sorption theory in estimating ϕ, along with other assumptions about particle concentrations in the atmosphere, the following estimates were made of the atmospheric lifetime of 5 hazardous compounds, assuming removal only by adsorption by particles: acrylonitrile, 8.6×10^9 days; vinylidene chloride, 5.3×10^{10} days; ethylene dichloride, 7.0×10^9 days; perchloroethylene, 1.8×10^9 days; and benzo(a)pyrene, 7.9 days. Among these compounds, it appears that the particle-sorption removal mechanism is significant only for benzo(a)pyrene.

TABLE 20.2 Estimated Atmospheric Half-Lives of Five Hazardous Chemicals, Assuming Removal Only by Dissolution in Raindrops

Compound	p_{is}, in *torr*	x_{is}	α_i	$\tau_{1/2}$, in years
Acrylonitrile	1.14×10^2	<1.0	$<9.05 \times 10^3$	<0.8
Benzo(a)pyrene	5.46×10^{-9}	2.8×10^{-11}	5.4×10^3	1.4
Ethylene dichloride	8.4×10^1	1.6×10^{-3}	1.9×10^1	3.9×10^2
Tetrachloroethylene	1.85×10^1	1.1×10^{-5}	6.1×10^{-1}	1.2×10^4
Vinylidene chloride	6.17×10^2	3.9×10^{-5}	6.5×10^{-2}	1.1×10^5

Dry deposition is a third possible means for physical removal of hazardous substances from the atmosphere. This process involves sorptive removal through contact with soil, water, or plants on the Earth's surface. It may be described in terms of a **deposition velocity**, v_d, given by the following formula:

$$v_d = \frac{\text{deposition flux}}{\text{concentration of contaminant in air}} \tag{20.7.1}$$

If the deposition flux is in units of $\mu g \times cm^{-2} \times sec^{-1}$ and the concentration is in $\mu g \times cm^{-3}$, v_d is in units of cm/sec. For several gases that have been investigated, v_d values range from 0.01 to 3 cm/sec, corresponding to residence times of 3 to 900 days. These values vary greatly with type of compound, type of surface, and weather conditions.

Investigations of the deposition velocities of methyl iodide and carbon tetrachloride produced v_d estimates of 10^{-3} cm/sec, which has also been assumed as a reasonable estimate for other organohalide compounds, such as perchloroethylene, dichloroethylene, and vinylidene chloride [15]. Assuming dry deposition as the sole removal mechanism, the estimated lifetime of organohalide compounds in the atmosphere would be about 25 years. This figure shows that dry deposition is an insignificant removal mechanism for most volatile hazardous organic compounds.

From the preceding discussion of physical removal processes, it may be concluded that, for a number of volatile organic compounds that are not very soluble in water, chemical removal processes must predominate. As discussed in Chapters 14 and 15, the most important of these processes is reaction with **hydroxyl radical**, $HO\cdot$, in the troposphere. In the case of compounds with a double bond, such as acrylonitrile or vinylidene chloride, reaction with ozone is also possible. Other oxidant species that might react with hazardous waste compounds in the troposphere and stratosphere include peroxyl radicals, $HOO\cdot$; alkylperoxyl radicals, $ROO\cdot$; NO_3; singlet molecular oxygen, $O_2(^1\Delta g)$; and the atomic oxygen species, $O(^3P)$ and $O(^1D)$.

Even though the concentration of $HO\cdot$ in the troposphere is normally lower than the concentrations of some other species that might react with volatile hazardous waste compounds, the reactivity of $HO\cdot$ is so high that the radical tends to initiate most of the reactions leading to the chemical removal of most refractory organic compounds from the atmosphere, particularly in the troposphere. As noted in Section 14.3, hydroxyl radical reacts with organic compounds containing hydrogen, R—H, by abstraction,

$$R{-}H + HO\cdot \rightarrow R\cdot + H_2O \tag{20.7.2}$$

and may react with those containing unsaturated bonds, R=R ′, by addition:

$$R{=}R' + HO\cdot \rightarrow R{-}\overset{\bullet}{\underset{\underset{H}{|}}{O}}{-}R' \tag{20.7.3}$$

In both cases, reactive free radicals are formed and undergo further reactions, leading to species that are nonvolatile or water-soluble, which are scavenged from the atmosphere by physical means. These scavengable species tend to be aldehydes, ketones, or acids. Halogenated organics may lose halogen atoms in the form of halo-oxy radicals and undergo further reactions to form scavengable species.

Reaction with **ozone** may be a significant pathway for the removal of unsaturated compounds from the atmosphere. Ozone adds across double bonds, leading to the formation of reactive species that undergo further reactions to form species that precipitate from the atmosphere (see Chapter 14). Rate constants for the reactions of unsaturated compounds with O_3 are typically only about 10^{-17} cm^3 per molecule per second, compared to approximately 10^{-11} cm^3 per molecule per second for reactions with HO $\cdot$. Despite its much lower reactivity level, ozone's concentration level in the troposphere is normally so much higher than that of HO $\cdot$ that ozonolysis mechanisms are competitive with those involving HO $\cdot$ for the removal of compounds with double bonds from the atmosphere. Although aromatic compounds also react with O_3, the rate is so slow compared to reaction with HO $\cdot$ that it is not normally a significant removal mechanism for these substances.

In general, reactions with species other than HO $\cdot$ or O_3 are not considered significant in the removal of hazardous organic waste compounds from the troposphere. Perhaps in some cases such reactions do contribute to a very slow removal of such contaminant compounds.

Photolytic transformations involve direct cleavage (photodissociation) of compounds by reactions with light:

$$\text{R—X} + h\nu \rightarrow \text{R}\cdot + \text{X}\cdot \qquad (20.7.4)$$

The extent of these reactions varies greatly with light intensity, quantum yields (chemical reactions per light quantum absorbed) and other factors. In order for photosynthesis to be an important process for its removal from the atmosphere, a molecule must have a **chromophore** (light-absorbing group) that absorbs light in a wavelength region of significant intensity in the impinging light spectrum. This requirement limits the importance of photolysis in this regard to removal of conjugated alkenes, carbonyl compounds, some halides, and some nitrogen compounds, particularly nitro compounds. However, these classes do include a number of the more important hazardous waste compounds (see Table 20.3).

In cases of extremely volatile and refractory hazardous waste compounds that escape destruction and removal in the troposphere, direct photolytic reactions predominate as removal mechanisms in the stratosphere, where the flux of highly energetic short-wavelength ultraviolet radiation is very intense. This process is most clearly illustrated by the stratospheric photodissociation of chlorofluorocarbon compounds, which are quite stable and resistant to removal mechanisms in the troposphere (see Section 12.17). In the stratosphere, photolysis can result in the removal of halide atoms from compounds through the action of ultraviolet light, as shown for the following reaction of chloroform:

$$\text{Cl—}\overset{\text{H}}{\underset{\text{Cl}}{\text{C}}}\text{—Cl} + h\nu \rightarrow \text{Cl—}\overset{\text{H}}{\underset{\text{Cl}}{\text{C}}}\cdot + \text{Cl}\cdot \qquad (20.7.5)$$

The highly reactive radical products undergo further reactions with species such as O atoms in the stratosphere (where M is an energy-absorbing third body)

$$\text{Cl—}\overset{\text{H}}{\underset{\text{Cl}}{\text{C}}}\cdot + \text{O} + \text{M} \rightarrow \text{Cl—}\overset{\text{H}}{\underset{\text{Cl}}{\text{C}}}\text{—O}\cdot + \text{M} \qquad (20.7.6)$$

to eventually form stable compounds that are scavenged from the atmosphere.

Table 20.3 summarizes estimates of residence times and possible fates of a number of hazardous chemicals that may get into the atmosphere. This table is based upon a study by the EPA [15]. Examination of the material in the table should provide some insights into the reactions and fates of hazardous chemicals in the atmosphere.

20.8 HAZARDOUS ORGANIC COMPOUNDS IN THE BIOSPHERE—BIODEGRADATION

Biological treatment can be used to eliminate many hazardous organic compounds [16]. In some cases the compounds are partially degraded to harmless substances. They may be completely oxidized to CO_2 under aerobic conditions or transformed to CO_2 and CH_4 under anaerobic conditions (see Section 5.26 for a discussion of the microbial degradation of pesticides). Anthropogenic compounds resist biodegradation much more strongly than do naturally occurring compounds. This is generally due to the absence of enzymes that can bring about an initial attack on the compound. Some structural groups on organic molecules are particularly likely to make organic compounds degradation-resistant. Branched carbon chains have already been mentioned in this regard (see Section 5.12). Other groups with inhibiting qualities include ether linkages, meta-substituted benzene rings, chlorine, amines, methoxy groups, sulfonates, and nitro groups.

Most studies of the biodegradation of refractory synthetic organic compounds have concentrated on aerobic, oxidative processes, which have usually been considered the most efficient and generally applicable. However, anaerobic processes have some advantages, such as the lack of any need for oxygen transfer and lower production of sludge. In addition, anaerobic processes bring about reduction of nitro groups, reductive dehalogenation, and reduction of sulfoxide groups, outlined below in the structure of Aramite, an insecticide:

$$\text{(4-}C(CH_3)_3\text{-}C_6H_4\text{)}-O-CH_2\overset{\displaystyle CH_3}{\overset{|}{C}}H-O-\overset{\displaystyle O}{\overset{\|}{S}}-O-CH_2CH_2Cl$$

A number of the physical and chemical characteristics of a compound are involved in its amenability to biodegradation. These include hydrophobicity, solubility, volatility, and octanol-water partition coefficient (see Section 6.15).

Several groups of microorganisms are capable of partial or complete degradation of hazardous organic compounds. Among the aerobic bacteria, those of the *Pseudomonas* family are the most widespread and most adaptable to the degradation of synthetic compounds. These bacteria degrade biphenyl, naphthalene, DDT, and many other compounds. Anaerobic bacteria are very fastidious, in that they require oxygen-free (anoxic) conditions and pE values of

---

[16] Kobayashi, H. and Rittman, B.E. 1982. Microbial removal of hazardous organic compounds. *Environmental Science and Technology* 16: 170A–181A.

TABLE 20.3 Estimated Residence Times and Possible Fates of Hazardous Waste Chemicals in the Atmosphere*

Compound	$k_{OH} \times 10^{12}$**	$k_{O_3} \times 10^{18}$**	*Photolysis probability*	*Physical removal probability*	*Residence time, days****	*Possible reaction products*
Acetaldehyde	16	—	Probable	Unlikely	0.03–0.7	H_2CO, CO_2
Acrolein	44	4	Probable	Unlikely	0.2	OCH—CHO, H_2CO, HCO_2H, CO_2
Acrylonitrile	2	≤0.05	—	Unlikely	5.6	H_2CO, HC(O)CN, HCO_2H, CN·
Allyl chloride	28	18.3	Possible	Unlikely	0.3	HCO_2H, H_2CO, $ClCH_2CHO$, chlorinated hydroxyl carbonyls, $ClCH_2CO_2H$
Benzo(a)pyrene	—	—	Possible	Probable	8	1-6-Quinone of benzo(a)pyrene
Benzyl chloride	3	0.004	Possible	Unlikely	3.9	ϕCHO, Cl·, chloromethylphenols, ring-cleavage products
Bis(chloromethyl)ether	4	—	Possible	Probable	0.02–2.9	HCl, H_2CO, chloromethylformate, ClHCO
Carbon tetrachloride	<0.001	—	—	Unlikely	>11,000	Cl_2CO, Cl·
Chlorobenzene	0.4	$<5 \times 10^{-5}$	Possible	Unlikely	28	Chlorophenols, ring-cleavage products
Chloroform	0.1	—	—	Unlikely	120	Cl_2CO, Cl·
Chloromethyl methyl ether	3	—	Possible	Probable	0.004–3.9	Decomposition products, chloromethyl and methyl formate, ClHCO

Chloroprene	46	8	Probable	Unlikely	0.3	H_2CO, $H_2C=CClCHO$, OHCCHO, ClCOCHO, $H_2CCHCClO$, chlorohydroxyl acids, aldehydes
o-, *m*-, *p*-cresol	55	0.6	—	Unlikely	0.2	Hydroxynitrotoluenes, ring-cleavage products
Dichlorobenzene	2.03	$\leq 5 \times 10^{-}$	Possible	Unlikely	39	Chlorinated phenols, ring-cleavage products, nitro compounds
Dioxane	3	—	—	Unlikely	3.9	$OHCOCH_2CH_2OCHO$, OHCOCHO, oxygenated formates
Dioxin	—	—	Probable	—	—	—
Epichlorhydrin (chloropropylene oxide)	2	—	Possible	Unlikely	5.8	H_2CO, OHCOCHO, $ClCH_2C(O)OHCO$
Ethylene dibromide	0.25	—	Possible	Unlikely	45	$Br\cdot$, $BrCH_2CHO$, H_2CO, BrHCO
Ethylene dichloride	0.22	—	Possible	Unlikely	53	ClHCHO, $H_2CClCOCl$, H_2CO, $H_2CClCHO$
Ethylene oxide	2	—	—	Unlikely	5.8	OHCOCHO
Formaldehyde	10	$<2 \times 10^{-5}$	Probable	Unlikely	0.1–1.2	CO, CO_2
Hexachlorocyclopentadiene	59	8	Probable	—	0.2	Cl_2CO, diacylchlorides, ketones, $Cl\cdot$

TABLE 20.3 *(continued)*

Compound	$k_{OH} \times 10^{12}$**	$k_{O_3} \times 10^{18}$**	*Photolysis probability*	*Physical removal probability*	*Residence time, days****	*Possible reaction products*
Maleic anhydride	60	160	Possible	Possible	0.1	CO_2, CO, acids, aldehydes and esters that should photolyze
Methyl chloride	0.14	—	Possible	Unlikely	83	Cl_2CO, CO, ClHCO, Cl·
Methyl chloroform	0.012	—	Possible	Unlikely	970	H_2CO, Cl_2CO, Cl·
Methyl iodide	0.004	—	Possible	Unlikely	2,900	H_2CO, I·, IHCO, CO
Nitrobenzene	0.06	$<5 \times 10^{-5}$	Possible	Unlikely	190	Nitrophenols, ring-cleavage products
2-Nitropropane	55	—	Possible	Unlikely	0.2	H_2CO, CH_3CHO
N-Nitrosodiethylamine	26	—	Probable	—	≤ 0.4	Photolysis products, aldehydes, nitramines
Nitrosoethylurea	13	—	Possible	—	≤ 0.9	Photolysis products, aldehydes, nitramines
Nitrosomethylurea	20	—	Possible	—	≤ 0.6	Photolysis products, aldehydes; nitramines
Nitrosomorpholine	28	—	Possible	—	≤ 0.4	Photolysis products, aldehydic ethers
Perchloroethylene	0.17	0.002	Possible	Unlikely	67	Cl_2CO, $Cl_2C(OH)COCl$, Cl·
Phenols	17	1	—	Possible	0.6	Dihydroxybenzenes, nitro phenols, ring-cleavage products
Phosgene	0	—	—	Possible	—	CO_2, Cl·, HCl
Polychlorinated biphenyls (PCB's)	< 1	5×10^{-5}	Possible	Unlikely	>11	Hydroxy PCB's, ring-cleavage products

Propylene oxide	1.3	—	—	Unlikely	> 8.9	$CH_3C(O)OCHO$, $CH_3C(O)CHO$, H_2CO $HC(O)OCHO$
Toulene	6	0.0001	—	Unlikely	1.9	Benzaldehyde, cresols, ring-cleavage products, nitro compounds
Trichloroethylene	2.2	0.006	Possible	Unlikely	5.2	Cl_2CO, $ClHCO$, CO, $Cl\cdot$
Vinylidene chloride	4	0.04	Possible	Unlikely	2.9	H_2CO, Cl_2CO, HCO_2H
o-, *m*-, *p*-Xylene	~16	~0.001	—	Unlikely	0.7	—

*From Reference 15.

**In units of cm^3 per molecule per second; most rate constants calculated theoretically. The term k_{OH} represents the rate constant for reaction with hydroxyl radical; k_{O_3} represents the rate constant for reaction with ozone.

***The shorter residence time (if a range is indicated) includes removal by photolysis.

less than −3.4 in order to survive, so they are difficult to study in the laboratory. Anaerobic bacteria **catabolize** (destructively metabolize, resulting in release of energy) biomass through hydrolytic processes, breaking down proteins, lipids, and saccharides. They are also known to reduce nitro compounds to amines, degrade nitrosamines, promote reductive dechlorination, reduce epoxide groups to olefins, and bring about the breakdown of aromatic structures. **Actinomycetes** are microorganisms that are morphologically similar to both bacteria and fungi. They are involved in the degradation of a variety of organic compounds, including degradation-resistant alkanes and lignocellulose. Other compounds attacked include pyridines, phenols, nonchlorinated aromatics, and chlorinated aromatics. Fungi are particularly noted for their ability to attack long-chain and complex hydrocarbons. Fungi are more successful than bacteria in the initial attack on PCB's. Phototrophic microorganisms, which include algae, photosynthetic bacteria, and cyanobacteria (blue-green algae), have two interesting aspects in terms of biodegradation capability: they tend to concentrate organophilic compounds in their lipid stores, and they can induce photochemical degradation of sorbed organic compounds. As an example of the action of photosynthetic bacteria on synthetic compounds, *Oscillatoria* can initiate the biodegradation of naphthalene by the attachment of —OH groups.

Practically all classes of synthetic organic compounds can be at least partially degraded by various microorganisms. These classes include nonhalogenated alkanes, halogenated alkanes (trichloroethane, dichloromethane), nonhalogenated aromatic compounds (benzene, naphthalene, benzo(a)pyrene), halogenated aromatic compounds (hexachlorobenzene, pentachlorophenol), phenols (phenol, cresols), polychlorinated biphenyls, phthalate esters, and pesticides (DDT, parathion).

20.9 HEALTH EFFECTS OF HAZARDOUS WASTES

The health effects of hazardous wastes in the environment should be studied to a greater extent than they have been previously. According to a report released in 1982 [17], there is a correlation between incidence rates of certain types of human cancers and the location of toxic waste dumps. This study was performed by the University of Medicine and Dentistry of New Jersey and concentrated on cancer in white males residing in the vicinity of hazardous chemical dumps in Middlesex, Hudson, and southeast Union counties in New Jersey. In these areas, mortality rates were "excessively higher" than the national average for esophagus cancer (25% excess), stomach cancer (50% excess), and colon-rectal cancer (35% excess). These findings do not *prove* that the toxic waste dumps are solely responsible for the excess occurrence of cancer. In addition to the waste dumps, the area studied is highly industrialized and has a high population density, which might be contributing factors. According to a 1972 study by the National Cancer Institute, New Jersey tends to have a higher-than-average occurrence of certain types of cancer; emissions from industrial plants, toxic waste dumps, and other factors may be involved.

---

[17] 1982. Tie between cancer deaths and toxic waste is found. *New York Times* 8 July 1982.

SUPPLEMENTARY REFERENCES

See references following Chapter 19.

QUESTIONS AND PROBLEMS

1. What is the lowest level of segregation of hazardous wastes and why is waste in this state particularly difficult to handle?
2. In terms of levels of segregation of wastes, how would you classify a mixture of methylene chloride, a very low-boiling, non-combustible organochlorine compound, and xylenes, which are relatively high-boiling aromatic hydrocarbons? What disposal and/or utilization techniques might be used with this mixture, knowing that methylene chloride is a useful solvent?
3. A dilute waste sludge is judged hazardous because of the presence of dissolved Cd^{2+} ion. Suggest an economical means of treating such a waste.
4. Match each waste constituent in the left column with the most desirable means of treatment in the right column.

(a) CrO_4^{2-}	(1) Lime
(b) Pb^{2+}	(2) Na_2S
(c) CN^-	(e) H_2SO_4
(d) sludge	(4) Oxidizing agent
(e) Dissolved NaOH	(5) Dewatering

5. What does an ideal fixative do?
6. Match each fixation procedure in the left column with the appropriate hazardous substance in the right column.

(a) Lime fixation	(1) Flue gas desulfurization sludge
(b) Thermoplastic techniques	(2) Pozzolanic material
(c) Self-cementing	(3) Urea-formaldehyde
(d) Organic polymers	(4) Bitumen or paraffin

7. What is used to isolate different classes of wastes in a secure hazardous-waste landfill?
8. What are some of the important processes determining the transformations and ultimate fates of hazardous chemical species in the hydrosphere?
9. According to a simple view of solution chemistry, a hazardous waste cation, such as Cd^{2+}, may be precipitated from solution as a simple salt (for example, $CdCO_3$). What are some alternatives for the removal of such a cation?
10. What part of the hydrosphere is most subject to long-term, largely irreversible contamination from the improper disposal of hazardous wastes in the environment?
11. Match each term in the left column with the appropriate description in the right column.

(a) Unsaturated zone	(1) Usually involves a saline water aquifer
(b) Landfill	(2) Determines limit of unsaturated zone
(c) Lagoon or pit	(3) Source of leachate
(d) Disposal well	(4) Area of relatively high waste attenuation
(e) Water table	(5) Source of leakage

12. Of the five "high-priority" atmospheric pollutants discussed in Section 20.7, which is most likely to be removed from the atmosphere by physical processes?
13. What structural features must a compound possess in order for direct photolysis to be a significant factor in its removal from the atmosphere?

ANSWERS TO ODD-NUMBERED QUESTIONS AND PROBLEMS

CHAPTER 2, PAGE 33

1. The oxygen in the air oxidizes soluble Fe^{2+} to acidic Fe^{3+}. The latter consumes alkalinity by the reaction,

$$Fe^{3+} + 3\ HCO_3^- \rightarrow Fe(OH)_3(s) + 3\ CO_2(g)$$

3. Phenolphthalein alkalinity $= \dfrac{V_p N}{V_s} \times 100$

5. 2.91 mg/L N; 12.9 mg/L NO_3^-

7. $[Fe^{3+}] = 1.00 \times 10^{-4}M$, $[FeOH^{2+}] = 3.99 \times 10^{-5}M$, $[Fe(OH)_2^+] = 9.92 \times 10^{-6}M$, $[H^+] = 2.22 \times 10^{-3}M$, pH = 2.65.

9. $[O_2(aq)] = 6.20 \times 10^{-4}M$

11. $[CO_2(aq)] = 4.49 \times 10^{-4}M$, $[HCO_3^-] = 2.00 \times 10^{-3}M$, $[CO_3^{2-}] = 9.38 \times 10^{-7}M$, $[OH^-] = 1.00 \times 10^{-7}M$.

13. $CaCO_3 + HCl \rightarrow Ca^{2+} + HCO_3^- + Cl^-$, hardness $= 1.00 \times 10^{-3}M$, alkalinity $= 1.00 \times 10^{-3}$ equivalents/liter for an HCl waste stream. For an HF waste stream: $CaCO_3 + HF \rightarrow Ca^{2+} + HCO_3^- + F^-$ until the solubility product of CaF_2 is exceeded, followed by $2\ CaCO_3 + 2\ HF \rightarrow Ca^{2+} + 2\ HCO_3^- + CaF_2(s)$. After $1.00 \times 10^{-3}M$ HF has come to equilibrium with $CaCO_3$, $[Ca^{2+}] = 6.25 \times 10^{-4}M$, alkalinity $= [HCO_3^-] = 1.00 \times 10^{-3}$ equivalents/liter, $[F^-] = 2.5 \times 10^{-4}M$.

15. To prevent accumulation of salts in the soil

17. (e)

19. (d)

21. 21.7%

CHAPTER 3, PAGE 56

1. *Acid-Base* / *Redox*

Acid-Base	*Redox*
$HOAC \rightleftharpoons H^+ + OAc^-$	$Fe^{2+} \rightleftharpoons Fe^{3+} + e^-$
$H_2O + H^+ \rightleftharpoons H_3O^+$	$H^+ + e^- \rightleftharpoons ½\ H_2(g)$
$HOAc + H_2O \rightleftharpoons H_3O^+ + OAc^-$	$Fe^{2+} + H^+ \rightleftharpoons Fe^{3+} + ½\ H_2(g)$

Given the Bronsted definition of acids, acid-base reactions involve the transfer of positively charged particles (protons, H^+) from acids (proton donors) to bases (proton acceptors). Redox reactions involve the transfer of negatively charged entities (electrons, e^-) from reducing agents (electron donors) to oxidizing agents (electron acceptors).

3. pE = 20.58 − pH

5. From the NO_3^--NH_4^+ half-reaction, pE = 6.15. From Equation 3.9.4, $\log P_{O_2} = -4x(20.75 - pE + \log[H^+]) = -30.4$, $P_{O_2} = 3.98 \times 10^{-31}$ atm.

7. pE = 14.58
9. pE = 14.5
11. P_{O_2} = 6.3×10^{-23} atm (in an actual system, such a low O_2 pressure would not be possible).
13. Because most half-reactions are not reversible
15. Decreasing pE with increasing depth

CHAPTER 4, PAGE 81

1. Copper is used as a "tracer metal" because of the relatively strong chelates that it forms. In some cases iron(III) might compete with copper(II) for the chelating agents, thus causing a negative interference.
3. By methods such as voltammetric reduction of free metal ions as compared to complex metal ions (see *voltammetry* in the index).
5. The equilibrium constant of the reaction,

$$Cu(OH)_2(s) + Y^{4-} \rightleftharpoons CuY^{2-} + 2\,OH^-$$

is 0.189. At pH 11, the ratio, $[CuY^{2-}]/[Y^{4-}]$, is 1.89×10^5, showing that the reaction tends strongly to the right.
7. pH = 6.10
9. $[PbT^-]/[CaT^-] = 0.39$
11. $[HT^{2-}] = 1.07 \times 10^{-2}M$
13. By shifting redox potentials and dissolving protective oxide coatings.
15. $[PbT^-]/[HT^{2-}] = 23.1$
17. Prevent removal of heavy metals with sludge
19. Chlorophyll
21. $[HT^{2-}] = 1.29 \times 10^{-9}M$
23. 0.0299

CHAPTER 5, PAGE 115

1. $CH_2CH_2CO_2H$
3. C
5. 42.9 g O_2
7. 2^{30} = 1,073,000,000 cells
9. The new bacteria have to become acclimated to the new medium. This may involve formation of a mutant strain that is better adapted to the medium.
11. Because bacterial enzymes are destroyed rather quickly beyond a certain temperature.
13. The autotrophic bacteria tend to be more complex because they are capable of the biochemical conversion of a wide range of materials, from very simple chemical species to complex biomolecules.
15. Virus
17. End of the lag phase and beginning of the log phase
19. (a) 1c + B; (b) 1 + E; (c) 3 + A; (d) 1 + A; (e) 1b + F.
21. To break down solid food material into components that can penetrate bacterial cell walls, thus allowing the digestion process to be completed inside the cell.
23. High organic and low nitrate and sulfate levels

CHAPTER 6, PAGE 144

1. 345 meq/100 g
3. (d)
5. (e)
7. By a flocculation process that probably involves flocculating agents produced by bacteria
9. 38.8 meq/100 g
11. It is the pH at which there is no *net* charge on a particle surface. There are positively and negatively charged functional groups, but their charges balance out.
13. CEC refers to the total equivalents of cationic material that can be bound per unit weight of ion-exchanging sediment, whereas ECS expresses levels of specific bound ions.
15. As coordinatively unsaturated Lewis acids

CHAPTER 7, PAGE 185

1. (b)
3. Radionuclides of "life elements", intermediate-length half-lives, alpha emitters
5. (e)
7. It forms an adherent precipitate, rather than a soluble complex, with Ca^{2+} ion, and it produces dangerously basic solutions.
9. 122 g
11. Soap precipitates from solution in the presence of Ca^{2+} or Mg^{2+} ions, and it is more biodegradable than ABS.
13. 315 cpm
15. (a) Cd, Pb, Hg; (b) CH_3HgCl, $(CH_3)_2Hg$; (c) DDT, methyl parathion, carbaryl.
17. It is ABS, which is poorly biodegradable, thus giving low BOD values.

CHAPTER 8, PAGE 223

1. 1.00 mole
3. Conversion to CO_2 and to settleable biomass
5. Reverse osmosis involves a sorbed pure water layer that selectively excludes solutes.
7. Advantages: high reliability, predictable behavior; disadvantages: high energy requirements, high chemical costs
9. 466,000 liters/day of air
11. Alkalinity
13. H^+ concentration
15. a-4, b-3, c-1, d-2

CHAPTER 9, PAGE 257

1. Color and specific conductance
3. Metals
5. (d)
7. $1.58 \times 10^{-4} M$ HCl
9. 400 μ/gm^3
11. $5.26 \times 10^{-4} M$
13. 2.88 mg/L
15. Hydrocarbons (organics)

CHAPTER 10, PAGE 284

1. $Fe_2O_3 + 6\,H^+ + 2\,e^- \rightarrow 2\,Fe^{2+} + 3\,H_2O$
   $MnO_2 + 4\,H^+ + 2\,e^- \rightarrow 2\,Mn^{2+} + 2\,H_2O$
3. $FeS_2 + \frac{15}{2}\,H_2O_2 \rightarrow Fe^{3+} + H^+ + 2\,SO_4^{2-} + 7\,H_2O$
5. 2,350 moles of H^+
7. (a)
9. 2 metric tons
11. Urea
13. From about 83 to about 123 meq/100 g
15. (1) by electrostatic attraction to positively charged acidic soil particles; (2) by ion-exchange of anions bound to soil mineral matter
17. Convert the area of an acre to cm^2 and perform the following calculation:

$$\frac{0.005\,\frac{\text{g lead}}{\text{g lime}} \times \frac{10 \times 10^6\ \text{g lime}}{\text{acre of soil}}}{\frac{\text{cm}^2}{\text{acre of soil}} \times 20\ \text{cm} \times \frac{2.0\ \text{g soil}}{\text{cm}^3}} \times 10^6 = \text{ppm Pb in soil}$$

19. Change solubilities and oxidation-reduction rates
21. Oxidation-reduction, acid-base reactions, precipitation, sorption, and biochemical degradation

CHAPTER 11, PAGE 320

1. Absorption of ultraviolet energy by O_3
3. Because of the absence of high-energy solar radiation at night
5. 1.03 kg
7. 0.1 atm
9. NO_2^*
11. 2.60%
13. 5.14×10^{16}
15. The stratosphere is high enough to be penetrated by energetic solar radiation that brings about chemical transformations, and it contains a high enough level of chemical species for their reactions to be significant.
17. The excited $O(^1D)$ atom may lose energy by collision with N_2 to form ground state $O(^1P)$, which in turn may react with O_2 to produce O_3.
19. * designates an excited state, whereas · designates an unpaired electron in a free radical.
21. Infrared light is not energetic enough to break chemical bonds or produce excited electronic states that would cause a species to be more reactive.

CHAPTER 12, PAGE 353

1. For all significant chemical species, the quantity coming into the reservoir per unit time is equal to the quantity leaving it or being consumed in it.
3. (c)
5. Dry limestone

7. 28.6 g
9. SO_2
11. 9,600 metric tons
13. 30 metric tons of SO_2
15. 150 million metric tons of $CaCO_3$
17. 21.2 km^3

CHAPTER 13, PAGE 378

1. 1.6×10^{10} electrons/cm^2
3. 0.3 cm/sec
5. 1.9 μm
7. Presence in the atmosphere of a non-volatile acid, particularly H_2SO_4.
9. 0.1 to 10 μm
11. Because of their higher surface-to-mass ratio
13. Carbonaceous material, metal oxides and glasses, water, dissolved ionic species, and ionic solids

CHAPTER 14, PAGE 401

1. (e)
3. SO_2
5. As part of a chain-terminating step
7. Oxidants
9. Higher excess air and no wall quench
11. Because the first must be operated under reducing conditions to eliminate NO, and the second must be operated under oxidizing conditions to remove CO and hydrocarbons
13. It is assumed that CO is essentially all from automotive sources and reacts only slowly, so that concentrations of other species may be compared to it. Neither of these assumptions may be totally correct.
15. Abstraction reaction with propane, addition reaction with propylene.
17. Both NO and NO_2 combine with free radicals, thus terminating the chain reactions that are involved in forming smog.
19. With methane the reaction is an abstraction reaction, $CH_4 + HO\cdot \rightarrow H_3C\cdot + H_2O$, whereas with ethylene, an addition reaction may occur:

$$H_2C{=}CH_2 + HO\cdot \rightarrow HO{-}CH_2{-}CH_2\cdot$$

CHAPTER 15, PAGE 425

1. a-2, b-1, c-3
3. The total concentrations should be lower due to mixing and photochemical reactions. The concentration of alkenes will be lower than that of the alkanes because of the greater photochemical reactivity of the alkenes.
5. Addition of $HO\cdot$ radical across the double bond
7. Carbonyl compounds, that is, the aldehydes and ketones
9. Ability to absorb near-ultraviolet light, leading to photochemical dissociation reactions

CHAPTER 16, PAGE 446

1. Use of a detector containing the same gas as that being determined
3. A very long path length is usually required and is often achieved through the use of multiple reflections in the cell.
5. Optical emission spectoscopy, multi-channel plasma emission, neutron activation analysis, and X-ray fluorescence are all possibilities.
7. 877 liters/min, 0.877 m^3/min
9. 3.82×10^3 liters, 0.382 m^3
11. Such methods give relative values that may still be helpful in establishing trends in pollutant levels, determining pollutant effects, and locating pollutant sources.
13. Primary standards are those deemed necessary to protect public health; secondary standards are designed to provide protection against known or expected adverse effects of air pollutants upon materials, vegetation, animals, and so on.
15. By the use of detectors selective for elements such as sulfur, nitrogen, and chlorine; by chemical separations, such as extraction of phenols into base

CHAPTER 17, PAGE 494

1. In order to obtain 1.5 million metric tons of pure copper per year from ore containing an averge of 0.7% copper, 214 million metric tons of copper ore per year would be required.
3. They are both aluminum silicate minerals.
5. Fuel consumption would be cut to 68% of current levels.
7. A diesel engine has a higher peak temperature which, according to the Carnot equation, gives a higher efficiency.
9. Coal consists largely of condensed aromatic structures.
11. Because the product contains nitrogen gas that cannot be removed
13. To prevent coal from caking
15. The term *resources* refers to quantities that are estimated to be ultimately available, whereas *reserves* are well-defined resources that can be produced profitably using existing technology.

CHAPTER 18, PAGE 533

1. The membrane controls which chemical species enter or leave a cell. By excluding some toxic substances the membrane can reduce their toxicity. Substances that disturb the function of a cell membrane may also be toxic.
3. Structure (primary, secondary, tertiary) and active sites
5. $$\text{enzyme}\begin{matrix}/\text{SH}\\ \backslash\text{SH}\end{matrix} + Pb^{2+} \rightarrow \text{enzyme}\begin{matrix}/\text{S}\backslash\\ \backslash\text{S}/\end{matrix}\text{Pb} + 2\,H^+$$

   This interaction is especially harmful when it occurs on an active site.
7. EDTA acts much like British Anti-Lewisite to remove heavy metals from the body as chelates. It is administered as the calcium salt so that it will not result in the net removal of essential calcium from the body.
9. Nitrite is an antidote for cyanide.

CHAPTER 19, PAGE 568

1. A skin disorder that can result from exposure to noxious chemicals
3. Oil and gas brine pits
5. Yes, on the basis of ignitability because they are "liable to cause fires through friction."
7. To define toxicity according to levels of toxic contaminants leached from the wastes following a standard leaching procedure.
9. Citrate anion is a heavy-metal chelating agent that tends to mobilize heavy metals as anionic chelates, which are poorly sorbed by solids.
11. a-3, b-1, c-2
13. a-4, b-5, c-3, d-6, e-1, f-7, g-2
15. a-2, b-4, c-5, d-3, e-1
17. The use of wet lime to scrub SO_2 from stack gas produces large quantities of sludge.

CHAPTER 20, PAGE 597

1. Mixture of organic and inorganic material in the form of dilute aqueous sludge. This type of waste is particularly difficult to handle, for the following reasons: it cannot be incinerated, it contains large quantities of water, and the isolation of individual constituents is very difficult.
3. Remove the Cd^{2+} by a process such as cementation and discharge the remaining harmless material to a publicly owned treatment works.
5. An ideal fixative makes toxic contaminants immoble and chemically non-reactive.
7. Cells divided by clay walls
9. Precipitation as a more complicated compound, such as a basic metal salt, coprecipitation as a minor constituent of some other compound, or sorption by the surface of another solid
11. a-4, b-3, c-5, d-1, e-2
13. The compound must have a chromophore, or light-absorbing chemical group.

INDEX